Biomass Burning and Global Change
Volume 1
Remote Sensing, Modeling and Inventory Development, and Biomass Burning in Africa

edited by Joel S. Levine

The MIT Press
Cambridge, Massachusetts
London, England

This book was set in Times Roman on the Monotype "Prism
Plus" PostScript Imagesetter by Asco Trade Typesetting Ltd.,
Hong Kong.
Editorial and production services were provided by Superscript
Editorial Production Services.
Printed and bound in the United States of America.

Library of Congress Cataloging-in-Publication Data

Biomass burning and global change / edited by Joel S. Levine.
      p.   cm.
    Includes bibliographical references and index.
    ISBN 0-262-12201-4 (vol. 1 : alk. paper). —
  ISBN 0-262-12202-2 (vol. 2 : alk. paper)
     1. Burning of land—Environmental aspects.
  2. Fuelwood—Burning—Environmental aspects.
  3. Climatic changes.   I. Levine, Joel S.
TD195.B56858   1996
628.5'3—dc20   96-35030
                            CIP

To the loving memory of my father, Hyman Levine,
for his interest, concern, and love of our planet
and the people who inhabit it.

Contents

Volume 1

Volume 2

83

84

Contributors

Mohammed Radzi bin Abas
University of Malaya
Kuala Lumpur, Malaysia

Brou Ahoua
INSET
Yamoussoukro, Ivory Coast

Thomas D. Akers
NASA Johnson Space Center
Houston, Texas

Plinio C. Alvalá
Instituto Nacional de Pesquisas Espaciais
São José dos Campos, São Paulo, Brazil

Vincent G. Ambrosia
NASA Ames Research Center
Moffett Field, California

Bruce E. Anderson
NASA Langley Research Center
Hampton, Virginia

Meinrat O. Andreae
Max Planck Institute for Chemistry
Mainz, Federal Republic of Germany

Harold J. Annegarn
University of the Witwatersrand
Johannesburg, Republic of South Africa

Paulo Artaxo
University of São Paulo
São Paulo, Brazil

Elliot Atlas
National Center for Atmospheric Research
Boulder, Colorado

Christopher H. Baisan
University of Arizona
Tucson, Arizona

Stephen P. Baker
U.S.D.A. Forest Service
Missoula, Montana

Donna C. Ballentine
University of Virginia
Charlottesville, Virginia

James A. Barber
Lockheed Martin Engineering and Sciences Company
Hampton, Virginia

Kimberly E. Baugh
University of Colorado
Boulder, Colorado

Terrence G. Bensel
Allegheny College
Meadville, Pennsylvania

Bibiana Bilbao
Instituto Venezolano de Investigaciones Científicas
Caracas, Venezuela

Donald R. Blake
University of California at Irvine
Irvine, California

Nicola J. Blake
University of California at Irvine
Irvine, California

Laura L. Bourgeau-Chavez
Environmental Research Institute of Michigan
Ann Arbor, Michigan

Kenneth D. Bowersox
NASA Johnson Space Center
Houston, Texas

James A. Brass
NASA Ames Research Center
Moffett Field, California

Guy Brasseur
National Center for Atmospheric Research
Boulder, Colorado

Delphine Brocard
Université Paul Sabatier
Toulouse, France

Edward V. Browell
NASA Langley Research Center
Hampton, Virginia

Sandra Brown
U.S. Environmental Protection Agency
Corvallis, Oregon

Jean-Michel Brustet
Laboratoire d'Aérologie
U.R.A.-CNRS
Toulouse, France

Roger A. Burke
U.S. Environmental Protection Agency
Athens, Georgia

Hélène Cachier
Centre National de la Recherche Scientifique
Commissariat à l'Énergie Atomique
Gif-sur-Yvette, France

Catherine F. Cahill
University of Washington
Seattle, Washington

Donald R. Cahoon, Jr.
NASA Langley Research Center
Hampton, Virginia

Meiqiu Cao
Research Center for Eco-Environmental Sciences
Chinese Academy of Sciences
Beijing, China

Karen L. Carleton
Physical Sciences Inc.
Andover, Massachusetts

Glen R. Cass
California Institute of Technology
Pasadena, California

John Vande Castle
University of Washington
Seattle, Washington

Sundar A. Christopher
South Dakota School of Mines and Technology
Rapid City, South Dakota

Catherine Chuang
Lawrence Livermore National Laboratory
Livermore, California

James S. Clark
Duke University
Durham, North Carolina

Wesley R. Cofer III
NASA Langley Research Center
Hampton, Virginia

James E. Collins Jr.
Science and Technology Corporation
Hampton, Virginia

Susan G. Conard
U.S.D.A. Forest Service
Riverside, California

Vickie S. Connors
NASA Langley Research Center
Hampton, Virginia

Garry D. Cook
CSIRO Tropical Ecosystems Research Centre
Darwin, Northern Territory, Australia

Craig E. Corrigan
Lawrence Berkeley National Laboratory
Berkeley, California

John A. Costulis
NASA Langley Research Center
Hampton, Virginia

Richard H. Couch
NASA Langley Research Center
Hampton, Virginia

Richard O. Covey
NASA Johnson Space Center
Houston, Texas

Bernard Cros
Université Paul Sabatier
Toulouse, France

Paul J. Crutzen
Max Planck Institute for Chemistry
Mainz, Federal Republic of Germany

Paul Cwielong
Centro de Investigación y Extensión Forestál Andino
Patagónico
Chubut, Argentina

Peter Daum
Brookhaven National Laboratory
Upton, New York

Ethan R. Davis
University of Colorado
Boulder, Colorado

Richard E. Davis
NASA Langley Research Center
Hampton, Virginia

Roseanne D. Diab
University of Natal
Durban, Republic of South Africa

Michel Dimbele
ORSTOM
Bangui, Central African Republic

Patrick K. Dowling
United States Air Force
MacDill Air Force Base
Tampa, Florida

Pete Dowty
University of Virginia
Charlottesville, Virginia

Francisco Echalar
Université Paris
Créteil, France

Thomas F. Eck
NASA Goddard Space Flight Center
Greenbelt, Maryland

Harold Eddleman
Lawrence Livermore National Laboratory
Livermore, California

Christopher D. Elvidge
NOAA National Geophysical Data Center
Boulder, Colorado

Gerd Esser
Justus-Liebig-University
Giessen, Federal Republic of Germany

Hugh Douglas Eva
Institute for Remote Sensing Applications
Joint Research Centre
Ispra, Italy

Agustin T. Ezcurra
University Pais Vasco
Vitoria, Spain

Philip M. Fearnside
National Institute for Research in the Amazon
(INPA)
Amazonas, Brazil

Zongwei Feng
Research Center for Eco-Environmental Sciences
Chinese Academy of Sciences
Beijing, People's Republic of China

Ronald J. Ferek
University of Washington
Seattle, Washington

Mike Flood
NASA Langley Research Center
Hampton, Virginia

Luke Flynn
University of Hawaii
Honolulu, Hawaii

Jacques Fontan
Laboratoire d'Aérologie
U.R.A.-CNS
Toulouse, France

Robert S. Fraser
NASA Goddard Space Flight Center
Greenbelt, Maryland

Nancy H. F. French
Environmental Research Institute of Michigan
Ann Arbor, Michigan

Amanda L. Frick
Environmental Research Institute of Michigan
Ann Arbor, Michigan

Valentin V. Furyaev
Russian Academy of Sciences
Krasnoyarsk, Russia

Paul A. Garn
NASA Langley Research Center
Hampton, Virginia

Michael Garstang
University of Virginia
Charlottesville, Virginia

Greg Gaston
U.S. Environmental Protection Agency
Corvallis, Oregon

Annie Gaudichet
Université Paris
Créteil, France

Chris D. Geron
U.S. Environmental Protection Agency
Research Triangle Park, North Carolina

William P. Gilhooly
University of Virginia
Charlottesville, Virginia

Josef Goergen
Cooperación Tecnica Argentino-Alemana
San Carlos de Bariloche, Argentina

Johann G. Goldammer
Freiburg University
Freiburg, Federal Republic of Germany

Barbara B. Gormsen
Science Applications International Corporation
Hampton, Virginia

Claire Granier
National Center for Atmospheric Research
Boulder, Colorado

Jean-Marie Grégoire
Institute for Remote Sensing Applications
Joint Research Centre
Ispra, Italy

David W.T. Griffith
University of Wollongong
New South Wales, Australia

Liane S. Guild
Oregon State University
Corvallis, Oregon

Wei Min Hao
U.S.D.A. Forest Service
Missoula, Montana

Geoffrey W. Harris
Max Planck Institute for Chemistry
Mainz, Federal Republic of Germany

Robert C. Harriss
NASA Headquarters
Washington, D.C.

Gary Hartley
Canadian Forest Service
Sault Ste. Marie, Ontario, Canada

Dean A. Hegg
University of Washington
Seattle, Washington

Günter Helas
Max Planck Institute for Chemistry
Mainz, Federal Republic of Germany

Michael R. Helfert
NASA Johnson Space Center
Houston, Texas

John A. Herring
University of Washington
Seattle, Washington

Robert G. Higgins
NASA Ames Research Center
Moffett Field, California

Lynn M. Hildemann
Stanford University
Stanford, California

Christine A. Hlavka
NASA Ames Research Center
Moffett Field, California

Peter V. Hobbs
University of Washington
Seattle, Washington

Jeffrey A. Hoffman
NASA Johnson Space Center
Houston, Texas

Johannes Hoffstadt
Justus-Liebig-University
Giessen, Federal Republic of Germany

Brent N. Holben
NASA Goddard Space Flight Center
Greenbelt, Maryland

Dale F. Hurst
University of Colorado
Boulder, Colorado

Galina A. Ivanova
Russian Academy of Sciences
Krasnoyarsk, Russia

Antony Jalink, Jr.
NASA Langley Research Center
Hampton, Virginia

Anthony C. Janetos
NASA Headquarters
Washington, D.C.

Russell D. Johnson
Environmental Research Institute of Michigan
Ann Arbor, Michigan

Tom Jones
NASA Johnson Space Center
Houston, Texas

Christopher O. Justice
University of Maryland
College Park, Maryland

Per Kållberg
European Centre for Medium Range Weather
Forecasting
Reading, England

Alexis D. Kambis
Science Systems and Applications, Inc.
Greenbelt, Maryland

Shin G. Kang
Physical Sciences Inc.
Andover, Massachusetts

Eric S. Kasischke
Environmental Research Institute of Michigan
Ann Arbor, Michigan

J. Boone Kauffman
Oregon State University
Corvallis, Oregon

Yoram J. Kaufman
NASA Goddard Space Flight Center
Greenbelt, Maryland

Ahmed Khedim
Forschungszentrum Jülich (KFA)
Jülich, Federal Republic of Germany

Eric A. Kihn
National Oceanic and Atmospheric Administration/
National Geophysical Data Center
Boulder, Colorado

Volker W.J.H. Kirchhoff
Instituto Nacional de Pesquisas Espaciais
São José dos Campos, São Paulo, Brazil

Yegor K. Kisilyakov
Russian Academy of Sciences
Krasnoyarsk, Russia

Richard Kleidman
Science Systems and Applications, Inc.
Greenbelt, Maryland

Donna V. Kliche
South Dakota School of Mines and Technology
Rapid City, South Dakota

Melanie A. Kneen
University of the Witwatersrand
Johannesburg, Republic of South Africa

Brigitte Koffi
Institute for Remote Sensing Applications
Joint Research Centre
Ispra, Italy

Thomas Konzelmann
Swiss Federal Institute of Technology
Zurich, Switzerland

Gudrun Koppen
University of Ghent
Ghent, Belgium

Ralf Koppmann
Forschungszentrum Jülich (KFA)
Jülich, Federal Republic of Germany

Georges Kouadio
École Nationale Supérieure
Abidjan, Ivory Coast

Herbert W. Kroehl
National Oceanic and Atmospheric Administration/
National Geophysical Data Center
Boulder, Colorado

Tom L. Kucsera
NASA Goddard Space Flight Center
Greenbelt, Maryland

Thomas A. J. Kuhlbusch
Max Planck Institute for Chemistry
Mainz, Federal Republic of Germany

Corinne Lacaux
Université Paul Sabatier
Toulouse, France

Jean-Pierre Lacaux
Université Paul Sabatier
Toulouse, France

Cong Lai
Institute of Environmental Medicine
New York University Medical Center
Tuxedo, New York

Carl Laterza
Brookhaven National Laboratory
Upton, New York

Krista K. Laursen
National Center for Atmospheric Research
Boulder, Colorado

Bruce D. Lawson
Canadian Forest Service
Victoria, British Columbia, Canada

Philippe Le Canut
Max Planck Institute for Chemistry
Mainz, Federal Republic of Germany

Joel S. Levine
NASA Langley Research Center
Hampton, Virginia

Chungcheng Li
Kunming Weather Bureau
Wuhua District, Kunming, People's Republic
of China

Rong R. Li
Science Systems and Applications, Inc.
Greenbelt, Maryland

Catherine Liousse
Centre des Faibles Radioactivités
Centre National de la Recherche Scientifique
Commissariat à l'Énergie Atomique
Gif-sur-Yvette, France

Dan A. Livingstone
Duke University
Durham, North Carolina

Robert N. Lockwood
U.S.D.A. Forest Service
Riverside, California

Kamlesh P. Lulla
NASA Johnson Space Center
Houston, Texas

James A. McAdoo
NASA Langley Research Center
Hampton, Virginia

Frank Mack
Justus-Liebig-University
Giessen, Federal Republic of Germany

Stephen A. Macko
University of Virginia
Charlottesville, Virginia

David McDougal
NASA Langley Research Center
Hampton, Virginia

Lisa M. McKenzie
University of Montana
Missoula, Montana

Donna P. McNamara
NASA Goddard Space Flight Center
Greenbelt, Maryland

Willy Maenhaut
University of Ghent
Ghent, Belgium

Amy E. Major
University of Georgia
Athens, Georgia

Jean Paul Malingreau
Institute for Remote Sensing Applications
Joint Research Centre
Ispra, Italy

Eino Mälkönen
The Finnish Forest Research Institute
Vantaa, Finland

Brian Markham
NASA Goddard Space Flight Center
Greenbelt, Maryland

Bice Martincigh
University of Natal
Durban, Republic of South Africa

J. Vanderlei Martins
University of São Paulo
São Paulo, Brazil

Monica A. Mazurek
Rutgers University
New Brunswick, New Jersey

E. Medina
Instituto Venezolano de Investigaciones Cientificas
Caracas, Venezuela

W. Paul Menzel
National Oceanic and Atmospheric Administration/
National Environmental Satellite Data and
Information Service
Madison, Wisconsin

William L. Miller
Dalhousie University
Halifax, Nova Scotia, Canada

Micheline Moula
Laboratoire d'Aérologie
Toulouse, France

Jean-François Muller
Belgian Institute for Space Aeronomy
Brussels, Belgium

F. Story Musgrave
NASA Johnson Space Center
Houston, Texas

J. David Nance
University of Washington
Seattle, Washington

Doreen Osowski Neil
NASA Langley Research Center
Hampton, Virginia

Leonard Newman
Brookhaven National Laboratory
Upton, New York

Claude Nicollier
NASA Johnson Space Center
Houston, Texas

Scott Nolf
Computer Sciences Corporation
Hampton, Virginia

Tihomir Novakov
Lawrence Berkeley National Laboratory
Berkeley, California

Gerald J. Olbu
U.S.D.A. Forest Service
Missoula, Montana

Maria A. de Oliveira
Instituto Nacional de Pesquisas
São José dos Campos
São Paulo, Brazil

Jennifer Richardson Olson
NASA Langley Research Center
Hampton, Virginia

Didier Orange
Laboratoire d'Hydrologie et de Géologie
Bangui, Central African Republic

I. Ortiz de Zárate
University Pais Vasco
Vitoria, Spain

Roger D. Ottmar
U.S.D.A. Forest Service
Seattle, Washington

Loyd W. Overbay
NASA Langley Research Center
Hampton, Virginia

Dirk A.B. Parsons
University of the Witwatersrand
Johannesburg, Republic of South Africa

Hamilton G. Pavão
Instituto Nacional de Pesquisas Espaciais
São José dos Campos
São Paulo, Brazil

Joyce E. Penner
University of Michigan
Ann Arbor, Michigan

João A. Pereira
Instituto de Pesquisas Espaciais
São José dos Campos
São Paulo, Brazil,

Marie-Hélène Pertuisot
Centre des Faibles Radioactivités
Centre National de la Recherche Scientifique
Commissariat à l'Energie Atomique
Gif-sur-Yvette, France

Jean Laurent Pfund
Projet FNS Tere-Tany
Antananarivo, Madagascar

Pham-Van-Dinh
Université Paul Sabatier
Lannemezan, France

Stuart J. Piketh
University of the Witwatersrand
Johannesburg, Republic of South Africa

Peter Pilewskie
NASA Ames Research Center
Moffett Field, California

Artemio Plana-Fattori
University of São Paulo
São Paulo, Brazil

James R. Plummer
U.S.D.A. Forest Service
Missoula, Montana

Elaine M. Prins
University of Wisconsin
Madison, Wisconsin

Lawrence F. Radke
National Center for Atmospheric Research
Boulder, Colorado

Wilson T. Rawlins
Physical Sciences Inc.
Andover, Massachusetts

Jon C. Regelbrugge
U.S.D.A. Forest Service
Riverside, California

Henry G. Reichle Jr.
North Carolina State University
Raleigh, North Carolina

Jeffrey S. Reid
University of Washington
Seattle, Washington

Lorraine A. Remer
Science Systems and Applications, Inc.
Greenbelt, Maryland

Aaron R. Rezendez
Stanford University
Stanford, California

Philip J. Riggan
U.S.D.A. Forest Service
Riverside, California

Don M. Robinson
NASA Langley Research Center
Hampton, Virginia

Norberto Rodriguez
Centro de Investigación y Extensión Forestál Andino
Patagónico
Chubut, Argentina

William A. Roettker
NASA Langley Research Center
Hampton, Virginia

Wolfgang F. Rogge
Florida International University
University Park, Florida

John L. Ross
University of Washington
Seattle, Washington

F. Sherwood Rowland
University of California at Irvine
Irvine, California

William F. Ruddiman
University of Virginia
Charlottesville, Virginia

Jochen Rudolph
Forschungszentrum Jülich (KFA)
Jülich, Federal Republic of Germany

Glen W. Sachse
NASA Langley Research Center
Hampton, Virginia

Washito A. Sasamoto
NASA Langley Research Center
Hampton, Virginia

Mary C. Scholes
University of the Witwatersrand
Johannesburg, Republic of South Africa

Robert J. Scholes
Forest Science and Technology (Forestek)
CSIR
Pretoria, Republic of South Africa

Daniel I. Sebacher
Science Applications International Corporation/
GATS
Hampton, Virginia

Roger Serpolay
University Blaise Pascal
Clement-Fd, France

Alberto W. Setzer
Instituto Nacional de Pesquisas Espaciais (INPE)
São José dos Campos,
São Paulo, Brazil

Ronald W. Shea
Oregon State University
Corvallis, Oregon

Gary Shelton
NASA Ames Research Center
Moffett Field, California

Robert T. Sherrill
NASA Langley Research Center
Hampton, Virginia

Bernd R.T. Simoneit
Oregon State University
Corvallis, Oregon

Ilya Slutsker
NASA Goddard Space Flight Center
Greenbelt, Maryland

Kelly D. Smith
NASA Langley Research Center
Hampton, Virginia

Irina Sokolik
NASA Ames Research Center
Moffett Field, California

David M. Sonnenfroh
Physical Sciences Inc.
Andover, Massachusetts

Darrell Wayne Sproles
Computer Sciences Corporation
Hampton, Virginia

Laurel J. Standley
Stroud Water Research Center
Avondale, Pennsylvania

Brian J. Stocks
Canadian Forest Service
Sault Ste. Marie, Ontario, Canada

Ronald A. Susott
U.S.D.A. Forest Service
Missoula, Montana

Wayne T. Swank
U.S.D.A. Forest Service
Otto, North Carolina

Robert J. Swap
University of Virginia
Charlottesville, Virginia

Thomas W. Swetnam
University of Arizona
Tucson, Arizona

Matthew A. Tarr
U.S. Environmental Protection Agency
Athens, Georgia

Anne M. Thompson
NASA Goddard Space Flight Center
Greenbelt, Maryland

Kathryn C. Thornton
NASA Johnson Space Center
Houston, Texas

Winston S. W. Trollope
University of Fort Hare
Alice, Republic of South Africa

Vaughan C. Turekian
University of Virginia
Charlottesville, Virginia

Peter D. Tyson
University of the Witwatersrand
Johannesburg, Republic of South Africa

Susan Ustin
University of California at Davis
Davis, California

Erik Valendik
Forest Fire Laboratory
Russian Academy of Sciences
Krasnoyarsk, Russia

Francisco P.J. Valero
Scripps Institution of Oceanography
University of California at San Diego
La Jolla, California

James M. Vose
U.S.D.A. Forest Service
Otto, North Carolina

John J. Walton
Lawrence Livermore National Laboratory
Livermore, California

Xiaoke Wang
Research Center for Eco-Environmental Sciences
Chinese Academy of Sciences
Beijing, People's Republic of China

Darold E. Ward
U.S.D.A. Forest Service
Missoula, Montana

Ross W. Wein
University of Alberta
Edmonton, Alberta, Canada

Karl-Friedrich Weiss
Max Planck Institute for Chemistry
Mainz, Federal Republic of Germany

Ray E. Weiss
University of Washington
Seattle, Washington

Ronald M. Welch
South Dakota School of Mines and Technology
Rapid City, South Dakota

Michael Welling
Max Planck Institute for Chemistry
Mainz, Federal Republic of Germany

Charles H. Whitlock
NASA Langley Research Center
Hampton, Virginia

Frank G. Wienhold
Max Planck Institute for Chemistry
Mainz, Federal Republic of Germany

Edward L. Winstead
Science Applications International Corporation/
GATS
Hampton, Virginia

Barbara E. Wyslouzil
Worcester Polytechnic Institute
Worcester, Massachusetts

Heng Yao
Research Center for Eco-Environmental Sciences
Chinese Academy of Sciences
Beijing, People's Republic of China

Véronique Yoboué
University of Abidjan
Abidjan, Ivory Coast

Thomas Zenker
Max Planck Institute for Chemistry
Mainz, Federal Republic of Germany

Richard G. Zepp
U.S. Environmental Protection Agency
Athens, Georgia

Ya-hui Zhuang
Research Center for Eco-Environmental Sciences
Chinese Academy of Sciences
Beijing, People's Republic of China

Introduction

Joel S. Levine

Our planet and global environment are witnessing the most profound changes in the brief history of the human species. Human activity is the major agent of those changes—depletion of stratospheric ozone, the threat of global warming, deforestation, acid precipitation, the extinction of species, and others that have not become apparent.

This statement is from the introduction to the report, *Global Change and Our Common Future*, published by the National Research Council in 1989. The processes of global change identified in the statement—depletion of stratospheric ozone, the threat of global warming, deforestation, acid precipitation, and the extinction of species—all have one thing in common: they are caused or significantly enhanced by biomass burning. Biomass burning is the burning of the world's living and dead vegetation, including grasslands, forests, and agricultural lands following the harvest for land clearing and land-use change. Biomass burning is not restricted to one geographical region, but is rather a truly global phenomenon.

Biomass burning is a significant global source of gaseous and particulate emissions to the atmosphere. Gases produced by biomass burning include (1) greenhouse gases, carbon dioxide (CO_2), methane (CH_4), and nitrous oxide (N_2O), that lead to global warming, (2) chemically active gases, nitric oxide (NO), carbon monoxide (CO), methane, and hydrocarbons (NMHCs), which lead to the photochemical production of ozone (O_3) in the troposphere, and (3) methyl chloride (CH_3Cl), and methyl bromide (CH_3Br), which lead to the chemical destruction of ozone in the stratosphere (Levine 1985). Particulates produced by biomass burning perturb the transfer of incoming solar radiation through the troposphere, and, hence, impact climate. Recent estimates of atmospheric gases produced by biomass burning are summarized in table I.1 (see Chapter 27 by Andreae et al.). In addition to these direct effects on atmospheric composition and chemistry and climate, biomass burning perturbs other components and processes in the earth's system, including (1) the biogeochemical cycling of nitrogen (N_2O and NO) and carbon (CO_2, CO, and CH_4) gases

from the biosphere to the atmosphere, (2) water run-off and evaporation, and, hence, impacts the hydrological cycle, (3) the reflectivity and emissivity of the land, which in turn changes the radiative properties of the land and hence, impacts climate, and (4) the stability of ecosystems which in turn impacts biological diversity. For these and other reasons biomass burning is an important driver for global change (Levine et al. 1995).

The 84 contributions in these two volumes represent the latest scientific results and technological advances in biomass burning research. Biomass burning is global in scope and the contributors to this volume are also global in scope representing a total of 254 researchers from 20 countries. The chapters here were originally presented at the Chapman Conference on Biomass Burning and Global Change, 13–17 March 1995, in Williamsburg, the colonial capital of Virginia. The conference was organized and sponsored by the American Geophysical Union (AGU), the world's largest professional organization of earth scientists, with membership in excess of 25 000. The conference was coorganized and cosponsored by the National Aeronautics and Space Administration (NASA), the NASA Langley Research Center, the U.S. Environmental Protection Agency, the U.S. Department of Agriculture Forest Service, the International Geosphere-Biosphere Program, and the International Global Atmospheric Chemistry Project. The 1995 conference was the second AGU Chapman Conference on biomass burning. The first conference, the Chapman Conference on Biomass Burning: Atmospheric, Climatic, and Biospheric Implications, took place 19–23 March 1990, again in Williamsburg, Virginia. The conference volume of the same name, was published by The MIT Press in 1991 and contained 63 contributions written by 160 researchers representing 14 countries (Levine 1991). In the field of biomass burning and global change much has happened and much has been learned since the 1990 conference. These two volumes describe what we have learned and point out directions for future research to better assess the role of biomass burning as a driver for global change.

Table 1 "Best guess" estimates of gaseous and particulate emissions from global biomass burning and all anthropogenic sources (including biomass burning) (Assumptions: Total biomass burned = 8910 Tg/yr; total carbon burned = 4100 Tg C/yr) (Andreae et al. chapter 27)

Species	Biomass burning contribution (units are Tg of species per year)	All anthropogenic sources (units are Tg of species per year)	% Due to biomass burning
CO_2	13 500	33 700[a]	40.1
CO	680	1600[a]	42.5
CH_4	43	275[a]	15.6
NMHC	42	100[b]	42.0
H_2	16	40[c]	40.0
NO	21	70[a]	30.0
N_2O	1.3	5.5[a]	23.6
NH_3	6.7	57[d]	11.8
SO_2	4.8	160[e]	3.0
COS	0.21	0.38[f]	55.3
CH_3Cl	1.1	1.1?	100
CH_3Br	0.019	0.11[g]	17.3
Aerosols			
TPM[h]	90	390[e]	23.1
Carbon	60	90[e]	66.7

a. Houghton et al. 1995
b. Ehhalt et al. 1986
c. Warneck 1988
d. Schlesinger and Hartley 1992
e. Andreae 1995
f. Chin and Davis 1993
g. WMO/UNEP 1995
h. Total particulate matter

Volume 1 is divided into three parts: I. Remote Sensing: Global and Regional Scales, II. Modeling and Inventory Development: Global and Regional Scales, and III. Biomass Burning in Africa. Volume 2 contains parts IV. Biomass Burning in South America, V. Biomass Burning in Southeast Asia, VI. Biomass Burning in Temperate Ecosystems, VII. Biomass Burning in the Boreal Ecosystem, and VIII. Oil Fires in Kuwait. The chapters in parts I and II are more general in nature and refer to regional or global-scale phenomena or processes; the chapters in parts III to VIII are unique to burning in a particular ecosystem or geographical location.

Remote Sensing of Biomass Burning from Space

Perhaps the greatest single challenge to the scientific community studying biomass burning is to accurately assess the spatial and temporal distribution of burning over a given period of time, that is, weeks, months, or a year. Once the spatial and temporal distribution of burning in a particular ecosystem or geographical region is known, this information combined with information obtained during field experiments on the amount of biomass consumed during burning and the gaseous and particulate combustion emissions, can

provide reliable estimates of the amount of gaseous and particulate emissions released to the atmosphere during the burning event. This information can be used in developing global budgets of these gaseous and particulate species and to assess their impact on the composition and chemistry of the atmosphere, and on climate. However, the key to assessing the atmospheric and climatic impact of burning is the monitoring of the spatial and geographical distribution of burning. Remote sensing of fires from space provides the only opportunity for this.

A number of new techniques have been developed to study the spatial and temporal distribution of burning from space using existing satellite systems, including the nighttime low-light satellite images obtained with the Defense Meteorological Satellite Program (DMSP) Block 5 satellites (figure I.1), the Advanced Very High Resolution Radiometer (AVHRR) aboard the NOAA operational meteorological satellites (figure I.2), geostationary satellites, and astronaut photography from the Space Shuttle. The general use of these space-based platforms to study and monitor biomass burning is discussed in chapters 2 through 9. Applications of these techniques to biomass burning in specific ecosystems and geographical regions are addressed in chapters 21 and 22 for fires in Africa, chapters 53 and 54 for fires in Brazil, chapter 63 for fires in Southeast Asia, and chapters 75 through 77 for fires in the boreal system. A new NASA Mission to Planet Earth (MTPE) Program Office research initiative in land-cover and land-use change, of which biomass burning is an important process, is discussed in chapter 1. A dedicated small and inexpensive NASA satellite experiment, FireSat, to monitor the global distribution and frequency of biomass burning on a near daily basis is discussed in chapter 12. Biomass burning is a major global source of carbon monoxide in the troposphere. Measurements of tropospheric carbon monoxide obtained with the NASA Measurement of Air Pollution from Satellites (MAPS) aboard the Space Shuttle in 1984 and 1994 are discussed in chapters 10 and 11. Enhanced levels of tropospheric carbon monoxide were found to be associated with biomass burning in South America and Africa.

Modeling and Inventory Development: Global and Regional Scales

Theoretical modeling is a powerful tool to assess the impact of the gaseous and particulate emissions from biomass burning on regional and global atmospheric composition and chemistry. Part II contains chapters that address the biogeochemical and atmospheric

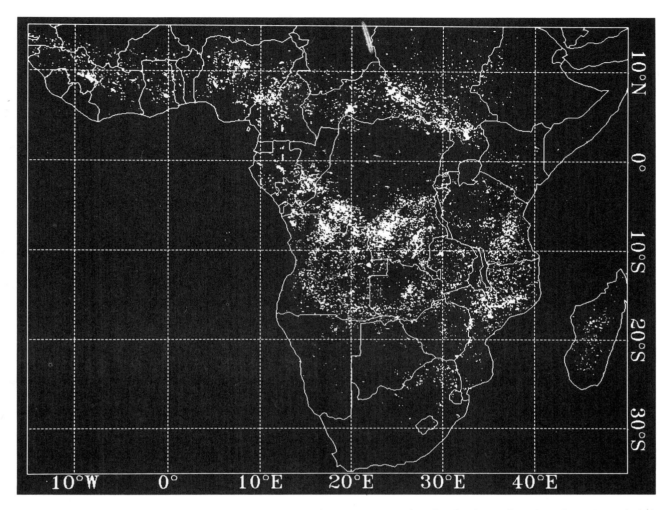

Figure I.1 Extensive burning of Africa's savannas mapped over entire year (1987) based on nighttime images from the Defense Meteorological Satellite Project (DMSP) (Cahoon et al. 1992)

transport modeling and the development of a biomass burning inventory on both global and regional scales. These chapters consider the biomass density for tropical forests, the global carbon dioxide and carbon cycles, and large-scale transport of biomass burning gases and particulates on the regional and global scale. Part II concludes with an approach to the development of an inventory of parameters of biomass burning, that is, spatial and temporal distribution of burning, biomass consumed during burning, and the gaseous and particulate emissions produced during burning.

Field Experiments

Africa and South America
Biomass burning field experiments are important to understand the burning characteristics of very diverse

ecosystems, including the amount of biomass consumed during burning, the gaseous and particulate combustion products produced, and the impact of burning on the biogeochemical cycling of gases from the soil to the atmosphere. Since the 1990 conference, a series of international and national field experiments have studied biomass burning in diverse ecosystems including the savannas of southern Africa and Brazil, the tropical rain forests of Brazil, the temperate forests of the United States Pacific Northwest and the boreal forests of Russia. The Biomass Burning Experiment: Impact on the Atmosphere and Biosphere (BIBEX), an activity of the International Global Atmospheric Chemistry (IGAC) Project, part of the International Geosphere-Biosphere Program (IGBP), has organized and coordinated biomass burning field experiments in the grasslands of southern Africa and Brazil, the tropical forests of Brazil, and the boreal forests of Russia

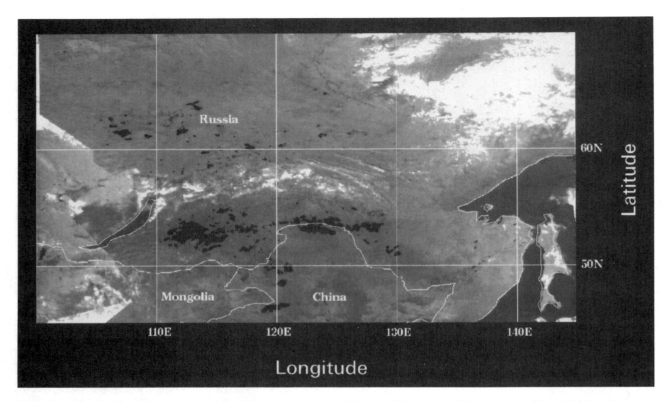

Figure I.2 Fire scars (burned areas) in northern China and southeastern Siberia in 1987 mapped with the Advanced Very High Resolution Radiometer (AVHRR) on the NOAA satellite (Cahoon et al. 1994)

and the former Soviet Union. NASA conducted a field experiment in the forests of the United States Pacific Northwest, the Smoke Cloud, and Radiation (SCAR) Experiment. In 1991, the government of Iraq conducted a large-scale burning experiment in Kuwait.

The first large, coordinated program conducted under BIBEX was the Southern Tropical Atlantic Regional Experiment (STARE). From August to October 1992, after two years of preparation, about 200 researchers participated in this field experiment, using eight aircraft and numerous ground-based measurement systems. STARE had two subcomponents, SAFARI-92 (Southern Africa Fire-Atmosphere Research Initiative-1992) and TRACE-A (Transport and Atmospheric Chemistry near the Equator-Atlantic). SAFARI-92 was planned, organized, and implemented by BIBEX; TRACE-A was planned, organized, and implemented by the NASA Global Tropospheric Experiment Program Office and coordinated with SAFARI-92.

The main objective of SAFARI-92 was the study of burning, fire ecology, and atmospheric chemistry over southern Africa (figure I.3). SAFARI-92 consisted of more than 150 scientists from 14 countries and in-

volved ground-based and airborne measurements. An important aspect of SAFARI-92 was a series of controlled or prescribed fires set at Kruger National Park in the northeast section of South Africa. The controlled fires ranged from more than one dozen small fires covering several acres to two large fires covering about 5000 acres. Ground-based and airborne measurements were obtained before, during, and after the Kruger National Park fires and included such parameters as the rate of fire spread, the amount of biomass consumed, the gaseous and particulate emissions, and the impact of these fires on the biogenic emissions of nitrogen and carbon gases from the soil to the atmosphere. During the SAFARI-92 experiment, measurements of ozone were obtained over Brazzaville, Congo, during a preliminary phase of the Experiment for Regional Sources and Sinks of Oxidants (EXPRESSO). Results from the SAFARI and EXPRESSO experiments form the largest number of contributions to these two volumes and are described in part III (chapters 21 through 46). Chapters 47 through 51 in part III deal with the general physics and chemistry of aerosols produced during biomass burning in Africa and in other locations.

Figure I.3 Burning of the savannas in Kruger National Park, South Africa, in September 1992 during SAFARI-92 (Photograph courtesy Joel S. Levine)

The main objective of TRACE-A was the exploration of the large-scale atmospheric composition of the region extending from Brazil over the South Atlantic to southern Africa. TRACE-A centered around airborne atmospheric chemistry and meteorological measurements obtained with the NASA DC-8 instrumented aircraft. The results of the TRACE-A experiment over Brazil are described in part IV (chapters 53 through 61). Studies on historic fires in Africa and South America are presented in chapters 52 and 62, respectively.

The Temperate Ecosystems

In September and October 1994, the NASA Smoke, Clouds and Radiation-C (SCAR-C) Project was conducted in the US Pacific Northwest. The experiment consisted of three prescribed fires. The measurements included the characterization of the nature and con-sumption rates of the fuels, fire behavior, and airborne in situ and remote-sensing measurements of the gaseous and particulate emissions. The SCAR-C results are summarized in chapters 65 through 67 in part VI. Other chapters in part VI include biomass burning measurements in other temperate ecosystems: in eastern North America (chapter 69), the southeastern United States (chapter 68), China (chapters 70 through 72), Spain (chapter 73), and Australia (chapter 74).

The Boreal Forests

Biomass burning can no longer be considered only a tropical phenomenon. Since 1990, there has been new interest in biomass burning in the world's boreal forests, particularly in the boreal forests of China, Siberia, Canada, and Alaska. Not long ago it was believed that on average, only about 4 million acres of boreal forest burned each year (Seiler and Crutzen 1980). At the

Figure I.4 Burning of the boreal forest during the Bor Forest Island Experiment in Krasnoyarsk, Russia, in July 1993 (Photograph courtesy Wesley R. Cofer)

1990 conference, it was reported that an average of about 20 million acres of boreal forests may burn each year, with significant year-to-year fluctuations (Stocks 1991). A recent study based on satellite measurements indicated that in 1987 about 36 million acres burned in the boreal forests of Asia alone (Cahoon et al. 1994). The estimates of biomass burning in boreal forests, which have increased by a factor of 10 over the last two decades highlight two things: the great geographical extent of biomass burning in the world's boreal forests and the power of satellite measurements to provide accurate information on the geographical extent of burning.

On 6 July 1993 BIBEX and the International Boreal Forest Research Association (IBFRA) organized a boreal forest biomass burn experiment, the Bor Forest Island Experiment, in Krasnoyarsk, Russia, the first phase of the Fire Research Campaign Asia-North (FIRESCAN) (figure I.4). Thirty researchers from eight countries studied fire behavior and the gaseous and particulate emissions from this fire. Measurements obtained during this experiment suggested that gases and particulates produced during biomass burning in the boreal forest may be directly transported to the upper troposphere and, perhaps, even into the lower

stratosphere. The very hot boreal forest fire may generate sufficient energy to permit very strong upward convective currents to traverse the entire vertical extent of the troposphere (see chapters 79 and 81).

The boreal system contains about 25% of the world's forests and is very susceptible to climate change. Calculations with General Circulation Models (GCMs) suggest that warming of the earth will result in warmer and drier boreal forests and, in turn, an enhanced frequency of fires. The release of large amounts of carbon dioxide by these fires will amplify the greenhouse warming. There is new interest in biomass burning in the world's boreal forest, which is evident by no fewer than seven chapters on this subject (chapters 75 through 81).

Biomass Burning, Methyl Bromide, and Stratospheric Ozone Depletion

The SAFARI, TRACE-A and Bor Forest Island Experiments led to an important scientific discovery—significant amounts of methyl bromide (CH_3Br) were produced during biomass burning (Mano and Andreae 1994 and chapter 56 by Blake et al.). Methyl bromide releases atomic bromine, which leads to the catalytic chemical destruction of stratospheric ozone, which is very similar to the catalytic destruction of stratospheric ozone by atomic chlorine. A major anthropogenic source of stratospheric chlorine is the photolysis of chlorofluorocarbons (CFCs), which were used mainly in aerosol sprays. The use of these CFCs is now banned by the Montreal Protocol. On an atom-to-atom basis, bromine is about 50 times more efficient in destroying stratospheric ozone than is chlorine.

Bromine is carried into the stratosphere in various forms such as halons and hydrocarbons, of which methyl bromide (CH_3Br) is the predominant form. The chemistry of bromine in the stratosphere is very interesting because of a chlorine-bromine synergism with respect to ozone depletion. A coupling reaction between BrO and ClO produces Br and Cl atoms, both of which react with ozone forming a catalytic cycle for ozone destruction (Wayne 1991):

$$BrO + ClO \rightarrow Br + Cl + O_2 \qquad (I.1)$$

$$Br + O_3 \rightarrow BrO + O_2 \qquad (I.2)$$

$$Cl + O_3 \rightarrow ClO + O_2 \qquad (I.3)$$

$$\text{Net Reaction: } 2O_3 \rightarrow 3O_2 \qquad (I.4)$$

There are no known efficient reservoirs for Br or BrO, in contrast to the case for Cl and ClO, because HBr and $BrONO_2$ are rapidly photolyzed. Other bromine reactions that impact the chemical destruction of stratosphere include:

$$BrO + BrO \rightarrow 2Br + O_2 \qquad (I.5)$$

There are two further channels for the BrO + ClO reaction:

$$BrO + ClO \rightarrow Br + OClO \qquad (I.6)$$

$$BrO + ClO \rightarrow BrCl + O_2 \qquad (I.7)$$

The branching ratios into the three channels for reactions (I.1), (I.6), and (I.7) are 0.45, 0.43, and 0.12, respectively, at room temperature (Wayne 1991). The discovery that methyl bromide is an important combustion product of biomass burning identified a previously unknown and very important connection between biomass burning and the chemical destruction of stratospheric ozone.

A few months after the conference, the Nobel Committee announced that the 1995 Nobel Prize in chemistry would be shared by three atmospheric scientists for their important contributions to our understanding of the chemical destruction of stratospheric ozone. The winners of the 1995 Nobel Prize in Chemistry were Paul J. Crutzen, Mario J. Molina, and F. Sherwood Rowland. Two of these winners, Crutzen and Rowland, participated in the conference and their contributions are included in this volume (Crutzen is an author of chapter 16 and Rowland as an author of chapter 56).

The Oil Fires in Kuwait

One of the consequences of Iraq's invasion of Kuwait was the ignition of more than 700 oil wells, storage tanks, and refineries in February 1991 (figure I.5). These fires initially consumed more than 600 barrels of oil per day. The combustion of this oil produced a tremendous flux of gaseous and particulate emissions into the atmosphere. While not biomass burning in its strictest definition (although in the very distant past, petroleum was indeed biomass!), the Kuwaiti oil fires presented an opportunity for many in the biomass burning community to assess the impact of this act of environmental terrorism on the local and regional atmosphere. The results of several studies are described in chapters 82 through 84.

Public Education and Outreach

As part of the public education and outreach aspect of the conference, a free public symposium, "Biomass Burning and Global Change" took place at the Williamsburg Lodge, the conference site, on Monday, 13 March 1995. The symposium speakers included M. O. Andreae (Max Planck Institute for Chemistry, Mainz, Germany), S. Brown (U.S. Environmental Protection Agency, Corvallis, Oregon), P. J. Crutzen (Max Planck Institute for Chemistry, Mainz, Germany), R. C. Harriss (Director, Science Division, NASA Mission to Planet Earth Program Office), J. S. Levine (NASA Langley Research Center, Hampton, Virginia), J. E. Penner (Lawrence Livermore National Laboratory, Livermore, California), R. G. Prinn (Chair, International Global Atmospheric Chemistry [IGAC] Project, Massachusetts Institute of Technology, Cambridge, Massachusetts), and F. S. Rowland (University of California, Irvine, California). The symposium audience consisted of science teachers and educators from all over Virginia. Public Television local affiliate, WHRO/TV in Norfolk, Virginia, video-taped the entire public symposium and, shortly thereafter, produced a one-hour program entitled "Biomass Burning and Global Change." This program was telecast across the United States on the Public Television Network on 25 April 1995, to coincide with the twenty-fifth annual celebration of Earth Day.

Figure I.5 Burning of Kuwaiti oil wells in July 1991 (Photograph courtesy Wesley R. Cofer).

Acknowledgments

It is a pleasure to acknowledge the assistance and support provided by members of the Conference Program Committee and session chairs: M. O. Andreae, D. R. Cahoon, J. S. Clark, W. R. Cofer, P. J. Crutzen, R. A. Delmas, J. Fishman, M. Garstang, J. G. Goldammer, P. V. Hobbs, J. P. Malingreau, J. E. Penner, R. G. Prinn, F. S. Rowland, A. W. Setzer, B. J. Stocks, A. M. Thompson, W. S. Trollope, D. E. Ward, and R. G. Zepp. Special thanks to R. J. Curran, SCAR-C Program Manager, NASA Mission to Planet Earth, and to J. L. McElroy, Global Change Research Program, U.S. Environmental Protection Agency, for their support of this publication.

Conclusion

Once again, it seems most appropriate to conclude the introduction to this volume with a quote from Stephen J. Pyne, fire historian at Arizona State University:

We are uniquely fire creatures on a uniquely fire planet, and through fire the destiny of humans has bound itself to the destiny of the planet.

References

Andreae, M. O. 1995. Climatic Effects of Changing Atmospheric Aerosol Levels. *World Survey of Climatology, Volume 16: Future Climates of the World* (A. Henderson-Sellers, Ed.), Elsevier, Amsterdam, Holland, pp. 341–392.

Cahoon, D. R., B. J. Stocks, J. S. Levine, W. R. Cofer, and K. P. O'Neill, 1992. Seasonal distribution of African savanna fires. *Nature*, 359, 812–815.

Cahoon, D. R., B. J. Stocks, J. S. Levine, W. R. Cofer, and J. M. Pierson, 1994. Satellite analysis of the severe 1987 forest fires in Northern China and southeastern Siberia. *Journal of Geophysical Research*, 99, 18,627–18,638.

Chin, M., and D. D. Davis, 1993. Global sources and sinks of OCS and CS_2 and their distributions. *Global Biogeochemical Cycles*, 7, 321–337.

Ehhalt, D. H., J. Rudolph, and U. Schmidt, 1986. On the importance of light hydrocarbons in multiphase atmospheric systems. In *Chemistry of Multiphase Atmospheric Systems* (W. Jaeschke, Ed.), Springer-Verlag Berlin/Heidelberg, pp. 321–350.

Houghton, J. T., L. G. Meira Filho, J. Bruce, H. Lee, B. A. Callandar, E. Haites, N. Harris, and K. Maskell, 1995. *Climate Change 1994: Radiative Forcing of Climate Change.* Cambridge University Press, Cambridge, 339 pages.

Levine, J. S. (Ed.), 1985. *The Photochemistry of Atmospheres: Earth, the Other Planets and Comets.* Academic Press, San Diego, California, 500 pages.

Levine, J. S. (Ed.), 1991. *Biomass Burning: Atmospheric, Climatic, and Biospheric Implications.* The MIT Press, Cambridge, Mass., 569 pages.

Levine, J. S., W. R. Cofer, D. R. Cahoon, and E. L. Winstead, 1995. Biomass burning: A driver for global change. *Environmental Science and Technology*, 29, 120A–125A.

Mano, S., and M. O. Andreae, 1994. Emission of methyl bromide from biomass burning. *Science*, 263, 1255–1257.

National Research Council. 1989. *Global Change and Our Common Future.* National Academy Press, Washington, D.C.

Schlesinger, W. H., and A. E. Hartley, 1992. A global budget for atmospheric NH_3. *Biogeochemistry*, 15, 191–211.

Seiler, W., and P. J. Crutzen, 1980. Estimates of gross and net fluxes of carbon between the biosphere and atmosphere from biomass burning. *Climate Change*, 2, 207–247.

Stocks, B. J., 1998. The extent and impact of forest fires in northern circumpolar countries. In *Global Biomass Burning: Atmospheric, Climatic, and Biospheric Implications* (J. S. Levine, Ed.), The MIT Press, Cambridge, Mass., pp. 197–202.

Warneck, P., 1988. *Chemistry of the Natural Atmosphere.* Academic Press, San Diego, 757 pages.

Wayne, R. P., 1991. *Chemistry of Atmospheres* (Second Edition). Oxford University Press, Oxford, 160–164.

WMO/UNEP, 1995. *Scientific Assessment of Ozone Depletion: 1994.* World Meteorological Organization Global Ozone Research and Monitoring Project Report No. 37, Geneva, Switzerland.

I

Remote Sensing: Global and Regional Scales

Mission to Planet Earth: Land-Cover/Land-Use Change Program

Anthony C. Janetos, Christopher O. Justice, and Robert C. Harriss

The ultimate vision is to develop the capability to perform repeated global inventories of land use and land cover from space, and to develop the scientific understanding and models necessary to evaluate the consequences of observed changes.

Accurately accounting for land-use and land-cover change with fine spatial and temporal resolution, as well as the underlying research to interpret it, will require a partnership of many scientific and natural resource management institutions around the world (Justice 1992). Although some of the measurements necessary can be made from space, the scientific knowledge to interpret and evaluate these observations cannot be obtained simply through remote sensing, and thus demands a partnership with in situ research. Remote-sensing technologies may also become part of the "green technology" revolution, which enables a more efficient use of natural resources through high-precision management of agriculture, forestry, and urban development. The primary role of the space agencies will be to develop methods to extend local and regional understanding to global scales. Space agencies around the world will need to coordinate efforts to help fulfill this role, which has its scientific underpinning articulated by the Global Climate Observing System (GCOS), the International Geosphere-Biosphere Programme (IGBP), and the World Climate Research Program (WCRP), and which could be implemented by the Committee on Earth Observation System (CEOS) agencies. NASA's role will be to develop methods and techniques, and to sponsor research that demonstrates the consequences of land-cover and land-use change, to establish methods to quantify the consequences and to develop the capabilities to explore alternative land-use and monitoring strategies. This research will provide scientific understanding and tools for future operational and commercial interests in land remote sensing.

Background

The underlying philosophy of the MTPE Land-Cover/Land-Use Change Program is to further the understanding of the consequences of land-use and land-cover change for continued provision of ecological goods and services.

Ecological goods and services refer to the suite of marketable materials produced in ecological systems, through either cultivation or harvesting of wild resources (e.g., rubber tapping), as well as the ecosystem-level processes of importance to human activities such as purification of water by wetlands and regulation of water flow by forested watersheds. The preservation of ecological goods and services depends on sustainable management of resources, and recognizes three facts:

1. Ecological processes controlling biogeochemical cycles and hydrologic processes result in both goods (i.e., food, fiber, etc.) and services (e.g., water purification, preservation of soil fertility);

2. Human influences that both transform land cover from one type to another and that change or intensify the management regime on lands clearly affect provision of goods and services regionally, and have the potential to affect them globally; and therefore

3. It is necessary to understand the human influence on these processes to determine the potential for continuing provision of resources and services for expanding human populations.

As with the other major processes of land-cover change, the extent and impacts of biomass burning at regional and global scales are only now being quantified through a combination of ground-based and airborne measurement campaigns and systematic satellite monitoring. These regional and global monitoring programs are starting to be coordinated and organized effectively through interagency collaboration within the United States and internationally through the IGBP. The scientific community recognizes that realistic models of the process and reliable estimates of the products of biomass burning are needed for comprehensive modeling of the earth's system. In addition, quantifying biomass burning and the associated emissions are a focus for the scientific assessments of the International Panel on Climate Change (IPCC) and are part of the land-use assessment for the national emission inventories of greenhouse gases currently being implemented by the signatory countries.

Biomass burning in many areas of the world is anthropogenic and was probably one of the first ways by which humans were able to alter large areas of the land surface. As with the other processes of land-cover and land-use change, understanding the human dimension is therefore an integral part of the research that is needed. Interagency collaboration will be needed to ensure a holistic approach and develop this relatively poorly understood aspect of this program

The National Aeronautics and Space Administration (NASA) recognizes the importance of coupling the new research efforts in land-use and land-cover change and existing research programs, platforms, and instruments in order to take best advantage of existing investments such as Asrar and Dozier (1994) and Sellers and Schimel (1993). At the same time, it is important to lay the plans for new campaigns and research topics, while exploring the opportunities for the development of new technologies and measurement capabilities. It will also be of critical importance to develop and show the links to operational classification and mapping programs within the US government agencies, and in the scientific and operational agencies around the world.

Program Elements

There are four underlying research dimensions to the NASA MTPE Land-Cover/Land-Use Change Program. A short description follows of the emphasis that NASA's program will have in each dimension.

Forcing Factors

These are the factors that drive changes in landscapes and the resultant impacts on the biogeochemical and hydrological cycles and energy and gas fluxes (figure 1.1). They can be broadly separated into two categories, which in reality may interact.

Climatic and Ecological Drivers Short- and long-term variability in weather, climate, and internal ecosystem dynamics, including, for example, variations in herbivore populations, drive some aspects of land-cover change and land use on decadal and multidecadal time scales (Graetz 1994) (figure 1.2). Extreme short-term climate conditions can act as a catalyst for land-cover change; for example, successive years of drought can change ecosystem composition as well as land use (Tucker et al. 1991; Le Houerou 1992). These drivers

must be taken into account in any attempt to understand current patterns of land cover and land use, and also must be taken into account in any attempts at projection. *The Land-Cover/Land-Use Change Program will rely primarily on other components of MTPE and national and international science agencies for the development of historical and climatic data sets.*

Socioeconomic Drivers This category includes the economic, social, and political factors that fuel the human activities responsible both for conversion of land cover and for intensification of management regimes (figure 1.3). Population dynamics and economic activity are clearly major factors in determining the distribution and intensity of land-cover change (Vitousek 1994; Sage 1994; Allen and Barnes 1985). Desires and pressures around the world for economic development, and the needs and options for increased food production and delivery must be quantified, understood, and ultimately modeled (Myer and Turner 1992; Wolman and Fournier 1987). Any attempt to understand current patterns of land cover and land use must explicitly take into account land-use history (Flint and Richards 1991). The MTPE program will be linked to other research on human dimensions of land-use change both within the United States and abroad (Stern et al. 1992). This coupling will enable a more complete understanding of human and physical factors in determining the extent, rate, and consequences of land-cover and land-use change. *The NASA program will investigate the human dimension processes directly when they can be coupled to observed recent changes in the landscape or regional predictive models.*

Responses and Consequences

Over the next several decades, the stresses of population pressure and economic development will occur largely in the developing world in the tropics and subtropics. Thus, NASA proposes to make the initial thrust of its effort in those parts of the world that are currently undergoing the most stress, where the stresses from human activities are sure to increase the most rapidly, and where major changes are already taking place. Two broad categories of response to the drivers of change are of particular importance to the NASA program.

Land-Cover Conversion *The primary NASA interest is to identify the current distribution of land-cover types, and to track their conversion to other types. Measurement of the rates of rapid conversion of forest cover to other types, as is occurring in the humid tropics, is of*

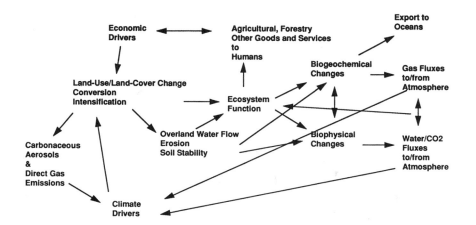

Figure 1.1 Interactions of land-cover/land-use change (Source: NASA)

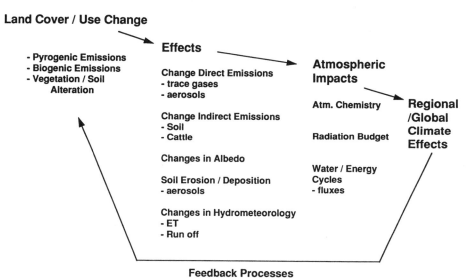

Figure 1.2 Land-cover change/climate interaction (Source: NASA)

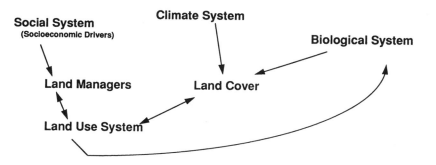

Figure 1.3 Land-cover/land-use human dimension interaction (Source: NASA)

particular interest to this program because of the links to trace gas emissions and sustainable development.

Data on the distribution and rates of change can be used in the development and verification of biogeochemical and biophysical models, and also in the analysis of the effect of spatial patterns and history of conversion on ecosystem processes and structure, on the effects on biodiversity at a variety of scales of analysis, and on the effects on the continued delivery of ecological services on local-landscape-regional scales (Skole 1994; Houghton 1991). These data will also be critical to identifying the types of land cover in which in situ process and ecosystem studies need to be performed in order to parameterize landscape-scale (and larger) ecosystem models, and to understand the current patterns of biogeochemical and biophysical functioning on the landscape (Vitousek and Matson 1992; Melilo 1994; Riebsame et al. 1994). Data on land cover and land-cover change are also needed to parameterize global climate models (Defries et al. 1995; Henderson Sellers 1992). Links to patterns and changes in tropospheric chemistry (Scholes et al. 1995) and possible changes in large-scale hydrometeorology can also be explored in this arena (Rogers 1994).

Land-Use Intensification *The primary interest of the NASA program will initially be in understanding the consequences of intensified management of agricultural and agroforestry and grazing systems, particularly in the tropics and subtropics, and being able to measure the longer-term, in situ degradation of forested ecosystems that occurs for example through imprudent forest management.*

These data can help understand the consequences of managing the end-states of land-cover change. Detecting in situ degradation of forest resources is difficult (Houghton and Hackler 1994). Current research in the Landsat Pathfinder suggests that while Landsat can be used to detect the presence of intensive logging,

additional methodological and technological research may be necessary to arrive at consistent and replicable regional estimates of long-term degradation. Similarly, detection of phenomena such as long-term change in the frequency of fire in forests or savannas will require both good historical data and also improvements in satellite measurement techniques (Justice et al. 1993).

There is the potential for increased emphasis on forestry and agriculture in temperate and boreal systems, as the implications of the Climate and Biodiversity Conventions for forests and agriculture become better understood by individual countries. There are strong opportunities here for the NASA program to develop interactions with a variety of national case studies, as well as with other U.S. programs in the land management agencies. These links will be especially important in building up the history of land management in particular case studies. The primary consideration of the location of the in situ research will be to have sites representative of the major land-cover types identified in the initial studies of land-cover conversion, as well as those areas undergoing intensification of management activities.

Modeling and Implications
We anticipate that an important contribution of the NASA MTPE Land-Cover/Land-Use Change Program will be the development of techniques to incorporate land-cover and land-use change into existing biogeochemical and biophysical models.

It will be important to be able to develop, parameterize, and evaluate models that are able to couple the biogeochemical and biophysical dynamics of the land surface and its interactions with the atmosphere (figure 1.4). The new emphasis from the perspective of this initiative will be the development of data sets and techniques to enable the models to use better representations of the actual land cover present, rather than

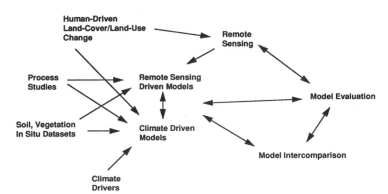

Figure 1.4 Links to Earth system modeling (Source: NASA)

potential natural vegetation, and to be able to represent changes in land cover, whether from the internal dynamics of the ecosystems, or from human-driven perturbations. Links to other process studies in tropospheric chemistry, aerosol radiative forcing, and hydrometeorology will need to be explored in order to gain a better understanding of the relationships among changes in land cover, the processes that drive those changes, and biogeochemical and physical changes in Earth's atmosphere and climate system.

Ultimately, it will be the ability to model systems undergoing land-use change that will provide tools for both scientists and decision makers to evaluate the potential consequences of different management practices, and to assess the consequences of policies that affect land-cover conversion. There will be close ties of the MTPE Land-Cover/Land-Use Change Program to the existing and developing efforts in Earth System Modeling, sponsored through MTPE Science and the NASA EOS Interdisciplinary Science teams. The project will also link to the modeling activities of the IGBP in the Global Analysis, Interpretation, and Modeling project (GAIM), the Global Terrestrial Ecosystem Core Project, and in particular to the GCTE Long-term Ecosystem Modeling Activity (LEMA) (IGBP 1992a).

Techniques and Methods

A part of this initiative will be devoted to developing, refining, and implementing techniques and methods in support of the science objectives of this program.

Techniques and methods for applying satellite data to address questions of land-cover and land-use change are currently being developed in various parts of the current NASA programs (Laporte et al. 1995; Running et al. 1994; Dobson et al. 1995). This new initiative will provide a focus for these activities. The use of high-resolution satellite data for quantifying land-cover change is by no means new; however, recent projects within the NASA Landsat Pathfinder Program have developed and applied a core of workable methods for regional tracking of land-cover conversion (Skole and Tucker 1993). This approach makes use of the global archives of high-resolution satellite data acquired over the last 20 years. Coarse resolution satellite time series data are also being used to develop improved land-cover classifications suitable for use in global and regional models (IGBP 1992c; Defries and Townshend 1994). More conventional multitemporal classification procedures are being augmented by improved decision-tree and neural network techniques (Heerman and Khazenie 1992; Michaelson et al. 1994;

Defries et al. 1995). Similarly, new techniques are being developed for automated change detection, for use on regional data sets (Lambin and Strahler 1994). Multiresolution satellite data methods are also being developed. For example, AVHRR and Landsat data are being combined to drive biomass burning trace gas and particulate emissions models for southern Africa (Scholes et al. 1995). One area in which methodological progress is urgently needed is the combination of physical and socioeconomic data in process and predictive modeling of land-cover change at a regional scale.

One of the technical and methodological challenges facing the remote-sensing research community is to decouple the diurnal, seasonal, and cyclical changes in the landscape from land-cover conversion. Similarly, experimental methods for land-surface characterization applied to local study sites need to be tested and applied in a more operational and regional context for model parameterization. A close coordination will be sought between the land-cover and land-use change initiative and the algorithm development and testing, which is being supported by the remote-sensing science component of the NASA Ecological Processes and Modeling Program.

In addition to utilizing data from the current and historic sensors, over the next several years, the contributions of TRMM, EOS AM-1, and Landsat-7 will become critical to the continued success of the proposed research program. These missions will be expected to provide quantitative, space-based data on land-surface characteristics, climatic regimes, and underlying vegetation dynamics around the world (Running et al. 1994; Irons et al. 1995). The MTPE Land-Cover/Land-Use Change Program will encourage the development of new algorithms pertinent for land-cover monitoring using data from these planned sensing systems. The acquisition of new digital elevation data for use by EOS AM-1 sensors, and the continued improvement of and access to digital elevation model (EDM) data around the world will also be necessary parts of the NASA program (Justice et al. 1995).

There are several opportunities available for infusing new technological development and exploration into the study of land-use change. An obvious candidate is the exploration of the use of high spectral resolution remote sensing to understand two aspects of dynamic land surfaces: their accurate classification into different cover types, and possibly the chemical composition of their plant canopies, including their photosynthetic pigment complements and nitrogen status.

There are other new technological capabilities to be explored, both in terms of potential new measurements and in terms of new capabilities to provide data. Ongoing projects in the Ecological Processes and Modeling Program that investigate the utility of synthetic aperture radar (SAR) for categorizing different land-cover types and for deriving biophysical parameters from landscapes will continue to be pursued, but will be refocused on issues of particular concern for understanding the consequences of land-use change. New technological capabilities for acquiring, managing, processing, disseminating, and electronically archiving data and information also should be investigated jointly with elements of the Earth Observing System Data and Information System (EOSDIS) (e.g., Pathfinders) in order to evaluate and develop the ability to get important data in the hands of both scientists and land managers quickly and easily.

Implementation

It is prudent to build on the investments that MTPE has made over the last several years. These investments have focused primarily on regions in which rapid change is currently happening, and which have implications for global changes in biogeochemistry and climate. The following targeted regions will form the initial focus of case studies that combine remote sensing and in situ methods (figure 1.5).

The Americas

Several process-based in situ studies will be initiated in conjunction with the planned international Large-Scale Biosphere-Atmosphere Campaign (LBA) and Inter-American Institute (IAI) activities. The most impor- tant initial campaign will be LBA, currently being planned by teams of scientists in Brazil, the United States, and Europe (Kirchoff 1994; Wofsy et al. 1994). The LBA activities will span the range of interests from large-scale hydrometeorology, to mesoscale atmospheric phenomena, to transect-based studies of ecological and atmospheric biogeochemistry. Its underlying rationale is to understand the functioning of the Amazon Basin as an integrated system, with a particular focus on understanding how the rapid land-use change of the past few decades has affected it. MTPE Land-Cover/Land-Use Change studies will focus primarily on the possibilities for and consequences of long-term management of altered landscapes in the Amazon Basin. A question of particular interest, for example, would be to investigate the possibility of managing the newly created matrix of disturbed and relatively undisturbed habitat to sustain productivity and soil fertility. Investigations will not be restricted to forests and pastures, but will also be encouraged to incorporate cerrado and agricultural areas.

Regional studies in Mexico and the United States will be encouraged to focus on the interactions of agriculture, water resources, and climate variability. Land-cover and land-use studies in North America will be closely coordinated with other U.S. agencies and nations through the U.S. Global Change Research Program and appropriate international organizations.

Southeast Asia

A base of scientific investigations must be built within Asian and Southeast Asian countries that will assess the consequences of forest conversion to agriculture and the long-term, in situ degradation of the forested landscape

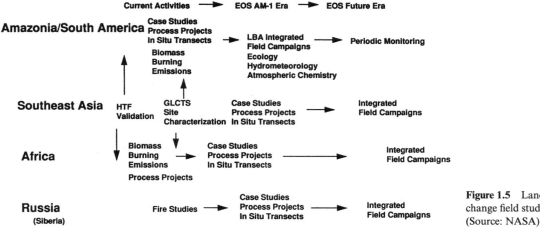

Figure 1.5 Land-cover/land-use change field studies and campaigns (Source: NASA)

and *agricultural intensification.* In addition, projects should be encouraged to investigate the interaction of agricultural areas, natural areas, and managed forests. These would include projects that investigate the capability of remote sensing to map changes, and to parameterize landscape-scale ecosystem models. The IGBP, LUCC (Land-Use and Land-Cover Change), and START (System for Training and Research) have developed a coordinated program within the region that will provide an important forum for international science and collaboration (IGBP 1992b). In particular the South East Asia Regional Committee for START (SARCS) has a strong land-use change component in its research program. One could anticipate that integrated campaigns will be planned for Southeast Asia, possibly focusing on areas with landscapes affected by selected logging, intensified agriculture, and urban expansion.

Southern and Central Africa

The land transformations currently going on in the dense humid forests, the seasonal woodlands, and savanna systems of Africa are of considerable interest to this program. A regional initiative as part of the NASA program would require initial planning efforts and a focused research investment. Both the global change research community and the natural resource management community are concerned with various aspects of land-cover and land-use change in this region. Large areas of land are currently undergoing transformations due to the pressure for agricultural land. The initial activity in this region will be to synthesize results from previous international, European, and South African sponsored campaigns in which NASA has been a participant, in order to understand which activities can be enhanced. This regional activity will be developed and implemented in conjunction with IGBP, LUCC, START, and the host countries (Andreae et al. 1994; Justice et al. 1995, IGBP 1994). Two new IGBP activities relevant to this program are getting underway in this region under the Kalahari Transect and the Miombo initiatives (Justice et al. 1994; IGBP 1995). A small number of studies currently under way in the NASA Ecological Processes and Modeling Program have been investigating processes along transects in this region and may be viewed as planning activities for the future.

Russia and the Countries of the Former Soviet Union

The primary MTPE efforts will be to take advantage of data sets and studies that have been developed within the context of the Earth Sciences Joint Working Group, and to coordinate activities with those of other agencies with particular interests in the boreal forest. There are at least three potential areas of interest: (1) to understand whether frequencies of major fires in the boreal region have increased over the past several decades, which has been hypothesized; (2) to be ready to evaluate the consequences for the global carbon cycle if large tracts of boreal forest are logged for timber as a result of recent concessions; (3) to conduct careful case studies of agricultural and forested regions, concentrating on the effects of past management practices; this may become possible because previously classified data sets are now becoming available.

Future Planning and Comprehensive Landscape Studies

In addition to the regional focus activities, a small part of the program resources will be devoted to conducting workshops and specialized, targeted analyses to coordinate appropriate activities within MTPE, among US agencies, and integrate the MTPE efforts with those of the IGBP and WCRP.

The four regions identified above present a range of land-use change scenarios from rapid conversion to long-term degradation. As the program develops, synthesis workshops will be held to provide comparisons among the different regional cases, providing insight into a spectrum of impacts and management questions. In addition to studies in the four mentioned regions, the initial program will undertake a series of workshops to examine the impacts of land-use changes and agricultural intensification in the United States; and incorporation of satellite-derived land-use change information into regional scale hydrological models will be investigated. Specific integrated studies to investigate the links between remotely sensed data and biogeochemical/biophysical processes will be planned in order to develop and parameterize models on landscape scales. Models on these scales are likely to be of great importance in developing the confidence in parameterizations for models that examine changes on regional and global scales, and/or the transient response of landscapes over long time periods. The implementation of GCIP in the Mississippi Basin of the United States provides one promising opportunity for such studies. MTPE will be especially interested in relationships with research programs that can provide a historical perspective on the importance of land-use history in explaining current patterns of ecosystem functioning in the United States.

A small number of studies will be supported to address requirements for future satellite missions, with

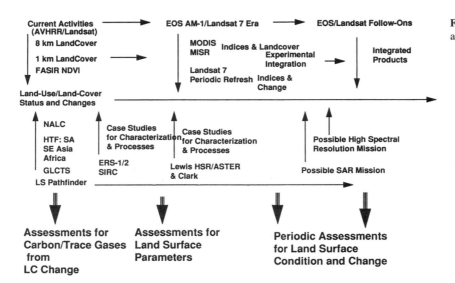

Figure 1.6 Satellite studies and assessments (Source: NASA)

outgrowths of existing technologies and with new technologies. A specific effort will be initiated to scope the scientific and operational effort in land-use change necessary to take full advantage of the expected data streams from TRMM, Landsat-7, EOS AM-1, and other appropriate international sensing systems (figure 1.6). Coordination will be sought with the various science working groups associated with these instruments to ensure that the observational and data requirements of this program will be met.

Links to NASA/NOAA Pathfinder and Other Data-Development Efforts within MTPE

The NASA EOS program has embarked on a series of Pathfinder projects aimed at making time series data from the current suite of Earth observation sensors available to the global change research community and preparing for the data-management challenges of EOS (Maiden and Greco 1994; James and Kalluri 1994). The links to the Pathfinder and other data-development efforts within MTPE are extremely important. Because of the proposed tropical/subtropical initial focus, the Landsat Pathfinder Humid Tropical Forest activity provides the major foundation of data on land-cover conversion for both rapid and slower, in situ changes. Thus, the science supported through the Land-Use Change Program builds on the observations and data sets provided and archived through Pathfinder. Funds set aside in EOSDIS for the purchase of commercial Thematic Mapper and SPOT data (and possibly data from other systems, such as IRS) will have the Land-Use Program as one of the major

clients, with the Pathfinders acting as intermediaries. Other NASA supported Pathfinder efforts, such as the GAC AVHRR Pathfinder, the IGBP Global 1 km project and the NASA IGBP Global Land-cover Mapping project, will also play important roles in global land-cover mapping and characterization (Townshend 1995). These latter activities are either included in or closely linked to the EOSDIS Version 0 activities at the Distributed Active Archive Centers (DAACs). Linkages will be established with EOSDIS concerning the availability and refinement of these data sets and their possible augmentation to meet the needs of the land-cover and land-use program.

Increased attention from this program will need to be given to the state of the global high-resolution satellite data archives, which are in various stages of maintenance. The Landsat Ground Station Operators Working Group (LGSOWG) may be the appropriate body with which the program can work towards the long-term security of these archives and the development of a coordinated acquisition strategy prior to the launch of Landsat-7.

Links among U.S. Agencies

At a time when agencies are reevaluating and reducing the scope of their programs and their external commitments, the coordination necessary for this program will be a particular challenge. Nevertheless, the benefits of interagency collaboration are obvious, and particular attention will be needed to secure the necessary mechanisms for effective collaboration. Interests of other agencies range from EPA's concerns about being able to account for changes in carbon flux on a country-by-country basis to concerns in the ecological research agencies for being able to map the occurrence of different land-cover types on Earth's surface and being able to track changes in their extent, in order to understand patterns of biological diversity. These discussions will be broadened to include the practical concerns of the land management agencies of being able to standardize a system of vegetation classification. The NASA program can play a major role in understanding the relationship between changes in broad aspects of landscape structure and the ways in which these landscapes function. We envision providing a link between the largely remote-sensing–based scientific community supported by NASA and the land managers of a variety of other federal agencies through the collaborations produced in the CENR process.

Outside of the scientific arena, MTPE is forming strong ties to U.S. agencies active in international development. Discussions with US AID (United States Agency for International Development) on a variety of projects have already begun, and there is some potential for stronger interactions in fields such as sustainable agriculture, agroforestry, and sustainable forest management in the developing world. For example, the new Central African Regional Program for the Environment (CARPE) being launched by US AID will provide focus for interagency collaboration on land-use change and forest management activities within that region.

Another area of potentially strong interaction is with the U.S. Country studies program of DOE (Department of Energy), EPA, and US AID, which helps developing countries develop and apply methodologies for calculating national greenhouse gas emissions, and studying climate change impacts, and possible mitigation strategies. Each of these topics requires both predictive ecosystem-level models and accurate assessments of land-cover and potential land-cover change.

Links to International Activities

Links to the international science community will come largely through the activities of the IGBP and WCRP projects that focus either directly on land-use change, or on the processes that are affected as a result. The IGBP in conjunction with CEOS is already working on a pilot project to understand the technical consequences of combining data from the suite of multispectral, high spatial resolution remote sensing instruments that currently are in orbit: SPOT, Landsat MSS, and TM, IRS. In addition, IGBP-DIS is currently embarked on developing the technical and logistic requirements for acquiring and processing a global, high spatial resolution land-cover data base, analogous to the 1-km AVHRR project. IGBP-DIS is also engaged in the development and validation of an improved global land-cover data base at 1-km resolution using the AVHRR data base.

The new joint HDP/IGBP Core Project on Land-Use and Cover Change (LUCC) is clearly of central importance to this program (Turner et al. 1993), as are some of the activities of the other Core Projects such as GCTE (Global Change and Terrestrial Ecology), IGAC (International Global Atmospheric Chemistry), BAHC (Biological Aspects of the Hydrologic Cycle) and possibly LOICZ (Land-Ocean Interactions in the Coastal Zone). For example the IGAC Biomass Burning Experiment (BIBEX), and the newly proposed IGBP Joint Core Project Transect Activities, will certainly be undertaking research of direct relevance to this program. The Framework Projects, GAIM, and IGBP-DIS are also implementing projects of direct relevance to this program such as the global wetlands project, the global soils data base, and the global fire data base. Similarly, there are component projects of the World Climate Research Program (WCRP) such as the Global Energy and Water Cycle Experiment GEWEX, in particular the GEWEX Continental Scale International Project (GCIP), and possibly the Global Ocean Atmosphere Land System (GOALS) activities that are relevant to this program.

The working relationship between the NASA Landsat Pathfinder and the ECE TREES Project provides a good example of how international programs can be combined to provide improved global land-cover assessments. Similarly, the NASA Landsat Pathfinder program provides a complementary activity to the UN/FAO Tropical Forest Assessment and AFRICOVER programs. The Alternatives to Slash and Burn (ASB) Project being undertaken by the Consultative Group of Institutes Agricultural Research (CGIAR) is currently seeking collaboration with the Landsat Pathfinder Program. Through these relationships links have been established with forest services and research groups in tropical countries. These kinds of mutually beneficial collaborations will be encouraged under the NASA land-use/land-cover change activity.

As in the United States, links must be established with the international development community. There has already been some success within the Landsat Pathfinder in developing connections with the World Bank's Global Environmental Facility (GEF). Similarly, the GEF is supporting the SARCS activity in Southeast Asia. Preliminary discussions with the UNDP (United Nations Development Programme) have also been initiated through NASA Headquarters.

Reporting

The Land-Cover/Land-Use Change Program will develop a series of reports that identify the consequences of land-use change in globally important areas of rapid change, and these reports will provide the scientific foundation for a new NASA series of reports on human influences on the biosphere.

These reports (figure 1.6) will go beyond the expected high number of scientific refereed papers from this program and will be targeted at high-level decision makers and the public. The reports will be along the lines of integrated assessments closely tying the science to improved resource and environmental management. The outline for this new series of reports will be developed as a high priority over the next year, along with a detailed plan for how information from NASA and other programs will be made available and used.

Conclusions

This chapter outlines a new initiative on land-cover and land-use change that provides an important research focus for applying existing satellite technologies and refining the plans for proposed sensing systems. The new MTPE Land-Cover/Land-Use Change Program is designed to address pressing questions associated with improved management of Earth's resources. Research undertaken by this program will serve the needs of both the global change and natural resource management communities. The program is shown as an integrating theme among various existing components of Mission to Planet Earth and to have a strong synergy with several research and development activities of other national agencies and international programs. In its implementation it will consist of a combination of remote sensing and in situ studies focused on tropical and subtropical regions. The results of the individual research projects from this program will be synthesized into a series of reports aimed at policy and resource management communities.

References

Andreae M. O., Fishman J., Garstang M., Goldammer J. G., Justice C. O., Levine J. S., Scholes R. J., Stocks B. J., and Thompson A. M. 1994. Biomass burning in the global environment: First results from the IGAC/BIBEX Field Campaign STARE/TRACE-A/SAFARI-

92. In Prinn R. (Ed). *Global Atmospheric-Biospheric Chemistry.* Plenum Press, New York, 83–101.

Asrar G., and Dozier J. 1994. EOS: Science strategy for the Earth Observing System. NASA, IBN 1-56396-198-9.

Defries R. S., and Townshend J. R. G. 1994. NDVI-Derived land-cover classifications at a global scale. *International Journal of Remote Sensing* 20(17):3567–3586.

DeFries R., Hansen M., and Townshend J. In press. Global discrimination of land-cover types from metrics derived from AVHRR Pathfinder data, *Remote Sensing of the Environment.*

DeFries R. S., Field C. B., Inez F., Justice C. O., Matson P. A., Matthews M., Mooney H. A., Potter C. S., Prentice K., Sellers P. J., Townshend J., Tucker C. J., Ustin S. L., and Vitousek, P. M. In press. Mapping the land surface for global atmosphere-biosphere models: Toward continuous distributions of vegetation's functional properties, *Journal of Geophysical Research-Atmospheres.*

Dobson M. C., Ulaby F. T., and Pierce L. E. 1995. Land-cover classification of terrain attributes using synthetic aperture radar. *Remote Sensing of the Environment* 51:19–214.

Graetz D. 1994. Grasslands. In Meyer B. L. II, and Turner W. B. (Eds). *Changes in Land-Use and Land-Cover: A Global Perspective.* Papers arising from the 1991 OEIS Global Change Institute, Cambridge University Press, 125–147.

Heerman P. D., and Khazenie, N. 1992. Classification of multi-spectral remote sensing data using a back-propagation neural network. *IEEE Transactions on Geoscience and Remote Sensing* 30:81–88.

Henderson Sellers A. 1992. PILPS (Project for Intercomparison Land Surface Parameterization Schemes) Workshop Report and Earth Science Plan, International GEWEX Project Office Publication Series #5, Washington, D.C.

Houghton R. A. 1991. Tropical deforestation and atmospheric carbon dioxide. *Climatic Change* 19:99–118.

Houghton R. A., and Hackler J. L. 1994. The net flux of carbon from deforestation and degradation in South and Southeast Asia. In Dale V. H. (Ed). *Effects of Land-Use Change on Atmospheric CO_2 Concentrations: South and Southeast Asia as a Case Study.* Springer-Verlag, New York, 301–327.

IGBP. 1992a. Global Change and Terrestrial Ecosystems: The Operational Plan. *IGBP Report* 21. IGBP Secretariat, Stockholm, 95 pp.

IGBP. 1992b. Report from the START Regional Meeting in Southeast Asia. *IGBP Report 22.* IGBP Secretariat, Stockholm.

IGBP. 1992c. Improved global data for land applications: A proposal for a new high-resolution data set. J. R. G. Townshend (Ed). *IGBP Report 20.* IGBP Secretariat, Stockholm.

IGBP. 1994. Africa and Global Change. Report from a meeting in Niamey, Niger, 23–27 November 1992. *IGBP Report 29.* IGBP Secretariat, Stockholm.

IGBP. 1995. The IGBP Terrestrial Transects. *IGBP Report* 34. IGBP Secretariat, Stockholm.

Irons J. R., Williams D. L., and Markham B. L. 1995. Landsat-7 ETM+ on-orbit calibration and data quality assessment. *Proc. 15th Annual International Geoscience and Remote Sens. Symp.* IGARSS'95. Firenze, Italy, 10–14 July 1995:1573–1575.

James M. E., and Kalluri S. N. V. 1994. The Pathfinder AVHRR land data set: An improved coarse resolution data set for terrestrial monitoring. *International Journal of Remote Sensing* 15(17):3347–3364.

Justice C. O. 1992. Satellite monitoring of tropical forests: A commentary on current status and institutional roles. *Proceedings of the International Space Year, World Forest Watch Conference*, São Jose Dos Campos, Brazil, CEC-EUR 14561 EN.

Justice C. O., Malingreau J. P., and Setzer A. 1993. Satellite remote sensing of fires: Potential and limitation. In Crutzen P. and J. Goldammer (Eds). *Fire in the Environment; Its Ecological, Climatic, and Atmospheric Chemical Importance*. John Wiley and Sons, Chichester.

Justice C. O., Scholes R., and Frost P. 1994. African savannas and the global atmosphere research agenda 1994–1998. (Eds). Proceedings of a joint IGBP START/IGAC/GCTE/GAIM/DIS workshop on African savannas, land-use, and global change: Interactions of climate productivity and emissions, Victoria Falls, Zimbabwe. *IGBP Report* 31. IGBP Secretariat, Stockholm.

Justice C. O., Bailey G. B., Maiden M. E., Rasool S. I., Strebel D. E., and Tarpley J. D. 1995. Recent data and information system initiatives for remotely sensed measurements of the land surface. *Remote Sensing of the Environment* 51:235–244.

Justice C. O., Kendall J. D., Dowty P. R., and Scholes R. J. In press. Satellite remote sensing of fires during the SAFARI campaign using NOAA-AVHRR data. *Journal of Geophysical Research*.

Kirchoff V. W. J. H. 1994. TAHBIS—a tropical atmosphere-hydrosphere-biosphere-integrated study in the Amazon. *Brazilian Journal of Geophysics (Revista Brasiliera de Geofísica)* 12(1):3–8.

Lambin E. F., and Strahler A. H. 1994. Change vector analysis in multitemporal space: A tool to detect and categorize land-cover change processes using high temporal resolution satellite data. *Remote Sensing of the Environment* 48:231–244.

Laporte N., Justice C. O., and Kendall J. 1995. Mapping the humid forests of Central Africa using NOAA-AVHRR data. *International Journal of Remote Sensing* 16(6):1127–1145.

Le Houerou, H. N. 1992. Climate change and desertification. *Journal of Rangeland Management* 42:183–201.

Maiden M. E., and Greco S. 1994. NASA's Pathfinder data set programme: Land surface parameters. *International Journal of Remote Sensing* 15(17):3347–3364.

Melilo J. M. 1994. Modelling land-atmosphere interactions. In Meyer and Turner (Eds). *Changes in Land-use and Land-cover: A Global Perspective*. Papers arising from the 1991 OEIS Global Change Institute, Cambridge University Press, 387–409.

Meyer W. B., and Turner B. L. II. 1992. Human population growth and global land-use/cover change. *Annu. Rev. Ecol. Syst.* 23:39–61.

Michaelsen, J., Schmiel, D. S., Friedl, M. A., Davis, F. W., and Dubayah, R. C. 1994. Regression tree analysis of satellite and terrain data to guide vegetation sampling and surveys. *Journal of Vegetation Science* 5:673–686.

Riebsame W. E., Parton W. J., Galvin K. A., Burke I. C., Bohren L., Young R., and Knop E. 1994. Integrated modeling of land-use and cover change. *Bioscience* 44(5):350–353.

Rogers P. 1994. Hydrology and water quality. In Meyer W. B. and Turner B. L. II, (Eds). *Changes in Land-Use and Land-Cover: A Global Perspective*. Papers arising from the 1991 OEIS Global Change Institute, Cambridge University Press, 231–257.

Running S. W., Loveland T. R., Pierce L. L., and Hunt E. R. Jr. 1994. A remote sensing based vegetation classification for global land-cover analysis. *Remote Sensing of the Environment* 51:39–48.

Sage C. 1994. Population and income. In Meyer W. B. and Turner B. L. II, (Eds). *Changes in Land-use and Land-cover: A Global Perspective*. Papers arising from the 1991 OEIS Global Change Institute, Cambridge University Press, 263–286.

Scholes R. J., Ward D., and Justice C. O. In press. Emissions of trace gases and aerosol particles due to vegetation burning in southern hemisphere Africa. *Journal of Geophysical Research*.

Sellers P., and Schimel D. 1993. Remote sensing of the land biosphere and biogeochemistry in the EOS era: Science priorities, methods and implementation—EOS land biosphere and biogeochemical panels. *Global and Planetary Change* 7:279–293.

Skole D. 1994. Data on global land-cover change. In Meyer W. B., and Turner B. L. II (Eds) *Changes in Land-Use and Land-Cover: A global perspective*. Papers arising from the 1991 OEIS Global Change Institute, Cambridge University Press, 437–471.

Skole D. L., and Tucker C. J. 1993. Tropical deforestation and habitat fragmentation in the Amazon: Satellite data from 1978 to 1988. *Science* 260:1905–1910.

Stern P. C., Druckman, D., and Young O. R. 1992. *Global Environmental Change: Understanding the Human Dimensions*. National Academy Press, Washington, D.C.

Townshend J. R. G. 1995. Global data sets for land applications from the Advanced High Resolution Radiometer: An introduction. *International Journal of Remote Sensing* 15(17):3319–3332.

Tucker C. J. Dregne H. E., and Newcomb W. W. 1991. Expansion and contraction of the Sahara Desert. *Science* 253:299–301.

Turner B. L. II, Moss R. H., and Skole D. L. 1993. Relating land-use and global land-cover change. *IGBP Report* 24, IGBP Secretariat, Stockholm.

Vitousek P., 1994. Beyond global warming ecology and global change. *Ecology* 75(7):1861–1876.

Vitousek P. M., and Matson P. A. 1992. Agriculture, the global nitrogen cycle and trace gas flux. In R. S. Oremalns (Ed). *The Biogeochemistry of Global Change: Radiative Trace Gases*. Chapman and Hall, New York, 193–208.

Ward D. E. In press. Modeling combustion efficiency and emission factors for selected savanna ecosystems. *Journal of Geophysical Research*.

Wofsy S., Harriss R., Skole D., and Kirchoff V. W. J. H. 1994. Amazon biochemistry and atmospheric chemistry experiment (AMBIACE). *Brazilian Journal of Geophysics (Revista Brasiliera de Geofísica)* 12(1):9–28.

Wolman and Fournier (Eds). 1987. *Land Transformation in Agriculture, SCOPE 32*, John Wiley and Sons, Chichester, U.K.

Developing a Global Vegetation Fire Monitoring System for Global Change Studies: A Framework

Jean Paul Malingreau and Jean-Marie Grégoire

The global significance of biomass burning in natural and managed ecosystems is now widely recognized (Crutzen and Andreae 1990) and the need to better characterize its impact on vegetation and the atmosphere has been identified as a priority in global change studies (IGAC 1994). Linkages among vegetation fires and atmospheric chemistry and aerosols are most relevant to the global warming and climate change debate and are understandably at the forefront of biomass burning-related studies. It is important to remember, however, that fires are closely associated with human activities and that their pattern of distribution is largely, albeit not entirely, related to land-use practices. This means that vegetation fires may follow relatively fixed patterns but may also appear with renewed intensity in regions where and during periods in which burning is not common. The association with land use also means that burning may cease to be of concern under new and prescribed land-management practices. In any case, quantitative predictions about global change will be much improved by the ability to predict future state and conditions of the land cover (Turner et al. 1994). Through new conjunctions between enhanced human activities, alterations in vegetation conditions, and climate change the role of fire in global geochemistry and land cover could thus be evolving in yet to be identified directions. The purpose of a global fire-monitoring system is to better assess the role of biomass burning as a potentially significant element of a largely human-controlled chain of events leading to large-scale changes in the earth's system.

While fire in ecosystems is a well-known process that has been studied intensively for decades (see reviews in Mueller-Dombois 1981; Goldammer 1990; Levine 1991), new perspectives are now being brought to the foreground. They are driven first by the relatively recent need to address ecological questions at scales hitherto little explored. Ecosystem properties or values emerging from their functioning at regional to continental and even global scales are now being intensively analyzed. Fire is one of the generic processes that contribute to ecosystem distribution, character-

istics, evolution or, in some case, rapid transformation. Although burning is concentrated in some areas of the world and during discrete periods of time, the fact that it is almost ubiquitous and can even appear in unsuspected areas (Malingreau et al. 1986; Leighton and Wirawan 1987) or during unusual periods of time make mandatory a global approach to its assessment. Given the complexity of biomass burning situations there is a need to develop an analytical framework that can lead to the right interpretation of detected events in terms of relevance to global issues. Capabilities offered by remote-sensing techniques to assess global burning (Kaufman et al. 1988; Malingreau 1991; Cahoon et al. 1992; Justice and Dowty 1994; Chuvieco and Pilar Martin 1994; Koffi et al. 1995) render even more urgent a correct evaluation of the increasingly available data. This chapter proposes a general framework for such analysis.

Global Perspective on Biomass Burning

The impact of biomass burning on atmospheric chemistry and biogeochemical cycles has been widely exposed in recent years (Hao and Liu 1994; Crutzen and Andreae 1990); it is through interactions with the atmosphere that biomass burning acquires its truly global character. Changes or trends in emission, transport, chemical transformation and, eventually, deposition of biomass burning gases are thus important to assess. The problem is first to correctly identify and measure possible changes in the magnitude and characteristics of the emission sources. Uncertainty affects various terms of the equation usually employed in the estimates (Hao and Liu 1994): general values only are used for describing the combustible material; the actual occurrence of fire in given ecosystems is, in most cases, still assumed (the comprehensive information on West Africa cited by Menaut et al. 1991 dates back to Phillips et al. 1965). Inaccuracies and compounding errors can thus quickly appear in the overall budget calculations. In such a context predictive capabilities cannot be improved unless significant progress is made in the collection of quantitative information on processes driving the global biomass burning phenomena. The basic requirements are spatially referenced and temporally related data on vegetation conditions, occurrence and type of fire, and amount of burned biomass (Malingreau et al. 1993).

The anthropogenic dimension of biomass burning is most relevant in the framework of transformations that are currently affecting large tracts of land. Deforestation, agricultural extension, savannization, fuel wood and charcoal burning are all related to changing food and wood consumption requirements, themselves linked to population growth, evolving economic policies, or new environmental management practices. Fire is being used as a land-management tool in vegetation clearing, nutrient recycling, and in forest management (prescribed fires). By affecting the nature and state of the land cover, humankind is also changing the distribution of fire-prone ecosystems; this is the case in the maintenance of savannas in Africa (or *cerrado* in South America and *alang-alang* in Asia) carved in the moist or semideciduous forest domain. The picture painted here must, in addition, be examined in the framework of changing climate scenarios, which would lead to enhanced stress in vegetation, reduced soil water reserves, and new fuel accumulation patterns in the ecosystems.

Fire can therefore be considered one of the focal points of the multiple relationships between humans and the environment. Changes in fire patterns can be taken as indicators of change in land-use patterns and overall environmental conditions. Land-use practices and their concomitant use of fire are at the forefront of a series of issues related to sustainable development and resource management. The need for a global analysis of biomass burning is thus justified beyond its relevance to and impacts on atmospheric chemistry. Although these issues are usually more regional or local, shifts in land-use policies or trends can involve very large areas. This complex web of relationships among fire, climate, human activities, and vegetation is the background for our elaboration of a global fire assessment.

Scale and Information Requirements

There are various ways to produce a global picture of environmental processes. The most common has been to extrapolate point or casual observations to larger areas to which findings are conjectured to apply. Statistical analysis can support the approach by stratifying the domain of interest and helping in the selection of sampling points (data on burn frequencies and efficiencies are typically collected in such manner). A second approach is to aggregate published or official data; this is possible when a systematic reporting system exists to support the compilation. The Siberian seasonal fire data are annually produced by compiling fire district reports; they are, however, unreliable (Shvidenko and Nilsson 1994). Finally, a global picture can be attempted by using a systematic and comprehensive observing system such as the one provided by the National Oceanic and Atmospheric Administration (NOAA) satellite (see chapter 3 by Setzer and Malingreau) or the Defense Meteorological Satellite Program (DMSP) system (Cahoon et al. 1994).

Whatever the case, it is important to recognize that due to inconsistencies in the data base and the difficulties in normalizing the information, a global picture cannot be derived from the sum of local assessments. Conversely, a picture acquired at a truly global scale (exhaustive in terms of territory and period covered) will perforce be based on a sampling procedure determined by the nature and timing of the measurement. Because of the elusive nature of burning, the effect of the temporal sampling singularly complicates the assessments (Malingreau 1992).

The largest uncertainties in the current global biomass burning picture relate to the occurrence of fires and to the amount of biomass consumed. Fires trigger the process, but their spatial versatility and their elusive nature make it difficult to generalize; truly global fire activity assessments have yet to be produced. Various approaches to the evaluation of emissions from burned vegetation have been proposed (Ward and Hardy 1991; Delmas et al. 1991b). Yet, while vegetation biomass can to some extent be modelled or measured the burned fraction of that biomass is more difficult to assess. Furthermore, the area component—that is, the surface where fire has been active at one or another level of intensity—is equally difficult to measure. Indirect clues given by information on the associated land-use practices, type and condition of vegetation, timing and type of burn can be useful but, again, cannot be systematically collected.

Attempts at developing a global picture of biomass burning should not hide the fact that, as as rule, the interest in fire events usually concentrates on regional and local impacts. Regional issues are somewhat more tractable and can give rise to international recognition of problems associated with biomass burning. The degradation of air quality in Southeast Asia is typically a regional feature because it is the burning in Sumatra and Kalimantan (Indonesia) that negatively affects the air quality in Singapore and Malaysia, to the point of creating health problems and impeding air traffic. As described by Koffi et al. (1995a), regional transport mechanisms determine the deposition in remote areas of biogenic components emitted by intensive fire activity (as in the case of Central Africa). Increasingly, evaluations of the national contribution to atmospheric biogenic gases are being integrated in a regional framework, which can then be used in international treaty negotiations. Biomass burning across national boundaries will undoubtedly assume a growing importance as pressure to control the phenomena increases. In many ways, regional perspectives on biomass burning patterns and calendars may in the future dominate the range of interests linked to the overall issue of burning vegetation.

Local monitoring and assessments of fire activity and impact are of more immediate concern in terms of fire prediction, firefighting, and overall land-cover management for prevention. They will not be considered here, although it is recognized that most of the current knowledge on fire characteristics and impacts are derived from such locally accumulated experience (Albini 1984).

It is becoming clear that the relevance of fire events in one part of the world or another or at a given period of time or another is always evaluated with respect to a particular concern or purpose. Thus, although all fires contribute, even temporally, to the chemical composition of the atmosphere, some are more significant than others. Although all fires destroy vegetation, some are more damaging than others. Although all fires contribute to the global picture, some events are more significant than others. A few examples demonstrate this selectivity in interests: Mediterranean fires raise intense local and regional concerns because of their economic impact but are usually not examined in priority in the framework of global change. The 1982–1983 and 1992 fires in Kalimantan raised some local concern and a lot of regional interest and, because they took place in essentially fire-free ecosystems, they are considered most significant in terms of changes in tropical biomass burning patterns. Early and late fires in savannas are viewed differently when considered from an agricultural or atmospheric point of view. Changing patterns in boreal forest fires may reveal self-amplifying coupling to global warming (Cofer et al. 1990).

Framework and Typology

The development of a global biomass burning analysis framework is thus faced at the outset with a fundamental question: must a comprehensive fire-monitoring system take into account all the biomass burning events taking place in terrestrial ecosystems or only "relevant" events? Should an observation system be universal or attuned to a specific type of biomass burning situation? Various alternatives can be envisaged:

• The monitoring system is designed to account for all biomass burning events at the surface of the earth at all times.

• The system addresses only events that are "significant" in the perspective of global change (i.e., large fires, large biogenic emission sources, exceptional fire situations, critical ecoystems, etc.).

• The system collects information on any biomass burning event detected by a given sensor or survey (this is essentially a logistically or technologically driven approach).

The first alternative can only be achieved at great cost and through the systematic compilation of a wide variety of data; it is essentially impractical and is not considered here. The question of addressing only significant events requires that a series of criteria be identified for detection and evaluation. Only when the question "significant for what and for whom?" has been answered can a strategy for data collection be designed. A typology of biomass burning that relies on an analysis of the direct agents of fire and of its impacts is proposed in the following section to guide this approach. If adopted, the purpose of a global assessment would then be to focus only on fires that would have direct relevance to, say, atmospheric chemistry or landscape impacts.

Remote-sensing techniques have been instrumental in bringing about a realization of the global importance of biomass burning in the world biomes (Robinson 1992; Malingreau et al. 1986; Cahoon et al. 1992; Belward et al. 1995; see chapter 21 by Koffi et al. and chapter 3 by Setzer and Malingreau). Satellite observations have brought to Earth very dramatic views of fire in natural and human-derived ecosystems. They have demonstrated that the process is widespread and that it can have very large dimensions. As spectacular as they are, the remote-sensing views have led us to erroneously believe that such an approach will automatically produce objective and relevant data. Yet, the method is far from satisfactory. Indeed, observations of Earth from space are blind to the relative importance of detected events. Thus if a thermal or light-sensitive sensor does not differentiate a small straw-burning fire and a deforestation fire one must make sure the data are not confused with respect to their significance from a global perspective. The effects of the satellite sampling on the representativity of the fire data must be correctly understood before such interpretation can be attempted. A brief review of those main sampling effects introduced is given in a following section.

A global framework for analysis could be structured around three possible typological approaches; these are related to the fire regime, causation agents, or fire impacts.

Fire Regime

Fire regime has been used to characterize the general dynamics of fire in ecosystems. It is related to the frequency of fire ($1/n$ years), the date of occurrence, the area affected, and the interannual variability of those characteristics. Thus biomass burning patterns can be used to stratify the world into fire zones where burning has assumed historical significance. This probability of occurrence is essentially associated with the vegetation burns and there is currently enough knowledge to propose such a practical stratification. Goldammer (1991) uses the present distribution of vegetation formations such as deciduous and monsoon forests, tropical pine forest, and tropical savannas to spatially characterize fire regimes (for example, "regular short-term interval fires"). Every region at the surface of the earth is thus attributed an empirically determined probability of burning. This probability has naturally led to the priority examination of African savannas, boreal forests, and the Mediterrannean Basin, and northeast and eastern Australian bush. However useful it is in a first-pass analysis, a frequency classification must be considered nominal because it does not take into consideration the interannual variability in both the magnitude of burning and in its calendar. In such context a generalized fire regime will not adequately describe a highly dynamic situation.

Typology of Causation Agents

In a global perspective it may be of interest to rapidly focus on the suite of causation agents that lead to changing biomass burning situations. The approach proceeds along a typology that provides as the main point of entry the actual use to which fire is put on land. This intended (or not intended) use then determines in a first pass the relevant characteristics of biomass burning in the global picture. For example, if grassland management foresees the use of fire as a means of regenerating a palatable grass cover and if this practice is relatively stable in terms of area concerned, then burning has relatively little interest for

global change per se. A systematic analysis of trends in land-use practices including the role of fire could be organized along the following lines:

I. Human-induced fires
   A. Land management
      1. Land clearing: shifting cultivation, agricultural expansion, deforestation
      2. Land maintenance, bush control, weed burning
      3. Agricultural residue burning: straw, stubble
   B. Forest management
      1. Prescribed fires
   C. Grassland management (early and late fires can be separated)
      1. Grazing (grass regeneration)
      2. Hunting
      3. Others
   D. Accidental, arson fires (in relation to population density, political situation, poor management practices, etc.).
II. Natural fires
   A. In grassland
   B. In forest

Fire Impacts

Using the impact of biomass burning as the point of entry in the stratification is the most direct way to proceed to establish the global framework. If global change is indeed the objective of the analysis, fire events that significantly contribute to a relevant Earth's system attribute will receive priority. A fire will thus be of interest in the assessment if and only if it is relevant to a particular and preset issue.

Impact on Atmospheric Chemistry Thresholds of magnitude are empirically determined based on the intensity of the burning, the area covered, the amount of biomass burned, and the relative importance of fire in the geochemical cycle of a particular ecosystem. If change is truly the focus of interest, then only data from areas where biomass burning is on the increase or decrease should be included. This may mean that large areas of African savannas where intense burning has been common during historical periods may not be included in the analysis. The difficulty is to assess whether there is a trend in the burning intensity and distribution associated with contemporary pressures on land. The transition areas between savanna and forest and the rain forest biome itself are more likely to represent areas of interest in this respect. The analysis of time series of satellite data may shed some light on the interannual variablity of the process (Belward et al. 1995; see chapter 21 by Koffi et al.).

Economic Impact Economic losses due to fire are, as a rule, highly localized and, despite the spectacular and often dramatic nature of those highly publicized events, they have so far not been considered in a global framework. Australia, California, the Mediterranean Basin, and parts of the southeastern Andes are areas of highly damaging fires. Regional economic impacts have been brought to light by the fires in Sumatra and Borneo, which have created harsh living conditions and impeded air traffic in the Malaysian Peninsula and Singapore. The economic impact of burning in Siberian forests will surface as a prime issue as attempts are made to develop new exploitation plans for the management of their vast resources.

Economic gain due to the use of fire is of course directly associated with agricultural and land-clearing practices. An entry in our stratification could thus be directly related to practices that are made possible and even benefit from fire. Again it is the change in those practices that will be of interest in the global picture, as well as trends such as those associated with agricultural intensification or decreases in return period of land cultivation.

Ecological Impact Burning areas could be stratified according to the role fire plays in ecosystem maintenance or destruction. This criterion would separate most savanna and temperate/boreal forests (fire-prone ecosystems) from areas where the occurrence of fire irremediably and drastically changes the characteristics of the land cover. This latter category includes all areas where the regrowth of a tree cover has recently been prevented by the use of fire. Examples are the *alang-alang* areas in Southeast Asia, parts of the Brazilian cerrado, or the "savannes incluses" of the Central African Basin. "Catastrophic" fire events could fit here in the stratification framework. Unexpected and large-scale fires are indeed important to detect because they could indicate changes in the nature of the relationships among population, vegetation, and climate. The capability to detect and interpret catastrophic fires in a systematic way and on a global scale requires, however, a powerful information system that can at the same time identify the "normal" and "abnormal" events and provide a base for comparison. Current information technologies are able to support the approach but their operational use may still be too costly for global fire monitoring.

The three-pronged typology presented above leads to a multiple-entry classification of fire areas; the classes are not exclusive and it is clear that several criteria are to be used to derive the global framework. Figure 2.1

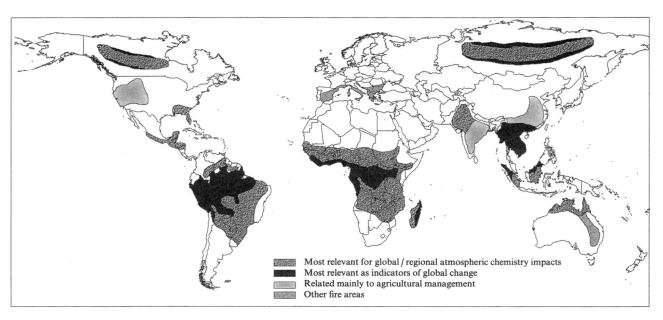

Figure 2.1 A stratification of vegetation fire areas according to a set of criteria retained in the proposed framework

is a first attempt at identifying a series of priority areas with respect to a few of those stratification criteria. A strict and well-documented procedure along the lines proposed above should now be used to support the systematic analysis of the processes and impacts associated with biomass burning around the world. The approach may also help in designing better research strategies focusing on the most relevant areas and issues.

The Role of Satellite Observations in Global Fire Assessment Observations

Earth observation satellites provide an essential support to the global monitoring of biomass burning. The Advanced Very High Resolution Radiometer (AVHRR) instrument of the NOAA satellite series and its capability to detect elevated heat sources at the surface of the earth has been intensively exploited in many parts of the world during the last decade (several papers in this volume specifically refer to this particular approach). Despite technical and operational limitations, the AVHRR has played a major role in alerting the scientific community to the global scale of biomass burning (Matson and Dozier 1981; Malingreau et al. 1985; Setzer and Pereira 1991; Grégoire 1993). The question of accurately assessing resolvable fire events using the thermal channel of the AVHRR has been extensively reviewed by Robinson (1991). However, a series of fundamental issues remain little examined in the analysis of such data; they pertain not only to the spatial sampling associated with the geometry and resolution of the instrument but, more important maybe, to the effect of the temporal sampling. This point has caught our attention because it seems to introduce the largest error term in current satellite-based assessments.

Orbiting satellites acquire data at specific and relatively constant times of day for any suborbital point. The overall ecosystem dynamics will determine whether such frequency and timing appropriately represent a very temporary event such as biomass burning (figure 2.2). The question of sampling the diurnal cycle of fire activity has been relatively little examined with respect to satellite assessments. Atmospheric conditions (wind and air humidity), fuel moisture, and local practices all determine the scheduling of field activities. Satellite overpasses occur at very specific times during the day and most remote-sensing analyses use one of those passes only as a reference. For the AVHRR, the afternoon pass (nominally 1430 local time but drifting to later in the afternoon during the satellite's lifetime) is often retained. Morning or night passes are rarely

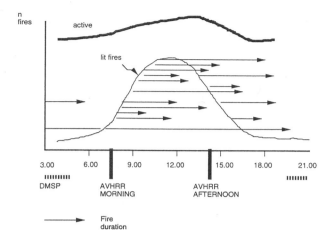

Figure 2.2 Graphic description of fire dynamics in vegetation expressed in terms of fire starting time and duration (horizontal arrows), number of fires lit at a particular time (thin curve), and total number of active fires (thick curve on top). The time of observation by various satellites is indicated for reference.

considered except for ad hoc campaigns. Given the diurnal burning cycle and the relatively well understood resolution capabilities of the AVHRR sensor, the AVHRR thermal fire detection equation summarizes the situation:

$$n_{fires,period\ t} = n_{fires\ detected} + n_{fires\ undetected},\ and$$

$$n_{fires\ undetected} = n_{fires\ <\ resolution} + n_{fires\ outside\ temporal\ window}$$
$$+ n_{fires\ cloudy\ areas\ gaps\ in\ coverage}.$$

To accept AVHRR fire assessment at face value one must assume that few subresolution fires were present in the scene, that the 1430 hr overpass has picked up most of the active fires for that day, and that extrapolation to cloudy areas and missing days has been made. Anyone familiar with biomass burning situations in the field will quickly realize that the first two assumptions are not always valid. Extrapolation to cloudy areas is, furthermore, very rarely performed.

Much remains to be done to properly evaluate the sampling effect introduced by the instrumental constraints. Similar questions can be raised with respect to the AVHRR Global Area Coverage (GAC) data; however, due to the fact that more than 10 years of daily data are now available in this format, it has been demonstrated that the highly sampled GAC product can provide a good description of fire calendars and, on a continental scale, of the position of the larger fire areas (Belward et al. 1995). These are, in any event, the only data available to determine historical burning patterns over large areas as demonstrated by Koffi et al. (1995).

The nighttime detection capabilities of the DMSP satellite suffer from similar weaknesses. A further drawback is introduced by the fact that at night many, if not most, of the agricultural fires are put out and that the representativity of the sample may be even poorer than with the AVHRR. The DMSP visible light fire detection equation:

$$n_{\text{fires period p}} = n_{\text{light sources}} - n_{\text{city industr. lights}}$$
$$+ n_{\text{fires outside temporal window}} + n_{\text{fires cloudy areas, gaps}}.$$

Given the above restrictions one must therefore consider satellite data as "a" sample of fire distribution in the region under observation. The exact value of the sample remains unknown. The AVHRR is one of the main sources of information on biomass burning at the scale of interest in global change analysis and, in the absence of dedicated instruments, work must continue to refine its utilization. Indirect methods to evaluate the information must be sought. They rely mainly on ancillary information which can lead to a better appreciation of the nature and impact of the detected fires. An important recommendation of a 1992 Dahlem Workshop specifically dedicated to biomass burning was to integrate the available data into a Fire Information System, which supports the interpretation (Crutzen and Goldammer 1993).

A Global Fire Product

The information necessary to put into application the framework described here is still piecemeal and needs to be more systematically collected. The proposed shortcut is essentially practical because it starts with the production of a fire product based on a given set of observations collected by satellite in an essentially nonpurposive manner. The product is then submitted to screening with reference to the framework described here. This global exercise has been initiated in the framework of the FIRE Project (JRC 1993) in collaboration with the Earth Observation System Company and the Natural Resources Institute of the United Kingdom (EOS 1995).

In order to derive a truly global fire product one must access a data base that satisfies a minimum set of conditions: it must be geographically global, obtained in a systematic manner, and similar everywhere with an acceptable frequency of acquisition and for a minimum period of time. The AVHRR data set developed in the framework of the land-cover mapping of the

International Geosphere-Biosphere Program (IGBP) verifies some of these conditions (Townshend et al. 1992). It contains the AVHRR afternoon acquisitions (all spectral channels) for most of the earth's surface for the period of April 1992 to September 1994. An additional requisite is that a detection method be available that is as automatic as possible yet still sensitive to local conditions. The work presented here uses a contextual fire-detection algorithm applied to the entire data set for a given period. Finally, a reference information system must be available to evaluate the data. This is essentially the one presented in this chapter.

Data from seven consecutive days (11–17 January 1993) were extracted from the NOAA-AVHRR High Resolution Picture Transmission/Local Area Coverage (HRPT/LAC) data set assembled for the Global Land Cover Project. The raw data received from a network of stations were ingested and stitched together to produce orbital segments by the EROS Data Centre (Eidenshink and Faundeen 1994). Fourteen orbits are needed to cover the entire land masses of the earth.

The detection technique applied to the AVHRR data set relies on a "contextual" algorithm (Prins and Menzel 1992; Justice and Dowty 1994) and runs in two phases. The first makes a selection of the pixels in a window that could be fires by applying a multithreshold algorithm but with a range of thresholds wide enough to keep all the possible fire pixels in the selection. The second pass leads to a final allocation of fire points by using automatically derived information about the immediate neighborhood of the potential fire pixels. The main advantage of this method is that the thresholds are set "automatically." This maintains the procedure constantly sensitive to regional and seasonal differences. This is difficult to do when setting local thresholds for each area. Furthermore, the use of a single algorithm that automatically determines its local parameters singularly reduces operator intervention—an important element in the preparation of any global product. Preliminary work on validation of the global products has shown that the contextual method of fire detection is as effective as the conventional one, as illustrated in figure 2.3 (after Flasse and Eva 1995).

Three products are prepared in the global AVHRR fire-processing chain: level I contains the geographical coordinates of the detected fires in a given orbital segment, level II contains the same data for the entire globe and for a given day, and level III contains a periodic synthesis. Figure 2.4 presents the level III product for the 11–17 January 1993 period. The display of

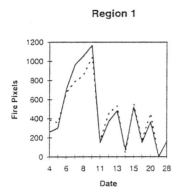

Region 1

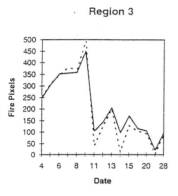

Region 3

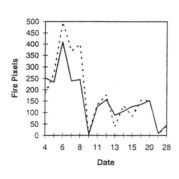

Region 5

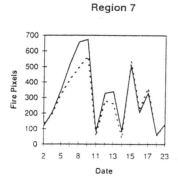

Region 7

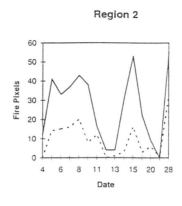

Region 2

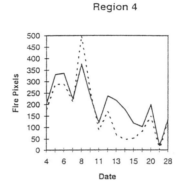

Region 4

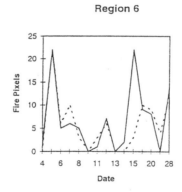

Region 6

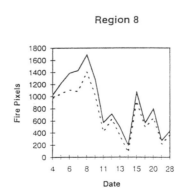

Region 8

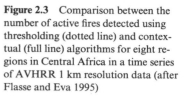

Figure 2.3 Comparison between the number of active fires detected using thresholding (dotted line) and contextual (full line) algorithms for eight regions in Central Africa in a time series of AVHRR 1 km resolution data (after Flasse and Eva 1995)

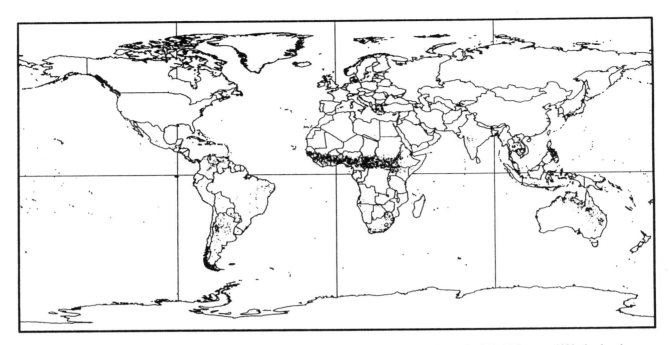

Figure 2.4 Global pattern of fire activity derived from the AVHRR global land data set for the week of 11–17 January 1993; the data have been processed by EOS, Ltd., and NRI UK.

active fires for that period is illustrative of the potential of a global product; it can be compared to the simple stratification proposed in figure 2.1. It must be noted, however, that the same restrictions apply to this product as to other regional assessments (Robinson 1992). The influence of the time of observation (nominally 1430 local time but later for the period of time selected here) is particularly crucial in a global product. One can indeed ask if it is relevant to compare fire data acquired around the same local time over savanna and tropical or boreal forests where burning practices and diurnal cycles are widely different. The representativity of the sample is clearly ecosystem dependent and is difficult to ascertain. When longer time series produced in this standard manner become available, it will be possible to produce a global "standard" picture very much in a manner suggested by Koffi et al. (1995) for the AVHRR GAC (Global Area Coverage) product. Changes and trends will then be sought and analyzed in terms of driving variables.

The global product presented here lends itself to a series of further investigations that will require adequate tools. As indicated, the support of a powerful geographical information system is crucial at this point. Ancillary spatial information and specific analytical tools must come to support both the production and analysis phases of the global product. A Vegeta-

tion Fire Information System (Malingreau et al. 1992) is an essential component of the approach.

Conclusions

Given the large diversity of biomass burning conditions around the globe and the wide variety of interests associated with a global analysis, it has been suggested that the use of a framework can enhance the relevance and effectiveness of the data collection and interpretation process. A series of criteria for asserting whether a particular fire situation is of interest in a defined context have been presented. A stratification of vegetation fires is proposed, which can guide the elaboration of an analytical framework. A global product derived from the AVHRR daily passes over all land masses of the earth is proposed as a starting point for this analysis. Limitations have been noted. Although the instrument is not ideal for such observations, it can lead to the preparation of longer time series of fire patterns which, by virtue of their consistency, can support trend and change analyses. The manipulation and processing of the very large data sets contained in the current global land-cover product (the total volume of the IGBP AVHRR land product is about 5 terabytes for the 1992–1994 period) currently precludes rapid progress in the derivation of a permanent global fire-monitoring system. Financial resources are not yet available in the international scientific community to conduct such a large-scale operation.

References

Albini, F. A. 1984. Wildland fires. American Scientist 72:590–597.

Belward, A., A. Hollifield, and M. James. 1995. The potential of the NASA GAC Pathfinder product for the creation of global thematic data sets: the case of biomass burning patterns. Int. Journ. of Remote Sensing, 16, (11):2089–2097.

Cachier, H. 1992. Biomass burning sources.In Encyclopedia of the Earth System Science. 1. Academic Press. pp. 377–385.

Cahoon, D. R., B. J. Stocks, J. S. Levine, W. R. Cofer and K. P. O'Neill. 1992. Seasonal distribution of African savanna fires. Nature. 359:812–815.

Chuvieco, E. and E. Pilar Martin. 1994. Global fire mapping and fire danger estimation using AVHRR images. Photogramm. Eng. and Remote Sensing 60(5):563–570.

Cofer, W. R. III, J. S. Levine, E. L. Winstead and B. J. Stocks. 1990. Gaseous emissions from Canadian boreal forest fires. Atmos. Environ (24A):1653–1659.

Crutzen, P. J. and M. O Andreae. 1990. Biomass burning in the tropics: impact on atmospheric chemistry and biogeochemical cycles. Science 250:1669–1673.

Crutzen, P. J. and J. G. Goldammer, (Eds). 1993. Fire in the Environment: The Ecological, Atmospheric and Climatic Importance of Vegetation Fires. John Wiley and Sons New York.

Delmas, R., P. Loudjani, A. Podaire and J. C. Menaut. 1991a. Biomass burning in Africa: an assessment of the annually burned biomass. In Levine, J. S. (Ed). Global Biomass Burning: Atmospheric, Climatic, and Biospheric Implications. MIT Press. Cambridge, Mass.

Delmas, R., A. Marenco, J. P. Tathy, B. Cros and J. G. R. Baudet. 1991b. Sources and sinks of methane in Africa. CH_4 emissions from biomass burning. Journ. of Geophys. Res. 96:7287–7299.

Eidenshink, J. C. and J. L. Faundeen. 1994. The 1 km AVHRR global data set: first stages in implementation. Int. J. of Remote Sensing, 15(17):3443–3462.

EOS. 1995. Global Vegetation Fire Product. Final Report Contract 1044-94-09-FIEP ISP GB.

Flasse, S. and H. Eva. 1995. A comparison of two methods for regional active fire detection in the Central African Republic using NOAA-AVHRR LAC data (in prep.).

Goldammer, J. G. 1991. Tropical wild-land fires and global changes: prehistoric evidence, present fire regimes and future trends. In Levine, J. S. (Ed). 1991. Global Biomass Burning. Atmospheric, Climatic, and Biospheric Implications. MIT Press. Cambridge, Mass. pp. 83–91.

Goldammer, J. G. (Ed). 1990. Fire in tropical biotas: ecosystems processes and global challenges. Ecological Studies Vol. 84. Springer-Verlag, Berlin.

Grégoire, J-M. 1993. Description quantitative des régimes de feux en zone soudanienne de l'Afrique de l'Ouest. Sécheresse 4:37–45.

Hao, W. M. and M. H. Liu. 1994. Spatial and temporal distribution of tropical biomass burning. Global Geochem. Cycles. 8(4):495–503.

IGAC. 1994. The International Global Atmospheric Chemistry Project. An Overview. IGAC Project Office. Princeton.

JRC. 1993. FIRE-Fire in Global Resource and Environmental Monitoring: A project of the Commission of the European Communities. JRC Public. SPI. 93.26. pp. 10. Ispra, Italy.

Justice C. O. and P. Dowty (Eds). 1994. IGBP DIS Satellite Fire Detection Algorithm Workshop Report. IGBP-DIS Working Paper No. 9, Univ. Paris VI. France.

Kauffman, J. P. and C. Uhl. 1990. Interactions of anthropogenic activities, fire, and rain forests in the Amazon Basin. In Goldammer J. G. (Ed). Fire in tropical biotas: ecosystems processes and global challenges. Ecological Studies Vol. 84. Springer Verlag, Berlin.

Kaufman, Y., C. J. Tucker and I. Fung. 1988. Remote sensing of biomass burning in the tropics. Journ. of Geophys. Reasearch 95:9927–9934.

Koffi, B., J-M. Grégoire, G. Mahé and J-P. Lacaux. 1995a. Remote sensing of bush fire dynamics in Central Africa from 1984 to 1988: analysis in relation to vegetation and pluviometric regional patterns. Atmosph. Res. 39:179–200.

Koffi, B., J-M. Grégoire and H. D. Eva. 1995b. Satellite monitoring of vegetation fires on a multiannual basis at continental scale in Africa. This volume.

Levine, J. S. (Ed). 1991. Global Biomass Burning: Atmospheric, Climatic, and Biospheric Implications. MIT Press. Cambridge, Mass.

Malingreau, J. P., G. Stephens and L. Fellows. 1985. Remote sensing of forest fires: Kalimantan and North Borneo in 1982–83. Ambio 14(6):314–315.

Malingreau, J. P. 1990. The contribution of remote sensing to the global monitoring of fires in tropical and subtropical ecosystems. In Goldammer J. G. (Ed). Fire in Tropical Biotas. Ecosystems Processes and Global Challenges. Ecological Studies Vol. 84. Springer-Verlag, Berlin.

Malingreau, J. P., F. A. Albini, M. O. Andreae, J. S. Levine, J. M. Lobert, T. A. Kuhlbusch, L. Radke, A. Setzer, P. M. Vitousek, D. E. Ward and J. Warnatz. 1993. Group Report: Quantification of fire characteristics from local to global scales. In Crutzen, P. J. and J. G. Goldammer (Eds). Fire in the Environment: The Ecological Atmospheric, and Climatic Importance of Vegetation Fires. John Wiley and Sons. pp. 329–343.

Matson, M. and J. Dozier. 1981. Identification of subresolution high temperature sources using a thermal infrared sensor. Photogramm. Eng. and Remote Sensing. 47(9):1311–1318.

Menaut J-C., L. Abadie, F. Lavenu, P. Loudjani and A. Podaireo. 1991. Biomass burning in west African savannas. In Levine, J. S. (Ed). 1991. Global Biomass Burning: Atmospheric, Climatic, and Biospheric Implications. MIT Press. Cambridge, Mass. pp. 133–142.

Mueller-Dombois D. 1981. Fire in tropical ecosystems. In Fires, Regimes and Ecosystem Properties. Proc. Conf. General Tehnical Report WO-26. U.S. Dept. of Agric. Forest Service. pp. 137–176.

Philipps, J. 1965. Fire—A master and Servant: Its Influence in the Bioclimatic Regions of Trans-Saharan Africa. Proceedings of the Tall Timbers Fire Ecology Conf 4:7–109.

Prinsm, E. M. and W. P. Menzel. 1992. Geostationary satellite detection of biomass burning in South Americs. Int. Journ. of Remote Sensing, 13(15):2783–2799.

Robinson, J. M. 1991. Fire from space: global fire evaluation using infrared remote sensing. Int. J. of Remote Sensing 12:3–24.

Shvidenko, A. and S. Nilsson. 1994. What do we know about the Siberian Forests? Ambio 23(7):396–404.

Townshend, J. R. G. (Ed). 1992. Improved global data for land applications—a proposal for a new high resolution data set. IGBP Global Change Report No. 20, Stockholm.

Turner B. L. II, W. B. Meyer, and D. L. Skole. 1991. Global land-use/land-cover change: towards an integrated study. Ambio 23, 91–95.

Ward, D. E. and C. C. Hardy. 1991. Smoke emissions from wildland fires. Environ. Int. 17:117–134.

AVHRR Monitoring of Vegetation Fires in the Tropics: Toward the Development of a Global Product

Alberto W. Setzer and Jean Paul Malingreau

Fire has accompanied and played important roles in the development of the vegetation of our planet. Since its control by humans about half a million years ago, it has been a major tool in hunting, forest conversion, agriculture, and pasture renewal, just to mention a few uses related to vegetation changes. Since the pioneer work of Crutzen et al. (1979) environmental effects resulting from vegetation fires have increasingly become a subject of scientific interest. With tropical deforestation taking place at unprecedented rates, they rapidly changed into a worldwide scientific and public environmental concern. Two recent publications (Levine 1991; Crutzen and Goldammer 1993) contain hundreds of references that extensively document the current importance of biomass burning.

Of particular interest is the estimate of the location and extent of vegetation burning on the planet, which has yet to be made. Supposedly, 60% of the pantropical savannas in the world, about 15×10^6 km^2, may be affected by fire every year (Goldammer 1993). Tropical deforestation, normally attained through the use of fire, was estimated at 15.4×10^4 km^2 for the last decade (Singh 1993); unknown immense areas of boreal forests are also destroyed by fires. On a global basis, carbon emissions from biomass burning could account for 30–80% of the fossil fuel burning rate of 5.7×10^{15} g of carbon per year, with significant effects in biogeochemical cycles and possibly also on the climate (Crutzen and Andreae 1990).

Vegetation fires occur on all continents (e.g., Andreae 1993) and difficulties in their detection with conventional ways for most of the world makes systematic remote sensing from space the only possibility for their comprehensive study. Because of its daily coverage, low (1.1 km) resolution images of the Advanced Very High Resolution Radiometer (AVHRR) aboard the National Oceanographic and Atmospheric Administration (NOAA) series satellites (Kidwell 1991) have been used in thermal detection of active fires in daytime since the early work of Matson and Dozier (1981) pointed to this possibility. An AVHRR-based real-time operational program for firefighting and monitoring has existed for many years on a regional scale (Setzer and Pereira 1991a) and an AVHRR global fire product is under consideration (IGBP 1992) following Malingreau (1990) and Malingreau et al. (1993). However, little validation of AVHRR fire detection algorithms is found in the literature; different authors use different methods and commonly refer to fires detected and mapped without any field verification. More recently Setzer et al. (1994) describe AVHRR responses to two large forest fires in Amazonia, and Belward et al. (1993) to five fires in West African savannas; in opposition to algorithms using a multichannel approach, these data sets indicate that AVHRR's channel 3 (3.55–3.93 μm) alone is enough and the best to detect active fires.

In the following text we present evidence from 330 cases of vegetation fires in NOAA-11/AVHRR images of three continents. Of significant importance, these fires were detected in daytime and not nighttime images, representing the most needed and difficult conditions for their detection. General statistics of the results are given but we concentrate on the main practical problems of fire detection with AVHRR, suggesting techniques toward a global fire product.

Validation of Fires in AVHRR Images

Validation of satellite detection of vegetation fires must rely on known active fires during image acquisition. Prescribed fires or any fires identified during a satellite overpass provide such information; however, due to logistical problems these techniques are limited. The works of Belward et al. (1993) and Setzer et al. (1994), the only known published examples of actual field validation of satellite detection of active fires, reflect the limitations of this method. Another possibility of AVHRR fire validation is cross-comparison with images from high-resolution satellites like the Landsat Thematic Mapper-TM. Pereira Jr. and Setzer (1996) examined new fire scars in a set of three time-consecutive Landsat-TM frames for the same savanna area in relation to the active fires detected with AVHRR during the same two periods of 16 days for the same area. In the period with best results, 26% of the fires detected by AVHRR could not be verified in the TM images, possibly because they occurred in very short and sparse grasses or because of highly reflective soils or of regrowth caused by rains; 43% of the TM fire scars had no corresponding active fire in the AVHRR image, presumably because these fires were not active during AVHRR imaging or were covered by clouds. Such limitations and the difficulties and the cost of comparing TM and AVHRR images on a worldwide basis limit this approach to a research scale.

In the present work an active vegetation fire in AVHRR images was retained only if it was at the origin of a smoke plume detectable in channel 1 (0.55–0.68 μm) and if simultaneously hot "fire pixels" in channel 3 (3.55–3.92 μm) existed at the same place. Plumes had to show a conical/bending shape typical of fire emissions, with the vortex over the "fire pixels." Fire pixels were those pixels in channel 3 with low digital counts (DN), below 60, which in the inverted scale of channel 3 corresponds to the highest temperature end of the scale. Channel 3 was preferred in relation to other channels because it is the most sensitive to thermal emissions from fires. Smoke plumes were visually detected through digital enhancement of channel 1 images, where they present higher reflectivity in relation to the other AVHRR channels (Pereira and Setzer 1993). The enhancement used was linear stretching, with settings that varied from image to image and according to the region of the images analyzed because of differences in solar illumination, satellite viewing geometry, atmospheric opacity, and background reflectance. Fire pixels were selected using a simple thresholding. Vegetation fires, henceforward also called "fire events," or simply "fires," were selected by overlapping the thresholded channel-3 image with that of channels 1 and applying different linear stretching until plumes could be associated with fire pixels. All processing was done with raw 10-bit resolution uncorrected images. Only images showing at least 10 such independent cases of fire were used in the analyses, and these images were selected after processing hundreds of NOAA-11/AVHRR images of the archives of the Monitoring of Tropical Vegetation Group-MTV at the Joint Research Centre (JRC), Ispra, Italy. Most images presented fewer cases of fires or had no clear, unmistakable association between smoke plumes and fire pixels.

Only AVHRR images of a single satellite, NOAA-11 in this case, were used to avoid the introduction of problems resulting from change of sensors. However,

Table 3.1 33 AVHRR/NOAA-11 images used in the study for validation of fires in tropical vegetation

	Dates of AVHRR images	Equatorial crossing	Area covered	Latitudes of fires	Longitudes of fires	Original Ecosystems
S	22-Aug -1989	49 W	5N-24S	5-10 S	46-65 W	Fo, TrFo,
	23-Aug -1989	46 W	2N-24S	8-12 S	57-62 W	Fo
A	26-Aug -1989	64 W	0-24S	5-13 S	60-70 W	Fo
M	1-Sep -1989	48 W	2-20S	5-15 S	45-50 W	TrFo, Wo
E	3-Aug -1990	49 W	1-24S	7-23 S	49-57 W	Fo, TrFo
R	4-Aug -1990	46 W	0-24S	7-17 S	45 -54 W	Fo, TrFo
I	11-Aug -1990	52 W	2-24S	11-24 S	49-57 W	Fo, Wo
C	10-Sep -1990	46 W	2-24S	18-21 S	46-55 W	Wo
A	14-Aug -1991	67 W	10-45S	10-24 S	53-64 W	Fo, Wo
	15-Aug -1991	64 W	10-45S	10-24 S	53-63 W	Fo, Wo
W	4-Jan -1989	10 E	2-16N	6-12 N	1 E-12 W	Fo, TrFo, Wo
E	1-Feb -1989	15 E	2-17N	7-12 N	3-14 W	TrFo, Wo
S	4-Feb -1989	12 E	2-17N	6-11 N	5-14 W	Fo, TrFo, Wo
T	21-Feb -1989	15 E	2-16N	5-11 N	5-9.5 W	TrFo, Wo
	23-Dec -1990	07 E	6-21N	6-9 N	4-9 W	Fo, TrFo, Wo
A	26-Dec -1990	02 E	4-19N	7-10 N	1 E-8 W	TrFo, Wo
F	27-Dec -1990	04 E	6-21N	8-12 N	4-1 E	TrFo, Wo
R	28-Dec -1990	07 E	8-22N	9-10 N	2-4 W	Tr Fo , Wo
I	30-Dec -1990	12 E	6-20N	7-9 N	7-9 E	TrFo, Wo
C	31-Dec -1990	10 E	6-21N	9-11 N	3-1 E	Wo
A	2-Dec -1991	03 W	4-19N	7-9 N	7-9 W	TrFo, Wo
	1-Jan -1992	09 E	2-16N	7-11 N	4-14 W	Tr Fo., Wo
	23-Mar -1990	117 E	0-35N	20-25 N	90-95 E	Fo,Wo
	24-Mar -1990	120 E	0-25N	20-25 N	95-105 E	Fo,Wo
S	29-Mar -1990	106 E	7-23N	12-19 N	96-101 E	Fo, Wo
E	30-Mar -1990	109 E	0-25N	15-20 N	97-103 E	Fo, Wo
	17-Apr -1990	109 E	10-15S	12-15 S	130-135 E	Wo
A	25-Mar -1991	102 E	5-25N	10-20 N	98-109 E	Fo, Wo
S	27-Mar -1991	108 E	5-30N	18-22 N	97-104 E	Wo
I	28-Mar -1991	111 E	8-35N	11-22 N	93-108 E	Fo, Wo
A	29-Mar -1991	113 E	15-35N	19-28 N	97-102 E	Wo
	5-Apr -1991	105 E	5-35N	15-20 N	96-100 E	Fo, Wo
	21-Apr -1991	101 E	5-35N	22-28 N	102-108 E	Fo, Wo

images of different dates along the life of the sensor were used to investigate sensor variations with time. The images concentrated on three regions on different continents where biomass burning is a common feature, and where different types of vegetation are burned. The 33 NOAA-11 images used and the areas covered are listed in table 3.1. Fires in South America included those related to forest conversion in southern Amazonia and in the Pantanal area, and to pasture renewal and agriculture in the savannas/cerrados of central Brazil. In West Africa the fire cases were of forest clearing and savanna burning. For Indochina, in Southeast Asia, fires represented forest conversion, diverse agricultural uses, and savanna burning. The latitudes of the 330 fires analyzed ranged from 28°N to 25°S, thus covering the tropical belt (see table 3.1 for a division of ranges by continent). Being in the north and south hemispheres, the regions studied have their dry, fire season in different periods of the year, and provided diverse sun-target-satellite geometry, thus representing an assorted collection of fire cases and conditions in diverse tropical vegetation ecosystems. For the 330 cases of fires selected, the digital counts of the pixels for a window with 15 lines and 15 columns in the five AVHRR channels centered around each fire were printed and used in the analysis given in the sections below. In some of the windows other fires also existed besides the selected fire; they amounted to 294 additional cases and were not eliminated in the analysis in order to make the windows represent typical operational conditions.

Digital Counts or Temperatures?

Digital counts (DNs)in the raw AVHRR images can be converted to albedo values in the case of channels 1 and 2 using prelaunch calibration coefficients. For channels 3, 4, and 5 DN conversion is made either to radiances or to temperatures using on-board calibration values available at each image scan line; these values are measured from a stable blackbody and from space (Kidwell 1991). The saturation temperature for channels 3, 4, and 5 is about 320°K, which is adequate to the temperature ranges of oceans and clouds, the primary targets for which AVHRR was designed; the minimum interval between temperature values is ∼0.1°K. According to Wien's displacement law, targets with temperatures in the range of vegetation fires, from 400°C to 700°C, have the maximum of emission from 4.3 μm to 3 μm, resulting in most of the energy concentrated in the band of channel 3 (3.55 μm–3.93 μm). For this reason, and also based on energy emissions from fires in channels 3 and 4 (see following section), our considerations about the use of DNs or temperature/radiances will refer only to channel 3, where the thermal signal from fires is stronger.

Fire pixels in channel 3, regardless of the size of the fire event and of the concentration of biomass burned, are normally not saturated, as shown below (see also Belward et al. 1993; Pereira and Setzer 1993; and Setzer et al. 1994). This fact is against theoretical calculations based on emitted thermal energy which indicate that even a small fire with ∼30 m × ∼30 m should saturate channel 3 (Robinson 1991). In the 330 cases of fires here analyzed together with the 294 additional cases in the windows, which amounted to 3094 fire pixels, just 54 pixels (1.75%) in 43 cases showed a zero DN, the nominal saturation value of channel 3. So far, the only explanation for this contradictory situation has been proposed by Setzer and Verstraete (1994), who assert that an engineering design problem exists in the on-board processing of the output signal from the channel-3 sensor. According to their hypothesis, signals much beyond the saturation limit of the sensor are indicated with the same values as those of targets below the saturation limit. Figure 3.1 shows the sensor curve proposed to explain why very hot targets like fires or very bright ones like sunglint on water do not saturate channel 3. Regardless of the reasons for this nonsaturation problem, a conceptual question ex-

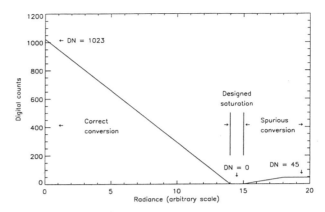

Figure 3.1 Possible explanation for the nonsaturation of channel 3 by fires and highly reflective targets, due to a spurious signal conversion (after Setzer and Verstraete 1993)

ists: what is the meaning of assigning a temperature of ∼310°K, therefore below the saturation limit of ∼320°K of the sensor, to a target known to have at least some 500°K and which should have saturated the sensor? Our view is that there is obviously a major fault in the conversion of DNs to temperature in the case of fires. With the use of temperatures or radiances instead of digital counts this error possibly increases because the causes and effects of the error are not taken into account in the radiance equations that rely on the DNs. For instance, if the curve suggested in figure 3.1 is correct, the higher the temperature of a fire the lower will be the calculated temperature using current calibration coefficients.

Another strong point in favor of using DNs instead of calculated temperatures or radiances to characterize fire pixels in channel 3 is that fire pixels have DNs about one order of magnitude different from surrounding pixels—see Pereira and Setzer (1993). Any classification algorithm to select fire pixels, based or not on surrounding background values, will be less effective when AVHRR temperatures or radiances are used because the differences are only of a few degrees K.

After Setzer and Verstraete (1994), NOAA modified the AVHRR on NOAA-14 prior to its launch in December 1994, and as a result, hot pixels, such as fires, tend to present full-saturation levels on channel 3 of this sensor. From a fire detection point of view, this change further increased the difficulties in discriminating fires on satellite overpasses during the afternoon.

Which AVHRR Channels?

Detection of active fires in day or nighttime AVHRR images relies on their thermal emissions and therefore only channels 3, 4, and 5 should be considered for this

Table 3.2 Summary of AVHRR digital count values (DNs) for 330 vegetation fires

	channel 2		channel 3		channel 4		channel 5	
	fires	window	fires	window	fires	window	fires	window
averages	156.1	159.9	34.4	414.9	332.1	347.5	290.6	304.4
stand.deviat.	32.0	26.8	8.8	89.5	49.2	42.8	55.9	50.1
maximum	362.0	283.0	50.0	600.0	602.0	575.0	596.0	573.0
minimum	93.5	109.0	1.0	114.0	200.0	227.0	148.0	182.0

purpose. Channels 1 and 2 in daytime images are useful only to see smoke plumes, or to detect fire scars in areas already burned. Nighttime images in the visible part of the spectrum have been used to detect fires on images of the military Defense Meteorological Satellite Program (DMSP) satellites (Cahoon et al. 1992), but the results are contradictory (Langaas 1993) as a possible result of diurnal patterns of burning practices and the difficulty in distinguishing between fires and artificial lights. Table 3.2 summarizes the response of AVHRR channels 2–5 in DNs for the 330 cases of independent vegetation fires selected and also for the corresponding image windows of 15×15 pixels surrounding and including each fire; the total number of fire pixels was 3094, and the total number of pixels in the windows was 74 250. For channel 3 the average values of the fires and the windows were 34.4 and 414.9, respectively, resulting in a ratio of $12:1$ among them. For channels 2, 4, and 5, the averages of the DNs for the fires and the windows were relatively close.

The average response of channel 3 is still very distinct from that of the windows when the standard deviations of the DNs of the fires are considered: 34.4 ± 8.8 for fires and 414.9 ± 89.5 for the windows (see table 3.2). For channels 2, 4, and 5, the standard deviations of the DNs for both fires and windows indicate no separability at all between fires and their surroundings. Also noticeable in Table 3.2 is that the range of DNs for fires in channel 3 was from 0 to 50, a very small one in comparison to those found in the other channels; the DN average ranges for fires in channels 2, 4, and 5 were 93.5–362, 200–602, and 148–596, respectively. A similar pattern was also observed when the individual fire pixels of each fire event were considered. Also important to note is that the DN range of fire pixels in channel 3, contrary to what occurs in other channels, is not associated to other natural targets, except for highly reflective soils or sun glint. Presenting all data for the 3094 fire pixels, or even to the 330 fire events would require space not available within the scope of this publication, and readers interested in the individual cases should contact the authors.

These observations agree with results from theoretical calculations based on the Planck equation of emissivity as function of wavelengths and temperatures. For the range of channel 3 (3.55 μm–3.93 μm) and assuming emissivity 1, a background surface at 30°C emits close to 0.6 W/m^2 while a fire at 500°C emits 1360 W/m^2, therefore some 2300 times more energy per unit area. For the range of channel 4 (10.35 μm–11.28 μm) the energy emissions at the same temperatures are, respectively, 30 W/m^2 and 510 W/m^2, with the fire emitting 17 times more energy than the background. On the other hand, the area in a pixel actually on fire (i.e., the fire front) is normally small compared to the total pixel area and this situation imposes a limiting effect for fire detection with channel 4. For example, a large fire with a hypothetical continuous front of 500 m by 5 m, or 2500 m^2, occupies only 0.2% of a pixel at nadir with diameter 1260 m. Weighing the emissions in relation to their areas for channel 4, the fire emits the negligible amount of 3.3% of the total energy of the pixel—1.25 million W against 37.3 million W for the rest of the pixel; at off-nadir angles this small percentage is reduced even further. The same situation for channel 3 results in the fire emitting 82% of the total energy of the pixel—3.4 million W against 0.746 million W for the rest of the pixel. Such values together with the curve proposed in figure 3.1 therefore explain why only channel 3 clearly detects the signal from fires, even from small ones. Considering in addition that fires in channel 4 are detected in a very wide range of DNs (200–602), its use in fire detection algorithms is not advisable; the same reasoning is also applicable to channel 5.

Figure 3.2 illustrates these problems by showing DNs cross-sections through two fires; in one case (figure 3.2a) channel 4 responds to the signal on channel 3, but in the other case (figure 3.2b) the response is the opposite, with channel 4 indicating lower temperature (higher DNs) at the fire pixels than at the nonfire pixels (the DN original scales for channels 3, 4, and 5 are inverted, so that higher counts represent lower temperatures). These fires belong to the same image of 27 March 1991 and are shown in the upper part of the photo of figure 3.5; their distance in columns was 212. Figure 3.3 shows the DNs for the windows surrounding the same two fires which are located in their center. Fire pixels in channel 3 are shown shaded, as well as the corresponding pixels in the other channels; other shaded areas off the center of the window are other fires existing in the same windows. In both cases the difference between fire and nonfire pixels in channel 3 is easily seen as more than one order of magnitude.

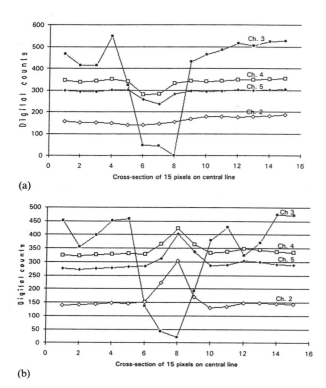

(a)

(b)

Figure 3.2 The DN cross-section through two fires in AVHRR channels 2–5. In (a) channels 3, 4, and 5 respond similarly to the fire, but in (b) the response of channels 4 and 5 is inverse. See figure 3.5 for the location of the two fires.

Channel 4 shows a significant reduction in DNs only for the first case but even there the minimum DN observed, 167, was only half of the surrounding nonfire pixels. The relatively high values in channels 1 and 2 in the center of the windows of the two cases result from the high reflectivity of the smoke plumes. The reason for the lower temperatures of channels 4 and 5 in the fire of figures 3.2b and 3.3b is probably the presence of a much larger smoke plume (see figure 3.5, fires A and B). In both fires the wind blows the plumes to the east and the AVHRR is also to the east. Therefore, with a larger and denser column of smoke in fire B than in fire A, the AVHRR channels 4 and 5 had more difficulty detecting fire B than A, and actually measured characteristics of the plume and not of the fire. The curve and data of channel 2 for fire B (see figures 3.2b and 3.3b) also shows a strong increase in reflectance from the smoke plume (scale not inverted in this channel) corroborating this interpretation. Channel 3, on the other hand, penetrated smoke to a much larger extent and was not attenuated in this case.

Furthermore, the relation between point temperatures over land and the surface brightness temperature indicated by channels 4 and 5 is not a straightforward one; in many cases, no correlation will even be found (Mansor and Cracknell 1992). This relation can only be estimated if the emissivity of the surface is known and by applying atmospheric correction algorithms based on actual atmospheric profiles of water vapor, aerosols, and temperature for the site of interest at the time of image acquisition—currently an impossible task in terms of available data for tropical regions and of computational needs for a global/regional fire product. Therefore, our suggestion is that channel 3 should be used to detect fires at daytime despite the possible sensor problem shown in figure 3.1 and also of the warning of Kidwell (1991) in the NOAA-series user's guide, page 3-2: "Users should be aware that AVHRR channel 3 data on each TIROS-N series spacecraft have been very noisy due to a spacecraft problem and may be unusable, especially when the satellite is in daylight."

Limitations of Channel 3

Many limitations should be considered when using AVHRR's channel 3 for fire detection. Some of them, already known (Setzer 1993), are: fires not active during the satellite overpass, fire fronts smaller than ~50 m, clouds in the fire-satellite line-of-sight, below canopy fires, and solar reflection. The problem of solar reflection and other limitations are discussed in the following subsections.

Pixel Geometry

The AVHRR resolution is generally referred to as 1.1 km at nadir, and degrading toward the image edges. However, the use of channel 3 for the detection of fires, which are small in relation to the satellite resolution and have a signal one order of magnitude higher than surrounding targets, needs a more thorough analysis. To start with, the instantaneous field-of-view (IFOV) of channel 3 is 0.00151 radians, larger than for other channels; channels 1, 2, 4, and 5 have IFOVs of 0.00139, 0.00141, 0.00141, and 0.00130 radians, respectively (Kidwell 1991). At nadir, considering the satellite at its nominal altitude of 833 km, this corresponds to a circle with a diameter of 1.26 km covering an area of 1.24 km². At the image edge the pixel becomes an ellipse with axes of 2.66 km in the along-track direction and 7.25 km in the along-scan direction, and with an area of 15.14 km². The interval between any AVHRR consecutive pixels in the same line is 0.0009443 radians, obtained from the scan range of 0.967 radians (55.4 degrees) divided by 1024, the

Channel 01

y \ x	691	692	693	694	695	696	697	698	699	700	701	702	703	704	705
1676	141	137	132	130	131	129	128	128	128	128	129	127	125	124	128
1677	136	132	129	126	129	126	125	126	128	126	128	124	122	144	168
1678	131	130	128	128	126	124	124	125	125	128	127	128	135	160	178
1679	127	128	130	129	124	122	123	122	123	125	141	175	187	203	190
1680	129	126	125	123	123	123	122	128	137	142	173	184	177	199	241
1681	129	128	124	123	124	125	125	155	192	181	192	219	220	217	356
1682	124	123	124	125	126	128	162	198	205	218	213	199	204	270	348
1683	121	122	131	133	128	131	145	146	143	142	136	125	121	120	133
1684	123	140	151	134	121	126	129	122	121	123	122	118	117	117	116
1685	127	158	158	129	116	120	122	120	117	114	114	115	117	122	133
1686	126	130	128	122	121	121	119	117	118	116	114	115	128	154	172
1687	129	128	125	124	123	121	121	118	116	116	115	119	147	164	144
1688	130	128	124	123	123	123	122	118	118	118	118	122	130	120	116
1689	129	128	126	124	124	124	125	122	118	119	119	117	116	115	116
1690	128	127	126	125	127	127	124	121	121	120	118	116	116	116	117

Channel 02

y \ x	691	692	693	694	695	696	697	698	699	700	701	702	703	704	705
1676	154	161	161	160	161	159	156	151	150	150	154	156	158	160	162
1677	153	164	165	161	162	153	150	150	150	148	146	159	164	168	179
1678	165	169	167	161	152	145	145	147	153	149	155	166	171	178	184
1679	160	163	163	156	153	155	156	160	160	157	174	190	191	193	184
1680	160	168	160	152	159	161	163	169	168	153	167	181	179	194	225
1681	158	161	158	155	161	163	166	185	195	176	181	207	208	209	340
1682	154	152	152	156	156	156	173	193	198	209	208	201	211	283	355
1683	157	152	152	148	141	140	145	155	168	178	179	175	179	182	188
1684	155	154	153	140	130	131	147	161	163	160	160	166	174	179	180
1685	163	170	161	153	156	160	158	167	165	160	161	168	174	176	180
1686	163	160	156	162	167	164	159	166	165	163	164	162	167	179	180
1687	160	158	159	164	167	164	164	162	160	163	163	164	169	170	167
1688	157	157	160	164	166	168	167	159	156	160	163	160	160	170	169
1689	157	160	160	166	168	167	163	160	160	164	166	168	169	169	168
1690	156	158	160	160	160	161	163	160	157	158	164	168	167	166	169

Channel 03

y \ x	691	692	693	694	695	696	697	698	699	700	701	702	703	704	705
1676	413	449	481	509	492	500	507	485	476	442	419	427	464	469	446
1677	401	472	484	521	493	485	483	456	440	388	277	400	479	441	418
1678	472	477	495	497	489	481	432	403	422	392	385	454	435	443	495
1679	471	478	443	457	496	504	482	456	448	419	418	445	460	466	478
1680	439	492	470	471	476	495	492	467	431	352	331	425	509	530	512
1681	451	449	468	455	435	423	454	464	491	454	481	502	539	529	372
1682	482	478	446	438	405	104	31	208	475	526	555	545	519	451	351
1683	468	415	414	549	325	46	45	7	430	464	485	517	504	524	526
1684	249	73	289	678	45	46	46	39	445	445	434	491	508	514	525
1685	45	46	45	171	293	1	43	372	490	515	502	494	485	500	526
1686	46	46	45	127	403	385	421	463	483	504	516	416	34	39	468
1687	21	34	104	373	457	459	459	485	505	498	637	43	46	46	16
1688	388	372	409	460	473	491	513	526	505	587	517	46	46	45	461
1689	460	469	449	477	500	480	483	504	498	494	440	348	352	449	511
1690	471	481	491	482	472	470	488	495	467	466	486	477	485	507	516

Channel 04

y \ x	691	692	693	694	695	696	697	698	699	700	701	702	703	704	705
1676	331	334	340	345	342	345	344	336	329	320	312	317	326	323	320
1677	326	333	342	348	343	340	334	325	319	303	292	313	328	322	326
1678	333	337	341	342	340	334	321	315	315	308	311	321	321	329	345
1679	333	335	330	335	342	340	332	324	321	313	317	330	340	348	355
1680	328	340	338	336	338	341	337	330	321	307	314	338	357	371	398
1681	330	332	335	332	331	333	336	342	350	347	347	371	379	400	478
1682	341	337	329	328	327	319	319	339	361	371	375	369	369	410	450
1683	345	339	342	350	341	280	282	331	343	340	344	349	348	351	355
1684	338	337	357	366	313	167	205	324	341	333	332	341	345	347	349
1685	275	151	296	348	347	330	317	339	349	350	343	342	343	348	354
1686	247	80	236	335	339	336	340	344	349	346	344	337	335	357	349
1687	344	338	336	338	340	340	341	348	349	340	335	308	249	291	349
1688	339	337	337	340	342	347	350	351	346	336	323	250	224	316	345
1689	336	336	335	340	344	341	342	344	342	343	342	339	340	347	348
1690	336	338	339	339	336	337	339	340	334	336	339	339	342	347	347

Figure 3.3 The AVHRR channels 1–4 DN values for windows of two fires located at their centers, but with inverse response in channels 3 and 4. Figures *a* and *b* contain, respectively, the two cross-sections of Figure 3.2. Shaded areas correspond to the fire pixels in channel 3.

(a)

Channel 01

y \ x	479	480	481	482	483	484	485	486	487	488	489	490	491	492	493
1704	101	99	99	101	102	104	108	112	117	123	122	131	157	165	154
1705	104	105	107	109	107	109	112	115	117	123	128	141	158	162	158
1706	109	113	114	114	108	109	115	117	115	130	147	144	146	161	163
1707	113	114	113	110	110	114	121	129	134	149	169	159	153	159	161
1708	112	112	110	111	116	124	129	144	173	176	166	156	158	159	161
1709	108	116	117	117	117	139	153	169	186	171	157	158	160	154	154
1710	101	106	113	115	123	160	201	202	157	144	146	156	158	152	148
1711	103	105	108	109	115	138	225	317	180	133	144	166	160	152	151
1712	105	113	114	116	121	130	146	187	161	145	159	187	163	157	168
1713	105	108	112	116	120	122	123	125	125	135	160	164	157	162	163
1714	110	110	110	109	110	113	113	113	114	117	126	137	153	161	155
1715	114	113	111	109	110	112	115	117	117	119	128	161	185	168	144
1716	114	114	114	111	110	112	113	119	120	120	125	179	289	255	193
1717	112	113	115	113	118	125	118	120	120	121	128	158	186	156	142
1718	112	116	116	115	124	132	123	127	136	135	130	130	132	131	134

Channel 02

y \ x	479	480	481	482	483	484	485	486	487	488	489	490	491	492	493
1704	129	136	141	137	139	144	140	143	141	141	136	137	150	160	156
1705	130	135	139	130	129	140	144	148	145	147	141	141	151	155	157
1706	133	136	140	135	130	140	146	150	148	152	150	144	143	148	151
1707	141	140	136	132	128	139	150	150	149	156	160	149	143	143	147
1708	139	141	141	145	144	148	155	155	163	163	155	146	143	145	149
1709	146	151	153	158	149	146	151	157	166	156	146	144	147	146	148
1710	146	151	153	153	149	160	187	180	141	133	133	139	147	146	144
1711	140	142	143	148	145	153	224	304	171	130	135	148	149	145	144
1712	138	138	138	144	150	152	163	196	162	148	159	177	153	148	157
1713	132	136	144	146	149	152	149	145	145	157	173	167	152	156	158
1714	139	140	141	136	137	141	144	142	141	148	155	151	152	160	156
1715	139	133	136	138	138	139	135	136	141	145	151	170	181	169	150
1716	135	133	137	140	142	143	137	138	141	144	145	193	287	253	198
1717	127	133	139	142	144	144	136	137	136	140	151	178	202	174	160
1718	123	134	138	139	141	144	137	139	145	147	146	147	152	152	154

Channel 03

y \ x	479	480	481	482	483	484	485	486	487	488	489	490	491	492	493
1704	326	329	376	375	465	462	442	411	404	396	389	357	368	416	448
1705	350	328	276	247	414	441	373	378	444	422	184	118	370	436	470
1706	349	273	115	215	417	311	38	18	443	444	128	61	392	404	433
1707	386	334	299	330	404	481	98	45	20	454	394	281	443	457	457
1708	335	208	326	420	464	310	432	412	466	472	479	474	482	501	482
1709	401	14	193	506	28	45	42	40	43	323	405	376	473	476	493
1710	446	416	447	434	44	45	43	43	44	120	461	468	466	454	475
1711	451	355	399	452	459	137	42	20	193	380	427	325	371	474	471
1712	400	331	447	461	464	242	169	27	44	119	390	261	257	461	455
1713	438	442	451	475	491	434	156	43	43	359	265	201	345	462	481
1714	403	405	447	435	438	358	373	380	383	362	297	334	366	416	476
1715	434	424	440	445	368	360	411	408	381	395	378	348	372	442	490
1716	392	408	416	427	386	402	391	421	405	410	432	352	228	271	352
1717	348	390	425	442	259	308	401	400	323	394	428	409	395	462	471
1718	281	319	352	422	302	363	411	412	397	369	379	280	256	427	472

Channel 04

y \ x	479	480	481	482	483	484	485	486	487	488	489	490	491	492	493
1704	291	300	304	310	324	321	313	306	303	303	298	294	304	317	325
1705	291	291	289	290	315	319	313	312	313	311	304	302	307	321	330
1706	300	293	288	296	318	323	312	316	320	319	315	311	312	318	325
1707	302	295	296	311	326	332	323	317	327	330	332	330	330	333	332
1708	304	295	305	326	338	334	336	341	347	346	343	340	342	343	337
1709	313	298	312	332	326	313	330	334	332	344	345	344	346	342	338
1710	317	313	319	329	323	325	362	362	323	334	343	345	344	337	334
1711	325	321	326	328	331	329	366	422	363	332	338	349	344	337	336
1712	324	326	329	330	330	325	328	340	314	314	337	353	343	337	343
1713	324	325	326	332	333	323	311	298	294	308	325	330	330	334	340
1714	328	329	324	321	324	320	316	307	301	301	312	317	321	332	340
1715	320	320	323	322	318	315	313	309	302	305	315	345	366	357	347
1716	307	312	317	324	321	312	310	313	309	309	314	353	423	414	369
1717	295	306	319	326	321	316	315	314	309	310	320	345	362	350	337
1718	282	296	315	323	318	318	316	315	311	307	313	324	332	333	332

(b)

Table 3.3 PIxel geometry for AVHRR channel 3 with IFOV = 0.0015rd, satellite altitude of 833 km, and earth radius of 6378 km

Scan Angle (degrees)	Alongtrack Diameter (km)	Alongscan Diameter (km)	Area of Pixel (km2)	Alongscan Pixel dist. (km)	Sat.Pixel Distance (km)	Sat.Px.Cn. Angle (degrees)
0.0	1.26	1.26	1.24	0.79	833.00	180.00
10.0	1.28	1.31	1.31	0.82	847.60	168.68
20.0	1.35	1.46	1.55	0.92	894.28	157.25
30.0	1.49	1.80	2.10	1.13	983.81	145.58
40.0	1.73	2.51	3.40	1.57	1142.82	133.39
45.0	1.91	3.18	4.78	1.99	1267.55	126.92
50.0	2.18	4.36	7.49	2.73	1446.90	119.99
55.4	2.66	7.25	15.14	4.54	1760.82	111.46

Table 3.4 Range of digital counts for AVHRR/NOAA-11 channel-3 fire pixels in 80 cases of vegetation fires (from Setzer and Malingreau 1993)

Date	Region	Case:	1	2	3	4	5	6	7	8	9	10
4-Feb-89	W-Africa	max.	41	41	41	36	39	41	36	41	41	37
		min.	1	5	39	28	39	36	36	10	12	0
		N.Pix	11	11	4	2	2	9	1	9	19	5
26-Aug-89	S-America	max.	43	43	41	43	43	44	43	43	43	42
		min.	25	6	38	3	1	5	4	3	7	11
		N.Pix	14	13	2	4	14	29	8	18	7	4
30-Mar-90	SE-Asia	max.	44	43	43	44	44	45	45	44	44	43
		min.	42	41	4	23	20	23	27	43	0	42
		N.Pix	6	4	11	10	6	12	8	2	4	2
11-Aug-90	S-America	max.	42	48	44	45	45	45	45	45	45	45
		min.	7	0	41	44	41	35	18	38	7	25
		N.Pix	4	23	3	3	5	7	7	15	10	10
31-Dec-90	W-Africa	max.	44	30	44	1	45	35	42	45	45	45
		min.	40	30	44	1	6	19	29	5	20	43
		N.Pix	2	1	2	1	4	2	2	7	5	2
28-Mar-91	SE-Asia	max.	45	26	46	42	45	46	46	46	46	46
		min.	44	19	42	22	44	45	15	43	43	45
		N.Pix	2	2	7	4	2	4	8	6	5	4
15-Aug-91	S-America	max.	47	47	47	47	47	47	47	47	47	47
		min.	37	2	11	33	33	12	2	3	30	22
		N.Pix	8	16	18	8	11	12	56	11	21	12
2-Dec-91	W-Africa	max.	45	47	45	37	45	46	46	47	45	46
		min.	27	17	45	26	42	25	4	41	42	37
		N.Pix	4	11	1	2	2	5	7	8	2	6

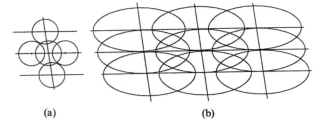

(a)   (b)

Figure 3.4 Overlap of neighboring channel-3 pixels (*a*) close to nadir and (*b*) at 50° off-nadir. Note that the distance between scan lines remains constant and that the distance between pixels increases proportionally to the along-scan diameter of the pixels; on the other hand, the increase in the along-track diameter causes major pixel overlaps. Pixels drawn in proportion to their dimensions.

number of pixels sampled in the range. At nadir this results in an along-scan distance of 0.79 km between centers of adjacent pixels in the same line; at the image edge, with a satellite-to-pixel distance of 1761 km, this distance becomes 4.54 km. Table 3.3 shows other values for intermediate scan angles. The distance between consecutive scan lines is 1.079 km, assuming the earth to be a sphere with radius 6378 km, the satellite orbit a circle at an inclination of 98 degrees, and 360 scans per minute; this distance is constant in any part of the image.

Therefore, neighboring pixels overlap in any part of the image. At nadir, 51.85% of a channel 3 pixel is also covered by the two neighboring pixels in the same scan line, and an additional 11.47% by the pixels in the neighboring scan lines (see figure 3.4*a*). So, at the minimum no-overlap condition, 36.84% of a pixel is not covered by adjacent pixels. Towards the edge of the image, as shown in figure 3.4*b*, only the percentage overlap from pixels in neighboring lines increases because in the along-scan direction the pixel size increases together with the distance between pixels; in the across-track direction, the distance between pixels, that is between scan lines, remains constant but the pixel across-track diameter increases with its distance from the satellite. At the image edge, almost 100% of a

pixel is covered by adjacent pixels and even skipping every other line will still cause some along-track overlap.

The effect of adjacent pixel overlap in channel 3 plays a complex role in fire detection. Depending on its position in the area covered by a pixel, a fire front of ~ 200 m at nadir can be detected simultaneously in up to three neighboring pixels, each one with 1.24 km². At the image edge this same extreme situation can result in the same fire being detected in four neighboring pixels, each with 15.14 km². Reduction of this overlap effect could be tried using image deconvolution techniques, but the AVHRR's system modulation transfer functions (MTF) are not available (Breaker 1990) and this possibility remains beyond the scope of the present work.

Degradation of the Sensor

Degradation of the AVHRR channels 1 and 2 was reported for NOAA-11 (Holben et al. 1990), and for NOAAs 7 and 9 (Kaufman and Holben 1993). In the case of channel 3, the existence of on-board calibration data reduces the effects of any sensor response change with time when conversion of radiances or temperatures is applied to the raw data; possibly for this reason no studies of channel 3 variations from other authors are found. Within our view that the use of raw DNs should be preferred for fire detection in relation to radiances and temperatures, any temporal variation in channel 3 will influence detection algorithms. Table 3.4, from Setzer and Malingreau (1993), shows the maximum and minimum pixel DNs for 80 cases of vegetation fires in 8 images representing almost three years in the life of of NOAA-11. It shows that the maximum DN of fire pixels increases with time, and varied from 41 after the satellite's launch in February 1989, to 46–47 in December 1991. A summary of table 3.4, excluding the DN 48 of the August 1990 image, yields the following temporal variation of channel 3 fire detection limit (Setzer and Malingreau 1993):

Date	DN max
Feb. 1989	41
Aug. 1989	43–44
Mar. 1990	44–45
Aug. 1990	44–45
Dec. 1990	44–45
Mar. 1991	45–46
Aug. 1991	47
Dec. 1991	46–47

Unpublished data from the AVHRR fire detection program in Brazil shows that this DN limit went up further to 48 in August 1992 and to 52 in August 1993. According to this same program, similar variations were observed in past years for the AVHRR channel 3 on-board NOAA-9, thus suggesting that channel-3 degradation with time seems a common feature in the AVHRR series. Therefore, if a fire detection algorithm is entirely or partially based on channel-3 thresholding, the DN value of fire definition has to be checked and adjusted a few times per year. New DN threshold limits for fires are easily found from a histogram of a channel-3 image containing many confirmed fires, that is, with associated well-defined smoke plumes; the DN in the histogram where a sharp peak occurs close to the saturation hot end-of-scale, immediately followed by a discontinuity towards the other end, marks this limit.

Sunglint

Reflections of solar light in water surfaces and highly reflective soils can impose a strong limitation on AVHRR fire detection because they result in the same DNs in channel 3 as fires do (Setzer and Verstraete 1994). Energy from either fires or sunglint reach the satellite at an intensity with the magnitude of $10e^{-9}$ W/m^2. Therefore, channel 3 has a similar sensitivity to reflected light and to fires, being even more sensitive to sunglint than channels 1 and 2. Sunglint occurs in oceans, rivers, lakes, rice paddies, and fish ponds, among others, whenever the sun is close to the plane of the AVHRR swath. The earth stripe parallel to the satellite trajectory which is subject to sunglint is on the order of several hundred kilometers wide by many thousand kilometers long and is found in the left half of images during afternoon ascending orbits. Figure 3.5 illustrates this effect by showing in light gray the Gulf of Siam entirely imaged with sunglint. Because the sun-earth-satellite geometry changes seasonally the sunglint problem is not constant in AVHRR images. For example, the dry/fire season in the central part of South America starts around June when the sun is over the Northern Hemisphere reaching the solstice, and until the equinox of late September no sunglint interferes with fire detection; the same conditions should also prevail for the fire season in South Africa. On the other hand, the detection of fires in Indochina during late March and April at the end or the dry/fire season in that region is strongly impaired by sunglint because at that time the sun is in the Southern Hemisphere, very close to the equator. For all purposes, if a full-year fire product is considered, sunglint will occur at all tropical regions during about one month.

Reflective Soils

Reflective soils also cause very low DNs in channel-3 images, and in many cases the same values associated with fires. Of the total solar energy emitted, only 0.32% is in the range of AVHRR's channel 3; compared to the maximum of 2074 W/m^2-μm at 0.48 μm, only 12 W/m^2-μm reach the earth's surface in the range of channel 3 with zenithal illumination. The reflectance of barren soils, according to the only two references found for channel 3 (Hovis 1966; Suits 1989), vary from 0.13 to more than 0.4. For these values the amount of reflected solar energy reaching the satellite will be of the same order as that from fires and sunglints ($10 \times e^{-9}$ W/m^2), which explains the confusion among these three targets in daytime AVHRR images. Vast areas can be subject to this reflection problem, like all the north of Africa and the Sahel zone in late winter (Grégoire et al. 1993). In the other regions analyzed, this problem was observed only in smaller scales: in the savannas of east central Brazil during October, and in the savannas of northern Indochina and in the Philippines in April. This phenomenon was noticed to be cyclical, occurring and changing its place according to the sun's position like in the case of sunglint from water. Figure 3.5 shows an area of sunglint from land, located close to the east limit of the Gulf of Bengal, and not too distant from an area of major fire activity in the seasonal forest and upland agriculture areas of South China, North

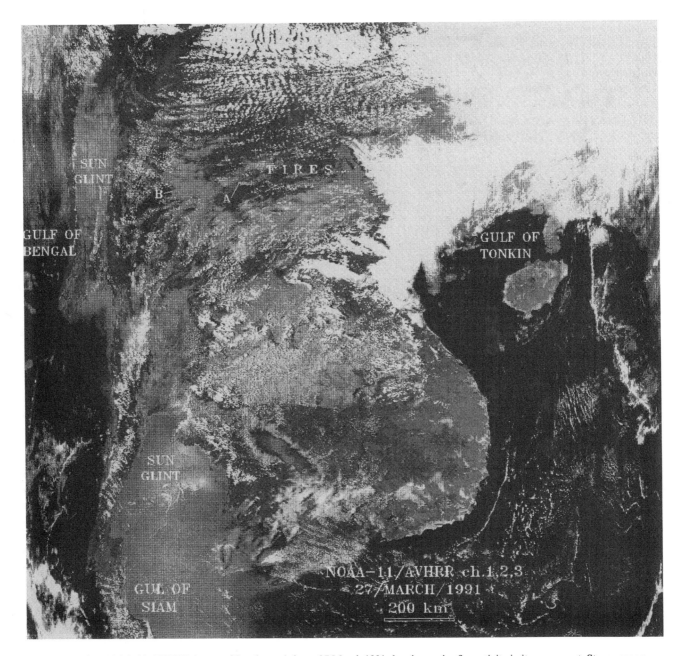

Figure 3.5 The NOAA-11 AVHRR image of Southeast Asia on 27 March 1991 showing major fire activity in its upper part. Strong ocean sunglint occurs in the Gulf of Siam, and land reflection is found east of the Gulf of Bengal. Fires A and B at the origin of large smoke plumes are described in the text.

Vietnam, North Laos, Northeast Birmania, and East Myanmar.

A/D Bit Conversion Problems

Another point to be considered in the detection of fires when channel-3 DNs are used results from imperfect analog-to-digital conversion (A/D) of signals on-board AVHRR. A close examination of a histogram of any AVHRR image of any channel shows peaks and dips occurring at regular intervals in the scale of DN values. These irregularities result from a bias of exactly one DN count in the allocation of DNs in the analog signal of the sensor. In the case of NOAA-11 we observed that this effect is particularly noticed in DN multiples of 8 and 32 in the scale of 0 to 1023, indicating that bits 5 and 7 (where bit 10 is the less significant one) in the A/D are the most deficient ones. As a consequence, for example, an AVHRR histogram will show a discontinuity in which DN 8 shows an excessive number of pixels while DN 7 shows a small number of pixels in relation to the overall trend of the histogram at this range. In this case, many pixels whose analog signal (voltage) nominally corresponds to DN 7 are considered as having the signal of DN 8. The consequence in terms of radiometric resolution is an occasional error of one DN, which is actually the expected precision limit of the AVHRR. In terms of fire detection this implies a faster "degradation" of the sensor when the threshold limit reaches DNs 47 and 55.

Eliminating False Channel-3 Fires

Figure 3.5 is a good example of a critical and real situation where sunglint in an AVHRR image must be eliminated in order to allow proper identification of active fires; without any screening most of the Gulf of Siam and parts of land in the upper left corner of the image would be erroneously classified as fires in channel 3 together with the many hundreds of real vegetation fires also present in the image. The most simple and practical automatic way to minimize sun reflection in channel-3 fire detection is the use of the latest of possible multiple and consecutive AVHRR overpasses that cover the region subject to sunglint or to soil reflection. With this technique an area in the west half of an image with specular reflection will be imaged at a very large scan angle towards the east in the next orbit. At equatorial latitudes this possibility is restricted only to regions at off-nadir scan angles larger than 44°, and just for 3 out of every 9 days—the orbital repetition cycle of the NOAA-series satellites. At latitudes of 30°

multiple coverage can be extended to scan angles larger than 38° during 4 out of every 9 days. A good illustration of these repetition patterns in consecutive orbits is found in Goward et al. (1991), while Gutman (1991) shows the 9-day periodicity of scan angles for a single location. Orbital prediction programs combined with calculations of sun-target-satellite geometry can be used to exclude sunglint areas from the fire-detection procedures in areas subject to sun reflection that can not be re-imaged in no-sunglint conditions.

Another possibility to exclude sunglint is the use of masks following contours of continents and main rivers. However, this is restricted only to large bodies of water; in addition, maps and navigation imprecisions will always cause a considerable amount of errors along even the main bodies of water. A third option is the combination of multiple AVHRR channel data. In most cases channels 4 and 5 will show no response at all for fires. Their use to determine if sun reflection is occurring is also very limited because both fires and reflection can occur in a background of relatively high or low temperatures. For example, in the same image of South America sunglint is found in the high peaks and dry salt lakes of the Andes mountains in a below-freezing temperature background and also in a 30°C environment in Brazilian cerrados. If thin clouds or smoke cover the region these channels may not penetrate them when channel 3 does and the temperature indication they will produce will not make sense in terms of fire detection. Channels 1 and 2 are probably the best ones to eliminate sun reflection. The key in this option is to select the correct thresholding value because it will vary with the sun-pixel-satellite geometry; bidirectional reflectivity equations may be thus used and account for variations of solar illumination. As in channels 4 and 5 thin clouds and smoke plumes over fires will cause errors because in this case either channel 1 or 2 will consider a fire as sunglint.

A last possibility to identify sun reflection can be considered. Fires are usually limited to a few contiguous pixels, while sun reflection is associated with hundreds or even thousands of contiguous pixels; for the total of 624 cases of fires in this study (330 cases selected and 294 others in the windows), the average was 4.96 fire pixels per fire. Therefore a simple algorithm that checks the number of pixels in the spatial distribution of fire pixels in channel 3 could eliminate sun reflections in most cases.

A global product of AVHRR full-resolution images involves very large amounts of data processing since one such composite image has ~1.4 gigabytes per channel. Having in mind the shortest possible computing demand for a global fire product, our initial tentative suggestion for detecting real fires and eliminating false fires in channel 3 caused by sun reflection is basically the following:

- Mask out oceans, seas, and large lakes in the processing of images.

- If a region is subject to sun reflection, try the use of the consecutive orbit for that region.

- Use simple thresholding to select hot/fire pixels in channel 3.

- For each fire pixel detected in channel 3, check the corresponding pixel value in channel 1 to eliminate strong glint cases; if the value in channel 1 is above a specific (bright) threshold for that part of the image, reject the pixel as a fire pixel.

- If the pixel value in channel 1 is below the specific (dark) threshold, check the spatial distribution of fire pixels in channel 3; for less than ~ 20 contiguous pixels, accept the fire event; otherwise reject it as a fire.

And finally, an effect that reduces sunglint and is provided by orbital variations of the afternoon pass of NOAA satellites must be mentioned. Although called sun-synchronous the equatorial crossing time of these satellites varies gradually from their launching dates, resulting in a drift, or delay, of about 20 minutes per year. The resulting change in the sun-earth-satellite geometry with a lower sun in the sky significantly shifts the sunglint toward the left edge of the images, improving fire detection in the more central part of the image.

Border Effects

Usually AVHRR analyses discard information from pixels beyond a scan angle of ± 30 degrees to avoid the use of large pixels and to minimize atmospheric optical effects. This is not necessary in the case of fire detection using channel 3. The size of a pixel does not interfere with fire detection because the energy emitted even by small fires (fire front larger than ~ 50 m) is enough to reach the fire detection limit regardless of the pixel size. At the 3.75-μm band range atmospheric transmittance is about 90% (LaRocca 1989), higher than for any other part of the spectrum (in the visible and near-infrared parts it is $\sim 60\%$). This makes channel 3 less sensitive to atmospheric attenuation, even at large-scan angles with longer slant distances (for the limit scan angle of 55.4° the air mass is ~ 2.7 times the zenith air mass). It is also much less sensitive to haze by

a factor of 30–200 times compared to channel 1 in the 0.64-μm band (Kaufman and Remer 1994). This last characteristic is of particular importance in off-nadir fire detection since regions subject to intense biomass burning are covered by dense smoke palls of millions of km^2 (see Andreae et al. 1988; Helfert and Lulla 1990; Setzer and Pereira 1991b). Many of the 330 cases of fires analyzed here were purposely selected very close to the image borders in order to find out if they presented any particular radiometric or spatial patterns. Most of them were in the column range of 0 to 100 in relation to the borders at scan angles larger than 50°, with an extreme case at column 5 and 55.13° off-nadir. One of the cases had just one fire pixel associated, but most of the others presented a distribution more elongated in the across-track direction, as expected from the considerations presented above. As in central parts of the images smoke plumes in the edges of the images also had channel-3 fire pixels at their origin within the same DN limit as for the rest of the image, thus indicating that channel-3 fire detection can be extended to the full image. The only constraint is the loss of geographical precision in the location of fires because of the larger size of pixels at the image edges.

Geometric Corrections

Because of the NOAA satellite's low altitude of ~ 830 km and the radiometer scanner wide-angle coverage of $\pm 55°$, the off-nadir geometry of pixels is distorted, becoming an ellipse of 2.7 km $\times$ 7.3 km at the image edges for channel 3. AVHRR geometrically corrected images are referenced to a base of pixels with constant size forcing the correction algorithm to repeat and interpolate original pixel values to obtain corrected pixels. Fire fronts are usually restricted to very small areas relative to the AVHRR resolution and most fires are indicated by a few contiguous pixels. Therefore, after geometric correction the number of fire pixels in a particular fire event will differ from that in the original image. This effect is illustrated in table 3.5, where the same 10 fire events in one AVHRR image were compared in their raw and geometrically corrected forms. The table shows that for large off-nadir angles, as in the last two cases, the number of fire pixels can double in the corrected image. In terms of ground-surface equivalence of fire pixels both types of images present similar results, but for characterizing individual fires the corrected image introduces additional difficulties. For example, one isolated fire pixel in a raw image indicates a fire front between ~ 50 m

Table 3.5 Effect of geometric correction in the number of fire pixels for the AVHRR image of 15 August 1991. Fires are listed in sequence of off-nadir angles.

	Raw image			Geometr.corrected image		
Off-nadir angle	No.Fire pixels	Maxim. DN	Minim. DN	No.Fire pixels	Maxim. DN	Minim. DN
7.5	16	47	2	13	47	2
9.0	8	47	37	6	47	37
16.4	18	47	11	17	47	11
18.2	57	47	2	53	47	2
21.7	21	47	30	18	47	30
22.9	12	47	22	11	47	22
23.8	11	47	3	9	47	3
28.2	12	47	12	13	47	12
48.1	11	47	33	23	47	33
48.3	8	47	33	20	47	33

and a few hundred meters long, most likely from just one fire. If in the corrected image two or more fire pixels are produced for the same fire, the same interpretation about its size cannot be made; in this case the possibility of two or more independent but close fire events has to be considered. This difficulty increases for mosaics composed of many corrected AVHRR images, causing random modifications in the number of fire pixels. For fire pixels close to nadir the effect of correction may actually reduce the number of fire pixels, as seen in table 3.5. In this case, for fires associated with few pixels the risk exists that important information like the minimum or maximum value of the fire pixels is lost in the transformation of images. Our suggestion towards a global composite product is that raw unprocessed images be used to reduce loss or modification of information, and the number of fire pixels in grid cells be updated on a daily basis if possible. If the objective is the real-time combat of fires, then the geographical coordinates of fire pixels in the raw images should be used.

8-Bit or 10-Bit Images?

The AVHRR images are transmitted by the NOAA-series satellite and received by ground stations on a 10-bit/1024-level radiometric scale. However, some image-processing systems operate on an 8-bit/256-level scale configuration either to speed up processing or to fit internal software and hardware requirements; as a result AVHRR image conversion to 8-bit resolution is common. This conversion is usually achieved by dropping the two less significant bits in the 10-bit data, or by dividing 10-bit values by 4 and giving the result as a round number. As shown above, fires are better detected in channel 3, up to a very specific and precise count level, which changes in the satellite's life. If the conversion to 8-bit is made, this distinction becomes more difficult and errors in the selection of fire pixels

may occur. For example, if the threshold detection is 0–46 in the 0–1023 scale, it will correspond to the range 0–11 on a 0–256 scale when the two least significant bits are dropped or to the range 0–12 if division by 4 with rounding is used. In the first case the eleventh level includes original counts 44–47, and in the second case the twelfth level includes the counts 46–49. The upper limit in both cases does not correspond to the exact fire threshold detection and may include values not associated with fires. Therefore, 10- to 8-bit conversion should be avoided in AVHRR fire detection whenever possible.

Another problem may occur in the 10- to 8-bit conversion, and is caused by badly designed software, as was the case in a system used worldwide that we tested at the start of the present work. Unsound as it may seem, in this case the conversion is done by dropping the two most significant bits instead of the two less significant ones. One of the symptoms caused by this error is that badly converted channel-3 images present many pixels at the cold saturation extreme of the scale with $DN = 255$; unfortunately, publications in the literature contain this mistake.

Conclusions

The use of NOAA/AVHRR 1.1 km resolution images for fire detection in different tropical ecosystems of the world and in diverse imaging conditions was analyzed in the perspective of developing a global fire product. The principal contribution of this analysis is the recommendation of the use of channel 3 as the main channel for fire detection anywhere and at any time; in particular, its DNs should be used instead of any derived radiative parameter information. Although this channel presents inaccurate response to fires due to a sensor engineering problem, it is this very problem that fortuitously allows a clear identification of fire spectral signatures. The main constraint in its use at daytime is solar reflection from exposed reflective soils and from water surfaces; to minimize it the combined use of channel 1 and of a spatial analysis of fire pixels is recommended. Because the energy seen by the satellite in fires or reflections has about the same magnitude, the resulting misidentification of targets is expected to occur in satellites with sensors of higher resolution operating in the 3–4 μm solar spectrum region.

Other limitations of channel 3 in fire detection presented and analyzed were excessive overlap of neighboring pixels causing repetition of fire detection; degradation of the sensor along the years, which require updates in fire detection thresholds; and slight misreading of one DN in the on-board analog-to-digital conversion. In a global fire product, these limitations have to be considered together with already known ones such as: fires not active during satellite overpasses, fire fronts smaller than ~ 50 m, clouds in the fire–satellite line of sight, and fires not reaching the canopy. Combined effects of these fire-related factors have been reviewed and evaluated in Malingreau (1990).

Notwithstanding many limitations in the use of high-resolution AVHRR images, a global fire product is attainable and its production should be started at the earliest possible time. No other source of data in the next years will provide regular and consistent world-wide detection of fires.

Acknowledgments

We acknowledge the support of the Brazilian Scientific Council (CNPq) which made the development of this work possible through grant no. 200602/79-9. A. S. Belward, J. M. Grégoire, and P. Kennedy were very helpful in discussions of fire detection and their location in the AVHRR images; G. de Souza assisted with image-processing hardware and software.

References

Andreae, M. O. 1993. Global distribution of fires seen from space. *EOS Transact.*, *AGU*, 74(12):129, 135.

Andreae, M. O., Browell, E. V., Garstang, M., Gregory, G. L., Harris, R. C., Hill, G. F., Jacob, D. J., Pereira, M. C., Sachse, G. W., Setzer, A. W., Silva Dias, P. L., Talbot, R. W., Torres, A. L., Wofsy, S. C. 1988. Biomass Burning Emissions and Associated Haze Layers Over Amazonia. *J. Geophy. R.*, 93(D2):1509–1527.

Belward, A. S., Grégoire, J.-M., D'Souza, G., Trigg, S., Hawkes, M., Brustet, J.-M., Serca, D, Tireford, J.-L., Charlot, J.-M., and Vuattoux, R. 1993. In-situ, real-time fire detection using NOAA/AVHRR data. *Proc.*, *6th AVHRR User's Conf.*, Belgirate, Italy, 28 June–2 July; Eumetsat/Joint Research Centre 333–339.

Breaker, L. C. 1990. Estimating and removing sensor-induced correlation from advanced very high resolution radiometer satellite data. *J. Geophys. Res.*, 95(C6):9701–9711.

Cahoon, D. R., Stocks, B. J., Levine, J. S., Cofer, W. R., O'Neill, K. P. 1992. Seasonal distribution of African savanna fires. *Nature*, 359:812–815.

Crutzen, P. J., Heidt, L. E., Kranec, J. P., Pollock, W. H., and Seiler, W. 1979. Biomass burning as a source of atmospheric gases CO, H_2, N_2O, NO, CH_3Cl, and COS. *Nature*, 282:253–256.

Crutzen, P. J. and Andreae, M. O. 1990. Biomass burning in the tropics: impact on atmospheric chemistry and biogeochemical cycles. *Nature*, 250:1669–1678.

Crutzen, P. J. and Goldammer, J. G. 1993. *Fire in the Environment: The Ecological, Atmospheric, and Climatic Importance of Vegetation Fires*. Chichester, John Wiley, 497 pp.

Goldammer, J. G. 1993. Historical biogeography of fire: tropical and subtropical. In *Fire in the Environment: Its Ecological, Climatic, and Atmospheric Chemical Importance*, Eds. P. J. Crutzen and J. G. Goldammer, chap. 15, pp. 297–314. New York, John Wiley.

Goward, S. N. Markham, B., Dye, D. G., Dulaney, W., and Yang, J. 1991. Normalized difference vegetation index measurements from the advanced very high resolution radiometer. *Remote Sens. Environ.*, 35:257–277.

Grégoire, J. M., Belward, A. S., and Kennedy, P. 1993. Dynamiques de saturation du signal dans la bande 3 du senseur AVHRR: Handicap majeur ou source d'information pour la surveillance de l'environnement en milieu soudanoguinéen d'Áfrique de l'Ouest? *Int. J. Remote Sensing*, 14 (11):2079–2095.

Gutman, G. G. 1991. Vegetation indices from AVHRR: an update and future prospects. *Remote Sens. Environ.*, 35:121–136.

Helfert, M. R. and Lulla, K. P. 1990. Mapping continental-scale biomass burning and smoke palls over the Amazon Basin as observed from the Space Shuttle. *Photogram. Eng.*, 56(10):1367–1373.

Holben, B. N., Kaufman, Y. J., and Kendall, J. D., 1990. NOAA-11 AVHRR visible and near-IR inflight calibration, *Int. J. Remote Sensing*, 11, 1511–1519.

Hovis, W. A. Jr. 1966. Infrared spectral reflectance of some common minerals. *Applied Optics*, 5(2):245–248.

IGBP. 1992. Improved global data for land applications: a proposal for a new high resolution data set. Ed. J. R. G. Townsend, *International Geosphere and Biosphere Program—IGBP*, Report 20, Stockholm.

Kaufman, Y. J. and Remer, L. A. 1994. Detection of forests using mid-IR reflectance: an application of aerosol studies. IEEE *J. Geosc. and Rem. Sens.*, 32:672–683.

Kidwell, K. B. 1991. *NOAA Polar Orbiter Data User's Guide*. NOAA/NESDIS, Washington, D.C.

Langaas, S. 1993. Diurnal cycles in savanna fires. *Nature*, 363, 120.

LaRocca, A. J. 1989. Atmospheric absorption. In *The Infrared Handbook*, Eds. W. L. Wolfe and G. J. Zissis, pp. 5.1–5.132 (Ann Arbor, MI; E.R.I.M.).

Levine, J. S. 1991. *Global Biomass Burning: Atmospheric, Climate, and Biospheric Implications*. MIT Press, Cambridge, Mass., 569 pp.

Malingreau, J. P. 1990. The contribution of remote sensing to the global monitoring of fires in tropical and subtropical ecosystems. In *Fire in the Tropical Biota*, Ed. J. G. Goldammer, chap. 15, pp. 337–370. *Ecosystem Processes and Global Challenges, Ecological Studies*. Springer-Verlag, Berlin-Heidelberg.

Malingreau, J. P., Albini, F. A. Andreae, M. O., Brown, S., Levine, J., Lobert, J. M., Kuhlbush, T. A., Radke, L., Setzer, A., Vitousek, P. M., Ward, D. E. and Warnatz, J. 1993. Group report: quantification of fire characteristics from local to global scales. In *Fire in the Environment: Its Ecological, Climatic, and Atmospheric Chemical Importance*, Ed. P. J. Crutzen and J. G. Goldammer, chap. 19, pp. 327–343. New York, John Wiley.

Mansor, S. B., and Cracknell, A. P. 1992. Land surface temperature from NOAA-9 AVHRR data. *Proceedings, 18th Annual Conf. of the Remote Sensing Society*. Ed. A. P. Cracknell and R. A. Vaughan. The Remote Sensing Society. University of Dundee, 15–17 Sept. 1992; pp. 274–286.

Matson, M. and Dozier, J. 1981. Identification of subresolution high temperature sources using thermal IR sensor. *Photogram. Engin. and Remote Sensing*, 47:1311–1318.

Pereira, Jr., A. C, and Setzer, A. W. 1996. Comparison of fire detection in savannas using AVHRR's ch. 3 and TM images, *Int. J. Remote Sens.*, 17(10)1925–1937.

Pereira, M. C. and Setzer, A. W. 1993. Spectral characteristics of deforestation fires in NOAA/AVHRR images. *Int. J. Remote Sensing*, 14(3):583–597.

Robinson, J. M. 1991. Fire from space: global fire evaluation using infrared remote sensing. *Int. J. Remote Sensing*, 12(1):3–24.

Setzer, A. W. 1993. Operational monitoring of fires in Brazil. *Internat. Forest Fire News*, 9:8–11.

Setzer, A. W. and Pereira, M. C. 1991a. Operational detection of fires in Brazil with NOAA/AVHRR. *24th. Internat. Symp. on Remote Sensing of the Environment*, Rio de Janeiro, Brazil (Ann Arbor, E.R.I.M.), 469–482.

Setzer, A. W. and Pereira, M. C. 1991b. Amazonia biomass burning in 1987 and an estimate of their tropospheric emissions. *Ambio*, 20(1):19–22.

Setzer, A. W. and Malingreau, J. P. 1993. Temporal variation in the detection limit of fires in AVHRR's ch. 3. *Proc. 6th AVHRR User's Conf.*, Belgirate, Italy, 28 June–2 July, Eumetsat/Joint Research Centre, pp. 575–579.

Setzer, A. W. and Verstraete, M. M. 1994. Fire and glint in AVHRR's ch. 3: a possible reason for the nonsaturation mystery. *Internat. J. Remote Sensing*, 15(3):711–718.

Setzer, A. W., Pereira Jr., A. C. and Pereira, M. C. 1994. Satellite studies of biomass burning in Amazonia: some practical aspects. *Remote Sensing Reviews*, 10:91–103.

Singh, K. D. 1993. The 1990 tropical forest resources assessment. *Unasylva/FAO*, 44(174):10–19.

Suits, G. H. 1989. Natural Sources. In *The Infrared Handbook*, Ed. W. L. Wolfe and G. J. Zissis, MI, pp. 3.1–3.154 (Ann Arbor, E.R.I.M.).

The Simulation of AVHRR Data for the Evaluation of Fire-Detection Techniques

Pete Dowty

Since the work of Dozier (1981), a number of studies have demonstrated the ability of the National Oceanic and Atmospheric Administration/Advanced Very High Resolution Radiometer (NOAA AVHRR) instrument to detect terrestrial vegetation fires. This work has been reviewed elsewhere (Robinson 1991; Justice and Dowty 1994). There are important limitations with this technique, which have also been discussed in the literature (Robinson 1991; Justice et al. 1993). These studies suggest that there are significant, but unknown, uncertainties associated with fire distributions derived from satellite data. Even as the limitations become more widely recognized, however, there has been considerable interest in utilizing remotely sensed fire distributions in emissions inventories and atmospheric transport studies (Thompson et al. 1994; Scholes et al. 1994). Most work in this area has utilized data from the NOAA AVHRR instrument because of its moderate temporal and spatial resolution (twice daily and 1.1 km at nadir, respectively). It is anticipated that this interest will only increase given the plans to maintain AVHRR coverage into the next century, the recent availability of the International Geosphere-Biosphere Program–Data and Information Systems (IGBP-DIS) 1-km global AVHRR data set, as well as the existing regional archives of AVHRR that have not been fully exploited. In addition, the launch of the Moderate-Resolution Imaging Spectrometer (MODIS) instrument aboard the Earth Observing System (EOS AM) platform later this decade will make data available for global fire-detection studies that will benefit from techniques developed with AVHRR data.

Although many studies have presented fire-detection algorithms for application to AVHRR data, few have attempted to validate the results (Flannigan and Vonder Haar 1986; Setzer and Pereira 1991). Since algorithm behavior varies across gradients in land cover and environmental conditions, validation of a global algorithm requires a large number of such validation exercises. As an alternative to such a massive effort, this study explores the use of theoretical models in the validation of fire-detection algorithms. A model that simulates AVHRR data in the presence of fire could be used to evaluate the performance of algorithms in a range of scenarios that might be encountered in a global study. The validation of such a model, however, requires simultaneous observation of a fire by aerial or ground reconnaissance and the AVHRR. The Block 56 burn during the SAFARI-92 experiment discussed here is the first fire known to this study that meets this criterion.

This chapter first describes a model that simulates AVHRR data and its sensitivity to various input parameters. This work builds on an earlier study describing the initial modeling effort (Dowty 1993). Then image data from the Block 56 burn is used to access the validity of the simulated data. Finally, the detection limits of three fire-detection algorithms are evaluated using simulated data.

Model Description

The model represents each pixel by up to three elements of uniform temperature. These elements represent background, active fire, and warm areas adjacent to active fire that are above background temperatures due to recent burning but are far below combustion temperatures. The area of the warm element is directly proportional to the fire area. The proportionality constant, *wfactor*, is specified as a fixed input parameter. Presumably *wfactor* is related primarily to fuel type and residence time of the flame front.

The MODTRAN model (Berk et al. 1989) is used to calculate atmospheric and solar contributions to the infrared signal associated with each pixel element. MODTRAN operates at a spectral resolution of 1 cm^{-1} through up to 100 km of the atmosphere divided into a maximum of 34 uniform layers. Six model atmospheres are included with MODTRAN which are defined by temperature, pressure, water vapor, and other trace gas profiles. Alternatively, a user may construct an atmosphere by including these profiles as model input.

For this application, MODTRAN is run in spectral increments of 1 cm^{-1} through the infrared bands corresponding to AVHRR channels 3 (3.8 μ), 4 (10.8 μ), and 5 (11.7 μ). In order to maintain flexibility, and in particular to allow model inversion, the results of the complete MODTRAN calculations are not used. Instead, the MODTRAN code has been modified to generate a file that includes parameters for each spectral increment. These parameters include atmospheric transmissivity from surface to satellite, transmissivity seen by the direct solar beam from top-of-atmosphere to surface to satellite, downwelling and upwelling atmospheric radiance, top-of-atmosphere solar radiance, coefficients for multiple and single scattering, and downwelling diffuse solar radiance. A separate parameter file is generated for each scenario of atmosphere and solar-satellite geometry that is considered.

The model calculates the thermal emission from each pixel element using Planck's law and the assumption that emissivity is uniform within each band or channel. The parameters generated by MODTRAN are used to calculate solar and atmospheric influences, including the portion of downwelling radiances reflected under the assumption of Lambertian surfaces with uniform reflectances within bands. In order to calculate the total radiance incident on the AVHRR detectors, the radiance from each pixel element is weighted by area and this pixel radiance is then integrated across wave number after weighting each term by the channel response function, ϕ_v. The channel response function weights are interpolated from the 60-point curves provided by NOAA for the NOAA-11/AVHRR (Brown 1988). For example, the calculated radiance in channel 3, L_3, is then:

$$L_3 = \frac{\sum_v \phi_v \cdot [p \cdot L_f + p \cdot wfactor \cdot L_w + (1-p-p \cdot wfactor) \cdot L_b]dv}{\sum_v \phi_v \cdot dv}$$

where p is the proportion of the pixel occupied by active fire and L_f, L_w, and L_b are the radiances from the fire, warm, and background elements, respectively.

If the MODTRAN calculations are considered part of an integrated model, then the model can be represented functionally as:

$$L_3(T_3) = f_3(p, T_f, T_w, T_b, \varepsilon_{f3}, \varepsilon_{w3}, \varepsilon_{b3}, atm)$$
$$L_4(T_4) = f_4(p, T_f, T_w, T_b, \varepsilon_{f4}, \varepsilon_{w4}, \varepsilon_{b4}, atm)$$
$$L_5(T_5) = f_5(p, T_f, T_w, T_b, \varepsilon_{f5}, \varepsilon_{w5}, \varepsilon_{b5}, atm)$$

where T_i is the brightness temperature in channel i, *atm* represents the parameters produced by MODTRAN that are used to characterize the atmosphere and solar geometry, and ε_{ij} represents the emissivity of fire, warm, and background areas ($i = f$, w, and b, respectively) in channel j. When the model is inverted, brightness temperature becomes a model input and

either p or T_f can be predicted. The channel 3 equation, for example, can be arranged to predict p:

$$p = f_3(T_e, p, T_f, T_w, T_b, \varepsilon_{f3}, \varepsilon_{w3}, \varepsilon_{b3}, atm)$$

In general, both p and T_f are considered unknowns and the equation from a second channel is used to solve them simultaneously.

Sensitivity Study

The sensitivity of the model to the various input parameters was explored using a single-factor analysis. Each parameter was varied above and below a central value as all other parameters were held at their central values. The values used are shown in table 4.1 and were chosen to represent reasonably wide ranges that might be encountered in data from diverse environments. Three atmospheres were used from the MODTRAN program: mid-latitude winter (ML win), U.S. Standard (US Std), and mid-latitude summer (ML sum). Water vapor content of these profiles is shown as total column precipitable (ppt) water in table 4.1. The low, mid, and high values of fire area as a proportion of a pixel (p) correspond to 300, 600, and 1200 m^2, respectively, of active fire in a nadir pixel. The resultant ranges in model output, (T_3, T_4, and T_5) associated with these ranges in input parameters are shown in

Table 4.1 Low, mid, and high values of parameters examined in sensitivity study

Parameter	Low	Mid	High
ε_{b3}	0.85	0.95	1.0
ε_{b4}	0.85	0.95	1.0
ε_{b5}	0.85	0.95	1.0
ε_{f3}	0.85	0.95	1.0
ε_{f4}	0.85	0.95	1.0
ε_{f5}	0.85	0.95	1.0
ε_{w3}	0.85	0.95	1.0
ε_{w4}	0.85	0.95	1.0
ε_{w5}	0.85	0.95	1.0
P	0.00025	0.0005	0.001
T_f	500	800	1100
T_w	350	400	450
T_b	295	305	315
wfactor	0.0	1.0	2.0
Scan angle	0	25	50
Water vapor	ML win	US Std	ML sum
(ppt. water)	(0.6 cm)	(1.1 cm)	(2.3 cm)

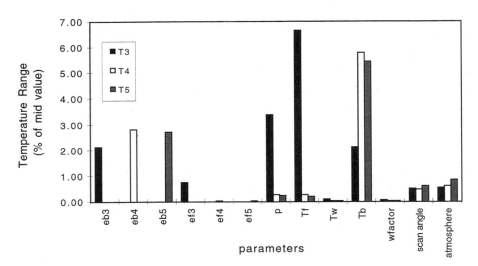

Figure 4.1 Effects of perturbation of input parameters on simulated brightness temperatures

figure 4.1. The effects of varying emissivity of the warm area were negligible and these are not shown.

The fire temperature had a strong effect on the channel 3 response and the background temperature had a strong effect on the response in all channels but especially in channels 3 and 4. Also the response in each channel was affected by the background emissivity of the channel. Scan angle and atmosphere had weaker effects in all three channels. Fire emissivity had a weak effect in channel 3 but was negligible in the other channels. Parameters associated with the warm area had only a minor effect on the output.

These results show the importance of environmental gradients relative to variations in fire characteristics. The latter dominate the channel 3 response while the environmental parameters dominate the responses in channels 4 and 5. This distinction is the basis for most fire-detection algorithms. However, both background emissivity in channel 3 and background temperature had a significant effect on the channel 3 response. These results also suggest that when a simulation is based on an actual field scenario where input parameters are derived from physical measurements with associated uncertainties, rather than a hypothetical scenario, emissivities of fire and the warm area can be roughly approximated with little loss of accuracy in the model output. The uncertainties associated with the other parameters have a more significant impact on the simulated brightness temperatures. Given the results in figure 4.1, the difficulty in determining landscape-scale emissivity and background temperature, and the simplistic model treatment of active flame, it is reasonable to assume that the uncertainty in model output is significantly better than 5% when physical

measurements with their own inherent uncertainties are used as input. It is not unrealistic to assume uncertainty in model output may be in the range of 1% with accurate input data. It is important to note that the ranges used for the input parameters were chosen to span broad gradients that could be encountered for these parameters. The envelope of uncertainty about measured quantities should be much less.

Model Validation

Approach

Data from an experimental fire during the SAFARI-92 experiment were used to quantitatively evaluate the performance of the model. Available data included a detailed map of active fire at the time of the NOAA-11 satellite overpass and a nearly contemporaneous characterization of the atmosphere. The fire was in Block 56 of the Kruger National Park in South Africa on 18 September, 1992. The most direct approach would have been to compare actual AVHRR data to simulated data on a pixel-by-pixel basis. Model input would be derived from detailed field characterization of surface and atmospheric conditions in the presence of a fire. Unfortunately, no field data was available to realistically estimate a spatially integrated emissivity or background temperature near Block 56. In addition this was a large fire extending across multiple pixels and the uncertainties in georeferencing the imagery made a pixel-by-pixel comparison impractical.

Instead, a different approach was used that inverted the model to estimate fire size given the actual AVHRR data as input and compared this estimate with one derived from the reconnaissance data. First, the pixels that contain fire, referred to as "fire pixels," were manually identified by their distinctive channel 3 response. Then the data from adjacent, background pixels were used to estimate surface emissivity and background temperature. This latter step is now described in more detail. For given atmospheric conditions, the radiative equations for channels 4 and 5 are (showing unknown independent variables only):

$$L_4(T_4) = f_4(T_b, \varepsilon_{b4}) \qquad L_5(T_5) = f_5(T_b, \varepsilon_{b5})$$

These equations do not have a unique solution because they contain three unknowns. We proceed by assuming that background emissivity is the same in both channels 4 and 5, denoted by ε_{b45}. It is then possible to solve for T_b and ε_{b45}. The channel 3 equation then has only one unknown and can be solved for ε_{b3}:

$$L_3(T_3) = f_3(T_b, \varepsilon_{b3})$$

This procedure of calculating emissivities and background temperature was applied to pixels surrounding each of the fire pixels after filtering for cloud and fire. The results from a minimum of five pixels were averaged to determine these values.

The radiative equations for channels 3 and 4 were then applied to the fire pixels and solved for the only remaining unknown quantities, fire area and temperature:

$$L_3(T_3) = f_3(p, T_f) \qquad L_4(T_4) = f_4(p, T_f)$$

In order to do this, the emissivity of the fire was held at 0.95 in all three channels and the emissivity of the warm area was set equal to the background. These were necessary simplifications that ignored possible variation due to soot content and depth of the flame in the former case, and effects of ash and charcoal in the latter. However, given the results of the sensitivity study these assumptions should have had a relatively minor effect on model output. The intent was to validate the model through the comparison of the total (multipixel) predicted fire area to the data from the aerial reconnaissance. However, the preliminary results reported here suggest that additional refinement of this technique is necessary and that it is premature to make this direct comparison.

Block 56 is irregularly shaped with the longest dimension spanning approximately 7 km and the perpendicular dimension spanning 4 km. At the time of the NOAA-11 satellite overpass at 1542 South African Standard Time (SAST) the linear distance of active firefront, both backfire and headfire, exceeded 10 km. Water vapor and temperature profiles up to 1 km were obtained from a tethersonde released at 1723 SAST

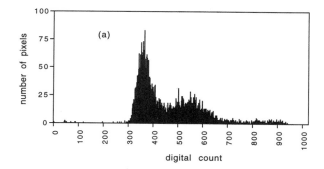

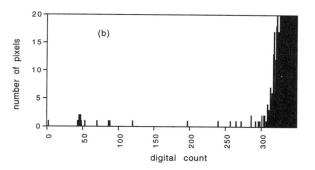

Figure 4.2 Raw data from region around Block 56: (a) full 10-bit range, (b) detail of hot pixels

at Pretoriuskop less than 5 km from Block 56. Above 1 km, data were used from a rawinsonde launched at 1330 SAST from Pietersburg approximately 250 km northwest of the burn site. These data were provided by Eskom. Total precipitable water in the column was 2.3 cm.

Pixels that were considered to contain fire were manually identified by examining the histogram of raw digital counts in channel 3 in a block of 7080 pixels surrounding the fire site (figure 4.2). The raw data is inverted in this channel so that lower counts represent higher brightness temperatures. A small number of hot pixels were quite distinct from the bulk of the distribution. A digital count of 300 was chosen as a threshold to distinguish these pixels, resulting in the identification of 22 fire pixels.

Results

The calculated emissivities fell in a range between 0.94 and 1.1. The values greater than 1.0 indicate a limitation in either the model itself or the input data. The wide range suggests that the AVHRR data are quite noisy since there is nothing in the local environment to indicate that surface emissivities vary so widely. In order to proceed with the calculations, the emissivities were constrained to values less than or equal to 1.0.

The calculated background temperatures were between 304 and 315°K with over half of the values between 305 and 307°K. The wide range here may in part be attributable to signal contamination from adjacent fire pixels that raised the apparent temperature but not enough for these pixels to be filtered out as fire pixels themselves.

The calculations of fire size and temperature produced reasonable values for all but two of the 22 fire pixels. For these two pixels, there was no solution within physically meaningful limits. Again, this indicates a limitation in either the data or the model. The validity of assuming uniform emissivity in channels 4 and 5 was evaluated using the technique of Becker and Li (1990). This is based on solving the radiative equations for channels 4 and 5 independently. The equations were inverted to calculate surface temperature for a range of emissivities for pixels that do not contain fire. If the emissivities were truly the same in each channel then the equations would predict the same surface temperature for that emissivity. This technique was applied to a number of pixels surrounding the Block 56 fire. The results from most of the pixels followed the pattern of figure 4.3a suggesting that $\varepsilon_4 > \varepsilon_5$ (Becker and Li 1990) and that the assumption of uniform emissivity in the two channels is invalid. The pattern in figure 4.3b was also observed, however, suggesting that $\varepsilon_4 = \varepsilon_5 = 1.0$ in those cases. These results are subject to the validity of model parameters and the accuracy of representation of the atmosphere by MODTRAN.

The entire set of calculations was then repeated 11 times while constraining $(\varepsilon_{b4} - \varepsilon_{b5})$ to values ranging from -0.01 to 0.03. There was no solution for two of the fire pixels while $(\varepsilon_{b4} - \varepsilon_{b5})$ was between -0.001 and 0.005 and outside of this range the number of fire pixels for which there was no solution increased rapidly. This result is insufficient to draw firm conclusions about the value of $(\varepsilon_{b4} - \varepsilon_{b5})$ but does not preclude the possibility that $\varepsilon_{b4} > \varepsilon_{b5}$. It does, however, suggest that some other input parameter contributes to the lack of solution for at least two of the fire pixels. It is therefore premature to use model output in an absolute sense, although the theoretical completeness of the model suggests that the model may still accurately reflect functional relationships between model input and output.

A sensitivity analysis of the inverted model was done in order to explore the lack of solutions for two of the Block 56 fire pixels. First, the model was used to simulate data for pixels both with and without the presence of fire. The mid-parameter values of the earlier sensitivity analysis were used as input (table 4.1). Again

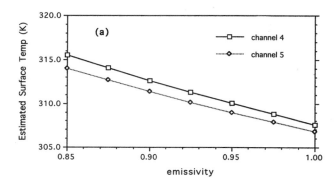

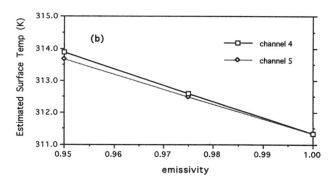

Figure 4.3 Estimated surface temperatures versus emissivity for two pixels from the Block 56 area: (a) indicates that $\varepsilon_4 > \varepsilon_5$; (b) indicates that $\varepsilon_4 = \varepsilon_5$

a single-factor analysis was used to examine the sensitivity of the inverted model to perturbations in four input parameters and three of the calculated background parameters. The scheme is shown in table 4.2. All values shown represent perturbations about the parameter values.

The levels of perturbation were chosen to represent reasonable uncertainties in an analysis such as that of the Block 56 burn. For T_3 and T_4, these perturbations are applied only to data from the pixel containing active fire, not the background pixel, and can be taken to represent noise in these data. Previous estimates in the uncertainty associated with AVHRR brightness temperatures have been as high as 3°K (Cooper and Asrar 1989) and channel 3 data is said to sometimes be rendered useless by noise (Kidwell 1991). These are therefore conservative estimates. The resultant range in the output variables is shown in table 4.3.

These results clearly indicate a critical sensitivity to uncertainties in the calculated parameters, ε_{b45} and T_b. Given the assumptions in the model and its use in an inverted mode, significant errors in the calculation of

Table 4.2 Perturbations of input parameters for sensitivity study of inverted model

	Parameter	Perturbation
Input parameters	ε_{f3}	± 0.15
	scan angle	$\pm 5°$
	T_3	$\pm 1°K$
	T_4	$\pm 1°K$
Calculated params	ε_{b3}	± 0.15
	ε_{b45}	± 0.15
	T_b	$\pm 2°K$

Table 4.3 Effects of perturbation of input parameters on the inverted model

	Range in output parameter (% of mid value)	
Parameter	p (fire area)	T_f
ε_{f3}	11	6
scan angle	0	0.01
T_3	9	5
T_4	(−)[a]	112
ε_{b3}	28	15
ε_{b45}	(+)[a]	(+)[a]
T_b	(+)[a]	(+)[a]

a. The model did not reach a solution within physically meaningful limits due to the positive (+) or negative (−) perturbation of the input parameter.

ε_{b45} and T_b can be expected and may explain the failure of the model to reach a solution. These results do not allow a firm quantitative estimate of model accuracy. However, the theoretical detail in the model suggests that accuracy is substantially improved over the Dozier model. Also, comparison of simulated and actual background data for the Block 56 scenario with rough estimates of input parameters suggest that simulated brightness temperatures are accurate to at least 6°K ($\sim 2\%$).

Algorithm Evaluation

Approach

The performance of three fire-detection algorithms was explored by using simulated AVHRR data to delineate detection limits. The effects of both environmental and algorithm parameters on these detection limits were investigated. The atmospheric profiles from the Block 56 fire were used for all data simulations. The mid-level values in table 4.1 were used for the

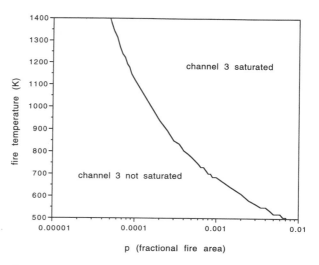

Figure 4.4 Channel 3 saturation curve derived from simulated data. Model parameters are described in text.

other parameters unless otherwise specified. Simulated temperatures were constrained to values less than or equal to 321°K to represent detector saturation.

Existing algorithms can be placed into four groups depending on whether they utilize raw digital counts (Setzer and Pereira 1991; Pereira and Setzer 1993), fixed thresholds (Flannigan and Vonder Haar 1986; Kaufman et al. 1990; Kennedy et al. 1994), Dozier model calculations (Lee and Tag 1990; Langaas 1993), or spatial analysis (Prins and Menzel 1994; Justice et al. 1994). In this study, three algorithms were studied: a fixed-threshold algorithm (Kaufman et al. 1990), a Dozier-based algorithm (Lee and Tag 1990), and an algorithm that utilizes spatial analysis (Justice et al. 1994). These are referred to as the Kaufman, Lee&Tag, and Goddard algorithms, respectively. It is also possible to evaluate algorithms using raw digital counts if an inverted calibration equation is used to convert simulated brightness temperatures into raw digital data, but that was not done for the current study.

The detection limits for each of the algorithms are displayed on axes of fire temperature versus fire area (as a fraction of the pixel area). For the parameters noted above, figure 4.4 shows the channel 3 saturation curve, delineating saturating from nonsaturating fires, on these axes. This curve will vary for different scenarios of atmosphere, surface characteristics, and solar-satellite geometry.

The Algorithms

The Kaufman algorithm was developed for the detection of deforestation fires in Brazil (Kaufman et al. 1990). It tests for the existence of three conditions:

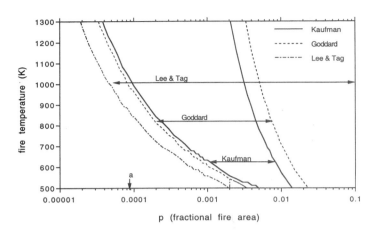

Figure 4.5 Detection limits for three algorithms for a background temperature of 315°K and other parameters noted in the text. Regions resulting in detection are indicated with arrows. Point *a* is referred to in the text.

1. $T_3 \geq 316°K$,

2. $\Delta T_{34} \geq 10°K$, and

3. $T_4 > 250°K$,

where T_3 and T_4 are brightness temperatures in channels 3 and 4 and ΔT_{34} is the difference between these two channels ($T_3 - T_4$). The first test ensures that the pixel is hot. The second requires a significant enhancement of the channel 3 signal relative to that in channel 4, which is less sensitive to high-temperature radiative sources. The last test screens out potential cloud pixels that are cool in channel 4 but may appear hot in channel 3 due to reflection of solar radiation.

The Lee&Tag algorithm was developed for application to nighttime data (Lee and Tag 1990). It applies a variable threshold to the channel 3 brightness based on calculations made with the Dozier model. A background temperature, T_{bm}, is estimated using data from adjacent pixels and the McClain split-window calculation (McClain et al. 1985). Deviations of background data from T_{bm} are used to correct the channel 3 and channel 4 brightness temperatures of the central pixel being tested. The Dozier model is then used to calculate channel 3 and channel 4 brightness temperatures for a range of fire sizes using T_{bm} and a user-specified threshold fire temperature. One fire size will lead to a match between the calculated response in channel 4 and the actual channel 4 data being tested. The associated calculated channel 3 value is applied as a threshold to the actual channel 3 data.

The Goddard algorithm was originally developed for savanna fires in southern Africa but is intended to have wider applicability (Justice et al. 1994). It consists primarily of four tests:

1. $T_3 \geq 316°K$,

2. $T_4 \geq 278°K$,

3. $T_3 \geq T_4$, and

4. $\Delta T_{34} > \Delta T_{b34} + 2 \cdot \sigma_{\Delta Tb34}$,

where ΔT_{b34} is the average difference between the responses in channels 3 and 4 in a background window around the pixel being tested and $\sigma_{\Delta Tb34}$ is the standard deviation of ΔT_{b34} in this same window. For the purposes of this study, the parameters ΔT_{b34} and $\sigma_{\Delta Tb34}$ were derived from actual NOAA-11 AVHRR data in the vicinity of the Block 56 burn.

Detection Limits

Figure 4.5 shows the detection limits for the three algorithms for a background temperature of 315°K and the other parameters mentioned above. The Lee&Tag algorithm was applied with a threshold fire temperature of 800°K. The resultant detection envelopes lie between the upper and lower limits or, in the case of the Lee&Tag algorithm, above the single detection limit. The extent of each detection envelope is indicated. The roughness of the plotted limits is an artifact primarily of the fineness of fire area–fire temperature matrix used. Below the lower limits, the enhancement of the channel 3 signal is too subtle to be detected by the algorithm. Above the upper limit, saturation constrains the response in channel 3 while enhancement in channel 4 leads to a small difference between the two channels and a detection does not occur. The exception is the Lee&Tag algorithm, which does not have an upper detection limit because saturation in channel 3 is a sufficient condition for detection.

It is surprising that the Lee&Tag detection limit follows the shape of the limit for the other two algorithms. The Lee&Tag algorithm is designed to detect fires with temperatures above a fixed threshold, which would result in a detection limit parallel to the x-axis running through the selected threshold fire temperature. In fact, the detection limit falls on the specified threshold fire temperature (800°K) only at a fire area indicated by point a in figure 4.5. For areas less than point a, the response in channels 3 and 4 in the simulated data is significantly below that predicted by the Dozier model. For a given fire size a fire hotter than the specified threshold is required to trigger a detection. For fire sizes greater than point a, the enhancement in channel 3 relative to channel 4 is greater in the simulated data than in the Dozier model to an extent that a positive detection occurs at fire temperatures below the specified threshold. The discrepancy between the model used to simulate the data and the Dozier model also leads to the sharp change in the detection limit near $p = 0.002$. It was found that this occurs where the simulated data approaches the Dozier-model predictions internal to the Lee&Tag algorithm.

Variations in background temperature were found to have a large impact on the detection limits of all three algorithms. Figure 4.6 shows detection limits for the three algorithms at background temperatures of 350°K, 315°K, and 320°K. Other parameters are the same as for the data used in figure 4.5. In most cases, an increase in background temperature leads to an expansion of sensitivity at the lower end and a contraction at the upper end of the detection envelope. For both the Kaufman and Goddard algorithms, the $T_3 \geq 316$°K test is limiting at $T_b = 305$°K. At higher background temperatures the threshold tests become critical. For these algorithms the effect at the upper detection limit is more pronounced than at the lower limit. This effect is due to saturation in channel 3 occurring with smaller and cooler fires as background temperature increases. There is a fairly extreme loss in sensitivity at the upper detection limit for both of the algorithms as background temperature approaches 320°K. The detection limit of the Lee&Tag algorithm behaves similarly to the lower detection limits of the other algorithms in response to changes in background temperature. The Lee&Tag algorithm is seen to have a significantly larger detection envelope that is slightly less stable across the gradient in background temperature as compared to the other algorithms. The detection limit at $T_b = 350$°K near $p = 0.0003$ also displays the erratic behavior noted earlier.

The effects of variation in background emissivities in all three channels was also explored for the Goddard algorithm. In general the effect on the detection limits was much less than the effects of variation in background temperature. Changes from 0.85 to 1.0 in ε_{b3} led to a moderate shift of both upper and lower de-

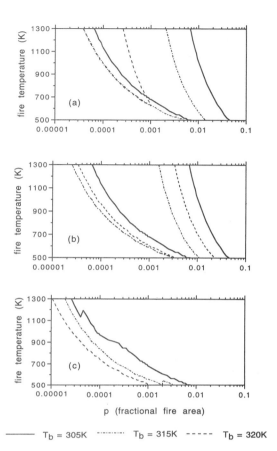

Figure 4.6 Effect of background temperature on detection limits of (*a*) Kaufman, (*b*) Goddard, and (*c*) Lee&Tag algorithms. Background temperatures of 305°K, 315°K, and 320°K are shown. Other parameters are noted in the text.

tection limits in the direction of hotter and larger fires. Only a small effect was observed due to changes in ε_{b4}, and changes in ε_{b5} had no effect at all.

Finally, the effects of algorithm parameters on detection limits were explored. For the Kaufman algorithm, varying the threshold for ΔT_{34} (originally at 10°K) from 5°K to 15°K led to a moderate contraction of the detection envelope at the upper end. In contrast, there was no effect on the lower limit because for the scenario used the requirement that $T_3 \geq 316$°K determined the detection limit. For the Goddard algorithm, the standard deviation multiplier (originally at 2) was changed to 3. This resulted in only a very small contraction of the upper limit of the detection envelope and no change at the lower limit. The main algorithm parameter of the Lee&Tag algorithm, the threshold fire temperature, was varied from 600°K to 1000°K resulting in a substantial contraction of the detection envelope.

Discussion

Algorithm performance may be evaluated from several perspectives. For application to large areas (continental, global scales) a range of fire regimes can be expected and the detection envelope should be as broad as possible to ensure sensitivity throughout this range. The results of this study, however, show that (1) limited signal strength associated with smaller, cooler fires and (2) detector saturation significantly restricts the extent of the detection envelope for each of the algorithms examined. There are fires with characteristics that are likely to occur frequently in continental or global studies that would be missed by each of the algorithms. It is also important for the detection limits to be as stable as possible across gradients in both surface and atmospheric parameters. The results show that detection limits vary strongly with background temperature and to a lesser extent background emissivities. This clearly has the potential to introduce a systematic bias into distributions of detected fires over large areas where gradients in these parameters occur. A third goal for algorithm performance is the absence of false detections. This study did not consider false detections and it is likely that the advantage of a wider detection envelope will have a trade-off of a higher incidence of false detections. The lack of an upper detection limit for the Lee&Tag algorithm is probably an example of how the high incidence of false detections would make the high sensitivity impractical. This algorithm, however, was developed for nighttime application where channel 3 saturation alone would be a better indicator of the presence of fire. Still, to fully evaluate these algorithms, the occurrence of false detections must be considered. Conceivably, background variability could be included in the model and the occurrence of false detections could be evaluated on a statistical basis. The variance and covariance structure determined from actual AVHRR images could be used to parameterize this model extension.

The importance of the extent and stability of the detection limits is closely linked to the fire regime under observation. The acceptability of a set of detection limits is closely related to whether the distribution of fire characteristics lies mainly inside or outside of the detection envelope or near the limits, in which case detection depends largely on environmental parameters. The importance of the latter case can potentially be minimized by varying algorithm parameters to counteract the effect of variation in environmental parameters. For instance, as background temperature increases, tending to widen the Lee&Tag detection limits, the threshold fire temperature could be increased, resulting in a tendency to reduce the detection envelope.

Conclusions

The study demonstrates the potential contribution of modeling to the evaluation and development of fire-detection algorithms. There are clear differences between the three algorithms examined in terms of the extent and stability of their detection envelopes. This means that not only will performance vary among algorithms, but also that the performance of any given algorithm will vary across environmental gradients, particularly gradients in background temperature. This will introduce systematic bias into distributions of detected fires that span such gradients. The results of the algorithm evaluation, however, suggest that if these environmental gradients are known, either through independent data or from the satellite data itself, it may be possible to adjust algorithm parameters to counteract the change in algorithm performance over the environmental gradient. This possibility requires further evaluation and may not be practical over large areas due to the sharp increase in processing requirements.

Given the recent opportunities to produce global fire distributions from 1 km AVHRR data, it is important that the behavior of fire-detection algorithms be better understood. The use of simulated AVHRR data is the ideal way to study algorithm behavior and to further develop the algorithms themselves. This approach also allows for explicit characterization of the limitations of algorithm sensitivity and the stability of detection envelopes. It is critical that this information be disseminated with satellite-derived fire distributions for the user to better understand the limitations of the technique. It is also important to put these limitations in the context of other issues such as cloud cover, pixel overlap, irregular channel 3 behavior (Setzer and Verstraete 1994), and the limited temporal sampling of the AVHRR.

The evaluation of the simulation model itself with data from the SAFARI-92 experiment shows that more work is needed in order to satisfactorily validate the model. The sensitivity of the inverted model to both background temperature and the $\varepsilon_{b4} = \varepsilon_{b5}$ assumption suggest that error in these calculated parameters prevented the model from reaching a solution for two of the 22 fire pixels associated with the Block 56 burn. Future field work that utilizes controlled burns in an attempt to evaluate fire detection with AVHRR should include components that would be helpful in further evaluating the simulation model. This requires detailed fire mapping at the time of the satellite overpass and measurements that could be used to calculate spatially integrated estimates of the temperature of the background as well as the emissivities in the three infrared bands of the AVHRR.

The addition of a stochastic generator of background variability would greatly enhance the usefulness of the simulation model. Detection algorithms could then be evaluated and tuned with respect to the occurrence of false detections. This is a critical aspect of algorithm performance. The extent of the Lee&Tag detection envelope in figures 4.5 and 4.6 is broader than the other two algorithms but it is likely that this is associated with an increase in false detections. The results of this study, therefore, do not allow for a clear ranking of algorithm performance.

This study represents a modeling effort that clearly needs further development to reach its full potential. However, the model has been successfully used to demonstrate the extent and stability of three fire-detection algorithms representing different approaches in algorithm development. The validation exercise with the SAFARI-92 experimental fire points to serious shortcomings with the model or, perhaps more likely, an incompatibility between the quality of data required by the model and that available in AVHRR imagery. If AVHRR-derived fire distributions are to be used with a higher level of confidence or in a more quantitative fashion, it is absolutely necessary to pursue a modeling approach.

Acknowledgments

The author thanks Brian Stocks for helpful discussions and a description and map of the Block 56 fire and Gerhard Held of Eskom for tethersonde and rawinsonde data. Assistance from Nick Zambatis, Kruger Park, during the field work and in obtaining maps of Kruger Park is also gratefully acknowledged. This work was funded by a NASA grant awarded through the Goddard Space Flight Center under the Graduate Student Researcher's Program.

References

Becker, F. and Z.-L. Li. 1990. Towards local split window method over land surfaces. *Int. J. Remote Sensing*, 11(3), 369–393.

Berk, A., L. S. Bernstein, and D. C. Robertson. 1989. MODTRAN: A Moderate Resolution Model for LOWTRAN 7. Final Report GL-TR-89-0122, Geophysics Laboratory, Air Force Systems Command, Hanscom AFB, Mass.

Brown, S. 1988. Updated Appendix B for NOAA-H/11 to *Data Extraction and Calibration of TIROS-N/NOAA Radiometers*. L. Lauritson, G. J. Nelson, and F. W. Porto, 1979, NOAA Technical Memorandum NESS 107, NOAA, Washington, D.C., 58 pp.

Cooper, D. I. and G. Asrar. 1989. Evaluating atmospheric correction models for retrieving surface temperatures from the AVHRR over a tallgrass prairie. *Remote Sens. Environ.*, 27, 93–102.

Dowty, P. R. 1993. A Theoretical Investigation of Fire Detection with AVHRR Data, M.S. Thesis, University of Virginia, Charlottesville.

Dozier, J., 1981. A method for satellite identification of surface temperature fields of subpixel resolution. *Remote Sens. Environ.*, 11, 221–229.

Flannigan, M. D. and T. H. Vonder Haar. 1986. Forest fire monitoring using NOAA satellite AVHRR. *Can. J. For. Res.*, 16, 975–982.

Justice, C. and P. Dowty (Eds.). 1994. Technical Report of the IGBP-DIS Satellite Fire Detection Algorithm Workshop. 25–26 Feb. 1993, NASA-GSFC, Greenbelt, Md, IGBP, Paris.

Justice, C. O., Kendall, J. D., Dowty, P. R., and Scholes, R. J. 1994. Satellite sensing of fires during the SAFARI campaign using NOAA-AVHRR data. *J Geophys. Res.* (in press).

Justice, C. O., J.-P. Malingreau, and A. W. Setzer. 1993. Satellite Remote Sensing of Fires: Potential and Limitations. In *Fire in the Environment: The Ecological, Atmospheric, and Climatic Importance of Vegetation Fires*, edited by P. J. Crutzen and J. G. Goldammer, John Wiley, New York, pp. 77–88.

Kaufman, Y. J., C. J. Tucker, and I. Fung. 1990. Remote sensing of biomass burning in the tropics. *J. Geophys. Res.*, 95(D7), 9927–9939.

Kennedy, P. J., A. Belward, and J.-M. Grégoire. 1994. An improved approach to fire monitoring in West Africa using AVHRR data. *Int. J. Remote Sens.*, 15(11), 2235–2255.

Kidwell, K. B. 1991. *NOAA Polar Orbiter Data User's Guide (TIROS-N, NOAA-6, NOAA-7, NOAA-8, NOAA-9, NOAA-10, NOAA-11, NOAA-12)*, NOAA, Washington, D.C., 301 pp.

Langaas, S. 1993. A parameterised bispectral model for savanna fire detection using AVHRR night images. *Int. J. Remote Sens.*, 14(12), 2245–2262.

Lee, T. F. and P. M. Tag. 1990. Improved detection of hotspots using the AVHRR 3.7-μm channel, *Bull. Amer. Meteor. Soc.*, 71(12), 1722–1730.

Matson, M., S. R. Schneider, B. Aldridge, and B. Satchwell. 1984. Fire Detection Using the NOAA-Series Satellites. NOAA Technical Report NESDIS 7, Washington, D.C., 34pp.

McClain, E. P., W. G. Pichel, and C. C. Walton. 1985. Comparative performance of AVHRR-based multichannel sea surface temperatures. *J. Geophys. Res.*, 90, 11587–11601.

Pereira, M. C. and A. W. Setzer. 1993. Spectral characteristics of deforestation fires in NOAA/AVHRR images. *Int. J. Remote Sens.*, 14(3), 583–597.

Prins, E. M. and W. P. Menzel. 1994. Trends in South American biomass burning detected with the GOES VAS from 1983–1991. *J. Geophy. Res.* 99(D8), 16719–16735.

Robinson, J. M. 1991. Fire from space: Global fire evaluation using infrared remote sensing. *Int. J. Remote Sens.*, 12(1), 3–24.

Scholes, R. J., D. E. Ward, and C. O. Justice. 1994. Emissions of trace gases and aerosol particles due to vegetation burning in southern-hemisphere Africa. *J Geophys. Res.* (in press).

Setzer, A. W. and M. C. Pereira. 1991. Operational Detection of Fires in Brazil with NOAA-AVHRR, Presented at 24th International Symposium on Remote Sensing of Environment, 27–31 May 1991, Rio de Janeiro.

Setzer, A. W. and M. M. Verstraete. 1994. Fire and glint in AVHRR's channel 3: A possible reason for the nonsaturation mystery. *Int J. Remote Sens.*, 15(3), 711–718.

Thompson, A. M., D. P. McNamara, K. E. Pickering, R. D. Hudson, J. Kim, T. Kucsera, M. R. Schoeberl, C. O. Justice, and J. D. Kendall, SAFARI/TRACE-A Science Teams. 1994. Ozone over southern Africa, Brazil, and the Atlantic during 1992. IGAC/ STARE/SAFARI/TRACE-A Missions, CACGP, Japan.

AVHRR and ERBE Investigations of Biomass Burning in the Tropics

Sundar A. Christopher, Donna V. Kliche, and Ronald M. Welch

Biomass burning is considered a major source of trace gas species and aerosol particles (Crutzen et al. 1979; Logan et al. 1981) which play a vital role in tropospheric chemistry and climate (Crutzen and Andreae 1990). Anthropogenic biomass burning has expanded drastically in the last 15 years, due to increased deforestation practices in Brazil's Amazon Basin, as well as to clearing land for shifting cultivation in South America, Southeast Asia, and Africa.

Biomass burning produces large amounts of carbon dioxide, carbon monoxide, water, hydrocarbons, nitrous oxides, and smoke particles (Crutzen et al. 1979). Recent estimates have shown that about 114 Tg of smoke is produced yearly in the tropics through biomass burning (Penner et al. 1992). These smoke particles affect boundary layer cloud microphysics by increasing the amount of available cloud condensation nuclei (CCN) (Radke 1989), and by decreasing the cloud droplet sizes (Kaufman and Nakajima 1993). These particles can scatter the incoming solar radiation, thereby having a cooling effect on climate, or can modify the shortwave reflective properties of clouds by acting as CCN (Charlson et al. 1987). This indirect effect of increasing cloud albedo may be as large in magnitude (but opposite in sign) as the greenhouse effect due to a doubling of CO_2 (Coakley et al. 1987). On the other hand, the graphitic carbon released during biomass burning can also increase the absorption of solar radiation by the atmosphere and clouds. Therefore, smoke aerosols represent an important regional climate variable. Smoke can also be transported hundreds of kilometers away from the source and can therefore affect the climates of regions adjacent to burning areas.

The detection of aerosols over water is relatively straightforward because of the large contrast between water and atmospheric aerosols. The detection of aerosols over land is often more difficult due to the high albedo of the underlying background. Textures (second-order statistics) offer the possibility of measuring the spatial distribution of gray levels in the image by the use of statistical measures such as the disorder or homogeneity of the scene, contrast between gray values at a given distance, and so on. In this study, a new technique based on a combination of spectral and textural measures is used for aerosol detection over land. Collocated Earth Radiation Budget Experiment (ERBE) measurements are used to determine the shortwave (SW), longwave (LW), and net radiative forcing of aerosols produced due to biomass burning.

Data

In this study, a total of 44 selected satellite images over South America from 1985, 1986, and 1987 are analyzed during the biomass burning season (July through October). The Advanced Very High Resolution Radiometer Local Area Coverage (AVHRR LAC) data are used to accurately detect fires and smoke. The nominal spatial resolution is about 1.1 km at nadir, which is adequate for smoke and fire detection (Kaufman and Nakajima 1993). The area of study is between 5°S to 20°S and 45°W to 65°W, which includes the province of Rondônia, where thousands of fires are encountered each year. Twenty-five National Oceanic and Atmospheric Administration (NOAA-9), AVHRR LAC images from 1987 are used to develop the smoke-detection algorithm (Kaufman, personal communication 1993). The ERBE instantaneous scanner data from NOAA-9, which has a nominal spatial resolution of about 40 km, is used to examine the top-of-the-atmosphere (TOA) SW and LW fluxes.

Methodology

Fire and Smoke Detection

Fire detection is performed based on the method described by Kaufman et al. (1990). The mid-IR (3.7 μm-channel 3, "T_3") and the infrared channels (11 μm-channel 4, "T_4") of the AVHRR data are used to monitor the number of fires (Kaufman et al. 1990). The first condition ($T_3 \geq = 316°K$) requires that a pixel be close to the 320°K saturation level. The second condition ($T_3 \geq T_4 + 10°K$) requires that the radiative temperature in channel 3 be much larger than that of channel 4, in order to ensure that the pixel is not a warm surface; and the third condition ($T_4 > 250°K$) overrules the possibility of saturating the pixel from highly reflective clouds.

In order to accurately detect smoke over land, combinations of spectral and textural measures are used. The spectral combinations that were examined include AVHRR channels 2/1, $(1 - 4)/(1 + 4)$, $1 - 2$, $4 - 5$, $3 - 4$, $(3 - 4)/(3 + 4)$, and so on. Based on the Gray Level Difference Vector (GLDV) approach (see Welch et al. 1988), several textural features were calculated for a group of 9×9 AVHRR pixels, which include (1) contrast, (2) local homogeneity, (3) angular second moment, (4) entropy, (5) mean, (6) difference cluster shade, and (7) difference cluster prominence. A detailed description can be found in Haralick and Shapiro (1992).

For smoke detection, the best combination was obtained with channel 1 in red, channel $(1 - 4)/(1 + 4)$ in green, and the texture "mean" of the channel $(1 - 4)/(1 + 4)$ in blue (the 3-band overlay is not shown). However, figure 5.1 shows an example of the spectral and textural features calculated for one of the analyzed images. The size of the image is about 200 km^2. Figure 5.1a shows channel 1 for this image where the dense smoke over fires appears highly reflective. Figure 5.1b shows the same area, but for the channel $(1 - 4)/(1 + 4)$. Here the dense smoke is brighter than the background, but less bright than the clouds. Figure 5.1c is an example of the calculated texture MEAN (channel $(1 - 4)/(1 + 4)$). The textural measure highlights the boundaries of the smoke clearly.

Collocation of AVHRR with ERBE Data

After identifying the smoke produced due to biomass burning, the ERBE instantaneous scanner data are used to study the effects of smoke on the TOA fluxes. The ERBE data consist of observations of solar constant, the reflected shortwave radiation, and the earth-atmosphere emitted longwave radiation at the TOA (Barkstrom et al. 1989). The ERBE spatial resolution is about 40 km, while the AVHRR data has a nominal spatial resolution of 1.1 km, in the LAC form.

In this study, a collocation procedure between ERBE and AVHRR LAC data, similar to that described in Li and Leighton (1991) is used. The collocation analysis is performed only for the 1985 and 1986 data, since ERBE data from NOAA-9 are not available for 1987. First, the center of the ERBE pixel that is closest to the AVHRR pixel is identified. Once

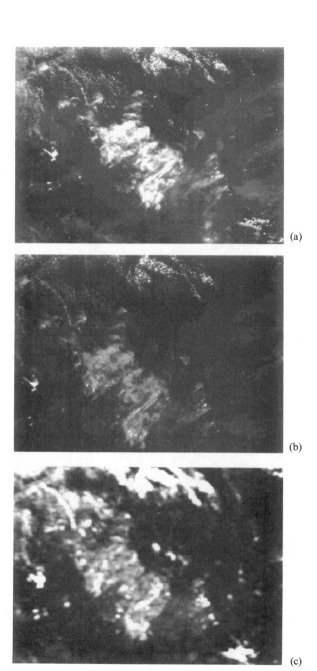

(a)

(b)

(c)

Figure 5.1 The AVHRR LAC image over Rondônia Province, Brazil, 3 September 1985: (*a*) channel 1; (*b*) channel $(1 - 4)/(1 + 4)$; (*c*) texture MEAN channel $(1 - 4)/(1 + 4)$.

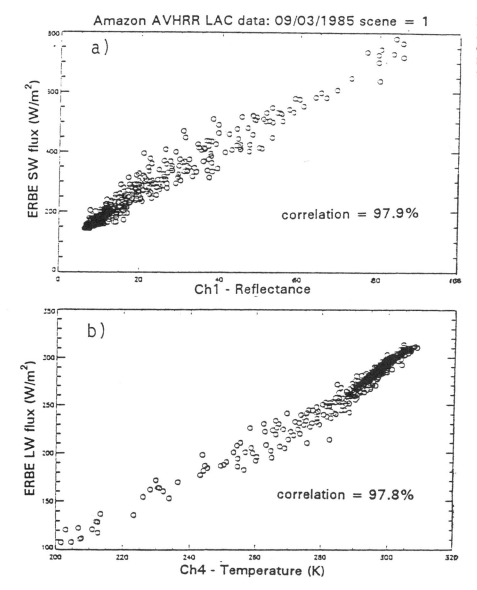

Figure 5.2 Collocation between AVHRR and ERBE: (*a*) AVHRR channel 1 reflectances versus ERBE SW fluxes; (*b*) AVHRR channel-4 temperatures versus ERBE LW fluxes.

the center pixel has been identified, a group of 37×37 pixels centered on the closest pixel is assumed to correspond to an ERBE pixel. Figure 5.2 shows an example of the correlation between AVHRR channel 1 reflectances versus REBE SW fluxes and AVHRR channel 4 temperatures versus ERBE LW fluxes. The correlation coefficients are 97%, which ensures that the data have been properly calibrated, navigated, and collocated.

Results

Understanding the influence of clouds on the earth radiation budget is extremely important for under-standing the earth's climate. The difference in the radiative heating between a column of clear and cloudy air represents the cloud radiative forcing. Similarly, the aerosol radiative forcing is calculated by subtracting the "smoky" air from the column of clear air. The shortwave (SWARF) and longwave (LWARF) aerosol forcing is defined as

$$SWARF = F_O(\alpha_{clr} - \alpha_{aer})$$

$$LWARF = LW_{clr} - LW_{aer}$$

where

F_O = incoming solar flux
α_{clr} = clear sky albedo

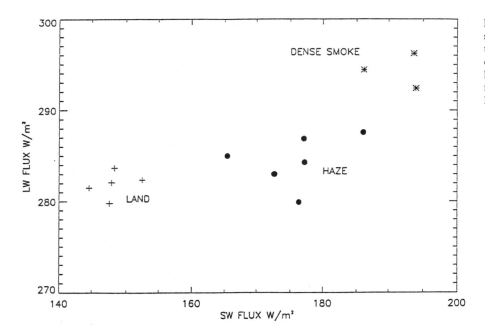

Figure 5.3 Samples of land, dense smoke, and diffuse smoke (hazelike) taken for one of the selected images of September 1985. For the samples having 1 to 10 fire pixels classified as fires, the LW TOA fluxes are slightly larger than for land.

α_{aer} = aerosol sky albedo
LW_{clr} = clear sky LW flux
LW_{aer} = aerosol sky LW flux.

Figure 5.3 shows the scatter plot between ERBE SW flux and LW flux for selected samples from one image over Rondônia. Three classes are defined: (1) land, (2) haze (diffuse smoke), and (3) dense smoke. The difference in SW flux values between dense smoke and clear land is about 40 W/m^2, whereas the difference between the diffuse smoke and clear land is about 25 W/m^2. The diffuse smoke (hazelike) class refers to the case when the smoke is transported, by winds, away from the source, therefore being optically thinner than the dense smoke over fires. The corresponding LW TOA flux values for the dense smoke class are larger than for land because of the large number of fires below, which increases the overall temperature of the sample. Diffuse smoke samples are taken downwind from the fire areas and had either no fire pixels or fewer than 10 pixels classified as fires. Therefore the LW TOA flux values for these samples are comparable with land values.

The SWARF and LWARF are calculated as described in the formulas above. For the dense smoke class, the mean SWARF is −46.8 W/m^2, and the mean LWARF is 9.9 W/m^2. For the diffuse smoke (hazelike) class, the mean SWARF is −25.9 W/m^2, and the mean LWARF is 8.9 W/m^2. Based on this data, we then calculated the instantaneous net radiative forcing. Therefore, the net radiative forcing of dense smoke is about −36.9 W/m^2 and of diffuse smoke (hazelike) is about −17 W/m^2.

Conclusions

Recent estimates show that about 114 Tg of smoke is produced yearly in the tropics through biomass burning (Penner et al. 1992). These aerosols can scatter the incoming solar radiation and can affect the reflective properties of the clouds, therefore having a cooling effect on climate. The aerosol particles represent an important regional climate variable. While detection for aerosols over water is relatively straightforward, the detection of aerosols over land is more difficult due to high albedo of the background. In this study, a new technique based on a combination of spectral and textural measures is developed for detecting smoke produced from biomass burning. The AVHRR channel 1 and $(1-4)/(1+4)$ ratio along with the "mean" textural measure provide the best results. The direct radiative effects of these aerosols are examined using broadband scanner measurements from the Earth Radiation Budget Experiment. Preliminary results show that the instantaneous net radiative forcing of aerosols is −37 W/m^2.

Acknowledgments

This research was supported under NASA Grant NAGW-3740 managed by Dr. Robert J. Curran. We thank Dan Baldwin for the navigation software, and NASA Langley for the ERBE S8 data. Special thanks to Dr. Yoram Kaufman for the 1987 NOAA-9 images and Joie Robinson for typing the manuscript.

References

Barkstrom, B. R., E. Harrison, G. Smith, R. Green, J. Kebler, R. Cess, and the ERBE Science Team. 1989. Earth Radiation Budget Experiment (ERBE) archival and April 1985 results, *Bull. Amer. Meteor. Soc.*, 70, 1254–1262.

Charlson, R. J., J. E. Lovelock, M. O. Andreae, and S. G. Warren. 1987. Oceanic phytoplankton, atmospheric sulphur, cloud albedo and climate, *Nature*, 326, 655–661.

Coakley, J. A., R. L. Bernstein, and P. A. Durkee. 1987. Effect of ship-stack effluents on cloud reflectivity, *Science*, 237, 1020–1022.

Crutzen, P. J., and M. O. Andreae. 1990. Biomass burning in the tropics: Impact on atmospheric chemistry and biogeochemical cycles, *Science*, 250, 1669–1678.

Crutzen, P. J., L. E. Heidt, J. P. Krasnec, W. H. Pollock, and W. Seiler. 1979. Biomass burning as a source of atmospheric gases, *Nature*, 282, 253–256.

Haralick, R. M., and L. G. Shapiro. 1992. *Computer and Robot Vision. Vol. I.* Addison-Wesley, Reading, Mass., 453–494.

Kaufman, Y. J., and T. Nakajima. 1993. Effect of Amazon smoke on cloud microphysics and albedo-analysis from satellite imagery, *J. Appl. Meteor.*, 32, 729–744.

Kaufman, Y. J., C. J. Tucker, and I. Fung. 1990. Remote sensing of biomass burning in the tropics, *J. Geophys. Res.*, 95, 9927–9939.

Li, Z., and H. G. Leighton. 1991. Scene identification and its effect on cloud radiative forcing in the Arctic, *J. Geophys. Res.*, 96, 9175–9188.

Logan, J. A., M. J. Prather, S. C. Wofsy, and M. B. McElroy. 1981. Tropospheric chemistry: a global perspective, *J. Geophys. Res.*, 86, 7210–7254.

Penner, J. E., R. E. Dickinson, and C. A. O'Neill. 1992. Effects of aerosol from biomass burning on the global radiation budget, *Science*, 256, 1432–1433.

Radke, L. F. 1989. Airborne observations of cloud microphysics modified by anthropogenic forcing; paper presented at *Symposium on Atmospheric Chemistry and Global Climate*, Amer. Meteor. Soc., 29 Jan.–3 Feb., Anaheim, Calif.

Welch, R. M., S. K. Sengupta, and D. W. Chen. 1988. Cloud field classification based upon high spatial resolution textural features, Part I: Gray level co-occurrence matrix approach, *J. Geophys. Res.*, 93, 12663–12681.

Monitoring Biomass Burning with the New Generation of Geostationary Satellites

W. Paul Menzel and Elaine M. Prins

With the launch of the Geostationary Operational Environmental Satellite-8 (GOES-8) in April 1994, a new capability for monitoring diurnal biomass burning activities in North and South America was introduced. The higher spatial resolution, greater radiometric sensitivity, and improved navigation offer many advantages over the GOES-7. The 1994 biomass burning season in South America and the United States provided many opportunities for the GOES-8 imager to detect fires and track smoke/aerosol transport regimes.

Although the GOES-8 was not yet operational during the 1994 burning season, several sample data sets were obtained in North and South America. These data sets were used to develop the GOES-8 Automated Biomass Burning Algorithm (ABBA), which was operational for the 1995 burning season in South America. In September 1994, the GOES-8 observed smoke palls in South America similar to those seen in August and September of 1988; they extended throughout the Amazon Basin and south into Bolivia, Paraguay, Uruguay, and Argentina. As in previous years, most of the burning occurred along the perimeter of the Amazon Basin and throughout the cerrado regions of southern Brazil and Bolivia. The diurnal signature was clearly evident with maximum burning occurring between 1500 and 1800 UTC. Comparisons with GOES-7 imagery reveal much more detail in the GOES-8 imagery, including surface features and individual fire activity. Throughout the wildfire season in North America only the largest fires were evident in GOES-7 imagery. Fires that displayed a strong brightness temperature signal in relation to the non-fire background temperatures in the GOES-8 data often showed no elevated signal in the GOES-7 data. For the first time it is possible to monitor diurnal variability in wildfire activity throughout the Western Hemisphere.

The GOES-8: The New Generation of Geostationary Satellites

On 13 April 1994, the first of the National Oceanic and Atmospheric Administration (NOAA) next generation of geostationary satellites, GOES-8, was launched. All components of the GOES-8 system are new or greatly improved: (1) the satellite is Earth oriented to improve instrument performance; (2) sounding and imaging operations are now performed by different and separate instruments; (3) a five-band multispectral radiometer with higher spatial resolution improves imaging capabilities; (4) a sounder with higher radiometric sensitivity enables operational temperature and moisture profile retrieval from geostationary altitude for the first time; (5) a different data format is used to retransmit raw data to direct receive users; and (6) a new ground data processing system handles the high volume of data and distributes advanced products to a variety of users (Menzel and Purdom 1994).

The GOES-8 imager has a five-band multispectral capability with 10-bit precision and high spatial resolution: (a) 0.52–0.72 μm (visible) at 1 km useful for cloud, pollution, and haze detection, and severe storm identification; (b) 3.78–4.03 μm (shortwave infrared window) at 4 km useful for identification of fog at night, discriminating between water clouds and snow or ice clouds during the daytime, detecting fires and volcanoes, and nighttime determination of sea surface temperature; (c) 6.47–7.02 μm (upper-level water vapor) at 8 km useful for estimating regions of mid-level moisture advection and drying and tracking mid-level atmospheric motions; (d) 10.2–11.2 μm (longwave infrared window) at 4 km familiar to most users for cloud-drift winds, severe storm identification, and location of heavy rainfall; and (e) 11.5–12.5 μm (infrared window more sensitive to water vapor) at 4 km useful for identification of low-level moisture, determination of sea surface temperature, and detection of airborne dust and volcanic ash. Onboard calibration provides brightness temperatures with $0.3°$K relative precision; in-flight determinations of noise levels indicate reduction by 2 to 10 times over those from GOES-7. Table 6.1 indicates the in-flight noise performance of the GOES-8 imager and compares it with the GOES-7 and Meteosat-5 performances.

Table 6.1 Comparison of GOES-8, GOES-7, and Meteosat-5 imagers

Band	Bit depth			Resolution (km)			Noise		
	G-8	G-7	M-5	G-8	G-7	M-5	G-8	G-7	M-5
								(counts)	
Visible	10	6	8	1	1	2.5	3	1	1
Infrared								(°C at 300°K)	
3.9 μm	10	10	NA[a]	4	16	NA	0.23	0.25	NA
10.7 μm	10	10	8	4	8	5	0.14	0.15	0.20
12.0 μm	10	10	NA	4	16	NA	0.26	0.40	NA
								(°C at 230°K)	
6.7 μm	10	10	8	8	16	5	0.22	1.00	0.40

a. NA indicates data were not available.

The improved performance is most notable in the visible band, where 10-bit data from silicon detectors shows much improved low light sensitivity and detector-to-detector consistency. The infrared window channels needed for detection of fires have been improved considerably; the short(long)-wave window has four (two) times the spatial resolution while maintaining similar radiometric performance. The water vapor band performance of the GOES-8 is also improved by an order of magnitude over the GOES-7 (twice the spatial resolution and one fifth the noise). The high spatial resolution and the good signal to noise of the imager data make it very useful at satellite viewing angles up to 75 degrees; GOES-8 images over Hudson Bay and near the Arctic Circle reveal details not seen in GOES-7 images (Purdom 1995).

These enhanced capabilities of the GOES-8 imager make it much more useful for detecting and characterizing biomass burning, not only in the Amazon Basin of South America, but also in the temperate and boreal forests of the United States and Canada. The following section will present some examples from 1994.

Detection of Biomass Burning with the GOES-8

During the past decade, the utility of geostationary remote sensing to monitor trends in biomass burning in South America has been demonstrated with the Visible Infrared Spin Scan Radiometer Atmospheric Sounder (VAS) on board GOES-4 through GOES-7 (Prins and Menzel 1994). A strong diurnal signature in the biomass burning was found; the burning peaks between 1500 and 1800 UTC in the early afternoon. Typically the burning area detected at peak periods is two to five times greater than that observed at other times. Trends in the burning detected with an auto-mated biomass burning algorithm (ABBA) from 1983 to 1991 were reported; locations of fires in the selva spread to three times as big a region in 1991 compared to 1983 (285 000 km^2 versus 85 000 km^2, respectively). The GOES-7 noise (about 0.25°C in the 3.9-μm band and 0.15°C in the 10.7 μm band) for 13.8 km fields of view (FOVs) suggests that the smallest fire with a temperature of 450°K that can be detected is 0.03 km^2; the smallest fires detected with the GOES-VAS ABBA were about 0.1 km^2. GOES-8 noise (roughly the same as GOES-7) for 4 km FOVs enable 450°K fires of 0.002 km^2 to be detected; a 0.02 km^2 fire was detected with an initial GOES-8 ABBA in South America in the fall of 1994.

The improved capability to detect individual fires is demonstrated in figure 6.1. The GOES-8 visible detection of smoke and haze associated with biomass burning in South America for 6 September 1994 is shown in figure 6.1a; the curved black outline indicates the extent of the smoke pall. Figure 6.1b shows a close-up of the GOES-8 3.9-μm image for one of the source regions (outlined in the black rectangle in figure 6.1a) where the burning occurred the previous day; hundreds of individual fires are evident as dark hot spots. Although a direct comparison of GOES-8 and GOES-7 in South America during the 1994 burning season was not possible due to the westerly location of the GOES-7 instrument, figure 6.1c and 6.1d show similar data from GOES-7 for 30 August 1988. General comparisons demonstrate the ability to see much more detail with the GOES-8 data, including land features and localized individual fire activity. Because the GOES-8 oversamples the 1-km visible and the 4-km infrared fields of view by a factor of 1.75, the aspect ratio of the GOES-8 images is stretched in the east-west direction.

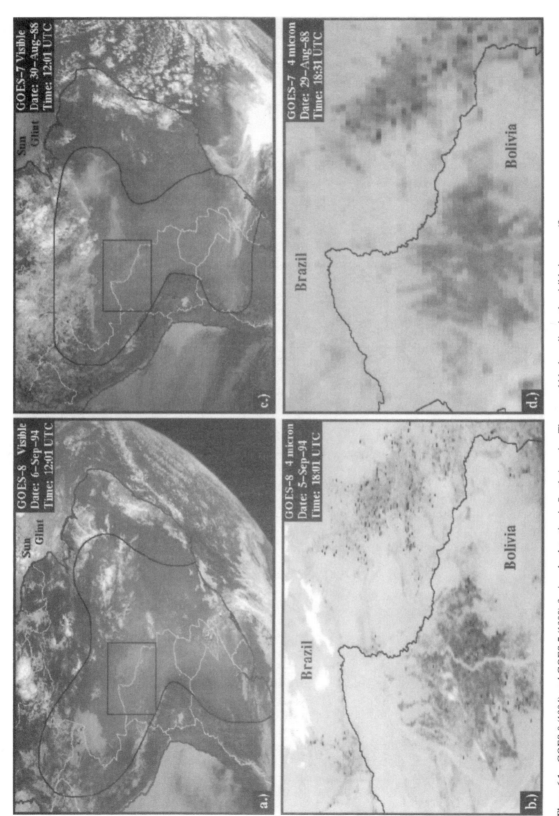

Figure 6.1 GOES-8 (1994) and GOES-7 (1988) fire/smoke detection in South America. The curved black outline in the visible imagery (figure 6.1*a*, *c*) identifies the extent of smoke associated with biomass burning activity throughout Brazil, Bolivia, Paraguay, and Argentina. The dark hot spots in the 4-μm imagery (figure 6.1*b*, *d*) indicate active fires along the border between Brazil and Bolivia that contributed to the smoke palls outlined in the visible imagery.

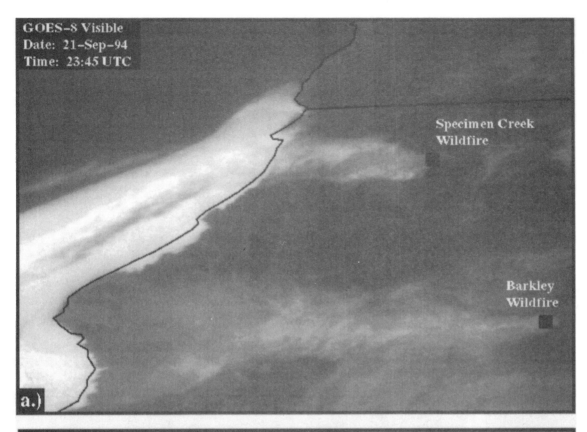

Figure 6.2 During the SCAR-C experiment, smoke transport associated with wildfire activity in Northern California was observed in GOES-8 (figure 6.2*a*) and GOES-7 (figure 6.2*b*) visible imagery. The GOES-8 imagery clearly identifies the smoke plume extending over coastal marine stratus. The smoke is not so easily identified in the GOES-7 imagery.

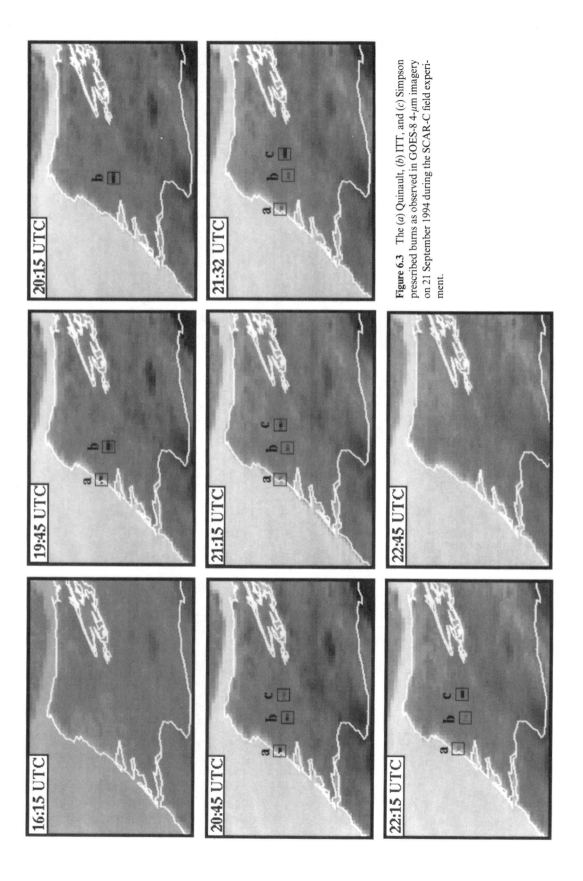

Figure 6.3 The (a) Quinault, (b) ITT, and (c) Simpson prescribed burns as observed in GOES-8 4-μm imagery on 21 September 1994 during the SCAR-C field experiment.

The GOES-8 visible data at 10 bits can be enhanced to show detail never possible before in the clouds and haze associated with biomass burning. Figure 6.2 shows an example of smoke and haze from two wild-fires transported over marine stratus clouds off the coast of northern California on 21 September 1994. The GOES-8 visible image (figure 6.2a) clearly shows the haze over the marine stratus and traces it back to the source fires; the smoke is not so easily identified in the GOES-7 visible image (figure 6.2b). The ability to distinguish haze from cloud will greatly facilitate estimates of aerosol transport associated with biomass burning.

The diurnal variability of burning associated with a given fire is demonstrated in the GOES-8 imagery in figure 6.3. Several prescribed burns were initated on 21 September 1994 in Washington in association with the Smoke Clouds and Radiation (SCAR-C) experiment, including the Quinault fire (48 acres, 47:19°N, 124:16°W), the Simpson fire (95 acres, 47:12°N, 123:30°W), and the ITT fire (97 acres, 47:08°N, 123:38°W). The Quinault fire consisted of approximately 5000 tons of red cedar debris ignited by the US Forest Service (USFS) at approximately 1810 UTC. More than 21 acres were flaming at 1830 UTC; approximately 30 acres remained in the smoldering phase at 2015 UTC; and less than 10 acres were smoldering at 2200 UTC (Ottmar 1994; see Hobbs et al., chapter 66 this volume). The first available GOES-8 shortwave window image at 1945 UTC clearly shows burning at Quinault corresponding to peak heat release rates computed by the USFS. At 2015 UTC the GOES-8 did not detect elevated 4 μm brightness temperature for the Quinault site; however at 2045 UTC the fire reappears in the GOES-8 image and remains until 2215 UTC. The GOES-8 ABBA estimates of area burning and average fire temperature are compared with the ground estimates of flaming and smoldering acres in table 6.2. Because the ABBA infers uniform background radiance from neighboring clear sky pixels, these estimates are somewhat hindered by the coastal location of Quinault, where background radiation for each GOES-8 fire pixel comes from a combination of ocean and land (Prins and Menzel 1994). The relatively good agreement between GOES-8 and ground truth estimates is very encouraging; the estimate of the size of the fires is within 20% on the average at any given time. The GOES-8 also shows the enhanced capability over the GOES-7, which did not detect the fire after 2045 UTC.

Another indication of the diurnal variation of the biomass burning in South America is evident in the application of the ABBA to GOES-8 observations

Table 6.2 GOES-8 and ground estimates of the intensity and extent of the Quinault, Washington, controlled burn on 21 September 1994. Note that 1 acre equals 0.004 km². Ground data courtesy of Roger Ottmar of the U.S. Forest Service Seattle Forestry Science Laboratory, Pacific Northwest Research Station.

	Ground observations		GOES-8 estimates	
Times (UTC)	Flaming (acres)	Smoldering (acres)	Total area (acres)	Temperature (°K)
1800	0	0	NA[a]	NA
1815	2	0	NA	NA
1830	21	0	NA	NA
1845	23	7	NA	NA
1900	22	12	NA	NA
1915	21	21	NA	NA
1932	15	24	NA	NA
1945	15	21	40	602
2000	7	26	NA	NA
2015		29	No elevated signal in G-8 data	
2030		23	NA	NA
2045		20	27	626
2100		18	NA	NA
2115		18	16	597
2132		13	17	586
2145		11	NA	NA
2200		10	NA	NA
2215		8	Fire barely detectable in G-8 data	
2230		7	NA	NA
2245		6	Fire not detected in G-8 data	
2300		5	Fire not detected in G-8 data	

a. NA indicates data were not available.

every three hours during the week of 5–11 September 1994. Figure 6.4a shows the study area over Brazil, Bolivia, and Peru; the 3.9-μm image indicates considerable fire activity at 1800 UTC on 5 September (figure 6.4b). Figure 6.4c indicates fire locations as observed with the ABBA. Temperatures and burning areas were estimated for about 1600 fires (in black); another 500 (in white) were detected in cloudy regions, but the ABBA could not find an adequate estimate of clear sky radiances; and fewer than 50 fires (in gray) saturated the detector so that the temperature for the field of view was reported at the maximum value of 335°K even though a hotter temperature is probable. Thus at a single time (1800 UTC, 5 September 1994), the GOES-8 detected roughly 2100 fires in the study region of South America (figure 6.4a). The diurnal variation of

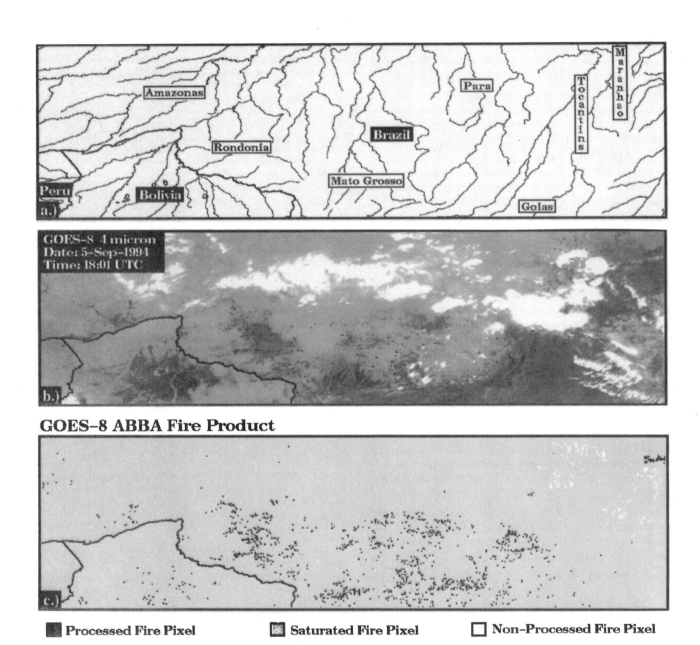

GOES-8 ABBA Fire Product

■ **Processed Fire Pixel**　　▨ **Saturated Fire Pixel**　　□ **Non-Processed Fire Pixel**

Figure 6.4　Application of the GOES-8 ABBA in South America: (*a*) The study area includes portions of Brazil, Bolivia, and Peru. (*b*) The 4-μm imagery shows considerable fire activity at 1800 UTC on 5 September 1994. (*c*) More than 2100 fires were identified by the GOES-8 ABBA for this time period.

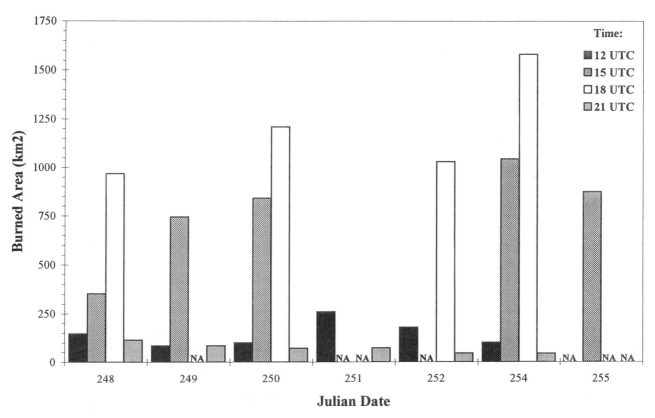

Figure 6.5 Diurnal variation in biomass burning in South America as observed with the GOES-8 ABBA. GOES-8 ABBA results show that during the week of 5–11 September 1994 peak burning occurred near 1800 UTC. (NA indicates data were not available.)

the biomass burning in the study region for the week is shown in figure 6.5. Peak burning of 1000 to 1500 km^2 occurred at 1800 UTC; burning at 1200 and 2100 UTC is down by an order of magnitude; and burning at 1500 UTC is reduced on the average by one third of the peak value. This differs somewhat from Prins and Menzel (1994), who reported that the GOES-7 detected a preference for burning at 1500 UTC in 1983.

Conclusions

The GOES-8 is demonstrating enhanced capability to detect fires with infrared window data at 4 km resolution with 0.2°C noise. The GOES-8 is detecting smaller fires and more of them than the GOES-7; the GOES-8 has already detected fires smaller than 5 acres. The GOES-8 operational schedule enables more frequent observations; routine imager coverage (in all five spectral bands) is scheduled every 30 minutes over South America and every 15 minutes over the continental United States. This enhanced capability can

be used to assist with early detection of fires in remote regions of North and South America.

Estimation of the areal extent and temperature of a fire with the automated biomass burning algorithm (ABBA) shows good early results in comparison to controlled burns of known characteristics in North America. With the ABBA, the GOES-8 continues to indicate a strong variation in the burning as a function of the time of day in South America; a preference for burning between 1500 and 1800 UTC indicated by the GOES-5 in 1983 is found again for one week in 1994.

The GOES-8 appears to be well suited to continue and improve the geostationary remote sensing capability that provides information on the diurnal signature in biomass burning, investigates aerosol transport, and monitors trends in biomass burning for the next decade throughout the Western Hemisphere.

References

Menzel, W. P. and J. F. W. Purdom. 1994. Introducing GOES-I: The first of a new generation of geostationary operational environmental satellites. *Bull. Amer. Meteor. Soc.*, vol. 75, no. 5, 757–781.

Ottmar, Roger. 1994. Personal Communication. United States Forest Service, Seattle Forestry Science Laboratory, Pacific Northwest Research Station, Seattle, Washington.

Prins, E. M., and W. P. Menzel. 1994. Trends in South American biomass burning detected with the GOES visible infrared spin scan radiometer atmospheric sounder from 1983 to 1991. *J. Geophysical Res.*, 99, 16 719–16 735.

Purdom, J. F. W. 1995. Observations in polar regions using GOES-8 imagery. Preprints, *Geophysics and the Environment, International Union of Geodesy and Geophysics, XXI General Assembly*, Boulder, Colo. 2–14, July 1995, American Geophysical Union, Washington, D.C., A265.

Investigation of Biomass Burning and Aerosol Loading and Transport Utilizing Geostationary Satellite Data

Elaine M. Prins and W. Paul Menzel

Recent modeling and analysis efforts have suggested that the direct and indirect radiative effects of aerosols from global biomass burning may play a major role in the radiative balance of the earth and are an important factor in calculations of climate change. One of the most active burning regions is in South America and is associated with deforestation in the Amazon and various agricultural burning practices throughout the continent. The GOES VAS series of satellites, operational since 1981, and the new generation of GOES satellites, introduced with the launch of GOES-8 in the spring of 1994, offer the unique opportunity to monitor trends in biomass burning activities and associated aerosol loading and transport regimes.

The GOES Automated Biomass Burning Algorithm (ABBA) provides information concerning the location, temperature, and size of subpixel fires. An automated GOES multispectral thresholding algorithm is being developed to catalogue the areal extent and transport of aerosols associated with biomass burning. GOES visible and infrared (4, 11, and 12 μm) data are used to distinguish smoke/aerosols associated with biomass burning from other multi-level clouds and low-level moisture. The visible, 4-, and 11-μm bands distinguish haze from clouds, while the 11- and 12-μm bands are primarily used to distinguish haze from low-level moisture and semitransparent cirrus. Furthermore, the prevailing circulation and transport of aerosols in South America can be estimated in half-hourly GOES visible and infrared images by tracking the displacement of smoke, aerosols, and adjacent clouds. A preliminary analysis of GOES satellite imagery collected over South America during the dry seasons of 1983, 1988, 1989, 1991, and 1994 revealed numerous examples of aerosol transport associated with biomass burning in the selva and cerrado. Three major transport regimes were identified with smoke palls extending thousands of kilometers from the emission sources.

Monitoring Biomass Burning and Aerosol Transport with Geostationary Satellites

For nearly 15 years the Geostationary Operational Environmental Satellite (GOES) has provided half-hourly visible and multispectral infrared coverage of the Western Hemisphere. Although the GOES is primarily a weather satellite, it offers the unique opportunity to monitor changes in the earth's surface and subsequent interactions with the atmosphere. Previous studies have shown the ability to use GOES visible and infrared data to monitor biomass burning activities and associated smoke/aerosol transport in South America (Menzel et al. 1991; Prins and Menzel 1992; 1994). Figure 7.1 provides an illustrative example. The black markers in figure 7.1a represent more than 1000 pixels identified by the GOES Automated Biomass Burning Algorithm (ABBA) as containing fire activity at 1830 UTC on 24 August 1988. The visible image shows the smoke pall associated with the burning on the following morning. The smoke pall five days later (29 August 1988) shown in figure 7.1b represents one of the largest ever documented in South America extending over 5 million km^2. It covers the area east of the Andes from the equator to 30°S and out over the Atlantic Ocean at its southern boundary extending to the edge of the satellite image. The smoke represents a combination of emissions from burning in the selva (tropical forest) and cerrado (grasslands) regions as well as wildfires in the Andes Mountains. The wind barbs in figure 7.1c indicate anticyclonic flow with the Andes acting to deflect the smoke southward where westerlies transport it over the Atlantic Ocean. The cloud and smoke drift winds give an indication of the height of the aerosols as well as the transport regime. The black wind barbs indicate winds above 400 mb, white wind barbs depict winds between 400 and 699 mb, and gray wind barbs indicate winds below 700 mb. Using multispectral GOES VAS data, it was possible to estimate the height of the aerosols in relation to adjacent clouds. In some regions the smoke

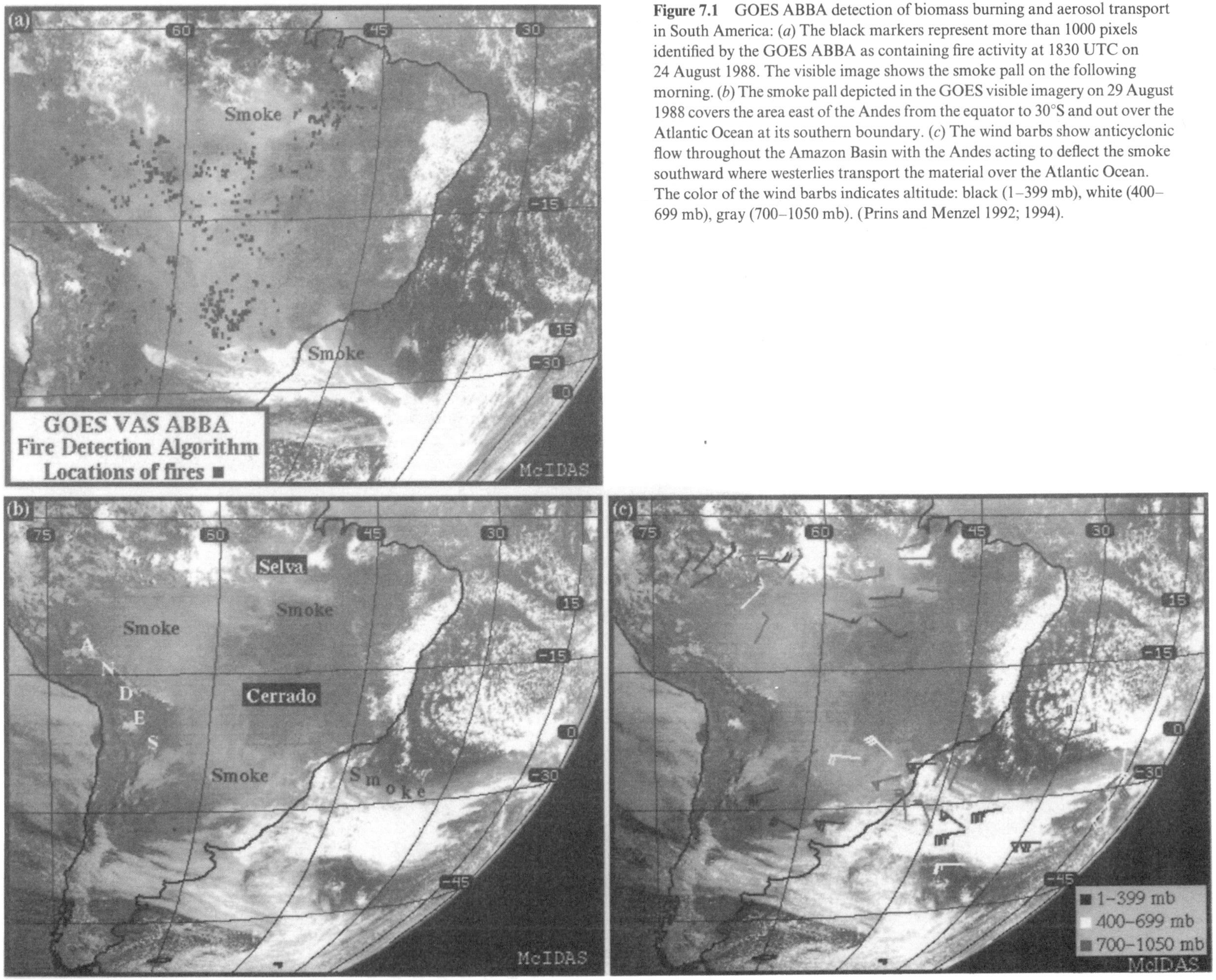

Figure 7.1 GOES ABBA detection of biomass burning and aerosol transport in South America: (*a*) The black markers represent more than 1000 pixels identified by the GOES ABBA as containing fire activity at 1830 UTC on 24 August 1988. The visible image shows the smoke pall on the following morning. (*b*) The smoke pall depicted in the GOES visible imagery on 29 August 1988 covers the area east of the Andes from the equator to 30°S and out over the Atlantic Ocean at its southern boundary. (*c*) The wind barbs show anticyclonic flow throughout the Amazon Basin with the Andes acting to deflect the smoke southward where westerlies transport the material over the Atlantic Ocean. The color of the wind barbs indicates altitude: black (1–399 mb), white (400–699 mb), gray (700–1050 mb). (Prins and Menzel 1992; 1994).

pall was estimated to extend above 4 km and possibly as high as 10 km.

The GOES archive offers the opportunity to survey the extent of burning in South America and to analyze trends in biomass burning from similar instrument platforms with a fixed algorithm. The GOES ABBA was developed at the Cooperative Institute for Meteorological Satellite Studies (CIMSS) to determine trends in fire activity in South America over the past decade utilizing the GOES Visible Infrared Spin Scan Radiometer (VISSR) Atmospheric Sounder (VAS) archive. The GOES ABBA is a dynamic multispectral thresholding algorithm that uses the 4- and 11-μm bands to locate fire pixels and provide estimates of subpixel fire activity and mean fire temperature. The automated algorithm was applied to a study area extending from 5°S to 15°S and from 45°W to 70°W every day for two weeks at the peak of the burning season in South America in 1983, 1988, 1989, and 1991 in an effort to estimate the areal extent of burning in South America during the past decade and to provide additional insight into the diurnal signature in satellite detection of biomass burning activities. The study regime included biomass burning associated with deforestation activities in the selva as well as grassland management and agricultural applications in the cerrado and mixed zone along the perimeter of the Amazon. Throughout the study area the GOES VAS ABBA detected twice as much daily burning during the two weeks in 1991 than in 1983. Fire activity nearly doubled in the selva and mixed regions, and tripled in the cerrado. Furthermore, the locations of the fires in the selva spread to a region in 1991 three times bigger than that of 1983. The region containing evidence of fire activity was roughly 85 000 km² in 1983 and increased to 285 000 km² in 1991. From 1988 to 1991 biomass burning rates associated with deforestation in previously undisturbed primary forests seemed to have leveled off, while fire activity in the mixed and cerrado regions increased. A detailed description of the algorithm and analyses can be found in Prins and Menzel (1994). The new series of GOES satellites, with improved capabilities for monitoring biomass burning and aerosol transport, enable continued monitoring of both short- and long-term trends in biomass burning throughout the Western Hemisphere.

Smoke/Aerosol Transport Regimes in South America

In order to assess the possible roles of global biomass burning in climate change and more directly in atmospheric chemistry and radiative transfer processes, the extent of global burning and resulting aerosol transport regimes must be determined. The majority of all biomass burning occurs in the equatorial and subtropical regions of Africa and South America. A review of GOES visible and infrared satellite imagery over South America during the dry seasons of 1983, 1988, 1989, 1991, and 1994 revealed numerous examples of aerosol transport associated with biomass burning in the selva and cerrado. This offers insights into the extent and transport of smoke/aerosol in South America under different meteorological conditions. From sequences of half-hourly visible and infrared data, it was possible to isolate various transport regimes for smoke/aerosols and adjacent clouds.

During the dry season the majority of the images depicted anticyclonic flow throughout the Amazon Basin, where easterly winds in the northern portion of the Amazon Basin transport the aerosols westward and the Andes Mountains deflect the smoke to the south-southeast. Typically the smoke pall is limited to within the continent east of the Andes by a synoptic disturbance over the south Atlantic Ocean, which extends along the eastern coast of South America (figure 7.2, *track a*). In 1983, 1988, and 1994 there were examples of anticyclonic flow throughout the Amazon Basin which resulted in transport out over the Atlantic Ocean near 30°S (figure 7.2, *track b*). This transport regime is often associated with a stalled frontal boundary located south of the Amazon Basin that extends out over the ocean and acts as a barrier at the southern extent of the smoke pall allowing for aerosol transport over the ocean. In 1991 and 1994 there were several episodes during which the smoke continued southward along the Andes Mountains into Bolivia, Paraguay, and Argentina instead of being channeled east toward the Atlantic Ocean (figure 7.2, *track c*). In 1991 this was characterized by a lack of cyclonic activity off the coast of Argentina that might otherwise act to channel the emissions east. On several occasions the smoke plume extended south along the front range of the Andes into the southern half of Argentina at 40–50°S. Another transport mechanism evident in all five years consists of flow from the Amazon Basin into the selva regions in the northwestern portion of the Brazilian Amazon and extending into Peru and Colombia (figure 7.2, *track d*). This is often accompanied by a large-scale convective complex associated with a frontal boundary extending from the front range of the Andes Mountains and across the southern and eastern sections of the Amazon Basin. Most other transport mechanisms are combinations of the above scenarios. Only a few examples of smoke/aerosol transport

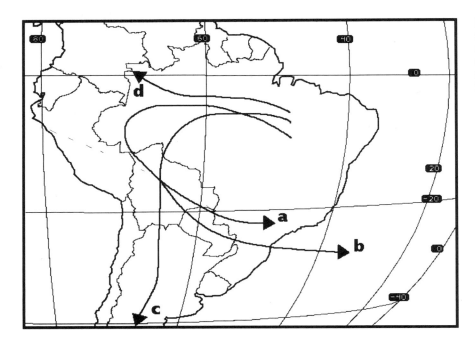

Figure 7.2 The four labeled trajectories (*a*, *b*, *c*, and *d*) represent the most common aerosol transport regimes observed in GOES satellite imagery collected during the dry seasons in 1983, 1988, 1989, 1991, and 1994.

were evident in the GOES imagery for 1989. The 1989 burning season was abnormally wet, and the region was covered by clouds, obscuring possible fire activity and associated smoke transport. These examples provide an initial perspective on the extent of aerosol coverage during the dry season in Brazil and typical transport regimes; more quantitative large-scale satellite analyses are needed to better characterize this phenomenon.

GOES Multispectral Smoke/Aerosol Detection

In order to document quantitatively the extent of smoke/aerosols associated with biomass burning using satellite imagery, it is necessary to be able to distinguish between multi-level clouds, low-level moisture, and smoke. This can be accomplished with multispectral techniques. Other studies have used NOAA AVHRR channels 1 (0.58–0.68 μm), 2(0.72–1.10 μm), and 4 (10.5–11.5 μm) in false color composites to distinguish between smoke and clouds and to monitor long-range transport of smoke associated with forest fires in northwestern Canada and China (Chung and Le 1984; Cahoon et al. 1991). When considering only the visible (channel 1) and near-infrared images (channel 2), it is often difficult to separate the smoke from the clouds, especially semitransparent cirrus. Smoke-covered regions are not readily identified in longwave infrared images either. Because smoke aerosols are composed primarily of particles in the sub-

micron range, they do not have the effect on outgoing longwave radiation that clouds do, except in situations where the smoke concentration is extremely large. Furthermore, absorption and emission processes occur at temperatures that are very similar to the surface (Lenoble 1991). In a study of four forest fires in North America and Asia, smoke plumes were easy to detect in the visible imagery, but were not seen in the infrared. Smoke plumes were associated with noticeable daytime cooling, but no nighttime effects were found (Robock 1991).

Fishman et al. (1986) give examples of active fires and aged smoke over South America as seen in GOES visible data in 1980. By utilizing both the visible and infrared window (11 μm) data, Fishman et al. (1986) suggest ways to discriminate between clouds and smoke. Additional work by Prins and Menzel (1992) demonstrates the ability to use a sequence of high temporal GOES imagery to determine the motion and extent of aerosols associated with biomass burning activity. The temporal resolution of the GOES data makes it possible to select the optimum time of the day to monitor the smoke. Although the smoke is often evident throughout the day, the best time to see the smoke is in the morning near 1200 UTC. At this time of day the low solar elevation angle produces the maximum scatter of incoming visible radiation, which makes it easier to detect the smoke. In addition, during the morning smoke is fairly stratified in distinct, highly concentrated layers. By the afternoon it is often mixed throughout the boundary layer due to convective instability and is not so readily seen in the visible imagery (Andreae 1988).

Recent work has shown the feasibility of using GOES visible and infrared (4, 11, and 12 μm) data in an automated multispectral thresholding scheme to distinguish smoke/aerosols from other multi-level clouds and low-level moisture. In visible imagery smoke often appears as a milky gray haze, but cirrus, low-level moisture, fog and other multi-level clouds can display the same signature. Some of these clouds can be distinguished from smoke/aerosol by incorporating a combination of the 4- and 11-μm bands. By utilizing brightness-temperature limits in both bands and considering the 4- minus 11-μm difference, it is often possible to screen out various cloud types. Since the best conditions for monitoring smoke/aerosols in South America are coincident with a low solar zenith angle, the effect of reflected solar radiation in the 3.9-μm region is minimized. A large positive 4- minus 11-μm brightness-temperature difference is often indicative of cirrus and a negative difference on the order of 4°C can indicate stratus. The split-window (11 and 12 μm) data can be used in conjunction with the visible and 3.9-μm data to create a more robust aerosol detection algorithm. The split-window channels have been used extensively to detect low-level moisture and the presence of clouds. They were designed to use differential water vapor absorption across this part of the spectrum to estimate the amount of water vapor in the lower atmosphere; the 12-μm region is more sensitive to water vapor absorption than the 11-μm region. For low visible brightness reflectance values and surface emissivity corrected brightness-temperature differences greater than 5°C, one can assume substantial amounts of low-level moisture. Semitransparent cirrus, which looks very similar to haze in the visible imagery, often displays a relatively cold 11-μm brightness temperature and can display a large difference (11-minus 12-μm) in brightness temperature due to the effective emissivity differences in the 11- and 12-μm regions (Inoue 1985).

The smoke/aerosol episode observed in GOES-7 VAS imagery from 24 August to 29 August 1988 consists of a unique blend of large-scale aerosol transport and multi-level cloud activity, including semitransparent cirrus and low-level stratus. Data collected during this week are being used to develop a GOES automated multispectral aerosol identification algorithm. Figure 7.3 illustrates an example from 1231 UTC on 29 August 1988.

Figure 7.3a is a visible image that has been enhanced to show only those pixels which fall within a visible reflectance brightness range (counts of 60–100 on a scale of 0–255) often associated with haze. The white area represents pixels that do not fall within the specified range and do not have a signature indicative of smoke. The light gray shading represents higher brightness counts more commonly associated with clouds, and dark gray shading is more indicative of surface reflectance. Comparisons with corresponding unenhanced visible imagery shown in figure 7.1b clearly indicate that there are many cloud-contaminated pixels that have not been screened out by this single-band approach (e.g., the stratus deck off the Pacific Coast of South America).

Figure 7.3b is a composite image of the 4- minus 11-μm bands. Only pixels that had brightness temperature values greater than 285°K in both bands were included in the differencing. The resulting enhanced composite indicates all those pixels which exhibited a brightness temperature difference greater than −4°C and less than +20°C in shades of light to dark gray, respectively. All other pixels are shown in white. The upper limit on the range was set at +20°C to avoid eliminating any smoke-covered fire pixels that might be present in the study area. Analysis of the 4- minus 11-μm composite indicates that this test successfully eliminated the stratus deck and altocumulus clouds over the Pacific Ocean west of the continent. Convective activity in the northern portion of the Amazon Basin and a stratus/fog deck on the eastern coast of Brazil were screened out. A variety of multi-level cloud pixels associated with synoptic scale disturbances over the south Atlantic were also eliminated.

Figure 7.3c is a composite image of the 11- minus 12-μm bands. Those pixels with 11-μm brightness temperature greater than 285°K, 12-μm brightness temperatures greater than 280°K, and 11- minus 12-μm brightness temperature differences between −4°C and +6°C are indicated in shades of gray. The white area indicates pixels that were eliminated by this test. This composite is fairly similar to the 4- minus 11-μm composite. The 11- minus 12-μm composite screens out low-level moisture in the northwestern portion of the Amazon Basin and in the northeastern portion of Brazil as well as thin cirrus and isolated cumulus activity over the Atlantic Ocean near the equator and 25°W.

The composite GOES smoke detection product shown in figure 7.3d is the result of combining each of the three tests outlined in table 7.1. The portion of the

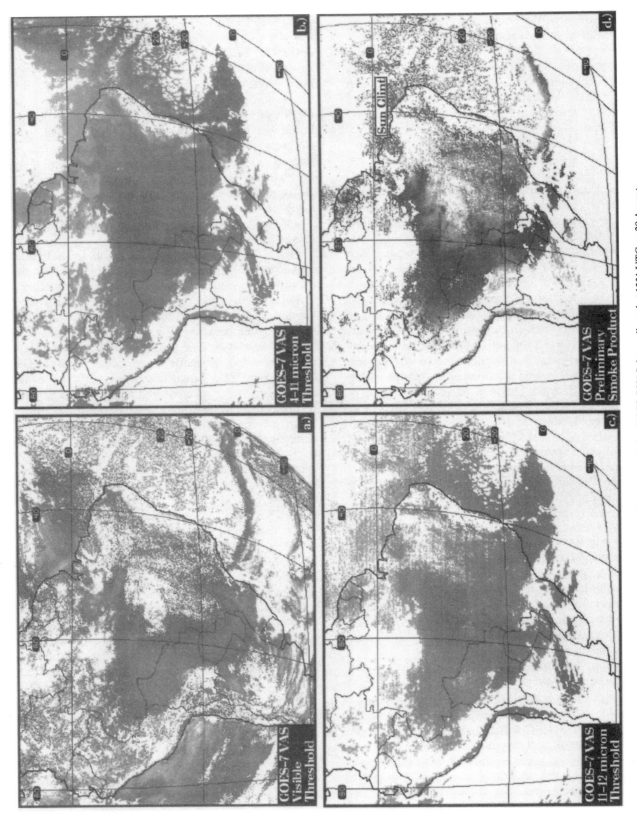

Figure 7.3 The GOES multispectral smoke/aerosol detection algorithm was applied to GOES-7 VAS data collected at 1231 UTC on 29 August 1988. Figure 7.3*a*, *b*, and *c* represent the composites obtained from applying each of the three thresholding tests outlined in table 7.1. The final composite (figure 7.3*d*) shows the result of combining the three tests.

Table 7.1 GOES multispectral smoke/aerosol detection algorithm

Test	Visible counts	4 μm Temp (°K)	11 μm Temp (°K)	12 μm Temp (°K)	4 − 11μm Temp (°K)	11 − 12μm Temp (°K)	Comments
1	60 ↔ 100						Limits the brightness count range to values indicative of haze
2		> 285	> 285		−4 ↔ +20		Screens for opaque clouds, cirrus, and stratus
3			> 285	> 280		−4 ↔ +6	Screens for low-level moisture, opaque clouds, and semi-transparent cirrus

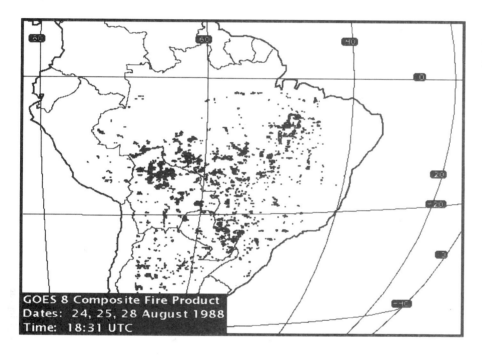

Figure 7.4 The GOES ABBA composite fire product shows the locations of fire activity (in black) for 24, 25, and 28 August 1988.

image with a white enhancement represents pixels that were eliminated by at least one of the threshold tests. The dark gray regions generally indicate areas of heavy smoke cover and the light gray enhancement represents minimal or no smoke coverage. Comparison with the visible imagery in Figure 7.1*b* is encouraging. The multispectral GOES smoke product clearly identifies the general outline of the smoke pall throughout the Amazon Basin and was able to delineate the aerosol plume from multi-level clouds over the Atlantic Ocean near 30°S. Figure 7.4 indicates the locations of fire activity (in black) as detected by the GOES ABBA on three of the most active burning days during this smoke episode, verifying that the aerosol detected with the multispectral GOES algorithm is actually associated with biomass burning activities throughout the region. The speckled region over the Atlantic Ocean in figure 7.3*d* clearly indicates problems distinguishing aerosols from isolated cumulus activity and identifying smoke in the vicinity of thin cirrus and along gradients between clear and cloud-contaminated pixels. Furthermore, solar zenith angle considerations must be incorporated to distinguish aerosols from sunglint.

Conclusions

The GOES series provides a means to monitor trends in biomass burning and aerosol production and transport in South America and throughout the Western Hemisphere over many years. The GOES ABBA was developed to provide diurnal information concerning the location, temperature, and size of subpixel fires in South America. Applications of the GOES ABBA in

South America from 1983 to 1991 have demonstrated the ability to document long-term trends in fire activity. Qualitative analyses of imagery collected during five biomass burning seasons in South America revealed many examples of large-scale smoke transport extending over several million km^2. Four major transport regimes were identified. Recent investigations have shown the feasibility of incorporating GOES visible and infrared (4, 11, and 12 μm) data into an automated multispectral algorithm capable of distinguishing smoke from multi-level clouds and low-level moisture in South America in an effort to identify, catalogue, and monitor aerosol loading and transport.

References

Andreae, M. O., E. V. Browell, M. Garstang, G. L. Gregory, R. C. Harriss, G. F. Hill, D. J. Jacob, M. C. Pereira, G. W. Sachse, A. W. Setzer, P. L. Silva Dias, R. W. Talbot, A. L. Torres, and S. C. Wofsy. 1988. Biomass burning emissions and associated haze layers over Amazonia. *J. Geophysical Res.*, 93, 1509–1527.

Cahoon, Jr., D. R., J. S. Levine, W. R. Cofer III, J. E. Miller, P. Minnis, G. M. Tennille, T. W. Yip, B. J. Stocks, and P. W. Heck. 1991. The great Chinese fire of 1987: A view from space. *Global Biomass Burning*, 61–66, J. S. Levine (Ed.). Cambridge, Mass.: The MIT Press.

Chung, Y. S., and H. V. Le. 1984. Detection of forest fire smoke plumes by satellite imagery. *Atmos. Environment*, 18, 2143–2151.

Fishman, J., P. Minnis, and H. G. Reichle Jr. 1986. Use of satellite data to study tropospheric ozone in the tropics. *J. Geophysical Res.*, 91, 14451–14465.

Inoue, T. 1985. On the temperature and effective emissivity determination of semitransparent cirrus clouds by bi-spectral measurements in the 10-micron window region. *J. Meteorol. Soc. Jpn.*, 63, 88–99, 1985.

Lenoble, J. 1991. The particulate matter from biomass burning: A tutorial and critical review of its radiative impact. *Global Biomass Burning*, 381–386, J. S. Levine (Ed.). Cambridge, Mass.: The MIT Press.

Menzel, W. P., E. C. Cutrim, and E. M. Prins. 1991. Geostationary satellite estimation of biomass burning in Amazonia during BASE-A. *Global Biomass Burning*, 41–46, J. S. Levine (Ed.). Cambridge, Mass.: The MIT Press.

Prins, E. M., and W. P. Menzel. 1992. Geostationary satellite detection of biomass burning in South America. *Int. J. Remote Sens.*, 13, 2783–2799.

Prins, E. M., and W. P. Menzel. 1994. Trends in South American biomass burning detected with the GOES visible infrared spin scan radiometer from 1983 to 1991. *J. Geophysical Res.*, 99, 16 719–16 735.

Robock, A. 1991. Surface cooling due to smoke from biomass burning. *Global Biomass Burning*, 463–476, J. S. Levine (Ed.). Cambridge, Mass.: The MIT Press.

Algorithm for the Retrieval of Fire Pixels from DMSP Operational Linescan System Data

Christopher D. Elvidge, Herbert W. Kroehl, Eric A. Kihn, Kimberly E. Baugh, Ethan R. Davis, and Wei Min Hao

Traveling in the Himalayan foothills near Darjeeling, India, in May 1848, British naturalist Joseph Dalton Hooker (1855) wrote that

fires, invisible by day, are seen raging all around. ... At night we were literally surrounded by them; some smouldering, ... others fitfully bursting forth, whilst others again stalked along with a steadily increasing and enlarging flame, shooting out great tongues of fire, which spared nothing as they advanced with irresistible might.

While numerous accounts such as this provide vivid descriptions of fires, recordings of such events are too few and contain insufficient detail to provide a basis for estimating the impact of biomass burning on greenhouse gas emissions or biodiversity. Daily satellite observations provide the only technical means for deriving the systematic global database on fire locations, dates, and sizes required to meet current scientific requirements.

In reviewing the literature, the earliest report we were able to find describing the observation of fire using a satellite sensor acquiring daily, global earth observations occurred when Croft (1973) described observing fires at night in Africa using "photographs" generated from the Defense Meteorological Satellite Program (DMSP) Operational Linescan System (OLS) visible band data. Croft (1973; 1979) was later able to use digital OLS data to observe fires, city lights, and gas flares. The first systematic inventory of fires with OLS data was accomplished by Cahoon et al. (1992) who manually digitized fire points from film produced from nighttime OLS orbits over Africa.

Since 1972, the U.S. Department of Defense (DOD) has maintained at least two DMSP platforms carrying OLS sensors in earth orbit. Because of the large volume of data and restrictions on access to the data, a digital archive for DMSP-OLS data was not established until 1992. A film archive established in 1974 at the National Oceanic and Atmospheric Administration (NOAA) National Geophysical Data Center holds analog data from approximately 1.7 million OLS orbits acquired in the 20 years prior to 1992.

With support from the Strategic Environmental Research and Development Program (SERDP), we have initiated the development of the algorithms and data bases required for the detection of fires at night using digital OLS data. In this chapter we describe the algorithms we have developed and provide preliminary results on fire and associated greenhouse gas emission in Southern Hemisphere Africa.

The DMSP System: Sensors and Archive

The DMSP maintains a constellation of two satellites in sun-synchronous, near-polar orbit at altitudes of approximately 833 km, an inclination of 98.8°, and an orbital period of 102 minutes. One satellite is in a dawn-dusk orbit, the second in a day-night orbit. The DMSP platforms are three axis stabilized, with roll, pitch, and yaw variations kept to within ±0.01°. This stability is unique compared to other polar orbiting systems such as Landsat or the NOAA Polar Orbiting Environmental Satellites. The currently orbiting DMSP satellites include F-12 with day-night overpasses at ~954 and 2154 local time, F-13 with dawn-dusk overpasses at ~604 and 1804 local time.

The NOAA National Geophysical Data Center (NGDC) serves as DOD's archive of data for the DMSP sensors. The U.S. Air Force Global Weather Central sends DMSP data tapes to NGDC daily. The DMSP archive was established in March of 1992, and began receiving data on a daily basis in September of 1992 and has operated continuously since that time. At NGDC the DMSP data are decompressed, deinterleaved into separate files for each sensor, and geolocated. NGDC has developed capabilities to recover DMSP data that have been corrupted by switched bits.

The NOAA NGDC DMSP archives are unique in that data from all of the DMSP sensors are available. Several other programs receive subsets of the DMSP data stream. The DMSP suite of sensors making observation of the earth and atmosphere includes four sensors (table 8.1).

Table 8.1 Defense meteorological satellite program (DMSP) sensor characteristics

Sensor	Wavelengths/Frequencies		Resolution	Swath width
OLS	VIS = 0.58 to 0.91 μm		Fine = 0.56 km	3000 km
	TIR = 10.5 to 12.6 μm		Smooth = 2.7 km	
SSM/I Microwave radiometer	19 GHz	H & V	70 × 45 km	1400 km
	22 GHz	V	60 × 40 km	
	37 GHz	H & V	38 × 30 km	
	85 GHz	H & V	16 × 14 km	
SSM/T1 Temperature sounder	Seven from 50 to 60 GHz		174 km	1400 km
SSM/T2 Water vapor sounder	91.5, 150, 183 GHz		46 to 120 km	1400 km

The Operational Linescan System (OLS) is an oscillating scan radiometer designed for cloud imaging with two spectral bands (VIS and TIR) and a swath of 3600 km. There are two spatial resolution modes in which data can be acquired. The full-resolution data, having nominal spatial resolution of 0.56 km, is referred to as "fine." On-board averaging of 5 × 5 blocks of fine data produces "smooth" data with a nominal spatial resolution of 2.7 km. Most of the data received by NOAA NGDC is in the smooth spatial resolution mode. The VIS bandpass straddles the visible and near-infrared (VNIR) portion of the spectrum with a full-width-half-maximum (FWHM) of 0.58–0.91 μm. The TIR band has a FWHM of 10.3–12.9 μm. The TIR band is calibrated using an on-board blackbody source and views of deep space to provide 8-bit data with a temperature range of 190 to 310°K, ideal for detecting and characterizing clouds. The wide swath widths provide for global coverage four times a day: dawn, day, dusk, night. The OLS produces 85% of the incoming DMSP data. NGDC recently improved its geolocation accuracy for the OLS data to ± one pixel using physically based orbital mechanics and terrain correction algorithms. A set of algorithms has been developed for retrieving cloud cover, cloud type, and cloud height from OLS data (e.g., Gustafson et al. 1994). The OLS has been included on all recent DMSP satellites (F-10, F-11, F-12, and F-13).

The other three sensors are passive microwave systems with coarse spatial resolution. The Special Sensor Microwave Imager (SSM/I) is a seven-channel, conical scanning radiometer, which measures Earth's emitted radiation in four thermal microwave frequencies. The 22.2-GHz channel is vertically polarized. Both horizontal and vertical polarization channels are acquired for the other three frequencies. A wide range of environmental products have been developed from SSM/I data (Hollinger 1989 and 1991), including ocean surface wind speed, ice area coverage, ice age, ice edge location, precipitation over water, precip-

itation over land, cloud amount, cloud water, integrated water vapor, land surface temperature, land surface type, soil moisture, and snow water content. The Special Sensor Microwave Temperature Sounder (SSM/T-1) is a seven-channel passive microwave sounder that measures atmospheric emissions in the 50 to 60 Ghz oxygen band at seven cross-track scan positions for use in constructing vertical profiles of atmospheric temperature. The Special Sensor Microwave Water Vapor Profiler (SSM/T-2) is a five-channel passive microwave sounder that measures atmospheric emissions from 91 to 183 GHz. The SSM/T-2 data are used to derive vertical profiles of water vapor, including relative and specific humidity and water vapor mass (Boucher et al. 1992).

Fire Detection with DMSP OLS Data

The algorithm we have developed for detection of fires in OLS data relies on detecting areas of active visible and near-infrared emission on the planet surface at night, when solar illumination is absent. The VIS band signal is intensified at night using a photomultiplier tube (PMT), making it possible to detect faint VNIR emission sources. The PMT system was implemented to facilitate the detection of clouds at night using the visible band. With sunlight eliminated, the light intensification results in a unique data set in which city lights, gas flares, and fires can be observed.

One adverse effect of the light intensification is that the system is quite sensitive to scattered sunlight. Under certain geometric conditions, the OLS telescope is illuminated by sunlight. Scattering of sunlight off the end of the telescope into the optical path results in visible band detector saturation, a condition referred to as glare (figure 8.l). The exact shape and orbital position of the glare changes through the year, but is generally confined to the western side of nighttime orbits in the 40 to 60° latitude range (north and south). Because of the substantial orbital overlap in this lat-

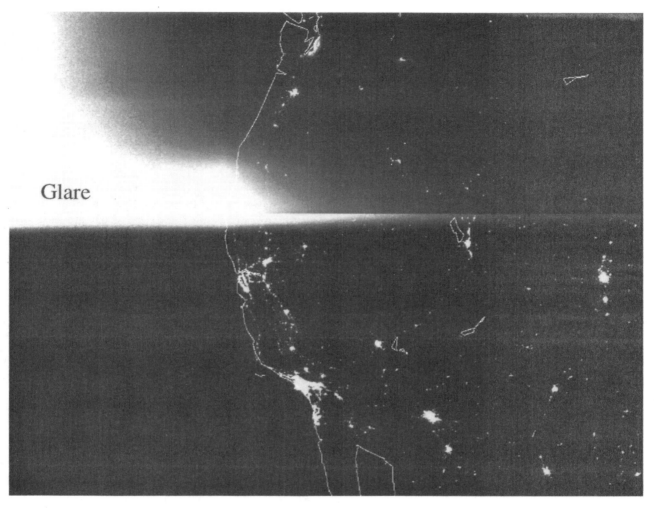

Figure 8.1 DMSP OLS nighttime image of the western United States showing glare (orbit F1017618, 16 April 1994)

itude range, it is still possible to acquire nearly global coverages of glare-free data each night.

Only the largest fires produce enough thermal emission in the 10–12 μm-region to be detected with the TIR band in smooth spatial resolution mode. At first this may seem counterintuitive to have better success at fire detection in the visible-NIR region than in the thermal region. Several factors contribute to the superior performance of the nighttime VIS data over the TIR data in smooth spatial resolution mode. First, the TIR band saturates at 310°K. Typical surface temperature backgrounds are in the 270 to 290°K range. Given the lack of contrast between the background and the 310°K limit, it takes a large number of saturated fine pixels to yield a discernible temperature anomaly in a smooth TIR pixel. Another factor is the expansion of the Instantaneous Field of View (IFOV),

which occurs when the VIS band signal is intensified with the PMT. The IFOV of the individual fine pixels under PMT operating conditions is approximately 1.74 km. The IFOV of the resulting smooth VIS band data at night is 3.98 km (center to center the smooth pixels are ~2.7 km apart). There is substantial overlap between adjacent VIS band fine pixels. As a result, a subpixel-sized fire bright enough to saturate a fine pixel in the VIS band will generally be observed in at least two and can potentially be observed in up to six of the 25 pixels that get averaged into a smooth pixel (Kihn this volume). The result is a variable amount of "double-counting" of fires visible in the nighttime. Because of the smaller IFOV of the TIR fine pixels, there is little overlap between adjacent pixels, and subpixel fires are generally observed only once. Finally, as described by Planck's law, the overall spectral emission of a blackbody increases as temperature increases and there is a strong shift towards emission at shorter wavelengths. An examination of Planck curves of spectral emission for blackbodies ranging from 300 to 1000° K indicates that the VIS band (0.58–0.91 μm) is in a better spectral position for detecting hot materials than the TIR band (10–12 μm). A preliminary examination of fine spatial resolution TIR data suggests that fires can be detected in either daytime or nighttime, but that many of the fire pixels will be saturated at 310° K.

Components of the Nighttime Fire-Detection Algorithm

Suborbiting
The OLS orbits are visually inspected to identify usable orbital segments. Suborbits are created for nighttime orbital segments over land areas. The suborbiting reduces the data volume that must be processed and is used to exclude features such as auroras, which are not relevant to the detection of fires.

Glare Removal
The OLS images are preprocessed to remove glare. Glare is detected when a 20 × 20 block of pixels is encountered with all pixels having saturated digital number (DN) counts of 63. The detection of the saturated block of pixels initiates an expanding search for all adjacent pixels with DN counts of 45 or greater. The DN counts for these pixels are set to zero. The cell size of 20 × 20 for glare detection was selected to avoid mistaking large cities as glare and to accommodate variations in glare size and shape. An example of OLS glare removal is shown in figure 8.2.

Identification of VNIR Emission Sources
Because brightness variations occur within and between orbits, it is not possible to set a single digital number (DN) threshold for identifying VNIR emission sources. We have developed software for detecting VNIR emission sources (lights) in nighttime OLS data using thresholds established based on the local background. This "light-picking" algorithm operates using background DN statistics generated from 50 × 50 pixel cells. Lights are identified in the central 20 × 20 pixel block inside the 50 × 50 pixel block used to generate the background statistics. Processing of an image proceeds by tiling the results from adjacent 20 × 20 pixel cells.

The nested configuration of the 20 × 20 cell inside of the 50 × 50 cell was designed to provide rapid processing of suborbits and to accommodate changes in background brightness. Identifying the local threshold for a 20 × 20 cell is 400 times faster than processing a new threshold for each individual pixel. This 20 × 20 pixel size matches the cell size used for glare detection. There is 60% overlap between the 50 × 50 pixel outer cells used to generate background statistics. This results in smooth transitions in threshold levels, avoiding threshold disparities between adjacent 20 × 20 pixel cells.

The distribution of DN values in each 50 × 50 pixel cell is analyzed to identify the set of pixels for use as the local background. VIS band DN values range from 0 to 63. Zeroes are missing data and 63s are saturated pixels. The lower limit of the background is taken to be DN = 1. The upper limit of the background is determined using a frequency distribution of pixel counts versus DN values (figure 8.3). Starting from DN = 63 and working down, the frequency distribution is analyzed to detect the first DN value where five consecutive DN bins have greater than 0.4% of the total pixel counts. For a full 50 × 50 cell, this corresponds to a search threshold of 10 pixels per DN bin. Once the upper limit of the background is selected, the mean and standard deviation of the background pixel set is calculated. Pixels containing VIS band emission sources are then identified using a threshold set at the DN value of the background mean plus three standard deviations. Figures 8.4–8.6 show examples of the light picker results applied to nighttime OLS imagery of Africa. Figure 8.7 shows the locations of lights that were detected in an OLS suborbit covering a large part of southern hemisphere Africa on 18 September 1992.

Lightning Removal
If an OLS linescan tracks across a cloud illuminated by lightning, a linear feature is produced that can be detected using the light-picking algorithm described above. Typically the lightning is observed in one or two scan lines, but can occur in up to three or four scan lines if multiple lightning strikes continue to illuminate the same cloud mass. We identify and remove lightning features from the images produced by the light-picking algorithm by testing the length versus width of all lights detected in cloud areas. Clouds are identified using the TIR band.

Geolocation
Our geolocation algorithm operates in the forward mode, projecting the center point of each pixel onto Earth's surface. The geolocation algorithm estimates the latitude and longitude of pixel centers based on the geodetic subtrack of the satellite orbit, satellite alti-

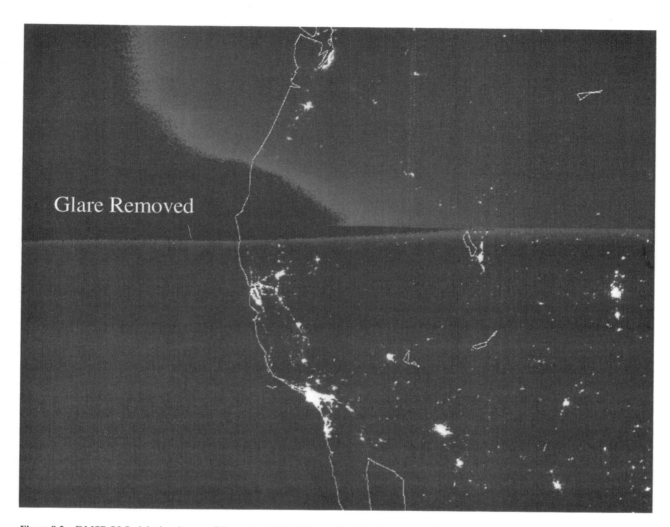

Figure 8.2 DMSP OLS nighttime image of the western United States showing glare removed (compare to figure 8.1) (orbit F1017618, 16 April 1994)

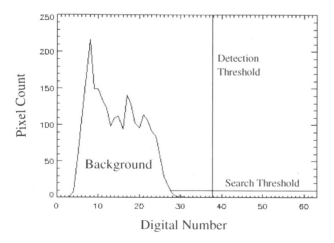

Figure 8.3 Histogram frequency distribution analysis to set local threshold for DMSP OLS light detection

tude, sensor model, an Earth sea-level model, and digital terrain data. The geodetic subtrack of each orbit is modeled using daily radar bevel vector sightings of the satellite (provided by U.S. Naval Space Command) as input into an Air Force orbital mechanics model that calculates the satellite position every 0.4208 seconds. The satellite heading is estimated by computing the tangent to the orbital subtrack. Satellite attitude is stabilized using four gyroscopes (three axis stabilization), a star mapper, Earth limb sensor, and a solar detector. The sensor model was developed based on the instrument mounting, scan characteristics, and the integration times for A/D conversion. We have used an oblate ellipsoid model of sea level and have used Terrain Base (Row and Hastings 1995) as a source of digital terrain elevations.

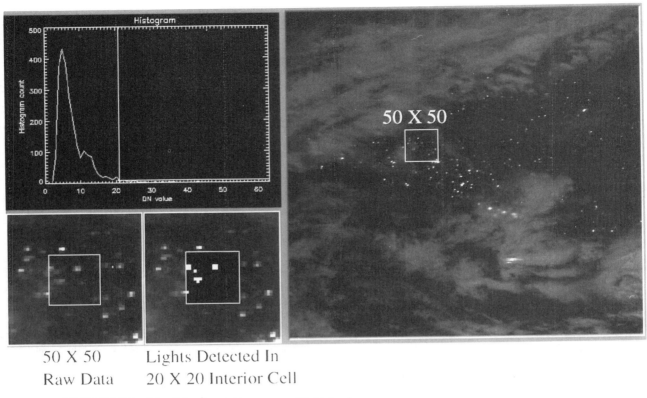

Figure 8.4   DMSP OLS light-picker lights detected in presence of light cloud cover

Identification of Stable Lights

Image time series analysis is used to identify stable lights produced by cities, towns, and industrial facilities. Heavy cloud cover blocks the detection of VNIR emission sources and light cloud cover tends to diffuse lights present on Earth's surface, making them appear larger than their actual sizes. In developing a stable lights data set we establish a reference grid and process large numbers of orbits, classifying pixels into one of three categories: (1) cloud, (2) cloud-free with no VNIR emission source, and (3) cloud-free with a VNIR emission source. Following classification, pixels are geolocated and resampled into the reference grid. The time series analysis is accomplished by running a counter for each of the three classes for each cell in the reference grid. In this way it is possible to identify which grid cells contain stable lights and also which areas need additional observations due to cloud cover. The stable lights for the southern hemisphere Africa derived from 96 orbits of data are shown in figure 8.8.

We have compared the OLS stable lights against the populated place lines from the Defense Mapping Agency (DMA) Digital Chart of the World (DCW) for several regions. The DCW-populated place lines were digitized from DMA Operational Navigation Charts (ONC) and are widely used as a source of geographic information on the global distribution of cities. The comparison was made to determine if it would be possible to use the DCW populated place lines as a surrogate for OLS stable lights. While the OLS stable lights covered each of the DCW populated place line features, a large number of OLS stable lights were not identified with DCW-populated place lines (see figure 8.9). These results indicate that it would not be possible to use the DCW-populated place lines in place of the empirical determination of OLS stable lights using the time series analysis.

Identification of Fires

Fires are identified as VNIR emission sources that are not associated with either stable lights or lightning. Lightning is screened out of the incoming data stream immediately after the light-picking algorithm is applied. Stable lights are masked out of the incoming data after the lights have been geolocated and resampled to the same reference grid as the stable lights data set. The algorithm that removes the stable lights identifies all pixels that occur in or directly adjacent to

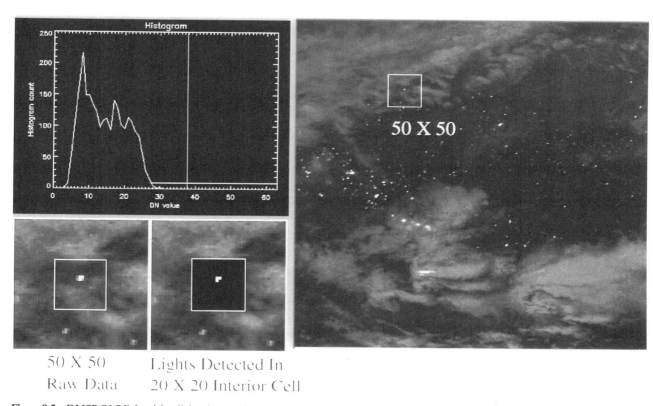

Figure 8.5 DMSP OLS light-picker lights detected in presence of heavy cloud cover

the known stable light locations and sets their DN values to zero. The remaining pixels, which contain ephemeral VIS band emission sources, are taken to be fires. Figure 8.10 shows the fires that were detected from the 18 September 1992 orbit of southern hemisphere Africa shown in figure 8.7 after masking out of stable lights (figure 8.8) and lightning. By compiling burn observations from adjacent orbits over specific time intervals it is possible to assemble continental-scale depictions of the spatial and temporal patterns of biomass burning. Figure 8.11 shows the cumulative burn observations from a three night, six orbit period in southern hemisphere Africa, covering 18–20 September 1992.

Calibration of Burn Area

Because the OLS sensor can detect fires that are sub-pixel in size, it cannot be assumed that the entire area of a pixel with a fire has been burnt. In estimating the cumulative burn area during a fire season it is reasonable to assume that an area only burns once, despite the fact that fires may be observed for multiple days in the same location. In order to calibrate burn areas it is necessary to have another source of data. Field sur-

veys are generally only feasible to validate burn areas for a small number of fires in a limited area. Procedures have been developed for using Landsat data to develop regionally applicable calibrations of burn area for Advanced Very High Resolution Radiometer (AVHRR) fire observations (Scholes et al. 1996). However, the acquisition of Landsat-style data in many parts of the world remains problematic, the data costs are high, and the processing requirements are intensive.

Early results obtained from simultaneously acquired nighttime OLS data in smooth and fine spatial resolution modes suggests that it may be feasible to calibrate OLS burn areas based on a temporal subsampling of OLS fine data. Fine-resolution OLS data can be acquired using a direct-readout ground-receiving station or by using the on-board tape recorders. The smooth resolution data continues to be produced even when the on-board tape recorders are used to acquire fine-resolution data.

Figure 8.12 shows a nighttime OLS data of fires in Sudan on 7 December 1994 in both the smooth and fine spatial resolution modes. With the increased spatial resolution of the fine data, it should be possible to

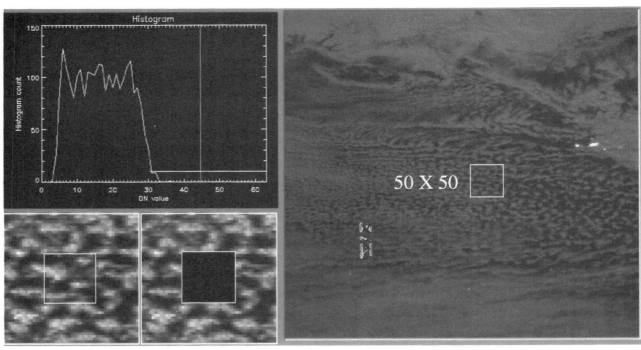

Figure 8.6 DMSP OLS light picker applied to moon-illuminated cloud cover with no VNIR emission sources detected

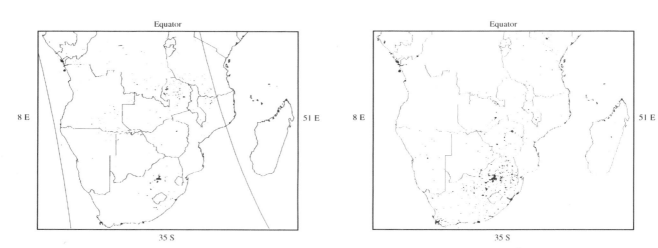

Figure 8.7 Lights (VNIR emission sources) detected in southern hemisphere Africa from a single DMSP OLS orbit 18 September 1992

Figure 8.8 Stable lights in southern hemisphere Africa, derived from 96 orbits of nighttime DMSP OLS data

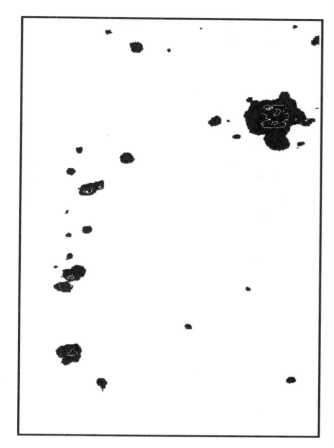

Figure 8.9 DMSP OLS stable lights for an area surrounding Harare, Zimbabwe. Populated place lines from the Digital Chart of the World (DCW) are shown in white

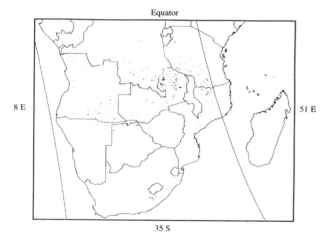

Figure 8.10 Nighttime fires detected in southern hemisphere Africa on a single orbit acquired on 18 September 1992

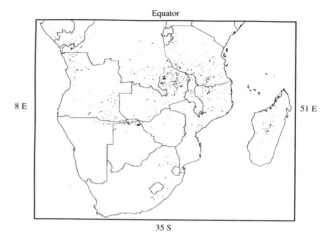

Figure 8.11 Nighttime fires detected in southern hemisphere Africa during six orbits acquired over three nights, 18–20 September 1992

make an improved estimate of the area of active fire. By using a temporal subsample of fine-resolution data, it may be possible to develop regionally valid burn area estimates that could be applied to nightly data acquired in the smooth spatial resolution mode.

Applications

Estimation of Trace Gas and Particulate Matter Emissions

One of the major applications of satellite-based biomass burning observations is to estimate trace gas and particulate emissions. To demonstrate the feasibility of using DMSP OLS fire observations to estimate the emissions of important trace gases and particulate matter from biomass burning, the USDA Forest Service's Intermountain Fire Sciences Laboratory used the OLS fire observations presented in figure 8.11 and estimated the resulting carbon dioxide emissions on a one-degree grid. Because we have not developed a calibration-to-burn area for the figure 8.11 fire observations, the carbon dioxide emission modeling was performed assuming that the entire area of each fire pixel was burnt. The results shown in figure 8.13 are thus overestimates of the actual emissions during the three-day period. More realistic estimates could be obtained with calibrated burn areas.

Biomass fires in southern hemisphere Africa are mostly used for shifting cultivation, deforestation, and savanna burning in this region. Other sources of biomass fires are fuel-wood use and clearing of agricultural residues. The ecosystems consist of tropical

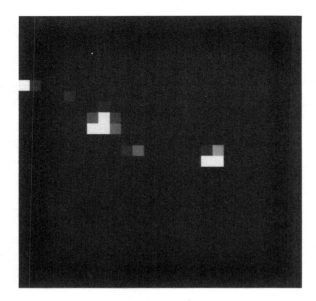 

Figure 8.12 Smooth-resolution (2.7 km pixels), left, versus fine-resolution (0.56-km pixels), right, DMSP OLS data of fires in southern Sudan, 7 December 1994

evergreen forests, semi-evergreen forests, secondary forests, deciduous forests, woodland and grassland savannas, and arid savannas. The trace gases for which emissions can be estimated from the OLS fire observations include CO_2, CO, CH_4, C_2-C_6 alkanes, alkenes, and alkynes, aromatic compounds (benzene, toluene, and xylenes), alcohols, aldehydes, ketones, organic acids, NO_x, NH_3, N_2O, and cyanogen compounds. The particulate matter that can be estimated includes total particulate matter and particles less than 2.5 mm in diameter.

The amount (T) of a trace gas or aerosol particles X emitted from biomass burning per unit time in each pixel is estimated using the following equation:

$$T = M * EFX_a = A * B * \alpha * (EFX_f * P_f + EFX_s * P_s)$$

where

T = amount of X produced from fires per unit time
M = amount of biomass burned per unit time
EFX_a = weighted-average emission factor of X
A = area burned per unit time
B = above-ground biomass density
α = fraction of above-ground biomass burned
EFX_f = emission factor of X during the flaming phase
EFX_s = emission factor of X during the smoldering phase
P_f = fraction of biomass burned during the flaming phase
P_s = fraction of biomass burned during the smoldering phase.

Note that the emission factors are a function of combustion efficiency:

$EFX_f = f(Ce_f)$ CE_f : combustion efficiency during the flaming phase

$EFX_s = f(Ce_s)$ CE_s : combustion efficiency during the smoldering phase

Further, combustion efficiency is a function of factors related to meteorological and fuel conditions at the time of burning:

CE_f or $CE_s = f$ (meteorological conditions; characteristics of biomass)

    $= f$ (wind velocity, temperature, humidity; amount, components, moisture and elemental composition of biomass)

Currently many of the above factors are estimated based on coarse resolution vegetation maps, statistical estimates of aboveground biomass density, and estimates of emission factors and generalized flaming-versus-smoldering combustion behaviors (Hao and Ward 1993; Hao and Liu 1994). The suite of DMSP microwave sensors have potential for estimating global precipitation that could be used with OLS surface temperature estimates as input into vegetation growth models to make dynamic estimates of fuel loads and fuel conditions. The SSM/T2 water vapor profiles and SSM/T1 air-temperature profiles could be used to estimate the near-surface relative humidity,

Equator

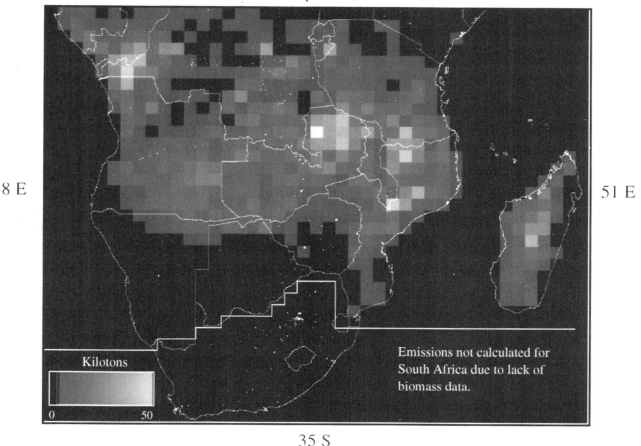

8 E

51 E

Kilotons

Emissions not calculated for
South Africa due to lack of
biomass data.

0   50

35 S

Figure 8.13 Carbon dioxide emissions from biomass burning in southern Africa based on DMSP OLS fire observations 18–22 September 1992

a key factor required to estimate the combustion efficiency.

Resource Management

The U.S. government spends approximately $500 million in fighting wildfires each year. Firefighters have long recognized that early detection of fires is key to cost-effective fire suppression and fire management. Fires not detected early have time to spread and become unmanageable and costly to fight, and result in great losses in lives, property, and resources. Satellite sensors have a proven capability for the detection of fires; however, the infrastructure required to transmit fire locations observed by satellites to local entities engaged in firefighting/fire management has not been widely developed.

The requirement for early detection of wildland fires goes beyond the protection of lives and property. Global conservation of biodiversity in the face of ex-

panding human populations will require individual countries to manage and often restrict biomass burning. Satellite observations of fires can be used to evaluate the effectiveness of government programs to reduce the frequency or seasonal timing of biomass burning. As an example, in 1994 the government of Madagascar issued a "no-burn" edict in an attempt to reduce widespread biomass burning. How effective was the edict? Figure 8.14 shows the nighttime fires detected with DMSP OLS data for 20 September 1992 and 1994. Although fires were not eliminated in 1994, there was a substantial reduction in the number of fires relative to 1992.

Conclusions

We present the first digital algorithm for nighttime fire detection with data from the DMSP OLS, a set of preliminary results and examples of applications that

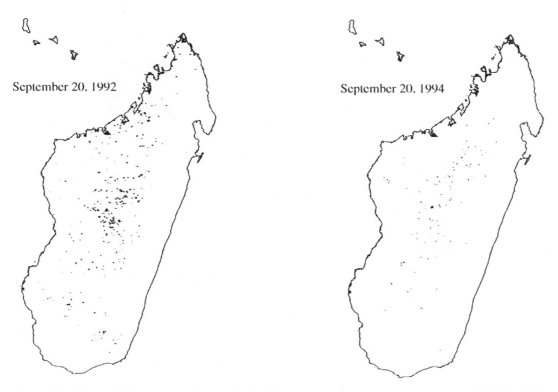

September 20, 1992   September 20, 1994

Figure 8.14 Effect of 1994 government "no-burn" edict in Madagascar on nighttime fires observed with DMSP OLS data

can be made with the fire observations. The algorithm we have developed is still being refined and needs to be thoroughly tested. The early results indicate that it would be feasible to produce a systematic global survey of fires for the years 1992–1995 using the DMSP archives held at NOAA NGDC.

In addition to fire monitoring, the nighttime OLS data can be used to map the distribution of stable visible-NIR emission sources present on Earth's surface. A global stable lights data set may be of considerable value for use as a global urban land-cover class, filling a gap in the current land-cover mapping capabilities of sensors such as the NOAA AVHRR. The global stable lights data set would also be useful in refining the data sets on the distribution of population and the consumption of energy.

The DMSP OLS has several advantages over currently available civil sensors for the daily monitoring of fires: (1) The positional stability of the OLS platform permits high accuracy geolocation of fires. (2) OLS provides superior spatial resolution (2.7-km pixels in smooth data and 0.56-km pixels in fine data). (3) Global coverages can be obtained without reliance on a network of receiving stations. (4) The nighttime

VIS band intensification makes it possible to detect faint VNIR emission sources on Earth's surface. (5) Because the OLS smooth data is a true average of 25 fine pixels, it is feasible to develop regionally valid smooth data burn area calibrations based on a temporal subsample of fine data. (6) The DMSP series is expected to continue until 2005–2006 without interruption, thus providing the potential for 15 years of continuous global fire observations.

Current efforts at NOAA NGDC are focusing on validating DMSP OLS fire observations through participation in international biomass burning experiments, intercomparison of fire-monitoring capabilities with other sensors, development of procedures for estimating burn area in 2.7-km resolution smooth OLS data with temporal subsamples of 0.56-km fine data, and installation of data lines for reception of near real-time OLS data for use in early detection of fires.

Acknowledgment

This project was supported by the Strategic Environmental Research and Development Program (SERDP).

References

Boucher, D., B. Thomas, and V. J. Falcone. 1992. Performance of the DMSP Special Sensor Microwave Humidity Profiles: Preliminary Results. In *Proceedings of the First NMC/NESDIS/DOD Conference on DMSP Retrieval Products*, R. Isaacs, E. Kalnay, G. Ohring, R. McClatchey (Eds.), PL-TR-92-2191, 17–24.

Cahoon, D. R., Jr., B. J. Stocks, J. S. Levine, W S. Cofer III, and K P. O'Neill. 1992. Seasonal distribution of African savanna fires. *Nature*, 359, 812–815.

Croft, T. A. 1973. Burning waste gas in oil fields. *Nature*, 245, 375–376.

Croft, T. A. 1979. The brightness of lights on Earth at night, digitally recorded by DMSP satellite. Stanford Research Institute Final Report Prepared for the US Geological Survey, Palo Alto.

Gustafson, G. B., R. G. Isaacs, R. P. d'Entremont, J. M. Sparrow, T. M. Hamill, C. Grassotti, D. W. Johnson, C. P. Sarkisian, D. C. Peduzzi, B. T. Pearson, V. D. Jakabhazy, J. S. Belfiore, and A. S. Lisa. 1994. Support of Environmental Requirements for Cloud Analysis and Archive (SERCAA): Algorithm Descriptions. PL-TR-94-2114, Scientific Report #2, Phillips Laboratory, Directorate of Geophysics, AFMC, Hanscom AFB, Mass., 01731–3010.

Hao, W. M. and M.-H. Liu. 1994. Spatial and temporal distribution of tropical biomass burning. *Global Geochemical Cycles*, 8, 495–503.

Hao, W. M. and D. E. Ward. 1993. Methane production from global biomass burning. *Journal of Geophysical Research*, 98, no. 20, 657–661.

Hollinger, J. 1989. DMSP Special Sensor Microwave Imager Calibration/Validation. Naval Research Laboratory, Final Report Volume I, Washington, D. C.

Hollinger, J. 1991. DMSP Special Sensor Microwave Imager Calibration/Validation. Naval Research Laboratory, Final Report Volume II, Washington, D. C.

Hooker, J. D. 1855. *Himalayan Journals*, John Murray, London.

Row, L. W. III and A. D. Hastings. 1995. TerrainBase Worldwide Digital Terrain Data, Documentation Manual and CD-ROM. NOAA National Geophysical Data Center, Boulder, Colo., NGDC Publication KGRD 30, 185 pp.

Scholes, R. J., D. E. Ward, and C. O. Justice. 1996. Emissions of trace gases and aerosol particles due to vegetation burning in southern-hemisphere Africa. *Journal of Geophysical Research*, in Press.

Forest Fire Detection from DMSP Operational Linescan System (OLS) Imagery

Eric A. Kihn

Biomass burning is a major global source of greenhouse gas emissions and reactive trace gases (Levine 1990). Because of the interannual variability and the wide distribution of fires, satellite monitoring of biomass burning is key to tracking the global carbon cycle. The spatial and temporal distribution of biomass burning is required to model the greenhouse emissions. To date only approximations exist for areas burned and quantities of biomass consumed on an extended global basis; no direct observations have yet been compiled. It is therefore important to document the ability of relevant spacecraft to both detect and estimate the area of surface fires.

A large body of work has been dedicated to this type of research with regard to the National Oceanic and Atmospheric Administration Advanced Very High Resolution Radiometer (NOAA AVHRR) sensor. Beginning with Dozier (1981) the ability of AVHRR to detect small high-temperature surface features, such as fires, has since been well established. For example Flannigan (1985), used the AVHRR sensor to observe more than 300 fires in Canada and compile statistics regarding detection and area estimation for a nine-day observation period. Matson et al. (1984) examined active fire-detection methods using the NOAA-N satellites. Work such as that done by Kaufman et al. (1990) describes fire-detection campaigns for a specific region using AVHRR. Lee and Tag (1990) use the 3.7-μm combined with the 10.8-μm channel to detect small but hot surface features such as fires and conclude that the method they developed allows for reliable AVHRR detection of hotspots. Even the detection of postfire "burn scar" observations such as in Cahoon et al. (1992) have been completed for the AVHRR instrument. Although it has been established since Croft (1973) that the OLS instrument can detect fires in the nighttime imagery, no comprehensive study has been performed to document what size fires the OLS instrument can detect or what percentage of fires for a particular period it was able to observe.

With both the Operational Linescan System (OLS) and AVHRR instruments because of the relatively large pixel size (0.55 or 2.7 km and 1 or 4 km, respectively), it is crucial to be able to estimate area for subpixel-sized fires. Dozier (1981) describes a method for the AVHRR instrument. This method uses the calculated response of channels 3 and 4, (i.e., 3.55–3.93 μm and 10.5–11.5 μm, respectively) and the observed brightness temperature in each channel to determine both a target temperature and the percentage of the pixel occupied by the high temperature source. Similar subpixel burn area estimation methods have yet to be developed for the DMSP OLS sensor. Therefore this chapter presents a similar method for OLS imagery and tests it against data for the Colorado fires of 1–10 July 1994 and Oregon's Wallowa-Whitman Forest fires of 24 July–3 Sept. using the 2.7-km resolution nighttime imagery. The results of this survey are then used to estimate the accuracy of OLS fire observations for detection, area estimation, and geolocation.

The Oregon and Colorado areas and times were chosen for several reasons. First, there were no fewer than 10 fires, of area greater than 200 acres, active in each area during the period, and the U.S. Forest Service was able to provide accurate records of location and eventual burn extent. Second, the OLS data were acquired daily throughout the period. In addition the weather conditions were favorable, with only one fire site being continuously obscured by cloud during its active period.

Methodology

This research was done using data recorded by the F-10 Defense Meteorological Satellite Program (DMSP) OLS sensor. The DMSP mission is to record meteorological, oceanographic, and solar-geophysical data in support of Department of Defense (DOD) operational requirements. The DMSP program currently operates three satellites in sun-synchronous, low altitude, polar orbits. The orbital period is 101 minutes and the inclination is 99°. The complement of instruments on DMSP satellites include the Operational Linescan

System (OLS) imager and other meteorological and solar-geophysical instruments, notably microwave imagers and sounders (see chapter 8 by Elvidge et al.).

The OLS instrument measures visible/near-infrared (VNIR) = (0.4–1.1 μm) and thermal infrared (TIR) = (10–13 μm) radiances. The TIR detector is a two-segment Mercury-Cadmium-Telluride photoconductive detector cooled to a temperature of about 110°K. Spectral bandpass is set by an optical filter on the detector germanium window for passing the 10.2- to 12.8-mm range and for rejection of water vapor, CO_2, and ozone effects. The nighttime VNIR is collected by a Photo Multiplier Tube (PMT), which is a cesiated GaAs (gallium arsenide) photocathode, image dissector type. The OLS sensor has the capacity to make observations with 0.55-km spatial resolution ($\sim$80 acres per pixel) called "fine." However, the data is typically reduced to 2.7-km resolution ($\sim$2,000 acres per pixel) called "smooth" resolution by the on-board. The smoothing is accomplished by analog averaging 5 of the fine data cells in the along-scan direction. Five of these cells are then averaged digitally in the along-track direction to produce a pixel with the desired 2.7-km resolution. The final TIR data values reported range from 190 to 310°K, and the VNIR data are reported as relative counts ranging from 0–63. In the case of the nighttime visible data the PMT has a spatial resolution of $\sim$1.75 km and therefore its final smooth pixel value is actually a running average over neighboring fine cells. This study concentrates on the OLS smooth data from satellite F-10, which makes a pass near midnight local time.

The OLS nighttime imagery records sources of visible and near-infrared emissions including city lights, aurora, fires, oil gas flaring, moonlit clouds, and lightning. As such it was necessary to develop a method to distinguish fires from the other light sources. In this study the method chosen was a manual comparison of images; however, a similar automated method has since been developed (see chapter 8 by Elvidge et al.). The manual method used involved first choosing images on five cloudless, moonless nights covering all of the region of interest. The composite of these five images then became the "Light Map" for the region of interest. Next every image for the appropriate period was compared against the Light Map and with each other. Those pixels that appeared newly bright in the observed images were considered fire candidates. Next the newly bright pixels were followed from night to night to ensure that they again went dark, as would be expected for a fire. Those pixels remaining were screened to eliminate possible lightning observations.

This proved very easy to accomplish because there were in general very few clouds for the period and lighting has a very discernible pattern in the OLS data. Finally a dark pixel count was established. A dark count is the pixel value assigned to nonilluminated pixels; typically this number varies between 1 and 4 counts. The dark count value is the final filter used to define pixels as bright or nonbright. A pixel passing through the above filters is finally considered bright if its value is greater than 3* (dark count) for that orbit. This definition does eliminate some pixels that appear visibly bright on the image, thereby it serves to lessen the possibility of a false fire-detection error. The pixels selected by this process are then classified as "fire pixels."

The final list of fire pixels was compared to a list of fires active during the period as obtained from the U.S. Forest Service. Because of the limits of the current geolocation method, a particular fire was considered successfully identified if the location of a fire pixel was within 2 km of the reported fire's location. Once identified, supporting information was obtained for each fire pixel. This included the corresponding TIR data and a background temperature of adjacent nonfire pixels.

Second, an attempt was made to identify each fire from its thermal signature. In this case the method was to look at the thermal value of "visible fire pixels" selected as above, and compare it with the thermal average of surrounding nonfire pixels (thermal background). If a pixel that was visibly bright had a thermal value greater than the thermal nonfire average plus one standard deviation it was considered a "thermal fire pixel." This process, while it had the advantage that it was simple to effect, also had the problem that pixels which were not actively on fire but which recently had been were used in computing the average for the thermal background. The effect of this was to raise both the background average and the standard deviation. However, since this should lead to fewer thermal identifications, not false ones, it was considered a satisfactory method for this experiment.

Once fire pixels of each type were identified, the next step was to create an area estimate from this information. The first method applied was a simple pixel correspondence method. In this method, it was assumed that all of an observed fire pixel was actually on fire. Then the total area equals the number of fire pixels observed multiplied by 2000 acres per smooth pixel.

As expected, this method could lead to gross overestimates of area (see Results), particularly when using the visible fire pixels. The thermal fire pixels, while somewhat more reasonable, were also subject to the same problems with overestimation. Therefore, it was necessary to develop a "subpixel" estimation technique for DMSP smooth OLS imagery.

Unfortunately, because of the frequency range of the OLS VNIR and TIR bands it is not possible to apply the method of J. Dozier (1981) to the observed data directly.

The TIR sensor simply saturates too soon to give any results when used with the VNIR sensor. However, we were able to calculate the upwelling radiance sensed by the thermal radiometer as in Dozier, and from this develop an alternative subpixel method. From Dozier the upwelling radiance sensed is given by

$$L(T) := \frac{\frac{1}{\pi} \int_{\lambda 1}^{\lambda 2} \frac{\varepsilon \cdot c1 \cdot \lambda^{-5} \cdot \Phi}{\exp\left(\frac{c2}{\lambda \cdot T}\right) - 1} \, d\lambda}{\int_{\lambda 1}^{\lambda 2} \Phi \, d\lambda}$$

where

$c1$ = first Planck constant
F = spectral response function of sensor
$c2$ = second Planck constant
λ = wavelength
e = emissivity
T = temperature
$\lambda 1$ = minimum sensor λ
$\lambda 2$ = maximum sensor λ

Some simplifying assumptions were applied before evaluating the integral. Emissivity (e) was assumed constant with respect to 1 $e = .9$,. F was taken as the linear fit to the actual response curve as shown in figure 9.1. Also the earth was assumed to be a perfect blackbody radiator.

With these assumptions it was then possible to numerically evaluate the integral. The result is displayed in figure 9.2.

Now as a simple model it was assumed that an observed fine *fire* pixel is composed of two temperature values: T_t, the target or fire temperature and T_b, the background or natural temperature. The total radiance sensed by the thermal detector becomes: $L_{\text{tot}} = L(T_t) + L(T_b)$.

The final step was to use the fact that each smooth pixel value is actually the average of 25 observed fine pixel values. The L_{tot} derived applies to each individual fine pixel which then is averaged, as described above, to form the smooth temperature value that is reported.

According to Albini (1980) flaming combustion requires a temperature of $\sim 600°$K and Vines (1981)

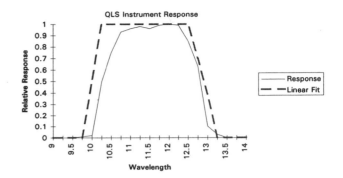

Figure 9.1 Comparison of actual and assumed OLS instrument response curves

reports that thick flames normally burn at $\sim 1300°$K. So we take as a reasonable assumption for our target fire temperature $1000°$K. With this in mind, we can see that nearly all fine pixels with an active fire in them will be saturated. This is because the OLS thermal channel saturates at $310°$K. This implies a fine pixel evenly at temperature $310°$K will just saturate the sensor. Now if the area of a pixel (A_f) is at a temperature of $1000°$K, its contribution to the total radiance detected is equivalent to 23.7 times that of an equal area at $310°$K (see figure 9.2). This implies that if as little as 4.2% of a fine pixel is actually on fire at the time of observation the sensor will be saturated and the value reported for that pixel will be $310°$K. Even if it is assumed that the fire is at the minimum temperature of $600°$K reported above, similar calculations show the total area necessary for saturation is only 12.5%. Therefore, we may reasonably assume any fine pixels containing an active fire are reported as saturated.

With that assumption it becomes a matter of simple algebra to obtain a subpixel estimate for each smooth pixel. First, assume that the nonfire fine pixels are at the calculated background temperature. The area is derived from the number of fine pixels (X), the measured brightness temperature (T_s), and the background brightness temperature (T_b) as

$$T_s = 310°\text{K} * X + (25 - X) * T_b.$$

Then after solving for X simply multiply by 80 acres per fine pixel to obtain the area in acres observed to contain active fire. This effectively reduces our counting unit from a 2000-acre increment to 80; while, of course, this can still lead to an overestimate of fire extent, the effect is greatly reduced as shown below.

One final consideration for the method is the observational overlap of the nighttime visible fine pixel observations. Unlike the daytime VNIR imagery and

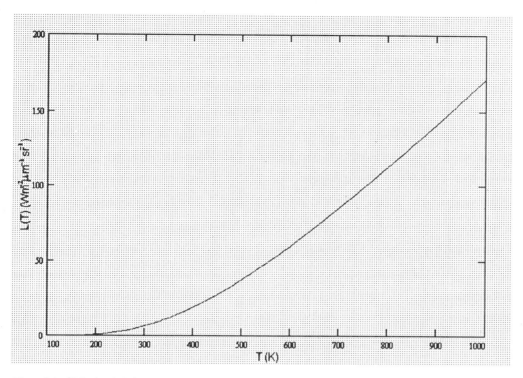

Figure 9.2 Calculated IR instrument response, upwelling radiance versus target temperature

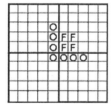

Figure 9.3 Theoretical fire distribution over an observed OLS "fine" grid

Table 9.1 Oregon and Colorado active fires

Fire name	Active dates	Total area
Baldy	24–31 July	223
Deep Creek	31 July–23 Aug	300
Starvation	31 July–23 Aug	675
Triangle Mt.	1–5 Aug	512
Thorn Creek	2–23 Aug	1210
Rapid River	2–23 Aug	6754
Alder Creek	28 Aug–3 Sep	5700
Cayouse	28 Aug–3 Sep	258
Vance	28 Aug–3 Sep	470
Hollow Log	28 Aug–3 Sep	708
Hourglass	1–8 July	1275
Elk	4–7 July	950
Pyramid Rock	4–8 July	1050
Spud Patch	4–9 July	1087
Wake	4–8 July	3460
Burn Canyon	5–9 July	200
Mitchell Lake	5–9 July	270
South Canyon	5–9 July	2430
Rabbit Mount.	6–9 July	630

all TIR imagery, which assigns a ~0.5-km area at the surface into each fine pixel, the PMT used for nighttime visible images actually assigns a ~1.75-km area at the surface into each fine pixel. The observed value taken by the PMT is then interpreted as representing a 0.5-km pixel, as with the daytime imagery, and is assigned into the data stream as such. This means that each fine pixel value reported for the nighttime visible imagery is actually a measurement of the photons emitted from an adjacent group of the 0.5-km pixels as measured at the surface. This leads to a situation in which the smooth values, computed from the fine data, report observed values for light sources outside their representative geographical range and tends to make

Table 9.2 Colorado fire observation results

Fire name	Date detected	Total acres actual	Pixels detected (VIS)	Pixels detected (IR)	Subpixel estimate
Hourglass	1 July	1275	6	1	320
Elk	4 July	950	4	2	320
Spud Patch	4 July	1087	4	1	640
Wake	4 July	3460	4	3	1600
Pyramid Rock	5 July	1050	1	1	160
Burn Canyon	5 July	200	3	3	480
Mitchell Lake	—	270			0
South Canyon	6 July	2430	3	2	240
Rabbit Mount.	6 July	630	3	2	160
Total:		11 352			3920

Table 9.3 Oregon fire observation results

Fire name	Date detected	Total acres actual	Pixels detected (VIS)	Pixels detected (IR)	Subpixel estimate
Baldy	25 July	223	2	1	80
Deep Creek	2 Aug	300	3	2	160
Starvation	2 Aug	675	3	1	240
Triangle Mt.	1 Aug	512	4	2	240
Thorn Creek	3 Aug	1210	10	4	480
Rapid River	2 Aug	6754	12	6	1920
Alder Creek	30 Aug	5700	12	4	1520
Cayouse	29 Aug	258	4	1	160
Vance	29 Aug	470	7	1	80
Hollow Log	28 Aug	708	6	2	400
TOTAL		16 810			5280

small surface lights such as a fire "blow up." An example of such a phenomenon can be seen below.

Consider the case of four adjacent smooth pixels, one of which contains an active fire designated F burning in the indicated fine pixels with no other light sources present, as in figure 9.3. Those fine pixels designated O will, because of the overlap described, detect the fire and their nonzero resultant value will be averaged into the appropriate smooth pixel. The net effect is that all four smooth cells will report light detected though only one actually contains a fire. To account for this, only pixels showing both a visible fire detection and a TIR value greater than the background average plus 1.5 standard deviations will be considered when applying the above-described area estimation algorithm.

Results

Nineteen fires were used as possible targets. Table 9.1 gives the complete list of the fires and their respective areas as reported by the U.S. Forest Service (USFS). In all cases, area refers to the total acres consumed by the fire and active dates are given by the USFS standards.

Every fire listed in table 9.1 except for Mitchell Lake was detected during its active period; Mitchell Lake remained under heavy cloud cover during its active burn phase. Of the fires detected, OLS typically recorded a shorter active period than the USFS report. This is likely because the Forest Service's definition of an "active" fire differs from the OLS observational requirements. However, observe that for this selection of fires a very reasonable lower bound for total area is produced by the method described above.

It should be noted that all of the fires observed during this study were of the Temperate Forest type, containing at least 30% tree cover of mixed type. Since the GIBBP will be used to monitor other biomass sources, notably grassland fires, additional study will be required for a complete ground truth validation. In tables 9.2 and 9.3 *date detected* refers to the date the fire was first observed, *pixels detected* refers to the total number of unique pixels observed over the fire's entire active period. Finally, *subpixel estimate* refers to the sum of the estimates for each night of observation.

It is clear from the tables that the method underestimates total area significantly. This is as expected since the estimate it produces is an estimate of active fire area at the time of each DMSP overflight. Since this study used only the F-10 spacecraft and the nighttime pass, only one observation was made in each 24-hr period.

Conclusions

• The DMSP OLS nighttime visible imagery will be a reliable detection system for temperate-zone forest fires of at least 200 total acres.

• The subpixel method described above can produce a reasonable instantaneous estimate for active fire area. This estimate can serve as a lower bound for the DMSP OLS fire-detection limits.

• The DMSP OLS will be susceptible to environmental effects such as cloud and smoke as outlined in Robinson (1991).

• In order to provide an accurate estimate of global biomass burning using OLS data, it will be necessary to develop a method to estimate total burn area. The method will need to be time flexible so that no one orbit is required to document the burn. This will most likely be detection of burn scar using 0.5-km resolution data at known fire locations.

References

Albini, F. A. 1980. Thermochemical Properties of Flame Gases from Fine Wildland Fuels. USDA Forest Service Research Paper INT-243, USDA, Intermountain Forest and Range Experiment Station, Forest Service, Ogden, Utah.

Cahoon, D. R., Jr., B. J. Stocks, J. S. Levine, W. R. Cofer III, and C. C. Chung. 1992. Evaluation of a Technique for Satellite-Derived Area Estimation of Forest Fires, *Journal of Geophysical Research*, 97, 3805–3814.

Croft, T. A. 1973. Burning Waste Gas in Oil Fields. *Nature*, 245, 375–376.

Dozier, J. 1981. A Method for Satellite Identification of Surface Temperature Fields of Subpixel Resolution. *Remote Sensing of Environment*, 11, 221–229.

Flannigan, M. D. 1985. Masters Thesis, Colorado State University, Collins, Colo.

Kaufman, Y. J., J. T. Compton, and I. Fung. 1990. Remote Sensing of Biomass Burning in the Tropics. *Journal of Geophysical Research*, 95, 9927–9939.

Lee, T. F., and P. M. Tag. 1990. Improved Detection of Hotspots using the AVHRR 3.7-mm Channel, *Bulletin of the American Meteorological Society*, 72, 1722–1730.

Levine, J. S. 1990. Global Biomass Burning: Atmospheric, Climatic, and Biospheric Implications. *EOS*, 71, 1075–1077.

Matson, M., S. R. Schneider, B. Aldridge, and B. Satchwell. 1984. Fire Detection Using the NOAA-Series Satellites. NOAA Tech. Mem. NESDIS 7, Silver Spring, Maryland.

Robinson, J. M. 1991. Fire from Space: Global Fire Evaluation Using Infrared Remote Sensing. *Int. J. Remote Sensing*, 12, 3–24.

Vines, R. G. 1981. Physics and Chemistry of Rural Fires, Fire and the Australian Biota, 129–151.

Refined Analysis of MAPS 1984 Global Carbon Monoxide Measurements

Doreen Osowski Neil and Barbara B. Gormsen

Carbon monoxide (CO) plays an important role in the chemistry of the troposphere (Seiler 1974). Logan et al. (1981) identify four sources of atmospheric CO: combustion of fossil fuel, biomass burning, and photochemical oxidation of methane and nonmethane hydrocarbons. The primary CO removal mechanism is reaction with the hydroxyl radical (OH); the primary sink of tropospheric OH is reaction with CO. The presence of CO therefore affects the oxidizing capacity of the troposphere.

The atmospheric CO mixing ratio varies widely in space and time. In the 1980s, the Measurement of Air Pollution from Satellites (MAPS) experiment addressed the question of CO presence in the atmosphere on a global scale. Using the technique of gas filter correlation radiometry from a space platform, the MAPS experiment obtained measurements of tropospheric CO in November 1981 and October 1984 on board the Space Shuttle (Reichle et al. 1986; Reichle et al. 1990). These brief snapshots demonstrated the feasibility of the technique for measuring trace gases in the troposphere, and provided some of the earliest measurements indicating that biomass burning is a significant global source of CO. With the loss of the Space Shuttle Challenger in 1986, the MAPS measurement program entered a hiatus. In 1992, MAPS returned to the Space Shuttle manifest, and we began this review of MAPS data retrieval techniques in preparation for upcoming missions. These new MAPS missions subsequently occurred in April and October 1994, and data from the new missions will be discussed elsewhere.

In the present work, we briefly describe the MAPS data retrieval process, and present algorithm revisions and new instrument characterization. As the primary focus of this chapter, we then report the effects of these characterizations and revisions of the algorithms upon the 1984 MAPS flight data.

Description of MAPS Data Retrieval

The MAPS data retrieval process is outlined in figure 10.1. This process is conceptually the same as the process discussed in Reichle et al. (1990). Significant changes made by the present work occur in the physics of the instrument model and the radiative transfer calculations.

The signal voltages generated by the MAPS instrument on orbit are converted to top-of-the-atmosphere (TOA) infrared radiance by means of a calibration. The vertical temperature structure through the lower atmosphere is constructed for each data point from standard meteorological products for the time of measurement, and weighted by the MAPS signal function discussed in Reichle et al. (1986) to generate the signal function weighted temperature, T_{sf}.

These measurements of the infrared absorption and the temperature state of the atmosphere are then matched to the radiance and temperature of model atmospheres, to infer the amount of both CO and nitrous oxide (N_2O) in the atmosphere at the time of measurement. The matching between measurement and model is based on the parameter T_{sf}, and is accomplished by means of regression coefficients. The regression coefficients are generated using the matrix inversion technique defined by Wallio et al. (1983).

In preparing to retrieve CO and N_2O mixing ratios, we limit check various instrument parameters to ensure consistency with the instrument model calculations, and iterate the instrument model to maximize its applicability to measured conditions. After retrieval, we remove data with clouds in the field of view ("cloud filtering") to provide CO and N_2O mixing ratio data, which represent the averaged values in the column from the surface to low Earth orbit.

Spectral Response Characterization

In the present work, we conducted an end-to-end spectral characterization of the MAPS instrument as a function of instrument operating temperature. The location of the spectral passband in the instrument model is crucial to the calculation of the modeled top-of-the-atmosphere radiance detected by the MAPS instrument on orbit. In previous work, the spectral

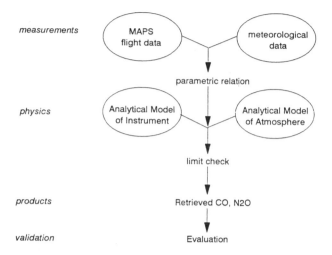

Figure 10.1 General description of the MAPS data-retrieval process

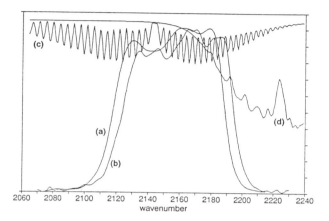

Figure 10.2 Location of MAPS passband for analytical models: (*a*) MAPS passband used by Reichle et al. (1990) for data retrieval, measured at room temperature; (*b*) passband used in present work, measured at 280°K, a temperature representative of a significant portion of the 1984 MAPS flight data; (*c*) atmospheric transmission of CO (Rothman et al. 1992); (*d*) Atmospheric transmission of N_2O (Rothman et al. 1992)

passband location used for these calculations was incorrect.

The end-to-end spectral transmission properties of the MAPS instrument were measured at room temperature in 1979. These properties are dominated by the optical band-pass filter and by the PbSe detectors. During the review of data from the first two MAPS flights in 1981 and 1984, some data anomalies were attributed to shifts in the instrument transmission function due to changes in the operating temperature of the MAPS instrument. However, this effect was not characterized by either analysis or test, and was not included in the data retrieval process.

In 1992, as part of the preparation of the MAPS instrument for new Shuttle flights, we designed a test to characterize the MAPS instrument spectral transmission as a function of instrument operating temperature. This spectral characterization test was performed for instrument operating temperatures from 0 to 25°C in 5-degree increments. The spectral resolution for the entire system (from monochromater through detector and data acquisition) was better than 1 cm^{-1}. The characterization provided sufficient data to construct a temperature-dependent instrument transmission function for temperatures within the MAPS operating envelope (0–15°C). Test results conform to the general properties of infrared multilayer coatings, as discussed by Baker and Yen (1967). In general, the MAPS spectral passband moves 0.275 cm^{-1} K^{-1} over the temperature range from 0 to 15°C, and widens less than 1 cm^{-1}.

A comparison of 1979 and 1992 spectral characterizations is displayed in figure 10.2. The MAPS transmission measured at room temperature in 1979 was used in the 1984 flight data retrieval reported by Reichle et al. (1990). The transmission at 7°C was calculated from the new spectral characterization, and represents the correct operating temperature of a significant portion of the 1984 MAPS flight data. We observe that the spectral passband was misplaced by 6 cm^{-1} (on a passband that is 70 cm^{-1} wide at the half power points) when the room temperature passband is used to represent the instrument operating at 7°C. The atmospheric transmission of CO and N_2O in this spectral range are also indicated for reference.

This characterization applies only to thermally stable instrument configurations. MAPS required the Space Shuttle to provide a constant temperature environment during on-orbit operations, although the flights in 1981 and 1984 did not fully comply with this requirement. Dynamic thermal conditions introduce large uncertainties in the data-retrieval process, and they are not included in the present analysis.

Calibration

The MAPS instrument is calibrated in the laboratory to define the relationship between its signal voltages (V, dV, dV') and radiance. Calibrated radiance data are combined with a line-by-line radiative transfer calculation to define the relationship between the signal channels and the variation in mixing ratio of CO and N_2O. The line-by-line correlation to atmospheric absorption spectra reveals the amount of the gas of interest in the field of view.

Because trace gas contributions to radiance are calculated quantities, the new knowledge of spectral lines from the 1992 revised HITRAN data base (Rothman et al. 1992), and of the MAPS instrument passband from our recent spectral characterization, provided the basis for deriving a new and better calibration.

For calibration, we operated the MAPS instrument in a vacuum chamber at constant instrument temperature to simulate operation on orbit with thermal control from the Shuttle's coolant loop. We introduced a series of gas mixtures into a gas cell that filled the instrument's field of view to simulate Earth's atmosphere. This cell was located immediately external to the MAPS instrument, also within the vacuum chamber. A secondary standard blackbody produced radiance to simulate Earth's surface emission. This radiance passed through the external (atmosphere-simulating) gas cell and entered the MAPS instrument.

Because the instrument's characteristics change with its own operating temperature, the calibration was conducted at a series of stable operating temperatures that span the expected on-orbit operating range. We allowed approximately 12 hours for thermal stabilization before conducting any testing at a given operating temperature. Calibration test procedures and data are contained in a NASA test report (NASA 1993).

Calibration equations form part of the instrument model. In our review of the instrument model, we also identified and corrected errors in the calculations of optical transmission through the instrument, errors in the calculation of radiance produced by on-board calibration blackbodies, and errors in the calculations of signal handling which had been correct for earlier configurations but which were in error for the existing MAPS instrument.

One channel on the MAPS instrument provides a simple infrared radiometer measurement of the top-of-the-atmosphere radiance in the 4.6-μm region. These data represents the most direct relationship between instrument signals and absolutely known conditions (calibration blackbody temperature viewed in geometry identical to on-orbit measurements). In figure 10.3, we present the distribution of observed top-of-the-atmosphere equivalent temperatures for the October 1984 flight data. These data were obtained in low earth orbit ($\sim$120 nautical miles altitude), between 57° south latitude and 57° north latitude, over land and water with cloud systems, during six days and nights in October 1984. These data define real instrument operating ranges for atmospheric measurements and simulations.

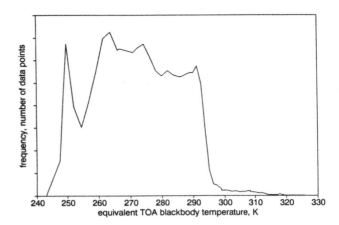

Figure 10.3 Top-of-the-atmosphere equivalent temperature as measured by MAPS on orbit in October 1984

We applied the new calibration to the existing 1984 MAPS data in this study, after analytically accounting for some configuration differences between the MAPS instrument that obtained the 1984 flight data and the MAPS instrument that we calibrated in 1992. In addition to providing high-quality calibration data for analysis, the 1992 calibration also verified that the MAPS instrument remained fully operational in preparation for its 1994 Space Shuttle flights.

Regression Coefficients

Eleven atmospheric models were defined for MAPS data retrievals in earlier work (Reichle et al. 1986; Reichle et al. 1990). We solved the nadir-viewing atmospheric radiative transfer equation for each modeled atmosphere over a range of underlying surface temperatures, and for a range of mixing ratios of the gas that MAPS measured (CO, N_2O), producing a library of top-of-the-atmosphere radiances. We fit sets of regression coefficients to this library of radiances so that we could interpolate between the CO and N_2O mixing ratios that are represented in the models. For the present work, we recalculated the top-of-the-atmosphere radiance for all 11 models using the correct instrument spectral passband, the recent spectral line parameters, and the revised instrument model, which were discussed in previous sections of this chapter.

We used the resulting accurate top-of-the-atmosphere radiances to calculate new sets of regression coefficients. These new regressions were applied to the existing 1984 MAPS flight data to retrieve CO and N_2O mixing ratios. Further discussion of the development of regression coefficients and of the limits and

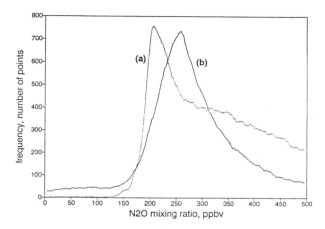

Figure 10.4 Comparison of N$_2$O frequency distributions: (*a*) N$_2$O frequency distribution from Reichle et al. (1990); (*b*) N$_2$O frequency distribution from present work. Present work peaks closer to expected value of 301 ppbv.

restrictions levied on the data during retrieval is beyond the scope of this chapter, and will be provided in a future manuscript.

Effect on Retrieval of Nitrous Oxide

We applied the new regression coefficient sets to a selected subset of the 1984 MAPS flight data and studied the effect on retrieved N$_2$O. We selected a subset of the 1984 MAPS flight data that was obtained at constant operating temperature, in keeping with the assumptions that went into the retrieval algorithms. We identified the extreme values as well as the mean and standard deviation of all the signal voltage channels and the instrument's internal temperature measurements. The variability of the internal temperature measurements is much less than 1°C, so we concluded that these measurements indicated thermally stable instrument operation. The specific mean values of these internal temperature measurements were used in the model radiance calculations, and consequently the model radiance calculations apply only to the conditions of the 1984 MAPS flight data.

We studied the frequency distribution of N$_2$O mixing ratios in the data set, presented in figure 10.4 for the original published data and the selected subset of 1984 MAPS flight data, which has been reprocessed in this work. This frequency distribution includes retrieved N$_2$O for all data points, including data over clouds. From the complete distribution, we verified that retrieved values of N$_2$O fall within the range of mixing ratios considered in the models; we identified the mean value of N$_2$O mixing ratio as a check against

the known value; and we observed some differences between the published data and the refined analysis.

For the data of figure 10.4, we observed that the mean of the selected subset of MAPS 1984 flight data is 257 ppbv with the corrections of the present work, compared to a value of 210 ppbv previously published. The accepted value for tropospheric N$_2$O in 1984 is 301 ppbv (Weiss 1981). The corrections of the present work remove a significant portion of the 30% error in the data set, but an error of 14% in the mean value remains in the reprocessed data.

We observe a significant change in the N$_2$O distribution on the high mixing ratio side of the peak value. The published data (curve *a* in figure 10.4) features a sharp departure from gaussian shape at about 260 ppbv; this feature was attributed to the presence of clouds in the field of view by Reichle et al. (1990). The refined analysis (curve *b* of figure 10.4) produces a distribution without a sharp departure from gaussian on the high mixing ratio side of the peak, although the distribution widens considerably as mixing ratio increases. Therefore, the definition of the presence of clouds must be reevaluated. Such an evaluation would be accomplished by reviewing the MAPS on-board infrared photography of daylit scenes, and comparing the actual clouds in the field of view with retrieved N$_2$O mixing ratios to set an objective threshold value for cloud contamination of the scene. A comprehensive cloud study based on the unique resource of MAPS on-board photography has not been conducted.

We also observe a significant change in the N$_2$O distribution on the low mixing ratio side of the peak value. The published data (curve *a* in figure 10.4) features a steeply rising curve beginning about three standard deviations below the peak N$_2$O value. The refined analysis (curve *b* of figure 10.4) indicates a low-level continuum of points with low values of retrieved N$_2$O mixing ratios. In addition, between two and three standard deviations below the peak N$_2$O value, we find that the distribution is wider than expected. Since the MAPS instrument photographs all daylit scenes, we consulted these space-based infrared photographs to determine the cause of these features. In the photographs, we observe large bright spots, almost always off water surfaces (ocean). We conclude that the low N$_2$O mixing ratios are related to sunglint. The literature indicates that sunglint has been evaluated as a microwave phenomenon (for example, Fett 1993) and in the visible portion of the spectrum (for example, Fett and Issacs 1979), but the electromagnetic properties of sunglint near 4.6 μm have not been presented. If the most significant property in the infrared is polarization, we speculate that this effect will be reduced in MAPS data from the 1994 missions, because of a hardware modification that was made to the MAPS instrument between the 1984 and 1994 flights.

Since N_2O is well mixed in the troposphere, the value of N_2O retrieved should be the same at all geographic locations. However, our studies of the retrieved N_2O for the published 1984 data set indicate many regions where retrieved N_2O varied far from the mean. We used this information to guide our development of regression coefficients, and to identify errors in the radiative transfer calculations, as we discussed earlier in this chapter. The N_2O values retrieved in our refined analysis show substantially less geographic variation.

Finally, we observe that the reprocessed data contain nearly as many points (area below the curve in figure 10.4) as the original data, although the reprocessing was performed on a subset of the original data. This suggests that the refined retrieval process produces a greater number of N_2O measurements within the range that was modeled.

Effects on the Data Set

We studied the global CO distribution pattern over the two-dimensional map of the planet to evaluate the effects of these changes in the MAPS data retrieval process on the reported data set.

Figure 10.5 presents the global CO distribution pattern for reprocessed 1984 MAPS flight data in 5° latitude by 5° longitude resolution. (MAPS obtains data once per second with a 20×20-km surface footprint. All acceptable second-by-second data points within the $5 \times 5°$ square are averaged for the value shown. Some squares represent the average of hundreds of data points; some represent the average of as few as six data points.) The previously published data indicate high levels of CO over biomass burning regions in South America and Africa, and lowest levels in the Pacific Ocean near the equator within 40 degrees either side of the dateline. The reprocessed data of figure 10.5 confirm these general features, but indicate higher CO values in the biomass burning regions, and stronger transport in the Atlantic between the continental biomass burning regions. In the reprocessed data, CO transport is also indicated from Africa into the Indian Ocean and past Australia. In addition, in the reprocessed data, we observe elevated CO over the industrialized regions of the globe (North America, Europe, and Japan). These regions showed almost no CO enhancement in the previously published data.

However, the CO values are still low compared to single-altitude aircraft measurements of the time (Reichle et al. 1990, p. 9852).

Some $5 \times 5°$ squares contain data in the original distribution, but do not contain data in the reprocessed distribution, and vice versa. These additions or deletions are due to new data values, which were retained or discarded differently through the limit-checking and cloud-filtering processes that were applied to the individual (second-by-second) data points. These individual data points were averaged for figure 10.5, and represent real changes in the retrieved data. From this result, we find that the refined retrieval process does not simply scale the values previously reported.

These revisions, corrections, and additions to the MAPS data-retrieval process have had a profound effect on the data. We identified and used the N_2O measurement as an objective and self-consistent assessment of error. Our refinement of the data-retrieval process removed more than half of the observed error from the quantitative measurement. The instrument and analytical models now feature significantly corrected and enhanced physics. We expect to improve the retrieval further with work discussed in the next section.

Future Work

While our work focused on a thorough review and revision of the analytical model and the instrument data, we did not address other important areas that remain for future work in the MAPS retrieval process.

First, the parameter that links the revised analytical models and the flight data should be reviewed. The parameter T_{sf}, established by Reichle et al. (1986) and used in this work, has the effect of minimizing atmospheric variations because of the integration through the column. Because of this heavy smoothing, few of the atmospheric models are used in actual data reduction (4 of the 11 models are used in more than 90% of the data). The same atmospheric model is used throughout the Tropics, over ocean, rain forest, and the Sahara, suggesting that at a minimum the effects of water vapor are not properly treated. Other techniques for linking the flight data and the analytical models, investigated in the early 1980s, may be more successful now that the radiative transfer calculations in both the instrument and the atmosphere have been substantially improved by this work.

Additionally, the implementation of signal function weighting in the analytical models should be further investigated. The instrument signal function, calculated from the instrument model viewing the U.S.

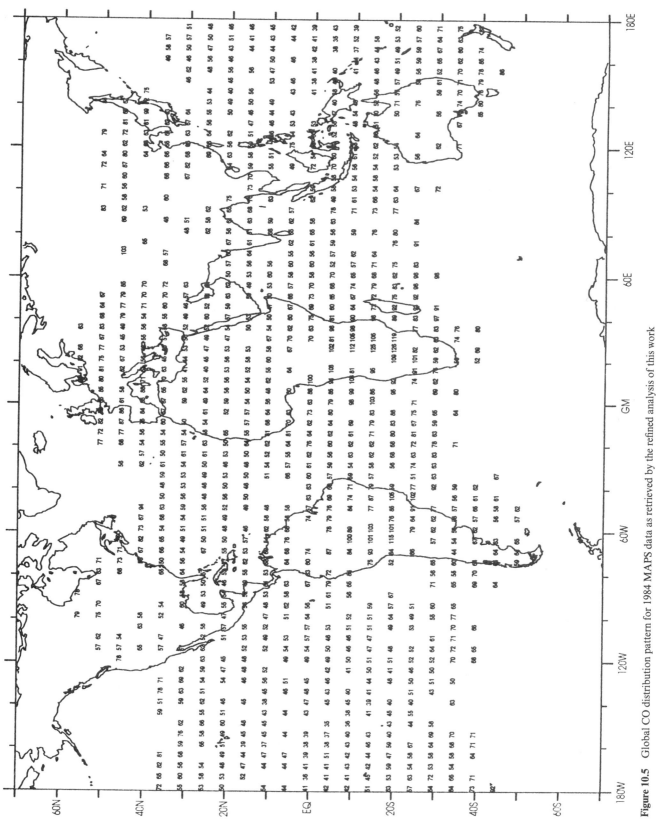

Figure 10.5 Global CO distribution pattern for 1984 MAPS data as retrieved by the refined analysis of this work

Standard Atmosphere, is applied to all data regardless of the actual atmospheric conditions. High-quality measurements of CO profiles with detailed knowledge of local atmospheric structure are now available from aircraft measurement programs (for example, Harriss et al. 1994), and provide a basis for checking the seasonal and geographic applicability of the signal function. Such a validation would be particularly important for data retrieval of the MAPS 1994 flight data from two different seasons.

Finally, it is important to recognize that the process described in this work is not the only nor even the best way to retrieve correlation filter radiometer data. Our ground rules in conducting this work required us to perform the retrieval in conceptually the same manner as previously established. However, during this work, we identified promising techniques that may totally or partially replace this process. Radiative transfer codes and atmospheric models are standard tools in the field; there is an advantage in using tools that have been subject to widespread scrutiny in the research community.

Acknowledgments

We thank Jack Fishman, who provided support crucial to the publication of this work. Without his intervention, this record of our work would not exist.

This work was funded by NASA's Office of Mission to Planet Earth, Flight Systems Division.

References

Baker, M. L., and V. L. Yen. 1967. Effects of the variation of angle of incidence and temperature on infrared filter characteristics, *Applied Optics*, 6, 1343–1351.

Fett, R. W. 1993. Kamishak gap wind as depicted in DMSP OLS and SSM/I data; final journal article, *International Journal of Remote Sensing*, 14, 403–423.

Fett, R. W., and R. Issacs. 1979. Concerning causes of anomalous gray shades in DMSP visible imagery, *Journal of Applied Meteorology*, 18, 1340–1351.

Harriss, R. C., G. W. Sachse, J. E. Collins, L. Wade, R. W. Talbot, E. V. Browell, L. A. Barrie, K. B. Bartlett, and G. F. Hill. 1994. Carbon monoxide and methane over Canada: July–August 1990, *J. Geophys. Res.*, 99, 1659–1670.

Logan, J. A., M. J. Prather, S. C. Wofsy, and M. B. McElroy. 1981. Tropospheric chemistry: A global perspective, *J. Geophys. Res.*, 88, 7210–7254.

NASA. 1993. MAPS TR-07-01-01 Space Flight Instrument Gas Response Test Report, *Report No. MAPS TR-07-01-01*, unpaginated, NASA Langley Research Center, Hampton, Va.

Reichle, H. G., Jr., V. S. Connors, J. A. Holland, W. D. Hypes, H. A. Wallio, J. C. Casas, B. B. Gormsen, M. S. Saylor, and W. D.

Hesketh. 1984. Middle and upper tropospheric carbon monoxide mixing ratios as measured by a satellite-borne remote sensor during November 1981, *J. Geophys. Res.*, 91, 10 865–10 887.

Reichle, H. G., Jr., V. S. Connors, J. A. Holland, R. T. Sherrill, H. A. Wallio, J. C. Casas, E. P. Condon, B. B. Gormsen, and W. Seiler. 1990. The distribution of middle tropospheric carbon monoxide during Early October 1984, *J. Geophys. Res.*, 95, 9845–9856.

Rothman, L. S., R. R. Gamache, R. H. Tipping, C. P. Rinsland, M. A. H. Smith, D. C. Benner, V. M. Devi, J. M. Flaud, C. Camy-Peyret, A. Perrin, et al. 1992. The HITRAN molecular data base—Editions of 1991 and 1992, *J. Quant. Spectroscopy and Radiative Transfer*, 48, 469–507.

Seiler, W. 1974. The cycle of atmospheric CO, *Tellus*, 26, 116.

Wallio, H. A., J. C. Casas, B. B. Gormsen, H. G. Reichle, Jr., and M. S. Saylor. 1983. Carbon monoxide mixing ratio inference from gas filter radiometer data, *Applied Optics*, 22, 749–754.

Weiss, R. F. 1981. The temporal and spatial distribution of tropospheric nitrous oxide, *J. Geophys. Res.*, 86, 7185.

Global Distribution of Biomass Burning and Carbon Monoxide in the Middle Troposphere during Early April and October 1994

Vickie S. Connors, Mike Flood, Tom Jones, Barbara Gormsen, Scott Nolf, and Henry G. Reichle, Jr.

Carbon monoxide is a colorless, odorless, and tasteless gas that is produced as a result of the partial oxidation of carbon compounds. Carbon monoxide (CO) is now recognized as being of great importance in the global-scale chemistry of the troposphere. Some evidence has been reported that CO may have a more direct radiative effect on the atmosphere than previously thought (Evans and Puckrin 1995). The largest sink for the hydroxyl radical (Logan et al. 1987), CO has both natural and anthropogenic sources. The principal anthropogenic sources are the combustion of fossil fuels, industrial processes like oil refineries and coke ovens, and the burning of biomass. The principal natural sources of CO are the oxidation of hydrocarbons like isoprene and methane. A comprehensive review of the sources and emissions of atmospheric carbon monoxide is given in Badr and Probert (1994b). The status of current CO measurements, techniques, and future measurement plans is summarized in Novelli and Rossen (1995).

With an intermediate lifetime in the troposphere ranging from a few weeks to a few months depending on the season and the location, CO can be transported long distances from its sources (Newell et al. 1988). The relationship between the distribution of the CO sources and sinks, the strength of those source and sink mechanisms, and the long-range transport of CO is responsible for the complex, non-uniform distribution that has been reported (Reichle et al. 1986; Reichle et al. 1990). The overall patterns in the global distribution of CO cannot be determined from either aircraft- or ground-based measurement systems; satellite-based instruments provide the best perspective and sampling frequency to make such global measurements.

The Measurement of Air Pollution from Satellites (MAPS) experiment was flown on board the space shuttle *Endeavor* during April and October 1994. During the two 10-day Space Radar Laboratory flights, extensive observations of the planet by both the STS-59 and STS-68 astronaut crews recorded the nature and extent of fires and smoke plumes within 57° of the equator in the form of visual reports and photographs. The accuracy of the space-based CO data was determined by comparisons with the airborne inter-calibrated measurements over North America and Australia, following a similar technique as used in Connors et al. (1991). This chapter will present an overview of both the global distribution of CO mixing ratios and the burning conditions during the two space shuttle missions. The distribution of burning events and the distribution of CO mixing ratios during October 1994 will be shown to be closely coupled.

Distribution of Carbon Monoxide Mixing Ratios

The measurements of carbon monoxide in the troposphere were made only from either ground-based or aircraft-based platforms until 1981. At the time, the data indicated that the CO mixing ratio was higher in the Northern Hemisphere than in the Southern Hemisphere, and that it was higher in the mixing layer than in the free troposphere. Generally, the average CO mixing ratio was twice as high in the Northern Hemisphere, about 120 ppbv, than in the Southern Hemisphere, about 60 ppbv. The seasonal cycle in the CO mixing ratio exhibits a maximum in local spring and a minimum in the local fall; springtime values tend to be approximately 50% higher than the values in the fall (Seiler et al. 1984; Fraser et al. 1986; Marenco 1988). Badr and Probert (1994a) and Levine (1991) provide the most comprehensive summaries of CO (and other gaseous) measurements.

The pre-1981 understanding of the global distribution of CO in the troposphere was substantively altered as a result of the first two space shuttle flights of the MAPS experiment. In November 1981, the MAPS measurements revealed maximum values in the Tropics between western South America and central Africa, rather than at northern mid-latitudes. The CO minimum was located over the central tropical Pacific Ocean. Patchy regions of high CO mixing ratios were located over portions of Europe and China. The transports of the surface-generated CO to the free troposphere was deduced to be the result of thunderstorm

Figure 11.1   The Space Radar Laboratory-2 payload over the mouth of the Amazon River on 7 October 1994

development or advancing cold fronts (Reichle et al. 1986; Newell et al. 1988). The October 1984 flight of the MAPS experiment occurred near the end of the burning season of Brazil and southern Africa. As before, the CO distribution showed strong gradients with both latitude and longitude. Very high mixing ratios were measured over and downwind from South America and southern Africa. Astronauts' visual observations and photography significantly contributed to the identification of both the locations of, and strengths for the CO emissions from the southern fires. The maximum amounts of CO were located over the tropical and mid-latitude regions of the Southern Hemisphere (Reichle et al. 1990). The measured latitudinal gradient of CO mixing ratio during the Southern Hemisphere spring was directly opposite the earlier view!

The MAPS experiment flew on the space shuttle *Endeavor* 9–19 April 1994 on STS-59 and 30 September–11 October 1994 on STS-68 as a primary instrument in the Space Radar Laboratory (SRL) payload. A view of the SRL-2 payload in operation over the mouth of the Amazon River on 7 October 1994 (Roll 267 Frame 008) is shown in figure 11.1. These two highly successful flights provided exceptional thermal control for the payload, enabling very

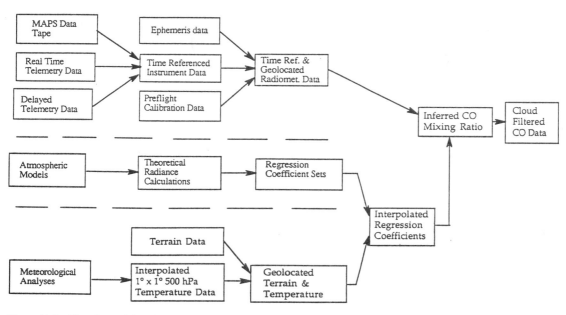

Figure 11.2 Flowchart of the 1994 MAPS data reduction scheme

high quality data collection. Because telemetry capability had been added to the MAPS instrument following the 1984 flight, preliminary analyses of CO were produced during each flight. These preliminary data sets (Reichle 1994; Connors 1994), each occurring in opposing seasons, were in striking contrast to each other. The April 1994 data showed that the maximum values of CO mixing ratios were located at high northern latitudes, with decreasing amounts towards the south. The variations with longitude were generally small at all latitudes. In contrast, the October 1994 preliminary analyses bore similarities to the October 1984 data in that the highest CO amounts were located over the tropical regions of the Southern Hemisphere with marked longitudinal variations.

Upon the completion of these two flights, the MAPS instrument data, both telemetered and recorded, were merged into a single, time-tagged and geolocated Level 0 data set. Then, the instrument data were processed following the general approach described in Reichle et al. (1986; 1990). A more detailed discussion of the data-processing scheme will be reported elsewhere; only changes to the overall process will be mentioned here. Figure 11.2 shows the logic flow for each of the major paths of the 1994 MAPS data-reduction process. As a result of analytical studies, the 500 hPa atmospheric temperature was determined to be the best temperature representation for the MAPS instrument. The 500 hPa temperature analyses were provided by the European Center for Medium Range

Weather Forecasts. The sets of regressions coefficients that were used are an improved set from what had been reported previously; the correct parameterization of the instrument filter and calibration data have been included both in the data reduction and in the atmospheric modelling. The radiative transfer calculations have been performed to match up the instrument characterization with the current situation in regard to both CO and N_2O. Much of the land surface and all of the ocean can be represented by an emissivity of 0.98 at 4.6 μm (Salisbury and D'Aria 1994). Only during high sun conditions over dry, barren soil or desert will the global approximation be inappropriate for the MAPS data reduction. For the case of the April and October 1994 MAPS flights, only 2.5% of the data were not represented by this condition; these data have been excluded from the data set.

The north-south gradients of CO during April and October are shown in figures 11.3 and 11.4, respectively. As suggested by the preliminary analyses, the April gradient is smooth with only small variations as represented in the one sigma gray shading. The solid line represents the zonal average, which ranges from a high in the northern high latitudes of about 120 ppbv, with a sharp transition through the tropical regions, and generally uniform amounts of about 50–60 ppbv over the Southern Hemisphere. Figure 11.4 shows quite a different view in October with the reversal in the gradient in the tropical regions. The variability across the latitude bands is greatest where the CO reaches the

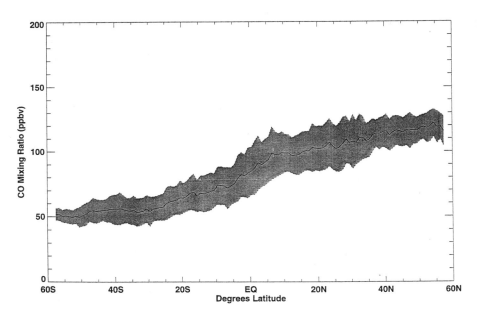

Figure 11.3 MAPS CO zonal averages during 9–19 April 1994

maximum amounts (about 125 ppbv) between 10°N and 35°S. Notice that the background amount in the high southern latitudes has increased to nearly 70 ppbv.

Comparisons between the MAPS CO data and aircraft survey flights over North America and profile flights near Australia made at nearly the same time were used to determine that the MAPS data are within 20% of the independent airborne measurements. When the space shuttle and aircraft pass over the same area at nearly the same time, the comparisons between the space-based and airborne measurements are within 10% of each other. More detailed data validation studies are ongoing and the results will be reported elsewhere.

Astronauts' Observations

The crews of both Space Radar Laboratory flights participated as observers for the MAPS experiment team. Their prime role was to keep the space shuttle *Endeavor* operating smoothly as a stable, accurately pointing science platform for the MAPS instrument. The instrument needed to be nadir-oriented while taking data and kept away from the unfiltered brilliance of the sun. Careful attitude planning was imperative because MAPS was not the only instrument in the cargo bay. The synthetic aperture radars, SIR-C from the Jet Propulsion Laboratory and X-SAR from Germany and Italy, required alignment with respect

to the target sites at proselyted angles and times. For example, the STS-68 crew on SRL-2 maneuvered *Endeavor* about 470 times in 11 days to accomplish the radar science objectives and still keep MAPS on target.

The second role that the astronaut crew performed for the MAPS experiment was to report all sightings of fires and smoke to the MAPS mission operations team during the flight. These verbal crew reports were plotted by the MAPS team and correlated with the preliminary CO analyses to link regions of maximum CO amounts with possible combustion sources on the ground. During the April 1994 flight, the STS-59 crew noted industrial smoke plumes and pollution palls over major cities of the Northern Hemisphere, particularly in Russia and the Far East. Small biomass fires were sighted in South America and Africa, but they were dwarfed by smoke plumes and palls in Burma and the East Indies. The night half of each orbit was very useful for detecting fires, whose orange outlines glowed brightly against the dark face of Earth below *Endeavor*'s windows. The astronauts reported everything from flaring natural gas fires in South America's oil fields, to slowly creeping fires along the Congo and Ivory Coast.

During the October 1994 flight, the industrial smoke plumes were still evident in north central Asia and eastern China, but the largest biomass fires had shifted to the East Indies and Australia. Borneo and New Guineau were wrapped in a shroud of haze and smoke

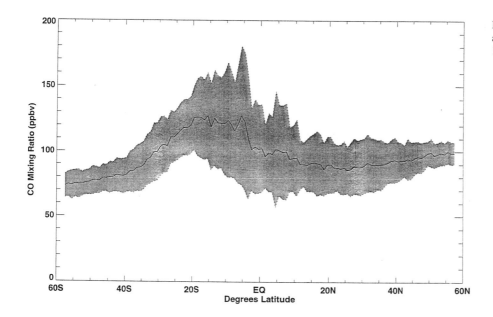

Figure 11.4 MAPS CO zonal averages during 30 September–11 October 1994

Table 11.1 Online access information for STS-59 and STS-68 photographs of Earth

Mission	Directory access	Contents
9–19 April 1994 STS-59		
	http://eol.jsc.nasa.gov/sseop.html	Introductory page
	http://ersaf.jsc.nasa.gov/sn5.html	Clickable map with .jpeg files
30 September–11 October 1994 STS-68		
	http://images.jsc.nasa.gov/html/home.htm	Contains .jpeg files

from dozens of fires caused by tropical forest clearing. Africa was undergoing the annual burning of the savanna grasslands in Mozambique and neighboring countries. The location and extent of the fires in Africa were similar to those analyzed by Cahoon et al. (1994) for 1986 and 1987. Australia, suffering from a long drought, was peppered with extensive range and forest fires, from the Cape York peninsula all the way to the eastern states of Queensland and New South Wales. South America was just emerging from its long burning season in the Amazon.

In addition to the verbal reports, both crews documented many of the CO sources with false-color infrared and color-visible photography. The crews photographed the MAPS correlative measurement sites repeatedly to record the regional environmental conditions on each mission day. In all, both Space Radar Laboratory crews combined to capture about 24 000 frames of Earth imagery for the payload experiments. The scanned photographs and brief descriptions of each from these two shuttle missions are electronically accessible to the public through MOSAIC in two directories (table 11.1).

Combined with images from the MAPS flight camera, the crew's hand-held and video-camera photography provided a unique data set characterizing the surface and atmospheric processes that contribute to the interpretation of the remotely sensed data from the radars and the MAPS instrument.

The astronauts recorded fires, smoke plumes, thunderstorms, weather systems, floods, and pollution palls with a synoptic view generally unavailable by any other system. After each flight, their debriefings expanded and complemented the spaceborne measurements. The highly successful collaboration between the payload experiment teams and the astronaut crews required regular communication among the groups; accomplished through training sessions, visits to the sites, participation in science team meetings, and ongoing active dialogue, the experience of the STS-59 and STS-68 missions should be regarded as models for future Earth-observing missions.

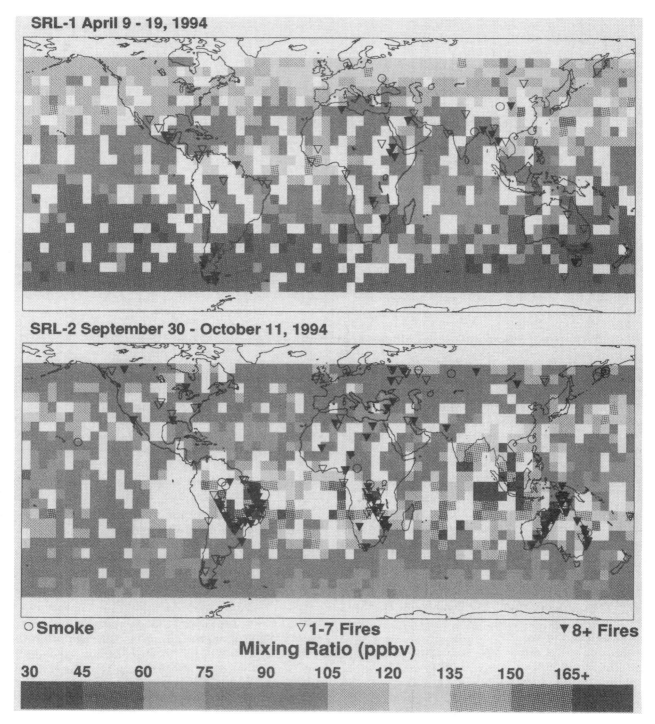

Figure 11.5 MAPS 5° × 5° averaged carbon monoxide mixing ratios and astronauts' observations of fires during (*top*) April and (*bottom*) October 1994

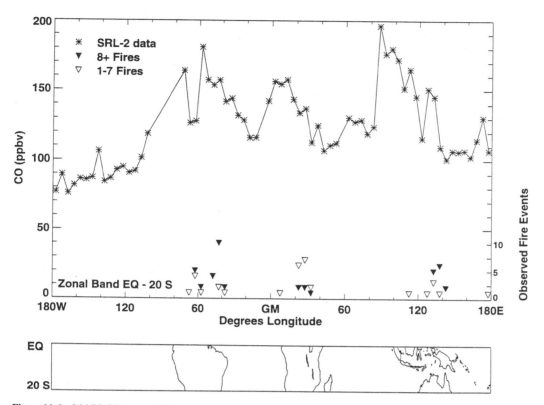

Figure 11.6 MAPS CO averaged between equator and 20°S and astronaut-observed fires 30 September–11 October 1994

Distribution of Fires and Smoke

The astronauts' observations for both flights were transmitted in near real-time from space and recorded, by hand, on the ground during both missions. The time-tagged observations were matched with the position of the nadir-location of the shuttle. An intensity index was developed to represent the extent and prevalence of the fires. All observations of smoke are plotted as open circles, reports of 1–7 fires are shown as open triangles, and more than 8 fires reported are shown as solid triangles. Figure 11.5 shows the distribution of the reported smoke and fires superposed on the MAPS 5° × 5° averaged CO distributions for April and October 1994. Please note that some of the fire reports may actually be the result of industrial burning in southern Argentina and over northern Africa and northern Europe.

Clearly, the widespread burning ongoing during the October 1994 flight and the concomitant high values in the vicinity of, and downstream from, the burning events are compelling. This relationship can be explored further by examining a transect from 180°W to 180°E between the equator and 20°S. In figure 11.6,

the averaged CO mixing ratios are shown as asterisks while the locations of the astronaut-reported fires are plotted as open and filled triangles below the CO transect. Generally, the CO maxima occur over each of the continental regions where the fire reports are the highest. A hint of downwind transport away from the Australia burning events is shown over 80°E–100°E where the peak is shifted away from the land area.

Conclusions

The MAPS experiment flew on the space shuttle *Endeavor* 9–19 April 1994 and 30 September–11 October 1994. During these two Space Radar Laboratory flights, extensive observations of the planet by both the STS-59 and STS-68 astronaut crews recorded the nature and extent of fires and smoke plumes within 57° of the equator in the form of visual reports and photographs. These unique observations contribute significantly to our interpretation and understanding of the relationship between the surface and atmospheric conditions that relate to the CO mixing ratios in the free troposphere. A strong relationship between the distribution of biomass burning, especially during the October 1994 flight, and the enhancement of CO mixing ratios near and downstream from the areas under the influence of biomass burning exists. Marked seasonal changes are shown in both the global distribution of CO mixing ratios and the distribution of burning regions during the two space shuttle missions.

Acknowledgments

The authors would like to recognize the incredible efforts of the STS-59 and STS-68 astronaut crews in thoroughly photographing and describing the conditions of Earth and the atmosphere during our two missions. We thank the NASA Johnson Space Center, Space Shuttle Earth Observations Project, in particular Dr. David Amsbury and Cynthia Evans for their support before, during, and after the missions. We appreciate the tireless efforts of the MAPS Mission Operations Team and the MAPS Correlative Measurement Team in producing the high-quality, multiple-platform, global CO data sets. We thank the NASA Langley Research Center Film Recording Laboratory for their gracious access to equipment during the preparation of this chapter.

References

Badr, O., and S. D. Probert. 1994a. Carbon monoxide concentration in the Earth's atmosphere, *Applied Energy*, 49, 99–143.

Badr, O., and S. D. Probert. 1994b. Sources of atmospheric carbon monoxide, *Applied Energy*, 49, 145–195.

Cahoon, D. R. Jr. 1994. The extent of burning in African savannas, *Adv. Space Res.*, 14, 447–454.

Connors, V. 1994. The grand global carbon monoxide (CO) experiment of 1994, *AGU EOS Transactions*, 1994 Fall Meeting, 75, 128.

Connors, V., D. Cahoon, H. Reichle, and H. Scheel. 1991. Comparisons between carbon monoxide measurements from spaceborne and airborne platforms, *Canadian Journal of Physics*, 69, Special Space Science Issue, 1, 128–1, 137.

Evans, W., and E. Puckrin. 1995. An observation of the greenhouse radiation associated with carbon monoxide, *Geophysical Research Letters*, 22, 925–928.

Fraser, P., P. Hyson, R. Rasmusson, A. Crawford, and M. Khalil. 1986. Methane, Carbon Monoxide, and Methyl-Chloroform in the Southern Hemisphere, *J. Atmos. Chem.*, 4, 3–42.

Levine, Joel S., Ed. 1991. *Global Biomass Burning*, MIT Press, Cambridge, Mass. 569 pp.

Logan, J., M. Prather, S. Wofsy, and M. McElroy. 1981. Tropospheric chemistry: A global perspective, *J. Geophys. Res.*, 86, 7210–7254.

Marenco, A. 1986. Variations of CO and O_3 in the troposphere: Evidence of O_3 photochemistry, *Atmos. Environ.*, 20, 911–918.

Newell, R., S. Shipley, V. Connors, and H. Reichle. 1988. Regional studies of potential carbon monoxide sources based on space shuttle and aircraft measurements, *J. Atmos. Chem.*, 6, 61–81.

Novelli, P., and R. Rossen, Eds. 1995. Report of the Meeting of WMO Experts on Carbon Monoxide, Boulder, Colo. 07–10 February 1994, World Meteorological Organization, Report No. 8.

Reichle, H., V. Connors, J. Holland, W. Hypes, H. Wallio, J. Casas, B. Gormsen, M. Saylor, and W. Hesketh. 1986. Middle and upper tropospheric carbon monoxide mixing ratios as measured by a satellite-borne remote sensor during November 1981, *J. Geophys. Res.*, 91, 10 865–10 887.

Reichle, H., V. Connors, J. Holland, R. Sherrill, H. Wallio, J. Casas, E. Condon, B. Gormsen, and W. Seiler. 1990. The distribution of middle tropospheric carbon monoxide during early October 1984, *J. Geophys. Res.* 95, 9845–9856.

Reichle, H. 1994. Spaceborne tropospheric carbon monoxide, past and future, *AGU EOS Transactions*, 1994 Fall Meeting, 75, 128.

Salisbury, J., and D. D'Aria. 1994. Emissivity of terrestrial materials in the 3 to 5 μm atmospheric window, *Remote Sensing of Environment*, 47, 345–361.

Seiler, W., H. Giehl, E. Brunke, and E. Halliday. 1984. The seasonality of CO abundance in the Southern Hemisphere, *Tellus*, 36B, 219–231.

FireSat and the Global Monitoring of Biomass Burning

Joel S. Levine, Donald R. Cahoon, Jr., John A. Costulis,
Richard H. Couch, Richard E. Davis, Paul A. Garn, Antony Jalink, Jr.,
James A. McAdoo, Don M. Robinson, William A. Roettker,
Washito A. Sasamoto, Robert T. Sherrill, and Kelly D. Smith

The report *Global Change and Our Common Future* published by the National Research Council (1989) concluded (p.1):

Our planet and global environment are witnessing the most profound changes in the brief history of the human species. Human activity is the major agent of those changes—depletion of stratospheric ozone, the threat of global warming, deforestation, acid precipitation, the extinction of species, and others that have not become apparent.

All five global environmental changes identified above have one important thing in common—they are all caused by biomass burning.

The burning of the world's living and dead biomass for land clearing and land-use change is a significant global source of atmospheric gases and particulates that impact the chemistry of the troposphere and stratosphere and the climate of our planet (Levine 1991; Levine et al. 1995). In addition to the production of significant amounts of gases and particulates to the atmosphere, biomass burning impacts Earth's atmosphere/biosphere system through the following processes: (1) the reflectivity and emissivity of the land and hence the global energy budget of out planet, (2) water run off and evaporation and the global hydrological cycle, and (3) the biogeochemical cycling of compounds from the biosphere to the atmosphere (Levine 1991). Biomass burning also impacts the stability of ecosystems and leads to the extinction of species (National Academy of Sciences 1990).

Biomass burning is not restricted to one country or to one region. Biomass burning is a regular feature of the world's tropical, temperate, and boreal forests, the savanna grasslands, and agricultural fields following the harvest. Biomass burning is a regular feature in the tropical forests in Brazil, Indonesia, Colombia, Ivory Coast, Thailand, Laos, Nigeria, Philippines, Burma, and Peru, the temperate forests of the United States and Europe, and the boreal forests of Siberia, China, Canada, and Alaska, the savanna grasslands of Africa, and the agricultural lands of the United States and Europe. It is generally believed that the vast majority of this burning is human-initiated (>90%) and that biomass burning has increased significantly over the last 100 years (Houghton 1991; Hao and Liu 1994).

Biomass burning is a significant global source of the following atmospheric gases: (1) greenhouse gases, carbon dioxide (CO_2), and methane (CH_4), that lead to global warming, (2) chemically active gases, nitric oxide (NO), carbon monoxide (CO), and hydrocarbons (HC), which lead to the photochemical production of ozone (O_3) in the troposphere. In addition, NO leads to the chemical production of nitric acid (HNO_3), the fastest growing component of acid precipitation, and (3) methyl bromide (CH_3Br), a major atmospheric source of bromide, which leads to the photochemical destruction of ozone in the stratosphere (Andreae 1991; Mano and Andreae 1994). Biomass burning is also a significant global source of atmospheric aerosols, which impact the transfer of incoming solar radiation through the atmosphere and hence impact both global climate and tropospheric chemistry (Penner et al. 1991).

To quantify the role and importance of biomass burning as a global source of gases and particulates to the atmosphere, information is needed on the global strength of biomass burning as a source of these environmentally significant compounds. To provide the needed information, international biomass burning research activities have been initiated over the last few years. The International Global Atmospheric Chemistry (IGAC) Project, a core activity of the International Geosphere-Biosphere Program (IGBP) has initiated two biomass burning research activities—The Biomass Burning Experiment (BIBEX) and the Global Emissions Inventory on Biomass Burning, part of the Global Emissions Inventory Activity (GEIA). In addition, over the last few years, international biomass burn field experiments have taken place in diverse ecosystems, including the South African Fire-Atmosphere Research Initiative (SAFARI-92), a ground-based and airborne measurement program in the savannas of southern Africa (Andreae et al. 1994), the Transport and Atmospheric Chemistry near the Equator-Atlantic (TRACE-A), an airborne mission over Brazil and southern Africa (Andreae et al. 1994), the Dynamics and Chemistry of the Atmosphere in Equatorial Forest (DECAFE) in the forests of equatorial Africa, the Biomass Burning Airborne and Spacecraft Experiment-Amazonia (BASE-A) in the Brazilian Amazon, and the Fire Research Campaign Asia-North (FIRESCAN) in the boreal forests of Siberia. Smaller biomass burn experiments have taken place in the chaparral ecosystem in the San Dimas Experimental Forest in southern California, in the wetlands in the Merritt Island National Wildlife Refuge, Kennedy Space Center, Florida, in the tropical rain forests of the Yucatán Peninsula, Mexico, and in the boreal forests of Canada.

In the last five years, biomass burning and its global environmental impacts were the subject of two international conferences sponsored by the American Geophysical Union—The Chapman Conference on Global Biomass Burning: Atmospheric, Climatic, and Biospheric Implications (March 1990) and the Chapman Conference on Biomass Burning and Global Change (March 1995).

More than a decade ago, Seiler and Crutzen (1980) showed that as a first approximation, the total amount of biomass burned (M) (in units of grams of dry biomass material per year) in a particular ecosystem may be given by the following equation:

$$M = A \times B \times a \times b \qquad (12.1)$$

where A is the total land area burned annually (m^2/year), B is the average biomass material per unit area in the particular ecosystem (grams of dry biomass material per m^2), a is the fraction of the aboveground biomass material relative to the total average biomass B, and b is the burning efficiency of the aboveground biomass. Parameter A, the area burned during a fire in a particular ecosystem, is the major uncertainty in solving equation (12.1). Parameters B, a, and b in equation (12.1) have been determined during a series of international and national biomass burn field experiments including SAFARI-92, DECAFE, BASE-A, and FIRESCAN, and the smaller scale experiments in California, Florida, Mexico, and Canada.

Once M is known, the total mass of carbon [$M(C)$] in grams released to the atmosphere during biomass burning may be calculated using the following equation:

$$M(C) = 0.45 \, M \qquad (12.2)$$

Since about 45% of biomass by weight is carbon (the remainder of the biomass weight is due to water (about

50%) with smaller amounts of nitrogen (about 1%), and still smaller amounts of sulfur, chlorine, and bromine). The carbon released to the atmosphere during biomass burning takes the form of several gaseous and particulate compounds, including CO_2, CO, CH_4, nonmethane hydrocarbons (NMHCs), and particulate carbon. The ratio of any carbon compound (such as CO, CH_4, NMHC) or nitrogen compound (such as NO, or N_2O) to carbon dioxide produced in biomass burning can be determined by knowledge of the emission ratio (ER). The emission ratio is the amount of any compound X produced during biomass burning normalized with respect to the amount of CO_2 produced during biomass burning. The emission ratio is usually normalized with respect to CO_2 because CO_2 is the overwhelming carbon species produced during biomass burning and it is a relatively easy gas to measure. The emission ratio is defined as

$$ER = \Delta X / \Delta CO_2 \qquad (12.3)$$

where ΔX and ΔCO_2 are the concentrations of the species X and CO_2 produced by biomass burning and are equal to ($X^* - X$) and ($CO_2^* - CO_2$) where X^* and CO_2^* are the measured concentrations in the biomass burn smoke plume and X and CO_2 are the background (out-of-plume) atmospheric concentration of the species. Information on the emission ratios for various gaseous and particulate compounds produced during biomass burning has been obtained during the series of international and national biomass burn field experiments including SAFARI-92, TRACE-A, DECAFE, BASE-A, and FIRESCAN, and the smaller scale experiments in California, Florida, Mexico, and Canada.

Over the past five years, as a result of a series of national and international biomass burn field experiments, we have gathered considerable information on all of the parameters shown in equations (12.1) to (12.3), including B, a, b and ER.

As previously noted, the major uncertainty in our current understanding of the global impact of biomass burning concerns the parameter A in equation (12.1), the total land area burned annually. The Global Emissions Inventory on Biomass Burning, a research activity of the IGBP/IGAC/GEIA, held an open meeting during the Chapman Conference on Biomass Burning and Global Change in March 1995 to discuss the uncertainties and unknowns in the development of a global emissions inventory on biomass burning to better assess the environmental impact of global biomass burning. There was unanimous agreement that the major uncertainty was accurate information on the spatial and temporal distribution of burning in the world's ecosystems.

At the present time, estimates of A are based on one of two sources of information—statistics of the occurrence of fires in each country tabulated and compiled by that country and collected and disseminated by the United Nations Food and Agricultural Organization (FAO) on an annual basis, and on certain satellite measurements, such as measurements obtained with the Advanced Very High Resolution Radiometer (AVHRR) on the NOAA polar orbiting operational meteorological satellites. The satellite systems used to deduce information about biomass burning were originally developed for other purposes and are not ideal to deduce information about the area consumed during burning. Each existing satellite system has its own set of unique problems that make it difficult to determine the needed information on the spatial and temporal variation of biomass burning. Problems include very poor or no coverage of the high latitude boreal forests, poor temporal and spatial coverage of the burned areas in the tropical and temperate regions, and low saturation temperature for the thermal channels, which makes detection of fires ambiguous.

The question of the rate of deforestation and the use of satellite measurements to deduce global burning was addressed in a National Academy of Sciences (1990) report (p.117):

The rates of deforestation vary widely, mainly because countries use different survey procedures.... and because satellite images of the entire globe are expensive and difficult to analyze. Simply put, we do not have reliable and up-to-date information on how much of the earth's surface is covered by forests and how fast it is being cut down.

It is ironic that there has never been a dedicated space experiment to study the geographical and temporal distribution of biomass burning. To determine the spatial and temporal distribution of biomass burning on our planet, the NASA Langley Research Center has proposed a dedicated satellite experiment, FireSat.

FireSat Science Mission and Requirements

A few very broad design goals have served as the underlying basis for the FireSat design and enhanced the science mission objectives. In order to make FireSat a smaller and less expensive instrument when compared to conventional imaging instruments, the latest technological advances have been carefully examined to determine their suitability for space-based application, to assess cost, and to minimize risk. From the outset, FireSat has been intended to incorporate new technology and provide insights and lessons that would be useful for the design of future Earth imaging systems. To minimize risk, the new technology chosen will facilitate an instrument design that requires no moving parts for operation. In addition to hardware advances, the FireSat design has the ability to provide a calibrated dataset for Earth interdisciplinary studies. FireSat will also provide a cloud-filtering algorithm and a real-time downlink to ground stations within the spacecraft's horizon. Cloud-filtering the data will reduce the data volume that requires archiving and, by only retaining the desired clear-sky data set, shorten computer processing time of the archived imagery.

With these fundamental instrument design goals always in mind, the FireSat instrument concept will meet the goals of a well-defined science mission. This mission is to develop a continuous global mapping of vegetation fires in forests (tropical, temperate, and boreal), grasslands, and agricultural fields. Further, after the analysis of the data set provided by FireSat, there will be the ability to better assess the environmental impact of fires on the atmosphere, climate, and land to determine the couplings between fire and global change. Given the established philosophy of this top-level definition of the science mission, the science mission has been further detailed in a top-down fashion in order to arrive at those aspects of the science mission that will drive the FireSat instrument design. The FireSat science mission will need to be about five years in length in order to gain better insight into the interannual variations of fire. During the five-year mission, FireSat data would be reduced to (1) map active vegetation fires globally and determine fire-front temperatures, (2) map the geographical extent of global vegetation fires, (3) monitor vegetation stress for indicating fire susceptibility, (4) estimate the gaseous and aerosol emissions of the fires released to the atmosphere, and (5) identify aerosol source regions for comparison with ground- and space-based radiation measurements.

Given the definition of the science mission above, a set of first-order requirements that drive the instrument design has been determined. In order to map active fires and gage the fire intensity along a firefront, the saturation of the thermal channel used for active fire detection should be at a temperature of at least 1000°K. Further, in order to spatially depict an active firefront and to accurately assess the area that has been burned, the spatial resolution of the instrument at nadir should be no greater than 500×500 m. In this case, the spatial resolution is defined as the actual

Table 12.1 Science products

Product	Archived format	Wavelengths (μm)	Unit of measure
Active fires	Geolocated fire pixels	3.7, 8.5	Temp. (K)
Burn scars (clear sky)	Geolocated burned area pixels and regional area burned estimates	0.5, 1.0, 2.2, 8.5	Area > 25 km^2 (5%)
Burn scars (through smoke aerosol)	Map recently burned areas	3.7, 8.5	Area > 25 km^2 (5%)
Smoke aerosol	Geolocated smoke-filled pixels	0.5, 1.0, 2.2, 8.5	Over-water optical depth
Vegetation changes	Geographical map	1.0, 1.65, 2.2	TBD
Surface temperature	Geographical map	8.5	Temp. (K)
Cloud masking	Bitmap	0.5, 1.0, 1.38, 8.5	NA[a]

a. Not applicable.

surface area contributing energy to an individual observation including all instrument effects that would broaden the instantaneous field of view (IFOV) geometric footprint. In order to maximize usefulness to the scientific community and provide a quantitative means for evaluating the data, the data set will be calibrated. Calibration will be evaluated over time in order to characterize instrument degradation and stability, which can impact the science mission. The geographical range of fires on Earth is roughly between 60°S to 75°N latitude. Given the north/south extent of fires to be monitored and a desire to minimize the difference in solar/spacecraft geometry between successive days, a sun-synchronous orbit of 830 km has been selected for the FiresSat design studies. Since meteorological conditions are more conducive to early afternoon fire development, a trade-off has been made between observing fire activity at its diurnal peak and the potential of increasing cloud cover in the afternoon when selecting a mid-afternoon local overpass time. Even though afternoon cumulus fields may obscure much of the ground, smoke aerosol can still likely be identified in the cloud gaps and permit establishing time histories of regional fire episodes. For the geographical mapping of fire activity and for co-registering images acquired from different orbital passes, the pointing knowledge of the instrument should be no worse than 250 m to assure that the imagery can be accurately mapped.

Only one of the science mission requirements has been relaxed somewhat. The requirement was that FireSat observe the entire planet at least twice daily. FireSat does only marginally worse than this by missing a narrow gap between successive orbits, in a 24-hour period, only in tropical regions. These gaps would be filled every other day. For the northern forest ecosystems there is at least four times daily coverage.

The reason for relaxing the requirement is to avoid imposing additional demands on the optical and detector designs when the gain in geographic coverage does not really merit the additional costs. The refined science requirement now specifies that the FireSat instrument has a 96-degree total field of view. Beyond this point the footprint size begins to grossly overlap with adjoining footprints by as much as 50%. The net result is blurring the imagery beyond the point that anything other than questionable interpretation of the data could result.

Based on the science goals, a fundamental set of science products has been outlined (table 12.1). In order to produce the specified science products, spectral bands have been chosen. The selection of the spectral bands has placed additional requirements on the FireSat instrument's conceptual design. Active fire detection will primarily be accomplished with a spectral channel centered at 3.7 μm. The 3.7-μm channel will have the high saturation of no less than 1000°K with sensitivity of no greater than 1°K so that small surface temperature changes can be detected through the smoke of active fires, which is indicative of burned surface. The 8.5-μm channel will be used in conjunction with the 3.7-μm channel for isolating active fires, but its prime role is to provide the surface temperature with peak efficiency over the range of land and ocean temperatures. The accurate detection of burned scars is accomplished by evaluating the spectral response in bands centered on 0.5, 1.0, 2.2, and 8.5 μm. The monitoring of smoke aerosol would use the same set of channels. Monitoring changes in vegetation would principally be through the combined use of bands centered at 1.0, 1.65, and 2.2 μm. A 1.38-μm channel has been added, and when used in combination with the other channels, will be used for masking clouds out of the imagery. In total, two infrared channels and five

Table 12.2 Comparisons of instruments

Platform	Resolution at nadir[a] (m)	Global coverage (days)	Thermal channel saturation for active fire detection[a] (°K)	Spatial coverage	In-flight cal.	Cloud-filtered archive
FireSat	**260**	2	**1100**	Global	Yes	Yes
MODIS	500/**1000**	1	**500**	Global	Yes	No
AVHRR	**1000**	1	320	Global	No	No
DMSP	600(2700)	1	No channel	Global	No	No
Landsat	30	14	No channel	Global	No	No
GOES	1000 (**4000**)	.02	**335**	Hemispheric	Yes	No

a. Bold values pertain to active fire detection.

visible-to-near infrared channels have been specified to meet the science objectives and are included in the FireSat instrument conceptual design.

This is a good point to assess how the FireSat design concept compares to other Earth imaging instruments, both operational and under development. The FireSat instrument, with its focused mission of fire detection and monitoring, exceeds the resolution (the IFOV footprint), and the thermal saturation for detecting active fires is above that of any other instrument reviewed. A summary of this comparison is given in table 12.2. The bold-valued figures are the characteristics than pertain to active fire detection. Other instruments saturate below 500°K in the active fire detection band (3.7 μm) and the IFOV footprint of the other instruments is, at a minimum, 1000 m for active fire detection. The Defense Meteorological Satellite Program (DMSP) Operational Linescan System (OLS) instrument and Landsat have been included in the comparison even though they contain no channels for active fire detection. The DMSP OLS instrument can be used to assess fire activity at nighttime by monitoring the low-light emissions in the visible part of the spectrum. Landsat has been included because its spectral band selection offers the potential for monitoring and assessing the areal extent of burned vegetation. Even though Landsat can be used to monitor burn area, the repeat coverage of every 14 days and very high spatial resolution are not practical for continual continental scale monitoring. The FireSat instrument concept has been developed by making several trade-offs in regard to spatial resolution, repeat coverage, and costs. As demonstrated by this comparison, the FireSat instrument has been optimized for global fire monitoring and is unique among the instruments.

In addition to defining the channel selection, it is necessary to further define the minimum acceptable precision of each of the channels that will meet the science mission requirements. Multiple-scattering radi-

ative transfer calculations were made while varying the solar zenith angle, the viewing angle, the atmospheric aerosol type and loading, the atmospheric model, and target characteristics. From these radiative simulations a reasonable value for the maximum and minimum detectable radiance was estimated for each of the spectral channels. Also, the smallest radiance change required to achieve the science mission goals was derived. As a result of these simulations, the quantization requirements were specified for each channel. In practically all of the channels, with the exception of 2.2 μm and 8.5 μm, 10-bit data is required. For the 2.2-μm and 8.5-μm channels, the required quantization is nine and seven bits, respectively.

Radiance Modeling for FireSat

A modeling and simulation effort, utilizing the MODerate resolution atmospheric TRANsmission model (MODTRAN) and other atmospheric radiative transfer software, was carried out to calculate the levels of total radiance at the aperture of the FireSat instrument, as the first step in the end-to-end instrument design process. This effort consisted of the following steps:

1. Determining which regions of the electromagnetic spectrum, visible and infrared, should be used to glean information on fire properties, while reducing interference from the atmosphere to a minimum in what is fundamentally a surface-reconnaissance task.

2. Determining optimal bandpasses within the chosen spectral regions, to further minimize atmospheric effects (e.g., avoid prominent water vapor absorption lines) as much as possible, consistent with maintaining adequate signal levels for sensitive measurements. The findings of steps 1 and 2 resulted in the definition of the seven FireSat channels described in the previous section.

Table 12.3 Parameters modeled for each FireSat simulation scenario

Location (latitude/longitude)

Season, date

Model atmosphere (either built-in or from rawinsonde data): temperature, pressure and molecular constituent concentration profiles, from sea level to 100 km altitude

Atmospheric aerosol content/scattering phase functions, using built-in or user-provided models

Surface temperature and spectral emittance

Solar illumination and sensor view geometries

Sensor and terrain altitudes

Solar zenith angle

Nadir view angle

Relative azimuthal angle (between sensor line-of-sight and solar azimuth direction)

Cloud cover (type and altitude)

Smoke particle-size distribution and concentration

Smoke transmission

Fire temperature and emissions models

Flame radiance and transmittance

---

3. Determining through simulations the range of signal levels likely to be encountered, under the full anticipated range of climatic, seasonal, atmospheric, sun illumination, and viewing geometry conditions.

4. Determining the sensitivity of measurement to changes of surface temperature, reflectance, aerosol content, solar illumination and viewing geometry, and the presence of clouds.

5. Developing a fire radiance model for assessing the relative contributions of the flame- and hot-surface components of the fire signature.

6. Modeling the transmittance and scattering properties of fire-generated smoke.

To perform the simulations, the parameters in table 12.3 were specified for each FireSat scenario. The PCModTRAN model (ONTAR 1992), a PC computer-based commercially available implementation of the U.S. Air Force Phillips Laboratory's MODTRAN model, was the main simulation tool used in this study. It uses a two-parameter band model of atmospheric transmittance, with an as-fine-as 2 cm^{-1} spectral resolution over the spectral range from ultraviolet to microwave. Calculations of transmittance, as well as of single- or multiple-scattered atmospheric and surface radiance components, are readily carried out with this model, which is in widespread use. MODTRAN is a development of the long-used LOWTRAN atmospheric radiative transfer model (Kneizys et al. 1988), and offers increased accuracy and finer spectral resolution than the latter. Other models used in the current simulations included BACKSCAT (Hummel et al. 1992), and HITRAN-PC (Killinger and Wilcox 1992). BACKSCAT, also developed under Phillips Laboratory sponsorship, is normally used to model laser back-scattering, but was used here in developing a preliminary model of smoke transmittance. HITRAN-PC, a line-by-line gas radiative transfer model developed at the University of South Florida, was used in calculating fire flame transmittance and absorption.

The main emphasis in the simulation effort was in applying PCModTRAN to calculate, for each FireSat scenario, the total radiance at the top of the atmosphere. This quantity is the input signal to the instrument; its computation is the first step in the end-to-end signal-modeling process. Calculation of the total spectral radiance and its subcomponents was carried out for several scenarios and for all seven FireSat channels.

For ease in comparing signal levels among the channels all initial simulations, except where indicated in the discussion, were performed using the following conditions:

- a 1976 Standard Atmosphere, which is broadly representative of mid-latitude spring/fall climatic conditions
- a rural haze aerosol model with 23-km visibility at the surface and a Mie scattering function
- 830-km satellite altitude, viewing at nadir
- 45-degree solar zenith angle
- 2-cm^{-1} spectral resolution
- no clouds or smoke
- 280°K surface temperature

Table 12.4 is a summary of some results for the seven channels, for this simulation condition-in-common. The table lists the bandpass for each channel, the total radiance in this bandpass incident upon the instrument aperture, and the surface reflectance value used in the simulation. The surface reflectances used for each channel were modeled after those in Bowker et al. (1985). The bandpass-averaged clear-sky transmittance from sea level vertically to satellite altitude is also listed in the table. It is noted that this transmittance value varies from nearly zero for channel 3, the cloud-detection channel, to over 0.9, for channels 4 and 5.

Figures 12.1 through 12.6 are samples from the simulation results.

Table 12.4 Nominal radiance and surface reflectance values

Channel #	Bandpass (μm)	Radiance, W/(cm^2*Sr)	Surface reflectance	Transmittance (average)
1	0.45–0.55	5.727E-4	0.07	0.585
2	0.98–1.07	1.077E-4	0.07	0.855
3	1.35–1.40	3.652E-7	0.60	0.004
4	1.60–1.74	1.615E-4	0.25	0.909
5	2.10–2.30	6.102E-5	0.20	0.911
6	3.60–3.80	3.970E-6	0.02	0.848
7	8.00–9.00	5.952E-4	0.02	0.712

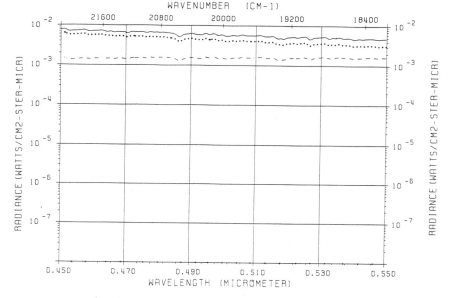

Solid line (____) total radiance

Dashed line (––––––) reflected radiance component

Dash-dot line (–·––·––·) thermal component: includes both surface thermal radiance and atmospheric thermal radiance contributions

Dotted line (.......) sum of single- and multiple-scattered radiance components

Figure 12.1 Total radiance and components calculated by PCMod TRAN for FireSat channel 1 (0.45–0.55 μm). Simulation conditions: 1976 Standard Atmosphere, nadir view, 23-km rural aerosol, 45-degree solar zenith angle, 0.07 surface reflectance, 280°K surface temperature. Reflected component dashed curve; atmospheric scattering component (dotted curve); thermal component—atmospheric plus surface—(dot-dashed curve, negligible for this case and lying on the x-axis). Total radiance (sum of components) (solid curve)

Except where indicated, all the figures use the same logarithmic ordinate range, extending from 1.0×10^{-8} to 1.0×10^{-2} spectral radiance units (W/(cm^2* Sr* μm)). Plotting all the components together is instructive for determining which components are dominant and which are negligible in a given wavelength region, and is very useful in deciding which parameters need to be varied, and which can be ignored in sensitivity studies.

Figure 12.1 is for FireSat channel 1 (0.45–0.55 μm). The results for this channel in the visible region show that most of the radiance from this low-reflectance scenario (typical of vegetated surfaces) stems from atmospheric scattering. The reflected component is secondary in importance, at only one half or less of the magnitude of the scattered component, and the thermal component is negligible.

Figure 12.2, for channel 3, the FireSat cirrus cloud detection channel (1.35–1.40 μm), shows a dramatically different result. Here, very strong absorption features (water vapor) are evident in the reflected component; this strong absorption is also obvious in table 12.4, where the average transmittance in the channel is seen to be only 0.004. With the thermal component still negligible in this band, the total radiance is almost totally dominated by the atmospheric-scattered component. Because of the very strong atmospheric absorption, the spectral radiance is only around 1.0×10^{-5} W/(cm^2*Sr*μm), two to three orders

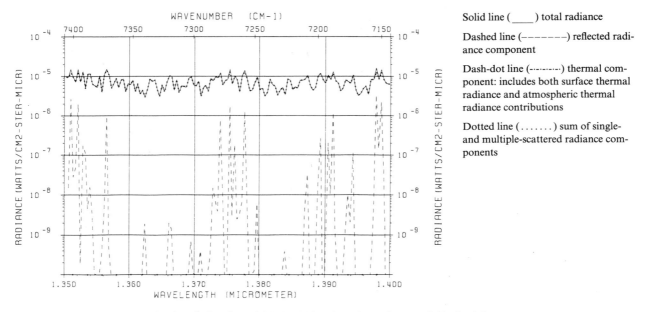

Figure 12.2 As in figure 12.1, but for FireSat channel 3 (1.35–1.40 μm); surface reflectance 0.60, cloud-free

of magnitude less than for channels 1 and 2. Figure 12.3 shows the result when a 1-km-thick cirrus cloud is inserted at 10 km altitude. In this case, the level of total spectral radiance is increased by one order of magnitude from that in Figure 12.2. Thus, the existence of a high value of radiance in a pixel in this channel means that high-altitude clouds are present. Simulations for thinner cirrus, not shown here, also show strong increases of radiance from the no-cirrus value. Therefore, channel 3 is shown to be a sensitive detector of high clouds' presence.

The radiance in the FireSat fire-detection channel (channel 6, 3.60–3.80 μm) was modeled under normal (280°K) and hot (600°K) surface-temperature conditions, both for cloud-free and overcast cloud scenarios. Figure 12.4 shows that, even at the baseline 280°K surface temperature condition, the thermal component dominates. The reflected component contributes about one third of, and the scattered only around 3% of the total radiance. Figure 12.5 models a cloud-free condition with a 600°K hot surface present, as in a low-intensity fire. In this case, the total radiance level is increased by around three orders of magnitude from that in figure 12.4; the thermal component dominates even more than in figure 12.4 (the ordinate maximum is increased to 1.0×10^{-1} radiance units for figure 12.5). Figure 12.6 simulates a stratus cloud deck overlying the 600°K hot surface. The total radiance in this cloud-obscured case is much reduced from that of the

cloud-free case (figure 12.5), but is about 80% above that of the normal temperature case (figure 12.4). Table 12.5 presents more detail on the effects of three cloud types (stratus, cumulus, and cirrus) and different surface reflectance values on the level of total radiance, for the common simulation condition. It shows that the level of radiance for the stratus-covered case is approximately equal to that for a cloud-free case with a 298°K surface temperature and 0.02 surface reflectance, or to that of a cloud-free case with a 280°K surface temperature and 0.11 surface reflectance. This result points out the need for a surface temperature estimate, and for a method of detecting clouds. The table shows that, as expected for thick clouds, the radiance is invariant with changes in underlying surface temperature and reflectance, for stratus and cumulus overcast conditions. Thus, surface temperature cannot be assessed through a stratus or cumulus cloud. However, the cirrus simulations do show a change with surface temperature and reflectance variability. This may indicate that it may be possible to detect hot surface through thin cirrus clouds. Both these findings again demonstrate the need for having a cloud-detection channel to indicate unambiguously when clouds are present.

Simulations for nighttime cloud-free viewing conditions were also carried out for all FireSat channels. The results, not shown here, demonstrate that for the thermal-sensitive channels 6 and 7 the surface thermal

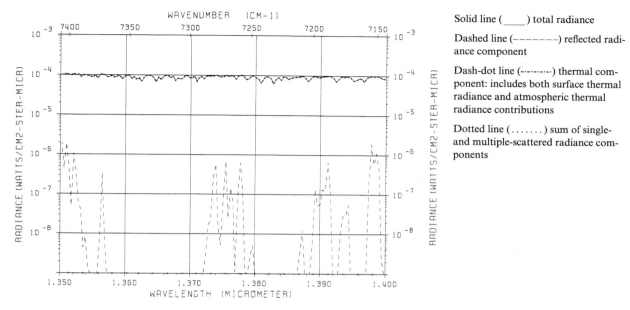

Figure 12.3 As in figure 12.2, but with 1-km-thick cirrus cloud present, with base at 10 km altitude

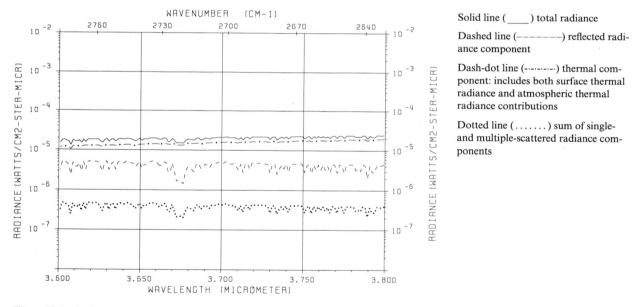

Figure 12.4 As for preceding figures, but for FireSat channel 6 (fire-detection channel, 3.60–3.80 μm); surface reflectance 0.02, surface temperature 280°K

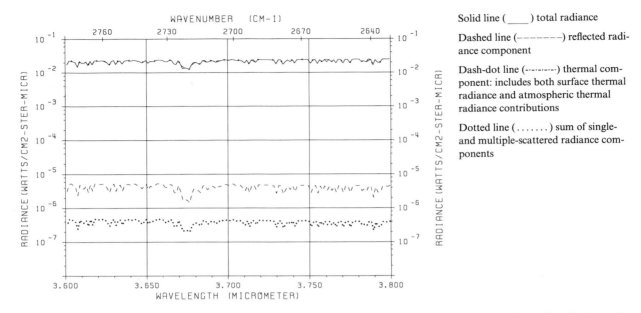

Figure 12.5 As for figure 12.4, but with 600°K hot surface present; surface reflectance 0.02. Note that total radiance signal is three orders of magnitude higher than for figure 12.4

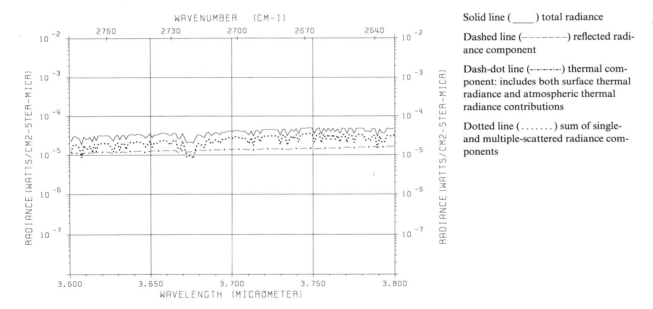

Figure 12.6 As for figure 12.5, but with MODTRAN default stratus deck present, with cloud base at 0.33, top at 1.0 km altitude

Table 12.5 Simulated total radiance levels[a] in FireSat channel 6[b], for various types[c] of overcast cloud cover conditions

T_{sfc},°K	Sfc. refl.	No cloud	Stratus	Cumulus	Cirrus
280	0.02	3.970E-6	7.339E-6	1.233E-5	4.900E-6
298	0.02	7.481E-6	7.339E-6	1.233E-5	7.942E-6
280	0.11	7.452E-6	7.339E-6	1.233E-5	7.799E-6
600	0.02	4.379E-3	7.339E-6	1.233E-5	3.795E-3
600	0.11	3.981E-3	7.339E-6	1.233E-5	3.450E-3

a. Total radiance units: $W/(cm^2 * Sr)$
b. Viewing conditions: Nadir, 45° solar zenith angle
Atmosphere: 1976 Standard, with 23-km visibility rural haze
Date: 3 April (Earth-solar distance = 1 a.u.)
c. Cloud models (per MODTRAN)
Stratus: base 0.33, tops 1.0 km
Cumulus: base 0.66, tops 2.0 km
Cirrus: 1.0 km thick, base at 10.0 km
Extinction coefficient: 0.14 km^{-1}

component so dominates the total radiance signatures that fire scenes can be detected readily under clear-sky nighttime conditions.

Calculations similar to those described above were performed to bound the range of expected total radiance signals. The total radiance values were also scaled to photon flux values for use in choosing detectors during the instrument design process. (Radiance is scaled to photon flux by multiplying the radiance value by 5.031447×10^{18} times the wavelength in μm.) Selected results are given in table 12.6, to give an indication of the range of signal levels to be encountered in nadir viewing. (All simulations used a 1976 Standard Atmosphere with a 23-km rural haze model.) The tabulation is in terms of condition simulated (surface reflectance for all channels except for channel 3, where cloud cover condition is described), surface temperature, and solar zenith angle (SZA).

While these results are adequate for assessing the major contributors to the total radiance, and for bounding the range of signals to be encountered in the FireSat channels for instrument design purposes, further study was deemed desirable to better understand the modeling of fires themselves. Therefore, modeling and simulation efforts were initiated in two areas—modeling the emission of fires, including that from flames as well as the hot surface, and modeling the transmittance of smoke. Preliminary results show that, for the cases simulated, the radiance from a fire scene stems predominantly from the heated surface; the flame radiance is small by comparison. Therefore, just simulating the radiance emitted by the fire-heated surface gives a good approximation (in smoke-free conditions) to the overall thermal signature, for

instrument design purposes. A preliminary smoke-transmittance model was also developed, using a smoke model given in BACKSCAT. Initial results show, as expected, that smoke transmittance varies widely with sensing wavelength, smoke particle size distribution, and concentration profile. It is planned to extend these simulations, using other smoke models that may be available, to additional transmittance and scattering studies, and to perform more detailed sensitivity studies in support of the FireSat design and data-interpretation process.

Systems Engineering Application to the FireSat Study

The systems engineering process has been applied from the inception of the FireSat study. The study process began with the establishment of the top-level goal for the FireSat instrument and continued with development of lower-level, supporting goals. These supporting goals were broken down further until system performance requirements were established. The criteria were met for inclusion in the list of performance requirements when a goal was deemed quantifiable, verifiable, and objective. This process developed a hierarchy of project goals that tied performance requirements directly to the top goal of the system. Technical as well as programmatic requirements were included.

The system performance requirements developed for FireSat included both constraints and performance measures. Constraints were defined as requirements whose numeric targets had to be obtained, but no premium was placed on exceeding the target values. Performance measures were those requirements in which more (or less) was considered to be better and would justify the expenditure of some other resource. Thus, the constraints were used to bound the problem and the performance measures to perform system-level trade-offs. Only after the development of these performance measures was consideration given to possible system architectures. System variations were constructed and each characterized by projections of the performance values for that option. Any option failing to meet established constraints was eliminated. This led to the most-preferred set of options currently under consideration.

FireSat Orbit Design and Coverage Analysis

Integral to the FireSat mission is the design of the orbit. There are three key factors in the design of the orbit: instantaneous spatial, long-term spatial, and

Table 12.6 Selected results from signal range modeling for FireSat

Channel	Condition (refl.)	SZA, deg.	Surface temperature (°K)	Radiance W/(cm²*Sr)	Photon flux/ (s*cm²*Sr)
1	Nominal (0.07)	45	280	5.73E-4	1.44E15
	Max. radiance (0.9)	0	280	4.91E-3	1.24E16
	Min. radiance (0.02)	45	280	4.19E-4	1.06E15
2	Nominal (0.07)	45	280	1.08E-4	5.58E14
	Max. radiance (0.9)	0	280	1.73E-3	8.95E15
	Min. radiance (0.05)	45	280	8.92E-5	4.62E14
3	Min. radiance (clear scene)	45	a	3.65E-7	2.54E12
	Max. radiance (1 km-thick cirrus cloud)	0	a	5.21E-6	3.62E13
	0.5 km-thick cirrus cloud	45	a	1.48E-6	1.03E13
4	Nominal (0.25)	45	280	1.62E-4	1.40E15
	Max. radiance (0.6)	0	280	5.54E-4	4.79E15
	Min. radiance (0.05)	45	280	3.46E-5	3.00E14
5	Nominal (0.20)	45	280	6.10E-5	6.74E14
	Max. radiance (0.40)	0	280	1.76E-4	1.94E15
	Min. radiance	45	280	6.76E-6	7.48E13
6	0.1 Refl.	45	260	5.45E-6	1.02E14
	0.1 Refl.	45	300	1.08E-5	2.01E14
	0.1 Refl.	45	340	3.30E-5	6.15E14
	0.1 Refl.	45	600	4.03E-3	7.49E16
	0.1 Refl.	45	800	2.04E-2	3.81E17
	0.1 Refl.	45	900	3.53E-2	6.57E17
	0.1 Refl.	45	1000	4.47E-2	1.02E18
7	0.1 Refl.	45	260	4.16E-4	1.78E16
	0.1 Refl.	45	280	5.70E-4	2.44E16
	0.1 Refl.	45	300	7.74E-4	3.31E16
	0.1 Refl.	45	320	1.04E-3	4.43E16
	0.1 Refl.	45	340	1.36E-3	5.80E16

a. Channel 3 (1.38 μm) is opaque to the surface.

periodic-temporal coverage. The instantaneous spatial coverage is a simple function of mission specifications and instrument performance. FireSat requires a spatial resolution size of less than 500 m. The physical limitations explained in the Instrument Concept section below yield an optimal altitude of 830 km. The chosen altitude is a trade between instrument performance (resolution and signal strength) and instantaneous spatial coverage.

The long-term monitoring mission of FireSat (globally between 75°N latitude and 60°S latitude) drives the minimum orbital inclination to greater than 60°. The long-term monitoring issues, along with the science request for constant sun angles (constant over a period of 30 to 70 orbits) drive the design orbit's inclination to a sun-synchronous inclination at an altitude of 830 km.

Another requirement is to be able to view as much of the area of regard (75–60° latitude) as possible while the ground is lighted and in a 24-hour period. The frequency with which fires occur escalates during a hemisphere's summer season. Choosing an ascending nodal crossing time (for an altitude of 830 km circular sun-synchronous orbit) of 1400 local time will afford a few more minutes of light on the downward pass, thus increasing the covered area during the prime fire season.

The FireSat coverage analysis presented here was done for an instrument with a full field of view (FOV) of 96.0°. Table 12.7 shows the initial spacecraft ephemeris used for all of the analysis presented here.

Table 12.7 FireSat ephemeris

Altitude (km)	830
Inclination (deg)	98.730
Eccentricity	0.0
Longitude of Asc Node (deg)	−149.65
True anomaly (deg)	0.0
Epoch (yyyy/ddd/hh : mm : ss)	1998/172/00 : 00 : 00
Sensor FOV (deg)	96.0
Minimum ground elev. (deg)	32.8755

The coverage of the design orbit is illustrated graphically in figures 12.7 through 12.9. Each of the figures is plotted on an equidistant cylindrical projection of the earth. For the purposes of the analysis herein, the term *lighted ground swath* refers to a satellite-to-sun angle of greater than 0.02 degrees, which in turn implies that at the subsatellite location the local sun angle is also a small positive number (after local sunrise). Figure 12.7 shows the entire earth (180 to −180° longitude and 90 to −90° latitude) along with FireSat's sensor swath for only the lighted portions of 24 hours worth of ephemeris. The swaths cover from approximately 60°S latitude northward to nearly over the northern pole and then descend to approximately 58°N latitude. There are 15 swaths in this figure. This includes the stunted swath which starts at the initial point of the ephemeris (see table 12.7). The gaps in the coverage are from −41 to 41° latitude in height by a maximum of approximately 5° in longitude at the equator. The coverage gaps have two striking results. First, the areas below and above −41 and 41° latitude respectively are covered 100% in a 24-hour period. This is significant because the number of viewing opportunities in the temperate-boreal areas of the globe is substantially less than those in the tropical regions (almost daily near the equator). Second, the gaps that appear in 24 hours of coverage are completely gone when the time limit for the evaluation is lengthened to 48 hours (because of nodal regression of the orbit). Also, the number of viewing opportunities ranges from two to eight above 41°N latitude.

The scale of figure 12.8 has increased; here the bounds of the plot are from −180 to −20° longitude and 0 to 75° latitude (again, only lighted ground swaths are plotted over a cylindrical projection of the earth). The continental United States is approximately centered in the plot. Two of the gaps in coverage extend into the United States. The relative sizes of the gaps are small with respect to the size of the United States. One can conclude that there is a low probability of any particular area of the United States being missed in any single day (24 hours). Furthermore, because full global coverage is obtained in two days (48 hours), no portion of the globe would be without coverage in that period. Also, from about 59°N latitude there are four opportunities to view any one location.

The last orbital analysis figure (figure 12.9) shows the ground swaths for a single spacecraft for 24 hours of ephemeris. The ground swaths shown in this figure are for both local daylight and local night. The coverage gaps are very small—only about 15° in latitude by 5° in longitude and occur between −30 and 30° latitude.

The orbit designed for FireSat (table 12.8) meets most of the requirements as set forth in the science section of the chapter. The areas where the orbit design falls short are only minor and are easily solved when a more realistic time value is used. The methodology is as follows: The near-infrared (NIR) and the middle-infrared (MIR) channels of the instrument do not require the presence of daylight. Therefore, for the purposes of fire detection and short-term monitoring of "hot spots," the solid 24-hour ground swath model can be considered valid. For long-term monitoring of "fire scars," the 48-hour lighted ground swath is more than sufficient.

FireSat Measurement Technique

FireSat obtains data by imaging the surface of Earth using a technique similar to pushbroom radiometry. In pushbroom radiometry, a linear array of detectors (an array with only one dimension) is moved over the surface to be imaged. Periodically, the detectors are sampled and the resulting data are stored. If the sampling occurs at the correct time interval, the data from sample to sample are contiguous and a map of the surface can be generated by reassembling the data contiguously. In FireSat, the linear arrays of the pushbroom radiometer are replaced by areal arrays, that is, arrays that have two dimensions and can obtain an image which is itself an area map of the surface scanned. Since the areal array has two dimensions, much more surface area is sampled in a single sample, and much more data is contained in a single sample. On the other hand, the time interval between samples is much greater because the spacecraft has to move farther to get to the proper location for the next sample. The net data rate is the same for both techniques, but the areal array allows time between samples to do any required on-board processing of the data.

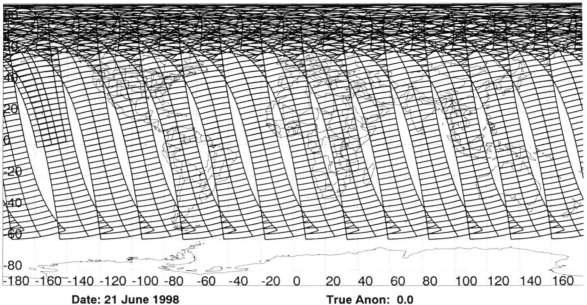

Date: 21 June 1998
Altitude: 830 km
Inclination: 98.730deg
Ecc: 0.0
Lon Asc: –149.65

True Anon: 0.0
Sensor:
Half Cone: 48.0 deg
Simulation Length: 24 hrs
Simulation Number: 1

Figure 12.7   FireSat single spacecraft lighted ground swath

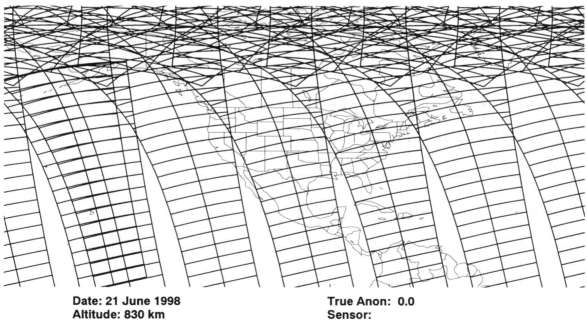

Date: 21 June 1998
Altitude: 830 km
Inclination: 98.730deg
Ecc: 0.0
Lon Asc: –149.65

True Anon: 0.0
Sensor:
Half Cone: 48.0 deg
Simulation Length: 24 hrs
Simulation Number: 3

Figure 12.8   FireSat single spacecraft lighted ground swath (North America)

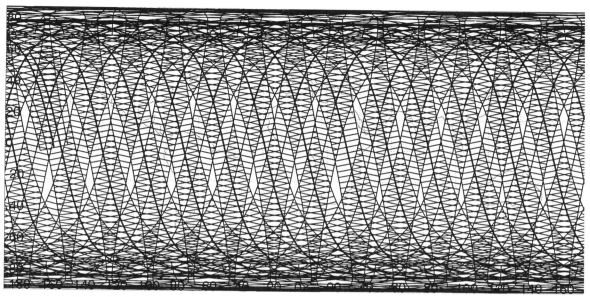

Date: 21 June 1998
Altitude: 830 km
Inclination: 98.730deg
Ecc: 0.0
Lon Asc: −149.65

True Anon: 0.0
Sensor:
Half Cone: 47.0 deg
Simulation Length: 24 hrs
Simulation Number: 4

Figure 12.9   FireSat single spacecraft 24-hour ground swath

The FireSat instrument is a hybrid instrument with both multichannel radiometer and imager characteristics. There are seven areal arrays (radiometer channels) covering a range of wavelengths from 0.5 μm to 8.5 μm. For optical simplicity the three longest wavelengths are imaged on one focal plane and the remaining four wavelengths share a second focal plane. The focal plane consists of an assembly of areal arrays arranged very precisely with respect to one another, called a mosaic, and appropriate spectral band-pass filters. The seven radiometer channels each have spectral bandwidths on the order of a few tenths of a μm. Each spectral bandwidth was chosen to maximize the signal content without undue interference from the intervening atmosphere. Each array contains the detector cells, or pixels, and the necessary readout electronics. The FireSat instrument composite field of view (FOV) is composed of images from three such mosaics concatenated electronically to form a single image. Some of the principal FireSat experiment parameters are shown in table 12.8.

FireSat Instrument Concept

The proposed baseline configuration for the FireSat instrument is shown in figure 12.10. The instrument is composed of two instrument modules and a cryogenic cooler. The cryogenic cooler is mounted to the longwave instrument module to provide cooling for the focal plane to a temperature of about 80°K. Each instrument module contains three telescopes with their associated focal plane arrays. The center telescope points toward nadir while the outboard telescopes point off nadir at an angle of 32°. The composite field of regard formed by concatenating the images from the three telescopes is approximately 96°.

As FireSat orbits Earth at an altitude of 830 km, the field of regard sweeps across the surface in three slightly overlapping swaths. The geometry of the swaths and the variation of the footprint size is shown in figure 12.11. The width of the center swath on the surface is 480 km and the outboard swaths are each 772 km wide, for a total swath width of 2024 km. These values take into account the curvature of the earth, which falls away 80.9 km at the outer limits of the field of regard from a plane tangent to Earth at nadir.

The instantaneous field of view (IFOV) is constant at 0.3173 milliradians, but the size of the footprint of each pixel on the surface of Earth varies with the viewing angle off-nadir. At the center of the composite field of view the footprint is smallest, about 263 m²,

Table 12.8 Parameters of the FireSat experiment

Altitude	830 km	Small-Sat mission; five years or more.
Data frequency	Daily revisit; 75°N–60°S	Sun-sync. Equator crossing: ∼ 1400 h ascending
Spatial resolution	500 × 500 m	Experiment critical parameter
Footprint	263 × 263 m	Needed for > 30% MTF response at 500 m
IFOV per pixel	0.32 × 0.32 mr	
Total Field of View	3 × 32°	
Aperture	30 mm	Driven by channel 7 diffraction limit
Optical speed	f/3	
Optical efficiency	66%	
Ch. 1 0.45–0.55 μm	Detect smoke + particulate	Range: = 1 to 100%, also see table 12.1
Ch. 2 0.98–1.07 μm	Map recently burned area	Range: = 1 to 100%, also see table 12.1
Ch. 3 1.35–1.40 μm	Cirrus detection	Range: = 1 to 100%, also see table 12.1
Ch. 4 1.60–1.72 μm	Vegetation moisture	Range: = 1 to 100%, also see table 12.1
Ch. 5 2.1–2.3 μm	Fire scars	Range: = 1 to 100%, also see table 12.1
Ch. 6 3.6–3.8 μm	Active fire + scar w.smoke	Range: T= 260–1100°K, also see table 12.1
Ch. 7 8.0–9.0 μm	Surface temperature	Range: T= 240–340°K, also see table 12.1
Split-focal plane FP-1: ch.1–5 FP-2: ch.6 and 7	SW channels are 7000 pix wide. Ch.2–5, use 7 arrays of 1000 pix each, ch.1 uses two arrays each 3500 pixels; ch.6 is 630 × 190 pixels; ch.7 is 380 × 190 pixels; ch.1 and 2 at 300°K; ch.3, 4, and 5 at 150–180°K; ch.6 and 7 at 80°K.	
Array dwell time	5.8 sec (40.6 km/7 km/sec)	integration time may be adjusted over wide range Smart Sensor: adaptive-gain or instant-IFC
Data rate	2 Mbits/sec	TRW [RIM] flight-qualified recorder is capable of 22 Gbyte, 80–160 Mb/sec
Data volume	3.6 Gbits, or < 1 Gbytes	2 Mbits/sec down-loaded every 30/min
Pointing knowledge	0.15 × 0.15 mrad	Pointing is equal to IFOV/2
Spacecraft jitter	$\mu \pm 0.5$ degree/sec (3Í)	Equals IFOV/dwell time, or 0.3 mr/30/msec

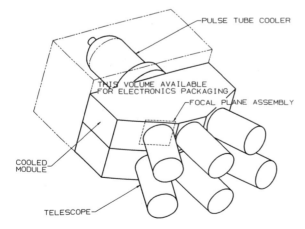

Figure 12.10 FireSat instrument configuration

but due to increasing slant range from the spacecraft to the surface with increasing look angle, the footprint size increases gradually, which is clearly shown in figure 12.11. Figure 12.12 shows the variation of the footprint multiplier with look angle, which increases to approximately 1.64 times the nadir value at 48° off nadir. In addition to experiencing a magnification due to increasing slant range, the footprint also becomes elongated due to the incidence angle being nonorthogonal to the surface. At 48° the footprint of a single pixel becomes roughly rectangular, measuring approximately 431 m × 795 m.

FireSat Optical Design

The telescopes used in the FireSat instrument are approximately 30 mm diameter, are f/3 speed, and are of lightweight design, each telescope weighing less

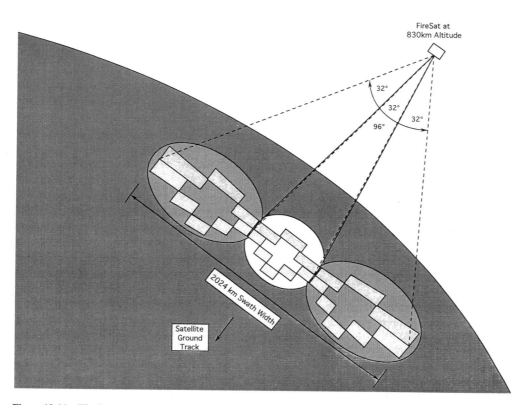

Figure 12.11 FireSat measurement geometry and effective footprint on Earth's surface

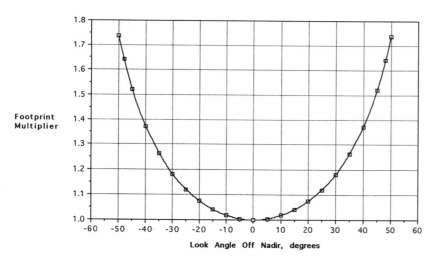

Figure 12.12 Footprint multiplier compared to look angle off-nadir

than 0.5 kg. Each telescope covers a 32° swath width with excellent spatial resolution. The optical performance, and radiation throughput, is optimized by using separate sets of telescopes for the visible/NIR and the MIR spectral regions. The multielement design is color-corrected over the wavelength range of interest. The spatial resolution of the visible/NIR telescopes is limited by aberrations, and their spatial resolution at Earth's surface is better than the 500×500 m spatial resolution science requirement. For the longer wavelength channels, diffraction effects predominate to degrade spatial resolution. Radiometer channel 7, which operates at 8.5 μm nominal wavelength, has an effective footprint of 500×500 m.

FireSat Radiometric Performance

The radiometric sensitivity of the detector channels in FireSat is determined by the time allowed for integration of photons from the scene that are incident on the pixels. The integration time required for the FireSat channels is on the order of milliseconds. Additional time is required for subsequent readout of the data from the arrays. The total time required is on the order of milliseconds; this time is called *one array data cycle time*. The arrays are sized so that the time it takes to move one array length along Earth's surface is significantly shorter than one array data cycle time. To this end the arrays are each 200 pixels in length, resulting in a 40-km-long footprint per array. Thus, at orbital speed, it takes 5.8 sec to move the array footprint a distance equal to its size. Note that each 5.8-sec period contains several array data cycle times. Thus, multiple data cycles are possible while the instrument travels the distance equal to the size of an array footprint. The FireSat experiment adds to its operational flexibility by using several different photon integration times per 5.8 sec period. Table 12.9 shows the predicted radiometric performance of channels 1 through 5. The table is a spreadsheet, showing several rows of information for these channels, culminating in signal-to-noise determination. The values shown in these rows bracket the instrument performance for the range of input radiance values expected for these channels. (The input radiance values are taken from table 12.6) Table 12.10 summarizes the performance of the 3.7 μm fire-detection channel. Table 12.10 shows parameters for a high-gain and a low-gain channel, which result from two integration times for the output of the same detector array. This use of two separate and different integration times per footprint sample has several advantages. A desirable very large dynamic measure-

ment range is possible without exceeding available detector full-well capacity. Also, the instrument can use analog-to-digital converters with a reasonable-to-implement 14-bit range. In fact, if three ranges are used (i.e., low-, medium-, and high-gain ranges), then the dynamic range for each channel allows the use of more common 12-bit analog-to-digital converters. Table 12.11 shows the performance of the 8.5-μm background measurement channel.

FireSat Radiometric Calibration

The biomass burning experiment requires the measurements in the 3.7- and 8.5-μm channels to be calibrated to $\pm 5\%$. The preflight primary calibration can draw on techniques that were proven for several previous space-based remote sensors such as LIMS, HALOE, and SABER.

In-flight radiometric calibration can be viewed as a check on both the offset (i.e., y-axis intercept) and the gain (i.e., slope) of the (straight-line) calibration curve. Typically the offset is checked by letting the sensor view space to provide a "zero-radiance" input to the instrument. Then, while still viewing space, a long-term stable source is input to check the instrument channel gain.

The in-flight calibration for the FireSat sensor presents special issues. The sensor is nadir-looking and to control reliability, weight, and cost, uses no moving optical parts. Thus no "space-look" can be obtained to check sensor output offset drifts and in-flight calibration must be performed while the sensor receives nonzero radiance from Earth. The FireSat instrument "smart sensor" capability, mentioned above, enables separating the instrument offset from Earth input radiance while viewing uniform Earth scenes that are slowly changing. It is worth noting that for the 3.7-μm fire-measuring channel the predicted offset due to total cavity radiation is less than 1%.

An in-flight check on the slope of the calibration curve (i.e., the monitoring of the end-to-end radiometer channel gain) can be accomplished using known radiance inputs to the instrument. FireSat uses diode lasers as sources for its in-flight calibration of the visible and NIR channels. For the longer wavelength channels the calibration sources will be thermo-electrically cooled, quaternary, type IV–VI diodes. Current research has identified candidate material systems for these diodes and commercially available devices are expected to be available in the near term (McCann 1995). The diode sources, in turn, are viewed by diode monitors to check the long-term source stability. The

Table 12.9 Predicted radiometric performance for FireSat channels 1 through 5

Channel	Condition[a]	Photon Flux/ $(S*cm^2*Sr)$	Photons[b]/ $(S*pixel)$	Photon/ pixel	Q.E., %	Signal e^{-1}	Dark noise, e^{-1}	ΣFPA noise, e^{-1}	Photon noise, e^{-1}	Total noise, e^{-1}	SNR
1	Nominal	1.44 E 15	4.10 E 7	1.64 E 5	25	4.10 E 4	4	16	203	204	201
	Max. Radiance	1.24 E 16	3.53 E 8	1.41 E 6	25	3.53 E 6	4	16	594	595	593
	Min. Radiance	1.06 E 15	3.02 E 5	1.21 E 5	25	3.02 E 4	4	16	174	175	173
2	Nominal	5.58 E 14	1.59 E 7	6.36 E 4	3.5	2.23 E 3	4	16	47	50	45
	Max. Radiance	8.95 E 15	2.55 E 8	1.02 E 6	3.5	3.57 E 4	4	16	189	190	188
	Min. Radiance	4.62 E 14	1.32 E 7	5.27 E 4	3.5	1.84 E 3	4	16	43	46	54
3	Clear Scene	2.54 E 12	7.24 E 4	2.90 E 2	60	1.74 E 2	173	176	13	177	1
	Max. Radiance	3.62 E 13	1.03 E 6	4.13 E 3	60	2.48 E 3	173	176	50	183	14
4	Nominal	1.40 E 15	3.99 E 7	1.60 E 5	65	1.04 E 5	173	176	323	368	283
	Max. Radiance	4.79 E 15	1.37 E 8	5.46 E 5	65	3.55 E 5	173	176	596	621	571
	Min. Radiance	3.00 E 14	8.55 E 6	3.42 E 4	65	2.22 E 4	173	176	149	231	96
5	Nominal	6.74 E 14	1.92 E 7	7.68 E 4	60	461. E 4	173	176	215	278	166
	Max. Radiance	1.94 E 15	5.53 E 7	2.21 E 5	60	1.33 E 5	173	176	365	405	328
	Min. Radiance	7.48 E 13	2.13 E 6	8.53 E 3	60	5.12 E 3	173	176	72	190	27

a. See Table 12.6 for definition of condition.
b. Phot/pix/sec: $[Wcm^{-2}sr^1]$. $[A\Omega\delta_{optics}]*[Ph/sec/watt] = [L]*[2.85\ E-8]*[5.0314\ E18*\lambda]$.
  Standard front-side illuminated silicon CCD at room temperature 27×27 pixels.
  NICMOS-type HgCdTe with switched FET multiplexer, at 150–180°K.
  Dynamic range of 9 bits is sufficient for all SW channels.
  Inherent CCD dynamic range = full-well electrons + noise electrons, or 175 000/20 = 8750, or ~13 bits.
  Integration time for channels 1 through 5 is 4 ms and Read-noise for channels 1 through 5 is assumed to be 15 electrons.

Table 12.10 Radiometric performance for FireSat channels 6a and 6b; bandpass: 3.6–3.8 μm

Scene, Temp/ε	L_{scene}, $Wcm^{-2}sr^{-1}$ a	$\Delta L/°K$, $Wcm^{-2}sr^{-1}$ a	Scene, e^{-1}/pixel b	Scene + cav, e^{-1}/pixel c	$\Delta e^{-1}/°K$, e^{-1}/pixel b	Noise, $(e^{-1}_{scene+cav})$, (1σ)	NEΔT, °Ks, (1σ)	SNR, (1σ)
High gain channel (260–500 kelvins)								
260/0.9	5.45E−6	5.00E−8	3.70E+4	1.01E+5	340	316	0.93	117
350/0.9	4.40E−5	1.27E−6	2.99E−5	3.62E+5	8636	602	0.070	497
500/0.9	1.10E−3	1.70E−5	7.48E+6	7.48E+6	115 600	2746	0.024	2734
Dynamic range 22 000 (14+ bits)								
Low gain channel (450–1100 kelvins)								
450/0.9	4.68E−4	8.90E−6	1.92E+4	1.96E+6	365	140	0.38	137
700/0.9	1.02E−2	8.00E−5	4.18E+5	4.18E+5	3280	647	0.20	646
1100/ 0.9	7.87E−2	2.60E−4	3.22E+6	3.22E+6	10 660	1794	0.13	1795
Dynamic range 8790 (13 bits)								

a. Scene input obtained from MODTRAN: (76 STD; 830 km alt; 45° SZA (3 PM); $\rho = 0.1$; visibility = 23 km.)
b. Integration time for high-gain channel, $\tau = 1.75$ msec; for low-gain channel, $\tau = 10.5\mu$ sec; also, $\delta_{optics} = 50\%$. So that for the high-gain channel: e^{-1}/pixel $= [Wcm^{-2}sr^{-1}] \cdot [A\Omega\delta_{optics}] \cdot [Ph/sec/Watt] \cdot [\xi] \cdot [t] = [L] \cdot [3.2E-7] \cdot [5.031E18*\lambda] \cdot [0.65] \cdot [1.75E-3] = [L] \cdot [6.8E+9]$. While, for the low-gain channel: e^{-1}/pixel $= [Wcm^{-2}sr^{-1}] \cdot [A\Omega\delta_{optics}] \cdot [Ph/sec/Watt] \cdot [\xi] \cdot [t] = [L] \cdot [3.2E-7] \cdot [5.031E18*\lambda] \cdot [0.65] \cdot [1.05E-5] = [L] \cdot [4.1E+7]$.
c. Cavity $e^{-1}_{(300K optics)} = [L] \cdot [A\Omega] \cdot [\varepsilon_{effective}] \cdot [Photons/sec/Watt] \cdot [\xi] \cdot [t] = [2.6E-5] \cdot [2.3E-6] \cdot [0.05] \cdot [5.031E18*\lambda] \cdot [0.65] \cdot [t] = 3.6E+7 \cdot [t]$ or, 6.3E+4 electrons and 378 electrons for high-gain and low-gain, respectively.
  The detector is assumed to be a NICMOS-type HgCdTe at 80°K and 27 μm pixels with switched MOS-FET multiplexer and a quantum efficiency of 0.65.
  The high-gain channel requires a detector $D^* = 2.85E+12$ cm · $Hz^{1/2} \cdot W^{-1}$; while $D^*_{BLIP} = 3.7E+12$ cm · $Hz^{1/2} \cdot W^{-1}$.
  The low-gain channel requires a detector $D^* = 2.05E+11$ cm · $Hz^{1/2} \cdot W^{-1}$; while $D^*_{BLIP} = 3.7E+12$ cm · $Hz^{1/2} \cdot W^{-1}$.
  Full-well: The high-gain channel MUX for a 500°K scene has 7.5E+6 electrons; while the low-gain channel MUX for a 1100°K scene has 3.25E+6 electrons.
  Array readout time: 0.4 sec (239 400 pixels transferred at a speed of 598 Kpix/sec); read noise less than 50 electrons.

Table 12.11 Radiometric performance for FireSat channels 6a and 6b; bandpass: 3.6–3.8 μm

Scene, Temp/ε	L_{scene}, Wcm^{-2}sr^{-1} a	$\Delta L/°K$, Wcm^{-2}sr^{-1} a	Scene, e^{-1}/pixel b	Scene + cav, e^{-1}/pixel c	Δe^{-1}/°K, e^{-1}/pixel b	Noise, (e$^{-1}_{scene+cav}$), (1σ)	NEΔT, °K, (1σ)	SNR, (1σ)
240/0.9	3.08E−4	4.5E−6	1.37E+6	2.77E+6	20 025	1664	0.083	823
260/0.9	4.16E−4	6.6E−6	1.85E+6	3.25E+6	29 370	1803	0.061	1026
300/0.9	7.74E−4	1.2E−5	3.44E+6	4.14E+6	53 400	2200	0.041	1564
340/0.9	1.36E−3	1.8E−5	6.05E+6	7.45E+6	80 100	2724	0.034	2216

Dynamic range 306 (i.e. 9 bits)

a. Scene input from MODTRAN: (76 STD; 830 km alt; 45° SZA (3 PM); $\rho = 0.1$; visibility = 23 km.)
b. e^{-1}/pixel = [Wcm^{-2}sr^{-1}] · [A$\Omega\delta_{optics}$] · [Ph/sec/Watt] · [ξ] · [t] = [L] · [3.2E−7] · [5.0314E18*λ] · [0.65] · [0.5E−3] = [L] · 4.45E+9; where, integration time, t = 0.5 msec; wavelength, λ = 8.5 μm; and quantum efficiency, ξ = 0.65.
c. Cavity e$^{-1}_{(300\,K\,optics)}$ = [L] · [AΩ] · [$\varepsilon_{effective}$] · [Photons/sec/Watt] · [ξ] · [t] = [9.5E−4] · [2.3E−6] · [0.05] · [5.0314E18*λ] · [ξ] · [t] = 1.58E+6 electrons.

The detector is assumed to be a NICMOS-type HgCdTe at 80°K and 27 μm pixels with switched MOS-FET multiplexer and a quantum efficiency of 0.65.
The channel requires a detector D* = 5.9E+10 cm · Hz$^{1/2}$ · W^{-1}; while D$^*_{BLIP}$ = 9.6E+11 cm · Hz$^{1/2}$ · W^{-1}.
Full-well on the MUX for a 340°K scene equals 7.5E+6 electrons.
Array readout time: 0.4 sec (144 400 pix transferred at a speed of 57 760 pix/sec); read noise less than 500 electrons.

output of one set of these sources illuminates the focal plane directly. These sources will be electronically controlled to provide two levels of input to their respective radiometer channels. The time response of the sources is very rapid; calibration can therefore be checked as often as desired. Another set of diodes will illuminate a row of pixels at the edge of each array, just outside the experiment field of regard. In a Jones method calibration (Jones 1960), the energy from these diodes will pass through the entire measurement channel optical system. This allows a check of the optics and detector response, assuming the average response of all array pixels changes proportionally with the average of the edge-row pixel responses.

FireSat Photodetector Arrays

Fortunately, the wavelength ranges of FireSat are coincident with other popular applications, including visible-NIR imaging (0.5 μm–1 μm), astronomical spectroscopy (1 μm–2.2 μm), terrestrial thermal imaging for military applications such as night vision systems for tanks (8 μm–9 μm), and other tactical military thermal imaging (3.7 μm). Accordingly mature photodetector and focal plane array technologies suitable for FireSat already exist. For the visible (0.5 μm) and first NIR (1 μm) channel, silicon charge-coupled device (CCD) technology, which allows near-photon counting capability, can provide focal plane arrays (FPAs) with small pixel sizes (as low as 9 μm). Such arrays are available in megapixel formats at relatively low cost, high yield, and high reliability due to the excellent material properties of silicon and the high level of

maturity of processing techniques, which is a result of the enormous commercial market for other devices based on metallic oxide semiconductors (MOS) such as computer memory chips (Sze 1988).

For the other wavelengths, the material problems are somewhat more difficult, because silicon does not respond to wavelengths longer than 1 μm or so. Because of the considerable importance of infrared (IR) detection, especially for military purposes, extensive resources have been devoted to development of IR imaging technology over the past 30 years. The most widely used and successfully developed material system is mercury cadmium telluride (HgCdTe). By changing the relative proportions of mercury and cadmium (in a stoichiometric sense), the wavelength response of these detectors can be tuned from NIR to FIR (Wolfe and Zississ 1978). This is particularly useful since photodetector leakage currents, which create noise, increase rapidly with the cut-off wavelength and temperature. Thus, a device with a cut-off wavelength as close as possible to the desired detection point will require less cooling for satisfactory operation than one with a cut-off wavelength that is greater than that of the useful channel. Furthermore, HgCdTe has several desirable material properties that lead to a lower dark current for a given cut-off wavelength than for indium antimonide (InSb), making it the material of choice for most shortwave infrared–medium-wave infrared (SWIR-MWIR) applications (DeWames et al. 1992).

Several important detector/FPA constraints and trade-offs have led to the current FireSat detector concept. First, satisfactory operation at the FireSat wavelengths can be obtained with temperatures achievable by thermoelectric cooling (185–200°K) in the visible to SWIR (0.5–2.20 μm) channels. The other two channels (3.7 μm and 8.5 μm) require cooling to 80°K or so, mandating an active cooler. Higher temperature detectors with background limited performance (BLIP) ability close to at relatively low background fluxes might alleviate this situation and make radiative cooling feasible, but this will require substantial technological advances.

Furthermore, FireSat requires coverage over a large field of view, which in turn requires very wide detector arrays. FPA size is driven by two constraints. First, although IR detector technology is fairly mature, limitations of fundamental material and device physics preclude fabrication of complicated, high-performance transistor structures that are needed to perform readout and signal-conditioning functions on materials such as HgCdTe (Barbe 1980). Thus, current state-of-the-art IR FPAs consist of a photodiode array fabricated on HgCdTe hybridized to a Si readout chip, or multiplexer (MUX), which performs the necessary signal processing (Jenkins 1987). Although highly developed, fabrication limitations prohibit die sizes of greater than approximately 2×2 cm (Kozlowski 1995). In addition, the extreme difficulties in processing IR photodetector materials such as HgCdTe place a limit on the size of the photodiode arrays themselves. This difference depends on the substrate on which the material is grown. For cut-off wavelengths of less than 5 μm or so, sapphire substrates are possible and FPA sizes of about 1.7×1.7 cm can be obtained. For longer wavelengths, however, cadmium zinc telluride (CdZnTe) must be used with a resulting size limit of 1×1 cm or so (Kozlowski 1995). The intersection of these limits, of course, determines the maximum available FPA size. For Si CCDs, a similar limit exists, but it is much larger, on the order of 3×3 cm or even slightly larger (Janesick and Elliot 1991).

The desire for as small an FPA as possible, driving the design to small pixel sizes, collides with the other limitation, that of full-well capacity. Both CCDs and hybridized IR FPAs work by integrating photo-generated charge, which is read out sequentially by an on-chip amplifier. The full-well capacity associated with each pixel, which is the maximum amount of charge that can be integrated without saturation, determines the maximum signal handling capacity of the FPA. To accommodate a large dynamic range, this should be as great as possible. Since full-well capacity scales with pixel area, one way to improve it is to use larger pixels. For a given number of pixels needed to accommodate a given field of view at a given resolu-

tion, this will lead to the requirement for a physically larger array. A careful trade-off study for FireSat has led to the choice of 27-μm pixels for the thermal channels and 60-μm pixels for the visible NIR channels. This in turn determines the number of arrays of a given size that must be used to obtain the desired coverage.

FireSat Focal Plane Cooling

The FireSat instrument is divided into two thermal zones. The short-wavelength detectors operate near the spacecraft's ambient temperature. The long-wavelength detectors require cooling to 80°K to achieve sufficient sensitivity. FireSat will use a mechanical cryogenic refrigerator to provide focal plane cooling below 80°K. Lightweight, low-power miniature refrigerators have been developed that provide 400+ milliwatts of cooling at 80°K. Innovative techniques such as fiber support technology (Batty et al. 1994) can be employed to thermally isolate the cooled focal plane from its surroundings, reducing the cooling requirements of the refrigerator.

FireSat Mechanical Design

The FireSat instrument's mechanical design is lightweight and simple in nature yet very functional. It is packaged in a single module containing all sensor mechanical, optical, and electronic subsystems. Structurally it consists of an aluminum monocoque structure designed primarily to provide and maintain the optical relationship required by the placement of the focal plane assemblies (FPAs) and telescopes, and to maintain the stringent thermal conditions resulting from the use of cooled FPAs and a mechanical cryocooler. In addition, the structure provides the load path required to withstand all launch loads and orbital thermal loads by interfacing with the spacecraft system. The instrument structure is also used to house and package many of the instrument's electronic subsystems. A sophisticated thermal design provides rigidity and thermal isolation for the FPAs and efficiently conducts heat from the FPAs to a spacelook radiator via a Stirling-cycle refrigerator, thus providing the optical alignment and its maintenance required by the six telescopes and their respective FPAs.

FireSat Thermal Control

The estimated heat load of the FireSat sensorcraft is 100 watts, of which 30 watts is produced by the mechanical cryocooler and its control electronics. Due to the small size of the instrument, much of its external surface area might have to be devoted to radiators for thermal control. Having the spacecraft bus handle the instrument's heat rejection may be a more effective alternative.

FireSat Spacecraft System

Given the FireSat instrument's mass and power estimates of 23 kg and 100 watts, respectively, and the orbital requirements of 830 km circular sun-synchronous at 98.7° inclination, a state-of-the art "'small-sat" bus design should surffice. The power, attitude control, and thermal control requirements appear to be reasonable and well within the capabilities of existing bus designs produced by several aerospace companies. Assuming a direct injection by a Pegasus-class launch vehicle, the mass fraction of the FireSat instrument is 12%, which is well within the range normally achieved. The cost of a bus will be from $5 to $15 million depending on redundancy requirements and manufacturer.

FireSat Launch Vehicle System

Two expendable launch vehicle systems are being considered for the FireSat mission. The first is a Pegasus-class vehicle capable of placing approximately 200 kg into a circular 830-km sun-synchronous orbit by direct injection. Alternatively, the use of an 185-km parking orbit and an integral propulsion system to perform a Hohmann transfer to an altitude of 830 km will increase the payload capability to more than 260 kg. There would be a 10- to 15-kg weight penalty required for the propulsion system. Launch systems in the LLV-1-class could provide higher performance (i.e., larger payloads or higher orbits). A direct injection of 290 kg to 830 km sun-synchronous can be accomplished. Alternatively, the use of a 185-km parking orbit and a Hohmann transfer to 830 km will deliver 388 kg to orbit.

New Technology for FireSat

The FireSat concept is based on available technology; the performance required from each individual component or subsystem has been demonstrated. However, there is significant challenge in combining these components and subsystems into a space-based sensor to provide the challenging FireSat science measurements. Additional technology development could significantly reduce the instrument development risk

and improve performance. We have identified several technological developments for the FireSat mission.

Benefit would be derived from the availability of highly integrated NIR-MIR detector readout multiplexers, for low-power compatible cryogenic temperature focal plane application. For maximum benefit this subsystem should simultaneously provide (1) on-chip correlated double sampling; (2) on-chip analog-to-digital converters with 14-bit dynamic range, capable of >1 MHz conversion speed; (3) on-chip smart-sensor-like adaptive sensor functions; (4) fast-readout (>1 MHz) with low readout noise (<20 electrons); and (5) full-well capacity in excess of 20 million electrons for a pixel size of approximately $25 \times 25 \mu m$.

Ideally, the large format FPAs (MUX plus detector) for FireSat would also have the following characteristics: (1) dimensions of 3×1 cm, or larger for channels in both the 3 to 5 μm and 8 to 10 μm range; (2) capability of high temperature, near-BLIP operation ($>110°K$) at low backgrounds and with extremely low drift characteristics; and (3) high-performance NIR and Mid-IR 'two-color' detector arrays.

For FireSat to benefit from its inherent Smart Imager capabilities the validation of the following techniques would be very useful: (1) multiple and adaptive gain concepts for full-well optimization; (2) demonstration of smart-sensor offset determination; and (3) for a high capability in-flight calibration subsystem of the 3.7- and 8.5-μm channels that used no moving parts, demonstration of quaternary IV–VI (e.g., PbSnSeTe) light-emitting diodes (LEDs) for operation at thermo-electric cooler temperatures.

FireSat development would also benefit from development of the following on-board data-processing technology: (1) boresighting of the seven radiometric channels by removing time delay between channel outputs; (2) automated cloud filter to remove most clouds for reduced data transmission; and (3) real-time ground station capability to automatically process FireSat data for fire monitoring.

The FireSat microspacecraft would benefit from improvements in several hardware disciplines including (1) lightweight, thermally stable structures; (2) low-power, high-speed digital electronics for data down-linking; (3) improved spacecraft attitude control system to provide IFOV pointing knowledge to 0.12 milliradians; (4) high-capacity, efficient, lightweight, reliable cryocooler compressor to allow added heat from smart-sensor, on-focal plane processing; and (5) for alternate methods of focal plane cooling, efficient high thermal load crysogenic temperature radiators.

References

Andreae, M. O. 1991. Biomass Burning: Its History, Use, and Distribution and Its Impact on Environmental Quality and Global Change. *Global Biomass Burning: Atmospheric, Climatic, and Biospheric Implications* (J. S. Levine, editor), The MIT Press, Cambridge, Mass., pp. 3–21.

Andreae, M. O., J. Fishman, M. Garstang, J. G. Goldammer, C. O. Justice, J. S. Levine, R. J. Scholes, B. J. Stocks, A. M. Thompson, and B. van Wilgen. 1994. Biomass Burning in the Global Environment: First Results from the IGAC/BIBEX Field Campaign SATARE/TRACE-A/SAFARI-92. *Global Atmospheric-Biospheric Chemistry* (R. G. Prinn, editor), Plenum Press, New York, pp. 83–101.

Barbe, D. F. (editor). 1980. *Charge-Coupled Devices*, Springer-Verlag, Berlin, pp. 126–128.

Batty, J. C., B. G. Williams, W. W. Burt, and W. A. Roettker. 1994. Cooling SABER: Interface, Performance, and Design Issues, *SPIE Proceedings* Vol. 2227-24, International Symposium on Optical Engineering in Aerospace Sensing, Orlando, Florida.

Bowker, D. E., R. E. Davis, D. L. Myrick, K. Stacy, and W. T. Jones. 1985. Spectral Reflectance of Natural Targets for Use in Remote Sensing Studies, *NASA Reference Publication RP-1139*, 184 pages.

DeWames, R. E., J. M. Arias, L. J. Kozlowski, and G. M. Williams. 1992. An Assessment of HgCdTe and GaAs/AlGaAs Technologies for LWIR Infrared Imagers, *Infrared Detectors: State of the Art*, Society of Photo-Optical Instrumentation Engineers (SPIE) Vol. 1735, pp. 2–16.

Hao, W. M., and M.-H. Liu. 1994. Spatial and Temporal Distribution of Tropical Biomass Burning. *Global Biogeochemical Cycles*, 8, pp. 495–503.

Houghton, R. A. 1991. Biomass Burning from the Perspective of the Global Carbon Cycle. *Global Biomass Burning: Atmospheric, Climatic, and Biospheric Implication* (J. S. Levine, Editor), The MIT Press, Cambridge, Mass., pp. 321–325.

Hummel, J. R., D. R. Longtin, N. L. DePiero, and R. J. Grasso. 1992. *Backscat LIDAR Simulation Version 3.0: Technical Documentation and Users Guide*. Scientific Report No. 3, SPARTA, Inc. Lexington, MA. (Available as report PL TR-92-2328 from Phillips Laboratory, Directorate of Geophysics, Air Force Materiel Command, Hanscom Air Force Base, Mass. 01731–3110).

Janesick, J. R., and T. Elliot. 1991. History and Advancement of Large Area Array Scientific CCD Imagers, *Astronomical Society of Pacific Conference Series '91*, Tucson, Ariz.

Jenkins, T. E. 1987. *Optical Sensing Techniques and Signal Processing*, Prentice-Hall, Englewood Cliffs, N.J., pp. 89–95.

Jones, R. Clark. 1960. *Proc. Infrared Information Symposium* (IRIS) 5(4), pp. 35.

Killinger, D. K., and W. E. Wilcox. 1992. *USF HITRAN-PC. Installation and Software User Manual.* (Commercial Version available from ONTAR Corp, North Andover, Mass. 01845–2000).

Kneizys, F. X., E. P. Shettle, L. W. Abreu, J. H. Chetwynd, G. P. Anderson, W. O. Gallery, J. E. A. Selby, and S. A. Clough. 1988. *User's Guide to LOWTRAN7*. (Available as Report AFGL-TR-88-0177 from Phillips Laboratory, Directorate of Geophysics, Air Force Materiel Command, Hanscom AFB, Mass. 01731–3110).

Kozlowski, J. T. 1995. Personal communication.

Levine, J. S. (editor). 1991. *Global Biomass Burning: Atmospheric, Climatic, and Biospheric Implications*. The MIT Press, Cambridge, Mass., 569 pages.

Levine, J. S., W. R. Cofer, III, D. R. Cahoon, and E. L. Winstead. 1995. Biomass Burning: A Driver for Global Change. *Environmental Science and Technology*, 29, pp. 120A–125A.

Mano, S., and M. O. Andreae. 1994. Emission of Methyl Bromide from Biomass Burning. *Science*, 263, pp. 1255–1257.

McCann, P. J., L. Li, J. E. Furneaux, and R. Wright. 1995. Optical Properties of Ternary and Quaternary IV-VI Semiconductor Layers on (100) Barium Fluoride Substrates, *Applied Physics Letters*, 66, 1355.

National Academy of Sciences. 1990. *One Earth, One Future: Our Changing Global Future*. National Academy Press, Washington, D.C., pp. 116–130.

National Research Council. 1989. *Global Change and Our Common Future*. National Academy Press, Washington, D.C., p. 1.

ONTAR, Inc. 1992. *PCModTRAN User's Manual*. Available from the ONTAR Corp., North Andover, Mass. 01845-2000.

Penner, J. E., M. M. Bradley, C. C. Chang, L. L. Edwards, and L. F. Radke. 1991. A Numerical Simulation of the Aerosol-Cloud Interactions and Atmospheric Dynamics of the Hardiman Township, Ontario, Prescribed Burn. *Global Biomass Burning: Atmospheric, Climatic, and Biospheric Implications* (J. S. Levine, editor), The MIT Press, Cambridge, Mass., pp. 420–426.

Seiler, W., and P. J. Crutzen. 1980. Estimates of Gross and Net Fluxes of Carbon between the Biosphere and Atmosphere from Biomass Burning. *Climatic Change*, 2, pp. 207–247.

Sze, S. M. 1988. *VLSI Technology*, 2nd edition. McGraw-Hill Book Company, New York, NY, pp. 1–7.

Wolfe, J., and G. Zississ (editors). 1978. *The Infrared Handbook*. The Infrared Information and Analysis (IRIA) Center, Environmental Research Institute of Michigan, pp. 11.91–11.93.

II

Modeling and Inventory Development: Global and Regional Scales

Estimates of Biomass Density for Tropical Forests

Sandra Brown and Greg Gaston

The role of tropical forests in global biogeochemical cycles, especially the carbon cycle and its relation to climate change, has heightened interest in estimating the biomass density of tropical forests. Forest biomass density provides estimates of the carbon pools in forest vegetation because about 50% of biomass is carbon. This pool is the potential amount of carbon, as carbon dioxide, that can be added to the atmosphere when the forest is cleared and/or burned. Attempts to estimate the biomass density of tropical forests have been made by the scientific community for use in models that assess the contribution of tropical deforestation and biomass burning to the increase in atmospheric carbon dioxide and other trace gases (Brown et al. 1989; Crutzen and Andreae 1990; Hall and Uhlig 1991; Houghton et al. 1987).

Estimates of the biomass density for many of the world's forests have been made. For example, a detailed summary of biomass density studies in tropical forests, from lowland to montane and from wet to very dry zones, was made by Brown and Lugo (1982). A later study by Olson et al. (1983) produced a global map of the biomass density of all ecosystem types, including disturbed and undisturbed forests, at a 0.5° × 0.5° grid-scale of resolution. These summaries of biomass density were based on ecological studies creating several problems with their use for global-scale analyses. Ecological studies are generally designed to characterize local forest structure and the study sites are usually not truly randomly located nor represent the population of interest (Brown and Lugo 1992). These type of studies are suitable for studying local forests but not for making inferences about larger populations (Brown et al. 1989). Furthermore, the total area covered by these studies is a very small fraction of the total forest area (e.g., less than 0.00001% for tropical forests; Brown and Lugo 1984).

A further problem with using biomass data from ecological studies for national to global analyses is the inherent bias of ecologists to adjust placement of plots based on the notion of what a mature forest should look like, that is, one with many large-diameter trees (Brown and Lugo 1992). The effect of adjusting plot placement to include large-diameter trees is to over-estimate biomass density of the forests because biomass per tree increases geometrically with increasing diameter. The result of this bias is to yield high biomass density estimates for forests (Brown et al. 1989). Thus data from ecological studies must be used with caution because they may not represent the biomass density of the forest over large areas.

Biomass density estimates for tropical forests have also been made by the Food and Agriculture Organization (FAO 1993) based on the FAO FORIS data base (Forest Resources Information System, a computerized data base) of volume over bark (VOB, commercial volume to a minimum tree diameter of 10 cm) often measured in forest inventories. On the positive side, VOB data from forest inventories are based on a large number of plots, generally collected from large sample areas using a planned sampling design from the population of interest. However, very few national or subnational inventories that report VOB have been done in the tropics. The compilation of the VOB data base by the FAO required much educated guesswork to produce estimates on a tropics-wide country-level basis. This approach is, therefore, of unknown reliability and any errors in VOB estimates were compounded during the conversion of these data to biomass density values. Clearly, new efforts to estimate biomass density more directly from forest inventory data are needed to provide more reliable data for national to global assessments of the quantity of forest resources.

This chapter summarizes the various approaches that we and colleagues have developed over the past decade or so for estimating biomass density of tropical forests, relying for the most part on forest inventory data and modeling in a geographic information system (GIS). Estimates of biomass density for a variety of tropical forests from different parts of the tropics are presented in tabular form and spatially distributed. We also discuss the factors that affect biomass density and show that it is not a static parameter but rather a moving target.

Definition of Biomass

A complete estimation of forest biomass density requires that the biomass of all forest components be estimated, including the above- and below-ground living mass of trees, shrubs, palms, saplings, other understory components, vines, epiphytes, and so on and the dead mass of fine and coarse litter. In this chapter we consider only the total amount of aboveground organic matter present in trees including leaves, twigs, branches, main bole, and bark, expressed as oven-dry tons per hectare (referred to as *biomass density*). For most forests or tree formations, biomass density estimates are based only on the biomass in trees with diameters greater than or equal to 10 cm, the usual minimum diameter measured in most inventories of closed forests. However, for forests or trees of smaller stature, such as those in the arid tropical zones, degraded forests, or secondary forests, the minimum diameter could be as small as 2.5 cm.

Most efforts on biomass estimation to date have generally focused on the aboveground tree component because it accounts for the greatest fraction of total biomass density and the methods are straightforward and generally do not pose too many logistical problems. However, a few estimates of these other components of tropical forests do exist, but they must be used with caution because the data base on which they are built is limited.

The amount of biomass in small-diameter trees, understory shrubs, vines, and herbaceous plants can be variable but is generally about 3–5% or less of the aboveground biomass of more mature forests (Jordan and Uhl 1978; Tanner 1980; Hegarty 1989; Lugo 1992). However, in secondary forests or disturbed forest, this fraction could be higher (e.g., up to 30%; Brown and Lugo 1990; Lugo 1992) depending on age of the secondary forest and openness of canopy. Palms are common in many tropical moist forests and they are also often ignored in forest inventories. Their contribution to total biomass density can be very variable, from almost 100% in pure palm forests to less than a few percent where they are a minor component of the forest (Brown and Lugo 1992).

The biomass of roots in tropical forests varies considerably among tropical forests depending mainly upon climate and soil characteristics (Brown and Lugo 1982; Sanford and Cuevas 1996). Root biomass is often expressed in relation to aboveround biomass, such as a root-to-shoot ratio (R/S ratio). From a recent review of the literature, R/S ratios for lowland to montane forests range from 0.04 to 0.85 (Sanford and Cuevas 1996). These estimates are based on only a few studies (about 30) and not all of them are consistent with respect to depth of sampling and whether all coarse roots were included.

The amount of dead plant material in a forest, or detritus, is composed of fine litter on the forest floor (leaves, fruits, flowers, twigs, bark fragments, branches less than 10 cm diameter, etc.), standing dead trees and lying dead wood greater than 10 cm diameter; the last two components are referred to as coarse woody debris (CWD). The biomass density of fine litter ranges from about 2 to 16 Mg/ha (average of 6 Mg/ha or less than 5% of aboveground biomass), with higher values generally in moist environments, although no clear trend is apparent in the data base (Brown and Lugo 1982). The amount of fine litter on the forest floor represents the balance between inputs from litterfall and outputs from decomposition, both of which vary widely across the tropics.

The amount of CWD in tropical forests is poorly quantified but extremely variable. It is potentially a large pool of organic carbon, perhaps accounting for an amount equivalent to 10 to more than 40% of the aboveground biomass of a forest (Saldarriaga et al. 1986; Uhl et al. 1988; Uhl and Kauffman 1990). Lack of data on this significant forest component obviously can lead to underestimates of the total amount of biomass in a forest.

It is clear from the above discussion that ignoring these other forest components can seriously underestimate the total biomass of a forest by an amount equivalent to about 70% or more of aboveground biomass. It is apparent that logistically and economically feasible methods and approaches must be developed to estimate this significant quantity of biomass, especially for improving estimates of terrestrial sources and sinks of carbon and other greenhouse gases.

Estimating Biomass Density from Inventory Data

Use of forest inventory data overcomes many of the problems in ecological studies as discussed above. Data from forest inventories are generally more abundant and are collected from large sample areas (subnational to national level) using a planned sampling method designed to represent the population of interest. However, inventories are not without their problems (Brown and Iverson 1992). Typical problems include:

• Inventories tend to be conducted in forests viewed as having commercial value, that is, closed forests, with little regard to the open, drier forests or woodlands.

• The minimum diameter of trees included in inventories is often greater than 10 cm, thus excluding smaller trees, which can account for more than 30% of the biomass (Gillespie et al. 1992).

• The maximum diameter class in stand tables is generally open ended, with trees greater than 80 to 90 cm in diameter often lumped into one class; the actual diameter distribution of these large trees significantly affects aboveground biomass density (Brown and Lugo 1992; Brown 1996).

• Not all tree species are included.

• Many of the inventories are from the 1960s, 1970s, or earlier and the forests often no longer exist or at least are not the same as they were at the time of the inventory.

Despite the above problems, many inventories are very useful for estimating biomass density of forests. During the last decade or so two main approaches for estimating the biomass density of forests based on existing forest inventory data have been developed. One uses existing volume estimates (m³ per ha), converted to biomass density (Mg/ha) using a variety of "tools" (Brown et al. 1989; Brown and Lugo 1992; Gillespie et al. 1992). A second approach directly estimates biomass density from the application of an appropriate allometric regression equation (biomass per tree as a function of diameter), selected on the basis of climate regime (dry, moist, or wet), to stand tables (number of trees per hectare in a given diameter class) often reported in forest inventories. The advantage of this second method is that it produces biomass estimates without having to make volume estimates and then to apply various expansion factors to account for non-commercial tree components. The disadvantage is that a fewer number of inventories report stand tables to small-diameter classes for all species, thus not all countries in the tropics are covered by these estimates.

Biomass Density Estimates

The above approaches have been used with inventories from many tropical Asian (9) and American (10) countries encompassing about 30 million hectares. The resulting estimates of aboveground biomass density for moist forests range from less than 50 Mg ha⁻¹ to more than 550 Mg ha⁻¹ (figure 13.1) with an arithmetic mean of 230 Mg ha⁻¹ for both tropical regions. In the wet zone of tropical America (mostly Panama), biomass density estimates range from less than 50 to about 300 Mg ha⁻¹, with an average of 150 Mg ha⁻¹. Forests in the wet zone tended to have lower biomass

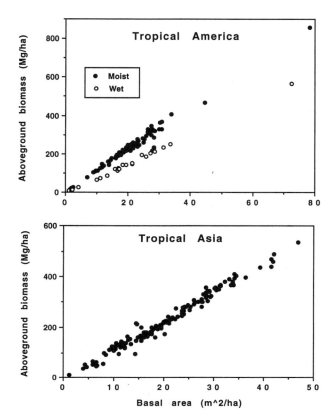

Figure 13.1 Aboveground biomass density estimates for forests of tropical America and Asia (from Brown 1996). The estimates are plotted against basal area as a way of showing the range of values; a high correlation is expected (see text)

densities for a given basal area, as has been shown before (Brown and Lugo 1982).

The range of biomass density estimates for moist tropical American forests is practically identical to that for moist tropical Asian forests (figure 13.1). As was the case for the tropical Asian forests (cf. Brown et al. 1991), many of the tropical American forests were identified as being disturbed (e.g., commercial harvesting, harvesting by indigenous communities, young to late secondary, shifting cultivation; Brown 1996).

Biomass density estimates for tropical moist forests of central and western Africa (Cameroon, Gabon, Côte d'Ivoire, and Ghana) based on inventories range between 187 and 378 Mg ha⁻¹ (ongoing research by Brown and Gaston). In the drier zones of West and East Africa where open forests or savanna woodlands dominate, biomass densities range from 22 to 196 Mg ha⁻¹. No inventory data for African moist forests available to date has produced biomass density estimates as high as those for tropical Asia or America, even though estimates from ecological

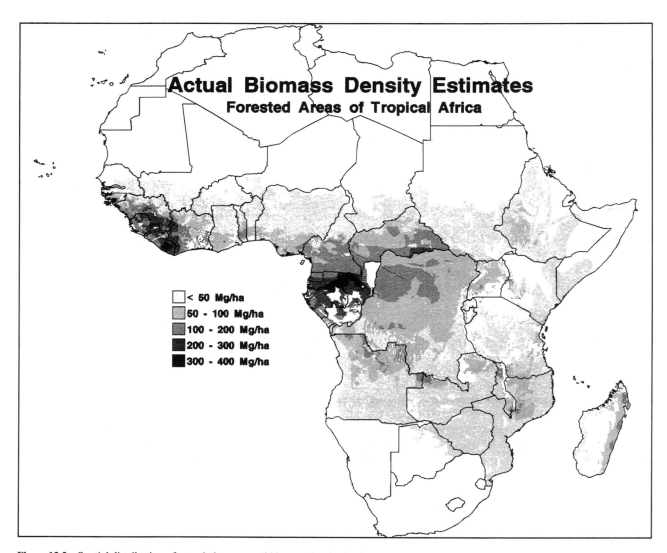

Figure 13.2 Spatial distribution of actual aboveground biomass density for forests and woodlands of tropical Africa from around 1980 (from Brown and Gaston 1995). This map is available as a ARC/INFO data base; a color version with 10 biomass density classes is available from authors.

studies show a similar range of values for all three tropical regions.

Estimating Biomass Density by Modeling in a GIS

Brown et al. (1993) and Iverson et al. (1994) developed a modeling approach using a GIS to produce spatial distributions of biomass densities for tropical forests. The method was developed to extend the few reliable, inventory-based biomass density estimates to regional scales in tropical Asia. The overall approach to making aboveground biomass density estimates was based on the assumption that the present-day distribution of

biomass is a result of a combination of the potential biomass density, based on prevailing climatic, edaphic, and geomorphologic conditions, and the cumulative impacts of human activities that reduce biomass. We have used this approach to generate biomass density estimates for forests and woodlands of tropical Africa (Brown and Gaston 1995, and ongoing research).

The modeling approach (described in detail in Iverson et al. 1994) first estimates potential biomass density by using a weighted overlay of input layers: precipitation, a climatic index, elevation and slope, and soil texture. Weighting factors were adjusted through an iterative process by comparing results to known localities (see Brown et al. 1993 and Iverson et al. 1994 for more details). The final iteration produced a raster grid with each pixel (5 km × 5 km) containing a potential biomass density (PBD) index ranging in value from about 40 to 100.

To calibrate the PBD indices into biomass density values required the assignment of biomass density estimates across the range of index values. The most critical values were those that identified the upper and lower biomass limits. A very limited set of ecological studies that gave biomass estimates for mature forests, woodlands, and wooded savannas were used to establish the upper and lower limits of biomass density (Brown and Gaston 1995). The process of establishing the linkage of PBD index values to biomass density relied heavily on prior field experience, experts in the area, and published information.

A variety of natural and anthropogenic factors reduce biomass in any system from its potential. Long-term human use has a dramatic effect on the density of biomass in forest ecosystems. Fuel-wood gathering, sanctioned and unsanctioned logging (Callister 1992), grazing, shifting cultivation, and anthropogenic burning all reduce the amount and density of biomass. Because these practices are continuing and ongoing as population pressure increases, the biomass density of forests becomes a "moving target." Past research has shown that population density is a good empirical indicator to quantify the long-term human impact on biomass density (cf. Brown et al. 1993). Using the methods described above, we estimated actual forest biomass density from the available forest inventories. The amount of biomass reduction as measured by the degradation index was calculated as the ratio of biomass density estimated from forest inventories to the modeled potential biomass density for the inventory location at the scale of a subnational unit or administrative unit such as a state. We then paired this degradation index to the population density of the subnational unit for the decade of the inventory and stratified the data base into two forest types: closed forest and open forest/woodland (Brown and Gaston 1995). We were able to identify only eight inventories for the whole of Africa for this step, four in the closed forest zone and four in the open forest/woodland zone.

We have shown that we can combine the African data base with a similar one for tropical Asia and develop statistically significant regression equations of degradation ratio versus population density for the closed forest and open forest/woodland zones (Brown and Gaston 1995, and ongoing research). We used these two regression equations with the population density map to produce a map of degradation ratios. The spatial distribution of "actual" biomass was the product of the potential biomass density map and the degradation ratio map. The estimates of actual biomass density were calculated pixel by pixel.

The spatial distribution of the actual biomass density for tropical African forests generally follows expected trends (figure 13.2). Because so few forest inventory data are available in the region, we were forced to use most of them to develop the degradation model. Only two inventories were not used, but were used for one step in the validation process. Results from a national forest inventory for the west African country of Guinea gave a weighted average biomass density estimate 135 Mg ha^{-1}. The weighted mean for this country from the modeling approach is 140 Mg ha^{-1}, almost equal to the measured estimate (Brown and Gaston 1995). Similarly for the wooded part of Mali, the inventory gave a range of 55 to 65 Mg ha^{-1} and the model gave a somewhat lower weighted mean of 45 Mg ha^{-1}. Furthermore, we used the process described in Brown et al. (1993) as a further check for our results. We used a reclassified map of the ecofloristic zones of Africa (something akin to a life zone map). Results of this step confirmed expected patterns. For example, actual biomass density decreased from about 300 Mg ha^{-1} in the lowland moist zone to 140, 60, and 20 Mg ha^{-1} in the lowland seasonal, lowland dry, and lowland very dry zones, respectively.

Highest estimates (>300 Mg ha^{-1}) are for dense humid forests located in parts of the west African countries of Liberia and Côte d'Ivoire and the central African countries of Congo, Equatorial Guinea, and Gabon (figure 13.2). Biomass densities decreased with increasing distance from these wetter areas to a low of less than 50 Mg ha^{-1} in the dry open woodlands of countries in the Sahel and East Africa. Area-weighted, country-level estimates of actual biomass density were also produced (table 13.1). Low coefficients of variation (CV) were obtained for countries with the highest biomass density estimates, suggesting a relatively homogeneous environment and lower population pressure (Brown and Gaston 1995, and ongoing research).

Conclusions

The biomass density of tropical forests is one of the most important variables that influence the magnitude of the terrestrial carbon flux and other trace gas fluxes. We have shown that a variety of tools are available to estimate biomass density at country to

Table 13.1 Mean area-weighted actual biomass density and coefficient of variation (CV) for forests of tropical Africa by country (from Brown and Gaston 1995)

Country	Actual (Mg ha^{-1})	CV (%)
Angola	73.3	81
Benin	58.0	53
Botswana	13.2	55
Burkino Faso	34.4	70
Burundi	42.7	49
Cameroon	217.4	54
CAR	199.6	44
Chad	42.8	58
Congo	343.6	29
Côte d'Ivoire	164.7	44
Equatorial Guinea	317.9	10
Ethiopia	51.5	100
Gabon	338.5	19
Gambia	29.2	55
Ghana	82.7	57
Guinea	139.6	58
Guinea Bissau	84.6	47
Kenya	33.0	80
Liberia	304.8	22
Madagascar	195.8	37
Malawi	47.1	65
Mali	44.9	57
Mozambique	57.3	75
Niger	8.6	50
Nigeria	49.0	88
Rwanda	33.7	40
Senegal	31.5	75
Sierra Leone	199.0	26
Somalia	12.5	40
Sudan	63.8	95
Tanzania	45.3	69
Togo	71.9	58
Uganda	102.2	43
Zaire	206.3	47
Zambia	46.8	85
Zimbabwe	13.6	71

regional scales, yet still capturing the heterogeneity of the environment. We have also suggested that estimates produced by these approaches are more suitable for regional-scale models because they are more representative of the larger landscape and attempt to encompass the human component. Finally, the GIS modeling approach has the advantage of producing biomass density maps that can be matched to similar ones produced by high-resolution satellite imagery that show the actual forest areas undergoing change.

Acknowledgments

Research reported on here was partially supported by a grant from the U.S. Department of Energy (DOE DEFGO2-90ER61081) to the University of Illinois (S. Brown, P.I.). The research reported in this chapter has also been funded in part by the U.S. Environmental Protection Agency. It has been subjected to the agency's administrative review, and it has been approved for publication as an EPA document.

References

Brown, S. 1996. Tropical forests and the global carbon cycle: estimating state and change in biomass density, in *The Role of Forest Ecosystems and Forest Management in the Global Carbon Cycle*, NATO Series, edited by M. Apps and D. Price, pp. 135–144, Springer-Verlag New York.

Brown, S. and G. Gaston, 1995. Use of forest inventories and geographic information systems to estimate biomass density of tropical forests: application to tropical Africa, *Journal of Environmental Monitoring and Assessment*, 38, 157–168.

Brown, S., A. J. R. Gillespie, and A. E. Lugo. 1989. Biomass estimation methods for tropical forests with applications to forest inventory data, *Forest Science*, 35, 881–902.

Brown, S., A. J. R. Gillespie, and A. E. Lugo. 1991. Biomass of tropical forests of South and Southeast Asia, *Canadian Journal of Forest Research*, 21, 111–117.

Brown, S. and L. R. Iverson. 1992. Biomass estimates for tropical forests, *World Resources Review*, 4, 366–384.

Brown, S., L. R. Iverson, A. Prasad, and D. Liu. 1993. Geographic distribution of carbon in biomass and soils of tropical Asian forests, *Geocarto International*, 8(4), 45–59.

Brown, S., L. Iverson, and A. E. Lugo. 1994. Land use and biomass changes of forests in Peninsular Malaysia during 1972–82: use of GIS analysis, in *Effects of Land Use Change on Atmospheric CO$_2$ Concentrations: Southeast Asia as a Case Study*, edited by V. H. Dale, Ch. 4, Springer-Verlag, New York.

Brown, S., and A. E. Lugo. 1982. The storage and production of organic matter in tropical forests and their role in the global carbon cycle, *Biotropica*, 14, 161–187.

Brown, S., and A. E. Lugo. 1984. Biomass of tropical forests: a new estimate based on volumes, *Science*, 223, 1290–1293.

Brown, S., and A. E. Lugo. 1990. Tropical secondary forests, *Journal of Tropical Ecology*, 6, 1–32.

Brown, S., and A. E. Lugo. 1992. Above-ground biomass estimates for tropical moist forests of the Brazilian Amazon, *Interciencia*, 17, 8–18.

Callister, D. J. 1992. *Illegal tropical timber trade: Asia-Pacific*, TRAFFIC International, Cambridge, UK.

Crutzen, P. J., and M. O. Andreae. 1990. Biomass burning in the tropics: impacts on atmospheric chemistry and biogeochemical cycles, *Science*, 250, 1669–1678.

Gillespie, A. J. R., S. Brown, and A. E. Lugo. 1992. Tropical forest biomass estimation from truncated stand tables, *Forest Ecology and Management*, 48, 69–88.

Food and Agriculture Organization. 1993. *Forest resources assessment 1990 tropical countries*, FAO Forestry Paper 112, Rome, Italy.

Hall, C. A. S., and J. Uhlig. 1991. Refining estimates of carbon released from tropical land-use change, *Canadian Journal of Forest Research*, 21, 118–131.

Hegarty, E. E. 1989. The climbers—lianas and vines, in *Tropical Rain Forest Ecosystems*, Ecosystems of the World 14B, edited by H. Lieth and M. J. A. Werger, pp. 339–354, Elsevier, New York.

Houghton, R. A., R. D. Boone, J. R. Frucci, J. E. Hobbie, J. M. Melillo, C. A. Palm, B. J. Peterson, G. R. Shaver, G. M. Woodwell, B. Moore, D. L. Skole, and N. Myers. 1987. The flux of carbon from terrestrial ecosystems to the atmosphere in 1980 due to changes in land use: geographic distribution of the global flux, *Tellus*, 39B, 122–139.

Iverson, L. R., S. Brown, A. Prasad, H. Mitasova, A. J. R. Gillespie, and A. E. Lugo. 1994. Use of GIS for estimating potential and actual forest biomass for continental South and Southeast Asia. In *Effects of Land Use Change on Atmospheric CO$_2$ Concentrations: Southeast Asia as a Case Study*, edited by V. H. Dale, Ch. 3, Springer-Verlag, New York.

Jordan, C. F., and C. Uhl. 1978. Biomass of a "tierra firme" forest of the Amazon Basin, *Oecologia Plantarum*, 13, 387–400.

Lugo, A. E. 1992. Comparison of tropical tree plantations with secondary forests of similar age, *Ecological Monographs*, 62, 1–41.

Olson, J. S., J. A. Watts, and L. J. Allison. 1983. *Carbon in live vegetation of major world ecosystems*, DOE/NBB-0037, National Technical Information Service, U.S. Department of Commerce, Springfield, Va.

Saldarriage, J. G., D. C. West, and M. L. Thorp. 1986. Forest succession in the Upper Rio Negro of Colombia and Venezuela, *Environmental Sciences Division Publication No. 2694, ORNL/TM9712*, Oak Ridge National Laboratory, Oak Ridge, Tenn.

Sanford, R. L., and E. Cuevas. 1996. Root growth and rhizosphere interactions in tropical forests, in *Tropical Forest Plant Ecophysiology*, edited by S. S. Mulkey, R. L. Chazdon, and A. P. Smith, Chapman Hall, New York.

Tanner, E. V. 1980. Studies on the biomass and productivity in a series of montane rain forests in Jamaica, *Journal of Ecology*, 68, 573–588.

Uhl, C., R. Buschbacher, and E. A. S. Serrao. 1988. Abandoned pastures in eastern Amazonia, 1. Pattern of plant succession, *Journal of Ecology* 76, 663–681.

Uhl, C., and J. B. Kauffman. 1990. Deforestation, fire susceptibility, and potential tree responses to fire in eastern Amazon, *Ecology*, 71, 437–449.

Land-Use Practices and Biomass Burning:
Impact on the Chemical Composition of the Atmosphere

Claire Granier, Wei Min Hao, Guy Brasseur, and
Jean-François Müller

Biomass burning results in the release of important amounts of gases and aerosols into the atmosphere. It has therefore a significant impact on the distribution of chemical species in the atmosphere (Crutzen et al. 1979; Andreae 1993; Hao and Ward 1993; Levine 1994). Biomass burning occurs mainly in the tropics, in relation to deforestation, shifting cultivation, fuelwood use, and clearance of agricultural residues (Hao and Liu 1994). The tropical deforestation rate is believed to have increased by 40% during the last decade (Hao et al. 1994). The use of wood to produce energy is currently increasing at a rate of 3% per year. Burning of savanna to produce forage and burning agricultural may also have increased recently.

This chapter evaluates the contribution of biomass burning to the emissions of several trace gases, including carbon monoxide (CO), methane (CH_4), nitrogen oxides (NO_x), and some nonmethane hydrocarbons (NMHCs). The impact of such emissions on the atmospheric distribution of these gases, as well as their impacts on the formation of reactive trace species, such as ozone (O_3) and the hydroxyl radical (OH), are assessed. The effect of sulfur compounds and aerosols emitted by biomass burning is not considered here.

The Three-Dimensional Model

The model used in this study is a three-dimensional chemical/transport model of the troposphere named IMAGES (Intermediate Model for the Annual and Global Evolution of Species). A detailed description of the model is given by Müller and Brasseur (1995). The horizontal resolution of the model is 5° in latitude and in longitude. The model includes 25 layers from the ground to 50 mbar, which are expressed in sigma-coordinates. Monthly average distributions of winds, surface pressure, temperature, water vapor, and clouds are specified, according to climatological analyses from ECMWF (European Center for Medium Weather Range Forecast) and from ISCCP (International Satellite Cloud Climatology Project). The transport of long-lived species is formulated through advection, diffusion, and convection. Advective transport is represented by a semilagrangian scheme (Smolarkiewicz and Rasch 1991).

The model simulates the evolution of 40 species, among which 20 are transported: these species include reactive compounds of oxygen, hydrogen, and nitrogen, as well as 6 nonmethane hydrocarbons (ethylene, ethane, propylene, isoprene, α-pinene, and a lumped species used as a surrogate for all other hydrocarbons). Approximately 120 chemical and 20 photodissociation reactions are considered. The emissions for chemical species by natural or anthropogenic sources have been compiled by Müller (1992), at a spatial resolution of 5° in latitude by 5° in longitude.

Distribution of Emission of Gases from Biomass Burning

The spatial and temporal distributions of the amount of biomass burned in tropical regions are taken from the data base established by Hao and Liu (1994). This data base provides the data for the amount of biomass burned monthly, at a 5°-by-5° resolution, for the reference year 1980. It accounts for the fires resulting from deforestation, shifting cultivation, savanna burning, fuel-wood use, and burning of agricultural residues. In nontropical areas, the amount of biomass burned is determined by Müller (1992), using statistics established by OECD (1989), USDA Forest Service (1986), and the United Nations (1988).

The amount of CO_2 emitted from biomass burning is first estimated assuming that the carbon content of biomass is about 50%, and that the fuel carbon is released as CO_2, CO, CH_4 and nonmethane hydrocarbons. The emission of each chemical species relative to CO_2 is then computed for each type of biomass fire (Hao and Ward 1993; Hao unpublished results). The emission ratios of CH_4, CO, or several nonmethane hydrocarbons (NMHCs) to CO_2 are given in table 14.1. It should be noted that in table 14.1, as well as in the rest of this chapter, "other" NMHCs do not

Table 14.1 Emission ratios to CO_2 from biomass burning (in percentages by volume)

	CO	CH_4	NO_x	C_2H_4	C_2H_6	C_3H_6	Other NMHCs
Tropical forest	12.4[a]	1.57[a]	0.21[b]	0.36[b]	0.116[b]	0.107[b]	0.16[b]
Temperate/boreal forest	14.94	1.05	0.21	0.24	0.078	$7.4\ 10^{-2}$	0.107
Savanna	4.17	0.24	0.074	0.05	$1.78\ 10^{-2}$	$1.63\ 10^{-2}$	$2.4\ 10^{-2}$
Fuel wood	9.89	2.49	0.065	0.57	0.184	0.169	0.25
Agriculture waste	8.70	0.44	0.19	0.1	$3.26\ 10^{-2}$	$2.99\ 10^{-2}$	$4.5\ 10^{-2}$

a. Data for CO and CH_4, revised from Hao and Ward (1993).
b. Data for other species from Hao et al. (1995) and Hao (Unpublished results).

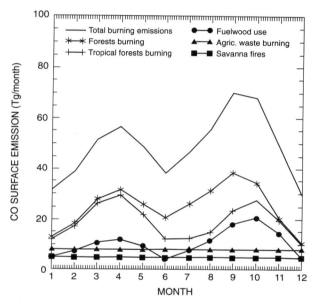

Figure 14.1 Seasonal variation of the CO emissions from tropical and total forests burning, savanna fires, fuel-wood, and agricultural waste burning, and from all sources of biomass burning, in Tg CO per month

include isoprene or terpenes, because the emission of these species from fires is insignificant.

Emissions of the trace gases vary significantly with season, following the seasonal variation of the total amount of biomass burned as shown by Hao and Liu (1994). Figure 14.1 shows the seasonal variation of the global amount of CO emitted by biomass burning and from each of the four types of biomass fires considered here. It should be noted that, although savanna burning represents the largest source of CO_2, its relative importance for the emissions of CO and of other trace gases is very low, due to the low emission ratios typical for this type of fire. Forest burning, more particularly tropical forests burning as shown in figure 14.1,

accounts for about half of the trace gas emissions of CO and other trace gases from biomass burning.

Table 14.2 presents the annual emissions of CO, CH_4, and NMHCs from various sources. It shows, for example, that biomass burning represents the largest source of CO (586 Tg/year, accounting for approximately 45% of the total surface source of CO). Biomass burning contributes significantly to the emissions of the NMHCs (45% in the case of ethylene). The contributions of biomass burning to the sources of CH_4 and NO_x are smaller, and represent only 8 and 13% of the source, respectively. The emissions of NO_x associated with biomass burning are, however, important because they take place mostly in nonindustrialized areas of the tropics.

The global emission estimates from biomass burning given in table 14.2 are lower than those presented in Müller (1992), since the emission factors used by Müller (1992) were generally higher and not dependent on the type of ecosystem. When compared to other estimates (Andreae 1993; Bonsang et al. 1994; Levine et al. 1994), the emission factors relative to CO_2 given here are significantly lower for savanna burning and slightly higher for forest fires than those used in previous studies, resulting in lower emissions of trace gases from biomass burning. However, the total emissions from biomass burning presented here are well within the range given by WMO/UNEP (1995) (i.e., 400–700 Tg/yr for CO, 20–80 Tg/yr for CH_4, 3–13 Tg N/yr for NO_x), but are lower than the WMO/UNEP estimate of 30–90 Tg C/yr for NMHCs.

Results of the Simulations

Several model simulations were performed to assess the importance of biomass burning on the global budget of several tropospheric species. Two cases were compared: the reference case, in which all sources of

Table 14.2 Annual emissions of trace gases

	CO (Tg CO/y)	CH$_4$ (Tg CH$_4$/y)	NO$_x$ (Tg N/y)	C$_2$H$_4$ (Tg C/y)	C$_2$H$_6$ (Tg C/y)	C$_3$H$_6$ (Tg C/y)	Other NMHCs (Tg C/y)
Forest burning	299	19	2.4	3.4	1.1	1	1.5
Tropics	233	17	2	2.9	0.9	0.9	1.3
Mid and high latitude	66	3	0.4	0.5	0.1	0.1	0.2
Savanna burning	128	4	1.1	0.7	0.2	0.2	0.3
Fuel-wood use	98	14	0.3	2.4	0.8	0.7	1.1
Agricultural waste	61	2	0.7	0.3	0.1	0.1	0.1
Total biomass burning	586	39	4.5	6.8	2.2	2.0	3.0
Anthropogenic emissions	383	131	23.4	2.1	3	0.6	16.9
Ocean + soils + other sources	330	300	6.6	6.2	0.8	3.7	35.1
Total emissions	1299	470	34.5	15.1	6.0	6.3	55

tropospheric species are included, and a simulation in which the biomass burning source is excluded. In order to better quantify the contribution of each burning source to the budget of the tropospheric chemical species, four additional simulations were performed, in which only one source was considered at a time.

Carbon Monoxide and Nonmethane Hydrocarbons

Since biomass burning represents almost half of the total CO source, as shown in table 14.2, such a source should have a large impact on the atmospheric CO distribution. Figure 14.2*a* shows the surface CO distribution in October, when all sources are included. The CO concentration reaches maxima in the industrialized regions and in the tropics. Figure 14.2*b* shows the change in the CO mixing ratio at the surface level (October) resulting from biomass burning emissions. The CO concentration increases at all latitudes. This increase is greater than 20% everywhere and reaches maximum values of 300% over the areas where biomass burning takes place. The corresponding change in the zonally averaged CO mixing ratio as a function of latitude and height is displayed in figure 14.2*c*. The largest changes in October (about 80%) take place in the southern tropics during the dry season. Large amounts of CO from biomass burning are efficiently transported to the upper troposphere by convection, and redistributed towards higher latitudes in the Southern Hemisphere. The CO increase is minimum in the 10–30°N latitude band, where other CO sources (i.e., biogenic and methane oxidation) dominate. The higher values calculated for the northern mid and high latitudes are due to temperate and boreal forest fires.

The global CO burden increases in October by about 45% as a result of biomass burning. A similar behavior is observed for all seasons, the increase in the tropospheric CO burden due to biomass burning varying from 30% in February to 45% in October, as a result of seasonal changes in the amount of biomass burned.

Simulations in which each biomass burning source is isolated show that, in South America, the increases in the CO mixing ratio associated with biomass burning result for about two thirds from forest fires, the remaining being mostly due to savanna fires. The CO increase in southern Africa is mostly due to savanna burning, which is most pronounced in this area during October, and the CO increase over equatorial Africa is mainly the result of forest burning. Forest fires and fuel-wood use are the main causes of the CO increase in southern Asia.

Table 14.3 provides the global budget of CO (annual mean) in the troposphere as calculated by the model, and the contribution of each source to the global burden of CO. These values show the importance of biomass burning for the determination of the CO tropospheric burden: about 26% of the tropospheric CO burden results from biomass burning. The tropospheric CO burden evaluated here is about 10% lower than in the previous estimate of Müller and Brasseur (1995) due to lower emissions from biomass burning, especially from the savanna fires.

Biomass burning represents 45%, 37%, 32%, and 5% of the total source of ethylene, ethane, propylene, and other NMHCs, respectively (see table 14.2). The global distribution of these compounds should there-

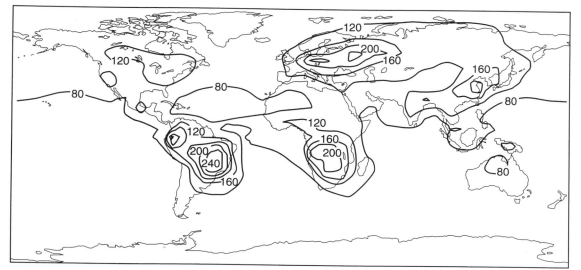

Contour Levels (ppbv) = (0, 40, 80, 120, 160, 200, 240, 280) (a)

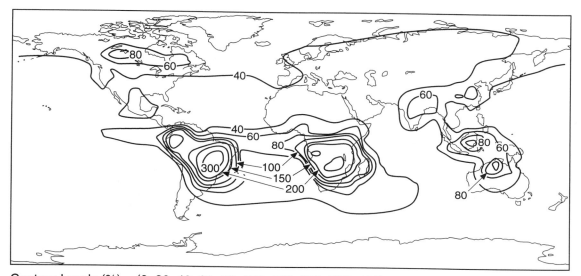

Contour Levels (%) = (0, 20, 40, 60, 80, 100, 150, 200, 300, 400) (b)

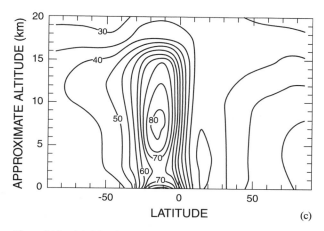

Figure 14.2 (*a*) CO mixing ratio distribution at the surface in October. (*b*) Change in the CO mixing ratio in October resulting from biomass burning emissions at the surface level and (*c*) as a function of latitude and altitude (zonal average)

Table 14.3 CO tropospheric budget (Tg CO/y)

Total emissions	1299 Tg/y
Emission from biomass burning	585 Tg/y
Photochemical production	946 Tg/y
Photochemical loss	2016 Tg/y
Dry deposition	229 Tg/y
CO burden—all sources	327 Tg
Forest burning	41 Tg
Savanna burning	15 Tg
Fuel-wood use	17 Tg
Agricultural waste	9 Tg
Total biomass burning	82 Tg

fore be largely affected by biomass burning. Figure 14.3 shows the change in the ethane (figure 14.3a) and ethylene (figure 14.3b) surface mixing ratios resulting from biomass burning in October. The mixing ratio of C_2H_6 increases by approximately a factor of 4 over biomass burning areas. As C_2H_6 has a relatively long lifetime (about 80 days), it can be transported to mid and high latitudes, like CO. This is not the case for C_2H_4, whose lifetime is on the order of 1 day. Hence, the additional ethylene concentration due to biomass burning remains near the source areas. Table 14.4 presents the percentage change in the tropospheric NMHCs burdens (annual mean) from a no-biomass burning case to a case in which each biomass burning source is considered separately. Forest fires and fuel-wood use have the greatest impact on the NMHCs surface mixing ratio.

Nitrogen Oxides

Biomass burning accounts for approximately 13% of the global NO_x (which represents the sum [NO] + [NO_2]) source (see table 14.2). Figure 14.4 represents the change in the NO_x concentration at the surface resulting from biomass burning. In contrast with CO, NO_x has a short lifetime, so that it is less efficiently transported. The observed increase in the NO_x mixing ratio is therefore located near the sources of biomass burning, where the NO_x increase can be very high. As for CO, increases in the NO_x concentration over South America and southern Africa are mostly due to forest and savanna burning, while changes over Asia result mainly from fuel-wood and agricultural waste burning. It should be noted that the concentration of other nitrogen compounds such as HNO_3, which have a longer lifetime than NO_x, or such as PAN, which is

affected by the increased NMHCs and NO_x emissions can also increase significantly at mid and high latitudes, away from the source areas.

A yearly averaged global budget for NO_x is given in table 14.5. Because the impact of biomass burning on the NO_x distribution is quite localized and remains significant only in the lowest layers of the atmosphere, biomass burning contributed to only 14% of the tropospheric NO_x burden in the model.

Ozone and the Hydroxyl Radical

As a consequence of the increase in the concentrations of the ozone precursor (i.e., CO, NMHCs, and NO_x), the ozone production is expected to increase as a result of biomass burning. This increase should affect the whole troposphere, due to vertical transport by advection and convection of the ozone precursors. The model predicts a 19% increase in the globally averaged net photochemical production of ozone (annual mean) resulting from biomass burning. This value is slightly higher than the value suggested by Bonsang et al. (1994), who evaluated the net ozone production in the troposphere from biomass burning in the tropics as being about 10% of the total ozone production. Such an increase leads to an enhanced concentration of ozone at the surface, as shown in figure 14.5a. Figure 14.5b displays the corresponding change in the vertical ozone distribution. The ozone concentration increases at all latitudes and altitudes by at least 5%, with maximum values around 100% in the vicinity of biomass burning areas.

The impact of biomass burning on the hydroxyl radical concentration is more complex (figure 14.6a and b). The increase in CO at all altitudes explains the calculated decrease of OH in the free troposphere and at mid latitudes. Near the source regions, however, the higher ozone concentrations result in an enhanced OH production. At the same time, the higher conversion of HO_2 to OH associated with the increase in nitrogen oxide concentration produces an increase (up to 10^6 cm^{-3} or 100%) in the OH density above the source areas.

The OH concentration determines the lifetime of most tropospheric species. When biomass burning is included in the model simulations, the OH concentration can either increase or decrease in different areas, so that the change in the globally averaged OH concentration is small. Consequently, the impact of biomass burning on the global lifetime of long-lived species is limited. For example, the yearly global average lifetime of methane changes from 9.66 years when biomass burning is ignored to 9.86 years when biomass

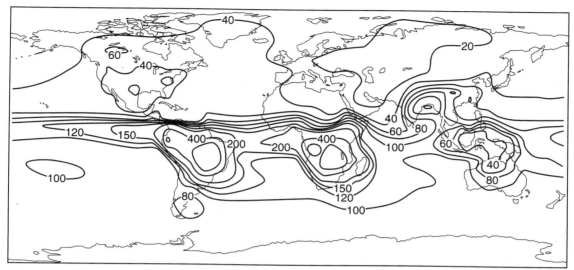

Contour Levels (%) = (0, 20, 40, 60, 80, 100, 120, 150, 200, 400, 600)   (a)

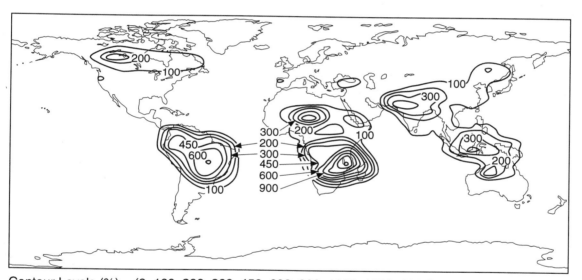

Contour Levels (%) = (0, 100, 200, 300, 450, 600, 900, 1200, 1500, 1800)   (b)

Figure 14.3   Change in the (*a*) ethane and (*b*) ethylene surface mixing ratios resulting from biomass burning in October

Table 14.4   Percentage increase in the NMHCs tropospheric burdens due to biomass burning averaged yearly

	C_2H_4	C_2H_6	C_3H_6	Other NMHCs
Forest burning	17%	23%	11%	2.8%
Savanna burning	3%	4.3%	1.5%	0.2%
Agriculture waste burning	3%	2.9%	1.4%	0.8%
Fuel-wood burning	21%	19%	12%	2.4%
Total	44%	49.2%	25.9%	6.2%

Table 14.5 NO_x tropospheric budget (Tg N/y)

Total emissions	34.5 Tg/y
Emission from biomass burning	4.5 Tg/y
Photochemical production	71.6 Tg/y
Photochemical loss	100 Tg/y
Dry deposition	4.5 Tg/y
NO_x burden	
All sources	0.52 Tg
Due to biomass burning	0.02 Tg

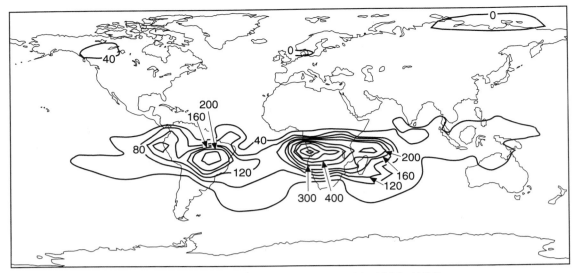

Contour Levels (%) = (0, 40, 80, 120, 160, 200, 300, 400, 500, 1000, 1500)

Figure 14.4 Change in the NO_x concentration at the surface resulting from biomass burning emissions in October

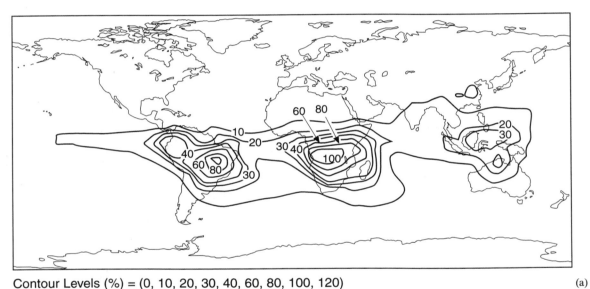

Contour Levels (%) = (0, 10, 20, 30, 40, 60, 80, 100, 120)

(a)

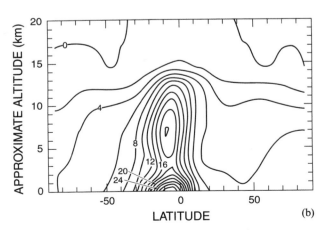

(b)

Figure 14.5 Change in the ozone mixing ratio in October resulting from biomass burning at (a) the surface level and (b) as a function of latitude and altitude (zonal average)

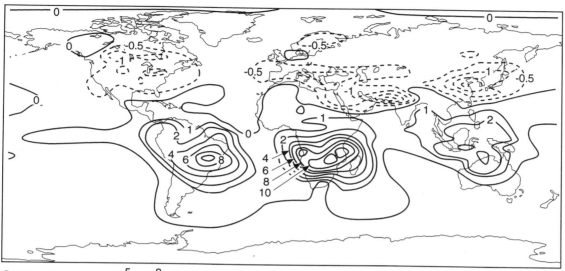

Contour Levels (in 10^5 cm^{-3}) = (-2, -1.5, -1, -0.5, 0, 1, 2, 4, 6, 8, 10, 12)   (a)

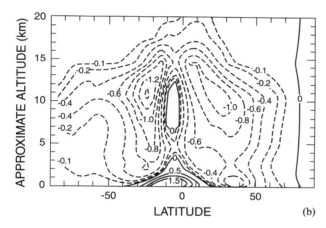

(b)

Figure 14.6   Change in the OH concentration in October resulting from biomass burning (*a*) at the surface level and (*b*) as a function of latitude and altitude (zonal average)

burning is included, which corresponds to a change of less than 2%.

Conclusions

This chapter confirms the large contribution of biomass burning to the emissions of CO, CH$_4$, NO$_x$, and nonmethane hydrocarbons: biomass burning represents about 45% of the surface source of CO and of ethylene, and 8 and 13% of the total source of CH$_4$ and NO$_x$, respectively. The impact of such sources on the chemical composition of the troposphere has been assessed with a three-dimensional model. It has been shown that trace gases emitted from biomass burning produce significant increase in the concentration

of several tropospheric compound concentrations on the global scale. As a result of biomass burning, ozone concentrations increase by up to 100% in the tropical burning areas, and by at least 5% at higher latitudes. The concentration of the hydroxyl radical, which determines to a large extent the oxidizing capacity of the atmosphere, either increases or decreases depending on the area, but its globally averaged concentration change is rather small.

Acknowledgments

The National Center for Atmospheric Research is sponsored by the National Science Foundation. Claire Granier is supported by NASA, under contract No.

W-18,531. Jean-François Müller is research assistant of the Belgian Fonds de la Recherche Scientifique. This work was partly supported by the Belgian Federal Office for Scientific, Technical and Cultural Affairs (O.S.T.C.).

References

Andreae, M. O. 1993. The influence of tropical biomass burning on climate and the atmospheric environment, in *Biogeochemistry of Global Change*, Ed. R. S. Oremland, Chapman and Hall, New York, pp. 113–150.

Bonsang, B., M. Kanakidou, and C. Boissard. 1994. Contribution of tropical biomass burning to the global budget of hydrocarbons, carbon monoxide, and tropospheric ozone, in *Non-CO$_2$ Greenhouse Gases*, Eds. J. van Ham et al., Kluwer Academic Publishers, Dordrecht, the Netherlands.

Crutzen, P. J., L. E. Heidt, J. P. Krasnec, W. H. Pollock, and W. Seiler. 1979. Biomass burning as a source of atmospheric trace gases, *Nature*, 282, 253–256.

Hao, W. M., and D. E. Ward. 1993. Methane production from global biomass burning, *J. Geophys. Res.*, 20 657–20 661.

Hao, W. M., and M.-H. Liu. 1994. Spatial distribution of tropical biomass burning in 1980 with 5° × 5° resolution, *Global Biogeochem. Cycles*, 8, 495–503.

Hao, W. M., M.-H. Liu, and D. E. Ward. 1994. Impact of deforestation on biomass burning in the tropics, in *Proceedings of Global Climate Change: Science, Policy, and Mitigation Strategies*, Ed. C. V. Mathai and G. Stensland, Air and Waste Management Association, 273–278.

Hao, W. M., D. E. Ward, G. Olbu, and S. P. Baker. 1995. Emissions of CO$_2$, CO, and hydrocarbons from fires in diverse African savanna ecosystems, *J. Geophys. Res.* special SAFARI issue, in press, 1996.

Levine, J., W. Cofer, and J. Pinto. 1994. Biomass burning, in *Atmospheric Methane: Source, Sinks, and Role in Global Change*, Ed. M. Khalil, NATO/ASI Series, Ch. 14, 299–313.

Müller, J. F., and G. Brasseur. 1995. IMAGES: A three-dimensional chemical transport model of the global troposphere, *J. Geophys. Res.*, June.

OECD. 1989. Organization for Economic Cooperation and Development/International Energy Agency, Quarterly oil statistics and energy balances, fourth quarter 1988, Paris.

Smolarkiewicz, P. K., and P. J. Rasch. 1991. Monotone advection on the sphere: An Eulerian versus semi-Lagrangian approach, *J. Atmos. Sci.*, 48, 793–810.

United Nations (UN). 1988. Statistiques de l'environnement en Europe et en Amérique du Nord, Recueil expérimental, *Normes et Etudes Statistiques*, vol. 39, New York.

USDA. 1986. United States Forest Service, An analysis of the timber situation of the United States, Appendix 3.

WMO/UNEP. 1995. *Scientific Assessment of Ozone Depletion: 1994.* World Meteorological Organization, Geneva, Switzerland.

Modeling the Influence of Vegetation Fires on the Global Carbon Cycle

Frank Mack, Johannes Hoffstadt, Gerd Esser, and Johann G. Goldammer

At present, annual global carbon emissions (gross fluxes) from vegetation fires and other phytomass (plant biomass) burning (e.g., fuel-wood and charcoal burning) are estimated at 4.08 Pg. This compares to global annual carbon emissions from fossil fuel burning of approximately 5 Pg (Andreae and Goldammer 1992; Andreae 1993). Thus, the contribution of vegetation fires to the total global pyrogenic carbon emissions is estimated to amount to approximately 40%. Savanna fires and the deforestation of dry, seasonal, and evergreen tropical forests contribute most to these emissions, whereas forest fires in temperate and boreal regions are of minor importance (Andreae 1993; Crutzen and Andreae 1990; Hao et al. 1990; Houghton et al. 1987; Olson et al. 1983).

While the release of carbon in various forms of radiatively active trace gases contributes to the "greenhouse effect," the formation of black carbon has a contrary effect because it constitutes a carbon sink. Black carbon (synonyms: *soot, elemental carbon*) is the chemically inert and biologically nondegradable fraction of carbon produced by fires. The magnitude of this potential carbon sink is unknown due to a lack of data dealing with the formation and the decomposition rate of black carbon. At present, Kuhlbusch and Crutzen (1995, chapter 16 this volume) assume an annual global formation of 50–270 Tg black carbon originating from vegetation fires.

With the exception of deserts and sparsely vegetated lands, fires are occurring in almost all biomes. However, reliable data on fire occurrence, (e.g., long-term statistics about average areas and phytomass burned per year and vegetation type) exist only in few places. Hence, the existing estimates of global pyrogenic emissions and of the formation of black carbon are not reliable.

In order to determine the pyrogenic carbon fluxes it is also important to know the proportions of phytomass, litter, and soil organic carbon at the total amount of biospheric carbon. How do vegetation fires influence the absolute amount of carbon and the shift of the relative proportions between the pools? In terms of vegetation science, one of the corresponding questions would be: will tropical savannas expand at the cost of forests?

Other questions are related to climate and particularly to climate anomalies, for example, how are interannual climate anomalies like El Niño-Southern Oscillation (ENSO) event affecting the number and the extent of vegetation fires as well as short-term anomalies of the atmospheric CO_2?

Such internal feedbacks can be handled in global models. Presuming a satisfactory adaptation at present conditions, trend extrapolations and scenarios are feasible in spite of dealing with a complex system characterized by internal feedbacks.

For the successful adaptation of a global fire model to a climate-driven global biospheric carbon cycle model, the following requirements for all grid elements needed to be matched:

- allocation of a biome for each grid element by a global biome model or a global vegetation map;

- information on whether a grid element is agriculturally used, including a clearing scenario for transition stages;

- prediction of the phytomass and its separation in a herbaceous and woody fraction and an above- and belowground fraction respectively; and

- prediction of the litter and its separation in a herbaceous and woody fraction and an above- and belowground fraction respectively.

With the exception of the last item, the High Resolution Biosphere Model (HRBM) fulfills these requirements (Esser et al. 1994).

Methods

Structure of the HRBM
The HRBM has been developed by the systems ecology group of the Institute for Plant Ecology at Giessen University (Germany). It succeeds the Osnabrück Biosphere Model (OBM) developed at the University

of Osnabrück (Esser 1990; 1987; 1986). The HRBM is a model of the terrestrial biosphere calculating the distribution (pools) and the turnover (fluxes) of carbon on a monthly scale for a $0.5 \times 0.5°$ grid (62 483 grid elements) of geographical latitude and longitude. It is driven by mean monthly data for precipitation, temperature, and cloud freeness, and by a soil fertility factor. The climate data are based on annual long-term means providing a constant climate situation (Leemans and Cramer 1991). In addition, a land-use scenario based on data of Richards et al. (1983) and Olson et al. (1983) is included. Grid elements are cleared or afforested following the numerical development of the population until 2050 as predicted by the Food and Agricultural Organization (FAO 1991).

The HRBM is characterized by a modular structure. The modules (or subroutines) deal with separate and different issues from other modules. All modules are adaptable to other carbon cycle models due to defined interfaces between the modules and the main program. Therefore, the HRBM was utilized for the results published in this chapter. For the first fire modeling approach by Mack (1994), the biome model of Prentice et al. (1992) was used, which is also incorporated as a subroutine in the HRBM.

The Module of Vegetation Fires in the HRBM

The fire module provides several fluxes which then can be included in a "generic" carbon cycle model. These fluxes fall into three classes, termed *burning* (*BL, burning loss*), *mortality* (*ML, mortality loss*), *and black carbon formation* (*CP, charcoal production*).

The burning fluxes transport carbon from the phytomass (*PH*) and litter (*L*) compartments to the atmosphere. Mortality fluxes add to the litter production from phytomass. Black carbon formation fluxes transport carbon from phytomass and litter to the black carbon pool.

All of these fluxes are assumed to be proportional to the corresponding source pool that is affected by fires. For instance, the flux "burning of aboveground herbaceous (*ha*) phytomass" is proportional to the amount of carbon in the aboveground herbaceous phytomass compartment: $PHBL_{ha} = \text{const} \times PH_{ha}$.

The calculation of the different proportionality factors is based on four coefficients, *burning probability* (*cburn*), *burning efficiency* (*cbef*), *burning mortality* (*cbmo*), *and the black carbon formation coefficient* (*cbch*).

The *burning probability* (*cburn*) in a given month determines the frequency of fires. Every month, this probability is used together with a random number

generator to decide whether the respective grid element will not be affected by fires, or will be totally burned in this month.

The *burning efficiency* (*cbef*) determines the amount of carbon emitted into the atmosphere by pyrolytic processes. The completeness of combustion is characterized by *cbef*. The *burning mortality* (*cbmo*) determines the amount of phytomass that dies during a fire without being transformed into CO_2 or black carbon. The *black carbon formation coefficient* (*cbch*) determines the amount of carbon converted to black carbon. In general, each coefficient depends on the material (i.e., it is different for herbaceous and woody material, and for phytomass and litter), and on the biome that dominates in the grid element.

Forming Groups of Biomes

Regarding the influence of the biome type on the various coefficients, it is possible to distinguish three major groups: forests (I); shrub formations (II); savannas, grasslands, and deserts (III). Forests are divided into four subgroups for the purpose of calculating mortality coefficients. Table 15.1 shows the biomes of the IIASA-biome model (Prentice et al. 1992) as they are assigned to the three groups.

Determining the Fluxes

Please note that all fluxes and coefficients are calculated for each grid element; those indices are omitted for the sake of readability. The four compartments (herbaceous/woody, above- /belowground) are denoted by the indices *ha, hb, wa, wb*.

All fluxes have a common structure. They contain a logarithmic factor which reflects the time-integrated interpretation of a coefficient, *coeff*, as the fraction of affected-to-total area of the respective grid element. If one takes only a single depleting flux F (e.g., burning loss) into account, a pool P loses carbon according to

$$dP/dt = -F = +\ln(1 - coeff) \cdot P$$

Integrated over one month, the resulting carbon amount is

$$P = P_0 \cdot (1 - coeff)$$

This is the fraction that corresponds to the area not affected.

Burning (*PHBL, LBL*)   Only aboveground pools are directly affected. There are four burning fluxes:

$$PHBL_{ha} = -\ln(1 - cbef_{PH,h}) \cdot PH_{ha}$$

$$PHBL_{wa} = -\ln(1 - cbef_{PH,w}) \cdot PH_{wa}$$

Table 15.1 IIASA-biomes grouped for burning characteristics

Forests (I)	Shrub formations (II)	Savannas, grasslands, and deserts (III)
Subgroup Ia Taiga Cold mixed forest Cool conifer forest Cool mixed forest	Tundra Broad-leaved evergreen forest/ warm mixed forest[c] Xerophytic woods/scrub	Ice/polar desert Cool grass/shrub Hot desert Semidesert
Subgroup Ib Cold deciduous forest Temperate deciduous forest		Warm gas/shrub Tropical dry forest/savanna
Subgroup Ic Broad-leaved evergreen forest/ warm mixed forest[a]		
Subgroup Id Burning tropical seasonal forest Tropical rain forest Cool grass/shrub[b]		

a. Only for calculating burning probability and for calculating burning efficiency.
b. Only for calculating burning probability.
c. Only for calculating dead fine fuel moisture content.

$$LBL_{ha} = -\ln(1 - cbef_{L,h}) \cdot L_{ha}$$

$$LBL_{wa} = -\ln(1 - cbef_{L,w}) \cdot L_{wa}$$

Mortality (*PHML*) Belowground herbaceous phytomass is not affected. It is assumed that the woody parts from both above- and below-ground die together, therefore they share a common coefficient. There are three mortality fluxes:

$$PHML_{ha} = -\ln(1 - cbmo_h) \cdot PH_{ha}$$

$$PHML_{wa} = -\ln(1 - cbmo_w) \cdot PH_{wa}$$

$$PHML_{wb} = -\ln(1 - cbmo_w) \cdot PH_{wb}$$

Black Carbon Formation (*PHCP, LCP*) Black carbon formation occurs only above ground. The *black carbon formation coefficient* (*cbch*) is determined by Kuhlbusch (1994) as that part of the charcoal which is produced by vegetation fires and which is biologically not decomposable. *cbch* is calculated separately for herbaceous and woody material.

There are four fluxes:

$$PHCP_{ha} = -\ln(1 - cbch_{PH,h}) \cdot PH_{ha}$$

$$PHCP_{wa} = -\ln(1 - cbch_{PH,w}) \cdot PH_{wa}$$

$$LCP_{ha} = -\ln(1 - cbch_{L,h}) \cdot L_{ha}$$

$$LCP_{wa} = -\ln(1 - cbch_{L,w}) \cdot L_{wa}$$

Determining the Coefficients
The four coefficients *cburn, cbef, cbmo,* and *cbch* are calculated using several auxiliary variables, namely the

humidity index (*hi*), the *diameter at breast height* (*dbh*), the *percentage of litter in the total above-ground herbaceous material* (*cur*—the "curing factor"), and the *dead fine fuel moisture content* (*fmc*). The first three are straightforward, but the fuel moisture content itself depends on two other numbers, the current values for temperature (t_f) and relative humidity (rh_f) during fires, or a short time before fires.

In forests, burning efficiency, mortality, and black carbon formation of woody phytomass drops with increasing diameter of the trees. The "affected fraction" of the wood, $\lambda(dbh)$, is a function of *dbh* which decreases linearly with increasing diameter. λ lies in the range 0 to 1 and is assumed to be 1 if not stated otherwise.

Auxiliary Variables The humidity index (*hi*) is an applied precipitation (P_p) to temperature (T) ratio by Gaussen (cit. by Kreeb 1983). It has been widely used by Walter and Lieth (1960) in their climate diagrams to distinguish between humid and arid months.

$$hi = \frac{P_p}{2mm} - \frac{T}{1°C} \tag{15.1}$$

As stated above, *cur* is the fraction of aboveground herbaceous litter in relation to the total aboveground herbaceous plant material [in %].

$$cur = \frac{L_{ha}}{L_{ha} + PH_{ha}} \cdot 100\% \tag{15.2}$$

The diameter at breast height (*dbh*) is used to determine the burning efficiency and the mortality coefficient in nontropical forests. The equation was empirically derived using the data published by Cannell (1982). *dbh* depends on the amount of herbaceous and woody phytomass.

$$dbh = 0.856 \cdot 10^{-4} \cdot PH^{0.857} \cdot \exp(-0.904 \cdot 10^{-5} \cdot PH) \tag{15.3}$$

The temperature during fires, t_f, is calculated from the monthly mean temperature:

$$t_f = 0.64 \cdot T + 14.7 \tag{15.4}$$

The relative humidity during fires, rh_f, is calculated from the humidity index *hi*, except for forest biomes, where an influence was not found. Therefore, rh_f is set to an empirically derived constant value in forests:

$$rh_f = \begin{cases} 0.5 \cdot hi + 40.5 & \text{everywhere, except} \\ 34.4 & \text{in forests} \end{cases} \tag{15.5}$$

Dead Fine Fuel Moisture Content   Fine fuel moisture content (*fmc*) of burned material is calculated from current values for temperature (t_f) and relative humidity (rh_f) during fires or a short time before fires and from the proportion of aboveground herbaceous litter at the total aboveground herbaceous plant material (*cur*). Different equations are used for different groups of biomes.

Forests (I)   Fine fuel moisture content is calculated following parts of the Canadian Forest Fire Weather Index System (CCFWIS) (van Wagner 1987):

$$fmc = 0.942 \cdot rh_f^{0.679} + 11 \cdot \exp((rh_f - 100)/10)$$
$$+ 0.18 \cdot (21.1 - t_f) \cdot (1 - \exp(-0.115 \cdot rh_f))$$

Shrub Formations (II)   Here, *fmc* is determined using Burgan's method to calculate fuel moisture content in fynbos (Burgan 1987):

$$fmc = \begin{cases} 0.03229 + 0.262577 \cdot rh_a \\ \quad - 0.0010404 \cdot t_a \cdot rh_a & \text{if } rh_a < 10 \\ 1.754402 + 0.160107 \cdot rh_a \\ \quad - 0.026612 \cdot t_a & \text{if } 10 \leq rh_a \leq 50 \\ 21.0606 - rh_a \\ \quad \cdot (0.00063 \cdot t_a + 0.0112) \\ \quad + 0.005565 \cdot rh_a^2 \\ \quad - 0.483199 \cdot rh_a & \text{otherwise} \end{cases}$$

t_a and rh_a are used to correct t_f and rh_f for the influence of radiation. Radiation is derived from the cloud freeness *clf*, ranging from 0 (complete cloud cover) to 1 (blue sky).

$$t_a = \begin{cases} t_f + 13.9 & \text{if } clf > 0.9 \\ t_f + 10.6 & \text{if } 0.9 \geq clf \geq 0.55 \\ t_f + 6.7 & \text{if } 0.55 > clf \geq 0.1 \\ t_f + 2.8 & \text{otherwise} \end{cases}$$

$$rh_a = \begin{cases} 0.75 \cdot rh_f & \text{if } clf > 0.9 \\ 0.83 \cdot rh_f & \text{if } 0.9 \geq clf \geq 0.55 \\ 0.92 \cdot rh_f & \text{if } 0.55 > clf \geq 0.1 \\ 1.00 \cdot rh_f & \text{otherwise} \end{cases}$$

Savannas, Grasslands, and Deserts (III)   *fmc* is calculated for grassland and desert biomes using the Australian Grassland Fire Danger Meter (GFDM) Mark 5:

$$fmc = \frac{97.7 + 4.06 \cdot rh_f}{t_f + 6.0} - 0.00854 \cdot rh_f + \frac{3000}{cur} - 30.0$$

Regarding the biome *tropical dry forest/savanna*, the equation was adapted by leaving out the proportion of above-ground herbaceous *litter* at the total aboveground herbaceous plant material (*cur*):

$$fmc = \frac{97.7 + 4.06 \cdot rh_f}{t_f + 6.0} - 0.00854 \cdot rh_f$$

Burning Probability   Burning probability (*cburn*) is derived from the humidity index. The less humid a region is, the higher is *cburn*. It is calculated for different groups of biomes.

The equations were empirically derived using data for fire cycles on a monthly time scale from Wein and Moore (1979) (transition zone between boreal and cool temperate forest in Nova Scotia, Canada) for forests (I), Brown et al. (1991) (fynbos) for shrub formations (II), Braithwaite and Estbergs (1985) as well as Lamotte et al. (1985 cit. by Menaut et al. 1991) and Menaut (pers. comm.) in savannas for savannas, grasslands, and deserts (III). The *fire cycle* is defined as the average time span needed to burn an area equal to the entire area of interest (Romme 1980). This approach assumes that *cburn* is the reciprocal value of the fire cycle (expressed in months).

Under certain conditions no fire is possible, *cburn* = 0 (see Mack 1994 for a detailed explanation):

if too cold: $it < 0°C$
if too humid: $hi > 50$
if fuel too wet: $fmc > 25\%$
                    (35% for biome group (III))
if fuel too sparse: fuel type and minimum amount
                    of above-ground plant material
                    depend on the biome group:

$$(I) \quad L_{ha} + L_{wa} < 45 \text{g} \cdot \text{m}^{-2}$$
$$(II) \quad PH_{ha} + PH_{wa} + L_{ha} + L_{wa}$$
$$< 180 \text{g} \cdot \text{m}^{-2}$$
$$(III) \quad PH_{ha} + PH_{wa} + L_{ha} + L_{wa}$$
$$< 45 \text{g} \cdot \text{m}^{-2}$$

Otherwise, *cburn* is determined according to

$$cburn = \begin{cases} 0.0058 \cdot \exp(-0.107 \cdot hi) & \text{for biome group (I)} \\ 0.00083 \cdot \exp(-0.117 \cdot hi) & \text{for biome group (II)} \\ 0.025 \cdot \exp(-0.081 \cdot hi) & \text{for biome group (III)} \end{cases}$$
$$(15.6)$$

Fuel comprises aboveground phytomass and aboveground litter pools with exception of the forests, where we assume that only litter contributes to the fuel.

Cleared Areas The burning probability for grid elements that are cleared is calculated by the use of a maximum function based on either humidity index *cburn_hi* or amount of litter *cburn_lw*. Similar to the previous equations, fire is believed to occur only if the mean monthly temperature is above 0°C. In contrast to the previous equations, fire is assumed to be independent from the calculated fuel moisture content since artificial drying is prevailing. Fire is excluded in agriculturally used areas.

$$cburn_{lw} = \min\{1, \max\{0, 0.214 \cdot \ln(0.023 \cdot L_{wa})\}\}$$
$$cburn_{hi} = \min\{1, \max\{0, 0.229 \cdot \exp(-0.049 \cdot hi)\}\}$$
$$cburn = \begin{cases} 0 & \text{if } T \le 0 \\ \max(cburn_{hi}, cburn_{lw}) & \text{otherwise} \end{cases}$$

The min/max expressions ensure that the *probabilities* cannot exceed the limits 0 and 1.

Burning Efficiency and Mortality Coefficient A simple, general formula is provided to calculate burning efficiency (*cbef*) and mortality coefficient (*cbmo*) for some pools and groups of biomes.

The general burning efficiency has been empirically derived from the dead fine fuel moisture content, *fmc*. The drier the fuel, the higher is the burning efficiency. Note that *cbef* ≥ 0.45.

$$cbef = 1.00 - \frac{0.55}{1 + \exp(5.24 - 0.76 \cdot fmc)} \qquad (15.7)$$

The *mortality coefficient* (*cbmo*) is calculated separately for herbaceous and woody material. The belowground woody phytomass pool is the only belowground pool directly affected by fires. It is assumed that the belowground parts die together with the aboveground parts. Therefore, similar values of *cbmo* for either

proportion of *PH* are assumed. In contrast, belowground herbaceous phytomass is *not* affected by vegetation fires.

The general mortality coefficient is based on the assumption that *all* carbon reached by fires (the fraction is given by λ) takes one of the three routes: into the air (*cbef*), into black carbon (*cbch*), and into litter (*cbmo*). Therefore,

$$cbef + cbch + cbmo = \lambda \quad \text{or} \quad cbmo = \lambda - cbef - cbch$$
$$(15.8)$$

Table 15.2 shows the cases to which this general formula and exceptions are applied.

Boreal and Cool Temperate Zones Dominated by Conifers (Ia) Based on the investigation by Hogan in a *Picea mariana-Cladonia alpestris* forest close to Schefferville, Canada (unpubl. cit. by Auclair 1983), one third of the aboveground woody phytomass is burned: $cbef_{const} = 0.33$.

The equations are based on investigations in *Pseudotsuga menziesii* and *Pinus banksiana* stands (Bergeron and Brisson 1990; Peterson and Arbaugh 1986, 1989). The affected fraction of the wood is given by

$$\lambda(dbh) = \min\{1, \max\{0, -2.68 \cdot dbh + 1.11\}\} \qquad (15.9)$$

The burning efficiency decreases from the maximum value (at $dbh < 0.042$) to zero (at $dbh > 0.413$):

$$cbef_{PH,w} = \lambda \cdot cbef_{const} \qquad (15.10)$$

One third of the affected phytomass is consumed by fire, and another (small) fraction is used to form black carbon. The rest of the affected phytomass dies according to equation (15.8).

Cold Deciduous Forest The equations are based on investigations in *Populus tremuloides* and *Populus tremuloides*/mixed hardwood stands (Alexander and Sando 1989; Quintillo et al. 1989). The affected fraction of the wood is given by

$$\lambda(dbh) = \min\{1, \max\{0, -4.79 \cdot dbh + 0.98\}\} \qquad (15.11)$$

leading again to a burning efficiency of

$$cbef_{PH,w} = \lambda \cdot cbef_{const} \qquad (15.12)$$

which reaches zero at $dbh = 0.204$. The mortality coefficient is determined according to the general rule, equation (15.8).

Burning Efficiency in All Other Forest Biomes The burning material consists mainly of herbaceous and litter material, since ground fires are prevailing (Albini

Table 15.2 Calculation of the coefficients related to burning, depending on biome groups (*gen.* means *general formulas*)

	Biome group I	Biome group II	Biome group III
$cbef_{PH,h}$	depends on dbh[a]	1.0[b]	*gen.*
$cbef_{PH,w}$	see text	*gen.*	$\begin{cases} 0.02^{c} & \text{Tropical dry forest/savanna} \\ gen. & \text{otherwise} \end{cases}$
$cbef_{L,h}$	gen.	*gen.*	*gen.*
$cbef_{L,w}$	gen.	*gen.*	0.25[d]
$cbmo_{ha}$	same as 'wa'[e]	0 (no litter)[f]	*gen. (all dead)*[g]
$cbmo_{wa}$	see text	*gen.*[h]	$\begin{cases} 0^{i} & \text{Tropical dry forest/savanna} \\ gen.^{j} & \text{otherwise} \end{cases}$
$cbmo_{hb}$		0[k]	
$cbmo_{wb}$		same as woody, above ground[l]	

a. Linear relations are assumed due to the available data providing relations between *dbh* and mortality coefficient:

$$cbef_{PH,h} = \max(0, -5.93 \cdot dbh + 1.0) \tag{15.17}$$

For temperate deciduous forests, $cbef_{PH,h}$ is set to 0 (Albini 1992).
b. Some investigations prove that the above ground herbaceous phytomass in shrub formations is totally consumed by vegetation fires (Cass et al. 1984; Griffin and Friedel 1984; van Wilgen 1982).
c. This is based on an investigation by Hopkins (1965) in the Okolomeji Forest, Nigeria.
d. Studies made by Frost (1985) resulted in lower burning efficiencies of woody litter in savannas compared to herbaceous litter. According to these studies, a value of 25% is estimated.
e. Due to lack of data, it is assumed that the mortality coefficient of herbaceous phytomass is identical with the mortality coefficient of woody phytomass. However, the restriction holds that the sum of burning efficiency, mortality coefficient, and black carbon formation coefficient may not exceed unity. Therefore, we have

$$cbmo_h = \begin{cases} 1 - (cbef_{PH,h} + cbch_{PH,h}) & \text{if } cbmo_w + cbef_{PH,h} + cbch_{PH,h} > 1 \\ cbmo_w & \text{otherwise} \end{cases} \tag{15.18}$$

f. Due to the complete combustion of herbaceous phytomass, $cbmo_h$ is set to 0.
g. This was proved by Bragg (1982) in the prairie (Nebraska).
h. See also van Wilgen 1982; Barro and Conrad 1991; Green 1981; Griffin and Friedel 1984; Kilgore 1973; Minnich 1983; Rutherford and Westfall 1986; Vogl and Schorr 1972; Wright et al. 1979.
i. The trees are adapted to frequent fires and therefore resistant.
j. Due to lack of data, $cbmo_w$ is assumed to be identical with $cbmo_w$ in shrub formations.
k. It is assumed that the belowground herbaceous phytomass is not affected by fires.
l. Due to lack of data, it is assumed that the belowground parts of a plant die together with the aboveground parts.

1992 for temperate deciduous forests). The amount of burned fine woody material is negligible compared to the largely unaffected stem wood of the trees. Therefore, $cbef_{PH,w}$ is set to zero.

Broad-leaved Evergreen Forest/Warm Mixed Forest (Ic) The equation is based on investigations in *Pinus spec.* stands (Storey and Merkel 1960; Lindenmuth 1960 cit. by Wright 1978).

$$cbmo_w = \begin{cases} 0 & \text{if } dbh > 0.191 \\ -2.56 \cdot dbh + 0.49 & \text{otherwise} \end{cases} \tag{15.13}$$

Tropical Rain and Tropical Seasonal Forests (Id) According to studies by Uhl and Kauffman (1990) at least 50% of the woody phytomass was transformed to litter.

$$cbmo_w = 0.5 \tag{15.14}$$

Black Carbon Formation Coefficient The general black carbon formation coefficient (*cbch*) increases with lower burning efficiency. A linear interdependence is assumed. *chmax* denotes the maximum black carbon formation coefficient (Comery et al. 1981; Fearnside 1991; Kuhlbusch 1994) and is set to 0.02. The formula is valid only for reasonably high values of efficiency.

$$cbch = chmax \cdot (1 - cbef), \quad cbef > 0.1 \tag{15.15}$$

A modified version of this formula is used in forest biomes (subgroups Ia to Ie): Regarding the herbaceous phytomass and the woody phytomass in forests, burning efficiencies less than 0.5 are common. In this case, a proportional relation between burning efficiency and black carbon formation coefficient is assumed to prevent unrealistically high black carbon formation coefficients with very low burning efficiencies.

$$cbch = chmax \cdot (0.5 - |cbef - 0.5|), \quad cbef > 0.1 \tag{15.16}$$

Results and Discussion

Global Results

The HRBM enlarged by the fire submodel computes mean annual carbon emissions of 4.14 $Pg \cdot yr^{-1}$ produced by vegetation fires (figure 15.1). Andreae (1993) and Olson (1981) estimated this carbon flux at 4.08 $Pg \cdot yr^{-1}$ and 4.98 $Pg \cdot yr^{-1}$, respectively. (The use of fuel wood of 0.88 $Pg \cdot yr^{-1}$ (Andreae 1993) and the burning of agricultural waste of 0.32 $Pg \cdot yr^{-1}$ (Andreae 1991) and 1.30 $Pg \cdot yr^{-1}$ (Olson 1981) were included in both cases.) The fact that grazing has not been included in the model may explain the high values of the model.

According to the model, 3.13 $Pg \cdot yr^{-1}$ carbon of the phytomass is globally transformed to litter, and 0.044 $Pg \cdot yr^{-1}$ black carbon is produced. Without including the burning of agricultural waste, the formation of black carbon is estimated by Kuhlbusch (1994) to range between 0.07 and 0.24 $Pg \cdot yr^{-1}$. This estimate is based on the same data the model relies on. The rather low model result derives from the model approach assuming a coefficient for the black carbon formation at the lower end of the possible range.

It is important to know whether formation and depletion of black carbon are in an equilibrium or whether natural fires alone or combined with an increasing number of human-caused fires may constitute a significant carbon sink. Houghton (1991) claims that black carbon cannot be accumulated for millions of years, but must be in a steady state. As a consequence, only disturbances of this equilibrium like the growing extent of fires may effect a net flux of black carbon.

The consideration of the impact of vegetation fires on the HRBM leads to a reduction of the total phytomass by 22.9%. At the end of 1980, the carbon content has been reduced from 583.4 Pg to 449.8 Pg. The herbaceous fraction comprises 20.1 Pg compared with 21.6 Pg omitting vegetation fires. The main reason for the difference is a dramatic reduction of the woody phytomass by 57.9% in tropical seasonal and tropical rain forests bordering tropical dry forests or savannas.

The carbon content of the litter pools dropped by 5.5%, the soil organic carbon content decreased by 16.1%. The litter pools comprise 84.4 Pg without fires and 80.1 Pg with fires, whereas the soil organic carbon content changed from 1239.2 Pg to 1039.9 Pg. Between 1860 and 1980, 5.8 Pg black carbon was produced. The total biospheric carbon content differs by 17.4%.

If burning processes are included in the model, the ratio between herbaceous and woody phytomass is affected. After four burns within 10 years, the herbaceous phytomass fraction of tropical seasonal forests may rise from 1.5% to 12%. This is similar to herbaceous phytomass fractions of tropical dry forests and savannas. In bush formations, the ratio may change from 8% to 27% after two burns within three years. On the other hand, the present model approach is not suitable for reflecting the decrease of woody phytomass in tropical dry forests and savannas. Annual burning increases the herbaceous phytomass fraction only from 13% to 15% regarding a grid element of the Ivory Coast. It is important to find a way to model aging and mortality of seedlings and saplings due to fire and other disturbance factors.

The present version of the biome model is not structured to react to such changes of the vegetation. Future biome models should be more dynamic. Biomes should be determined not only by climatic parameters but also by natural and anthropogenic disturbance regimes.

The atmospheric CO_2 content of the HRBM rose to 338.6 ppm in 1980 including the fire submodel and up to 342.5 ppm without fire. The consideration of burning processes forced a reduction of the phytomass in the model prerun resulting in a lower net release of carbon on cleared areas after 1860. According to the model, an increase of vegetation fires after 1860 due to climate change causes a rise of atmospheric CO_2. This is the result of model runs based on a variable instead of a constant climate. Looking at the seasonal cycle of the carbon emissions by vegetation fires, peaks are recorded in July and August (0.43 Pg), January (0.40 Pg), and February (0.38 Pg). Highest carbon sinks are in April (0.26 Pg) and in October (0.23 Pg).

The fire submodel has not yet included agricultural waste burning, effects of grazing, browsing, and trampling, as well as prescribed burning, impacts of fire exclusion, and shifting cultivation. The use of fuel wood is partly considered. Fuel wood not collected by people is considered as fuel available for free-burning fires.

Regional Results

The fire cycles for European and North American countries computed by the fire submodel are shorter than the actually known cycles. The model results were compared with data published annually by the United Nations Economic Commission for Europe (ECE) and the Food and Agriculture Organization of the United Nations (FAO) for both continents (ECE/FAO 1992; 1982). The fire control policies in the industrialized countries as well as landscape fragmentation prevent the spread of large fires. The fire cycle of the model for Germany is 136 times

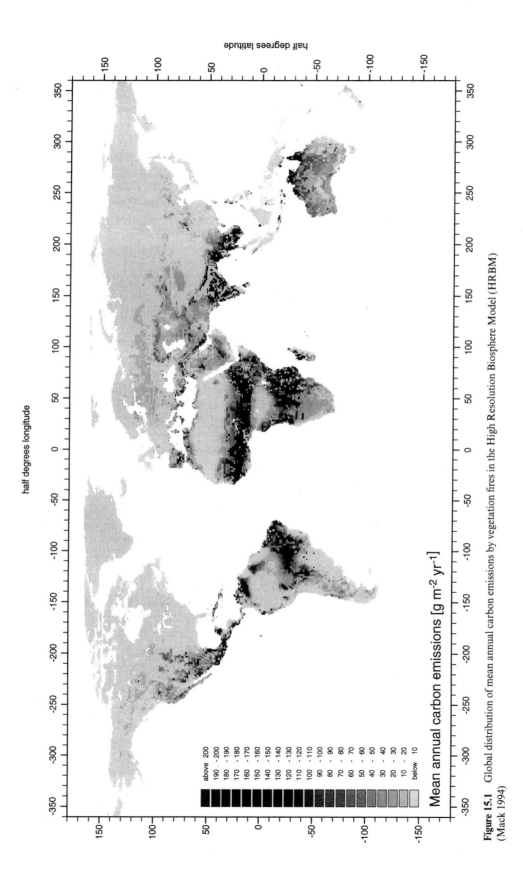

half degrees latitude

half degrees longitude

above	200
190	- 200
180	- 190
170	- 180
160	- 170
150	- 160
140	- 150
130	- 140
120	- 130
110	- 120
100	- 110
90	- 100
80	- 90
70	- 80
60	- 70
50	- 60
40	- 50
30	- 40
20	- 30
10	- 20
below	10

Mean annual carbon emissions [g m⁻² yr⁻¹]

Figure 15.1 Global distribution of mean annual carbon emissions by vegetation fires in the High Resolution Biosphere Model (HRBM) (Mack 1994)

shorter, for Sweden 63 times, and for the USA 38 times shorter as compared with statistical data from the 1980s. As a conclusion, the model provides more natural fire cycles than fire regimes of densely populated industrial countries. A comparison of known natural fire cycles and modeled fire cycles in Sweden confirms this: Zackrisson (1977) determined a natural fire cycle for Sweden of 102 years and Engelmark (1984) of 110 years, respectively. The model computes a fire cycle of 190.1 years exceeding the natural cycle only by a factor of 1.8.

The fire cycles of the Mediterranean countries range from 8.4 to 60.1 years. The southernmost country (Greece) shows the shortest and the northernmost country (France) the longest cycle. These model results match the majority of the available data providing 20 to 50 years for *natural* conditions. The fire cycles calculated by the model for countries with a high proportion of grasslands and savannas are close to most of the available data and estimates ranging between 1 and 5 years (Australia: 1.9 years; Ivory Coast: 3.4 years; India: 2.2 years; Tanzania: 1.8 years; Thailand: 2.5 years).

Model Sensitivity

The coefficients determined in the differential equations of the fire module are based on limited empirical data and many estimates as well as on assumptions by the author. Hence, the coefficients are rather uncertain. Thus, the sensitivity of the model results with regard to changes of the most important coefficients had to be tested by several model runs. The range of uncertainty was estimated referring to the data used. All model runs are characterized by the modification of one coefficient, whereas the others remained unchanged compared with the standard model run.

The doubling of the burning rate caused an increase of the carbon emissions by vegetation fires by 46% to 6.04 Pg. The mortality flux rose by 40% to 4.38 Pg and the black carbon formation by 38% to 0.061 Pg. At the same time, the herbaceous phytomass pools were reduced by 4.2% and the woody phytomass pools by 12.4% to 377.54 Pg carbon. Setting the burning efficiency generally to the maximum of 1 leads to carbon emissions of 7.70 Pg. This amount is 86% higher than the emissions of the standard model run. The herbaceous phytomass grew by 0.3% and the woody phytomass decreased by 5.1%. Another model run implying a general mortality rate of 1 showed an increase of the mortality flux by 116% to 6.78 Pg. This approach additionally caused a reduction of the woody phytomass by 12.2%.

Another model run was characterized by the prevention of the black carbon formation since 1860, the start of the main model run. Compared with the standard model run, the atmospheric CO_2 content rose by 1.1 ppm to 339.7 ppm. The standard model run calculated a loss of 5.8 Pg from the carbon cycle between 1860 and 1980. The atmosphere and the ocean contribute 40% each and the biosphere 20% or 1.16 Pg in 120 years.

Conclusions

Strictly speaking, the fire model does not represent natural fires. Most input data are strongly influenced by human activities.

Savannas and tropical dry forests account for most carbon emissions originating from vegetation fires, while temperate and boreal forests are of minor importance.

The integration of vegetation fires into the HRBM causes a reduction of phytomass stored in the biosphere. This is mainly due to a dramatic decrease of carbon stored in tropical seasonal forests.

This reduction of carbon stored in the biosphere provides less carbon caused by clearing for agricultural purposes resulting in a less accelerated increase of atmospheric CO_2.

Important data are lacking in terms of *monthly* values concerning fire frequency, burning efficiency, and mortality rate.

Model results regarding the fraction of vegetation fires according to emissions of CO_2 are now available. Therefore, emissions of other trace gases and aerosols can be derived using emission ratios.

The role of black carbon in the global carbon cycle is still uncertain due to a lack of experimental studies.

The effects of an expected climate change and the expected population increase on the change and the shift of biomes are not investigable with static biome models. The development of dynamic biome models including burning effects as a dominant factor in many regions should be focused on. This fire submodel provides basic ideas.

References

Albini, F. A. 1992. Dynamics and modeling of vegetation fires: observations, in *Fire in the Environment: Its Ecological, Climatic and Atmospheric Chemical Importance*, edited by P. J. Crutzen and J. G. Goldammer, pp. 39–52, John Wiley and Sons, Chichester.

Alexander, M. E., and R. W. Sando. 1989. Fire behavior and effects in aspen-northern hardwood stands, in *Proceedings of the 10th Conference on Fire and Forest Meteorology, April 17–21, 1989, Ottawa, Ontario*, Forestry Canada, Ottawa, Ontario.

Andreae, M. O. 1991. Biomass burning: its history, use, distribution and its impact on environmental quality and global climate, in *Global Biomass Burning: Atmospheric, Climatic, and Biospheric Implications*, edited by J. S. Levine, pp. 3–21, MIT Press, Cambridge, Mass., 569 pp.

Andreae, M. O. 1993. The influence of tropical biomass burning on climate and the atmospheric environment, in *Biogeochemistry of Global Change: Radiatively Active Trace Gases*, edited by R. S. Oremland, pp. 113–150. Chapman and Hall, New York.

Andreae, M. O., and J. G. Goldammer. 1992. Tropical wildland fires and other biomass burning: environmental impacts and implications for land-use and fire management, in *Conservation of West and Central African Rainforest*, edited by K. Cleaver, M. Munasinghe, M. Dyson, N. Egli, A. Peuker, and F. Wencelius, pp. 79–109, World Bank Environmental Paper Number 1, 353 pp.

Auclair, A. N. D. 1983. The role of fire in lichen-dominated tundra and forest-tundra, in *The Role of Fire in Northern Circumpolar Ecosystems*, edited by R. W. Wein and D. A. MacLean, pp. 235–252, John Wiley and Sons, New York.

Barro, S. C., and S. G. Conrad. 1991. Fire effects on California chaparral systems: an overview, *Environment International*, 17, 135–149.

Bergeron, Y., and J. Brisson. 1990. Fire regime in red pine stands at the northern limit of the species range, *Ecology*, 71, 1352–1364.

Bragg, T. B. 1982. Seasonal variations in fuel and fuel consumption by fires in a bluestem prairie, *Ecology*, 63, 7–11.

Braithwaite, R. W., and J. A. Estbergs. 1985. Fire patterns and woody vegetation trends in the Alligator Rivers region of northern Australia, in *Ecology and Management of the World's Savannas, Int. Savanna Symp. 1984*, edited by J. C. Tothill and J. J. Mott, pp. 359–364, Commonwealth Agricultural Bureau, Aust. Acad. Sci., Canberra.

Brown, P. J., P. T. Manders, and D. P. Bands. 1991. Prescribed burning as a conservation management practice: a case history from the Cederberg Mountains, Cape Province, South Africa, *Biological Conversation*, 56, 133–150.

Burgan, R. E. 1987. A comparison of procedures to estimate fine dead fuel moisture for fire behaviour predictions, *S. Afr. For. J.*, 142, 34–40.

Cannell, M. G. R. 1982. *World Forest Biomass and Primary Production Data*. Academic Press, London.

Cass, A., M. J. Savage, and F. M. Wallis. 1984. The effect of fire on soil and microclimate, in *Ecological Effects of Fire in South African Ecosystems*, edited by P. de V. Booysen and N. M. Tainton, pp. 311–325, Springer, Berlin-Heidelberg.

Comery, J. A., G. R. Fahnestock, and S. G. Tickford. 1981. Elemental carbon deposition and flux from prescribed burning on a longleaf pine site in Florida, Final report to National Center for Atmospheric Research, Boulder, Colo.

Conklin, H. C. 1957. Shifting cultivation and succession to grassland climax, *Proc. Ninth Pacific Sci. Congr.*, 7, 60–62.

Crutzen, P. J., and M. O. Andreae. 1990. Biomass burning in the tropics: Impact on atmospheric chemistry and biogeochemical cycles, *Science*, 250, 1669–1678.

ECE/FAO. 1982. Supplement 10 to Volume XXXIV of the Timber Bulletin for Europe, United Nations Economic Commission for Europe (Geneva)/Food and Agriculture Organization of the United Nations (Rome), ECE/FAO Agriculture and Timber Division, Geneva, 23 pp.

ECE/FAO. 1992. Forest Fire Statistics 1988–1990, United Nations Economic Commission for Europe (Geneva)/Food and Agriculture Organization of the United Nations (Rome), ECE/FAO Agriculture and Timber Division, Geneva, 23 pp.

Engelmark, O. 1984. Forest fires in the Muddus National Park (northern Sweden) during the past 600 years, *Can. J. Bot.*, 62, 893–898.

Esser, G. 1986. The carbon budget of the biosphere—structure and preliminary results of the Osnabrück Biosphere Model (in German with extended English summary), *Veröff. Naturf. Ges. zu Emden von 1814, New Series*, 7, 160 pp.

Esser, G. 1987. Sensitivity of global carbon pools and fluxes to human and potential climatic impacts, *Tellus*, 39B, 245–260.

Esser, G. 1990. Modelling global terrestrial sources and sinks of CO_2 with special reference to soil organic matter, in *Soils and the Greenhouse Effect*, edited by A. F. Bouwman, pp. 47–261, John Wiley and Sons, New York, 575 pp.

Esser, G. 1991. Interdisciplinary study on the contemporary carbon cycle: Seasonality of carbon fluxes and climate relevant parameters of the terrestial biosphere, Final report for the contract No. EV4C-0032-D(B), International Institute for Applied Systems Analysis, Laxenburg, Austria, (unpubl. draft).

Esser, G., J. Hoffstadt, F. Mack, and U. Wittenberg. 1994. High Resolution Biosphere Model, Documentation, Model version 3.00.00, *Mitteilungen aus dem Institut für Pflanzenökologie der Justus-Liebig-Universität Giessen, Heft* 2, 68 pp.

Fearnside, P. M. 1991. Greenhouse gas contributions from deforestation in Brazilian Amazonia, in *Global Biomass Burning: Atmospheric, Climatic, and Biospheric Implications*, edited by J. S. Levine, pp. 92–105, MIT Press, Cambridge, Mass. 569 pp.

FAO. 1991. Agrostat PC Statistical Data Base Vers.1.1., Rome.

Frost, P. G. H. 1985. Organic matter and nutrient dynamics in a broad-leaved African savanna, in *Ecology and Management of the World's Savannas, Int. Savanna Symp. 1984*, edited by J. C. Tothill and J. J. Mott, pp. 200–206, Commonwealth Agricultural Bureau, Aust. Acad. Sci., Canberra.

Green, L. R. 1981. Burning by prescription in chaparral. *UDSA For. Serv. Gen. Tech. Rep., PSW*-51, Pac. Southw. For. and Range Exp. Sta., Berkeley, Calif.

Griffin, G. F., and M. H. Friedel. 1984. Effects of fire on central Australian rangelands. I. Fire and fuel characteristics and changes in herbage and nutrients, *Aust. J. Ecol.*, 9, 381–393.

Hao, W. M., M. H. Liu, and P. J. Crutzen. 1990. Estimates of annual and regional release of CO_2 and other trace gases to the atmosphere from fires in the tropics, in *Fire in the Tropical Biota*, edited by J. G. Goldammer, pp. 440–462, Springer, Berlin-Heidelberg, 497 pp.

Hopkins, B. 1965. Observations on savanna burning in the Olokemeji Forest Reserve, Nigeria, *J. Appl. Ecol.*, 2, 367–381.

Houghton, R. A. 1991. Biomass burning from the perspective of the global carbon cycle, in *Global Biomass Burning: Atmospheric, Climatic, and Biospheric Implications*, edited by J. S. Levine, pp. 321–325, The MIT Press, Cambridge, Mass. 569 pp.

Houghton, R. A., R. D. Boone, R. Fruci, J. E. Hobbie, J. M. Melillo, C. A. Palm, B. J. Peterson, G. Shaver, G. M. Woodwell, B. Moore, D. L. Skole, and N. Myers. 1987. The flux of carbon from terrestrial ecosystems to the atmosphere in 1980 due to changes in land use: geographic distribution of the global flux, *Tellus*, 39B, 2–139.

Kilgore, B. M. 1973. The ecological role of fire in Sierran conifer forests: Its application to national park management. *Quaternary Research*, 3, 496–513.

Kreeb, K. 1983. Vegetationskunde: Methoden und Vegetationsformen unter Berücksichtigung ökosystemarer Aspekte, Ulmer, Stuttgart, 331 pp.

Kuhlbusch, T. A. 1994. Schwarzer Kohlenstoff aus Vegetationsbränden: eine Bestimmungsmethode und mögliche Auswirkungen auf den globalen Kohlenstoffzyklus, Dissertation, Johannes-Gutenberg-Universität Mainz, 88 pp.

Kuhlbusch, T. A., and P. J. Crutzen. 1995. A global estimate of black carbon in residues of vegetation fires representing a sink of atmospheric CO_2 and a source of O_2, *Global Biogeochem. Cycles*, (subm.).

Leemans, R., and W. P. Cramer. 1991. The IIASA database for mean monthly values of temperature, precipitation, and cloudiness on a global terrestrial grid, *IIASA Research Report, RR-91-18*, International Institute for Applied Systems Analysis, Laxenburg, Austria.

Mack, F. 1994. Zur Bedeutung von Vegetationsbränden für den globalen Kohlenstoffkreislauf. Dissertation, Institut für Pflanzenökologie der Justus-Liebig-Universität Giessen, 203 pp.

Menaut, J. C., L. Abbadie, F. Lavenu, P. Loudjani, and A. Podaire. 1991. Biomass burning in West African savannas, in *Global Biomass Burning: Atmospheric, Climatic, and Biospheric Implications*, edited by J. S. Levine, pp. 133–142, MIT Press, Cambridge, Mass.

Minnich, R. A. 1983. Fire mosaics in southern California and northern Baja California. *Science*, 219, 1287–1294.

Noble, I. R., G. A. V. Bary, and A. M. Gill. 1990. McArthur's fire-danger meters expressed as equations, *Aust. J. Ecol.*, 5, 201–203.

Olson, J. S. 1981. Carbon balance in relation to fire regimes. *USDA For. Serv. Gen. Tech. Rep.*, WO-26, Washington, D.C., 327–378.

Olson, J. S., J. A. Watts, and L. J. Allison. 1983. Carbon in live vegetation of major world ecosystems, ORNL-5862, Oak Ridge National Laboratory, Oak Ridge, Tenn.

Peterson, D. L., and M. J. Arbaugh. 1986. Postfire survival in Douglas-fir and lodgepole pine: comparing the effects of crown and bole damage. *Can. J. For. Res.*, 16, 1175–1179.

Peterson, D. L., and M. J. Arbaugh. 1989. Estimating postfire survival of Douglas-fir in the Cascade Range. *Can. J. For. Res.*, 19, 530–533.

Prentice, I. C., W. Cramer, S. P. Harrison, R. Leemans, R. A. Monserud, and A. M. Solomon. 1992. A global biome model based on plant physiology and dominance, soil properties and climate, *J. Biogeogr.*, 19, 117–134.

Quintillo, D., M. E. Alexander, and L. Ponto. 1989. Spring fires in a semi-mature trembling aspen stand, central Alberta, *Inf. Rep., NOR-X-000*, For. Can., North. For. Cent., Edmonton, Alberta.

Richards, J. F., J. S. Olson, and R. M. Rotty. 1983. Development of a data base for carbon dioxide releases resulting from conversion of land to agricultural uses, Institute for Energy Analysis, Oak Ridge Assoc. Universities, ORAU/IEA-82-10(M), ORNL/TM-8801.

Romme, W. H. 1980. Fire history terminology: Report of the Ad Hoc Committee, in Proceedings of the Fire History Workshop, October 20–24, 1980, Tucson, Arizona, pp. 35–137, *General Technical Report, RM-81*, Rocky Mountain Forest and Range Experiment Station, Forest Service, U.S. Department of Agriculture, 142 pp.

Rutherford, M. C., and R. H. Westfall. 1986. Biomes of Southern Africa—an objective categorization, *Memoirs of the Botanical Survey of South Africa, 54*, Botanical Research Institute, Department of Agriculture and Water Supply, Republic of South Africa.

Seavoy, R. E. 1975. The origin of tropical grasslands in Kalimantan, Indonesia, *J. Trop. Ecol.*, 40, 48–52.

Storey, T. G., and E. P. Merkel. 1960. Mortality in a longleaf-slash pine stand following a winter wildfire, *J. For.*, 58, 206–210.

Uhl, C., and J. B. Kauffman. 1990. Deforestation, fire susceptibility, and potential tree responses to fire in the eastern Amazon, *Ecology*, 71, 437–449.

Vogl, R. J., and P. K. Schorr. 1972. Fire and manzanita chaparral in the San Jacinto Mountains, California, *Ecology*, 53, 1179–1188.

van Wagner, C. E. 1987. Development and structure of the Canadian Forest Fire Weather Index System, *Canadian Forestry Service Forestry Technical Report*, 35, Ottawa, 37 pp.

Walter, H., and H. Lieth. 1960 ff. Klimadiagramm Weltatlas, Gustav Fischer Verlag, Jena.

van Wilgen, B. W. 1982. Some effects of post-fire age on the aboveground plant biomass of fynbos (macchia) vegetation in South Africa, *J. Ecol.*, 70, 217–225.

Wein, R. W., and J. M. Moore. 1979. Fire history and recent fire rotation periods in the Nova Scotia Acadian Forest, *Can. J. For. Res.*, 9, 166–178.

Wright, H. A. 1978. The effect of fire on vegetation in ponderosa pine forests: A state-of-the-art review, *Texas Tech. Univ. Range and Wildl, Information Series*, No. 2, Lubbock, Tex.

Wright, H. A., L. F. Neuenschwander, and C. M. Britton. 1979. The role and use of fire in sagebrush-grass and pinyon-juniper plant communities, *USDA For. Serv. Gen. Tech. Rep., INT-58*, Intermt. For. and Range Exp. Stn., Ogden, Utah.

Zackrisson, O. 1977. Influence of forest fires on the North Swedish boreal forest, *Oikos*, 29, 2–32.

Black Carbon, the Global Carbon Cycle, and Atmospheric Carbon Dioxide

Thomas A. J. Kuhlbusch and Paul J. Crutzen

Black carbon (often called charcoal,[1] soot, elemental carbon, etc.) is one of the ubiquitous materials circulating around the surface of the Earth. It is found in the air, soils, sediments, crustal rocks, meteorites, waters, and ices. Its universality is related to its refractory nature with respect to reactions with its surroundings and to its origin in burning processes, which are widespread. (Goldberg 1985, p. vii)

The term *black carbon* (BC) was first used by Novakov (1984) to describe the carbon fraction in combustion aerosols and particles which is of black color. It is used here to summarize a certain relatively inert form of carbon of blackish color, comprising a range of carbon compounds from highly polycyclic aromatic to elemental or graphitic carbon. BC is produced by any incomplete combustion process and is contained in the smoke or residues on the ground. It can be formed either via gas phase processes and thus be a fraction of soot (organic carbon + black carbon) or via pyrolysis in the residues of liquid or solid fuels. Fractions of the residues being carried by turbulence during flaming and smoldering combustion as well as soot may form the particulates in the smoke (figure 16.1). Because black carbon is formed by the combustion process it is absent in most fuels (except charcoal and coal) and uncharred residues. The term *thermally altered carbon* is used for the total residue comprising fine and coarse ash particles and the partially charred residue. The interest in black carbon is mainly threefold.

Black carbon emitted with smoke has been well investigated because of its optical and physicochemical properties. It is the main component in aerosols that strongly absorbs solar radiation. Despite being hydrophobic in nature, BC particles may act as cloud condensation nuclei (CCN) when they combine with other chemicals of CCN nature. Even though the latter process is still discussed, it is known that some smoke particles do act as CCN since clouds frequently form on smoke plumes (Andreae 1993). Both properties, absorbing solar light and acting as CCN, directly and indirectly influence the solar radiation budget and climate. BC formation is always accompanied by the formation of polyaromatic hydrocarbons, some of which are carcinogenic. Chang et al. (1982) suggest that atmospheric BC may act as a catalyst for various reactions.

The other two main points of interest in BC are based on its chemical and microbial inertness, which leads to residence times in soils, sediments, and ice of millions of years. Due to this inertness, BC and thermally altered organic matter are used by several scientists as a tracer in sediments and ice to investigate the history of fire and climate of different regions (Griffin and Goldberg 1975; Suman 1983; Herring 1977; Clark 1988; Chylek et al. 1995). The potential importance of the inertness of BC on the global carbon cycle, which is one focus of this chapter, was first pointed out by Seiler and Crutzen (1980). Since black carbon produced by vegetation fires either will not, or will very slowly decompose, this carbon will be sequestered from the short- to the long-term carbon pool and will thus represent a sink of biospheric carbon and of atmospheric carbon dioxide (CO_2).

Three attempts to estimate the quantities of thermally altered carbon in the residues of vegetation fires have so far been conducted (Fearnside et al. 1994; Fearnside et al. 1993; Comery et al. 1981). Fearnside et al. (1993) and Fearnside et al. (1994) sampled all blackened material (charcoal), but not the fine residue after a deforestation fire, dried it, and weighed it. Since this blackened material will still contain some degradable organic matter, not all of its carbon can be considered inert and thus a sink for the bioatmospheric carbon cycle. Comery et al. (1981) analyzed the residues of a prescribed long-leaf pine forest fire for organic and total carbon. The total carbon content was determined by burning the sample at 1650°C in oxygen and following quantification of the produced carbon dioxide. Organic carbon was determined by chemical oxidation with $Cr_2O_7^{-2}$ and following quantification of the excess $Cr_2O_7^{-2}$ by titration with standard $FeSO_4$ solution. "Elemental carbon" was then calculated by subtraction of organic from total carbon leading to an overestimation of black carbon

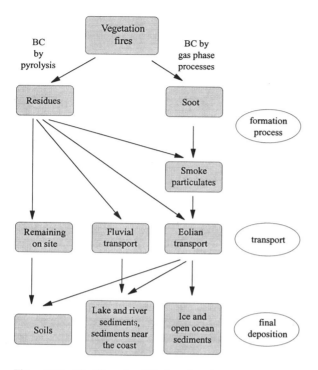

Figure 16.1 The life cycle of black carbon from formation to final reservoir

formation by neglecting the carbonate carbon, which was certainly present in those residues. The analyses of various savanna grass fire residues showed that the amount of carbonate carbon equaled about 35% of that of BC (Kuhlbusch et al. 1995). In some residues carbonate carbon even exceeded the amount of BC. Since the study by Comery et al. (1981) involved only a minor fraction of worldwide forest fires and due to their overestimation of "elemental carbon" (BC) formation, it will be disregarded.

A newly developed analytical method (Kuhlbusch 1995) was used to determine black carbon, two organic carbon fractions, and carbonate carbon in the residues of fires and is here briefly summarized. The results obtained with this method on the formation dependencies and on the quantitative production of black carbon by experimental and field vegetation fires will be presented and used to estimate the annual, global formation of black carbon.

Sampling and Analytical Method Determining Black Carbon

Residues obtained form a burning apparatus at the Max-Planck-Institut für Chemie (Lobert et al. 1991)

were taken from fires conducted with different vegetation types: pine needles, needle litter, savanna grasses (African and Venezuelan), straw, deciduous wood, and hay. The weight of the fuel and the residues was recorded. The residues consisted mainly of ash and contained hardly any uncharred or partially charred matter. A mill was used to pulverize both residue and fuel samples to an average particle size of 40 μm (maximum diameter 60 μm). Following this the samples were dried in an oven at 105°C for 24 hours and weighed prior to an elemental analysis to determine their carbon and hydrogen content. By subtracting the residual from the initial carbon load we derived the volatilized carbon (VC).

Residues sampled during a field experiment in Kruger National Park, South Africa (SAFARI-92), were obtained for comparison from six experimental sites that can be characterized as deciduous open-tree savanna. Details about the kind of vegetation, vegetation density, fire characteristics, and sampling procedures are given elsewhere (Shea et al. 1995; Trollope et al. 1995; Ward et al. 1995; Kuhlbusch et al. 1995) and are only briefly summarized here. By sampling the non-woody standing vegetation prior to the fire and the residues afterwards and by determining the dry weight and elemental contents of the samples after pulverization, we calculated the volatilized carbon (VC). The residues were partitioned during sampling into uncharred, partially charred (both manually sampled), and ash (sampled with a handheld vacuum cleaner; for discussion on sampling technique see Kuhlbusch et al. 1995). The ash was divided by sieving into ash1 (< 0.63 mm diameter) and ash2 (> 0.63 mm diameter) prior to the determination of dry weight and elemental contents.

This separation of the residue into various fractions was conducted to derive information on the dependency of the formation of BC on the degree of heat the residue was exposed to during the fire. Optical properties were used for differentiation and are introduced in the order of increasing exposure to heat: Uncharred residues did not show any visible alterations due to fire, partially charred residues showed the original structure and color as well as pyrolyzed parts. In contrast, the ash fraction sampled with the vacuum cleaner was totally blackened on the outside. This fraction was separated by sieving since ash particles above 0.63 mm (ash2) still had the original cell structure of the vegetation and did contain some less pyrolyzed organic matter. Ash1 was composed of black, unstructured, and thermally totally degraded matter.

These residues were then analyzed by a method developed by Kuhlbusch (1995), which is briefly summarized here. Acronyms used are BC: black carbon; BH: hydrogen associated with the black carbon; OC1 + OH1: organic carbon and hydrogen removed by the solvent extraction, 1st treatment step; OC2 + OH2: organic carbon and hydrogen removed by thermal treatment, 2nd treatment step; IC: inorganic carbon, carbonates; CE: carbon exposed to fire; VC: volatilized carbon by the fire, expressed in percentage of CE; TC and TH: total carbon and hydrogen content in the residue; TC1 + TH1: total carbon and hydrogen content in the sample after the solvent extraction; % dm: % dry matter in the untreated sample.

An elemental analysis determining the carbon and hydrogen content of the dried and pulverized residue sample is conducted to derive TC and TH. For the solvent extraction a weighed fraction of the sample is placed in a centrifuge tube and treated with different solvents [NaOH (1 molar), 70 mass-% HNO_3, 1 mass-% HCl, and twice distilled water]. A second elemental analysis is conducted after drying and weighing the sample to derive TC1 and TH1 enabling us to calculate the removed carbon and hydrogen (IC + OC1, OH1 respectively). This solvent extraction removes all inorganic carbon (IC) and some organic carbon (OC1), especially organic carbon not thermally altered (Kuhlbusch 1995).

To remove the residual organic carbon (OC2), weighed subsamples of the pretreated material are placed in a pure oxygen flow (ca. 500 ml min^{-1}) in an oven at 340°C for 2 hours. After the removal of all inorganic and organic carbon during the pretreatments, BC as well as BH (defined here as the hydrogen measured together with BC) are quantified by a third elemental analysis. The molar H/C ratio (BH/BC) determined in this last step has to be used with caution because some hydrogen is bonded to minerals, requiring a correction (see Kuhlbusch 1995). OC2 and OH2 respectively were quantified by subtraction of BC or BH from TC1 or TH1.

Kuhlbusch and Crutzen (1995) showed that BC can be defined by its molar H/C ratio and they postulated that the lower the H/C ratio of a carbon compound the more likely this compound is to be resistant to both chemical and microbial degradation processes. Thus black carbon was defined as a fire-derived, polycyclic aromatic to graphitic carbon fraction in smoke particulates and residues which is resistant to a treatment at 340°C in pure oxygen and has a molar H/C ratio of ≤ 0.2.

Results and Discussion

Formation Dependencies of Black Carbon in Vegetation Fire Residues

The results obtained from the experimental fires in the burning apparatus are summarized in table 16.1. In figure 16.2 we plotted BC as the percentage of TC in the residue versus the average molar emission ratio of CO/CO_2. The linear regression (r^2: 0.66; constant: 34.7 ± 4.6%; x-coefficient: −2.32 ± 0.45; degrees of freedom: 14) shown in figure 16.2 clearly demonstrates the dependence of the formation of BC on the burning conditions. Since the molar CO/CO_2 emission ratio is much lower during the flaming than during the smoldering phase, the negative slope of the regression line indicates that flaming rather than smoldering combustion is the major source of BC. Since woody fuels behave differently from non-woody fuels concerning BC production (Cadle and Groblicki 1982; Cachier et al. 1989; Kuhlbusch and Crutzen 1995), we did not include the results obtained from the deciduous wood fires in the previous and following correlation studies. Nevertheless the results from the deciduous wood residues were included in the global formation estimate.

Figure 16.3 shows the dependency of BC formation as a fraction of the residual carbon (TC) and the fire-exposed carbon (CE) versus the volatilized carbon (VC in % of CE). The curves shown in this figure were derived by fitting an equation developed by Kuhlbusch and Crutzen (1995) to data obtained from various residues (tables 16.1 and 16.2). Figure 16.3a demonstrates that the more carbon is volatilized, normally accompanied with higher combustion temperatures, the more important becomes BC in the residue, which agrees with the interpretation of figure 16.2. The fitted curve in figure 16.3b shows that the produced BC is enriched in the residue up to a certain VC ratio at which all exposed plant material, including BC, undergoes thermal degradation. The latter process becomes significant at about 90% VC. A comparison of the field data to the data obtained from the experimental fires in a burning apparatus reveals a wider scattering along the curve for the former. This is mainly due to the inhomogeneous residues in the field, where the amount of uncharred and partially charred residues were quite variable and influenced the determined BC/CE ratio (Kuhlbusch et al. 1995). Thus the residue composition (uncharred, partially charred, and ash residue) and the volatilization ratios for carbon (VC) for the various kinds of vegetation fires are indicators of the degree the residue was exposed to pyrolysis by the fire. Since the

Table 16.1 Black carbon content in the residues including the corresponding H/C ratios and the emissions of CO_2 by the fire

Biomass	BC (% dm)	BH (% dm)	H/C of BC (mol)	BC of TC (%)	BC of CE (%)	BC/CO_2 (%)	Type of fire[a]
Deciduous wood	49.88	0.435	0.10	62.65	2.81	3.42	backing
Deciduous wood	43.97	0.439	0.12	53.38	3.42	4.56	backing
Hay	1.01	0.022	0.28	12.98	0.14	0.17	backing
Hay	2.44	0.032	0.14	22.99	0.51	—	backing
Hay	3.11	0.048	0.18	28.02	1.01	—	backing
Hay	4.29	0.038	0.11	36.36	2.23	2.43	backing
Hay	3.48	0.049	0.17	17.49	1.76	2.87	backing
Needle litter	7.24	0.087	0.15	16.91	1.05	1.46	backing
Needle litter	1.72	0.028	0.20	5.63	0.97	1.96	heading
Needle litter (Phil.)	3.19	0.168	0.63	9.46	1.31	1.64	heading
Needle litter (Phil.)	0.74	0.189	3.09	2.87	0.61	1.30	heading
Pine needle	3.25	0.041	0.15	18.39	0.21	0.26	backing
Pine needle	4.79	0.051	0.13	29.21	0.47	0.55	backing
Pine needle	4.08	0.051	0.15	13.68	0.38	0.43	backing
Pine needle	5.60	0.078	0.17	23.12	1.41	2.00	backing
Savanna grass (Afr.)	3.58	0.074	0.25	37.24	0.63	0.80	backing
Savanna grass (Ven.)	4.21	0.073	0.21	28.91	0.93	1.49	backing
Savanna grass (Ven.)	3.84	0.084	0.26	24.73	1.06	1.92	backing
Savanna grass (Ven.)	4.01	0.101	0.30	22.19	1.00	1.60	heading
Straw	2.91	0.051	0.21	22.97	0.14	0.15	backing
Straw	2.13	0.032	0.19	17.73	0.22	0.30	heading
Straw	4.38	0.081	0.22	18.24	0.56	0.68	backing
Savanna grass (SAFARI-92)[c]	2.63	0.13[b]		11.97	1.00	1.23	heading
	2.11			8.33	0.94	1.22	backing

a. Data by Lobert 1989; backing, fire spreading against the wind; heading, fire spreading with the wind.
b. Average ratio given in Kuhlbusch et al. 1995.
c. Data from Kuhlbusch and Crutzen 1995 and Kuhlbusch et al. 1995.

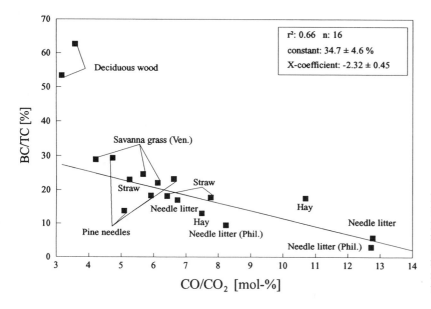

Figure 16.2 The correlation between black carbon formation, expressed in percentage of total carbon, and the molar CO/CO_2 emission ratio (from Kuhlbusch and Crutzen 1995). The negative slope indicates that flaming rather than smoldering combustion is the major source.

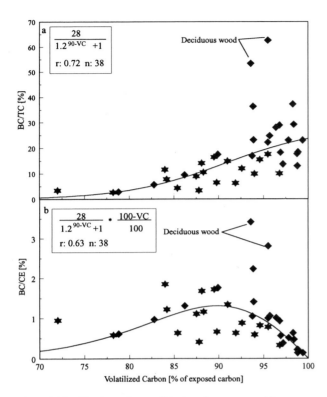

Figure 16.3 The dependence of black carbon expressed in percentage of residual carbon (TC) and carbon being exposed to fire (CE), on the volatilization ratio for carbon (VC). (data from Kuhlbusch and Crutzen 1995 ◆, and from Kuhlbusch et al. 1995 ★)

latter determines the amount of BC in residues, both the residue composition and VC are indicators of the formation of BC as well (further details in Kuhlbusch et al. 1995).

Production Ratios of Black Carbon for Various Kinds of Vegetation Fires

Most of the quantitative global vegetation fire estimates are based on burned biomass or volatilized carbon (Crutzen and Andreae 1990; Andreae 1993; Houghton 1991; Hao and Liu 1994) and can be converted to annual CO_2 release fluxes. Therefore, the global formation estimate in table 16.3 is based on the BC/C-CO_2 ratio (the amount of BC in the residue divided by the amount of carbon volatilized as CO_2). An overview of published BC/CO_2 ratios of BC in smoke and residues is given in figure 16.4 and demonstrates that the majority of BC remains on site after a fire. BC/CO_2 ratios obtained from experimental fires in the burning apparatus are summarized in table 16.1 and will be discussed for fuel wood, shifting agriculture, permanent deforestation, temperate and boreal

forest, agricultural waste, and savanna fires for use in the formation estimate.

The volatilization ratios for carbon in the two deciduous-wood fires (table 16.1) were 93.6% and 95.5%. Thus the resulting residues were taken as representative for fuel-wood burning, and a range from 3.4–4.6% for BC/CO_2 was obtained for this category. The kind of biomass and the fraction of aboveground biomass that is burned (30–45%; Hao and Liu 1994; Fearnside et al. 1993; Kasischke et al. 1995) are similar for shifting agriculture, permanent deforestation, and temperate and boreal forest fires. Therefore and because only two investigations concerning the formation of charcoal are available so far (Fearnside et al. 1993; Fearnside et al. 1994; both investigations were conducted on permanent deforestation fires), a single BC/CO_2 ratio will be used for these kinds of vegetation fires in the global formation estimate.

In the two investigations by Fearnside et al. the volatilization ratios of carbon were 27.6% (Fearnside et al. 1993) and 42% (Fearnside et al. 1994). A substantially higher volatilization ratio in the latter study was mainly due to the fuel composition. The fine fuel, mostly vines and litter, comprised about 8% of the aboveground carbon stock in the 1993 study and 25% in the 1994 study. Fearnside et al. (1994) concluded that the overall lower combustion efficiency determined in the 1993 study of 27.6% is more typical for the Amazonian burning. The formation ratios for charcoal in these two studies (excluding fine ash) were determined to be 2.7% (Fearnside et al. 1993) and 1.3% (Fearnside et al. 1994) of the preburn aboveground carbon stock. Charcoal contains degradable organic carbon as well as BC. The dependence of the formation of BC on the degree of heat to which the residue was exposed was previously shown. Therefore, and since the deciduous wood fire residues were similar to charcoal, we used the BC/TC ratios determined for wood fire residues (58% ± 5%, table 16.1) to estimate the amount of BC being produced in the deforestation fires. Thus we derived BC/CE ratios of 1.4–1.7% and 0.7–0.8%, respectively. Dividing these ratios by the volatilization ratios for carbon (0.28 and 0.42, respectively) and assuming that 90% of the volatilized carbon was emitted as CO_2 we obtained BC/CO_2 ratios of 5.6–6.8% and 1.9–2.1%, respectively. The differences in these two studies clearly show that there is a need for further investigations concerning BC formation by permanent deforestation, shifting agriculture, and temperate and boreal forest fires. For the global BC formation estimate for permanent deforestation, shifting agriculture, and temperate and boreal

Table 16.2 Elemental contents and amounts of the different fuel and residue fractions determined on 6 savanna sites in Kruger National Park, South Africa, during SAFARI-92 (from Kuhlbusch et al. 1995)

	C [% dm][a]	H [% dm][a]	H/C molar	% of fuel or residue	BC/TC [%]
Fuel	43.9 ± 1.4	5.7 ± 0.1	1.55 ± 0.02	6133 ± 2006 [kg/ha]	
Grass	44.0 ± 1.3	5.9 ± 0.2	1.61 ± 0.03	57.4 ± 17.5	
Litter	43.6 ± 3.4	5.4 ± 0.3	1.49 ± 0.07	41.4 ± 13.9	
Herbs	44.1 ± 4.7	5.6 ± 0.8	1.51 ± 0.08	1.2 ± 1.3	
Residue	26.4 ± 6.0	2.3 ± 0.7	1.01 ± 0.10	1055 ± 452 [kg/ha]	
Uncharred	44.1 ± 2.6	5.3 ± 0.7	1.42 ± 0.12	1.2 ± 2.4	
Partially charred	35.9 ± 8.3	3.6 ± 0.9	1.22 ± 0.09	36.5 ± 14.7	2.4 ± 1.5
Ash	21.9 ± 5.3	1.5 ± 0.5	0.83 ± 0.08	61.6 ± 25.8	16.7 ± 5.5
Ash 2 (>0.63 mm)	44.3 ± 2.7	3.2 ± 0.6	0.86 ± 0.14	14.8 ± 8.2	14.6 ± 8.0
Ash 1 (<0.63 mm)	14.8 ± 3.2	1.0 ± 0.2	0.81 ± 0.06	46.8 ± 17.6	17.0 ± 5.2
In comparison					
Rice straw[b] fuel	40.4 ± 0.1	5.4 ± 0.1	1.61 ± 0.05		
Rice straw[b] residue	13.2 ± 2.0	0.5 ± 0.1	0.47 ± 0.06		
Charcoal[c]	74.5	2.5	0.40		

a. Average and standard deviation of 20 sampling plots for the fuel and 18 plots for the residue.
b. Jenkins et al. 1991.
c. Lacaux et al. 1994.

Table 16.3 Global estimate of black carbon formation in residues based on carbon release estimates

Source	Carbon released (Tg C/a)				CO_2 released (Tg C/a)[a]	BC/CO_2 ratio (%)	Black carbon formed (Tg C/a)
	Crutzen and Andreae 1990	Andreae 1993	Houghton 1991	Hao and Liu 1994 (tropics only)			
Shifting agriculture	500–1000	—	500–700	590	450–900	2–6[b]	9–54
Permanent deforestation	200–700	570	400–1300	230	360–900	2–6[b]	7–54
Temperate and boreal forest	—	510	100–165	—	90–180	2–6[b]	2–11
Savanna fires	300–1600	1660	400–2400	1200	900–1600	1.0–2.0	9–32
Fuel wood	300–600	880	800	310	360–630	3.4–4.6	12–29
Agricultural waste	500–800	380	800	130	450–720	1.0–2.0	5–14
Total	1800–4700	4000	3000–6200	2430	2610–5130		44–194

a. Best estimate, based on carbon release data and assuming that 90% of the released carbon was emitted as CO_2.
b. Based on data by Fearnside et al. 1993, Fearnside et al. 1994, and this work.

forest fires we used the BC/CO_2 ratio range of 2–6%. This ratio is believed to be conservative because the amount of BC formed in the fine ash residue is unknown and not included.

The volatilization ratios for carbon by savanna fires and agricultural waste burning are around 80–90% (Seiler and Crutzen 1980; Andreae 1991; Menaut et al. 1991; Kuhlbusch et al. 1995). It is interesting to note here that the maximum formation of BC (figure 16.3) occurs around these volatilization ratios. Since fuel type and volatilization ratios are similar (both mainly non-woody vegetation), a single BC/CO_2 ratio is used for these kinds of fires. A volatilization ratio weighted average of the BC/CO_2 ratios for non-woody fuels presented in table 16.1 gives a range of 1.3–2.9% (VC range: 80–96%).

During a field study in southern Africa (SAFARI-92) BC was quantified in various residue fractions of savanna fires to compare the results with those obtained from the experimental fires in the burning apparatus. Six prescribed savanna fires were investigated during this campaign and the results concerning the fuel and residue composition are presented in tables 16.1 and 16.2 (for more details see Kuhlbusch et al. 1995). It is interesting to note the high carbon content in the A2 residue fraction (>0.63 mm diameter). Although A2 represented only about 15% of the total mass after the fire, it contained about 35% of the total amount of BC, 25% of the total amount of residual carbon, and about 22% of the total amount of residual nitrogen. Still, the largest pool of BC was in the A1

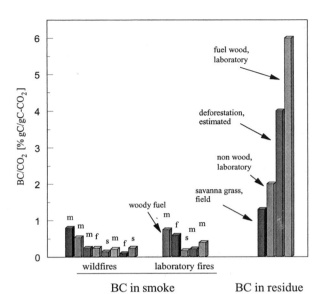

Figure 16.4 BC/CO_2 ratios determined in smoke and residues: f, sampled during flaming combustion; s, smoldering combustion; m, mixed combustion. Partially adapted values from emission factors (gC/kg dry matter) and converted with a factor of 2.5. (Turco et al. 1983; Crutzen et al. 1984; Patterson et al. 1986; Patterson et al. 1984; Andreae et al. 1988; Crutzen and Andreae 1990; Cachier et al. 1993; Turn et al. 1993; Kuhlbusch et al. 1995; Kuhlbusch and Crutzen 1995; Fearnside et al. 1993).

fraction with a particle size < 0.63 mm, representing about 52% of the total BC.

The average determined BC/CE ratio in these savanna fires was 0.96 ± 0.45 while $(88.4 \pm 6.2)\%$ of the exposed carbon was volatilized (Kuhlbusch et al. 1995). The same ratio obtained from the burning apparatus experiments was $(1.5 \pm 0.5)\%$ for volatilization ratios of carbon of 80–96%. With the assumption that 95% of the volatilized carbon was emitted as CO_2 we derive an average BC/CO_2 ratio for the savanna fires of $(1.2 \pm 0.5)\%$, being at the low end of the range derived from the burning apparatus experiments (1.3–2.9%). In combining those two ranges a BC/CO_2 ratio of 1–2% is used in the global formation estimate for savanna fires and agricultural waste burning.

Global Formation Estimate, Transport, and Reservoirs
The BC/CO_2 ratio is used in the estimate presented in table 16.3 because we believe it to be less dependent on the volatilization ratio than, for example, the BC/CE ratio due to the dependency that the less carbon is volatilized by a fire the less CO_2 is emitted, the less heat is produced and the less BC is produced. To derive the CO_2 release rates by the different kinds of vegetation fires we listed the four estimates of carbon release rates

in table 16.3 and used those to estimate the amount CO_2 released per year. Based on these estimates we obtained an annual formation rate of black carbon with the BC/CO_2 ratios discussed above of 44 to 194 Tg C (Tg = 10^{12}g). This estimate is 20–40% lower than a previous estimate of 76–241 Tg BC/a (Kuhlbusch and Crutzen 1995) This lower estimate is mainly due to results reported by Fearnside et al. (1994), which demonstrate the high variability in charcoal (BC) formation by deforestation fires. The latter study leads to BC/CO_2 ratios for shifting agriculture, permanent deforestation, and temperate and boreal forest fires of 2–6%, which is lower than the 5–7% used in Kuhlbusch and Crutzen (1995). Second, the BC/CO_2 ratio of 1.3–2.9% used by Kuhlbusch and Crutzen (1995) for savanna fires and agricultural waste was reduced to 1.0–2.0% in this estimate because of the lower formation ratios determined for savanna fires in the field study presented here and in Kuhlbusch et al. (1995).

Shifting agriculture and permanent deforestation have the strongest source strength and together account for 36–55%. Savanna fires contribute 16–20% and fuel wood 15–27%. Agricultural waste burning is of minor importance for the global budget of BC. To significantly improve the estimate of annual formation of black carbon, field measurements are needed, especially for shifting agriculture, permanent deforestation, and temperate and boreal forest fires.

In addition to the formation of BC in the residues, annually about 6 Tg BC are emitted into the atmosphere by vegetation fires (Liousse et al. 1995) leading to an annual total global black carbon formation of 50–200 Tg C. The major fraction of BC (> 80%) is thus contained in the residues. The transport and final reservoirs of BC particles can be quite different and are summarized in figure 16.1.

Black carbon residing on the site after the fire may remain or be transported off-site by rain or wind (figure 16.1). In the case of savanna fires, wind transport seems to be quite significant. Visual observation of a burned savanna site showed that within a few weeks a great proportion of the residue was transported off-site even in the absence of rain. This can be explained by the fact that a major fraction of the savanna fire residues were highly accessible to aeolian transport (table 16.2). Since 55% of the total amount of BC in savanna fires was determined to be in the fine ash1 residue, which exceeds the amount of BC emitted with the smoke, we suggest that fire residues may contribute significantly to atmospheric BC background concentrations. Such contributions may influence the solar radiation budget in certain regions. Additionally, about 38% of residual nitrogen was found to be in the ash1 fraction, and it would be transported along with the BC, thus perhaps contributing to the circulation of nutrients on regional scales. Besides the above-mentioned possible importance of transport of vegetation fire residues, more detailed knowledge concerning transport of BC is desirable and necessary to evaluate historic records of BC or fire-altered material in sediments, soils, and ice. Investigation including particle sizes in residues may reveal a relationship between analyzed sites, main biomass fuel, source region, and source strength.

Black carbon emissions with smoke can be deposited to soils, sediments, and ice. Its long residence time in these compartments can be deduced from its ubiquitous presence and age. Charcoal or BC is common in soils (Griffin and Goldberg 1975; Trumbore 1993) and charcoal samples from the Amazon were dated at up to 6300 years B.P. (Sanford et al. 1985). Sediment core analyses showed continuous records of BC in the North Pacific and of fusain (charcoal) in the North Atlantic dating from the Upper and Lower Cretaceous to today (Herring 1977, 1985; Summerhayes 1981). Lake sediment analyses disclosed continuous charcoal records over the last 10 000 years (Clark et al. 1989). Transport of BC to ice of the polar regions has been observed (Clarke and Noone 1985). The oldest ice record dates back about 14 000 years B.P. (Chylek et al. 1992).

Recently Suman et al. (1995) estimated the preindustrial annual flux of BC to marine sediments to be about 10 Tg BC/a, representing 5–20% of the total amount of BC produced by vegetation fires. Sanford et al. (1985) quantified charcoal in tierra firme soils in the Amazon by sieving the soil, sorting out charcoal particles > 0.5 mm, and determining their dry weight. From this investigation they estimated the amount of charcoal in the uppermost meter to range from 4.6 to 13.9 metric tons dry weight per hectare. More studies investigating the amount of BC in other soils and lake and river sediments would be desirable to be able to estimate annual fluxes. Thus the annual BC formation could be estimated independently and used for comparison.

The Impact of Black Carbon on the Global Carbon Cycle

Due to the inertness and stability of BC, carbon is sequestered from the short-term bioatmospheric to the long-term geological carbon cycle. This amounts to about 50–200 Tg C/a. This may, however, be an underestimate when evaluating the source strength of

black carbon and its impact on the global carbon cycle. We emphasize also that some organic matter being removed by the thermal pretreatment may be resistant to decomposition and be long-lived as well. Thus the carbon sink of 50–200 Tg BC/a could increase to about 100–400 Tg BC/a. More detailed investigations in the direction of degradation of thermally altered organic matter is essential for a better understanding of this sequestration process.

Part of the BC emanates from deforestation that results in net release of CO_2 to the atmosphere. However, a considerable amount of other types of fire occur without net CO_2 release. For instance, agricultural waste fires and savanna fires do not contribute net CO_2 emissions since the carbon being volatilized by the fire will be fixed again in plants after 1 to 2 years by photosynthesis. Altogether, BC formation from all kinds of fires reduces the net CO_2 release to the atmosphere by 50 to 200 Tg C/a, which is 2–18% of the net CO_2 emission by permanent deforestation estimated at 1100–2300 Tg C/a (Legget et al. 1992) and may constitute a significant fraction of the "missing carbon" in the anthropogenically disturbed global carbon balance (Detwiler and Hall 1988).

By sequestering carbon from the bioatmospheric carbon cycle, BC formation releases oxygen to the atmosphere. With a molar O/C ratio of 0.1 for BC (Kuhlbusch and Crutzen 1995), we derive an annual release of 0.95 mole O_2 per mole of BC formed. Thus today's BC formation releases annually 127–507 Tg O. The oldest documented record of charcoals, aged 340 million years, was found in coal from Pennsylvania (Warg and Traverse 1973). If only 10% of the present rate of vegetation fires were due to natural fires, then over a period of 340 million years about 8 times O_2 may have been released to the atmosphere by the formation of BC than the current atmospheric O_2 content. An estimate of oxygen released by the amount of organic carbon imbedded in sediments gives a value of 25 times the present atmospheric oxygen content (Warneck 1988). These numbers are only shown to indicate the potential importance of vegetation fires for the present atmospheric O_2 content.

Conclusions

Quantitative determination of BC in various vegetation fire residues has shown that BC is mainly a flaming phase product. The formation ratios and estimates presented here indicate that more than 80% of the total amount of BC produced by the fire remains on site and is not emitted with the smoke. It is possible that the BC remaining on site after the fire, especially the fraction in the fine ash residues, may contribute significantly to atmospheric BC background concentrations, thus influencing the solar radiation budget in certain regions, in addition to BC directly released to the atmosphere during the fire. Furthermore, transport of vegetation fire residues should be investigated concerning nutrient transport and in view of evaluating sediment core analyses to determine fire history. Based on BC measurements in various residues and global carbon release estimates by vegetation fires the annual formation of BC in residues was estimated to be 44–194 Tg BC worldwide. When the amount of BC being emitted by vegetation fires was added, a formation of 50–200 Tg BC/a was derived. In addition, a similar amount of microbially resistant C may have been removed by our pretreatment procedure. This carbon is sequestered from the short-term bioatmospheric to the long-term geological carbon cycle and represents 2–18% of the net CO_2 emissions by permanent deforestation. It may constitute a significant fraction of the "missing carbon" in the anthropogenically disturbed global carbon balance. By removing atmospheric CO_2 and releasing oxygen, fire may have had a geologically significant impact on the CO_2 and O_2 content of today's atmosphere.

Acknowledgments

The authors wish to thank R. Zepp for reviewing the first draft as well as C. Liousse and H. Cachier for sending us an early version of their paper.

Note

1. The term *charcoal* is used here for all material that was blackened by the combustion process, thus comprising organic and black carbon.

References

Andreae, M. O. 1993. The influence of tropical biomass burning on climate and the atmospheric environment, in *Biogeochemistry of Global Change: Radiative Active Trace Gases*, R. S. Oremland, (Ed.), pp. 113–150, Chapman and Hall, New York.

Andreae, M. O., E. V. Browell, M. Garstang, G. L. Gregory, R. C. Harriss, G. F. Hill, D. J. Jacob, M. C. Pereira, G. W. Sachse, A. W. Setzer, P. L. Silva Dias, R. W. Talbot, A. L. Torres, and S. C. Wofsy. 1988. Biomass-burning emissions and associated haze layers over Amazonia, *J. Geophys. Res.*, 93, 1509–1527.

Cachier, H., M. Bremond, and P. Buat-Ménard. 1989. Carbonaceous aerosols from different tropical biomass burning sources, *Nature*, 340, 371–373.

Cachier, H., C. Liousse, P. Buat-Ménard, and A. Gaudichet. 1993. Particulate content of savanna fire emissions, *J. Atmos. Chem.*

Cadle, S. H., and P. J. Groblicki. 1982. An evaluation of methods for the determination of organic and elemental carbon in particulate samples, in *Particulate Carbon: Atmospheric Lifecycle*, G. T. Wolff, and R. L. Klimisch, (Eds.), pp. 89–109, Plenum Press, New York-London.

Chang, S. G., R. Brodzinsky, L. A. Gundel, and T. Novakov. 1982. Chemical and catalytic properties of elemental carbon, in *Particulate Carbon: Atmospheric Lifecycle*, G. T. Wolff and R. L. Klimisch, (Eds.), pp. 159–181, Plenum Press, New York-London.

Chylek, P., B. Johnson, P. A. Damiano, K. C. Taylor, and P. Clement. 1995. Biomass burning record and black carbon in the GISP2 ice core, *Geophys. Res. Let.*, 22, 89–92.

Chylek, P., B. Johnson, and H. Wu. 1987. Black carbon concentration in Byrd station ice core: From 13 000 to 700 years before present, *Ann. Geophysicae*, 92, 9801–9009.

Clark, J. S. 1988. Stratigraphic charcoal analysis on petrographic thin sections: Application to the history in northwestern Minnesota, *Quaternary Research*, 30, 81–91.

Clark, J. S., J. Merkt, and H. Müller. 1989. Post-glacial fire, vegetation, and human history on the nothern Alpine forelands, southwestern Germany, *J. Ecology*, 77, 897–925.

Clarke, A. D., and K. J. Noone. 1985. Soot in the Arctic snowpack: A cause for perturbation in radiative transfer, *Atmos. Environ.*, 19, 1045–2053.

Comery, J. A., G. R. Fahnestock, and S. G. Pickford. 1981. Elemental carbon deposition and flux from prescribed burning on a longleaf pine site in Florida, in *Final Report to the National Centre for Atmospheric Research*, Boulder, Colo.

Cope, M. J., and W. G. Chaloner. 1980. Fossil charcoal evidence of past atmospheric composition, *Nature*, 283, 647–649.

Crutzen, P. J., and M. O. Andreae. 1990. Biomass burning in the tropics: Impact on atmospheric chemistry and biogeochemical cycles, *Science*, 250, 1669–1678.

Crutzen, P. J., I. E. Galbally, and C. Brühl. 1984. Atmospheric effects from post nuclear fires, *Climatic Change*, 6, 323–364.

Detwiler, R. P., and C. A. S. Hall. 1988. Tropical forests and the global carbon cycle, *Science*, 239, 42–47.

Fearnside, P. M., P. M. L. Graca, N. L. Filho, F. J. A. Rodrigues, and J. M. Robinson. 1994. Tropical forest burning in Brazilian Amazonia: Measurements of biomass, combustion efficiency, and charcoal formation at Altamira, Para, unpublished report.

Fearnside, P. M., L. Niwton Jr., and F. M. Fernandes. 1993. Rain forest burning and the global carbon budget: Biomass, combustion efficiency, and charcoal formation in the Brazilian Amazon. *J. Geophys. Res.* 98, 16 733–16 743.

Goldberg, E. D. 1985. *Black Carbon in the Environment*, John Wiley and Sons, New York.

Griffin, J. J., and E. D. Goldberg. 1975. The fluxes of elemental carbon in coastal marine sediments, *Limnol. Oceanogr.*, 20, 456–463.

Hao, W. M., and M.-H. Liu. 1994. Spatial and temporal distribution of tropical biomass burning, *Glob. Biogeochem. Cycles*, 8, 495–503.

Harrison, W. E. 1976. Laboratory graphitization of a modern estuarine kerogen, *Geochim. Cosmochim. Acta* 40, 247–248.

Herring, J. R. 1977. Charcoal fluxes into Cenozoic sediments of the North Pacific, Ph.D. Thesis, Scripps Institution of Oceanography, University of California, San Diego.

Herring, J. R. 1985. Charcoal fluxes into sediments of the North Pacific Ocean: The Cenozoic record of burning, in *The Carbon Cycle and Atmospheric CO₂: Natural Variations, Archean to Present*, pp. 419–442, E. T. Sundquist and W. S. Broecker (Eds.), American Geophysical Union: Washington, D.C.

Houghton, R. A. 1991. Biomass burning from the perspective of the global carbon cycle, in *Global Biomass Burning*, J. S. Levine (Ed.), pp. 321–325, MIT Press, Cambridge, Mass.

Jenkins, B. M., S. Q. Turn, R. B. Williams, D. P. Y. Chang, O. G. Raabe, J. Paskind, and S. Teague. 1991. Quantitative assessment of gaseous and condensed phase emissions from open burning of biomass in a combustion wind tunnel, in *Global Biomass Burning*, J. S. Levine (Ed.), pp. 305–317, MIT Press, Cambridge, Mass.

Kasischke, E. S., N. L. Christensen, and B. J. Stocks. 1995. Fire, global warming, and the carbon balance of boreal forests, *Ecological Appl.*, in press.

Kuhlbusch, T. A. J. 1995. A method determining black carbon in residues of vegetation fires, *Env. Sc. & Techn.*, 29, 2695–2702.

Kuhlbusch, T. A. J., M. O. Andreae, H. Cachier, J. G. Goldammer, J.-P. Lacaux, R. Shea, and P. J. Crutzen. 1995. Black carbon formation by savanna fires: measurements and implication on the global carbon cycle, *J. Geophys. Res.* special issue SAFARI-92, accepted.

Kuhlbusch, T. A. J., and P. J. Crutzen. 1995. Toward a global estimate of black carbon in the residues of vegetation fires representing a sink of atmospheric CO₂ and source of O₂, *Glob. Biogeochem. Cycles.* 9, No. 4, 491–501.

Lacaux, J.-P., D. Brocard, C. Lacaux, R. Delmas, A. Brou, V. Yoboué, and M. Koffi. 1994. Traditional charcoal making: an important source of atmospheric pollution in the African Tropics, *Atmos. Res.*, 35, 71–76.

Leggett, J., W. J. Pepper, and R. J. Swart. 1992. Emissions scenarios for the IPCC: an update, in *Climate Change 1992: The Supplementary Report to the IPCC Scientific Assessment*, J. T. Houghton, B. A. Callander, and S. K. Varney, (Eds.), pp 69–96, Cambridge University Press.

Liousse, C., J. E. Penner, J. J. Walton, H. Eddleman, C. Chuang, and H. Cachier. 1995. Modeling biomass burning aerosols, in *Biomass Burning and Global Change*, J. S. Levine (ed), submitted.

Lobert, J. M. 1989. Verbrennung Pflanzlicher Biomasse als Quelle Atmosphärischer Spurengase: Cyanoverbindungen, CO, CO₂, and NOₓ, Ph.D. Thesis, Johannes Gutenberg-Universität, Mainz.

Lobert, J. M., D. H. Scharffe, W. Hao, T. A. J. Kuhlbusch, R. Seuwen, P. Warneck, and P. J. Crutzen. 1991. Experimental evaluation of biomass burning emissions: Nitrogen and carbon containing compounds, in *Global Biomass Burning*, J. S. Levine, (Ed.), pp. 289–304, MIT Press, Cambridge.

Menaut, J.-C., L. Abbadie, F. Lavenu, P. Loudjani, and A. Podaire. 1991. Biomass burning in West African savannas, in *Global Biomass Burning*, J. S. Levine (Ed.), pp. 133–142, MIT Press, Cambridge, Mass.

Novakov, T. 1984. The role of soot and primary oxidants in atmospheric chemistry, *Sci. Total Environ.* 36, 1–10.

Patterson, E. M., and C. K. McMahon. 1984. Absorption characteristics of forest fire particulate matter, *Atmos. Environ.* 18, 2541–2551.

Patterson, E. M., C. K. McMahon, and D. E. Ward. 1986. Absorption properties and graphitic carbon emission factors of forest fire aerosol, *Geophys. Res. Lett.*, 13, 129–132.

Sanford, R. L., J. Saldarriaga, K. E. Clark, C. Uhl, and R. Herrera. 1985. Amazon rain-forest fires, *Science*, 227, 53–55.

Seiler, W., and P. J. Crutzen. 1980. Estimates of gross and net fluxes of carbon between the biosphere and the atmosphere from biomass burning, *Climatic Change*, 2, 207–247.

Shea, R. W., B. W. Shea, J. B. Kaufmann, D. E. Ward, C. Haskins, and M. C. Scholes. 1995. Fuel biomass and combustion factors associated with fires in savanna ecosystems of South Africa and Zambia, *J. Geophys. Res.*, special issue SAFARI-92.

Suman, D. O. 1983. Agricultural burning in Panama and Central America: Burning parameters and the coastal sedimentary record, thesis, University of California, San Diego.

Suman, D. O., T. A. J. Kuhlbusch, and B. Lim. 1995. Marine sediments: A reservoir for black carbon and their use as spatial and temporal records of combustion, in *Sediment Records of Biomass Burning and Global Change*, J. Clark, (Ed.), Springer.

Summerhayes, C. P. 1981. Organic facies of Middle Cretaceous black shales in deep North Atlantic, *Am. Assoc. Pet. Geol. Bull*, 65, 2364–2380.

Trollope, W. S. W., A. L. F. Potgieter, N. Zambatis, and L. A. Trollope. 1995. SAFARI-92: Characterization of biomass and fire behaviour in controlled burns in the Kruger National Park, *J. Geophys. Res.*, special issue SAFARI-92.

Trumbore, S. E. 1993. Comparison of carbon dynamics in tropical and temperate soils using radiocarbon measurements, *Glob. Biogeochem. Cycles*, 7, 275–290.

Turco, R. P., O. B. Toon, T. P. Ackerman, J. B. Pollack, and C. Sagan. 1983. Nuclear winter: Global consequences of multiple nuclear explosions, *Science*, 222, 1283–1292.

Turn, S. Q., B. M. Jenkins, J. S. Chow, L. C. Pritchett, D. Campbell, T. Cahill, and S. A. Whalen. 1993. Elemental characterization of particulate matter emitted from biomass burning: Wind tunnel derived source profiles for herbaceous and wood fuels, presented at the Fall Meeting, American Geophysical Union, San Francisco, Calif. 6–10, December.

Ward, D. E., W. M. Hao, R. A. Sussot, R. A. Babbit, R. W. Shea, J. B. Kauffman, and C. O. Justice. 1995. Effect of fuel composition on combustion efficiency and emission factors for African savanna ecosystems, *J. Geophys. Res.*, special issue SAFARI-92.

Warg, J. B., and A. Traverse. 1973. cited in M. J. Cope, and W. G. Chaloner (1980), *Geosci. Man*, 7, 39–46.

Warneck, P. 1988. *Chemistry of the Natural Atmosphere*, Academic Press, New York, p. 614.

Biomass Burning and the Production of Carbon Dioxide:
A Numerical Study

Alexis D. Kambis and Joel S. Levine

Recently the need to quantitatively assess the long-term climatic implications of global biomass burning has been asserted (Houghton 1991; Walker 1992). Based on existing data, it has been predicted that biomass burning is responsible for 20%–40% of the total anthropogenic CO_2 emitted into the atmosphere (Levine 1990).

This burning releases large amounts of CO_2 into the atmosphere, some of which is reincorporated into the terrestrial biosphere over the course of months to years. This reincorporation occurs during the period of regrowth following the annual burning of certain ecosystems, primarily grasslands. However, when an ecosystem, such as a tropical rain forest, is burned, it can no longer act as a sink for atmospheric CO_2. The CO_2 released is not reincorporated, since the rain forest is not replaced. This affects the global carbon cycle in two ways: (1) the carbon once stored in forest structural material must now be redistributed within the reservoirs of the global carbon cycle; and (2) an area once acting as a sink for atmospheric CO_2 no longer does so. Whether these disturbances will have prolonged effects upon the global carbon cycle is presently unknown.

The global carbon cycle regulates the amount of CO_2 remaining in the atmosphere. Because biomass burning affects the global carbon cycle in many ways, it is important to develop quantitative descriptions of how the carbon cycle will be affected and, in turn, how the atmospheric levels of CO_2 will respond.

A global carbon cycle model has been constructed with a realistic terrestrial biosphere and a realistic world ocean carbon cycle. The numerical model has a horizontal resolution of 5° by 5° latitude by longitude, seasonal time scales, and a maximum simulation period of approximately 15 years. The terrestrial biosphere is represented by a set of eight site-specific ecosystem models. The ocean carbon cycle model is embedded in an existing ocean general circulation model in order to give a realistic geographic distribution of ocean sources and sinks for CO_2. The ocean and biosphere modules are coupled with an atmo-spheric circulation module to form the complete model. The model takes input from data sets for fossil fuel combustion and biomass burning.

The model ultimately seeks the solution to the continuity equation for carbon dioxide in the atmosphere:

$$\frac{\delta C\rho}{\delta t} = -\nabla \cdot (\vec{V} C\rho) + \Phi + T_v C\rho$$

Where C represents the CO_2 concentration in a particular grid cell, ρ is the density of air, $\vec{V}$ is the horizontal wind velocity, T_v is the horizontal transport due to wind, and Φ is the corresponding source or sink at the surface beneath the grid box. The objective of each individual module is to compute Φ as a function of time, location, and local driving variables governing CO_2 exchange.

The Atmospheric Circulation Module

In addition to serving as a sink to CO_2, the atmosphere couples the oceans, terrestrial biosphere, and anthropogenic CO_2 sources. An accurate model of atmospheric transport is necessary due to the short time scales involved in the ocean and global terrestrial carbon cycles relative to the atmospheric transport times. Specifically, the time required for atmospheric transport is greater than the time required for atmosphere-ocean or atmosphere-biosphere exchange. That is, in a single model time step, the amount of carbon in an atmospheric grid box may be significantly altered by the sources and sinks for carbon at the surface of the earth.

The atmospheric circulation module consists of a set of prescribed wind fields and a numerical advection algorithm. Its purpose is to predict the global concentration field of CO_2, both as an output variable of the complete model and as an input variable for the ocean and biosphere modules. The atmospheric circulation module begins with some initial distribution of atmospheric CO_2 specified on a global grid at time t_i, and returns a final distribution at a later time, t_f. Then, after being computed by the appropriate module, the net CO_2 released (absorbed) is added (subtracted) from the appropriate grid location. Due to the sensitivity of the ocean and biosphere to atmospheric CO_2 concentration and to the variability of tropospheric transport time scales, it is critical to reproduce the atmospheric CO_2 distribution as accurately as possible. In addition to satisfying the above model requirements, an accurate transport scheme is of fundamental importance when attempting to reproduce observed values of atmospheric CO_2 for purposes of validation. In the present module, a set of wind field data obtained from the European Center for Medium-range Weather Forecasting (ECMWF) was used to drive the atmospheric advection scheme. The data were in the form of 0-hour and 12-hour observational analysis fields of horizontal winds. The atmospheric advection algorithm calculates the distribution of CO_2 on a computational grid in the bottom-most atmospheric layer defined in the data set. The horizontal grid resolution is 5° by 5° latitude by longitude, and the grid box height, measured in pressure coordinates, is from 1000 mb to 850 mb (0–1.5 km height), corresponding with the previously mentioned wind field data set. The advection algorithm solves the finite-difference form of the tracer transport equation by the method of conservation of second-order moments (Prather 1986).

The Terrestrial Biosphere Module

A common feature of the concentration of atmospheric CO_2 is a high-frequency oscillation superimposed on steady exponential growth. This high-frequency oscillation is due to seasonal photosynthesis/respiration processes of the terrestrial biosphere; the nearly monotonically increasing component is correlated with the release of CO_2 into the atmosphere from fossil fuel combustion until about 1987, when the fossil fuel consumption ceases to increase exponentially, while atmospheric CO_2 continues to increase.

The terrestrial biosphere module consists of a global ecosystem data base and eight site-specific ecosystem models which, taken together, represent more than 80% of the net atmospheric CO_2 exchange due to the world's ecosystems. The ecosystem data base specifies each ecosystem over a $1° \times 1°$ global grid. The criteria for ecosystem module selection were sufficiently stringent that only eight models were chosen from the literature. The two major criteria were (1) the models must have been verified, and (2) the models' output must depend on regional driving variables in such a way as to make extrapolation from site-specific CO_2 fluxes to regional CO_2 fluxes feasible. Other criteria, such as computational economy, were considered as well.

The cycling of carbon by the terrestrial biosphere is controlled by photosynthesis, respiration, and decomposition. These processes are different for each ecosystem, and the factors that affect them, namely temperature, rainfall, solar radiation, and wind speed, vary as functions of both time and geographic location. The functional dependence of the processes on the factors, however, is assumed to be independent of geographic location.

The terrestrial biosphere is not static. It has been observed to change in response to increased atmospheric CO_2 concentration (Acock and Allen 1991). The response to increased CO_2 is known as the CO_2 fertilization effect. The biosphere will also respond to changes in climatic driving variables such as temperature and rainfall. Changes in variables such as these are expected to accompany increased atmospheric CO_2. The response of the model biosphere is also strongly dependent on changes in external forcings such as temperature and rainfall.

A set of eight existing ecosystem models, which together represent more than 80% of the terrestrial biosphere, have been taken from the literature (King 1986). These models are integrated into a global terrestrial biosphere module with both spatial and seasonal variability. Spatial and seasonal variability are necessary in order to reproduce large-scale features of the atmospheric CO_2 distribution as well as being critical to predict model feedback effects, such as the response of the individual ecosystems to increasing levels of atmospheric CO_2, changing rainfall distributions, and changing temperature distributions.

The selected models have several advantages, which include realistic representation of seasonality and response to atmospheric CO_2 concentrations. They can also be used to represent changing ecosystem distributions and can be either improved or replaced individually without affecting the rest of the terrestrial biosphere module.

The Ocean Carbon Module

An existing ocean general circulation model (Cox 1984) is used to predict transport of carbon compounds within the ocean. This model is used both to advect the compounds relevant to the ocean carbon cycle and to determine values of temperature and salinity that affect the carbon chemistry within the earth's oceans. It is necessary to represent the physical ocean circulation explicitly, since the ocean biochemical processes controlling the cycling of carbon strongly depend on factors such as equatorial circulation, upwelling of deep water, and downwelling of surface water.

Because most of the ocean carbon chemistry takes place in the surface ocean and is on the time scale of years to decades, it is important that the ocean General Circulation Model (GCM) accurately resolve the upper ocean structure. The two critical components of upper ocean circulation are equatorial flow and mixed-layer processes. The latter is crucial to an ocean carbon cycle model. The ocean circulation module is built on the Geophysical Fluid Dynamics Laboratory's (GFDL) Modular Ocean Model (MOM). This model has been configured with a horizontal resolution of $4.5°$ latitude by $3.75°$ longitude, 12 vertical levels, realistic continental boundaries, and bottom topography. The maximum depth represented in the model is 5000 meters. The model is forced at the surface by temperatures, salinities, and wind stresses. This model has been used successfully to predict both steady state distributions of ^{14}C in the world ocean and distributions of bomb produced ^{14}C (Toggweiler et al. 1989). However, no attempts were made to include biochemical carbon processes represented in either work.

The Fossil Fuel Input

The effects of fossil fuel combustion are included in the model in the form of a global data set of CO_2 emissions. It has been determined that fossil fuel combustion is the primary source of anthropogenic CO_2 released into the atmosphere (Bowden 1992), thus, future atmospheric CO_2 levels as well as the state of the global carbon cycle will depend strongly on the future CO_2 emissions from the consumption of these fuels. In order to completely specify the inventory of CO_2 sources, scenarios for the use of fossil fuels must first be determined. Given a specific scenario, a data set of global CO_2 emissions can be generated and used as model input. For the present simulation, fossil fuel sources of CO_2 were assumed to be constant for the period 1990 to 2000. Seasonal variations in fossil fuel consumption were also neglected.

The Tropical Biomass Burning Input

Distribution and frequency of occurrence of tropical biomass burning are given on the $5° \times 5°$ global grid along with estimates of the CO_2 that is released during this burning for specific types in the form of a data set describing the type of ecosystem burned (grassland or forest), the months of peak burning, and the average CO_2 released per month during these months. A simple representation of the effects of deforestation associated with burning in the tropics is then incorporated into the model. An assessment can then be made of the impact of biomass burning and associated deforestation on the global carbon cycle and on atmospheric CO_2 distributions.

Global biomass burning is a significant source of radiatively active gases to the atmosphere, the most relevant to global climate change being CO_2 (Lashof 1989). The primary sources of biomass burning are the burning of forests for land clearing and the annual burning of grasslands. These data can then be converted into the CO_2 released during a particular biomass burning event. Carbon dioxide is also released during deforestation associated with biomass burning in certain geographic regions.

Seasonality of biomass burning can be inferred from data taken over the same location at different times (Hao and Liu 1994).

The data representing CO_2 release from deforestation comes directly from the biomass burning data. No attempt has been made to model the actual physical processes that control the exchange of CO_2 resulting from deforestation. Deforestation is therefore represented in the module by the following procedure: (1) determine the location and areal extent of the deforestation, (2) locate the grid box in the model where the deforestation has taken place and the corresponding ecosystem from the ecosystem database, (3) determine a CO_2 exchange weighting coefficient defined as the ratio of the areal extent of deforestation estimated in step one above to the area of the grid box where the deforestation has occurred, and (4) multiply the CO_2 exchanges calculated by the terrestrial biosphere module for the particular grid cell by the weighting function. That is, the terrestrial biosphere module takes the data from the deforestation module and applies it to the grid-scale CO_2 exchange function.

In the model, global biomass burning is represented as monthly average fluxes specified on the computational grid. The biomass burning data are combined with the deforestation data set previously generated from the biomass burning data to form a profile of biomass burning CO_2 emission over the globe. The data for global biomass burning used as model input are specified on a $5°$-$\times$-$5°$ grid extending from $35°N$ to $60°S$ and $110°W$ to $170°E$. The data set describes biomass burning in tropical America, tropical Asia, and tropical Africa. The type of burning is classified as either forest or grassland, and the CO_2 flux is given in terms of 10^{12} g CO_2 per month. Finally, the months of intensive burning are specified.

Module Coupling

The individual modules representing the atmosphere, ocean, and terrestrial biosphere, as well as biomass burning, deforestation, and fossil fuel combustion, are fully coupled with respect to CO_2. The modules representing the ocean and biosphere are sensitive to the CO_2 content of the atmosphere, thus they exhibit different patterns depending on the atmospheric CO_2 mixing ratio. Care was taken to ensure that each module functioned properly when isolated from the rest of the model. This was an important step towards validation of the complete model. When possible, results from the present implementation of the modules were compared with past results from the literature.

In the numerical computer model, the individual modules pass information through files on disk. The files describe either CO_2 concentrations, in the case of data passed to and from the advection module, or data base modifiers. The data base modifiers are generated by one module process and act on a data base that controls CO_2 fluxes within a separate module, as in the case of deforestation or biomass burning acting on the ecosystem data set used by the terrestrial biosphere module. The different modules comprising the model have different time scales associated with their own internal processes. For example, the time scale associated with the terrestrial biosphere module is taken to be one day, while the time step for the ocean carbon cycle is six hours, and the advection time step for the atmosphere is six hours.

Several model processes are unaffected by global carbon cycle processes—if feedback effects are ignored. Examples of such processes are ocean and atmospheric circulation and rainfall distribution. Still other variables are affected, such as atmospheric temperature, but the time for the effect to become noticeable is substantially longer than the model simulation time.

These observations can be used to drastically reduce the amount of computational resources necessary to evaluate a model scenario without substantially altering the model predictions. To this end, the modules representing atmospheric and oceanic general circulation are run beforehand, and the results are stored on disk for subsequent model runs.

If all these processes are considered from the perspective of the rate of error introduced by ignoring

certain processes and the time scales for certain feedback loops to have an effect, the model time horizon can be estimated to be 15 years. The time horizon is defined to be the amount of model time before the uncertainty introduced in state variables becomes as large as the variables themselves.

A flowchart represents the numerical model at the level of the individual modules in figure 17.1.

Validation

The Carbon Dioxide Information Analysis Center (CDIAC) publishes data sets gathered from the various CO_2 monitoring sites around the world. This data is in the form of monthly average atmospheric CO_2 concentrations. The monitoring sites are typically located well away from land in order to obtain CO_2 concentrations from relatively uncontaminated air. The primary features in this data are the seasonality and latitudinal variation of the atmospheric CO_2 concentration. Model predictions, along with measured values, are shown in figures 17.2*a* through 17.2*e*.

Results

Two model runs are used to investigate the effects of biomass burning on the atmospheric concentration of CO_2: (1) In the first run, Business-as-Usual (BaU) for the period 1990 to 2000. Biomass burning, fossil fuel combustion, and deforestation are allowed to increase at their present rates. (2) In the second run, biomass burning is halted; however, everything else is kept identical to the first run. The net contribution of biomass burning to the global average atmospheric level of CO_2 as a function of time is presented in figure 17.3. The global distribution of atmospheric CO_2 originating from biomass burning is shown in figure 17.4 as the difference in the CO_2 concentration fields for the two runs described above at the end of the 10-year simulation period.

Discussion

A global carbon cycle model has been constructed with a realistic terrestrial biosphere and a realistic world ocean carbon cycle. The numerical model has a horizontal resolution of 5° by 5° latitude by longitude, seasonal time scales, and a maximum simulation period of approximately 15 years. The terrestrial biosphere is represented by a set of eight site-specific ecosystem models. These models have several advantages, which include realistic representation of seasonality and

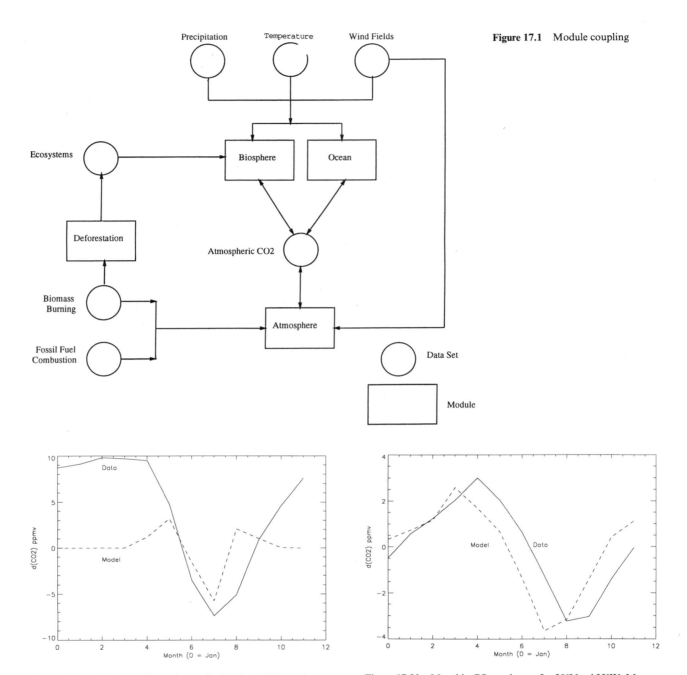

Figure 17.1 Module coupling

Figure 17.2a Monthly CO$_2$ exchange for 70°N × 155°W, Point Barrow

Figure 17.2b Monthly CO$_2$ exchange for 20°N × 155°W, Mauna Loa

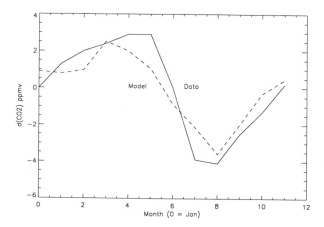

Figure 17.2c Monthly CO_2 exchange for $15°N \times 145°E$, Mariana Islands

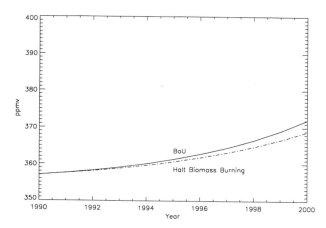

Figure 17.3 Average CO_2 concentration

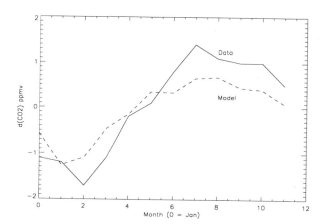

Figure 17.2d Monthly CO_2 exchange for $10°S \times 15°W$, Ascension Island

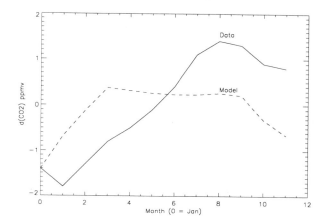

Figure 17.2e Monthly CO_2 exchange for $65°S \times 65°W$, Palmer Station

response to atmospheric CO_2 concentrations. These models can also be used to represent changing ecosystem distributions and can be either improved or replaced individually without affecting the rest of the terrestrial biosphere module. The ocean carbon cycle model is embedded in an existing ocean general circulation model in order to give a realistic geographic distribution of ocean sources and sinks for CO_2.

As seen in figure 17.3, the buildup of CO_2 in the atmosphere is somewhat rapid. This is due to the atmospheric module's underestimation of vertical mixing. The vertical mixing is presently driven by wind only, and to include other mixing schemes, such as moist convection, would require a considerably more complicated model atmosphere.

Important to note also is the relatively poor agreement of the model results with the measured values of atmospheric CO_2 in the high latitude regions (figures 17.2a through 17.2e). This is a result of poor representation of the ecosystem processes. In the mid latitudes, and particularly near the equator, however, the model agrees well with the measured values in both the amplitude and phase of atmospheric CO_2 concentrations. Also responsible for the disagreement between the model and data shown in figures 17.2a through 17.2e are limitations in the atmospheric model. Specifically, the atmospheric transport model exhibits little numerical diffusion, and underestimates vertical mixing, as has been previously discussed. These limitations give rise to the strong latitudinal gradient seen in figure 17.4. A final explanation of the discrepancy is that the model results are averaged over a $5° \times 5°$ area, where the measurement data represent a point somewhere in that grid cell.

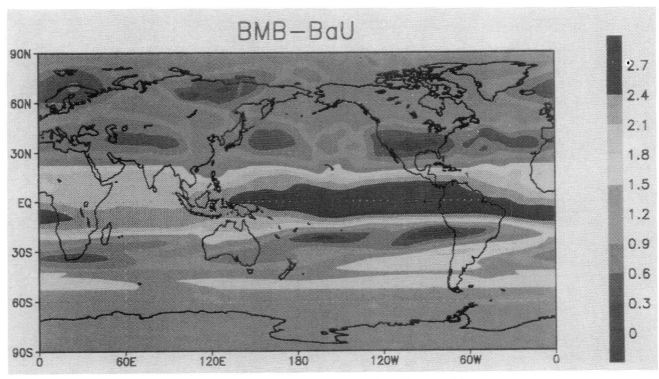

Figure 17.4 Net contribution of tropical biomass burning to atmospheric CO_2 after a 10-year model run

No net change in the distribution of sources and sinks for atmospheric CO_2 was observed for the model run from 1990 to 2000. This is due to the fact that observed changes in the ocean and biosphere take several decades to be detected, as in the case of CO_2 fertilization of the terrestrial biosphere and, in the case of the ocean sinks, the decadal exchange times for water flow between the surface and deep ocean.

It is concluded that atmospheric CO_2 concentrations will continue to rise well into the next century given any realistic scenario for either future fossil fuel use or biomass burning, keeping in mind that the model assumes the only net CO_2 released by biomass burning originates from burning of the tropical forests. Also excluded from the biomass burning sources of CO_2 is the burning of fuel wood and charcoal. This results in a 15% underestimate of CO_2 released. However, the modeling of the burning of fuel wood and charcoal was beyond the scope of this research.

The difficulties encountered in global carbon cycle modeling center primarily on the climate–carbon cycle feedbacks. Until a coupled global climate model is developed, the effects of global warming and associated climatic changes upon the carbon cycle must be treated as parameters. The other model limitations include processes such as the CO_2 released during

biotic changes immediately following deforestation, terrestrial ecosystem succession processes, river runoff processes depositing carbon into the ocean, and coastal ocean carbon processes. These limitations impose constraints on how far into the future predictions can be assumed to be accurate.

References

Acock, B. and L. Allen. 1991. Crop Response to Elevated Carbon Dioxide Concentrations. In *Direct Effects of Increasing Carbon Dioxide on Vegetation*. U. S. Dept. of Energy, Washington, D.C.

Bowden, T. A., R. J. Sepanski, I. W. Stoss, Eds. 1992. *Trends 91: A Compendium of Data on Climatic Change*, Publication CDIAC/ ORNL-46, Carbon Dioxide Information Analysis Center, Oak Ridge, Tenn.

Cox, M. 1984. GFDL Ocean Group Technical Report 1, Geophysical Fluid Dynamics Laboratory/NOAA, Princeton University, Princeton, N.J.

Hao, W. and M-H Liu. 1994. Spatial and temporal distribution of tropical biomass burning, *Global Biogeochem. Cycles*, 8 (4), 495–503.

Houghton, R. A. 1991. Biomass Burning from the Perspective of the Global Carbon Cycle. In *Global Biomass Burning: Atmospheric, Climatic, and Biospheric Implications* (Joel S. Levine, ed.), pp. 321–325, MIT Press, Cambridge, Mass.

King, A. and D. DeAngelis. 1986. Site-Specific Seasonal Models of Carbon Fluxes in Terrestrial Biomes, Environmental Sciences Division Publication No. 2515. Oak Ridge National Laboratory, Oak Ridge, Tenn.

Lashof, D. 1989. The dynamic greenhouse: feedback processes that may influence future concentrations of atmospheric trace gases and climate change, *Climatic Change*, 14, 213–242.

Levine, J. S. 1990. Global biomass burning: atmospheric, climatic, and biospheric implications, *Eos* 17, (37), 1075–1077.

Prather, M. 1986. Numerical advection by conservation of second-order moments, *J. Geophys. Res.* 91, 6671–6681.

Toggweiler, J., K. Dixon, K. Bryan. 1989. Simulations of radiocarbon in a coarse-resolution world ocean model, Part 2: distributions of bomb-produced carbon 14, *J. Geophysical Res.*, 94, *C6*, 8243–8264.

Walker, J. and J. Kasting. 1992. Effects of fuel and forest conservation on future levels of atmospheric carbon dioxide, *Palaeogeography, Palaeoclimatology, Palaeoecology (Global and Planetary Change Section)*, 97, (97).

A Study of the Mass Transport of Enhanced Continental Ozone in the Tropics and Its Impact over the Remote Southern Atlantic Ocean

Jennifer Richardson Olson

The recent finding that tropospheric ozone (O_3) over the southern tropical Atlantic Ocean exhibits a distinct seasonality with maximum values during the austral spring (figure 18.1) has led to several investigations aimed at identifying the potential mechanisms for this increase (e.g., Fishman et al. 1990, 1991; Krishnamurti et al. 1993, 1996; Jacob et al. 1996; Pickering et al. 1996; Thompson et al. 1996). Some of the major processes proposed to be at least partially responsible for the seasonal O_3 buildup are the downward transport of enhanced O_3 concentrations from the stratosphere or from the upper troposphere through divergent planetary east-west circulations (Walker circulations), and transport from middle altitudes through mechanisms such as frontal descent (Krishnamurti et al. 1993; Loring et al. 1996). A potential for in situ photochemical production also exists over the tropical Atlantic Basin, with O_3 precursors generated both by natural processes (e.g., biogenic soil or lightning sources of NO_x) and anthropogenic activities (e.g., biomass burning emissions of CO, hydrocarbons, and NO_x), followed by transport of these precursors into the remote Atlantic. It is well documented that a considerable amount of O_3 is generated in the lower troposphere over regions near large-scale fires (Delany et al. 1985; Chatfield and Delany 1990; Richardson et al. 1991; Cros et al. 1992; Kirchhoff et al. 1991; Lopez et al. 1992]. The mass transport of this enhanced O_3 into the Atlantic region is also likely to contribute to the observed O_3 maximum.

This study uses the Geophysical Fluid Dynamics Laboratory Global Chemical Transport Model (GFDL GCTM) to focus on the latter of these potential mechanisms. The approach is to force the model to mimic the expected seasonality of lower tropospheric O_3 over major burning regions and then examine the resulting transport of this seasonal O_3 enhancement. The major transport regimes that are likely to affect the tropical Atlantic during the primary austral burning season (July through October), the construction of the model simulation, and model results are discussed.

Major Transport Regimes into the South Atlantic

Examination of the meteorological climatology compiled by Oort (1983) for the southern tropical Atlantic region during the austral winter and spring (July through October; figures 18.2 through 18.4) shows that over its eastern portion, where O_3 concentrations appear to be largest, an upper tropospheric convergence and lower tropospheric divergence of winds indicate a region of subsidence above the marine boundary layer (see also Krishnamurti et al. 1993). Upper tropospheric horizontal wind speeds in this region are generally low (< 5 m s^{-1}) north of about 10°S. Within the boundary layer, persistent SE trade winds blow toward the equator with velocities near 10 to 15 m s^{-1}. Four major transport regimes from adjacent continents that are likely to affect the southern tropical Atlantic during the austral spring can be identified from studies of the climatological wind patterns (Krishnamurti et al. 1993; Richardson 1994; Thompson et al. 1996): (1) low-altitude tradewind advection from Africa, (2) low and middle tropospheric advection to the east off the South American east coast (roughly 15° to 30°S), (3) middle/upper tropospheric transport into the eastern tropical Atlantic from equatorial Africa, and (4) middle/upper tropospheric transport to the east across the Atlantic from northern South America.

Figure 18.2 shows the observed 950-mb climatological flow field over the South Atlantic Basin for July and October (Oort 1983). The quasi-permanent anticyclone over the subtropical Atlantic provides the driver for the persistent southeast trade winds that blow from the southwestern African continent toward the equatorial Atlantic. Near the mouth of the Congo River (10°S), a slight onshore, cyclonic curvature of the surface wind field is present along this coastal region year-round up to about 1 km above the surface (~ 900 mb) (Fontan et al. 1992). This "monsoonal" flow is somewhat strengthened during October with the development of a thermal low over the region of

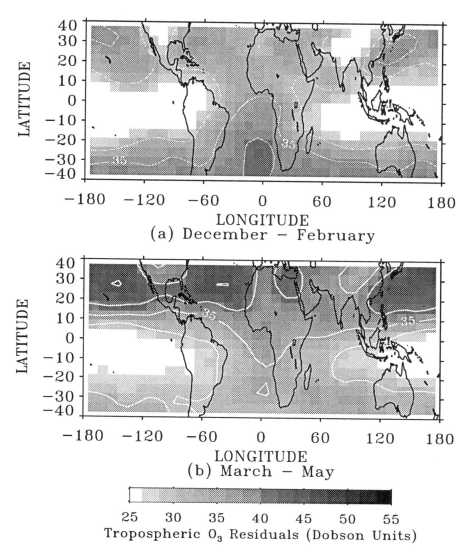

Figure 18.1 Seasonally averaged smoothed tropospheric ozone residuals (TOMS/SAGE method) in Dobson Units: (*a*) December–February, (*b*) March–May, (*c*) June–August, (*d*) September–November. Contours are at 5 Dobson Units. From Fishman et al. 1990

the Kalahari desert of southwestern Africa. At lower tropospheric altitudes above the monsoonal flow (700 mb, figure 18.3), air masses are freely transported out over the west African coast from the equator to 15°S into the southern Atlantic region. South of about 20°S at these altitudes, air masses follow an anti-cyclonic curvature and are swept into the strong east-ward flow back across southern Africa and toward the Indian Ocean.

An anticyclonic curvature of the winds over the eastern South American continent in the lower tropo-sphere (figure 18.3) produces onshore flow north of about 20°S with a westerly offshore flow dominant south of about 25°S (figure 18.3). Air masses from the lower and middle troposphere in the vicinity of the Amazon Basin are thus carried by the offshore flow into the western Atlantic (Richardson 1994).

At high altitudes during the austral winter and spring, an anticyclone just south of the equator over central Africa directs a circulation of air out over the eastern Atlantic with a continuation around the anti-cyclonic circulation back toward the east (figure 18.4). Persistent westerlies exist at these altitudes over north-ern South America. It may be expected that as convection begins to increase over central Africa and northeastern Brazil with the onset of the austral spring (October/November), lower tropospheric air trans-ported to higher altitudes via convection may be then advected into the Atlantic by the described upper tro-pospheric cirulation.

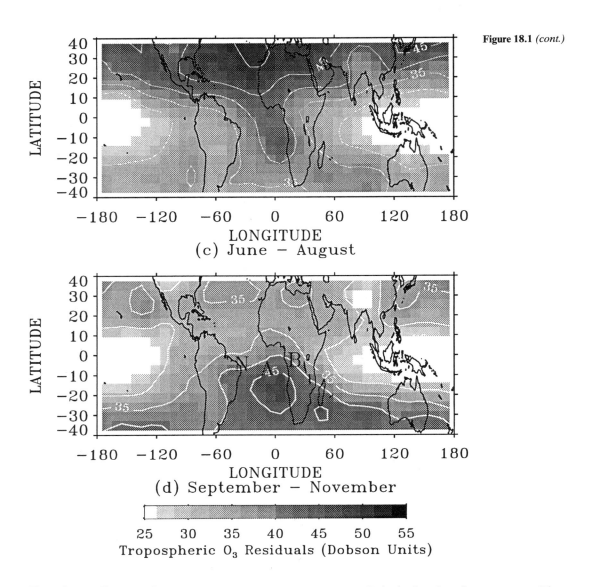

Figure 18.1 *(cont.)*

(c) June – August

(d) September – November

Tropospheric O$_3$ Residuals (Dobson Units)

Experiment Construction

The GFDL GCTM

The GFDL General Circulation Model (GCM) re-produces the general climatology of winds described above, including smaller-scale features such as the near-surface monsoonal flow over the southwestern African coast and overriding trade winds (Manabe et al. 1974; Manabe and Holloway 1975). The GCTM (Mahlman and Moxim 1978; Mahlman et al. 1980; Levy et al. 1982) is run from six-hour averages saved from the GCM wind fields with a 2.4° × 2.4° resolution and 11 vertical σ (terrain following) levels, 8 of which are in the tropical troposphere. The pressure (P_i) of a given level i is defined by: $P_i = \sigma_i \times P_o$, where

P_o is the local surface pressure. The convective vertical redistribution of model tracers is handled via a down-gradient diffusion which is increased when model parameters indicate atmospheric instability (see Levy et al. 1982 for details).

The purpose of this experiment is to examine the increase of ozone during the burning season; therefore, a model tracer is used that represents an O$_3$ enhancement over a background equal to the observed minimum season distribution (March through May for the region of interest). Assumptions are formulated to err on the side of maximizing the effect of this mass transport on the O$_3$ seasonality over the remote Atlantic and thus give a lower limit to the seasonality that must be a result of some other mechanism(s).

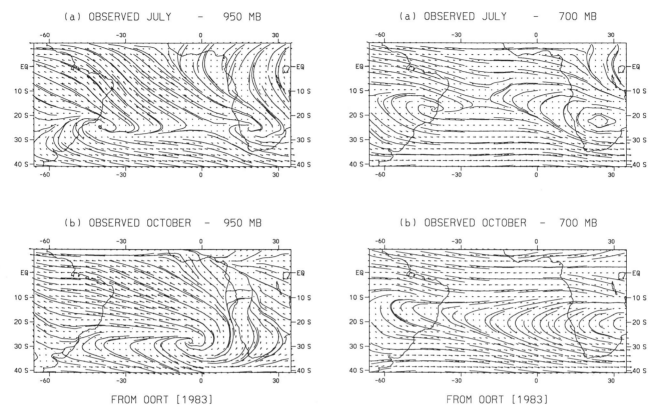

Figure 18.2 Observed climatological flow field at 950 mb from Oort (1983): (*a*) July, (*b*) October

Figure 18.3 Observed climatological flow field at 700 mb from Oort (1983): (*a*) July, (*b*) October

Seasonality of Lower Tropospheric Ozone over Continental Burning Regions

As discussed above, the approach of this study is to force the model tracer to mimic the expected seasonality of O_3 in the troposphere below $\sigma = 0.6$ (~ 600 mb) over the burning regions in tropical Africa and South America (shown in figure 18.5). To estimate this value, remote satellite information is first used to derive a climatological average of the regional tropospheric column amplitude (maximum minus minimum seasonal averages). Seasonal averages are used rather than monthly averages to assure that the amplitude is statistically robust (Olson et al. 1996). Multiyear in situ ozonesonde data sets representative of the various burning regions are then analyzed to determine the climatological percentage of this amplitude that is due to O_3 below the $\sigma = 0.6$ level.

Tropospheric Column Ozone The tropospheric column O_3 amount (called a "residual") is calculated by subtracting the stratospheric component of O_3 measured by the satellite-borne SAGE (Stratospheric Aerosol and Gas Experiment) and SAGE II instru-

ments from a near-concurrent total atmospheric amount measured by the TOMS (Total Ozone Mapping Spectrometer) instrument on the Nimbus-7 satellite (Fishman et al. 1990). Some recent studies point out potential errors in the TOMS data that may result in artificially elevated total ozone amounts where marine stratus are present (Thompson et al. 1993) or where TOMS may underestimate how much total ozone is present over land under the influence of a heavy pollution event generating very large amounts of ozone (Fishman et al. 1996a). However, a comparison of the O_3 seasonality from multiyear ozonesonde records at sites within and bordering the southern Atlantic region and over subtropical southern Africa has shown that the TOMS/SAGE residuals represent local climatological seasonal means to within 10% and seasonal amplitudes to within 15% (Olson et al. 1996). For the case of simplicity, throughout the rest of this chapter, the residuals will be referred to as observed, although strictly speaking, they are a derived quantity from calculations using two independent sets of observations.

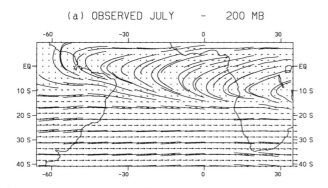

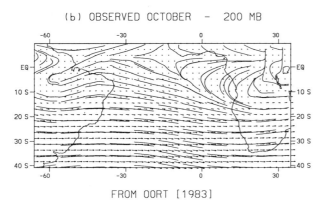

Figure 18.4 Observed climatological flow field at 200 mb from Oort (1983): (*a*) July, (*b*) October

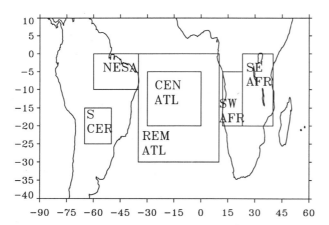

Figure 18.5 Boundaries for major regions discussed in this work. Major burning regions are NESA (Northeast South America), S CER (Southern Cerrado), SW AFR (Southwest Africa), and SE AFR (Southeast Africa). The remote Southern Atlantic is shown by REM ATL and the southern central Atlantic region is CEN ATL.

This climatological O_3 seasonality derived from the residuals averaged over each of the four major burning regions is shown in figure 18.6 and the associated quantitative seasonal amplitudes (maximum season minus minimum season) are listed in table 18.1. For the seasonal amplitude calculation, a period of minimum values of March through May is assumed for all regions. Seasonal maximum periods are region specific and are assigned as described in Olson et al. (1996) (see table 18.1).

Vertical Distribution of Seasonality The difficulty in the approach for this study then lies in the need to apply a vertical distribution to the amplitude provided by the residuals. Olson et al. (1996) show that there are distinct differences in the vertical distribution of the seasonal O_3 amplitude between the sites of Brazzaville, Congo (4°S, 15°E), and Natal, Brazil (6°S, 35°W). I assume here that the burning regions shown in figure 18.5 are homogeneous environments with pyrogenic emissions the driver of O_3 seasonality in the lower troposphere, and that the vertical distribution of the O_3 amplitude is similar throughout the individual regions. Thus, available ozonesonde data sets for sites within the various burning regions are analyzed to determine typical region-specific vertical distributions of the tropospheric amplitude.

Within the African regions, two stations with multiyear records of ozonesonde data were analyzed: Brazzaville, Congo, and Irene (Pretoria), South Africa (25°S, 27°E), located in the southwestern and southeastern African regions, respectively (Cros et al. 1992; Olson et al. 1996; Diab et al. 1996). At both of these sites, close to 55% of the seasonal tropospheric O_3 amplitude is due to contributions below $\sigma = 0.6$ (see table 18.2).

The coastal site of Natal, Brazil, is located in the northeast South American region (NESA) and has a continuous record of ozonesonde data dating back to 1978 (Kirchhoff et al. 1991, 1996). Only about 35% of the total amplitude is due to O_3 in the lower troposphere at this site (table 18.2). However, because Natal is located on the extreme eastern coast within the NESA region and is subjected to near-surface marine onshore flow, it is probably not representative of regions in the NESA nearer to large-scale fires.

Kirchhoff et al. (1996) present shorter-term (few months) records of ozonesonde profiles obtained over central Brazil during both the wet and dry seasons of 1992 as part of the TRACE-A (Transport and Chemistry near the Equator-Atlantic) campaign (Fishman et al. 1996b). Based on comparisons of the March

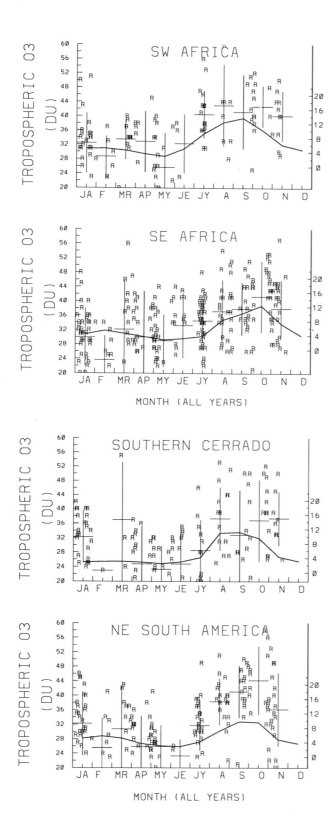

through May profiles at several sites within Brazil, Kirchhoff et al. (1996) conclude that ozone concentrations throughout the Amazon Basin and central Brazil are similar during this minimum season. Therefore, seasonal amplitudes are calculated here by subtracting this representative minimum season profile from burning season O_3 profiles. Porto Nacional (11°S, 48°W) is located within the NESA region in a cerrado environment near the Amazon Basin, and is subject to the effects of many local large-scale fires during the burning season. A seasonal enhancement of about 30 ppbv that is relatively consistent with altitude is calculated (Kirchhoff et al. 1996). At this site, about 45% of the column amplitude is located below $\sigma = 0.6$ (Table 18.2).

The lower tropospheric contribution from the Southern Cerrado site of Cuiaba, Brazil (16°S, 56°W), is calculated in a similar fashion. The September/ October 1992 average minus the representative wet season profile shows that 53% of the total column amplitude is in the lower troposphere (table 18.2) (Kirchhoff et al. 1996).

These percentages are assumed to be representative of the appropriate burning regions and are then applied to the regional average residual column amplitudes to derive an estimate of the regional-scale ozone enhancement in the lower troposphere (shown in table 18.1).

Model Tracer Parameterization

Tracer Source Distribution As discussed previously, I assume that the lower tropospheric seasonal O_3 enhancement over the four major burning regions neighboring the tropical Atlantic is driven by photochemical generation from biomass burning emissions. Therefore, the geographic and temporal distribution of the input model O_3 enhancement tracer must be related to the distribution of fire emissions. At present, however, a robust geographic and temporal distribution of fires remains undefined. Remote satellite measurements are currently being investigated as a method to estimate area burned and the temporal variation of the large-scale fires (Justice et al. 1985; Malingreau 1990; Men-

Figure 18.6 Seasonal variation of column O_3 over the major burning regions in Dobson Units. The bottom axis shows months, and all years of residuals are overlaid. Individual residual values ("observed") are shown with the Rs (left axis) and the horizontal lines and vertical bars show monthly averages and ± 1 standard deviation. The solid curve shows the model results (right axis). SW Africa, SE Africa, Southern Cerrado, NE South America.

Table 18.1 Seasonal amplitudes[a] of tropospheric ozone from residuals and from the pyrogenic O_3 enhancement model tracer

Burning region (maximum season)	Observed[b] column amplitude (Dobson units)	Observationally derived lower tropospheric amplitude[c] (Dobson Units) (% of column amplitude)	Model tracer in lower troposphere[c] (Dobson Units)
SW Africa (July–October)	12.1	6.7 (55%)	6.6
SE Africa (August–October)	8.8	4.8 (55%)	5.0
Southern Cerrado (August–October)	11.8	6.3 (53%)	6.0
Northeastern South America (August–October)	13.8	4.8–6.4 (35%–46%)	3.8

a. Seasonal amplitude is maximum season average (months shown in table) minus the March through May average.
b. Data labeled "observed" are O_3 tropospheric residual calculations as in Fishman et al. (1990).
c. "Lower troposphere" refers to the lowest four model levels; from the surface to $\sigma = 0.6$. This quantity is derived by applying the percentage of the seasonal amplitude from the lower troposphere listed in table 18.2 to the column amplitudes from the residuals.

Table 18.2 Climatological contribution to observed tropospheric ozone amplitude from the lower troposphere (surface to $\sigma = 0.6$)

Associated burning region (from figure 18.5)	Ozonesonde site	Calculated percentage of seasonal amplitude below $\sigma = 0.6$	Comments
SW AFR (southwest Africa)	Brazzaville, Congo 4°S, 15°E (Cros et al. 1992; Olson et al. 1996)	55%	(JASO)[a] minus (MAM)[b] data 1990–1992
SE AFR (southeast Africa)	Pretoria, South Africa 25°S, 27°E (Olson et al. 1996; Diab et al. 1996)	55%	(ASO)[c] minus (MAM)[b] data 1990–1993
NESA (northeast South America)	Natal, Brazil 6°S, 35°W (Kirchhoff et al. 1991; Olson et al. 1996)	35%	(ASO)[c] minus (MAM)[b] data 1978–1992 *Not likely affected by local fire emissions
	Natal, Brazil (Kirchhoff et al. 1996)	33%	(Sept–Oct 1992) minus (wet season profile from Goiania)[d] *Not likely affected by local fire emissions
NESA	Porto Nacional, Brazil 11°S, 48°W (Kirchhoff et al. 1996)	46%	(Sept–Oct 1992) minus (wet season profile from Goiania)[d]
S CER (Southern Cerrado)	Cuiaba, Brazil 16°S, 56°W (Kirchhoff et al. 1996)	53%	(Sept–Oct 1992) minus (wet season profile from Goiania)[d]

a. JASO = July, August, September, October average.
b. MAM = March, April, May average.
c. ASO = August, September, October average.
d. Minimum season O_3 profile throughout Brazil is assumed to be consistent with that measured at Goiania (16°S, 49°W) during April 1992 as discussed in Kirchhoff et al. (1996).

zel et al. 1991; Cahoon et al. 1992). Additionally, statistical methods are used to estimate fire distributions through analysis of maps of vegetation type, land-use trends and biomass densities (Seiler and Crutzen 1980; Crutzen and Andreae 1990; Hao et al. 1990). For this study, the distribution of emissions of CO_2 from deforestation, grass fires, and slash-and-burn agriculture published by Hao et al. (1990) is used as the relative geographic distribution of the source. The magnitude of the O_3 enhancement tracer is determined by scaling to these emissions to match observed lower tropospheric O_3 amplitudes.

The emissions reported in Hao et al. (1990) are in the form of mass emitted within $5° \times 5°$ grid boxes across the tropics, so a temporal distribution for burning must be prescribed. Prior studies have based fire emissions timing solely on rainfall statistics (Hao et al. 1990; Levy et al. 1991; Law and Pyle 1993). Hao et al. (1990), for example, assumed that fires occurred during the last five months of the dry season, with a fractional monthly distribution of (0.125, 0.25, 0.25, 0.25, 0.125). A temporal distribution based solely on precipitation data, however, may not produce realistic burning patterns. A timing scheme is used here which takes factors such as cultural practices and vegetation type into account, in addition to the local climatology of a region (Richardson 1994). During the Southern Hemisphere burning season, maximum emissions predicted by this temporal scheme are from August through October, about two months later into the season than that predicted from only precipitation (Hao et al. 1990). This is in better agreement with available regional observations of fires and combustive tracers such as CO, which indicate that the peak period for Southern Hemisphere fires is August through September (Kirchhoff et al. 1989; Müller 1992; Andreae 1993).

To approximate the heat-driven, near-surface turbulent mixing due to fires, the O_3 enhancement tracer is emitted from the model's lowest four levels ($\sigma = 0.99, 0.94, 0.835, 0.685$) with a mass distribution based on average profiles of CO obtained directly over burning cerrado regions in Brazil (Crutzen et al. 1985). This distribution, which places nearly 80% of the mass above $\sigma = 0.90$, allows "instantaneous mixing" of the tracer to levels above the surface layer.

Tracer Physical Characteristics The tracer is assumed to be photochemically inert to approximate conditions where there is equal photochemical production and destruction of O_3 in the remote troposphere. A second simulation was run with a tracer lifetime representative

of the photochemical O_3 loss due to the reaction of atomic oxygen with water vapor. Results from that simulation are not very different from the inert tracer simulation, and because this study reports an "upper limit" estimate for the effect of transported continental O_3, only the results from the inert tracer experiment are discussed here. A surface destruction is computed from a deposition rate of 1.0 cm s^{-1} over land and 0.1 cm s^{-1} over oceans, and gives the tracer a global lifetime on the order of 2 to 4 months. The model was initialized with near-zero concentrations and was run for 24 months, the last 12 of which are used in the analysis. Further integration shows that at this time, tracer concentrations have converged to a seasonal cycle equilibrium to within better than 5% at tropical latitudes.

Scale Factor To force the model tracer to mimic the magnitudes expected for the lower tropospheric amplitudes listed in table 18.1, a scale factor must be applied to relate the tracer to the input source magnitude of CO_2 in Hao et al. (1990). I used lower tropospheric amplitudes directly observed from in situ ozonesonde data to determine this scale factor because the use of a data set independent from the residual-based estimations shown in table 18.1 lends an additional constraint to the tracer parameterization. To most closely match the model amplitudes below the $\sigma = 0.6$ level at the appropriate grid boxes to those directly measured by ozonesondes at Brazzaville, Irene (Pretoria), and Porto Nacional, the model tracer requires a scale factor of ~ 18 (m-mole O_3 tracer)/(mole emitted CO_2). Interestingly, this value is well within the range of 4.8 to 40 (m-mole ΔO_3)/(mole ΔCO_2) for biomass burning plume enhancements measured by both field studies and in the laboratory reported by Andreae (1991).

Results

Comparison of Seasonality

Continental Burning Regions The first step in the model analysis is to examine how well this tracer parameterization reproduces the expected lower tropospheric enhancements over the burning regions. Figure 18.6 shows a comparison of the seasonality of the tropospheric O_3 residuals to that of the model tracer (integrated to 100 mb) (solid curve). The seasonal minimum (March through May) averages for the model results (right axis) and the residuals (left axis) were matched in order to compare the increases throughout the burning season. Qualitatively, the parameterization represents the shape of the O_3 seasonality quite well. A noteworthy point is that the period of enhanced concentrations appears slightly later in the year over southeastern Africa than over the southwestern portion of the continent. This is most likely due to the migration of the period of peak burning season across the African continent from the west coast (June through August) toward the east (July through October) (Cahoon et al. 1992).

Table 18.1 shows the quantitative values of the model amplitudes in the lower troposphere (below $\sigma = 0.6$) compared to the observationally derived lower tropospheric amplitudes. The model-predicted amplitudes in the lower troposphere over southwestern and southeastern Africa are 6.6 and 5.0 DU, which are in in good quantitative agreement (better than 5%) with observations. Similar to the African regions, the observationally derived lower tropospheric amplitude is reproduced over the South American Southern Cerrado region to within better than a 5% agreement. The results over northeastern South America, however, do not show as good a comparison. Only 28% of the observed column amplitude over this region (3.8 DU) is present in the model lower troposphere, compared to the expected 35 to 45% listed in table 18.2.

Tropical Southern Atlantic Ocean The observed column amplitude averaged over the Remote Southern Atlantic Ocean region (shown in figure 18.5; equator to 30°S, 35°W to 10°E) is 12.2 DU. Model results using the tracer parameterization described above predict that no more than 25% (~ 3 DU) of this observed amplitude may be explained by the mass transport of enhanced continental, lower tropospheric ozone (figure 18.7). If the estimated up to 15% differences in the residual calculations compared to in situ data are considered (Olson et al. 1996), the contribution is at most 33%. This leaves at least a 7 to 9 DU "deficit" averaged over the southern tropical Atlantic that must be explained by another mechanism.

Additional High-Altitude Source over Northeastern South America

The largest discrepancy between the model column seasonality over the continental regions and that observed is clearly over northeastern South America (figure 18.6d). This is because there is a greater impact from middle and upper tropospheric O_3 on the total column amplitude expected over northeastern South America as compared to the African sites (Olson et al. 1996).

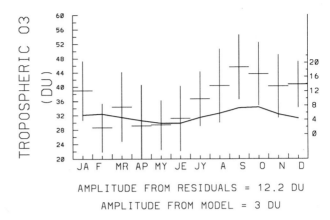

Figure 18.7 Seasonal variation of column O_3 over the remote Southern Atlantic region; Plot as for figure 18.6

There have been some investigations that speculate that the increase of convection over northern Brazil during the austral spring may contribute to an increase of O_3 in the upper troposphere over this region (Chatfield and Delany 1990; Pickering et al. 1991). The vertical redistribution of O_3 precursors results in a prolonged net production of O_3 at high altitudes, largely due to the transferral of nitrogen species into the free troposphere and away from surface and boundary layer sinks. Additionally, due to the nonlinear relation between O_3 production and NO_x concentration (Liu et al. 1987; Lin et al. 1988; Davis et al. 1996), the venting of highly concentrated pools of NO_x from the boundary layer into the free troposphere dilutes NO_x and increases the potential amount of O_3 generated per NO_x molecule.

Examination of the model gridpoints corresponding to Natal and Porto Nacional, Brazil, show that above 400 mb, the seasonal amplitude indicated by ozonesonde data at these sites is underpredicted by the model by about 20 ppbv. The large underprediction at these upper altitudes is likely to be largely due to this described upper-tropospheric photochemical generation of O_3, which is not represented in the model simulation. However, there is also the possibility that the model does not adequately represent the subgrid scale convective transport of the model enhancement tracer from the lower troposphere into higher altitudes.

To estimate the additional flux strength necessary to duplicate the observed upper tropospheric O_3 amplitudes over the northeastern South American region, a source is added to the model over this region between 400 mb and the tropopause from August through October. The flux required to reproduce the observed amplitudes is then determined to be on the order of

187
Olson

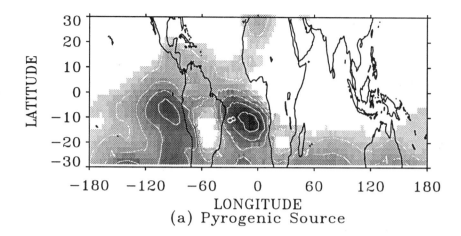

(a) Pyrogenic Source

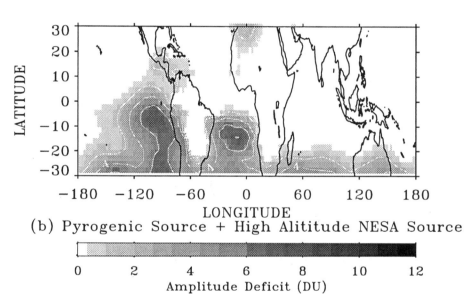

(b) Pyrogenic Source + High Alititude NESA Source

Amplitude Deficit (DU)

Figure 18.8 Amplitude deficit for MAM to ASO in Dobson Units. The amplitude deficit is calculated by subtracting the model-predicted seasonal amplitude from that observed. The observed values used to generate this plot are smoothed residuals (Fishman et al. 1990). (*a*) Deficit for model simulation of pyrogenic O_3 enhancement source only, (*b*) Deficit for model simulation of pyrogenic source plus the NESA upper tropospheric flux.

8.8 ppbv per day averaged throughout the three-month period. The mass transport of this upper tropospheric flux over northeastern South America results in a contribution to the observed O_3 amplitude over the southern Atlantic Ocean approximately equal to that calculated from the pyrogenic source (25% of the observed value). Therefore, about half of the observed amplitude may be explained by these two model sources. This indicates the importance of the transport regime discussed above due to the strong upper tropospheric westerlies blowing from over northeastern Brazil toward the central Atlantic in contributing to the O_3 column seasonality over the remote Atlantic.

Photochemical model calculations using measured chemical trace species in the tropical upper troposphere from the NASA TRACE-A mission during September and October 1992 have further indicated that the potential for a significant net O_3 production does exist at these upper altitudes (Jacob et al. 1996; Pickering et al. 1996; Thompson et al. 1996). Within cloud outflow regions over northeastern Brazil, the average net O_3 production was calculated by these investigators to be on the order of 3 to 4 ppbv d^{-1}. Although this production rate is only about half of the required upper tropospheric flux calculated here, photochemical calculations for upper tropospheric air masses obviously impacted by lightning generation of NO_x showed the potential for an initial net production rate of 7 to 8 ppbv d^{-1} (Pickering et al. 1996). Additionally, Pickering et al. (1996) and Thompson et al. (1996) present calculations that suggest photochemical production of O_3 in upper tropospheric air masses transported from South America may account for ~25% of the excess O_3 over the South Atlantic, which is quite similar to the contribution calculated from the additional upper tropospheric flux in this study.

Inclusion of this upper tropospheric source increases the tracer amplitude in the lower troposphere of northeastern South America by nearly 1 DU. When this is added to the results from the pyrogenic source, about a 4.5-DU amplitude is produced in that model region's lower troposphere. This is in much closer agreement with the lower end of the range of observationally derived values in table 18.1.

Ozone Deficit Calculation

Horizontal Distribution of the Deficit An amplitude "deficit" is defined here as the difference between the observed O_3 amplitude and that calculated by the model. In other words, the deficit is the portion of the observed amplitude that is not attributable to the sources used in the model. Figure 18.8 shows the deficit for the March–May (MAM) to August–October (ASO) amplitude across the tropics after including the pyrogenic source (figure 18.8*a*) and then adding the upper tropospheric source over NESA (figure 18.8*b*). Note that for the generation of this plot, the residual fields required a temporal and spatial smoothing in order to attain global coverage. The net effect from this smoothing is a suppression of the amplitudes, but the large-scale features remain unchanged. Therefore, this plot should be examined for general patterns of behavior rather than be used to determine the quantitative values of the deficit.

There are two centers of maximum deficit shown: one is located off the western coast of South America, and the second is found in the center of the Atlantic. The center of maximum deficit off the western South American coast is interesting in that it coincides with the strong subsidence branch driven by the Indian/Pacific Walker Circulation (Bjerknes 1959; Newell 1979). The rest of the deficit discussion, however, will focus on that over the southern Central Atlantic Region (defined here as 5° to 20°S and 30°W to the Greenwich Meridian; see figure 18.5).

Using the unsmoothed residuals, the observed column amplitude averaged over the central Atlantic regions is 16.5 DU. The pyrogenic source reproduces 3 DU of that amplitude, and the high-altitude flux over South America reproduces an additional 4 DU, resulting in a deficit of about 9.5 DU.

Vertical Distribution of the Ozone Deficit In order to calculate a vertical distribution of this deficit, a vertical distribution of the observed column amplitude is required. The residuals, however, only give a value for the tropospheric column amplitude. As described earlier for over the major burning regions, the fractional contribution of various altitude slabs to the tropospheric column amplitude is taken from ozonesonde data and applied to the residual column amplitude averaged over the region. In this case, the multiyear ozonesonde record obtained at Ascension Island (8°S, 15°W), which is located in the central Atlantic region, is used (Fishman et al. 1993; Olson et al. 1996).

The resulting vertical distribution of the deficit is calculated by subtracting the model-predicted amplitude from the observationally derived amplitude within model-equivalent altitude slabs. The model amplitudes (in DU) from both the pyrogenic and the upper tropospheric NESA sources are shown in figure 18.9*a* and the resulting deficit is shown in figure 18.9*b*. The deficit is shown in units of mixing ratio (ppbv) at each altitude along the right axis. These numbers show that the primary maximum in the deficit mixing ratio is seen at about 700 mb, with values slightly greater than 20 ppbv. There is also a significant deficit at upper altitudes, where there remains an underprediction on the order of 10 to 15 ppbv.

Upper Tropospheric Source over the Atlantic
The flux required to explain the central Atlantic upper tropospheric deficit shown in figure 18.9*b* is determined in a fashion similar to that described for the NESA region: that is, a source is put into the model from 400 mb to the tropopause from August through October. The flux strength required to remove the deficit shown in figure 18.9*b* is thus determined to be on the order of 4.8 ppbv d^{-1}.

Some mechanisms that may account for this additional flux could include eastward transport of upper tropospheric enhancements over the Southern Cerrado region that are not reproduced in this model simulation, and/or in situ photochemical production over the Atlantic (not simulated in this model study). Jacob et al. (1996) examined data obtained over the remote Atlantic during the TRACE-A campaign and determined that there is a net photochemical production of O_3 in the upper troposphere (8–12 km) over this region during the austral spring, with values averaging from around 2 to 4 ppbv d^{-1}. These values are a significant portion of the required additional flux calculated here.

Figure 18.10 shows the adjusted deficit over the central Atlantic when this upper tropospheric flux is added to the pyrogenic and the NESA upper tropospheric source simulations. While the deficit in the

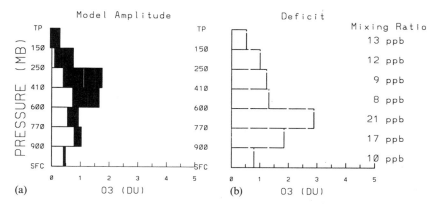

Figure 18.9 Vertical distribution of model amplitude and the amplitude deficit over the central Atlantic region (5° to 20°S, 30°W to Greenwich Meridian) in Dobson Units. (a) Model-predicted amplitude for the simulation using the pyrogenic O_3 enhancement source (white portion) and the NESA upper tropospheric source (shaded portion). Bars show the August through October average minus the March through May average for each model level. The equivalent pressure in mb for the model levels are shown on the left axis. (Note that the two lowest model levels are combined.) (b) Amplitude deficit, obtained by subtracting the model amplitude shown in (a) from the vertical distribution of the observed amplitude. The deficit is shown in mixing ratio units along the right axis.

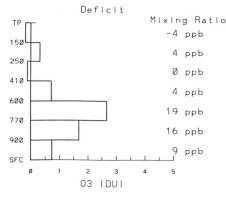

Figure 18.10 Vertical distribution of amplitude deficit in Dobson Units over the Central Atlantic region for model simulations of the pyrogenic source, and both the NESA and Central Atlantic upper tropospheric sources. The equivalent pressures in mb for the model levels are shown on the left axis. The deficit is shown in mixing ratio units along the right axis.

upper troposphere is essentially removed, the total column deficit in this region remains nearly 6 DU. There has been a slight reduction in the lower tropospheric deficit due to subsidence, but it is still quite large, with values near 20 ppbv. Alternately, as shown earlier, the lower tropospheric deficits over the continental burning regions are near zero (table 18.1).

The lower tropospheric deficit over the central Atlantic is therefore isolated from the local upper troposphere and from lower tropospheric source regions to the east and west. From this, it appears that an in situ production of O_3 in the lower troposphere over the central Atlantic region could be a significant source of the seasonal O_3 amplitude. However, while the in-transit net O_3 production in emission plumes at these low altitudes is limited to a time scale on the order of hours (Evans et al. 1974, 1977; Chatfield and Delany 1990; Jacob et al. 1996; Thompson et al. 1996), the transport time from the continents of Africa and South America into the central Atlantic is on the order of days.

There could be a delayed photochemical production of O_3 from biomass burning emissions in the central Atlantic region if a significant local regeneration of NO_x were to occur from subsidence and subsequent warming of nitrogen reservoir species such as PAN or HNO_4. Photochemical modeling calculations using observations obtained over the remote Atlantic during the TRACE-A campaign, however, show that in the lower troposphere (0–4 km), there is a strong net destruction of O_3 (Jacob et al. 1996; Thompson et al. 1996). Jacob et al. (1996) did calculate a net production at slightly higher altitudes (4–8 km) over this region, with values between 0 and 3 ppbv d^{-1}. The flux necessary to increase the model seasonal amplitude in the lower troposphere of this central Atlantic region to that observed is calculated here to be on the order of 7 to 8 ppb d^{-1}, more than 2.5 times the larger estimates of Jacob et al. (1996).

Further work is required to elucidate the mechanism that drives this lower tropospheric amplitude in the central Atlantic, including more photochemical analyses. Additionally, recent work has suggested that transport from middle latitude upper tropospheric regions via tropopause folds and descent in the vicinity of frontal boundaries could contribute to the seasonal buildup of lower tropospheric O_3 in the central Atlantic region (Loring et al. 1996).

Conclusions

Simulations using the GFDL GCTM suggest that no more than about 25% of the observed seasonal O_3 over the remote tropical southern Atlantic Ocean can be attributed to the mass transport of enhanced O_3 concentrations in the lower troposphere produced over biomass burning regions. An additional flux imposed in the upper troposphere (above 400 mb) over the northeast portion of South America with a strength of nearly 9 ppbv d^{-1} was required to duplicate amplitudes from ozonesonde records within that region. This additional flux increased the model-predicted seasonal amplitude over the tropical Atlantic to about half of what is observed.

The amount of the seasonal increase not accounted for by these two sources (amplitude deficit) was calculated, and the highest values are located over the central Atlantic and off the west coast of South America. The vertical distribution of the Atlantic deficit indicates that there are likely to be two additional fluxes required in this region: one located in the upper troposphere and one in the lower troposphere. The flux strength in the upper troposphere that is required to nullify the deficit at these altitudes is just less than 5 ppbv d^{-1}.

Measured concentrations of O_3 precursors in the middle and upper troposphere over the remote southern Atlantic Ocean and over northeastern Brazil during the austral spring show the likelihood of a significant photochemical source at high altitudes over the Atlantic Basin with rates that are not too far removed from the required flux strengths calculated here (Jacob et al. 1996; Pickering et al. 1996; Thompson et al. 1996). This study supports the conclusions from these analyses that it is possible, even likely, that photochemical processes at upper altitudes over the Atlantic Basin are sufficiently large to drive the O_3 seasonality at these altitudes.

When these upper tropospheric fluxes are included with the pyrogenic O_3 enhancement source in the model, a significant region of O_3 deficit equal to nearly 6 DU (about 20 ppbv) is isolated over the central Atlantic at lower altitudes. A flux of 7 to 8 ppb d^{-1} at ~ 700 mb is required through the burning season to account for the observed amplitude. This value is several times larger than prior estimates of photochemical production at these lower altitudes. The identification of the source of the lower tropospheric seasonality in the central Atlantic remains to be addressed through further photochemical and meteorological analyses in this complex region.

Acknowledgments

I thank H. Levy II and W. M. Moxim at the NOAA Geophysical Fluid Dynamics Laboratory in Princeton, New Jersey, both for helpful advice and for their assistance in the use of the GFDL GCTM. Jack Fishman and W. L. Chameides are also gratefully acknowledged for helpful discussions and comments. This work was funded by the NASA Mission to Planet Earth/Atmospheric Chemistry Modeling and Analysis Program.

References

Andreae, M. O. 1991. Biomass burning: Its history, use, and distribution, and its impact on environmental quality and global climate, in *Global Biomass Burning: Atmospheric, Climatic, and Biospheric Implications*, J. Levine, ed., MIT Press, Cambridge, Mass., 3–21.

Andreae, M. O. 1993. Global distribution of fires seen from space, *Eos Trans. AGU.* 74, 129, 135.

Bjerknes, J. 1969. Atmospheric teleconnections from the equatorial Pacific, *Mon. Weather Rev.*, 97, 163–172.

Cahoon, D., B. J. Stocks, J. S. Levine, W. R. Cofer III, and K. P. O'Neill. 1992. Seasonal distribution of African savanna fires, *Nature*, 359, 812–815.

Chatfield, R. B., and A. C. Delany. 1990. Convection links biomass burning to increased tropical ozone: However, models will tend to overpredict O_3, *J. Geophys. Res.*, 95, 18 473–18 488.

Cros, B., D. Nganga, A. Minga, J. Fishman, and V. Brackett. 1992. Distribution of tropospheric ozone at Brazzaville, Congo, determined from ozonesonde measurements, *J. Geophys. Res.*, 97, 12 869–12 875.

Crutzen, P. J., A. C. Delany, J. Greenberg, P. Haagensen, L. Heidt, R. Lueb, W. Pollock, W. Seiler, A. Wartburg, and P. Zimmerman. 1985. Tropospheric chemical composition measurements in Brazil during the dry season, *J. Atmos. Chem.*, 2, 233–256.

Crutzen, P. J., and M. O. Andreae. 1990. Biomass burning in the tropics: Impact on atmospheric chemistry and biogeochemical cycles, *Science*, 250, 1669–1678.

Davis, D. D., J. Crawford, G. Chen, W. Chameides, S. Liu, J. Bradshaw, S. Sandholm, G. Sachse, G. Gregory, B. Anderson, J. Barrick, A. Bachmeier, J. Collins, E. Browell, D. Blake, S. Rowland, Y. Kondo, H. Singh, R. Talbot, B. Heikes, J. Merrill, J. Rodriguez, and R. E. Newell. 1996. Assessment of ozone photochemistry in the western North Pacific as inferred from PEM-West A observations during the fall 1991, *J. Geophys. Res.*, 101, 2111–2134.

Delany, A. C., P. Haagensen, S. Walters, A. F. Wartburg, and P. J. Crutzen. 1985. Photochemically produced ozone in the emissions from large-scale tropical vegetation fires, *J. Geophys. Res.*, 90, 2425–2429.

Diab, R., A. Thompson, M. Zunckel, G. Coetzee, C. Archer, J. Combrink, F. Sokolic, G. Bodeker, D. McNamara, J. Fishman, and

D. Nganga. 1996. Vertical ozone distribution over southern Africa (and adjacent oceans) during SAFARI-92, *J. Geophys. Res.*, (in press).

Evans, L. F., N. K. King, D. R. Packham, and E. T. Stephens. 1974. Ozone measurements in smoke from forest fires, *Environ. Sci. Technol.*, 8, 75–76.

Evans, L. F., I. A. Weeks, A. J. Eccleston, and D. R. Packham. 1977. Photochemical ozone in smoke from prescribed burning of forests, *Environ. Sci. Technol.*, 11, 896–900.

Fishman, J., C. E. Watson, J. C. Larsen, and J. A. Logan. 1990. Distribution of tropospheric ozone determined from satellite data, *J. Geophys. Res.*, 95, 3599–3617.

Fishman, J., K. Fakhruzzaman, B. Cros, and D. Nganga. 1991. Identification of widespread pollution in the Southern Hemisphere deduced from satellite analyses, *Science*, 252, 1693–1696.

Fishman, J., J. Logan. P. E. Artaxo, H. Cachier, G. R. Carmichael, R. Dickinson, M. A. Fosberg, G. Helas, M. Kanakidou, J.-P. Lacaux, and F. Rohrer. 1993. Group Report: What is the impact of fires on atmospheric chemistry, climate, and biogeochemical cycles? in *Fire in the Environment: The Ecological, Atmospheric, and Climatic Importance of Vegetation Fires*, P. J. Crutzen and J. G. Goldammer, eds., John Wiley and Sons, West Sussex, England, 245–358.

Fishman. J., V. G. Brackett, E. V. Browell, and W. B. Grant. 1996a. Tropospheric ozone derived from TOMS/SBUV measurements during TRACE-A, *J. Geophys. Res.*, (in press).

Fishman, J., J. M. Hoell, Jr., R. D. Bendura, R. J. McNeal, and V. W. J. H. Kirchhoff, 1996b. NASA GTE TRACE-A experiment (September–October 1992): Overview, *J. Geophys. Res.*, (in press).

Fontan, J., A. Druilhet, B. Benech, R. Lyra, and B. Cros. 1992. The DECAFE experiments: Overview and meteorology, *J. Geophys. Res.*, 97, 6123–6136.

Hao, W. M., M. Liu, and P. J. Crutzen. 1990. Estimates of annual and regional releases of CO_2 and other trace gases to the atmosphere from fires in the tropics, based on the FAO statistics for the period 1975–1980, in *Fire in the Tropical Biota: Ecosystem Processes and Global Challenges*, J. G. Goldammer, ed., Springer-Verlag, Berlin, Heidelberg, 440–460.

Jacob, D. J., B. G. Heikes, S.-M. Fan, J. A. Logan, D. L. Mauzerall, J. D. Bradshaw, H. B. Singh, G. L. Gregory, R. W. Talbot, D. R. Blake, and G. W. Sachse. 1996. The origin of ozone and NO_x in the tropical troposphere: A photochemical analysis of aircraft observations over the South Atlantic Basin, *J. Geophys. Res.*, (in press).

Justice, C. O., J. R. G. Townshend, B. N. Holben, and C. J. Tucker. 1985. Analysis of the phenology of global vegetation using meteorological satellite data, *Int. J. Remote Sensing*, 6, 1271–1318.

Kirchhoff, V. W. J. H., A. W. Setzer, and M. C. Pereira. 1989. Biomass burning in Amazonia: Seasonal effects on atmospheric O_3 and CO, *Geophys. Res. Lett.*, 16, 469–472.

Kirchhoff, V. W. J. H., R. A. Barnes, and A. L. Torres. 1991. Ozone climatology at Natal, Brazil, from in situ ozonesonde data, *J. Geophys. Res.*, 96, 10 899–10 909.

Kirchhoff, V. W. J. H., J. R. Alves, F. R. daSilva, and J. Fishman. 1996. Observations of ozone concentrations in the Brazilian cerrado during the TRACE-A field expedition, *J. Geophys. Res.*, (in press).

Krishnamurti, T. N., H. E. Fuelberg, M. C. Sinha, D. Oosterhof, E. L. Bensman, and V. B. Kumar. 1993. The meteorological environment of the tropospheric ozone maximum over the tropical South Atlantic Ocean, *J. Geophys. Res.*, 98, 10 621–10 642.

Krishnamurti, T. N., M. C. Sinha, M. Kanamitsu, D. Oosterhof, H. Fuelberg, R. Chatfield, D. J. Jacob, and J. Logan. 1996. Passive tracer transports relevant to the TRACE-A experiment, *J. Geophys. Res.*, (in press).

Law, K. S., and J. A. Pyle. 1993. Modeling trace gas budgets in the troposphere: 2. CH_4 and CO, *J. Geophys Res.*, 98, 18 401–18 412.

Levy, H. II, W. J. Moxim, P. S. Kasibhatla, and J. A. Logan. 1991. The global impact of biomass burning on tropospheric reactive nitrogen, in *Global Biomass Burning: Atmospheric, Climatic, and Biospheric Implications*, J. Levine, ed., MIT Press, Cambridge, Mass., 363–369.

Levy, H. II. J. D. Mahlman, and W. J. Moxim. 1982. Tropospheric N_2O Variability, *J. Geophys, Res.*, 87, 3061–3080.

Lin, X., M. Trainer, and S. C. Liu. 1988. On the nonlinearity of the tropospheric ozone production, *J. Geophys, Res.*, 93, 15 879–15 888.

Liu, S. C., M. Trainer, F. C. Fehsenfeld, D. D. Parrish, E. J. Williams, D. W. Fahey, G. Hubler, and P. C. Murphy. 1987. Ozone production in the rural troposphere and implications for regional and global ozone distributions, *J. Geophys. Res.*, 92, 4191–4207.

Lopez, A., M. L. Huertas, and J. M. Lacome. 1992. Numerical simulation of the ozone chemistry observed over forested tropical areas during DECAFE experiments, *J. Geophys. Res.*, 97, 6149–6158.

Loring, R. O., Jr., H. Fuelberg, J. Fishman, and E. V. Browell. 1996. The role of middle latitude cyclones on the distribution of ozone over the South Atlantic Ocean, *J. Geophys. Res.*, (in press).

Mahlman, J. D., and W. J. Moxim. 1978. Tracer simulation using a global general circulation model: Results from a midlatitude instantaneous source experiment, *J. Atmos. Sci.*, 35, 1340–1374.

Mahlman, J. D., H. Levy, II, and W. J. Moxim. 1980. Three-dimensional tracer structure and behavior as simulated in two ozone precursor experiments, *J. Atmos. Sci.*, 37, 655–685.

Malingreau, J.-P. 1990. The contribution of remote sensing to the global monitoring of fires in tropical and subtropical ecosystems, in *Fire in the Tropical Biota: Ecosystem Processes and Global Challenges*, J. G. Goldammer, ed., Springer-Verlag, Berlin, Heidelberg, 337–369.

Manabe, S., D. G. Hahn, and J. L. Holloway, Jr. 1974. The seasonal variation of the tropical circulation as simulated by a global model of the atmosphere, *J. Atmos. Sci.*, 31, 43–83.

Manabe, S., and J. L. Holloway, Jr. 1975. The seasonal variation of the hydrologic cycle as simulated by a global model of the atmosphere, *J. Geophys. Res.*, 80, 1617, 1649.

Menzel, W. P., E. C. Cutrim, and E. M. Prins. 1991. Geostationary satellite estimation of biomass burning in Amazonia during BASE-A, in *Global Biomass Burning: Atmospheric, Climatic, and Biospheric Implications*, J. S. Levine, ed., MIT Press, Cambridge, Mass., 41–46.

Müller, J.-F. 1992. Geographical distribution and seasonal variation of surface emissions and deposition velocities of atmospheric trace gases, *J. Geophys. Res.*, 97, 3787–3804.

Newell, R. E. 1979. Climate and the ocean, *Amer. Sci.*, 67, 405–416.

Olson, J. R., J. Fishman, V. W. J. H. Kirchhoff, D. Nganga, and B. Cros. 1996. An analysis of the distribution of ozone over the southern Atlantic region, *J. Geophys. Res.*, (in press).

Oort, A. H. 1983. Global Atmospheric Circulation Statistics, 1958–1973, *NOAA Professional Paper* 14, National Oceanic and Atmospheric Administration Rockville, Md.

Pickering, K. E., A. M. Thompson, J. R. Scala,. W. Tao, J. Simpson, and M. Garstang. 1991. Photochemical ozone production in tropical squall line convection during NASA/GTE/ABLE2A, *J. Geophys. Res.*, 96, 3099–3114.

Pickering, K. E., A. M. Thompson, Y. Wang, W.-K. Tao, D. P. McNamara, V. W. J. H. Kirchhoff, B. G. Heikes, G. W. Sachse, J. D. Bradshaw, G. L. Gregory, D. R. Blake. 1996. Convective transport of biomass burning emissions over Brazil during TRACE-A, *J. Geophys. Res.*, (in press).

Richardson, J. L. 1994. *An investigation of large-scale tropical biomass burning and the impact of its emissions on atmospheric composition*, Ph.D. Thesis, Georgia Institute of Technology, Atlanta, Georgia.

Richardson, J. L., J. Fishman, and G. L. Gregory. 1991. Ozone budget over the Amazon: Regional effects from biomass-burning emissions, *J. Geophys. Res.*, 96, 13 073–13 087.

Seiler, W., and P. J. Crutzen. 1980. Estimates of gross and net fluxes of carbon between the biosphere and the atmosphere from biomass burning, *Climatic Change*, 2, 207–247.

Thompson, A. M., D. P. McNamara, K. E. Pickering, and R. D. McPeters. 1993. Effect of marine stratocumulus on TOMS Ozone, *J. Geophys. Res.*, 98, 23 051–23 057.

Thompson, A. M., K. E. Pickering, D. P. McNamara, M. R. Schoeberl, R. D. Hudson, J. H. Kim, E. V. Browell, V. W. J. H. Kirchhoff, and D. Nganga. 1996. Where did tropospheric ozone over southern Africa and the tropical Atlantic come from in October 1992? Insights from TOMS, GTE/TRACE-A and SAFARI-92, *J. Geophys. Res.*, (in press).

Biomass Burning Smoke in the Tropics: From Sources to Sinks

Krista K. Laursen and Lawrence F. Radke

For the past several years, extensive research has been directed toward the assessment of the contribution of biomass burning to global particulate emissions. In particular, the increasing rates of deforestation and changes in savanna land agricultural practices in South America and Africa have prompted researchers to more closely examine particulate emissions from burning in the tropics. Such studies of tropical biomass combustion have been undertaken with the dual goals of assessing the regional atmospheric impact of the burning and of estimating the contribution of burning in the tropics to the global emission of particles from biomass combustion. One recent estimate by Andreae (1991) placed the annual emission of total particulate matter from tropical biomass burning at 90 Tg, 86% of particulate emissions from global biomass burning and 6% of particulate emissions from all global emission sources. However, despite their important role in the development of a better understanding of global climate change and global ecosystem processes in general, such particulate-emissions estimates remain highly uncertain. The magnitude of this uncertainty is illustrated by the work of Penner et al. (1994), who estimate that the impact on the global radiation budget of aerosols produced by biomass combustion is to cause a net cooling of approximately 0.8 $W-m^{-2}$. However, Penner et al. also state that this estimate is uncertain by a factor of roughly 2.5.

In an effort to characterize both the combustion processes and resultant emissions from tropical biomass combustion, several studies have been carried out in Brazil. These independent research programs have involved ground-based measurements in the fire environment and at air quality monitoring stations (Ward et al. 1992; Artaxo et al. 1994), airborne studies of fire emissions and properties (Crutzen et al. 1985; Andreae et al. 1988; Ward et al. 1991; Riggan et al. 1993), and analyses of satellite data to determine fire locations and number and to estimate areas burned and quantities of emitted species (e.g., Kaufman et al. 1990; Menzel et al. 1991).

These various research programs have provided valuable information on the nature and extent of biomass combustion in Brazil and have furnished some useful data regarding the amount of particulate matter produced by such burning. However, work must still be done to characterize and more fully describe the processes governing the transport and removal of combustion-generated aerosols from the atmosphere in Brazil. This latter analytical step must be undertaken to determine if particulate emissions from biomass burning in Brazil are truly of global significance or if the impact of such emissions is one of purely regional significance.

In this chapter, we present a summary of some recent information regarding aerosol production from biomass burning in Brazil and discuss several ideas pertaining to smoke (i.e., aerosol) transport and lifetime. The purpose of these discussions is to outline some ideas pertaining to the fate of particles produced by biomass combustion in Brazil and in the tropics in general. We use these ideas as the framework for a set of guidelines for future research projects to be carried out in the tropics. These guidelines will, in our opinion, enable researchers to gain some valuable closure on the issue of particulate production and fate as pertaining to tropical biomass combustion and should make it possible to reduce, and possibly eliminate, some of the uncertainty surrounding the ultimate impact of biomass burning emissions on the atmosphere.

Smoke Production, Transport, and Removal: Background and Theory

Satellite image analyses have been applied to the study of global aerosol distribution patterns. Husar et al. (in press) have analyzed NOAA/AVHRR polar-orbiting satellite data and have developed a summary of the global oceanic aerosol pattern between July 1989 and June 1991. As is outlined by these authors, the dark ocean surface serves as a good background for the detection of atmospheric aerosol layers by satellite sensors. Aerosol plumes of continental origin are discussed in some detail by Husar et al., and the spatial and temporal distributions of these plumes are analyzed and presented. Figure 19.1, adapted from the work of Husar et al., shows mappings of the equivalent aerosol optical depth ($\times 1000$) for the periods June through August (figure 19.1a) and September through November (figure 19.1b) as derived from meteorological satellite data. As discussed by Husar et al., a conspicuous feature in both panels of figure 19.1 is the region of high aerosol optical depth off the west coast of Africa. The southwest African aerosol plume, separated from the west African aerosol plume in figure 19.1b, is, according to Husar et al., presumed to be attributable to biomass burning in equatorial Africa. Similarly, the region of high aerosol optical depth off the coasts of Mexico to Venezuela is, in the opinion of Husar et al., attributable to strong seasonal biomass burning in central America.

But a particularly conspicuous feature in figure 19.1 is the absence of a significant aerosol plume over the south Atlantic Ocean along the Brazilian coastline. In theory, such a plume should be present, given the high volume of biomass burning in Brazil and the normal airflow patterns over Brazil during the dry (burning) season. (Generally, the flow during this period is anticyclonic, resulting in winds blowing to the south and southeast as airstreams, first moving to the west, are deflected to the south and southeast by the Andes Mountain Range. See figure 19.2.)

The absence of a smoke plume off the coast of Brazil brings to light a critical point regarding the study of particulate emissions from biomass burning: the mechanisms controlling the transport and removal of the smoke particles must be examined and incorporated into any truly comprehensive study of the regional and global effects of biomass burning. While estimates of the particulate production term provide an important first-order idea of the impact of biomass combustion on the atmosphere, such estimates must be coupled with smoke transport and removal process studies in order to fully describe the fate of the smoke particles and to assess more accurately the environmental effects of the fire emissions.

The important role of atmospheric transport is outlined in detail by Garstang et al. (in press). As these authors discuss, transport pathways for particles produced by biomass combustion are determined by the location of the fire (or fires) and subsequent interaction of the smoke with atmospheric motions ranging in scale from convective (dry and moist) to mesosynoptic (cloud ensemble to storm) to planetary (subtropical anticyclones). On the scale of planetary and synoptic motions, semipermanent anticyclones in the tropics play a major role in determining both the structure of the atmosphere and the dominant pathways for vertical and horizontal transport of particulate emissions. As Garstang et al. discuss, large-scale subsidence within these anticyclones produces nearly clear skies and a dry atmosphere. This can result in inversions and stable layers in the atmosphere, the temporal and spatial persistence of which strongly influence the vertical and horizontal transport of particles produced by biomass combustion. The presence of multiple stratified atmospheric layers will act to suppress strong vertical transport and will, according to Garstang et al., concentrate fire emissions into horizontal atmospheric layers. Under these conditions, fire-generated particles and gaseous products can be transported quite rapidly over significant horizontal distances, possibly as great as 1000 km in as little as 10 hours.

Garstang et al. discuss the role of subtropical anticyclones in atmospheric transport primarily in the context of studies of fire emissions and transport carried out in south Africa. However, the presence of a strong seasonal anticyclone over Brazil (see figure 19.2 and Harriss et al. 1988) dictates that identical ideas and methods of analysis must be applied to the study of particulate emissions from biomass combustion in Brazil. Such an approach was, in fact, implemented by investigators during the Amazon Boundary Layer Experiment (ABLE 2A), during which particulate and gaseous emissions from biomass burning were investigated in conjunction with studies of atmospheric transport pathways (see Sachse et al. 1988). The work begun during ABLE 2A must be continued and expanded during future projects in Brazil to more accurately assess the fate, and subsequently the atmospheric impact, of combustion-generated aerosols.

In addition to the influence of anticyclonic flow patterns on aerosol transport, the impact of mesosynoptic and convective scale motions must also be considered. As is outlined by Garstang et al. (in press), large-scale motions such as those associated with tropical anticyclones can transport emissions to geographic regions in which dry or moist convective processes are active. These latter processes can result in the injection of emission products into layers of the atmosphere well above the planetary boundary layer. The work of Sachse et al. (1988) provides one example of the coupled influence of horizontal and vertical transport processes on the spatial distribution of fire emissions. In this paper, the authors present data illustrating how carbon monoxide (CO) produced by

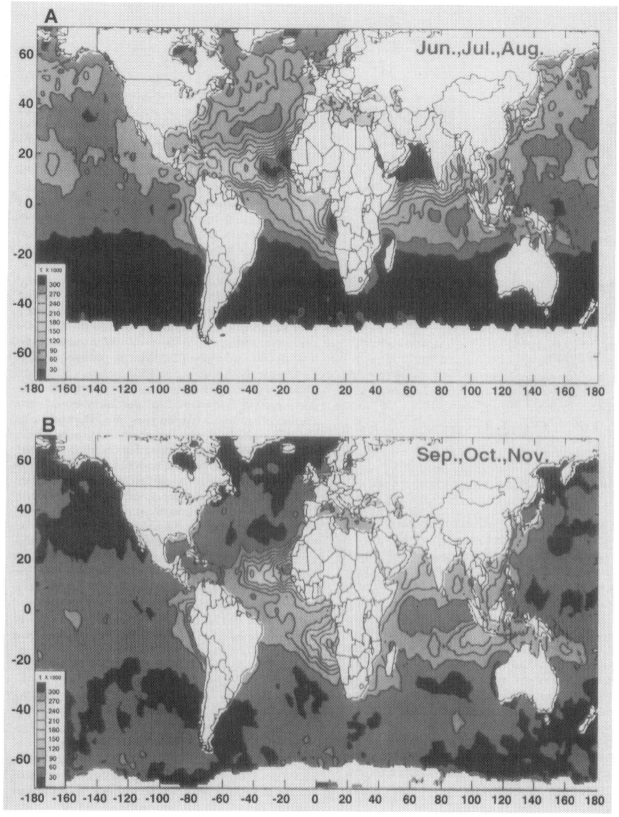

Figure 19.1 Equivalent aerosol optical depth (×1000) over the oceans for the periods (*a*) June to August and (*b*) September to November as derived from operational meteorological satellite data collected between July 1989 and June 1991 (adapted from Husar et al.)

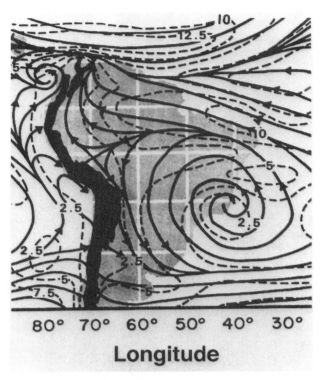

Figure 19.2 Streamlines (solid lines) and isotachs (dashed lines) over South America for July. The contours shown are for the 850-mb pressure level (adapted from Van de Boogaard 1977).

fires ~1000 km south of the Amazon Basin is transported into the basin and then, through a combination of convective mixing in the boundary layer and moist convection, is elevated to just below an inversion at an altitude between 3 and 3.5 km.

In the case of combustion-generated particles, the potential for transport into a region of moist convective activity raises the possibility for the activation of a critical mechanism affecting smoke lifetime, that of the scavenging and possible removal of a significant fraction of the aerosols by clouds and precipitation. Research has shown that particles produced by biomass combustion serve as very efficient cloud condensation nuclei (CCN) (Stith et al. 1981; Rogers et al. 1991). This high CCN efficiency indicates that particles produced by biomass fires may not only be transported in cloud systems, but may also be removed from the atmosphere in precipitation. (It should be noted that, in addition to being removed from the atmosphere as nuclei in raindrops, particles can also be removed as a result of impaction on falling raindrops, a process referred to as washout.)

A phenomenon related to the influence of moist convective activity on smoke lifetime is that of fire-induced convective activity. High-intensity fires occurring in regions of atmospheric instability can generate capping cumulus or cumulonimbus clouds, as has been documented by some investigators (Radke et al. 1991; T. Campos, personal communication). Precipitation from these clouds can remove combustion-generated aerosols from the atmosphere, although the exact scavenging efficiency of such clouds must still be quantified. (Preliminary work by Radke et al. [1991] yielded estimated scavenging efficiencies of approximately 50 to 80% from a study of Canadian fires.) Furthermore, it seems plausible that, in the case of fires of extremely high intensity, some of the smoke particles could be lofted to higher, possibly stratified, layers of the atmosphere in which horizontal transport over significant distances could take place. While the role of fire-induced convective activity in the transport and removal of smoke particles is probably not so significant as the influence of large-scale horizontal and local convective motions, it must, nevertheless, be considered in any attempt to adequately characterize the fate of smoke from biomass fires. Depending on the atmospheric conditions on the day a given fire occurs, sufficient convective activity could result from the fire to produce either a scavenging cloud and precipitation or to inject the particles into higher regions of the atmosphere in which transport over significant horizontal distances could occur.

Although almost certainly not so important as cloud and precipitation scavenging, particle coagulation within dense smoke plumes can also serve as an unexpectedly efficient smoke-removal mechanism. (See Radke et al. [1995] for a discussion of this process.) It is believed that such coagulation, which occurs as smoke plumes age, can be of a great enough magnitude to significantly increase the effective particle radii values retrieved from spectral radiometric data (see discussion in Westphal and Toon [1991]). Theoretically, such an increase in particle size as plumes age could lead to the removal, via sedimentation, of more particulate matter.

As the above discussion illustrates, the fate of particles produced by biomass combustion depends on a set of complex relationships between the geographical location of the fire or fires, the stability of the surrounding atmosphere, and the interaction of air motions of varying spatial and temporal scales with the smoke. Characterization of the smoke source term alone is not sufficient and must be combined with studies of the general meteorological environment and possible transport pathways in order to adequately address the critical issue of smoke lifetime. Indeed, it is only by combining particulate production estimates with analyses of the various transport and removal mechanisms that researchers will arrive at a better understanding of the impact of biomass burning particulate emissions on the atmosphere.

In the next two sections, we outline what we feel to be the central components (measurement and analytical) that should be incorporated into future studies of biomass burning in Brazil and in the tropics in general. We also give recommendations as to how such studies should be carried out. Our suggestions for measurement programs and experiment design plans are drawn directly from the various aerosol transport and removal phenomena discussed above and represent a direct application and extension of the principles discussed by Garstang et al. (in press) and the work already carried out by other investigators (e.g., Garstang et al. 1988; Sachse et al. 1988).

Suggested Components of Tropical Biomass Burning Studies

In our opinion, future studies of tropical biomass burning should incorporate three main components: airborne smoke measurements, satellite data analyses, and forecasting support. Executed individually, each of these components would yield some information needed to examine the production and fate of particles from tropical biomass combustion. However, it is the merging of all three elements in any study program undertaken in the tropics that will result in the type of comprehensive data set that is required to adequately assess the issue of smoke lifetime in the tropics.

Airborne Smoke Measurements

Several types of airborne measurements are critical for the study of the fate of aerosols generated by biomass combustion. Measurements we believe to be of the greatest importance are outlined below.

Microphysical Measurements Data regarding number concentrations of particles and particle size distributions must be collected at various locations within smoke plumes not only to quantify the amount of particulate matter produced by fires but also to assess how aerosol size distributions are altered as a result of transport processes (cf. Radke et al. 1995) and to aid in the assessment of the efficiency of smoke-removal mechanisms.

Particle Filter Collection and Analyses Such analyses are needed to determine the composition of combustion-generated aerosols and, subsequently, to deter-

mine if the composition is such that the aerosols can serve as condensation nuclei (CN) and CCN. Additionally, filter sample analyses can be used to detect the presence of chemical species which can serve as markers of particulate matter produced by biomass combustion (e.g., potassium and elemental carbon; see Ward and Hardy [1991] and Artaxo et al. [1994] for discussions of biomass burning tracer elements). The analysis of filter samples collected at various locations within smoke plumes can be used to identify smoke plumes from biomass burning and to track the movement of such plumes over large distances.

CN and CCN Measurements Although data on CN and CCN production by biomass combustion do exist, they have largely been obtained from studies of biomass fires in temperate and boreal regions (cf. Eagan et al. 1974; Stith et al. 1981; Radke et al. 1988). (One study of CCN production from tropical fires of which we are aware is that of Desalmand et al. [1985].) CN and CCN measurements must be made in the smoke from tropical biomass fires in order to quantify more exactly the production of CN and CCN in the tropics and better assess the potential for the removal of combustion-generated aerosols from the tropical atmosphere.

Lidar Measurements As has been demonstrated by several investigators, lidar is an extremely valuable tool for examining the horizontal and vertical structure of plumes and haze layers in the atmosphere (McCormick and Fuller 1975; Collis and Russell 1976) and for deriving more quantitative information such as the light-scattering and extinction characteristics and mass, area, and size of aerosols within these layers (Johnson and Uthe 1971; Uthe et al. 1982; Laursen et al. 1995). These various applications make lidar a critical component of any airborne program to be undertaken to study particulate emissions from tropical biomass burning. The work of Browell et al. (1988) during ABLE 2A provides a good example of the use of lidar to detect and map the structure and altitudes of aerosol layers produced by biomass combustion. Such information on the vertical stratification of smoke from fires is essential for examining how the smoke is transported in the atmosphere, and analyses such as those presented by Browell et al. should be carried out in future airborne projects in the tropics.

Spectral Radiometric Measurements Research by King (1987), Nakajima and King (1990), and Nakajima et al. (1991) has demonstrated the utility of scanning spectral radiometers for retrieving the optical depth and effective particle radius of stratocumulus clouds. The analytical methods proposed by these authors, which rely on the comparison of measured reflectance function values with computed reflectance function values to derive optical depth and effective radius, should be equally suitable for deriving the optical depth and particle size for biomass burning smoke plumes. The type of spatial information on optical depth and particle size that can be obtained from spectral radiometers (see figures in Nakajima et al. [1991] for examples) is needed to adequately assess the production and transport of particulate matter from biomass burning and dictates that spectral radiometric measurements should be incorporated into airborne studies of tropical biomass burning whenever possible.

The list of airborne measurements outlined above is not intended to be exhaustive, nor is it by any means inclusive of all types of measurements that should be made during airborne studies of tropical biomass combustion. Measurements of several trace gases (i.e., CO, CO_2, hydrocarbons, NO_x, N_2O, etc.) are equally vital components of airborne measurement programs. Indeed, CO, CO_2, and hydrocarbon measurements are essential for the determination of particulate and chemical emission factors using the carbon mass-balance technique. (See Ward et al. [1979], Nelson [1982], and Radke et al. [1988] for discussions of this method.) Furthermore, CO to CO_2 mass or volume mixing ratios can be used as tracers for biomass smoke due to the enhancement of these ratio values in plumes over normal ambient ratio values. The above discussion of specific airborne measurements is intended to highlight the types of aircraft instrumentation and measurements that are especially critical for determining the composition and quantity of particles produced by biomass combustion and for examining the transport and potential for removal from the atmosphere of these aerosols.

Satellite Data Analyses
Satellite imagery can be used to determine the location and temporal distribution of fires in the tropics (Brustet et al. 1991) and to detect and monitor the movement of smoke plumes from tropical biomass burning (Menzel et al. 1991). Additionally, recent research has shown that satellite data (specifically, NOAA AVHRR data) can be used to estimate the mass of emitted particles from tropical biomass burning (Kaufman et al. 1990).

All three applications, from detecting fires and smoke plumes to estimating particulate emissions, make satellite data analyses important components in

studies of tropical biomass combustion. Information on fire and smoke plume locations obtained from satellite images can be used to determine geographical regions in which airborne measurements should be made. The tracking of smoke plume movement using successive satellite images can provide not only information on the horizontal transport pathways for the smoke but can also be used to direct research aircraft to the appropriate geographical regions in which to make measurements in the smoke and characterize the nature and quantity of the particulate emissions downwind of the fires. Estimates of the emitted mass of particles derived from satellite data analyses can be compared with emission estimates derived from aircraft data, a process that will allow for the verification of satellite retrieval algorithms and will result in aircraft and satellite data sets that are internally consistent.

Forecasting Support
As is discussed by Garstang et al. (in press), meteorological conditions play dual roles in terms of the occurrence and environmental impact of biomass fires. Weather (i.e., dry) conditions produce the dry fuels that are needed for fires to start and burn sustainably, and meteorological conditions (atmospheric stability or instability, horizontal airflow patterns, dry and moist convective activity) dictate how the smoke from fires is transported through and removed from the atmosphere.

Forecasting must, of course, be employed to determine if the appropriate weather conditions exist for burning to occur. This type of analysis has long been carried out as a precursor to airborne missions to study biomass burning (i.e., to determine when such missions should occur) and has also been a critical component of prescribed and wildfire management programs. However, forecasting support should also be used to help in the assessment of the atmospheric impact of the smoke produced by biomass combustion. In the case of tropical biomass burning, forecasts should be used to examine the stability of the atmosphere in various geographical regions. As discussed above, stably stratified atmospheres suppress vertical smoke transport and concentrate fire emissions into horizontal layers which can then be transported over significant distances. Atmospheric stability analyses should be used to predict the potential for the formation of such stratified emission layers, and these predictions should then be used to dictate the type of flight patterns that are flown by aircraft to sample in the smoke plume and characterize the emissions. Analyses that reveal the presence of regions of atmospheric instability should also play a key role in studies of tropical biomass burning. As outlined above, fires that occur in such regions can generate capping cumulus or cumulonimbus clouds that may scavenge a significant fraction of the aerosols. Also, smoke may be transported into regions of atmospheric instability, and the resulting convective activity in such regions can result in the removal or enhanced vertical transport of particulate matter. Forecasting should be relied upon as an important first indicator of the potential for both fire-induced and normal environmental cumulus and cumulonimbus formation and the subsequent transport and removal of aerosols.

Synoptic analyses to determine regional- and continental-scale airflow patterns must also play a central role in the formulation and execution of tropical biomass burning studies. Knowledge regarding such airflow patterns is mandatory for determining how combustion-generated particulate matter will be transported horizontally, and advance (i.e., preflight) information on wind speeds and directions can be used to deploy and position research aircraft in the appropriate geographical areas for sampling in smoke plumes, and thereby characterizing the particulate matter, downwind of fires.

Recommended Experimental Approach

As stated above, it is our opinion that all three of the research components discussed should be incorporated into future studies of tropical biomass burning in order to arrive at a more accurate assessment of the fate of aerosols generated by tropical biomass burning. We advocate a comprehensive and process-oriented approach to studies of such biomass combustion, an approach that recognizes the complexity of the interactions between fire emissions and atmospheric motions of varying scales and that makes use of all analytical and measurement tools necessary for the study of the nature of the particulate emissions and their ultimate impact on the atmosphere. Several ideas were presented in the preceding section for how airborne measurements, satellite data analyses, and forecasting support should each be applied to the study of particulate emissions. In this section, we bring these ideas together and suggest how the three study components can best be merged for the formulation of more comprehensive biomass burning research programs.

In our opinion, the experimental approach most suitable for studying particulate emissions from trop-

ical biomass burning is one in which general, or larger-scale, information serves as a precursor to the collection of more detailed, smaller-scale information. By larger-scale information, we refer to information obtained from satellite imagery and from synoptic analyses. By smaller-scale information, we refer to in situ and remote sensing data obtained from research aircraft. At the beginning, and indeed throughout, a research program, satellite imagery should be used to locate areas in which a significant number of fires are burning and, as outlined above, to monitor the movement of smoke plumes whenever possible. Synoptic analyses should be carried out in parallel with satellite data analyses to provide further information on airflow directions. Next, regional forecasts and meteorological data analyses for regions in which fires are burning and for regions into which smoke may be transported should be carried out with the aim of assessing the stability of the atmosphere and hence the potential for convective activity. The information obtained on fire locations, direction of smoke transport, and atmospheric stability conditions can then be used to direct research aircraft to the appropriate geographical areas in which to execute detailed studies of the quantity and composition of the particulate matter and to examine the horizontal and vertical structure of smoke plumes.

Implicit in the study of the impact of atmospheric motions on smoke transport and removal is the need to employ a "source-to-sink" approach when performing airborne studies of particulate emissions. As stressed earlier in this chapter, it is not sufficient to simply characterize the particulate emission term at or near a fire or fires. Measurements must be made at several points downwind of fires in order to monitor and characterize changes in particulate composition and concentration. If possible, airborne measurements should be continued until a particulate "sink" (i.e., incorporation of particles into cloud systems, vertical transport due to strong convective activity, precipitation scavenging) is encountered, so that the magnitude of the sink can be estimated. In some cases, for example in Brazil, it may not be possible to track individual smoke plumes due to the fact that plumes from fires in various geographic regions often merge into large, composite smoke plumes. In this instance, airborne remote-sensing instrumentation and satellite sensors could play an especially critical role in the examination of particulate-removal mechanisms. For example, airborne spectral radiometric data and satellite data can be used to retrieve effective particle radius values for clouds and hence could be used to look for changes (i.e., decreases) in cloud radii values caused by the ingestion of particulate matter from biomass fires.

Conclusions

It is important to stress that the types of airborne measurements, satellite analyses, and forecasting support we have outlined here have each played a role in previous campaigns to study tropical biomass burning. As examples, we note the work carried out during ABLE 2A (e.g., Browell et al. 1988; Garstang et al. 1988; Harriss et al. 1988; Sachse et al. 1988) and during the Biomass Burning Airborne and Spaceborne Experiment in the Amazonas (BASE-A) (e.g., Ward et al. 1991; Kaufman et al. 1992). These and other studies of biomass burning have yielded important information on the quantity and composition of the aerosols produced by biomass combustion and have, via satellite images and synoptic analyses, generated some ideas as to how smoke in the tropics is transported and possibly removed from the atmosphere. Now, the task of building upon these previous research programs must be undertaken. Airborne, satellite, and forecasting approaches should be integrated into each future project in the tropics, and as many measurements as are practicable (i.e., CN and CCN measurements, particle filter analyses, lidar and spectral radiometric retrievals, etc.) should be carried out simultaneously during each project.

Finally, when planning and executing any program to study particulate emissions from tropical biomass burning, it should be remembered that the production and eventual removal of particulate matter is a process, a process governed by a complex set of relationships between the fire, its geographical location, and the surrounding small- and large-scale atmospheric motions. Consequently, a process-oriented approach should be applied to the study of tropical fire emissions where, as we have outlined here, the study of meteorological conditions and the actual fire emissions receive equal consideration and the complete lifecycle of the smoke, from source to sink, is studied and delineated. Adherence to these ideas and deployment and application of the various instruments and analytical methods we have outlined here should, in our opinion, enable researchers to work toward and achieve a more thorough understanding of the fate and ultimate environmental impact of particles generated by tropical biomass burning.

For purposes of relating the impact of biomass fire emissions to global issues, the experimental approach adopted by researchers must go beyond mere cataloging and inventorying of emissions. Scientists must now explicitly scale experimental protocols to scales much larger than the fire scale, indeed, to scales approaching the global, if the actual impact of biomass fire emissions is to be more accurately assessed.

Acknowledgments

We thank Teresa Campos (NCAR/RAF) for reviewing an earlier draft of this manuscript and Barbara Knowles for her work in preparing the figures. The National Center for Atmospheric Research is sponsored by the National Science Foundation.

References

Andreae, M. O. 1991. Biomass burning: Its history, use, and distribution and its impact on environmental quality and global climate, in *Global Biomass Burning: Atmospheric, Climatic, and Biospheric Implications*, edited by J. S. Levine, pp. 3–21, MIT Press, Cambridge, Mass.

Andreae, M. O., E. V. Browell, M. Garstang, G. L. Gregory, R. C. Harriss, G. F. Hill, D. J. Jacob, M. C. Pereira, G. W. Sachse, A. W. Setzer, P. L. Silva Dias, R. W. Talbot, A. L. Torres, and S. C. Wofsy. 1988. Biomass-burning emissions and associated haze layers over Amazonia, *J. Geophys. Res.*, 93, 1509–1527.

Artaxo, P., F. Gerab, M. A. Yamasoe, and J. V. Martins. 1994. Fine mode aerosol composition at three long-term atmospheric monitoring sites in the Amazon Basin, *J. Geophys. Res.*, 99, 22 857–22 868.

Browell, E. V., G. L. Gregory, R. C. Harriss, and V. W. J. H. Kirchhoff. 1988. Tropospheric ozone and aerosol distributions across the Amazon basin, *J. Geophys. Res.*, 93, 1431–1451.

Brustet, J. M., J. B. Vickos, J. Fontan, K. Manissadjan, A. Podaire, and F. Lavenu. 1991. Remote sensing of biomass burning in West Africa with NOAA-AVHRR, in *Global Biomass Burning: Atmospheric, Climatic, and Biospheric Implications*, edited by J. S. Levine, pp. 47–52, MIT Press, Cambridge, Mass.

Collis, R. T. H., and P. B. Russell. 1976. Lidar measurement of particles and gases by elastic backscattering and differential absorption, in *Laser Monitoring of the Atmosphere*, edited by E. D. Hinkley, pp. 71–151, Springer, Berlin.

Crutzen, P. J., A. C. Delany, J. Greenberg, P. Haagenson, L. Heidt, R. Lueb, W. Pollock, W. Seiler, A. Wartburg, and P. Zimmerman. 1985. Tropospheric chemical composition measurements in Brazil during the dry season, *J. Atmos. Chem.*, 2, 233–256.

Desalmand, F., R. Serpolay, and J. Podzimek. 1985. Some specific features of the aerosol particle concentrations during the dry season and during a bushfire event in West Africa, *Atmos. Environ.*, 19, 1535–1543.

Eagan, R. C., P. V. Hobbs, and L. F. Radke. 1974. Measurements of cloud condensation nuclei and cloud droplet size distributions in the vicinity of forest fires, *J. Appl. Meteorol.*, 13, 553–557.

Garstang, M., J. Scala, S. Greco, R. Harriss, S. Beck, E. Browell, G. Sachse, G. Gregory, G. Hill, J. Simpson, W. K. Tao, and A. Torres. 1988. Trace gas exchanges and convective transports over the Amazonian rain forest, *J. Geophys. Res.*, 93, 1528–1550.

Garstang, M., P. D. Tyson, H. Cachier, and L. Radke, 11–15 October, 1994, in press. Atmospheric transports of particulate and gaseous products by fires, to be published in *Sediment Records of Biomass Burning and Global Change*, proceedings book for the NATO Workshop on Biomass Fires and Climate, Algarve, Portugal.

Harriss, R. C., S. C. Wofsy, M. Garstang, E. V. Browell, L. C. B. Molion, R. J. McNeal, J. M. Hoell, Jr., R. J. Bendura, S. M. Beck, R. L. Navarro, J. T. Riley, and R. L. Snell, 1988. The Amazon Boundary Layer Experiment (ABLE 2A): Dry season 1985, *J. Geophys. Res.*, 93, 1351–1360.

Husar, R. B., L. L. Stowe, and J. M. Prospero, in press. Satellite sensing of tropospheric aerosols over the oceans with NOAA/AVHRR, submitted for publication in *J. Geophys. Res.-Atmospheres*.

Johnson, W. B., and E. E. Uthe. 1971. Lidar study of the Keystone stack plume, *Atmos. Environ.*, 5, 703–724.

Kaufman, Y. J., A. Setzer, D. Ward, D. Tanre, B. N. Holben, P. Menzel, M. C. Pereira, and R. Rasmussen. 1992. Biomass burning airborne and spaceborne experiment in the Amazonas (BASE-A), *J. Geophys. Res.*, 97, 14 581–14 599.

Kaufman, Y. J., C. J. Tucker, and I. Fung. 1990. Remote sensing of biomass burning in the tropics, *J. Geophys. Res.*, 95, 9927–9939.

King, M. D. 1987. Determination of the scaled optical thickness of clouds from reflected solar radiation measurements, *J. Atmos. Sci.*, 44, 1734–1751.

Laursen, K. K., D. G. Baumgardner, and B. M. Morley. 1995. Optical properties of the Kuwait oil fires smoke plume as determined using an airborne lidar system: preliminary results from 28 and 29 May 1991 case studies, *Atmos. Environ.*, 29, 951–958.

McCormick, M. P., and W. H. Fuller, Jr. 1975. Lidar measurements of two intense stratospheric dust layers, *Appl. Optics*, 14, 4–5.

Menzel, W. P., E. C. Cutrim, and E. M. Prins. 1991. Geostationary satellite estimation of biomass burning in Amazonia during BASE-A, in *Global Biomass Burning: Atmospheric, Climatic, and Biospheric Implications*, edited by J. S. Levine, pp. 41–46, MIT Press, Cambridge, Mass.

Nakajima, T., and M. D. King. 1990. Determination of the optical thickness and effective particle radius of clouds from reflected solar radiation measurements. Part I: Theory, *J. Atmos. Sci.*, 47, 1878–1893.

Nakajima, T., M. D. King, J. D. Spinhirne, and L. F. Radke. 1991. Determination of the optical thickness and effective particle radius of clouds from reflected solar radiation measurements. Part II: Marine stratocumulus observations, *J. Atmos. Sci.*, 48, 728–750.

Nelson, R. M., Jr. 1982. An evaluation of the carbon mass balance technique for estimating emission factors and fuel consumption in forest fires, *Res. Pap. SE-231*, 9 pp., Southeast For. Exp. Stn., U.S. Dep. Agric. For. Serv., Asheville, N.C.

Penner, J. E., R. J. Charlson, J. M. Hales, N. Laulainen, R. Leifer, T. Novakov, J. Ogren, L. F. Radke, E. E. Schwartz, and L. Travis, 1994. Quantifying and minimizing uncertainty of climate forcing by anthropogenic aerosols, *Bull. Amer. Meteor. Soc.*, 75, 375–400.

Radke, L. F., A. S. Ackerman, J. H. Lyons, D. A. Hegg, P. V. Hobbs, and J. E. Penner, 1995. Effects of aging on the smoke from a large forest fire, *Atmos. Res.*, 38, 315–332.

Radke, L. F., D. A. Hegg, P. V. Hobbs, J. D. Nance, J. H. Lyons, K. K. Laursen, R. E. Weiss, P. J. Riggan, and D. E. Ward. 1991. Particulate and trace gas emissions from large biomass fires in North America, in *Global Biomass Burning: Atmospheric, Climatic,*

and Biospheric Implications, edited by J. S. Levine, pp. 209–224, MIT Press, Cambridge, Mass.

Radke, L. F., D. A. Hegg, J. H. Lyons, C. A. Brock, P. V. Hobbs, R. Weiss, and R. Rasmussen. 1988. Airborne measurements on smokes from biomass burning, in *Aerosols and Climate*, edited by P. V. Hobbs and M. P. McCormick, pp. 411–422, A. Deepak Publishing Co., Hampton, Va.

Riggan, P. J., J. A. Brass, and R. N. Lockwood. 1993. Assessing fire emissions from tropical savanna and forests of central Brazil, *Photogramm. Eng. Remote Sens.*, 59, 1009–1015.

Rogers, C. F., J. G. Hudson, B. Zielinska, R. L. Tanner, J. Hallett, and J. G. Watson. 1991. Cloud condensation nuclei from biomass burning, in *Global Biomass Burning: Atmospheric, Climatic, and Biospheric Implications*, edited by J. S. Levine, pp. 431–438, MIT Press, Cambridge, Mass.

Sachse, G. W., R. C. Harriss, J. Fishman, G. F. Hill, and D. R. Cahoon. 1988. Carbon monoxide over the Amazon Basin during the 1985 dry season, *J. Geophys. Res.*, 93, 1422–1430.

Stith, J. L., L. F. Radke, and P. V. Hobbs. 1981. Particle emissions and the production of ozone and nitrogen oxides from the burning of forest slash, *Atmos. Environ.*, 15, 73–82.

Uthe, E. E., B. M. Morley, and N. B. Nielsen. 1982. Airborne lidar measurements of smoke plume distribution, vertical transmission, and particle size, *Appl. Optics*, 21, 460–463.

Van de Boogaard, H. 1977. The mean circulation of the tropical and subtropical atmosphere—July, NCAR Technical Note, NCAR/TN-118+STR, National Center for Atmospheric Research, Boulder, Colo.

Ward, D. E., and C. C. Hardy. 1991. Smoke emissions from wildland fires, *Environ. Int.*, 17, 117–134.

Ward, D. E., R. M. Nelson, and D. F. Adams. 1979. Forest fire smoke plume documentation, paper presented at the 77th Annual Meeting, Air Pollut. Control Assoc., Air and Waste Manage. Assoc., Pittsburgh, Pa.

Ward, D. E., A. W. Setzer, Y. J. Kaufman, and R. A. Rasmussen. 1991. Characteristics of smoke emissions from biomass fires of the Amazon region—BASE-A experiment, in *Global Biomass Burning: Atmospheric, Climatic, and Biospheric Implications*, edited by J. S. Levine, pp. 394–402, MIT Press, Cambridge, Mass.

Ward, D. E., R. A. Susott, J. B. Kauffman, R. E. Babbitt, D. L. Cummings, B. Dias, B. N. Holben, Y. J. Kaufman, R. A. Rasmussen, and A. W. Setzer. 1992. Smoke and fire characteristics for cerrado and deforestation burns in Brazil: BASE-B experiment, *J. Geophys. Res.*, 97, 14 601–14 619.

Westphal, D. L., and O. B. Toon. 1991. Simulations of microphysical, radiative, and dynamical processes in a continental-scale forest fire smoke plume, *J. Geophys. Res.*, 96, 22 379–22 400.

The Development of a Global Biomass Burning Inventory

Darrell Wayne Sproles

Over the last five years the importance of the burning of various ecosystems has been shown to be comparable to the use of fossil fuels on global atmospheric pollution. New imaging techniques permit the detection of ecosystem burning from satellites. Data collection with a number of other methods is also ongoing. Collecting this data into a global inventory on burning would be a valuable research tool for atmospheric scientists.

What Is the Biomass Burning Inventory?

The amount of data available to research the effect of global biomass burning is increasing. This data comes from satellite measurements of biomass burning, sampling of gases produced, and surveying of the amounts of biomass consumed. The Atmospheric Sciences Division at NASA Langley Research Center is interested in using this data as the starting point for detailed research on global biomass burning. A requirement of this research is to seek a methodology to facilitate exchange of data among researchers. A burning inventory is an approach considered for satisfying this requirement. A goal of the burning inventory would be to contribute to research by alleviating many data-management problems suffered by researchers.

Limited budgets are forcing innovative approaches and increased use of existing data. A burning inventory could maintain a history of biomass burning and the spatial and temporal distribution of biomass burning for exchange between researchers. Effort has been ongoing in this area to prove the feasibility of a global biomass burning inventory. This effort includes the development of a prototype data base with relational data base management systems to create the inventory. Figure 20.1 depicts a schema researched for the biomass data base model. Another area studied was in devising methods of distribution taking into consideration possible data storage mediums, computer platform availability, and so forth.

The data base divided the globe into $1° \times 1°$ grid cells. The data base schema allowed for static and reported data. Static data included coordinates, land areas, and ecosystem parameters. Ecosystem data was obtained from the National Center for Atmospheric Research (NCAR). The classes used with the global vegetation archives of Matthews (1983) were selected as the ecosystem types. This classification scheme employs 178 vegetation types using the UNESCO (1973) classification scheme and is summarized into 32 classes of vegetation (table 20.1). The variable data included information on date and location of burning, gases produced, amount of carbon consumed by a fire, and square area burned. The type of flexibility aspired for when modeling the data base is exemplified in the methodology for storing carbon consumed. *Burn coefficient* is the name of the attribute whose values represent carbon consumed. Multiple definitions are allowed for the values of burn coefficient. Each value in burn coefficient has an associated key in the attribute named *data category* that indicates how that specific value is defined.

Why Have a Biomass Burning Inventory?

A Biomass burning inventory would offer a means for assembling data from the many available sources. A burning inventory would provide researchers a single site to search and from which to obtain data for their analyses. Up-to-date vegetation information could be maintained as well. A distribution system would allow submittal and retrieval of this information. The final burning inventory's form, data formats, access, interface, and tools would be developed in response to the community it serves.

Who Would Use the Biomass Burning Inventory?

The targeted user of a biomass burning inventory would be the researchers who require data. The focus of the biomass burning might be natural, industrial, domestic fossil fuels, or agriculture.

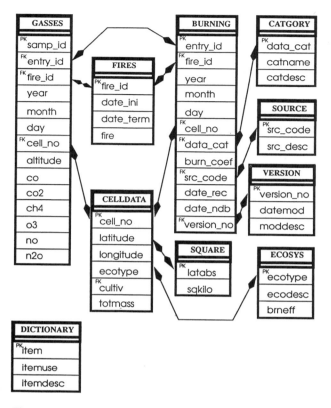

Figure 20.1 The graphical depiction of a biomass relational data base model. The relation's names are shown in the bold cell of each table and the attribute's names in each of the other cells. The primary and foreign keys are identified with the characters "PK" and "FK"; their relational connections are presented by the double-headed arrows.

An Example

As an example a model scenario will be set up in which a researcher would need the burning inventory. This simplistic yet illustrative example identifies a research objective and an approach to solving it.

Interest
Amount of CO_2 produced by savanna burning compared to CO_2 produced by forest burning.

Source
Distribution of CO_2 emissions from biomass burning in tropical Africa. Data published by Hao and Liu is obtained in paper form showing a $5° \times 5°$ grid (figure 20.2).

Processing
The data base was populated by distributing the data over the respective ecosystems it represents while

Table 20.1 The Matthews ecosystem classifications for a 1° by 1° resolution (from Matthews 1983)

eco_type	ecodesc
0	Water
1	Tropical evergreen rainforest, mangrove forest
2	Tropical/subtropical evergreen seasonal broadleaved forest
3	Subtropical evergreen rainforest
4	Temperate/subpolar evergreen rainforest
5	Temperate evergreen seasonal broadleaved forest, summer rain
6	Evergreen broadleaved sclerophyllous forest, winter rain
7	Tropical/subtropical evergreen needleleaved forest
8	Temperate subpolar evergreen needleleaved forest
9	Tropical/subtropical drought-deciduous forest
10	Cold-deciduous forest, with evergreens
11	Cold-deciduous forest, without evergreens
12	Xeromorphic forest/woodland
13	Evergreen broadleaved sclerophyllous woodland
14	Evergreen needleleaved woodland
15	Tropical/subtropical drought-deciduous forest
16	Cold-deciduous woodland
17	Evergreen broadleaved shrubland/thicket, evergreen dwarf shrubland
18	Evergreen needleleaved or microphyllous shrubland/thicket
19	Drought-deciduous shrubland/thicket
20	Cold-deciduous subalpine/subpolar shrubland, dwarf shrubland
21	Xeromorphic shrubland, dwarf shrubland
22	Arctic/alpine tundra, mossy bog
23	Tall/medium/short grassland with 10–14% woody tree cover
24	Tall/medium/short grassland with <10% woody tree or tuft-plant cover
25	Tall grassland with shrub cover
26	Tall grassland, no woody cover
27	Medium grassland, no woody cover
28	Short grassland, no woody cover
29	Meadow
30	Desert
31	Ice

Table 20.2 Data category designations specifying what the burn coefficient value represents

data_cat	catname	catdesc
0	Coutput	The calculated amount, in grams, of carbon material output in the corresponding $1° \times 1°$ geographical report site
1	MassBrnt	The calculated amount, in kilograms, of biomass materials burnt in the corresponding $1° \times 1°$ geographical report site
2	AreaBrnt	The area, in square kilometers, burnt in the corresponding $1° \times 1°$ geographical report site
3	PctBrn	The percentage of biomass meterials burnt in the corresponding $1°$ by $1°$ geographical report site

mapping it to a $1° \times 1°$ degree grid. Then the data base was used to query and select data for cells of a desired ecosystem. Further analysis was applied to the selected data both internal and external to the data base. These data output were displayed for visual comparison by a number of tools. Tools ranged from simple in-house software to advanced commercial package. Consideration was given to researchers who must maintain low cost, possibly limited to hardware such as personal computers and printers. However equal time is given to upper end capabilities.

Visualization

Product descriptions of the three levels of visualization performed are as follows:

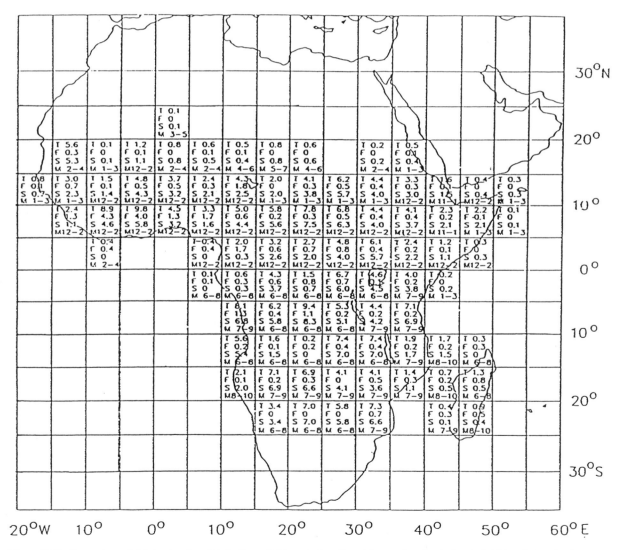

Figure 20.2 Example model data showing distribution of CO_2 emissions from biomass burning in tropical Africa during the dry season on a $5° \times 5°$ grid (from Hao and Liu 1994) T = total, F = forest; M = months of intensive burning (unit 10^{12} g $CO_2 - $ C month^{-1})

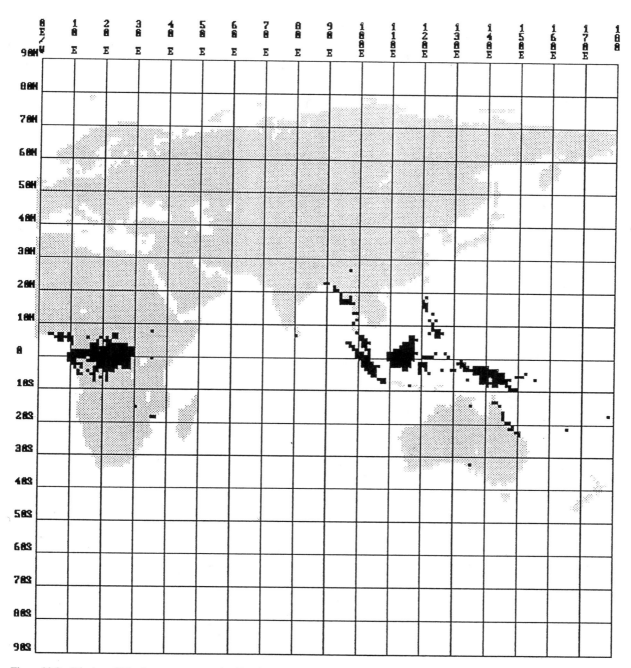

Figure 20.3 Display of Matthews ecosystem classification number 1, "Tropical evergreen rainforest, mangrove forest," produced from Postscript output of in-house software

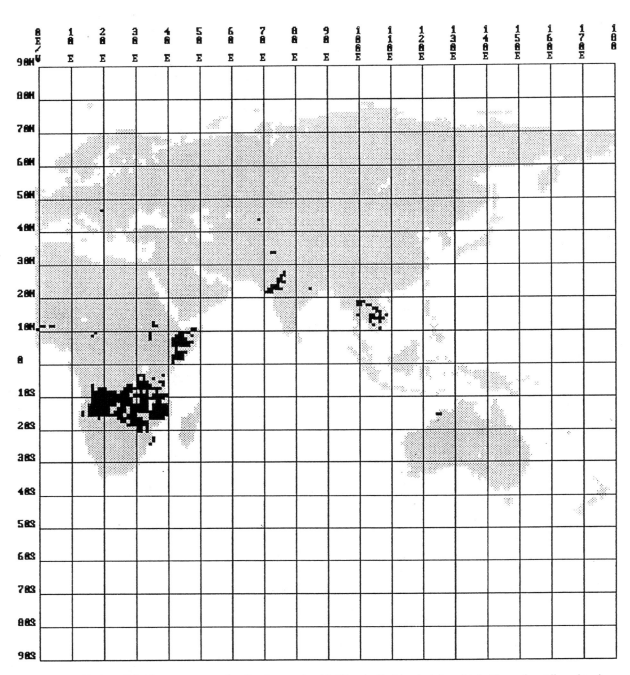

Figure 20.4 Display of Matthews ecosystem classification number 15, "Tropical/subtropical drought-deciduous forest," produced from Postscript output of in-house software

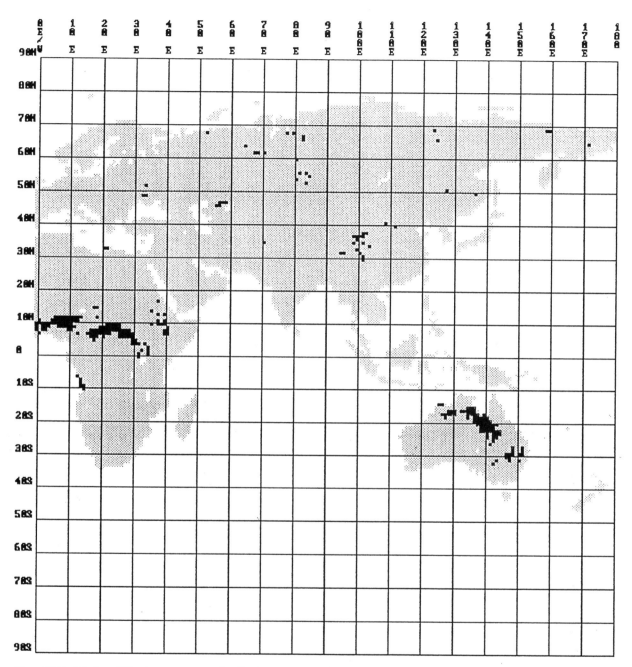

Figure 20.5 Display of Matthews ecosystem classification number 24, "Tall/medium/short grassland with <10% woody tree cover or tuft-plant cover," produced from Postscript output of in-house software

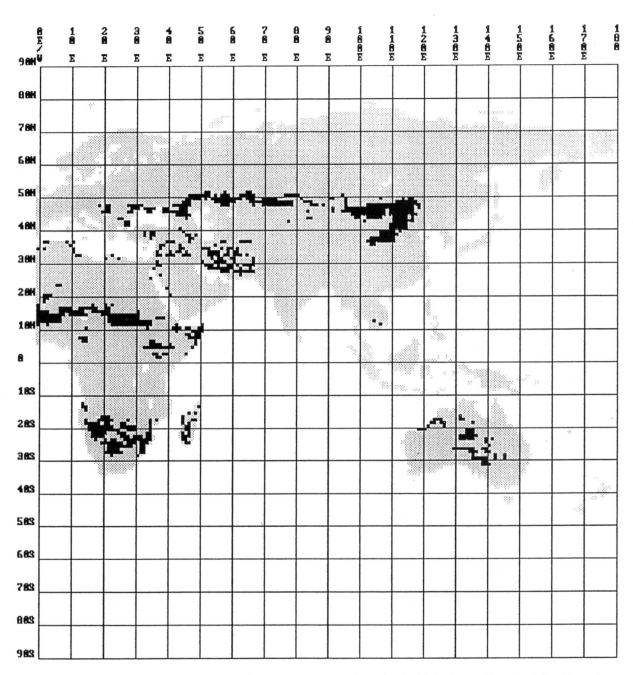

Figure 20.6 Display of Matthews ecosystem classification number 25, "Tall grassland, with shrub cover," produced from Postscript output of in-house software

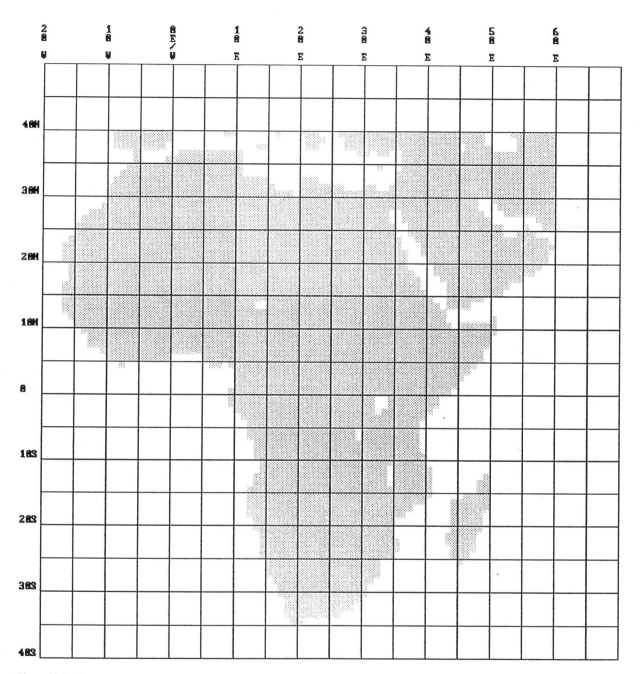

Figure 20.7 Sample of Postscript geographical mapping capability showing the African continent at a 1° × 1° resolution, produced by querying the data base for all cells of eco_types 1–31, excluding 0 (water), then filling in each cell in a gray scale

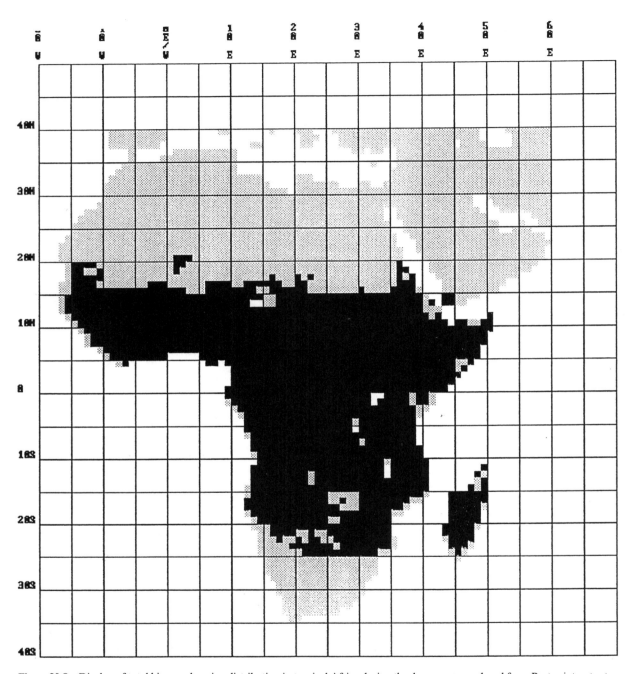

Figure 20.8 Display of total biomass burning distribution in tropical Africa during the dry season, produced from Postscript output of in-house software

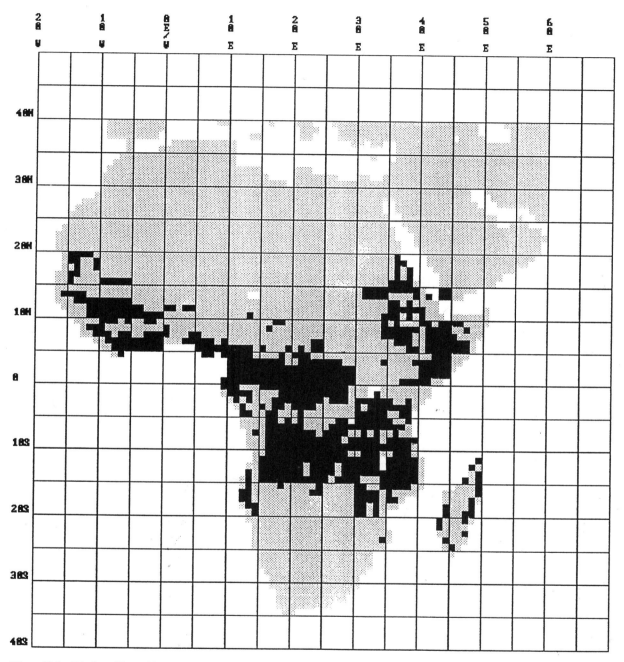

Figure 20.9 Display of forest biomass burning distribution in tropical Africa during the dry season, produced from Postscript output of in-house software

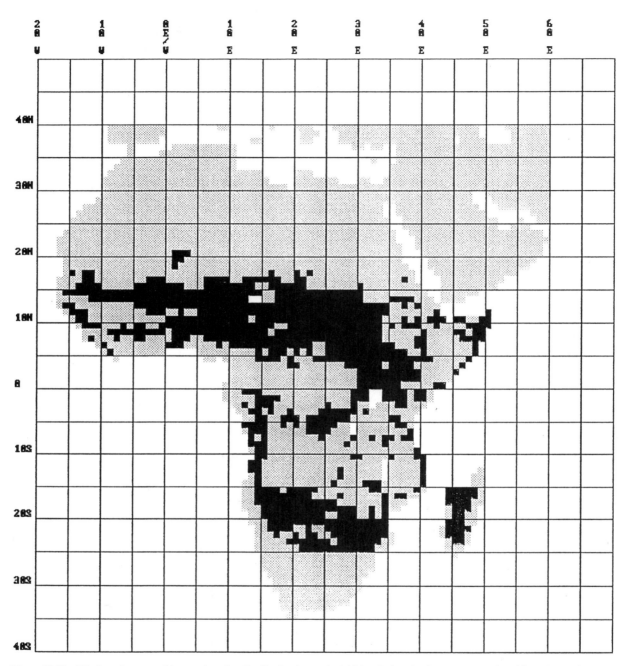

Figure 20.10 Display of savanna biomass burning distribution in tropical Africa during the dry season, produced from Postscript output of in-house software

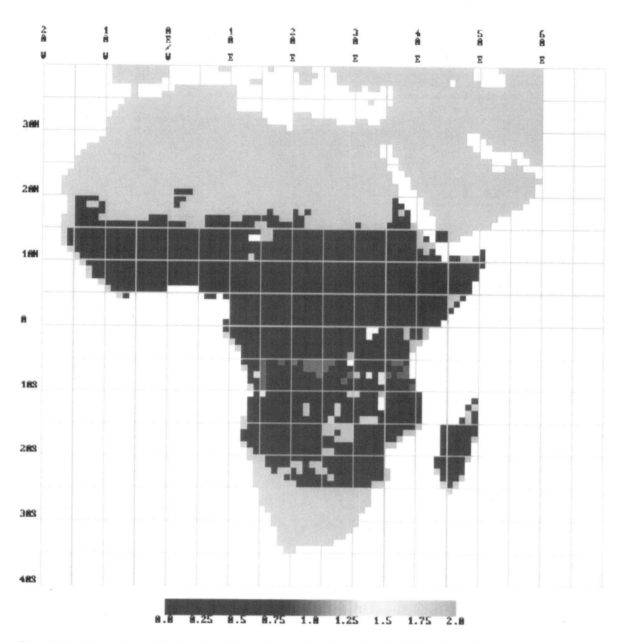

Figure 20.11 Color-enhanced display of total biomass burning distribution in tropical Africa during the dry season, produced from Postscript output of in-house software

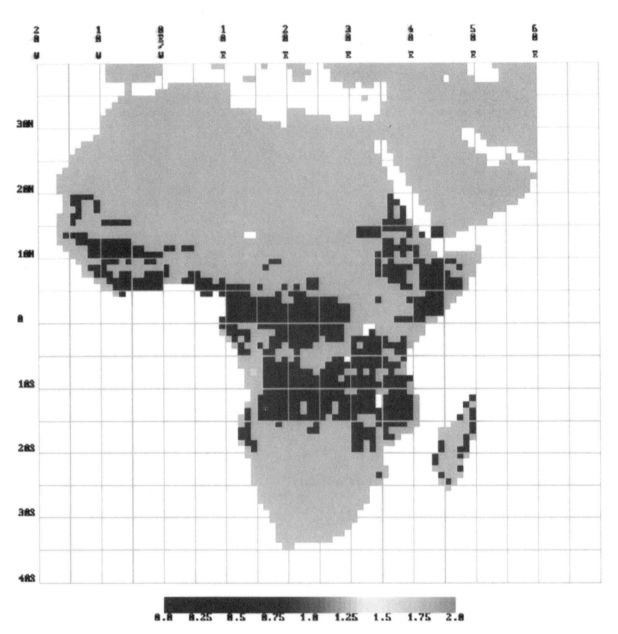

Figure 20.12 Color-enhanced display of forest biomass burning distribution in tropical Africa during the dry season, produced from Postscript output of in-house software

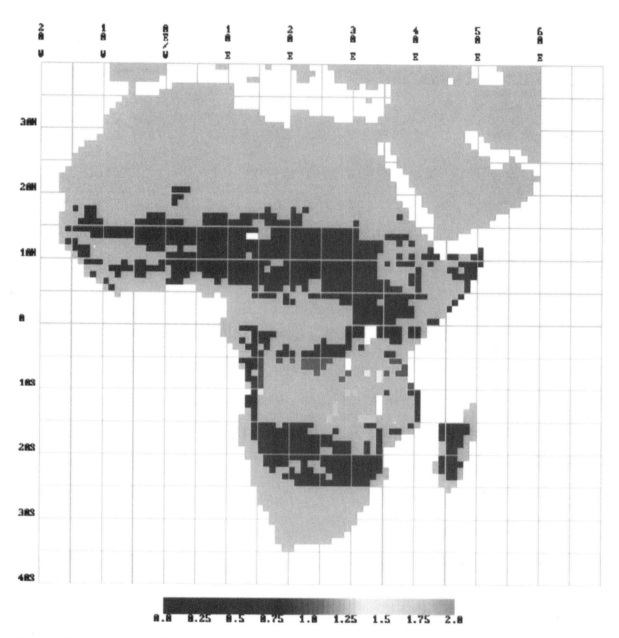

Figure 20.13 Color-enhanced display of savanna biomass burning distribution in tropical Africa during the dry season, produced from Postscript output of in-house software

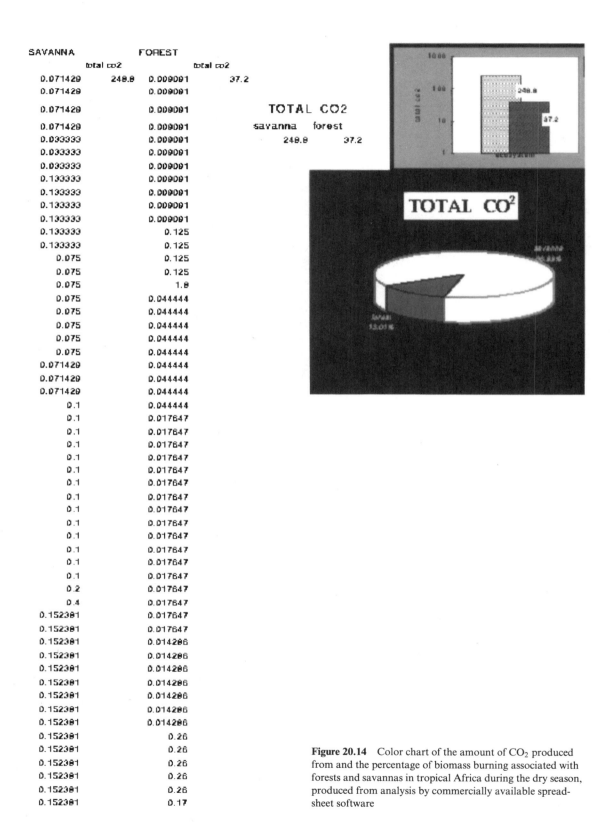

SAVANNA		FOREST	
total co2		total co2	
0.071429	248.8	0.009091	37.2
0.071429		0.009091	
0.071429		0.009091	
0.071429		0.009091	
0.033333		0.009091	
0.033333		0.009091	
0.033333		0.009091	
0.133333		0.009091	
0.133333		0.009091	
0.133333		0.009091	
0.133333		0.009091	
0.133333		0.125	
0.133333		0.125	
0.075		0.125	
0.075		0.125	
0.075		1.8	
0.075		0.044444	
0.075		0.044444	
0.075		0.044444	
0.075		0.044444	
0.075		0.044444	
0.071429		0.044444	
0.071429		0.044444	
0.071429		0.044444	
0.1		0.044444	
0.1		0.017647	
0.1		0.017647	
0.1		0.017647	
0.1		0.017647	
0.1		0.017647	
0.1		0.017647	
0.1		0.017647	
0.1		0.017647	
0.1		0.017647	
0.1		0.017647	
0.1		0.017647	
0.1		0.017647	
0.2		0.017647	
0.4		0.017647	
0.152381		0.017647	
0.152381		0.017647	
0.152381		0.014286	
0.152381		0.014286	
0.152381		0.014286	
0.152381		0.014286	
0.152381		0.014286	
0.152381		0.014286	
0.152381		0.26	
0.152381		0.26	
0.152381		0.26	
0.152381		0.26	
0.152381		0.26	
0.152381		0.17	

TOTAL CO2

savanna   forest

248.8   37.2

Figure 20.14   Color chart of the amount of CO_2 produced from and the percentage of biomass burning associated with forests and savannas in tropical Africa during the dry season, produced from analysis by commercially available spreadsheet software

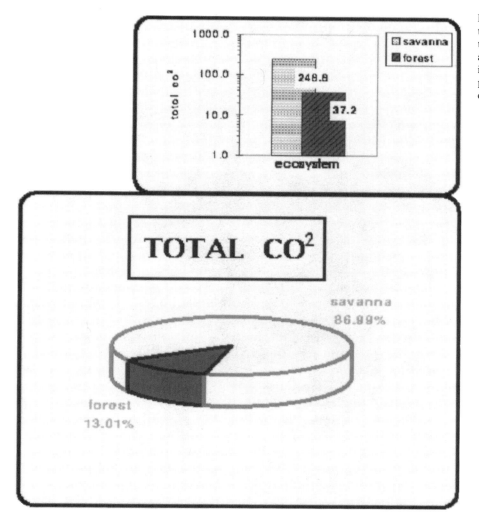

Figure 20.15 Enlarged color chart of the amount of CO_2 produced from and the percentage of biomass burning associated with forests and savannas in tropical Africa during the dry season, produced from analysis by commercially available spreadsheet software

1. In-house Postscript mapping software is used to produce figures 20.3 through 20.13. The maps in figures 20.3 through 20.6 show the expected locations of the Matthews (1983) classifications that could be grouped to explore savanna and forest burning. Figure 20.7 is a sample of a Postscript geographical mapping capability. The spatial distributions of biomass burning identified from the data base queries are displayed in figures 20.8 through 20.10. The previous three figures were externally processed adding color to enhance the images (figures 20.11 through 20.13).*

2. Widely available commercial spreadsheet software is used to bring in data from a query of the data base and begin some preliminary analysis. Figures 20.14 and 20.15 display the charts produced from this analysis from the spreadsheet.

3. High-level visual data analysis software and its command language was used to write a program for data display. This program read output from data base queries and scaled the values to a color bar for the purpose of again enhancing the images. The program then produced the graphical instructions to be sent to a color plotting device. Figures 20.16 through 20.19 are the results of this activity.

Model Extendibility

This is only one scenario of many that would be served by a burning inventory. However, varying data sets need to be provided and different methods of data acquisition discussed to model a useful data resource.

*Note: The figures in this chapter originally displaying color have been reproduced in black and white for this book. Thus the inclusion of the word "color" in the descriptive text and in the captions for these figures should be considered for referencing the process performed to produce the display and not as a description of the figure as reproduced for this book.

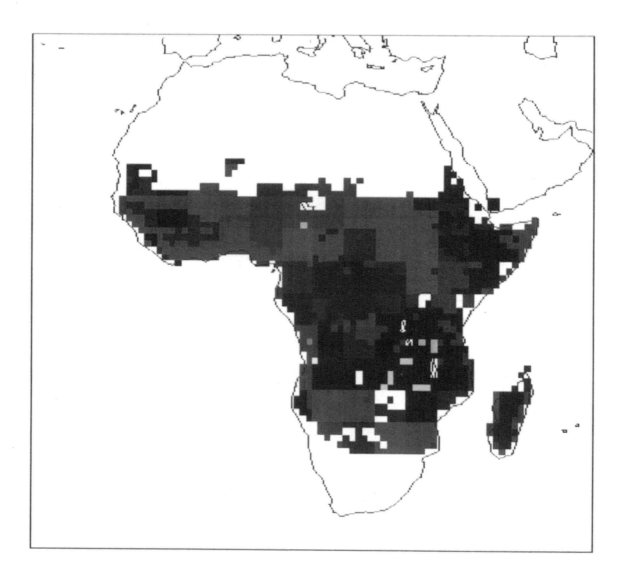

Figure 20.16 Color image of total biomass burning distribution in tropical Africa during the dry season, produced from output of visual data analysis software. The method of regridding the data from a 5° × 5° to a 1° × 1° format is the association method as described for figure 20.11.

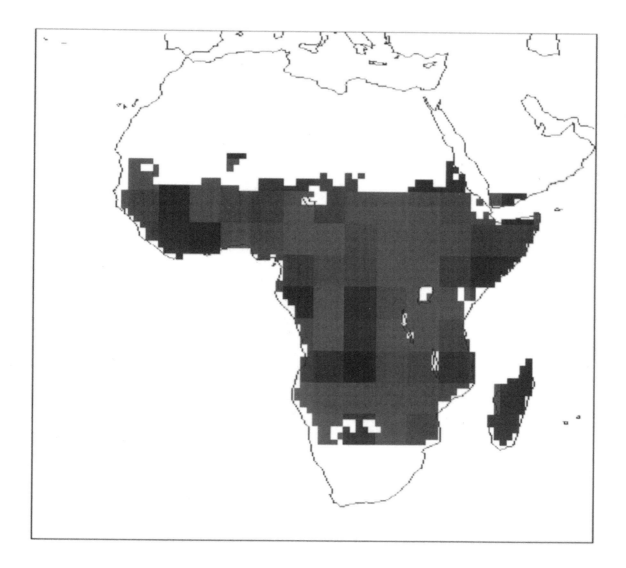

| 0.0 | 1.0 | 2.0 |

Figure 20.17 Color image of total biomass burning distribution in tropical Africa during the dry season, produced from output of visual data analysis software. The method of regridding the data from a $5° \times 5°$ to a $1° \times 1°$ format is to average the value of total CO_2 emissions over each of the 25, $1°$ cells of each $5°$ cell. This was deemed a less sophisticated method than that used in figures 20.11 and 20.16

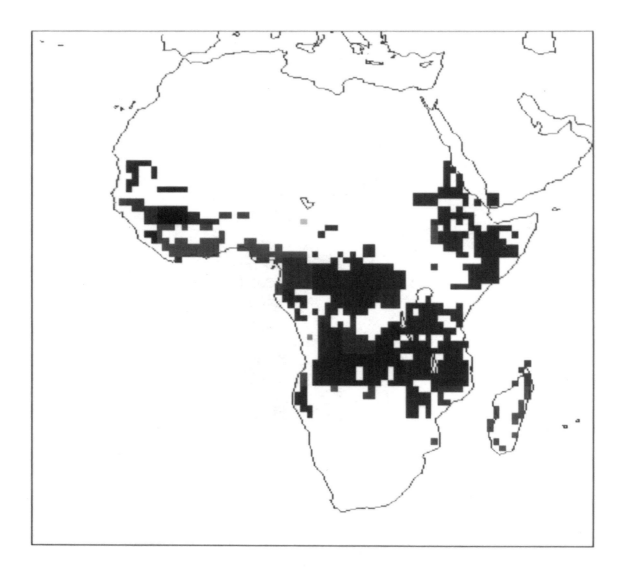

0.0   1.0   2.0

Figure 20.18   Color image of forest biomass burning distribution in tropical Africa during the dry season, produced from output of visual data analysis software. The association method was used to regrid the data.

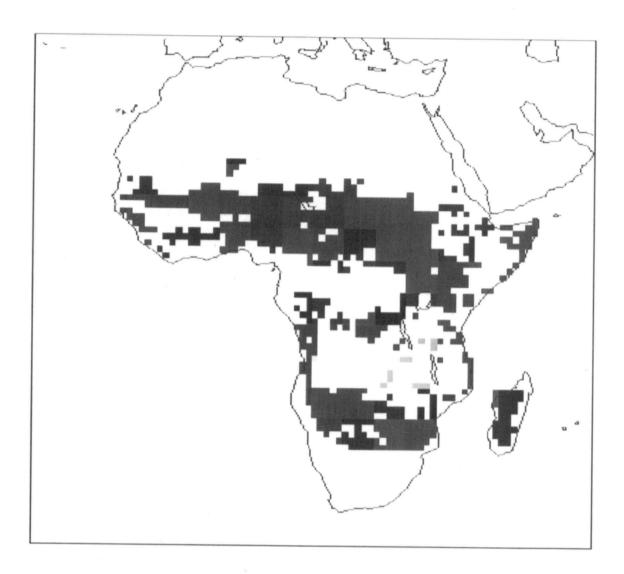

Figure 20.19 Color image of savanna biomass burning distribution in tropical Africa during the dry season, produced from output of visual data analysis software. The association method was used to regrid the data.

Where Would a Biomass Burning Inventory Reside?

Acquiring useful data in applicable formats is half of the task; developing a widely available distribution system is the other. The Langley Distributed Active Archive Center (DAAC), located at NASA Langley Research Center in Hampton, Virginia, is responsible for archiving and distributing NASA science data in the areas of radiation budget, clouds, aerosols, and tropospheric chemistry. The DAAC will also archive some of the data sets that result from the Earth Observing System (EOS) program and other elements of Mission to Planet Earth (MTPE).

The DAAC is a widely available and supported activity with resources in place to facilitate worldwide data distribution. A current activity is to assist the DAAC to develop and implement some dynamically accessible form of a global biomass burning inventory that the DAAC can accommodate with its capabilities. The goal is to interactively access the data for querying and polling to retrieve a user-defined data subset tailored to their requirements. In the interim all data received from the research community will be statically archived through the DAAC's current procedures, availability of data for exchange being the most pertinent issue.

Access to a Biomass Burning Inventory

The DAAC has developed an on-line computer system that allows the user to logon, search through the DAAC's data inventory, choose the desired data sets, and place an order. Data may be received either electronically (via FTP) or on media such as 4-mm tape, 8-mm tape, or CD-ROM (prepackaged data sets only).

Participation by Worldwide Research Communities

Feedback is important to the development of a final form for a global biomass burning inventory. A burning inventory must be designed with flexibility in mind; a data base is one method being researched to satisfy this requirement. The many sources of data, data formats, the manner in which data is acquired for inclusion, and distribution media are all variables in implementation. Developing an accessible tool requires input from the research communities to define these variables. The most essential detail of a successful implementation is undoubtedly selection and agreement on the data parameters required and the formats to use. The other requirement will be participation in submitting data to load such an inventory.

Process for Dialogue

What should be the process? It is hoped this chapter will initiate exchange in the research community directed at further development of this concept, a process that is driven by the researchers.

Conclusion

Design of a burning inventory is contingent on a comprehensive knowledge of the research requirements, thus data to be included. Design of the supporting software is contingent on knowledge of the source and format of the data. Therefore, the first objective must be to identify data requirements, sources, formats, and means of acquisition.

Acknowledgments

Sponsored by NASA Langley Research Center under the direction of Joel S. Levine with financial support from the Global Change Research Program of the US Environmental Protection Agency (EPA) under EPA Interagency Agreement DW-80936540-01-0.

References

Matthews, E. 1983. Global vegetation and land use: New high resolution data bases for Climate studies. *J. Clim. Appl. Meteor.*, 22, 474–487.

Henderson-Sellers. A., M. F. Wilson, G. Thomas, and R. E. Dickinson. 1986. Current Global Land-Surface Data Sets for Use in Climate-Related Studies. NCAR Technical Note NCAR/TN-272+STR, National Center for Atmospheric Research, Boulder, CO., 110 pp.

Hao, Wei Min, and Mei-Huey Liu. 1994. Spatial and temporal distribution of tropical biomass burning. *Global Biogeochemical Cycles*, 8, (4), 495–503.

Levine, J. S., 1991. Global Biomass Burning: Atmospheric, Climatic, and Biospheric Implications, MIT Press, Cambridge, Mass., 569 pp.

UNESCO. 1973. *International Classification and Mapping of Vegetation*, UNESCO, Paris.

III

Biomass Burning in Africa

Satellite Monitoring of Vegetation Fires on a Multiannual Basis at Continental Scale in Africa

Brigitte Koffi, Jean-Marie Grégoire, Hugh Douglas Eva

Biomass burning, owing to deforestation, shifting cultivation, savanna fires, fuel-wood use, and clearing of agricultural residues, is a phenomenon spread all over the tropical regions of the globe and is capable of causing large and long-lasting environmental changes such as atmospheric chemistry alterations, deforestation, and climate change (Crutzen and Andreae 1990; Lacaux et al. 1993). According to a recent study (Hao and Liu 1994), the African continent, where about two thirds of the biomass burning is due to savanna fires, contributes to 46% of biomass burnt in the tropics. Although many studies were carried out the last few years to characterize vegetation fire activity and its impact on environmental quality and climate, large uncertainties remain at continental and global scale, in both quantitative and spatiotemporal terms. There is therefore increased demand by the scientific community for systematic documentation of continental and global biomass burning, especially by scientists involved in fire emission, land-cover changes, and tropical tropospheric chemistry studies, in both experimental and modelling fields.

The present study has been carried out in the framework of the FIRE (Fire in global Resource and Environmental monitoring) project of the European Commission, whose objective is to define and to understand the geographical and temporal distributions of vegetation fires in tropical regions using remote-sensed data. It focuses on the monitoring of multiannual biomass burning activity for the African continent, using the Global Area Coverage (GAC) images, from the Advanced Very High Resolution Radiometer (AVHRR) of NOAA 7, 9, and 11 (National Oceanic and Atmospheric Administration) satellites (Kidwell 1990). Previous studies demonstrated that these 5-km resolution data do provide a good description of fire calendar and, on a continental scale, a good indicator of the locations of intense fire activity (Belward et al. 1994; Koffi et al. 1995). Indeed, the GAC data, available on a daily basis back to July 1981, is the only source of information actually usable to determine burning patterns in Africa, both on a continental scale and over long periods of time. The objectives of this study are (1) to map the spatial distribution of vegetation fires over the entire African continent, for a 10-year period (1981–1990), using AVHRR GAC imagery; (2) to analyse, subsequently, the temporal patterns of this distribution in order to determine the general trends of fire activity and of its spatiotemporal variability, on different time (10–daily to annual), and spatial (regional to continental) scales; (3) to improve the understanding of burning trends as a function of environmental (ecological, meteorological, anthropic, etc.) patterns, and finally (4) to assess the impact of biomass burning on the atmospheric chemistry of the tropics. With this last aim in view, great effort is taken to assemble and format the fire-related information, as derived from remote sensing, in consistent data sets adapted to the modelling of trace gases and particulate matter.

Fire Detection Method

The method used to detect vegetation fires in Africa, from AVHRR GAC daytime images, is a pixel-by-pixel multichannel threshold algorithm, taking into account four criteria: (1) channel 3 (3.55–3.93 μm) brightness temperature (B_{t3}) of a fire pixel has to be above a specified value; (2) B_{t3} must be greater than channel 4 (10.5–11.3 μm) brightness temperature (B_{t4}) plus a specified value; (3) B_{t4} itself must be above a specified value; and (4) channel 2 (0.72–1.10 μm) top of atmosphere reflectance (ToA 2) must be under a specified value.

The first three criteria have been proposed by Kaufman et al. (1988, 1989), and the fourth has been suggested by Kennedy et al. (1994). Many previous works using algorithms with these three or four criteria at regional scale have shown that universal threshold values do not exist to identify fire pixels. In fact, the spectral response of channel 3 and 4 by pixels containing active fires is especially strongly dependent both upon the ecological patterns and upon the period within the fire season (Grégoire et al. 1993; Belward et al. 1994; France et al. 1996; Kennedy et al. 1992, 1994; Koffi et al. 1995). In this study, the threshold values are therefore monthly defined from a previous statistical study of the spectral response of what we called "effective fire pixels," which are identified on the daily GAC images by the presence of associated visible smoke. A set of threshold values is so determined for each studied month, as follows: (1) $B_{t3} \geq B_{t3\,min\,1\%}$; (2) $B_{t4} \geq B_{t4\,min\,1\%}$; (3) $B_{t3} - B_{t4} \geq (B_{t3} - B_{t4})_{min\,1\%}$, and (4) $ToA2 \leq 10\%$, where $B_{t3\,min\,1\%}$, $B_{t4\,min\,1\%}$ and $(B_{t3} - B_{t4})_{min\,1\%}$ are the lower limit values of the confidence intervals (with a confidence level of 99%) of the "effective fire pixels" population mean of B_{t3} and B_{t4} and $B_{t3} - B_{t4}$, respectively.

The fire-detection method described above was applied to the daily GAC images from November 1984 to October 1989. The monthly averaged equatorial crossing times of these images, ranging from 1242 to 1524 GMT, are close to the peak of diurnal fire activity (See Moula et al. chapter 26, this volume). We computed B_{t3}, B_{t4}, and $B_{t3} - B_{t4}$ frequency distributions for the 776 effective fire pixels identified over this five-year period: Channel 3 brightness temperature shows a relatively small variation (mean = $318.6°K$; standard deviation = $1.8°K$), whereas B_{t4} (mean = $304.6°K$) and $B_{t3} - B_{t4}$ (mean = $14.0°K$) present a much higher standard deviation (6.0 and $5.3°K$, respectively). The statistical comparison between Northern and Southern Hemispheres (unpaired two group t-test with a 95% confidence level) of the mean values of the three above parameters does not show significant difference, over the whole of the five-year time series. However, a more detailed analysis of these frequency distributions according to the month and to the ecosystem provides significant differences for each of the three parameters (Koffi et al. 1995). Table 21.1 presents the threshold values determined for each of the 45 months from November 1984 through October 1989. The months of March, April, and May, which are known to be without significant fire activity at continental scale in Africa, are not processed. Behind several fire-detection tests, minima of $10°K$ and $14°K$ were imposed to $(B_{t3} - B_{t4})$ threshold value for NOAA 7, NOAA 9, and NOAA 11 satellite data, respectively.

Spatiotemporal Patterns of Vegetation Fires in Africa

From the daily fire maps, 10-daily, monthly, seasonal, and annual continental fire maps were produced and archived in a single data base, for the five-year time series processed. These composites were created in such a way that only pixels flagged as a fire on a single date are considered to be fire pixels.

As an example, figure 21.1 presents the distribution of fire pixels detected over Africa from November 1987 to October 1988, within selected biomes, as defined by UNESCO/White (1983). This example clearly shows the importance of fire activity in the African savanna of Northern and Southern Hemispheres. In the Northern Hemisphere, an almost unbroken belt of fire activity is observed along the savanna domain. In the Southern Hemisphere, the fire activity is the highest in the more humid savanna domain which surrounds the equatorial forest. The fire activity, decreases farther South, towards drier savanna domains. The rain forest of the Congo Basin appears to be squeezed by these two sub-continental fire patterns. Looking in detail at this map reveals the presence of fire pixels detected inside the equatorial forest domain (SW and NE), as defined by UNESCO/White (1983). These fire pixels are in fact located either in savanna (e.g., Congolian savanna), or in degraded forest (e.g., near the northern border) areas, which are not identified by UNESCO/White classification.

In figure 21.2, we present the total fire pixels detected, per degree square, for the months of June 1987, August 1987, October 1987, December 1987, January 1988, and February 1988. Although the spatial sampling of the AVHRR GAC data (Belward et al. 1994) results in the detection of only a small proportion of the active vegetation fires, these 5-km resolution data do provide a good spatiotemporal description of active fire dynamics, at subcontinental scale. From this data, one can therefore clearly observe the dynamics of the fire activity during the dry seasons of the Southern and Northern Hemispheres, which occur on average from June to October and November to February, respectively. In the south, one can observe a temporal shift from west to east, from the Atlantic coast to the coast of the Indian Ocean, whereas in the north the west-east fire belt moves from higher to lower latitudes, from the beginning to the end of the fire season. At the continental level, if we exclude March, April, and May, human-made fires are observed throughout the year. The month of October, with fire pixels detected north and south of the equator, appears to be a transition period between the fire seasons in Southern and Northern Hemispheres: it corresponds in fact to the end of the fire season in the south and the beginning in the north.

Table 21.1 Channel 3 ($B_{t3min1\%}$), channel 4 ($B_{t4min1\%}$) and channel 3 minus channel 4 (($B_{t3} - B_{t4})_{min1\%}$) brightness temperature thresholds, applied to the AVHRR GAC daily images, to identify fire pixels

Dry season	NOAA	NI	Equatorial crossing time	Threshold values $B_{t3} > (K)$	$B_{t4} > (K)$	$(B_{t3} - B_{t4}) > (K)$
Nov. 1984	7	5[a]	1459[b]	318	293	17
Dec. 1984	7	13[a]	1503[b]	318	298	17
Jan. 1985	7	26[a]	1504[b]	315	298	15
Feb. 1985	9	14[a]	1322[b]	319	311	10
Jun. 1985	9	23[a]	1326[b]	318	304	10
Jul. 1985	9	21[a]	1328[b]	317	299	13
Aug. 1985	9	23[a]	1329[b]	319	304	11
Sep. 1985	9	19[a]	1332[b]	319	305	10
Oct. 1985	9	17[a]	1334[b]	318	302	10
Nov. 1985	9	25[a]	1335[b]	318	300	11
Dec. 1985	9	29[a]	1334[b]	317	299	14
Jan. 1986	9	24[a]	1337[b]	317	305	10
Feb. 1986	9	22[a]	1337[b]	320	309	10
Jun. 1986	9	25[a]	1347[b]	317	303	11
Jul. 1986	9	22[a]	1349[b]	318	302	13
Aug. 1986	9	24[a]	1350[b]	318	303	11
Sep. 1986	9	23[a]	1355[b]	318	302	13
Oct. 1986	9	23[a]	1355[b]	317	297	15
Nov. 1986	9	21[a]	1358[b]	319	305	10
Dec. 1986	9	25[a]	1359[b]	316	301	12
Jan. 1987	9	27[a]	1402[b]	317	301	11
Feb. 1987	9	18[a]	1407[b]	314	297	13
Jun. 1987	9	29[a]	1417[b]	317	300	14
Jul. 1987	9	31[a]	1417[b]	317	300	15
Aug. 1987	9	28[a]	1422[b]	317	300	13
Sep. 1987	9	23[a]	1424[b]	318	302	13
Oct. 1987	9	29[a]	1427[b]	317	299	12
Nov. 1987	9	25[a]	1430[b]	319	301	10
Dec. 1987	9	27[a]	1437[b]	318	301	14
Jan. 1988	9	25[a]	1437[b]	317	297	15
Feb. 1988	9	26[a]	1432[b]	317	292	14
Jun. 1988	9	29[a]	1453[b]	314	293	17
Jul. 1988	9	22[a]	1459[b]	316	300	14
Aug. 1988	9	29[a]	1457[b]	317	298	15
Sep. 1988	9	22[a]	1524[b]	316	299	15
Oct. 1988	9	22[a]	1519[b]	318	300	12
Nov. 1988	11	30[a]	1239[b]	319	307	14
Dec. 1988	11	30[a]	1242[b]	318	303	14
Jan. 1989	11	27[a]	1241[b]	317	305	14
Feb. 1989	11	27[a]	1243[b]	319	307	14
Jun. 1989	11	28[a]	1255[b]	318	305	14
Jul. 1989	11	28[a]	1252[b]	318	306	14
Aug. 1989	11	23[a]	1255[b]	319	307	14
Sep. 1989	11	10[a]	1253[b]	320	308	14
Oct. 1989	11	21[a]	1257[b]	320	311	14

a. Number of available GAC images.
b. Monthly averaged equatorial crossings of the satellite (GMT).

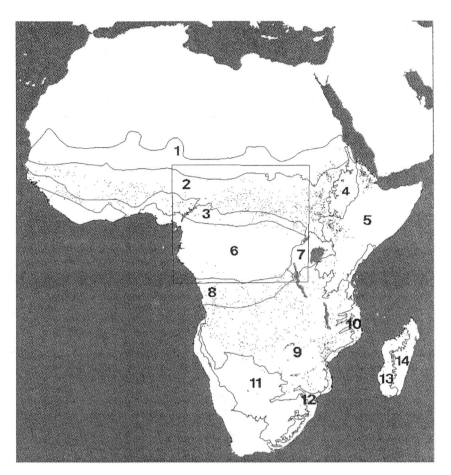

Figure 21.1 Distribution of fire pixels (dots) on continental Africa from November 1987 to October 1988, as derived from NOAA AVHRR GAC data (subsampling of the original data reduces the number of pixels by a factor of 4) within selected biomes as defined by UNESCO/White (1983): (1) Sahel, (2) Sudanian, (3) Guinea-Congolia/ Sudania, (4) Afromontane, (5) Somalia-Wasai, (6) Guineo-Congolian, (7) Lake Victoria, (8) Guineo-Congolia/ Zambesia, (9) Zambesian, (10) Zanzibar-Inhambane, (11) Kalahari-Highveld, (12) Tangaland-Pandoland, (13) West Malagasy, (14) East Malagasy. The analysis of fire dynamics in relation to environmental patterns presented here was performed for the region of Central Africa defined in this figure.

Understanding of Fire Dynamics in Relation to Environmental Patterns

The understanding of fire dynamics as a function of environmental patterns (ecological, meteorological, anthropic), requires the use of Geographical Information Systems (GIS) adapted to fire studies. Tools for spatiotemporal analysis are therefore being developed by the FIRE project to analyse burning patterns and dynamics in relation to different environmental factors, (e.g., land use, vegetation types, and condition, and meteorological patterns). The two examples presented in this section are drawn from a previous study (Koffi et al. 1995) focusing on fire dynamics in Central Africa, for the four 1984–1985 to 1987–1988 consecutive dry seasons of the Northern Hemisphere (November to February), using a similar detection method as the one previously described. The study area (8.1°E–31.4°E; 14.7°N–6.0°S) is composed of four vegetation zones, according to the UNESCO/White (1983) classification (figure 21.1): They are, from north to south, (1)

the Sahelian zone which corresponds to arid savanna; (2) the Sudanian zone which corresponds to a dry savanna domain; (3) the more humid Guinea-Congolia/ Sudania savanna zone, and (4) the Guineo-Congolian zone, corresponding approximatively to the equatorial forest area. It covers nine countries: Cameroon, Central African Republic, Gabon, Congo, Equatorial Guinea, southeastern Nigeria, southern Chad, southwestern Sudan, and northern Zaïre.

Terminology

The main terms referring to fire pixels used in this chapter have been defined by Kennedy et al. (1994) and Koffi et al. (1995) as follows. First, all the pixels identified as probable *fire pixels* represent a sample in time, that is, the number of fires and their distribution reflect the situation on the ground, at the time of the satellite overpass. Second, it is assumed that each fire pixel represents a single fire event, however, in reality, the 5-km resolution AVHRR GAC pixel may contain many fires. For a given region, the *burning season*

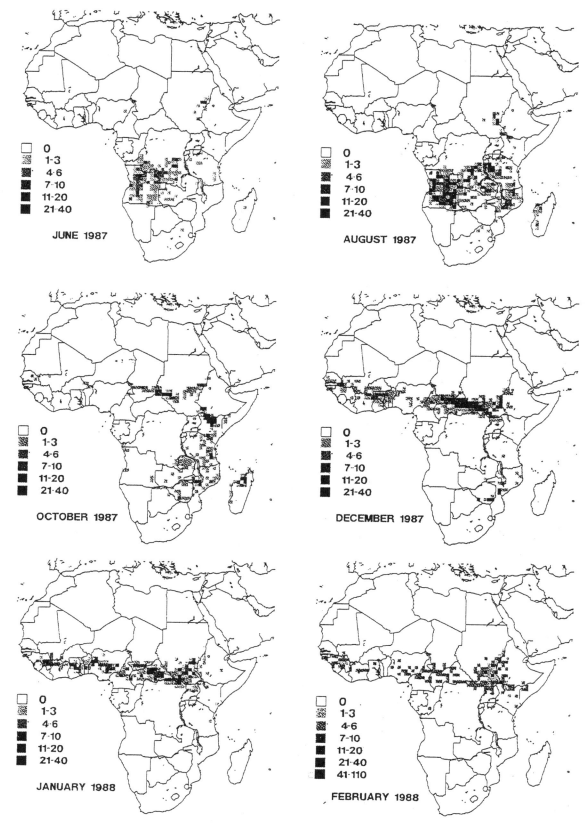

Figure 21.2 Number of fire pixels per degree square, as detected from the daily AVHRR GAC images for June 1987, August 1987, October 1987, December 1987, January 1988, and February 1988

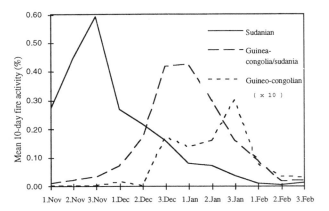

Figure 21.3 10-day fire activity (fire pixels/total pixels) for the 1987–1988 dry season of the Northern Hemisphere, per vegetation zone as defined by UNESCO/White (1983) within the study area presented in figure 21.1

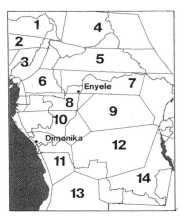

Figure 21.4 Experimental rain-forest sites (Enyélé and Dimonika) and pluviometric areas as defined by Mahé and Citeau (1993): (1) SAHEL, (2) JOS, (3) ADAMAOUA, (4) TCHAD, (5) RCA, (6) SANAGA, (7) OUBANGUI, (8) SANGHA, (9) CZAIRE, (10) BATEKE, (11) BRAZZA, (12) LOMANI, (13) ANGOLA, (14) KATANGA. Units 6 and 7 together form the SANOUB pluviometric region.

starts when the first fire pixels are detected and stops when no further fire pixels are in evidence; the *fire calendar* is the temporal distribution of fire pixels, for a given region, during the burning season; the *fire dynamics* refers to the spatial and temporal distribution of fire pixels; and the *fire activity* is the ratio (%) of fire pixels to the total number of pixels for a given area (in order to take into account the number of unavailable GAC daily images, the 10-daily and monthly counts are respectively brought to 10 and 30 days).

Fire in Central Africa, as a Function of the Vegetation Zone

In figure 21.3, the fire distribution and the fire calendar during the 1987–1988 dry season of the Northern Hemisphere are analyzed as a function of the main vegetation zones presented in figure 21.1. In order to compare the fire activity of the different vegetation zones over the fire season of the Northern Hemisphere, we calculated for each zone the mean values of the 10-daily fire activity: The fire activity appears to be mainly concentrated within the Sudanian and Guinea Congolia/Sudania savanna as defined in figure 21.1, that is, in relatively high biomass savanna, where the use of fire for agriculture, hunting, and cattle breeding is quite common. No fire pixel was detected in the arid savanna Sahelian zone. The fire activity observed within the Guineo-Congolian zone represents less than 5% of the fire pixels detected in the study area and takes place, indeed, in a transition zone between savanna and forest. It seems, in fact, according to recent studies, that the border between forest and the northern savanna domain, in the eastern part of the forest, is located somewhat more south than on the UNESCO/

White (1983) classification (Ehrlich and Lambin 1995; Janodet 1995). More work is needed, such as crossing land-cover and active fire data, to examine and monitor the deforestation processes in savanna-forest transition zones. Concerning the fire calendar, one can observe a shift of the fire activity from November in the dry savanna in the north to January in the more southern and more dense vegetation zones. This temporal shift in the start of the burning season of the Northern Hemisphere, already observed in figure 21.2, is associated with a shift in the end of the wet season, and can be related to the progressive decrease in water content of the vegetation, which occurs first in the Sahelian, then the Sudanian, and last in the Guinean savannas (Menault et al. 1991).

Fire Dynamics in Central Africa as a Function of Rainfall Patterns

In the same study area, we then looked for multiannual relationships between the fire activity, during the three dry seasons of 1985–1986, 1986–1987, and 1987–1988, and the rainfall. A description of the sources and processing of the monthly rainfall data base may be found in Koffi et al. (1995). The resulting three-year data base (March 1985 to February 1988) consists of monthly rainfall amounts for 105 stations, distributed over 14 pluviometric units, as defined by Mahé and Citeau (1993) and presented in figure 21.4.

Figure 21.5 shows the total fire activity detected in the TCHAD, JOS, RCA, SANOUB, and CSNIGADAM units, for each dry season (November to

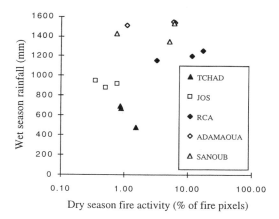

Figure 21.5 Seasonal fire activity for the 1985–1986, 1986–1987, and 1987–1988 dry seasons of the Northern Hemisphere (November to February) as a function of the previous annual amount of rainfall (March to February) for the TCHAD, JOS, RCA CSNIGADAM, and SANOUB pluviometric regions presented in figure 21.4 (Koffi et al. 1995)

February), as a function of the previous annual rainfall (March to February). One can observe in this figure that relatively dry units (TCHAD and JOS) have weaker fire activity than the relatively wet pluviometric areas (RCA, SANOUB, and CSNIGADAM). Although fire activity depends on many environmental factors (e.g., ecological, climatic, and anthropic), such a distribution may basically reflect the relation between the rainfall amount and the biomass density: a higher annual rainfall may result in a higher savanna biomass density (Loudjani 1988), and therefore, a higher fire activity. The interannual variability of the fire activity appears more pronounced for the wet southern pluviometric units than for the drier northern ones. This may in part be due to the fact that the vegetation of northern units is mainly herbaceous, whereas the wetter southern zones are more complex savannah, associating high grass, shrub, and woody strata: the degraded herbaceous savanna regions, where fires have been occurring for a longer time, may be more in equilibrium with the annual fire process than the humid southern ones, which are composed of degraded units and which are more often submitted to land-cover changes. A combined analysis of rainfall patterns and fire dynamics was performed; however it did not explain the interannual variability of the fire activity observed in Central Africa. It appeared that, together with an improved temporal resolution of the rainfall series (10-day interval), other factors such as dry season meteorological factors (e.g., occasional rainfall events, air humidity, and ITCZ position) must also be taken into account to investigate such results.

Assessment of the Impact of Fire Emissions on Tropospheric Chemistry

The effect of biomass burning on the tropospheric chemistry (impact on oxidant cycles, ozone pollution, greenhouse effect, acid deposition) is multifold. One of the goals of the FIRE project is to improve the understanding and the quantification of the impact of gaseous and particulate emissions from tropical vegetation fires, on the tropospheric chemistry. An essential part of achieving this goal is the characterization of spatiotemporal distribution of active fires, and the assessment of burned area and of burnt biomass. The example presented here combines the results of fire-detection studies of Africa, using AVHRR GAC data, with precipitation chemistry measurements in equatorial forests of the Congo. Mapping and spatiotemporal analysis tools developed by the FIRE project are used to analyse the rain chemistry at two Congolian forest sites of North and South Congo, in relation to the fire dynamics in Central Africa, during two annual experimental periods.

Precipitation Chemistry in African Equatorial Forests

As part of the DECAFE (Dynamique et Chimie de l'Atmosphère en Forêt Equatoriale) program (Fontan et al. 1992), precipitation chemistry measurements were carried out at two sampling sites in the equatorial forests of the Congo (figure 21.4): Dimonika in the south (4°N; 12°3 E) and Enyélé in the North (2°49 N; 18°1 E). The objectives of these studies were to determine the chemical characteristics of the precipitation and to estimate the seasonal influence of the various sources of gases and particles. The analysis of the chemical content and the physical characteristics of the rainfall events collected from November 1986 to October 1987 at Dimonika, and from June 1988 to June 1989 at Enyélé, clearly showed an important contamination of the troposphere during the dry season, by the vegetation fires from the Southern and Northern Hemispheres, respectively. This influence of biomass burning emissions, which induced higher concentrations for specific chemical species (e.g. NO_3^-, NH_4^+, H^+ and $C_2O_4^{2-}$) may be explained by the transport in the northeastern (November to May) and southeastern (June to October) trade winds of gases and particles produced by the savanna burning, north and south of the equator, respectively. These pollutants are then progressively scavenged by the precipitations in the equatorial belt (Cachier and Ducret 1991; Lacaux et al. 1991, 1992a, 1992b; Lefeivre 1993).

Table 21.2 Seasonal fire activity (% of fire pixels), as derived from AVHRR GAC imagery for the two annual experimental periods of rain chemistry

Experimental periods	Dry season	Pluviometric zones										
		Tchad 1	RCA 2	Sanaga 3	Oubangui 4	Sangha 5	Czaire 6	Bateke 7	Brazza 8	Lomani 9	Angola 10	Katanga 11
Nov. 1986–Feb. 1987	N. Hem.	1.1	7.9	1.3	1.8	0.0	0.0	0.0	0.8	0.0	0.0	0.0
Jun. 1987–Oct. 1987	S. Hem.	0.6	0.0	0.0	0.0	0.0	0.0	0.1	1.0	1.4	2.8	1.5
Jun. 1988–Oct. 1988	S. Hem.	0.2	0.0	0.0	0.0	0.0	0.0	0.1	1.0	0.9	1.5	1.4
Nov. 1988–Feb. 1989	N. Hem.	3.7	4.8	0.1	0.2	0.0	0.0	0.0	0.1	0.0	0.0	0.0

Comparison of Fire Activity and Precipitation Chemistry Measurements

In order to argue and complete these results, we analyzed the fire distribution in Central Africa, during the rain sampling periods, for the 11 pluviometric units situated in the path of the trade winds to the sampling sites (figure 21.4). Table 21.2 summarises the total fire activity (% of fire pixels) detected during the dry seasons of the Northern and Southern Hemispheres for both annual experimental periods. The periods of fire activity detected do correspond (1) to the November to February months for the pluviometric units located north of the equator (TCHAD, RCA, SANAGA, and OUBANGUI), and (2) to the June to October months for the ones located south of the equator (BATEKE, BRAZZA, LOMANI, ANGOLA, and KATANGA), that is, to the dry seasons of the Northern and Southern Hemispheres, respectively. The fire activity obtained in the TCHAD unit during the period June to October, is due in fact only to the months October 1987 and October 1988. It therefore corresponds to the very beginning of the fire season in the Northern Hemisphere. In the same way, the fire activity detected in the BRAZZA pluviometric zone for both periods of November to February corresponds indeed only to the November months, that is, to the very end of the fire season in the Southern Hemisphere. No fire pixel has been detected in SANGHA and CZAIRE units, which are completely inside the forest domain.

The 10-daily fire activities during the two experimental dry seasons presented in figures 21.6 *top* and 21.7 *top*, are compared to the rainfall nitrate concentration at Enyélé (figure 21.6 *bottom*) and Dimonika (figure 21.7 *bottom*), respectively.

North Congo If we consider the total of the four pluviometric areas (figure 21.6 *top*), the maximum of the fire activity, which occurs in December 1988, coincides with the maximum of the nitrate concentration measured in the Enyélé precipitation (figure 21.6 *bottom*).

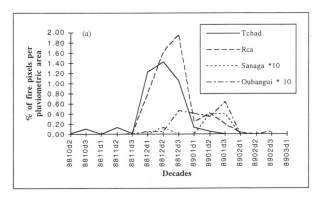

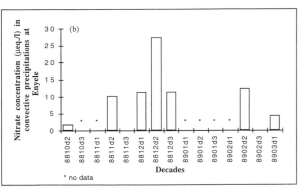

Figure 21.6 (*top*) Fire activity per pluviometric zone of the African Northern Hemisphere presented in figure 21.4, and (*bottom*) nitrate concentrations in convective precipitations of Enyélé (North Congo) during the 1988–1989 dry season

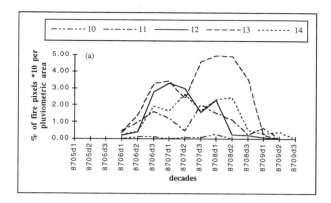

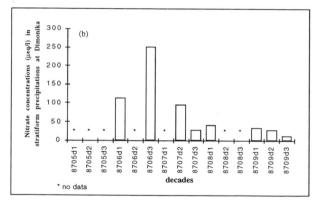

Figure 21.7 (*top*) Fire activity per pluviometric zone of the African Southern Hemisphere as defined in figure 21.4: (10) BATEKE; (11) BRAZZA; (12) LOMANI; (13) ANGOLA; (14) KATANGA, and (*bottom*) nitrate concentrations in stratiform precipitations of Dimonika (South Congo) during the 1987 dry season

According to the fire calendars, which show an important variability as a function of the pluviometric region, it seems that the contamination of the atmosphere in North Congo is mostly due to fire emissions from more northern savannah regions (TCHAD, RCA) which present the highest fire activity. Although they are nearer to the Enyélé sampling site, the SAN-AGA and OUBANGUI regions can't be considered as responsible for the increased rain concentration of nitrate observed in December 1988. These results are in good agreement with previous studies focusing on the DECAFE experiment in North Congo, which concluded a synoptic origin of the contamination of the troposphere by biomass burning emissions from the Northern Hemisphere (Cautenet and Lefeivre 1994; Fontan et al. 1992; Lefeivre 1993; Lopez et al. 1992), which are transported at an altitude of 1 to 3 km, by the northeastern trade winds.

South Congo The same type of analysis for the 1987 dry season of the Southern Hemisphere does not show such an obvious relationship between the temporal evolution of the rain concentration of nitrate at Dimonika (figure 21.7 *bottom*) and the temporal evolution of the fire activity in the southern pluviometric zones (figure 21.7 *top*). Contrary to North Congo, the contamination of precipitations in the South Congo does not seem to be linked to the biomass burning emissions from the regions of highest fire activity. Because of its temporal evolution, the fire activity detected, for instance, in ANGOLA region, cannot explain the nitrate concentration in Dimonika precipitation. One explanation could be that gases and particulate matter emitted by fires in the most southern savanna (e.g., ANGOLA area (figure 21.4)) are mostly carried by the southeastern trade winds more south of the experimental site, towards the Atlantic Ocean. Neither can the fire activity in LOMANI area, which presents during the whole fire season the highest correlation with the rain nitrate concentration at Dimonika, explain the maximum of nitrate concentrations observed for the third decade of June 1987, about one decade earlier than the maximum fire activity in this region. Such results also agree with precipitation chemistry studies (Lacaux et al. 1991, 1992a, 1992b; Lefeivre 1993) which showed an important contribution of local biomass burning emissions to the chemical content of Dimonika precipitation—local biomass burning which coincides with the first maximum of the fire activity observed in BRAZZA area, including the site of Dimonika. More work is needed to understand such relationships between regional patterns of fire distribution and point measurements of atmospheric chemistry.

With this objective in mind, the FIRE project, in collaboration with the CNR-IRRS of Milano (Brivio et al. 1994, 1995), is developing tools for spatial and temporal analysis of fire distribution. Some of these tools allow us to explore spatial characteristics of vegetation fire at regional and continental scales, to quantitatively describe these patterns on the basis of texture analysis techniques, and to relate them to some location of interest. Results are presented in the form of rose-diagram of frequency and length-distance, taking some reference point, such as an experimental site of tropospheric chemistry measurement. As an example, figure 21.8 shows the fire distribution for the third decade of June 1987, with reference point Dimonika (Koffe et al. 1994). Current work focuses on developing this analysis tool, with reference to a line or area. Such a tool will be especially useful in monitoring

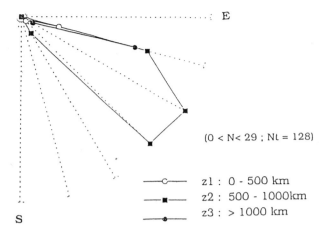

$(0 < N < 29 ; Nt = 128)$

- z1 : 0 - 500 km
- z2 : 500 - 1000 km
- z3 : > 1000 km

Figure 21.8 Number of fire pixels per angular sector, as derived from the AVHRR GAC images for the third decade of June 1987, with reference point Dimonika (located in figure 21.4): Nt = total number of fire pixels in the 90°SE sector; N = number of fire pixels per 15° angular sector

the fire pressure and deforestation processes, particularly in sensitive regions such as protected areas or equatorial forests.

Conclusions and Perspectives

Because of the spatial and temporal sampling, it is obvious that absolute quantitative information such as the number of fire events or the extent of burnt surfaces cannot be directly derived from AVHRR GAC data. On the other hand, data from this source do provide a good description and, related to other types of data, a good understanding of burning patterns at regional and continental scales—spatial patterns such as regional sets of fire activity as well as temporal patterns such as seasonality of biomass burning. Combined with atmospheric chemistry data, this information on fires can also be meaningfully interpreted in terms of their impact on atmospheric chemistry.

Current work is focusing on the completion of the continental fire maps for the 12 years of data (1981–1992) available in the GAC archive. The 10-day daily and monthly fire maps are available now for the five-year period from November 1984 to February 1989. The interpretation of the fire dynamics in relation to various environmental patterns (anthropic, ecological, meteorological) will be improved by current and foreseen developments in our Geographic Information System and spatiotemporal analysis tools. As far as impact of fire emissions is concerned, distribution of active fires and seasonality patterns are used as input for global modelling of biomass burning emissions

(black carbon, nitrogen compounds). Moreover, the FIRE project is also involved in the preparation phase of the future EXPRESSO (EXPeriment for Regional Sources and Sinks of Oxidants) interdisciplinary experiment, designed to investigate chemical processes in the tropical atmosphere. Finally, the FIRE project is also planning to participate in the IGBP project, called IGAC/DEBITS/AFRICA (MEDIAS- France, 1994) whose objective is to quantify wet and dry chemical depositions in Africa and to identify and simulate the parameters that control these fluxes.

These research activities related to satellite monitoring of active fires, using continental AVHRR GAC data, will be extended to other remote-sensing data and to other tropical regions. The integration of these data into a global fire information system, using a range of satellite and nonsatellite data sources, is also foreseen: this system, predominantly based on 1-km resolution data sets, will allow analysis of biomass burning impacts at scales ranging from the local or national level (Grégoire 1993) to global level (See Malingreau and Grégoire, chapter 2, this volume).

References

Belward, A. S., P. J. Kennedy, and J.-M. Grégoire. 1994. The limitations and potential of AVHRR GAC data for continental scale fire studies, *Int. J. of Remote Sensing*, 15(11): 2215–2234.

Brivio, P. A., I. Doria, J.-M. Grégoire, and B. Lefeivre. 1994. Use of the rose-diagram method for vegetation fire patterns analysis at regional scale in Africa, *14th International Conference CODATA'94: The Quest for a Healthier Environment*, Chambéry, France, 18–22 Sept.

Brivio, P. A., J.-M. Grégoire, B. Koffi, and G. Ober. 1995. An automatic clustering technique applied to the study of vegetation fires patterns distribution in the African continent, *IGARSS'95 conference*, Firenze, Italy, 10–14 July 1995.

Cachier, H., and J. Ducret. 1991. Influence of biomass burning on equatorial African rains, *Nature*, 352:228–230.

Cautenet, S., and B. Lefeivre. 1994. Contrasting behaviour of gas and aerosol scavenging in convective rain: A numerical and experimental study in the African equatorial forest, *J. Geophys. Res.*, 99 (D6): 13 013–13 024, June.

Crutzen, P. J., and M. O. Andreae. 1990. Biomass burning in the Tropics: Impact on atmospheric chemistry and biogeochemical cycles, *Sciences* 250:1669–1678.

Ehrlich, D., and E. Lambin. 1996. Broad-scale classification and interannual climatic variability, *Internal Journal of Remote Sensing* 17(5): 845–862.

Fontan, J., A. Druilhet, B. Bénech, and R. Lyra. 1992. The DECAFE experiments: Overview and meteorology, *J. Geophys. Res.*, 97: 6123–6136.

Franca, J. R. A., J. M. Brustet and J. Fontan. 1996. A multispectral remote sensing of biomass burning in West Africa, *Journal of Atmospheric Chemistry*, in press.

Grégoire, J.-M. 1993. Description quantitative des régimes de feu en zone soudanienne d'Afrique de l'Ouest, *Sècheresse*, 1 (4) mars, 37–45.

Grégoire, J.-M., A. S. Belward, and P. Kennedy. 1993. Dynamique de Saturation du signal dans la bande 3 du senseur AVHRR: Handicap majeur ou source d'information pour la surveillance de l'environnement en milieu soudano-guinéen d'Afrique de l'Ouest? *Int. J. Remote Sensing*, 14(11):2079–2095.

Hao, W. M., and M.-H. Liu. 1994. Spatial and temporal distribution of tropical biomass burning, *Global Biogeochemical Cycles*, 8(4):495–503.

Janodet, E. 1995. Stratification saisonnière de l'Afrique Continentale et cartographie fonctionnelle des écosystèmes forestiers africains. Rapport final Contrat CCR n°. 10 350–94.07 FIED ISPF p. 72 and annexes.

Kaufman, Y., C. J. Tucker, and I. Fung. 1988. Remote sensing of biomass burning—method, in *Proc. International Radiation Symposium*, Lille, France, 18–24. aug.

Kaufman, Y., C. J. Tucker, and I. Fung. 1989. Remote sensing of biomass burning in the tropics, *Advanced Space Research*, 9:265–268.

Kennedy, P., 1992. Biomass burning studies: The use of remote sensing, *Ecological Bulletins*, 42:133–148, Copenhagen.

Kennedy, P. J., A. S. Belward, and J.-M. Grégoire. 1994. An improved approach to fire monitoring in West Africa using AVHRR data, *Int. J. Remote Sensing*, 15(11):2235–2255.

Kidwell, K. B. 1990. Global vegetation index user's guide. Washington, DC: NOAA/NESDIS, Technical Report NOAA National Climatic Center.

Koffi, B., J.-M. Grégoire, P. A. Brivio, et J.-P. Lacaux. 1994. Télédétection satellitaire et chimie des pluies: Deux approches complémentaires pour l'étude des feux de végétation sur le contenu chimique de la troposphère en région tropicale, *Séminaire "Dépôts atmosphériques en Afrique,"* IGAC/DEBITS-MEDIAS, INSET Yamoussoukro, déc.

Koffi, B., J.-M. Grégoire, G. Mahé, and J.-P. Lacaux. 1995. Remote sensing of bush fire dynamics in Central Africa from 1984 to 1988: Analysis in relation to regional vegetation and pluviometric patterns, *Atmospheric Research*, 39, 179–200.

Lacaux, J.-P., H. Cachier, and R. Delmas. 1993. Biomass burning in Africa: An overview of its impact on atmospheric chemistry, in *Fire in the Environment*, ed. P. J. Crutzen and J. G. Goldammer, 159–191. J. Wiley and Sons Ltd, Chichester, England.

Lacaux, J.-P., R. Delmas, B. Cros, B. Lefeivre, and M. O. Andreae. 1991. Influence of biomass burning emissions on precipitation chemistry in the equatorial forests of Africa, In *Global Biomass Burning*, ed. J. S. Levine, pp. 167–174. MIT Press, Cambridge, Mass.

Lacaux, J.-P., R. Delmas, G. Kouadio, B. Cros, and M. O. Andreae. 1992a. Precipitation chemistry in the Mayombé Forest of equatorial Africa, *J. Geophys. Res.* (Special Issue DECAFE) 97:6195–6207.

Lacaux, J.-P., J. Loemba-Ndembi, B. Lefeivre, B. Cros, and R. Delmas. 1992b. Biogenic emissions and biomass burning influences on the chemistry of the fog water and stratiform precipitations in the African Equatorial Forest, *Atmos. Environ.* 26A, (4):541–551.

Lefeivre, B. 1993. Etude expérimentale et par modélisation des carctéristiques physiques et chimiques des précipitations collectées en forêt équatoriale africaine, *Thèse de doctorat* de l'Université Paul Sabatier, Toulouse, France, 308 pp.

Lopez, A., M. L. Huertas, and J. M. Lacome. 1992. A numerical simulation of the ozone chemistry observed over forested tropical areas during DECAFE experiments, *J. Geophys. Res.*, 97:6149–6158.

Loudjani, P. 1988. Cartographie de la production primaire des zones savanicoles d'Afrique de l'Ouest à partir de données satellitaires: comparaison avec des données de terrain. DEA diss., Univ. Paris XI.

Mahé, G., and J. Citeau. 1993. Interactions between the ocean, atmosphere and continent in Africa, related to the Atlantic monsoon flow: General pattern and the 1984 case study, *Veille Climatique Satellitaire*, 44:34–54.

Malingreau, J. P., and J.-M. Grégoire. 1997. Developing a global vegetation fire-monitoring system for global change studies: A framework (see chapter 2).

MEDIAS-France. 1994. Workshop on "Atmospheric Depositions in Africa" IGAC/DEBITS-MEDIAS-INSET Yamoussoukro (Côte d'Ivoire)—Dec. *La lettre de MEDIAS*, no. 4, p. 8, Feb.

Menault, J. C., L., Abbadie, F., Lavenu, P., Loudjani, and A. Podaire. 1991. Biomass burning in west African savannas. In *Global Biomass Burning: Atmospheric, Climatic, and Biospheric Implications* ed. J. S. Levine, pp. 133–142. MIT Press, Cambridge, Mass.

Moula, M., J. M. Brustet, H. Eva, J.-P. Lacaux, J.-M. Grégoire, and J. Fontan. 1997. Contribution of spread-fire model in the study of savannah fires (see chapter 24).

UNESCO/White, F. 1983. Vegetation map of Africa (Unesco/AETFAT/UNSO).

Use of the Earth Observation System in the Space Shuttle Program for Research and Documentation of Global Vegetation Fires: A Case Study from Madagascar

Johann G. Goldammer, Jean Laurent Pfund, Michael R. Helfert, Kamlesh P. Lulla, and STS-61 Mission Crew*

Madagascar is the world's fourth largest island with an area of about 587 000 km^2 and well known for biodiversity of landscapes, flora, and fauna and for its richness of endemic vegetation. According to White (1983) about 20% of Madagascar's flora is endemic. High landscape diversity is not only the result of topography and distinct climatic differentiation but also of human impacts. In the late Quaternary period, fires seem to have influenced the island's vegetation before the presumable arrival of humans on the island (Burney 1987). Paleo-Indonesians were the first immigrants to the island and practised swidden agriculture and pastoralism in which fire was an inevitable basic tool for clearing land and maintaining secondary grassland vegetation (Battistini and Verin 1972).

The original extent of Madagascar's rain forests has been estimated at 11.2×10^6 ha. Evaluation of historic aerial photographs and modern satellite imagery shows that between 1950 and 1985 ca. 111 000 hectares were deforested annually (Green and Sussman 1990). Today approximately one fourth of the forest formations of the prehuman era is left. Still today fire is the main tool in swidden agriculture and maintenance of secondary formations. No reliable statistical data on the extent of fires in forests, grasslands, and agricultural systems are available. Recent statistics compiled by the Forest Service (Direction des Eaux et Forêts) differentiate between savanna/grassland fires and forest fires. According to these data ca. 95% of all surface burned between 1987 and 1990 was in savanna-type ecosystems, averaging 440 000 ha yr^{-1}.

Study Area

The region of Beforona is in the humid zone of Madagascar and in the center of the fire observations by the STS-60 space shuttle mission. Meteorological data were collected in Beforona village (18°53′20″S, 48°38′30″E) which is in midst of the Terre-Tany project area, a Swiss-Madagascan research cooperation project in which different vegetation types are studied in function of distance (ecological gradient) to the primary forest (Zurbuchen 1993).

The dwellers of this region, the *Betsimisaraka*, are living at the forest edge. The cultural system based on fire cannot be, in this region, denominated as shifting cultivation in the strict sense. Primary forest is generally burned after selective exploitation, but most present burning activities are in secondary vegetation (locally called *savoka*; figure 22.1). The quantities of the dried plant biomass determine intensity and impacts of the fire: Aboveground biomass is ca 150–250 txha^{-1} of the forest already selectively exploited (2 to 5% of observed fires), and 25–50 txha^{-1} in young *savoka*. Savoka burning represents 95% of burned biomass in the study zone. Sowing is conducted directly after the fire. Cultivation and weeding follow until harvest between April and June. Harvest residuals are left on the fields for soil protection.

Environmental Consequences of Fire

After three or four fire cultivations on fallow lands the percentage of forest species is reduced drastically (ca. 80% of species are lost). The erosion rate is extremely high on steep and unprotected sites, particularly as a consequence of cyclones and landslides.

Long-term influence of fire has led to the formation of savannas on the ridges and tops of mountains, characterized by species like *Pteridium aquilinum, Dicranopteris linearis, Sticherus flagellaris*. Degraded lands are occupied by *Imperata cylindrica*.

The emission from forest and grassland burning add to the pan-African pyrogenic emission load in two ways (Garstang et al. 1995). Westerly zonal transport of air masses originating in the Indian Ocean region consist of maritime air not influenced by pyrogenic emissions. These air masses become affected by fire

* Richard O. Covey, Commander (USA), Kenneth D. Bowersox, Pilot (USA), F. Story Musgrave, Payload Commander (USA), Thomas D. Akers, Jeffrey A. Hoffman, Claude Nicollier, Kathryn C. Thornton, Mission Specialists (USA, Switzerland).

Figure 22.1 Fresh *tavy* (forest conversion burn) in primary mountainous rain forest

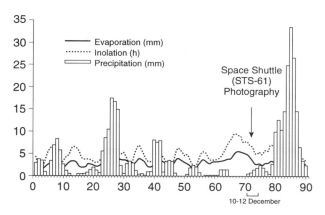

Figure 22.2 Weather observations (3-day average) in Beferona Village, Madagascar, during the period 1 October to 31 December 1993. Photographs were taken from the Space Shuttle STS-61 just before the end of the dry season.

Figure 22.3 View of Madagascar in the morning of 11 December 1993 at 1020 local time (0720 GMT). No burning activities can be observed at this time of day.

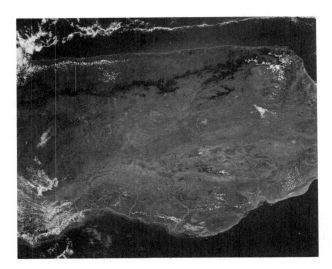

Figure 22.4 Close-up of the south-central part of Madagascar on the morning of 8 December 1993 at 0937 local time. Vegetation patterns are clearly visible (dark: mountain rain forest; light: secondary scrub, dry forest, and savanna)

Figure 22.5 Close-up (looking south) of the east coast of Madagascar during afternoon overpass of 9 December 1993 at 1700 (same overpass as in figure 22.7)

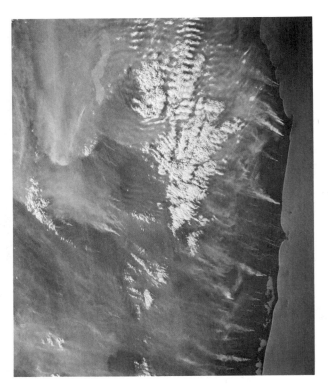

Figure 22.6 Close-up (at nadir) of the east coast of Madagascar during afternoon overpass of 9 December 1993 at 1700 local time (same overpass as in figure 22.7)

Figure 22.7 Looking south on Madagascar from spacecraft altitude 320 mm asl in the afternoon of 10 December 1993 at 1704 local time

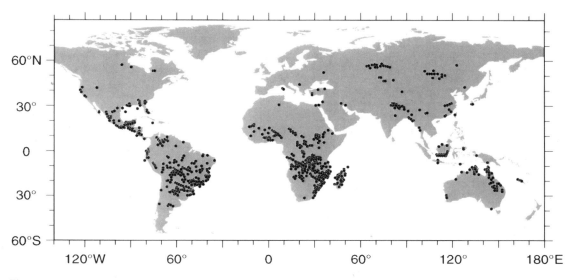

Figure 22.8 Global distribution of fires as seen from manned space flights. Dots are the center points of photographs showing smoke from vegetation fires. Source: Andreae 1993

emissions when crossing Madagascar and entering the continent.

The air masses may also be on recirculation mode, originating in the southern African subcontinent where they were already affected by pyrogenic products. These aged continental fire emissions are then refreshed with Madagascar's emissions and recirculated to the continent.

Documentation of Fire from Space

The use of the history of manned space missions to document vegetation fires has been described earlier (Helfert and Lulla 1990; Wood and Nelson 1991; Andreae 1993, figure 22.8). In this case study 163 orbits of the Space Transportation System 61 (STS-61) mission between 2 and 13 December 1993 were evaluated. STS-61 (Space Shuttle *Endeavor*) nominal altitude was 587 km, inclination 28.5°. The photographs were taken with a NASA-modified Hasselblad 500 EL/M, 70-MM (NASA 1994).

The observation period was at the end of the dry season in Madagascar during which the last cultivation fires had to be performed in order to take advantage of the upcoming rainy season (figure 22.2).

The photographs show a typical diurnal pattern of burning. In the mornings of the overpasses the whole island was completely smoke-free (figures 22.3 and 22.4). In the afternoon hours when fuels become drier and burning efficiency high enough to guarantee a sufficiently hot fire to clear the site, smoke palls

were abundant (figures 22.5–22.7). These photographs show the transport patterns of smoke, which is driven west towards the African continent (Mozambique, Tanzania).

Conclusions

Earlier observations of fires from space have proved useful in visualizing regional distribution patterns of fires and trajectories of smoke transport. The STS-61 observations for the first time were closely linked with ground observations in order to explain diurnal patterns of fire occurrence and pyrogenic emissions. Subsequent Space Shuttle missions in which large fire experiments were involved have failed so far. This is mainly due to lack of communication between remote experimental field sites and the NASA facility at Johnson Research Center which coordinates the Earth Observation Programme. Future experiments under the umbrella of IGAC/BIBEX should coordinate with Space Shuttle missions and with the permanently manned space station MIR.

References

Andreae, M. O. 1993. Global distribution of fires seen from space, *Eos Trans. AGU*, 74, 129–135.

Battistini, R., and P. Verin, 1972. Man and the environment in Madagascar, in *Biogeography and Ecology of Madagascar*, edited by R. Battistini and G. Richard-Vindard, pp. 311–336, Dr. W. Junk Publ., The Hague.

Burney, D. A. 1987. Late Quaternary stratigraphic charcoal records from Madagascar, *Quat. Res.*, 28, 274–280.

Garstang, M., P. D. Tyson, R. Swap, M. Edwards, P. Kållberg, and J. A. Lindesay. 1995. Horizontal and vertical transport of air over southern Africa. *J. Geophys. Res., Special SAFARI Issue* (in press).

Green, G. G., and R. W. Sussman. 1990. Deforestation history of the eastern rain forests of Madagascar from satellite images, *Science*, 248, 212–215.

Helfert, M., and K. P. Lulla. 1990. Mapping continental-scale biomass burning and smoke palls over the Amazon Basin as observed from the Space Shuttle, *Photogramm. Eng. Remote Sens.*, 56, 1367.

NASA. 1994. Catalog of Space Shuttle Earth observations hand-held photography. STS-61. NASA, JSC-25914-61. Houston, Texas, p. 160.

White, F. 1983. The vegetation of Africa. A descriptive memoir to accompany the UNESCO/AETFAT/UNSO vegetation map of Africa. UNESCO, *Natural Resources Research*, 20.

Wood, C. A., and R. Nelson. 1991. Astronaut observations of global biomass burning, in *Global Biomass Burning:Atmospheric, Climatic, and Biospheric Implications*, edited by J. S. Levine, pp. 29–40, MIT Press, Cambridge, Mass.

Zurbuchen, J. 1993. L'évolution du paysage entre Andevoranto et Analamazaotra au 19e siècle, *Terroirs et Ressources*, 1, Projet Terre-Tany, Antananarivo.

Chlorine and Bromine in the Biomass of Tropical and Temperate Ecosystems

Lisa M. McKenzie, Darold E. Ward, and Wei Min Hao

The photolysis of halogenated hydrocarbons in the stratosphere results in the release of ozone-destroying chlorine and bromine radicals (Molina and Rowland 1974; Wofsy 1975), with bromine being up to 100 times more destructive than chlorine (Solomon 1990). Two halogenated hydrocarbons, methyl chloride (CH_3Cl) and methyl bromide (CH_3Br), may be the largest reservoirs of atmospheric organic chlorine and bromine, respectively (Intergovernmental Panel on Climate Change 1990; Khalil 1993).

Biomass burning may be a major source of both CH_3Cl and CH_3Br. Several investigators have measured elevated levels of CH_3Cl in smoke plumes from biomass fires (Crutzen et al. 1979; Tassios and Packham 1985; Laursen, Hobbs, and Radke 1992; Andreae et al. 1995) and from controlled laboratory fires of biomass (Rasmussen et al. 1980; Reinhardt and Ward 1995; Lobert et al. 1991). Using these results and estimates for the total amount of biomass burned (Hao and Liu 1994; Crutzen and Andreae 1990), it has been estimated that biomass burning contributes up to 50% of total atmospheric CH_3Cl (Andreae et al. 1995). More recently, elevated levels of CH_3Br have been measured in smoke plumes from biomass fires (Manø and Andreae 1994; Andreae et al. 1995), and it has been estimated that up to 30% of atmospheric CH_3Br may result from biomass burning (Andreae et al. 1995). These estimates of total emissions of CH_3Cl and CH_3Br from biomass burning assume that emissions from one type of biomass are similar to emissions from another. This may not be a valid assumption.

Reinhardt and Ward (1995) have shown that the emission of CH_3Cl during a fire depends on the amount of chloride added to a fuel, fire intensity, and combustion efficiency. The emission ratio of CH_3Cl to carbon monoxide determined from smoke plumes over African savannas are about 10 times higher (Andreae et al. 1995) than those determined from smoke plumes over temperate forests in North America (Laursen, Hobbs, and Radke 1992). Methyl chloride emissions in laboratory fires vary over two orders of magnitude during smoldering combustion, depending on the fuel

being burned (Rasmussen et al. 1980). This evidence suggests that a relationship exists between chlorine and bromine concentrations in biomass and emissions of CH_3Cl and CH_3Br during burning. This relationship needs to be understood better to achieve improved estimates of CH_3Cl and CH_3Br emissions from biomass burning. An important step in gaining this understanding is determining the chlorine and bromine concentrations in various types of vegetation.

Bromine and chlorine are chiefly present in vegetation as chloride and bromide ions (Marschner 1986). Chlorine is an essential micronutrient in plants that is used in processes related to charge compensation, osmoregulation, and photosynthesis, and it accumulates in the chloroplasts (Marschner 1986). Bromine is not considered an essential nutrient but may replace chlorine when chlorine is deficient, albeit less effectively (Marschner 1986). Chlorine concentrations in vegetation vary from 15 to 27 000 mg kg^{-1} (Bowen 1979; Degroot 1989). Bromine concentrations may also vary but have rarely been measured. Some angiosperms contain approximately 15 mg kg^{-1} bromine (Bowen 1979). Chlorine and bromine concentrations may vary depending on chloroplast concentrations, as well as with proximity to large sources of chloride and bromide ions. For example, sea spray may deposit chloride and bromide ions on vegetation, and therefore coastal vegetation may have particularly high concentrations of chlorine and bromine. There may also be additional chlorine and bromine sources in duff, litter, and/or soil.

In this chapter, we present the results for chlorine and bromine from our analyses of several different types of fuels from temperate and tropical ecosystems. The influence of proximity to the Pacific Ocean on the chlorine and bromine concentrations in Douglas fir tissues has been examined. The aboveground biomass, percentage of fuel consumed, and chlorine and bromine concentrations from a Zambian savanna were used to estimate the contribution of each fuel type to the chlorine and bromine released to the atmosphere when the savanna was burned.

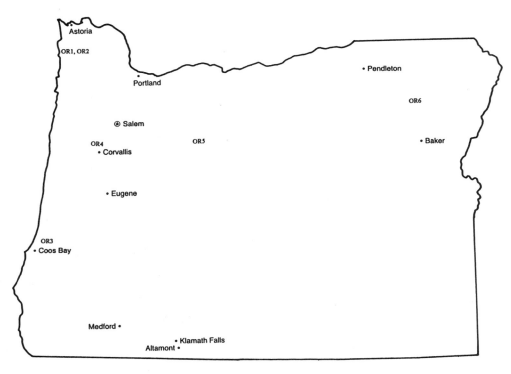

Figure 23.1 Sampling sites from temperate forests in Oregon, northwestern United States

Experimental Methods

Sample Collection

Samples of Douglas fir (*Pseudotsuga menziesii*) tissues were collected from six sites in Oregon (northwestern United States) shown on the map in figure 23.1 and listed in table 23.1. Only mature trees were sampled. Wood and bark were sampled at approximately 1.2 m from the base of the trees. Needles were collected at 2.1 to 6.1 m from the base of the trees, and the preceding year's growth was separated out for analysis. Forest floor beneath the trees was collected and separated into litter, duff, and humus layers when possible. The samples were ground in a Wiley mill to pass a 1-mm sieve and then stored in brown paper bags at room temperature. A soil sample was also collected at each site and stored in a clear glass vial. The sampling procedures for the Zambian and Brazilian samples have been described by Shea et al. (1995) and Kauffman, Cummings, and Ward (1994), respectively.

Chlorine and Bromine Extractions

Chlorine and bromine were extracted from the samples according to the method described by Gaines, Parker, and Gascho (1984), which is briefly described here.

Ground samples were oven dried at 70°C for two days. Two grams of a dried sample and 0.5 g of calcium oxide were weighed into a porcelain crucible. After the calcium oxide and sample were mixed thoroughly, 10 ml of Millipore water was stirred in to form a thin paste. This mixture was then ashed in a furnace at 550°C for three hours, resulting in all chlorine and bromine being in the inorganic forms of chloride and bromide. Fifteen milliliters of hot Millipore water was added to the resulting ash, which was then gently boiled for 30 minutes. The solution was filtered through #5 Whatman filter paper into a 25-ml volumetric flask and brought to volume with Millipore water. One method blank was prepared in the preceding manner, excluding the 2 g of sample, for each set of extractions.

Instrumental

The aqueous extracts were analyzed for chloride and bromide ions on a Dionex 2000i ion chromatograph equipped with a conductivity detector. Separation was achieved on an Ion Pac AS4A 4-by-250 mm column with an alkanol quaternary ammonium stationary phase and a carbonate-bicarbonate mobile phase.

Table 23.1 Codes for samples analyzed for chlorine and bromine

Sampling site	Wood	Foliage	Bark	Litter	Duff	Humus	Soil	Grass	Dicots	Ash
Oregon, moist temperate forest, 1.6 km from coast[a]	—	ORN1	ORB1	ORL1	ORD1	ORH1	ORS1	—	—	—
Oregon, moist temperate forest, 3.2 km from coast[a]	—	ORN2	ORB2	ORL2	ORD2	—	ORS2	—	—	—
Oregon, moist temperate forest, 16 km from coast[a]	ORW3	ORN3	ORB3	ORL3	ORD3	—	ORS3	—	—	—
Oregon, moist temperate forest, 60 km from coast[a]	ORW4	ORN4	ORB4	ORL4	—	—	ORS4	—	—	—
Oregon, dry temperate forest, 190 km from coast[a]	—	ORN5	ORB5	ORL5	ORD5	—	ORS5	—	—	—
Oregon, dry temperate forest, 440 km from coast[a]	ORW6	ORN6	ORB6	ORL6	ORD6	ORH6	ORS6	—	—	—
Zambia, Miombo (moist, wooded savanna)[b]	ZMW7	ZMF7	—	ZML7	—	—	—	ZMG7	ZMD7	ZMA7
Zambia, Dambo (moist savanna grassland)[b]	—	—	—	—	—	—	—	ZDG8	—	ZDA8
Brazil, tropical 1° forest[c]	BPW9	BPF9	—	BPL9	—	—	—	—	BPD9	BPA9
Brazil, tropical 2° forest[c]	BSW10	—	—	BSL10	—	—	—	—	BSD10	BSA10

a. All samples collected from Douglas fir.
b. Described in Shea et al. (1995).
c. Described in Kauffman, Cummings, and Ward (1994).

Quantitation

Chlorine and bromine concentrations were determined from four-point calibration curves using an external standard method. Standard solutions were prepared by dissolving potassium chloride and potassium bromide in Millipore water. All samples were extracted and analyzed in duplicate or triplicate. The method quantitation limits for chlorine and bromine were 0.38 mg kg^{-1} and 0.89 mg kg^{-1}, respectively.

Results and Discussion

Results from the chlorine and bromine analyses are presented in table 23.2 and figure 23.2. Both chlorine and bromine concentrations varied over two orders of magnitude. The highest chlorine concentration was in Zambian foliage, followed closely by Zambian dicots and grasses and Brazilian dicots. Temperate wood had the lowest chlorine concentration, which was an order of magnitude lower than tropical woods. The highest bromine concentrations were in Brazilian dicots from the secondary tropical forest and humus from the temperate forest nearest the coast. Temperate woods and bark and Zambian dambo grasses had the lowest bromine concentrations.

Molar ratios if Cl/Br are also presented in table 23.2. They varied over three orders of magnitude, with the highest ratio calculated from Zambian dambo grass and the lowest ratios from temperate humus collected nearest the coast. With the exception of the wood samples, Zambian savanna fuels had significantly higher molar Cl/Br ratios than Brazilian forest fuels and temperate forest fuels.

Chlorine and Bromine Profiles and Distance from Pacific Coast

A profile of chlorine concentration and distance from the Pacific Coast for the various components of Douglas fir is shown in figure 23.3. Chlorine concentrations in needles decreased sharply as coastal distance increased. Chlorine concentrations in bark, litter, and soil also decreased as coastal distance increased, but not as smoothly or as sharply. The wet deposition of chloride and the deposition of chloride normalized to annual rainfall, which were measured at similar sites in 1993 as part of the National Atmospheric Deposition Program (NADP 1993), also decreased with coastal distance (figure 23.4). Chlorine concentrations in wood did not vary with coastal distance. An analogous profile for bromine is shown in figure 23.5. Bromine concentrations decreased as coastal distance increased across all the tissue types. Bromine was not detected in Douglas fir wood.

The results clearly show that chlorine concentrates in needles, while bromine concentrates in soil. These separate fates for chlorine and bromine could be ex-

Table 23.2 Chlorine and bromine results

Sample[a]	Cl mg kg^{-1} (%RSD)	Br mg kg^{-1} (%RSD)	Molar Cl/Br
ORN1	940 (2.5)	13 (12)	160
ORN2	300 (9.8)	5.9 (13)	110
ORN3	230 (9.0)	4.7 (18)	110
ORN4	130 (8.1)	3.2 (13)	92
ORN5	120 (0.17)	1.9 (25)	140
ORN6	54 (0)	1.7 (26)	72
ORB1	51 (1.3)	3.0 (7.1)	38
ORB2	40 (16)	1.0 (29)	91
ORB3	47 (0)	<0.89[b]	>120
ORB4	52 (1.3)	<0.89	>130
ORB5	18 (8.9)	<0.89	>46
ORB6	16 (10)	<0.89	>40
ORL1	240 (0.04)	35 (1.6)	15
ORL2	200 (1.5)	20 (8.8)	23
ORL3	130 (3.4)	7.0 (0.55)	42
ORL4	130 (10)	6.7 (10)	44
ORL5	50 (5.9)	3.8 (0.08)	30
ORL6	67 (7.5)	3.3 (21)	46
ORD1	170 (1.6)	86 (0.77)	4.5
ORD2	230 (0.26)	38 (0.55)	13
ORD3	96 (5.9)	14 (3.8)	15
ORD5	36 (7.0)	2.6 (16)	31
ORD6	50 (17)	2.9 (6.4)	39
ORS1	140 (3.2)	100 (10)	3.5
ORS2	190 (7.2)	88 (0.5)	4.8
ORS3	41 (4.7)	17 (2.0)	5.4
ORS4	40 (8.9)	13 (23)	7.7
ORS5	<0.38[c]	3.1 (62)	<0.28
ORS6	18 (6.2)	1.4 (12)	30
ORH1	190 (1.5)	120 (0.87)	3.6
ORH6	15 (4.5)	14 (9.8)	24
ORW3	8.0 (5.8)	ND[d]	—
ORW4	10 (13)	ND	—
ORW6	9.0 (29)	ND	—
ZMW7	50 (11)	2.7 (18)	43
ZMF7	1600 (48)	11 (22)	320
ZML7	490 (47)	3.3 (31)	320
ZMD7	830 (9.3)	9.7 (8.0)	190
ZMG7	1000 (11)	4.7 (8.9)	490
ZMA7	760 (36)	7.5 (4.1)	230
ZDG8	710 (30)	<0.89	>1800
ZDA8	740 (1.5)	1.9 (0.51)	860
BPW9	90 (27)	2.7 (47)	93
BPF9	230 (32)	23 (41)	19
BPL9	100 (7.8)	12 (2.7)	20
BPD9	470 (45)	19 (25)	59
BPA9	520 (120)	42 (62)	48
BSW10	120 (40)	4.7 (74)	97
BSL10	66 (40)	20 (47)	11
BSD10	1200 (30)	130 (13)	20
BSA10	730 (18)	31 (20)	52

a. Sample codes described in table 23.1.
b. Detected but below method quantitation limit of 0.89 mg kg^{-1}.
c. Detected but below method quantitation limit of 0.38 mg kg^{-1}.
d. ND = not detected.

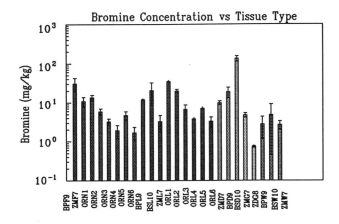

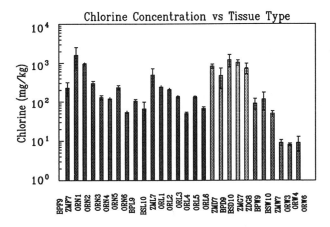

Figure 23.2 Bromine and chlorine concentrations versus tissue type. Codes on x-axis correspond to codes in table 23.1. The different shadings distinguish tissue types, for example, grasses have the lightest shadings.

plained as follows. Both chloride and bromide ions are deposited on trees and forest floor. Chloride ions in soil associate with the aqueous soil solution and are readily available for transport to the roots (Bowen 1979). Chloride ion, which is an essential micronutrient for plants, is selectively taken up both at the roots from the soil solution and at the stomata and accumulates in chloroplasts in needles (Marschner 1986; Peel 1974). Plants will take up as much chloride ion as is available, even to the point of toxicity (Lacasse and Rich 1964; Hofstra and Hall 1971). Bromide ions in soil are not as readily available for transport as are chloride ions (Bowen 1979). Relatively little bromide ion, which is not an essential nutrient, is taken up by the tree, and it therefore accumulates in the soil. Chlorine and bromine in bark are probably primarily due to direct deposition of chloride and bromide ions. Chlorine and bromine in wood are probably a result of

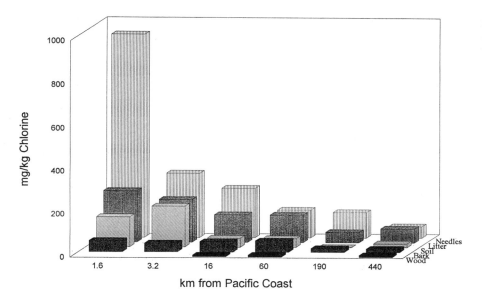

Figure 23.3 Chlorine concentrations in Douglas fir versus kilometers from the Pacific Coast

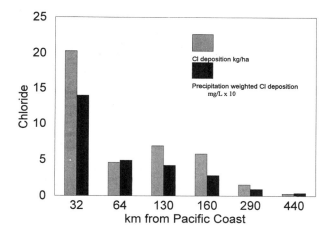

Figure 23.4 1993 annual mean chloride deposition concentrations (NADP 1993)

transport of chloride and bromide ions through the xylem (Peel 1974).

Theoretical Chlorine and Bromine from Zambian Fires
It is important to estimate chlorine and bromine emissions from African savanna fires, since most biomass burning occurs in the tropics, with African savanna burning accounting for 30% of the global total (Hao and Liu 1994). The aboveground biomass by fuel type and combustion factors (percent fuel burned) for each fuel type from Shea et al. (1995), along with the chlorine and bromine concentrations in table 23.2, were used to estimate how much chlorine and bromine were released to the atmosphere during fires of a Zambian miombo (wooded savannah) and a Zambian dambo

(savanna grassland) in 1992 (Shea et al. 1995). Figure 23.6 shows the chlorine and bromine released to the atmosphere during the fires from each type of fuel. The litter, the greatest component of the aboveground biomass, contributed 51% of the chlorine and 56% of the bromine released to the atmosphere during the miombo fire. Grasses contributed another 42% of the chlorine and 32% of the bromine emitted. The remainder of the chlorine and bromine was released from dicots, foliage, and wood. Even though the highest chlorine concentrations were in the foliage, its contribution to the chlorine released during the fire was minor due to its low contribution to the aboveground biomass. The dambo contained only grasses (Shea et al. 1995). Therefore, grasses were assumed to contribute 100% of the chlorine and bromine released.

Overall, about $2\,100\,000 \pm 730\,000$ mg of chlorine and 9500 ± 1800 mg of bromine were estimated to be released to the atmosphere for each hectare of miombo burned. About $1\,700\,000 \pm 530\,000$ mg ha^{-1} of chlorine and 1000 ± 94 mg ha^{-1} of bromine were estimated to be emitted from the dambo fire. The chlorine emissions are similar, with the ranges overlapping. The estimated amount of bromine released from the dambo fire is 10 times lower than that for the miombo fire. This is not surprising, since dambo grasses had significantly lower concentrations of bromine than miombo grasses and litter. Our estimated emissions are two times lower for chlorine and bromine than the lowest estimates from savanna fires in Kruger National Park, South Africa (Andreae et al. 1995). This is due to the lower chlorine and bromine concentrations

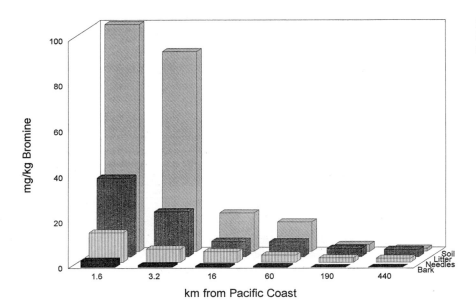

Figure 23.5 Bromine concentrations in Douglas fir versus kilometers from the Pacific Coast

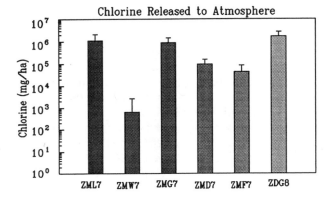

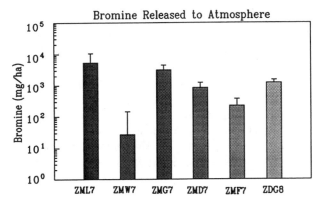

Figure 23.6 Chlorine and bromine releases to the atmosphere during fires in a Zambian miombo and dambo during 1992 (Shea et al. 1995)

in the aboveground biomass, as presented in table 23.3, in the Zambian fires.

We estimate that approximately 76% of fuel chlorine and 56% of fuel bromine were emitted to the atmosphere in the Zambian savanna fires. The estimates by Andreae et al. (1995) of 83% for chlorine and 81% for bromine fall within the high end of our ranges, which are presented in table 23.3. Estimated emissions of CH_3Cl and CH_3Br, based on 3% of the chlorine emitted in the form of CH_3Cl and 5% of the bromine emitted in the form of CH_3Br (Andreae et al. 1995), are also given in table 23.3.

Conclusions

Both chlorine and bromine concentrations varied over two orders of magnitude among fuels. The highest concentrations of chlorine were in foliage, grasses, and dicots. The highest concentrations of bromine were in tropical forest dicots and coastal temperate forest floor layers. Chlorine and bromine concentrations in Douglas fir tissues, except for wood, were dependent on the proximity of trees to the Pacific Ocean, a large source of chloride and bromide ions. Molar ratios of chlorine to bromine varied over three orders of magnitude. The Zambian fuels had the highest ratios.

The amount of chlorine and bromine released to the atmosphere during a fire depended on chlorine and bromine concentrations in the fuel burned, the contribution of the fuel to the aboveground biomass, and the percentage of fuel burned. These factors would also be

Table 23.3 Total chlorine and bromine for Zambian miombo and dambo

Site	Chlorine ± 1 SD	Bromine ± 1 SD
Zambian Miombo		
Aboveground biomass (mg ha^{-1})	3 300 000 ± 920 000	22 000 ± 4300
Fuel consumed (mg ha^{-1})	2 900 000 ± 1 000 000	17 000 ± 3300
Ash (mg ha^{-1})	760 000 ± 280 000	7500 ± 740
Percent released to atmosphere[a]	73 ± 43	56 ± 23
CH$_3$Cl[b] or CH$_3$Br[c] released (mg ha^{-1})	89 000 ± 39 000	560 ± 110
Zambian dambo		
Aboveground biomass (mg ha^{-1})	2 200 000 ± 670 000	2200 ± 170
Fuel consumed (mg ha^{-1})	2 200 000 ± 670 000	2200 ± 170
Ash (mg ha^{-1})	460 000 ± 65 000	1200 ± 230
Percent released to atmosphere[a]	79 ± 39	55 ± 14
CH$_3$Cl[b] or CH$_3$Br[c] released (mg ha^{-1})	72 000 ± 22 000	59 ± 6

a. Percent released to atmosphere = [(fuel consumed) − (ash)]/[fuel consumed].
b. Based on 3% of released chlorine in the form of CH$_3$Cl (Andreae et al. 1995).
c. Based on 5% of bromine released in the form of CH$_3$Br (Andreae et al. 1995).

expected to influence CH$_3$Cl and CH$_3$Br emissions. We are currently investigating the relationship between chlorine and bromine concentrations in fuels and CH$_3$Cl and CH$_3$Br emissions from laboratory fires.

Acknowledgments

We thank Dr. Geoff Richards for helpful discussions; Dr. Johnnie Moore and Ms. Lynn Biegelsen for their assistance with the I.C. analyses; Mr. Kirk Hash for assistance with sample preparation; and Dr. Rick Kelsey, Dr. Larry Byrant, Mr. Don Messerle, and Mr. Dave Messerle for their assistance in collecting the Douglas fir samples. Financial support from NASA's Global Change Research Program (4054-GC93-0134) is gratefully acknowledged.

References

Andreae, M. O., E. Atlas, G. W. Harris, G. Helas, A. Kock, R. Koppmann, W. Maenhaut, S. Manø, W. H. Pollock, J. Rudolph, D. Scharffe, G. Schebeske, and M. Welling. 1995. Methyl halide emissions from savanna fires in southern Africa, submitted to *J. Geophys. Res.*

Bowen, H. J. M. 1979. *Trace Elements in Biochemistry*, Academic Press, New York, pp. 49–121.

Crutzen, P. J., and M. O. Andreae. 1990. Biomass burning in the tropics: Impact on atmospheric chemistry and biogeochemical cycles, *Science*, 250, 1669–1678.

Crutzen, R. J., L. E. Heidt, J. P. Krasnec, W. H. Pollock, and W. Seiler. 1979. Biomass burning as a source of atmospheric gases, CO, H$_2$, N$_2$O, NO, CH$_3$Cl and COS, *Nature*, 282, 253–256.

Degroot, W. F. 1989. Methyl chloride as a gaseous tracer for wood burning? *Environ. Sci. Technol.*, 23, 252.

Gaines, T. P., M. B. Parker, and G. J. Gascho. 1984. Automated determination of chlorides in soil and plant tissues by sodium nitrate extraction, *Agronomy J.* 76, 371–374.

Hao, W. M., and M. H. Liu. 1994. Spatial and temporal distribution of tropical biomass burning, *Global Biogeochemical Cycles*, 8, 495–503.

Hofstra, G., and R. Hall. 1971. Injury on roadside trees: Leaf injury on pine and white cedar in relation to foliar levels of sodium and chloride, *Can. J. Bot.*, 49, 613–622.

Intergovernmental Panel on Climate Change, *Climate Change: The IPCC Scientific Assessment*, Cambridge University Press, London, 23–25, 1990.

Kauffman, J. B., D. L. Cummings, and D. E. Ward. 1994. Relationships of fire, biomass, and nutrient dynamics along a vegetation gradient in the Brazilian cerrado, *J Ecology*, 82, 519–531.

Khalil, M. A. K., R. A. Rasmussen, and R. Gunawardena 1993. Atmospheric methyl bromide: Trends and global mass balance, *J. Geophys. Res.*, 98, 2887–2896.

Lacasse, N. L., and A. E. Rich. 1964. Maple decline in New Hampshire, *Phytopathology*, 54, 1071–1075.

Laursen, K. K., P. V. Hobbs, and L. F. Radke. 1992. Some trace gas emissions from North American biomass fires with an assessment of regional and global fluxes from biomass burning, *J. Geophys. Res.*, 97, 20 687–20 701.

Lobert, J. M., D. H. Scharffe, W. M. Hao, T. A. Kuhlbusch, R. Seuwen, P. Warneck, and P. J. Crutzen. 1991. Experimental evaluation of biomass burning emissions: Nitrogen and carbon containing compounds, in *Global Biomass Burning: Atmospheric, Climatic and Biospheric Implications*, edited by J. S. Levine, pp. 289–304, MIT Press, Cambridge, Mass.

Manø, S., and M. O. Andreae. 1994. Emission of methyl bromide from biomass burning, *Science*, 263, 1255–1257.

Marschner, H. 1986. *Mineral Nutrition of Higher Plants*, Academic Press, New York, pp. 336–340.

Molina, M. J., and F. S. Rowland. 1974. Stratospheric sink for chlorofluoromethanes: Chlorine atomc-atalysed destruction of ozone, *Nature*, 249, 810–812.

Peel, A. J. 1974. *Transport of Nutrients in Plants*, John Wiley, New York, pp. 5–27, 111.

Rassmussen, R. A., L. E. Rassmussen, M. A. K. Khalil, and R. W. Dalluge. 1980. Concentration distribution of methyl chloride in the atmosphere, *J. Geophys. Res*, 85, 7350–7356.

Reinhardt, T. E., and D. E. Ward. 1995. Factors affecting methyl chloride emissions from forest biomass combustion, *Environ. Sci. Technol.*, 29, 825–832.

Shea, R. W., B. W. Shea, J. B. Kauffman, D. E. Ward, C. I. Haskins, and M. C. Scholes. 1995. Fuel biomass and combustion factors associated with fires in savanna ecosystems of South Africa and Zambia, *J. Geophys. Res.*, in press.

Solomon, S. 1990. Antarctic ozone: Progress towards a quantitative understanding, *Nature*, 347, 347–354.

Tassios, S., and D. R. Packham. 1985. The release of methyl chloride from biomass burning in Australia, *J. Air Pollut. Contr. Assoc.*, 35, 41–42.

U.S. Geological Survey, National Atmospheric Deposition Program, *NADP/NTN Annual Data Summary: Precipitation Chemistry in the United States 1993*.

Wofsy, S. C., M. B. McElroy, and Y. L. Yung. 1975. The chemistry of atmospheric bromine, *Geophys. Res. Let.*, 2, 215–218.

Carbon, Hydrogen, Nitrogen, and Thermogravimetric Analysis of Tropical Ecosystem Biomass

Ronald A. Susott, Gerald J. Olbu, Stephen P. Baker, Darold E. Ward,
J. Boone Kauffman, and Ronald W. Shea

Biomass fires in the tropics are a significant source of global smoke emissions and greenhouse gases. Levine (1991) provides a review of progress and current research interests in global biomass burning. In Brazil, the majority of the cerrado regions are burned every few years (Coutinho 1990). The burning of second-growth and primary tropical forest for land clearing in the Amazon also makes a sizeable contribution to global biomass burned (Fearnside 1990). African savanna burning also makes a considerable contribution to biomass burned in the tropics (Hao and Liu 1994). Little information is available on the elemental content of the complex mixture of fuels involved in biomass burning. This chapter provides a more detailed investigation of elemental carbon, hydrogen, and nitrogen (CHN) composition of typical fuel complexes burned in Africa and Brazil. The fuels analyzed in this study were from distinct ecosystems burned as part of BASE-B (Biomass Burning Airborne and Spaceborne Experiment—Brazil, 1990 to 1992) and SAFARI-92 (Southern African Fire-Atmosphere Research Initiative—1992) research programs to characterize smoke emissions. The CHN results are combined with reported area fuel loads and combustion factors (percent of fuel class burned) to calculate an overall CHN composition typical of the ecosystem. Ash content and pyrolytic char formation were also investigated by thermogravimetric analysis and related to elemental composition.

There has been considerable recent research to estimate total global area burned, the amount of biomass consumed, and the release of carbon as trace gases to the atmosphere (e.g., Hao et al. 1990; Hao and Ward 1993; Crutzen and Andreae 1990). Approximate or assumed constants are often used to calculate the carbon content of fuels to estimate carbon release to the atmosphere. Values of 45% (Crutzen and Andreae 1990; Seiler and Crutzen 1980) and 50% carbon content (Ward et al. 1991; Ward et al. 1992) are used most often, although actual measured values are also used (Lobert et al. 1991). The 50% carbon value is close to that calculated from Byram's (1959) approximate ratio

of atoms in wood, $C_6H_9O_4$. This agrees with the 50% carbon content of wood cited by Pettersen (1984) for an overall average. The 45% carbon value compares to the average of 46.3% used by Ewel et al. (1981) for ash-free plant tissue and is close to the value used by Lobert et al. (1991) of 45.5% carbon for an average mixture of selected bimoass fuels. These standard or consensus values are useful for comparing global estimates where other variables are more uncertain, but more accurate values are needed for calculations on individual well-defined ecosystems or fuel classifications.

Elemental CHN values are used to calculate emission factors for trace gases emitted from biomass burns. When biomass loads are known or can be estimated, the carbon content is needed to calculate total carbon present in the biomass or that released by burning (Olson, Watts, and Allison 1983). Alternatively, when fuel carbon content and the concentrations of major carbon-containing gases and particles are measured, carbon mass balance procedures are used to estimate the amount of fuel burned for emission factor calculations (Ward et al. 1991; Nelson 1982). Similar nitrogen mass balances are complicated by the production of elemental nitrogen (Lobert et al. 1991), which is not normally measured against the high nitrogen concentration of the atmosphere. The emission of oxides of nitrogen during pyrolysis has been found to be directly related to the fuel nitrogen content (Clements and McMahon 1980).

Oxygen stoichiometry during combustion can be calculated directly from the CHN data. This stoichiometry can be used to estimate useful fire parameters such as the fuel heat of combustion, the amount of air required for complete combustion, and flame temperatures (Susott, Shafizadeh, and Aanerud 1979). Equations are also available to estimate heats of combustion from elemental composition (Gnaiger and Bitterlich 1984; Patel and Erickson 1981).

Emission factors for particulate matter are inversely related to oxygen content of biomass (Ward 1990; Ward 1980). Biomass is approximately composed of carbon, hydrogen, nitrogen, oxygen, and mineral ash components. Ash content is readily measured by gravimetric techniques. If ash content is added to CHN contents, oxygen content can be calculated by difference (Susott et al. 1991). The oxygen estimate is fairly accurate, since other trace elements in the fuel are normally present in small concentrations (Pettersen 1984), and most are accounted for in the ash content.

Experimental Methods

Sampling

The ground-based studies in Brazil focused on the main ecosystems burned: cerrado, slashed tropical second-growth forest, slashed primary tropical forest, and pasture for livestock. Details of the areas burned and biomass loads are given by Kauffman, Cummings, and Ward (1994) and J. Kauffman et al. (manuscript in preparation, 1995). The cerrado vegetation is actually a continuum of plant communities from open grassland campo limpo to forest or cerradão, as described by Coutinho (1990). Campo sujo, campo cerrado, and cerrado senso stricto have increasingly more and larger shrubs and trees compared to the campo limpo grasslands. No cerrado areas were burned in this study, since fires are less common in these areas. This chapter reports on fuels from six cerrado burns near Brazilia, three slashed second- or third-growth forest burns near Jamari in Rondônia or near Maraba in Pará, four slashed primary forest burns near Jamari or Maraba, and three livestock pasture burns near Jamari.

In Africa, our burn study included semiarid and moist savanna ecosystems with a variety of grass and woody fuels. This chapter reports on nine semiarid savanna burns at Kruger National Park in South Africa, one semiarid treed savanna near Choma, Zambia, and two moist savanna woodlands and one moist dambo grassland in Kasanka National Park in Zambia. Details on the burn sites and sampling methods are described by Shea et al. (1995).

Fuels were partitioned into classes of grasses, litter, dicot seedlings, shrub leaves and stems (from plants < 2 m high), and woody debris. Woody debris was divided into size classes separated by diameter using the division of Deeming, Burgan, and Cohen (1977). Woody fuels respond to changes in humidity, temperature, or precipitation with a characteristic time lag. Woody diameters up to 0.64 cm will come to about two-thirds of their equilibrium moisture content in under 1 hour. Larger-diameter fuels take longer times

to respond. Samples for 1-hour (≤ 0.64 cm), 10-hour (0.64 to 2.54 cm), 100-hour (2.54 to 7.6 cm), and 1000-hour (7.6 to 20.3 cm) time lag woody debris were measured in this study. Several other fuel classifications were added, such as red slash foliage in Brazil's slashed primary forest and dung in Kruger National Park, where sizeable area loadings suggested they could be an important component of the fuels burned.

Each fuel class was sampled by combining 10 to 20 random grab samples representative of that class collected within the area to be burned. Five such combined samples were taken for each aboveground fuel class that made a sizeable contribution to the total fuel load burned. All samples were air dried and stored in plastic bags for transport to the laboratory. Samples were prepared for analysis by grinding in a Wiley mill to pass a 40-mesh screen (0.5 mm).

Elemental Analysis for Carbon, Hydrogen, and Nitrogen

A Perkin-Elmer 2400 elemental analyzer was used for CHN analysis. This analyzer burns the sample in an oxygen and helium mixture at 925°C and measures the CO_2, H_2O, and N_2 combustion products by gas chromatography. Approximately 2 mg of sample, weighed to 1 μg, was analyzed for each run, following manufacturer-suggested operating procedures. At least two replicate runs were averaged for each sample. Moisture contents were used to convert all elemental compositions to a dry-weight basis. Standard runs on acetanilide were used to calibrate the instrument. Measurements from standard runs interspersed with the fuel sample runs are summarized in table 24.1. Moisture contents for the 1990 Brazil samples were measured by thermogravimetric (TG) analysis. Moisture for all other samples was measured by oven drying for 24 hours at 100°C. The two methods for measuring moisture agreed to within 0.2% moisture. CHN values for three to five of the separate samples for a fuel class at a burn site were averaged.

Thermogravimetric Analysis

A Perkin-Elmer TGS-2 was used for TG analysis. In a typical experiment, a 5- to 6-mg sample was heated in nitrogen to 500°C at 20°C per minute. Moisture content was calculated on a dry-weight basis from the weight lost at 140°C. This moisture content was used to convert the elemental analysis data for 1990 Brazil samples to a dry-weight basis. Rapid heating for TG analysis requires 140°C to remove an amount of water comparable to 24-hour oven drying. Pyrolysis char was calculated from the weight remaining at 500°C.

Table 24.1 Carbon, hydrogen, and nitrogen standards analysis accuracy

	C% (SD)	H% (SD)	N% (SD)
Acetanilide theory	71.09	6.71	10.36
Africa 1992	71.10 (0.21)	6.78 (0.07)	10.35 (0.21)
Brazil 1990	70.99 (0.17)	6.62 (0.12)	10.33 (0.10)
Brazil 1991	71.11 (0.25)	6.82 (0.08)	10.35 (0.15)
Brazil 1992	71.14 (0.16)	6.80 (0.09)	10.36 (0.11)
NIST 1572 citrus leaves known	—	—	2.86
Measured	46.29 (0.69)	5.53 (0.26)	2.97 (0.07)
NIST 1575 needles known	—	—	1.20
Measured	53.81 (0.69)	6.91 (0.23)	1.22 (0.07)

After cooling from 500°C to 300°C, the sample was ashed by heating to 650°C in air. Ash content was calculated on a dry-weight basis from the weight remaining after ashing to 650°C. This ash content was used to calculate selected elemental composition data on an ash-free basis. Two or more replicate TG curves were averaged to form the final curve.

Results and Discussion

Table 24.1 summarizes CHN measurements made on an acetanilide standard during the analyses of biomass material and compares them to theoretical values for the standard. Measured carbon content agreed with theory to within 0.3% carbon. Hydrogen content agreed to within 0.2% hydrogen, and nitrogen content agreed to within 0.2% nitrogen. Table 24.1 also summarizes CHN results for National Institute of Standards and Technology (NIST) standard samples of citrus leaves and pine needles, with reported nitrogen contents closer to those of the biomass materials analyzed in this study. Measured nitrogen contents for the NIST samples were within 0.2% nitrogen of the reported values. Carbon and hydrogen contents of the NIST standards were not reported, but they are constant. Standard deviations for carbon and hydrogen contents for these NIST standards were two to three times those measured for the acetanilide standard. Variable moisture content could explain much of this added uncertainty, since only one moisture content was measured for the NIST standards (the acetanilide sample had negligible moisture content, while 8.0% moisture was found for the NIST standards).

Figures 24.1, 24.2, and 24.3 summarize the biomass carbon, hydrogen, and nitrogen contents (on a dry-weight basis), respectively. These figures include biomass from 13 fires in Africa (samples within a fuel class were combined for the first four Kruger fires) and 16 fires in Brazil. Fuel class data were combined for sev-

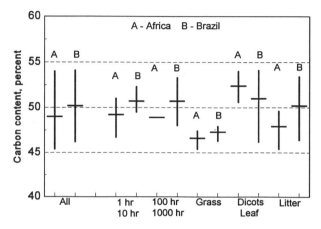

Figure 24.1 Carbon content by fuel class for biomass burned in Africa and Brazil

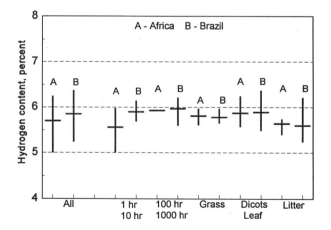

Figure 24.2 Hydrogen content by fuel class for biomass burned in Africa and Brazil

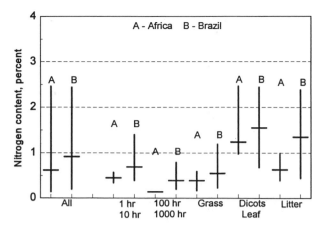

Figure 24.3 Nitrogen content by fuel class for biomass burned in Africa and Brazil

eral classes to simplify the presentation. The vertical bars in these figures represent the range of values for a fuel class, and the horizontal bars represent the average value. Our primary interest is in the carbon content to use in the carbon mass balance method for calculating emission factors of combustion products. The hydrogen and nitrogen data have been included to complete the data set for reference purposes. A more detailed set of CHN data is included in the appendix (tables 24.A1, 24.A2, 24.A3, and 24.A4).

All carbon contents for all fuel classes in figure 24.1 are within a range of 45% to 55% carbon. The Africa set ranges from 45.4% for grass and litter samples from the Kruger Block 56 fire, to 54.0% for a leaf sample from the Kruger Block 55 fire. The Brazil set carbon content ranges from 46.2% for a dicot sample from a second-growth forest burn in 1990, to 54.0% for a red slash (leaf) sample from a third-growth forest burn in 1992. The average for all Africa samples was 49.0% carbon compared to an average of 50.2% carbon for all Brazil samples. Grass samples had the lowest average carbon content, at 46.6% for Africa and 47.3% for Brazil. All classes of woody debris averaged close to 49% carbon for Africa samples and 51% for Brazil samples. Dicots and leaves made relatively minor contributions to the biomass burned at most sites but had somewhat higher average carbon content of 52.4% for Africa samples and 51.0% for Brazil samples. The carbon content of litter samples from Brazil averaged 50.2% with a wide range, from 46.4% for a 1992 primary forest sample to 53.4% for a cerrado senso stricto sample. There was less variation in the Africa litter samples, with an average of 47.9% carbon. The Africa samples averaged about 1% lower carbon content than

the Brazil set for most fuel classes, but there was a considerable range for most fuel classes.

Hydrogen content, in figure 24.2, averaged 5.7% hydrogen for all Africa samples and 5.8% for all Brazil samples. Hydrogen content ranged from 5.0% for 10-hour woody debris from a Zambia Miombo woodland site, to 6.4% for red slash leaves from a Brazil primary forest site. The litter fuel class has low average hydrogen along with the Africa 1-hour/10-hour combined woody debris, while the other woody classes had the highest average hydrogen content. Differences between fuel classes are small and may be influenced by uncorrected variations in the sample moisture content.

Nitrogen content was more variable, as shown in figure 24.3. Dicots, leaves, and litter classes generally had the highest nitrogen content (up to 2.46%), but a large variation in values was found. All Africa samples had an average 0.62% nitrogen content, and all Brazil samples averaged 0.92% nitrogen. These values are in general agreement with values reported by Dignon and Penner (1991).

Carbon composition for biomass reported here can be compared to estimates based on expected chemical composition. A biomass fuel can be divided into carbohydrate polymers, lignin polymers, and other extraneous materials that can be extracted with solvents. Susott, DeGroot, and Shafizadeh (1975) reported on this breakdown for a number of biomass materials. The lowest carbon content is expected for cellulose, with the formula $C_6H_{10}O_5$, or 44.4% carbon by mass. Simple monosaccharides such as glucose have carbon contents of 40%, but these constituents are only present in low percentages. Carbon content lower than that for glucose can only occur in more oxidized components, such as simple organic acids, but these acids make a small contribution to overall carbon content. The lignin fraction can have variable composition for biomass but generally has relatively high carbon content compared to other fractions. Sarkanen and Hergert (1971) report lignins from hardwoods at about 60% carbon. Extractives include hydrocarbons such as pinene, for example, with 88% carbon, as well as simple monosaccharides such as glucose with 40% carbon. Shafizadeh, Chin, and DeGroot (1977) reported carbon contents ranging from 46.2% to 76.8% for extracted mixtures from several foliage samples. The measured carbon content of biomass decreases as the inorganic ash fraction increases. Inorganic ash content in plants is also quite variable. Comparisons made on an ash-free basis can eliminate this source of variation in carbon contents.

Table 24.2 Elemental composition for fires in tropical ecosystem fuels based on combined analysis of all fuel classes weighted by amount burned (values on a dry-weight basis)

Fire ecosystem type and location	%C	%H	%N
Brazil 1990			
Cerrado campo limpo near Brasília	47.5	5.69	0.33
Cerrado campo sujo near Brasília	47.4	5.70	0.31
Cerrado campo cerrado near Brasília	49.5	5.80	0.55
Cerrado senso stricto near Brasília	50.5	5.62	0.60
Slashed second-growth forest near Jacunda, Pará	49.5	5.79	0.85
Slashed primary tropical moist forest near Jacunda, Pará	50.5	5.88	0.58
Brazil 1991			
Cerrado senso stricto (dense) near Brasília	51.8	5.82	0.83
Cerrado senso stricto (open) near Brasília	52.4	5.80	0.87
Pasture (12 years since primary deforestation) Jamari, Rondônia	48.1	5.76	0.56
Slashed third-growth forest near Maraba, Pará	49.5	5.76	0.80
Slashed primary tropical moist forest near Maraba, Pará	50.2	5.97	0.58
Brazil 1992			
Pasture (4 years since primary deforestation) Jamari, Rondônia	49.4	5.74	0.77
Pasture (10 years since primary deforestation) Jamari, Rondônia	49.1	5.83	0.40
Slashed third-growth forest near Jamari, Rondônia	50.7	5.90	0.70
Slashed primary tropical moist forest near Jamari, Rondônia	50.8	6.08	0.65
Slashed primary tropical moist forest near Jamari, Rondônia	51.0	6.05	0.70
South Africa 1992			
Savanna Kambeni 5 (late-winter biennial burns) Kruger National Park	48.0	5.78	0.55
Savanna Faai 1 (late-winter biennial burns) in Kruger National Park	48.0	5.77	0.59
Savanna Numbi 4 (late-winter biennial burns) in Kruger National Park	47.9	5.79	0.57
Savanna Shabeni 1 (late-winter biennial burns) in Kruger National Park	47.9	5.77	0.62
Savanna Kambeni 3 (early-spring biennial burns) in Kruger National Park	47.4	5.76	0.53
Savanna Shabeni 5 (early-spring biennial burns) in Kruger National Park	47.7	5.83	0.54
Savanna Block 55 (last burned in 1988) in Kruger National Park	47.2	5.80	0.55
Savanna Block 56 (last burned in 1990) in Kruger National Park	45.8	5.71	0.52
Savanna Kambeni E (last burned in 1954) in Kruger National Park	48.1	5.77	0.40
Zambia, Africa 1992			
Dambo grassland moist savanna near Kasanka National Park	46.5	5.73	0.18
Fallow Chitement moist savanna woodland near Kasanka National Park	47.9	5.58	0.43
Moist savanna Miombo woodland near Kasanka National Park	47.0	5.68	0.78
Semiarid savanna Miombo woodland near Choma	49.0	5.68	0.54

On an ash-free basis, the minimum carbon contents for biomass are expected to be higher than the 44% for carbohydrate polymers. A value of 50% carbon has been used to approximate wood materials (Byram 1959; Pettersen 1984). Other biomass materials have more extractives, lignin-like components, and hydrocarbons that should increase the carbon content compared to wood. A sample of rotten ponderosa pine wood, selected for a high lignin content, had a carbon content of 56% (Susott, DeGroot, and Shafizadeh 1975). Based on these considerations, most biomass materials are expected to fall in the range of 45% to 55%, on an ash-free basis. This range is consistent with the range for fuels in temperate forests (Vogt 1991). Isolated fuel components such as bark or small twigs may have higher carbon contents (Susott et al. 1991), but the composite grab sample collections used in the present study tend toward an average carbon content that should fall in the expected range. All the samples

in figure 24.1 fall into the 45% to 55% carbon range. Converting to an ash-free basis raises the average carbon contents (by up to 3% carbon based on table 24.A5).

The elemental analysis data (available in appendix tables 24.A1 to 24.A4) can be combined with area loads and combustion factors reported by Ward et al. (1992), J. Kauffman et al. (1994), and Shea et al. (1996) to calculate the elemental compositions of the fuel mixture that burned. Combustion factor is defined as the percent of the area fuel load that is burned. Elemental compositions for 29 sites burned from 1990 to 1992, weighted by the fraction of each fuel class burned, are given in table 24.2. The weighted carbon content varied from 45.8% carbon for a dry savanna in Kruger National Park to 52.4% for a cerrado senso stricto site near Brasília. Grass and litter made the largest contribution to the carbon contents in savanna areas, while woody debris greater than 2.54 cm diam-

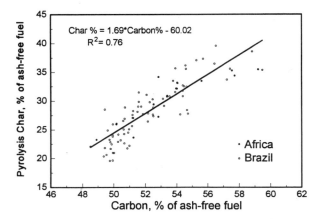

Figure 24.4 Pyrolysis char production as a function of carbon content for a variety of fuels burned in Africa and Brazil

eter and litter dominated the fuel burned in tropical forest and pasture sites in Brazil. The importance of litter and the variability of litter carbon content seen in figure 24.1 lead to much of the variability in carbon content seen in table 24.1.

Weighted carbon contents were generally in good agreement for sites in the same ecosystem type (where replicate burns were available). Slashed primary forests in Brazil had carbon contents that varied from 50.2% to 51.0%, pastures varied from 48.1% to 49.4% carbon, and eight savanna sites in South Africa varied from 47.2% to 48.1% carbon. A low carbon content was calculated for one savanna site (Block 56) due to a low carbon content of 45.4% measured for both grass and litter. Samples of the same fuel class were combined for the first four South Africa savanna fires in table 24.2. With the same carbon content for a fuel class, the weighted carbon contents only depend on the amount of each class burned in the four fires. A difference of only 0.1% carbon for these four sites indicates that variations due to fuel load and combustion factor were minor for these burns. The cerrado senso stricto areas burned near Brasília provide a contrasting case. One fire in 1990 and two fires in 1991 had weighted carbon contents of 50.5%, 51.8%, and 52.4%. The lowest value (50.5%) is mainly due to more grass (with a low carbon content) on the site burned in 1990. Even though the sites were similar and had the same fuel classes, the biomass load and combustion factor caused some difference in the weighted carbon content of the material burned.

An interesting correlation exists between the carbon content of biomass and the amount of char remaining after pyrolysis. Figure 24.4 contains a plot of char yield, measured by thermogravimetric analysis, versus carbon content. Both variables are on an ash-free basis to minimize effects of high ash contents for some of the samples. Ash contents of several litter samples were over 12%, one litter sample had 29.4% ash, and a dung sample had 23.4% ash (see appendix table 24.A4). Figure 24.4 shows that increased carbon content leads to increased char. The linear least squares fit had an R^2 of .76 and a slope of 1.69. Since the char is formed at the expense of volatile products that burn in the flames, higher carbon content may lead to lower flammability. If the char combustion is not complete, postfire residues would increase for fuels with high carbon content. Higher residue remaining at a site could compensate for higher fuel carbon content in determining the loss of carbon from a burn. The fuels analyzed by Susott, DeGroot, and Shafizadeh (1975) show the same trend over a wider range of carbon contents (from pure cellulose to decayed wood). As fuel carbon increases, the carbon released by burning could actually decrease if the residual ash or char is high in carbon. This could result from fuel components such as lignin with a high carbon content and a propensity to char during pyrolysis (Safizadeh 1984). The natural inorganic content of biomass fuels can also increase char formation by altering pyrolysis pathways (Philpot 1970), but this effect appears to be minimal for the fuels analyzed in this report. There is also a weak tendency for high char yield to be associated with higher nitrogen contents, but this trend is not as well defined as that in figure 24.4.

Conclusions

Detailed carbon, hydrogen, and nitrogen contents have been measured for tropical ecosystem biomass from experimental fires in Africa and Brazil. The consistent data set allows more accurate estimates of the elemental content to be made for a variety of ecosystem types if area loading of the major fuel classes is known or can be estimated. Elemental content, weighted by the proportion of the fuel class burned, provides a characteristic value for the site burned. Variations in carbon content appear to be both site specific and ecosystem specific. Carbon contents where grass dominated burning were about 2.5% carbon lower than areas with more woody debris and shrubs. Carbon content was close to 50% for the slashed forest areas of Brazil. Ash content should be considered when net carbon emission calculations are made. Fuel classes higher in carbon content tend to form more char residue in thermogravimetric experiments. If this

Table 24.A1 Elemental composition by fuel class for sites burned in African ecosystems in 1992 (values on a dry-weight basis with standard deviation in parentheses)

Fuel type	C% (SD)	H% (SD)	N% (SD)
Combined savanna fuels from Kambeni 5, Faai 1, Numbi 4, and Shabeni 1 (early-spring biennial burns) in Kruger National Park, South Africa			
1 hr (<0.64 cm)	49.7 (0.3)	5.57 (0.09)	0.53 (0.08)
10 hr (0.64 to 2.54 cm)	49.8 (0.4)	5.69 (0.04)	0.51 (0.08)
Grass	47.4 (0.7)	5.92 (0.10)	0.53 (0.11)
Leaf	52.3 (1.2)	5.58 (0.10)	1.04 (0.10)
Litter	48.2 (1.1)	5.65 (0.13)	0.63 (0.08)
Dung	46.7 (1.8)	5.44 (0.21)	1.60 (0.26)
Savanna Block 55 (last burned in 1988) in Kruger National Park, South Africa			
1 hr (<0.64 cm)	47.9 (0.5)	5.44 (0.14)	0.57 (0.07)
10 hr (0.64 to 2.54 cm)	49.1 (0.4)	5.41 (0.09)	0.47 (0.11)
Grass	46.6 (0.3)	5.96 (0.00)	0.57 (0.05)
Leaf	54.0 (1.4)	6.23 (0.05)	1.09 (0.03)
Litter	47.8 (0.5)	5.67 (0.08)	0.54 (0.08)
Dung	45.5 (3.6)	5.38 (0.45)	1.39 (0.40)
Savanna Block 56 (last burned in 1990) in Kruger National Park, South Africa			
1 hr (<0.64 cm)	49.8 (0.3)	5.67 (0.12)	0.42 (0.10)
10 hr (0.64 to 2.54 cm)	49.8 (0.1)	5.62 (0.13)	0.44 (0.13)
Grass	45.4 (0.7)	5.82 (0.06)	0.59 (0.19)
Leaf	50.6 (0.9)	5.67 (0.05)	2.46 (0.11)
Litter	45.4 (0.3)	5.53 (0.17)	0.50 (0.06)
Dung	46.2 (1.3)	5.35 (0.13)	1.58 (0.65)
Savanna Kambeni 3 (early-spring biennial burns) in Kruger National Park, South Africa			
1 hr (<0.64 cm)	49.8 (0.6)	5.52 (0.12)	0.56 (0.00)
10 hr (0.64 to 2.54 cm)	51.0 (0.2)	5.62 (0.00)	0.40 (0.09)
Grass	46.8 (0.1)	5.81 (0.07)	0.49 (0.02)
Leaf	53.7 (0.5)	6.08 (0.13)	1.16 (0.04)
Dung	46.1 (0.1)	5.62 (0.01)	1.57 (0.13)
Savanna Kambeni E (last burned in 1954) in Kruger National Park, South Africa			
1 hr (<0.64 cm)	48.7 (0.4)	5.97 (0.06)	0.39 (0.03)
10 hr (0.64 to 2.54 cm)	48.2 (0.2)	5.88 (0.10)	0.39 (0.17)
Grass	47.0 (0.5)	5.85 (0.12)	0.42 (0.08)
Leaf	52.7 (0.6)	5.91 (0.35)	1.01 (0.08)
Litter	48.4 (0.4)	5.72 (0.27)	0.39 (0.03)
Dung	42.8 (4.7)	5.50 (0.56)	1.20 (0.14)
Savanna Shabeni 5 (early-spring biennial burns) in Kruger National Park, South Africa			
1 hr (<0.64 cm)	48.4 (1.6)	5.66 (0.12)	0.45 (0.05)
10 hr (0.64 to 2.54 cm)	50.6 (0.1)	5.66 (0.22)	0.34 (0.06)
Grass	46.9 (0.6)	5.92 (0.04)	0.50 (0.08)
Leaf	51.2 (0.2)	5.71 (0.11)	1.17 (0.02)
Litter	48.6 (0.1)	5.73 (0.18)	0.59 (0.14)
Dung	48.0 (1.7)	5.74 (0.15)	1.64 (0.13)
Combined 100-hour fuel from all Kruger burn sites			
100 hr (2.54 to 7.62 cm)	48.9 (0.5)	5.93 (0.11)	0.14 (0.06)
Dambo grassland moist savanna woodland near Kasanka National Park, Zambia			
Grass	46.5 (0.5)	5.73 (0.04)	0.18 (0.02)
Fallow Chitemene moist savanna woodland near Kasanka National Park, Zambia			
Grass	47.4 (0.2)	5.78 (0.06)	0.17 (0.01)
Leaf	53.6 (0.5)	5.94 (0.15)	1.11 (0.05)
Litter	47.7 (0.3)	5.41 (0.10)	0.79 (0.11)

Table 24.A1 (continued)

Fuel type	C% (SD)	H% (SD)	N% (SD)
Moist savanna Miombo woodland near Kasanka National Park, Zambia			
1 hr (<0.64 cm)	48.9 (0.6)	5.25 (0.11)	0.50 (0.10)
10 hr (0.64 to 2.54 cm)	46.7 (4.8)	5.01 (0.45)	0.35 (0.07)
Grass	46.0 (0.3)	5.62 (0.05)	0.24 (0.01)
Leaf	52.0 (0.4)	5.93 (0.01)	0.98 (0.02)
Litter	47.3 (4.0)	5.70 (0.01)	0.99 (0.23)
Semiarid savanna Miombo woodland near Choma, Zambia			
1 hr (<0.64 cm)	50.5 (0.2)	5.61 (0.12)	0.47 (0.03)
10 hr (0.64 to 2.54 cm)	48.4 (1.3)	5.39 (0.30)	0.39 (0.10)
Grass	45.7 (0.3)	5.71 (0.08)	0.26 (0.05)
Leaf	51.2 (0.4)	5.76 (0.04)	1.10 (0.20)
Litter	49.6 (0.6)	5.72 (0.08)	0.63 (0.06)

Table 24.A2 Elemental composition by fuel class for sites burned in Brazilian ecosystems in 1990 (values on a dry-weight basis with standard deviation in parentheses)

Fuel type	C% (SD)	H% (SD)	N% (SD)
Cerrado campo limpo near Brasília			
Grass dead	47.4 (0.2)	5.66 (0.03)	0.25 (0.01)
Grass live	47.9 (0.6)	5.82 (0.07)	0.57 (0.03)
Cerrado campo sujo near Brasília			
Grass dead	47.1 (0.3)	5.69 (0.05)	0.23 (0.02)
Grass live	47.8 (0.2)	5.79 (0.04)	0.57 (0.05)
Dicots	49.0 (0.7)	5.71 (0.06)	0.78 (0.09)
Litter	50.3 (1.1)	5.59 (0.18)	0.80 (0.06)
Cerrado campo cerrado near Brasília			
Grass dead	47.7 (0.2)	5.75 (0.03)	0.30 (0.02)
Grass live	47.8 (0.4)	5.89 (0.07)	0.64 (0.04)
Dicots	51.8 (0.6)	5.93 (0.16)	0.77 (0.11)
Litter	52.1 (0.8)	5.73 (0.09)	0.86 (0.08)
Cerrado senso stricto near Brasília			
1 hr (<0.64 cm)	51.5 (0.3)	5.74 (0.06)	0.46 (0.03)
10 hr (0.64 to 2.54 cm)	51.9 (0.4)	5.70 (0.05)	0.39 (0.06)
Grass dead	47.1 (0.2)	5.66 (0.04)	0.31 (0.03)
Grass live	47.3 (0.2)	5.81 (0.05)	0.50 (0.07)
Dicots	52.0 (0.7)	5.97 (0.15)	0.76 (0.12)
Litter	52.2 (0.9)	5.45 (0.09)	0.78 (0.05)
Slashed primary tropical moist forest near Jacunda, Pará			
1 hr (<0.64 cm)	50.5 (0.4)	5.80 (0.07)	0.84 (0.16)
10 hr (0.64 to 2.54 cm)	50.2 (0.0)	5.90 (0.00)	0.62 (0.00)
100 hr (2.54 to 7.62 cm)	50.1 (0.2)	5.97 (0.04)	0.41 (0.07)
1000 hr (7.63 to 20.5 cm)	50.5 (0.2)	5.90 (0.05)	0.36 (0.06)
Leaf	51.2 (1.3)	5.69 (0.10)	2.12 (0.21)
Dicots	48.6 (0.3)	5.71 (0.09)	1.79 (0.25)
Litter	51.2 (0.2)	5.67 (0.07)	1.97 (0.14)
Slashed second-growth forest near Jacunda, Pará			
1 hr (<0.64 cm)	49.5 (0.4)	5.79 (0.04)	0.70 (0.10)
10 hr (0.64 to 2.54 cm)	50.3 (0.6)	5.85 (0.06)	0.74 (0.09)
100 hr (2.54 to 7.62 cm)	49.2 (0.3)	6.04 (0.04)	0.37 (0.04)
1000 hr (7.63 to 20.5 cm)	49.6 (0.3)	5.99 (0.04)	0.27 (0.04)
Leaf	50.0 (1.2)	5.50 (0.19)	2.21 (0.36)
Dicots	46.2 (1.2)	5.58 (0.15)	1.92 (0.29)
Litter	48.8 (0.5)	5.33 (0.07)	1.98 (0.23)

Table 24.A3 Elemental composition by fuel class for sites burned in Brazilian ecosystems in 1991 (values on a dry-weight basis with standard deviation in parentheses)

Fuel type	C% (SD)	H% (SD)	N% (SD)
Cerrado senso stricto (dense) near Brasília			
Grass	47.2 (0.3)	5.74 (0.05)	0.41 (0.06)
Leaf	51.8 (0.9)	6.07 (0.09)	1.07 (0.07)
Brush stems	53.6 (1.2)	5.71 (0.02)	0.68 (0.14)
Dicots	52.7 (0.2)	6.05 (0.07)	0.95 (0.13)
Litter	52.9 (1.4)	5.71 (0.19)	0.95 (0.10)
Cerrado senso stricto (open) near Brasília			
1 hr (<0.64 cm)	51.8 (0.3)	5.84 (0.05)	0.59 (0.07)
10 hr (0.64 to 2.54 cm)	52.3 (0.5)	5.74 (0.12)	0.56 (0.09)
100 hr (2.54 to 7.62 cm)	51.4 (0.6)	5.94 (0.12)	0.29 (0.06)
Grass	47.4 (0.4)	5.72 (0.06)	0.35 (0.02)
Leaf	52.9 (1.1)	5.90 (0.08)	1.20 (0.28)
Dicots	52.3 (1.0)	5.98 (0.11)	0.94 (0.22)
Litter	53.4 (0.7)	5.77 (0.06)	1.06 (0.17)
Pasture (12 years since primary deforestation) near Maraba, Pará			
1 hr (<0.64 cm)	49.8 (0.3)	5.82 (0.04)	0.82 (0.13)
10 hr (0.64 to 2.54 cm)	50.0 (0.2)	6.13 (0.05)	0.53 (0.06)
100 hr (2.54 to 7.62 cm)	50.2 (0.4)	5.98 (0.04)	0.27 (0.04)
1000 hr (7.63 to 20.5 cm)	51.0 (0.3)	5.89 (0.09)	0.32 (0.04)
1000 hr (7.63 to 20.5 cm)	48.0 (0.0)	5.61 (0.04)	0.50 (0.12)
Grass	46.3 (0.2)	5.85 (0.04)	1.19 (0.19)
Dicots	49.3 (0.4)	6.05 (0.09)	1.40 (0.31)
Litter	46.7 (1.1)	5.63 (0.11)	0.52 (0.09)
Slashed primary tropical moist forest near Maraba, Pará			
1 hr (<0.64 cm)	49.7 (0.2)	5.81 (0.09)	1.00 (0.18)
10 hr (0.64 to 2.54 cm)	50.9 (0.3)	5.93 (0.10)	0.81 (0.14)
100 hr (2.54 to 7.62 cm)	49.9 (0.2)	6.15 (0.05)	0.47 (0.04)
1000 hr (7.63 to 20.5 cm)	50.2 (0.1)	6.05 (0.08)	0.35 (0.05)
1000 hr (7.63 to 20.5 cm)	50.7 (1.1)	5.63 (0.21)	0.63 (0.10)
Leaf	52.4 (0.5)	5.98 (0.06)	1.12 (0.03)
Dicots	49.8 (0.9)	6.30 (0.14)	2.44 (0.19)
Red slash	51.6 (1.0)	6.19 (0.32)	2.36 (0.24)
Litter	48.3 (2.2)	5.49 (0.25)	1.76 (0.20)
Slashed third-growth forest near Maraba, Pará			
1 hr (<0.64 cm)	49.9 (0.4)	5.76 (0.06)	0.71 (0.09)
10 hr (0.64 to 2.54 cm)	49.9 (0.2)	6.00 (0.05)	0.62 (0.05)
100 hr (2.54 to 7.62 cm)	49.7 (0.8)	5.98 (0.10)	0.32 (0.06)
1000 hr (7.63 to 20.5 cm)	50.2 (0.3)	6.06 (0.07)	0.20 (0.09)
Red slash	51.6 (0.3)	5.76 (0.11)	1.98 (0.23)
Litter	47.4 (3.2)	5.25 (0.37)	1.31 (0.11)

Table 24.A4 Elemental composition by fuel class for sites burned in Brazilian ecosystems in 1992 (values on a dry-weight basis with standard deviation in parentheses)

Fuel type	C% (SD)	H% (SD)	N% (SD)
Slashed third-growth forest near Jamari, Rondônia			
1 hr (<0.64 cm)	51.5 (0.0)	5.99 (0.09)	0.59 (0.07)
10 hr (0.64 to 2.54 cm)	50.8 (0.2)	6.10 (0.02)	0.39 (0.11)
100 hr (2.54 to 7.62 cm)	49.9 (0.8)	6.20 (0.03)	0.20 (0.01)
1000 hr (7.63 to 20.5 cm)	51.4 (0.6)	6.09 (0.09)	0.23 (0.04)
1000 hr (7.63 to 20.5 cm)	53.3 (0.6)	6.03 (0.07)	0.40 (0.20)
Dicots	48.1 (0.0)	5.55 (0.00)	2.00 (0.00)
Red slash	54.1 (0.7)	6.00 (0.04)	1.82 (0.23)
Litter	49.1 (2.2)	5.31 (0.13)	1.72 (0.30)
Slashed primary tropical moist forest near Jamari, Rondônia			
1 hr (<0.64 cm)	50.7 (0.5)	6.01 (0.02)	1.40 (0.02)
1000 hr (7.63 to 20.5 cm)	51.1 (1.6)	5.74 (0.08)	0.57 (0.01)
Red slash	51.9 (1.7)	6.36 (0.48)	2.00 (0.25)
Litter	51.9 (0.3)	5.87 (0.16)	2.38 (0.23)
Slashed primary tropical moist forest near Jamari, Rondônia			
1 hr (<0.64 cm)	51.2 (0.8)	5.89 (0.16)	1.10 (0.10)
10 hr (0.64 to 2.54 cm)	50.4 (0.7)	6.00 (0.08)	0.64 (0.02)
100 hr (2.54 to 7.62 cm)	50.8 (0.5)	6.06 (0.07)	0.52 (0.19)
1000 hr (7.63 to 20.5 cm)	50.5 (0.6)	6.16 (0.14)	0.28 (0.02)
Leaf	51.5 (0.8)	5.88 (0.45)	2.08 (0.17)
Dicots	49.7 (0.0)	5.79 (0.00)	2.17 (0.00)
Litter	53.2 (1.5)	6.20 (0.15)	1.74 (0.24)
Pasture (4 years since primary deforestation) near Jamari, Rondônia			
1 hr (<0.64 cm)	51.6 (0.5)	6.00 (0.11)	0.57 (0.09)
10 hr (0.64 to 2.54 cm)	50.8 (0.9)	6.11 (0.06)	0.46 (0.07)
100 hr (2.54 to 7.62 cm)	51.9 (0.3)	6.12 (0.12)	0.53 (0.22)
1000 hr (7.63 to 20.5 cm)	51.1 (1.1)	5.85 (0.17)	0.47 (0.10)
1000 hr (7.63 to 20.5 cm)	51.8 (0.6)	6.07 (0.11)	0.30 (0.02)
Grass	47.6 (0.9)	5.96 (0.06)	0.98 (0.19)
Dicots	52.2 (1.9)	5.82 (0.03)	1.77 (0.16)
Litter	46.5 (1.1)	5.36 (0.11)	1.30 (0.44)
Pasture (10 years since primary deforestation) near Jamari, Rondônia			
1000 hr (7.63 to 20.5 cm)	52.2 (0.4)	6.03 (0.10)	0.28 (0.07)
Grass	46.5 (0.8)	5.82 (0.05)	0.87 (0.32)
Dicots	50.5 (0.1)	5.80 (0.01)	1.02 (0.19)
Litter	46.4 (1.5)	5.63 (0.14)	0.44 (0.17)

char formation is similar in full-scale biomass fires, the net release of carbon from fires may be fairly constant for all the fuel classes.

Appendix

This appendix contains additional detailed data on the CHN analysis for fuel samples from Africa and Brazil. Tables 24.A1, 24.A2, 24.A3, and 24.A4 list the CHN values for each fuel class sampled at each fire site for the 1992 Africa, 1990 Brazil, 1991 Brazil, and 1992 Brazil fires, respectively. Analysis results from three to five samples within a fuel class at a fire site were averaged. Standard deviations are listed in parentheses.

All CHN values in these tables were calculated on a dry-weight basis using either the thermogravimetric or oven-dry methods described in the Experimental Methods section.

Table 24.A5 summarizes thermogravimetric analysis results on samples from the Africa and Brazil sets. Samples were selected to have a range of carbon contents in order to examine the effects of inorganic constituents on the carbon analysis results. The listed values are for individual fuel samples, rather than the average of several samples within a fuel class as described for table 24.A1 to 24.A4. The residue remaining after heating in air to 650°C by the thermogravimetric method described in the Experimental Methods section was assumed to be inorganic ash. Organic materials are assumed to be lost from the

Table 24.A5 Carbon content and thermogravimetric analysis results on both an ash-included and ash-free basis (all values on a dry-weight basis)

Sample number	Fuel type	Ash-included basis			Ash-free basis	
		C%	TG char %	TG ash %	C%	TG char %
Brazil 1990						
975	Live grass	47.2	26.6	5.3	49.9	22.5
977	Dead grass	47.6	24.8	4.8	50.0	21.0
955	Live grass	48.0	30.9	5.4	50.7	27.0
956	Live grass	46.9	30.8	6.5	50.2	26.0
921	Leaf litter	50.1	36.4	7.1	53.9	31.6
919	Leaf litter	52.0	31.4	4.9	54.7	27.8
982	Dead grass	47.2	24.4	4.7	49.5	20.7
980	Dead grass	46.5	25.1	5.9	49.4	20.4
962	Live grass	47.9	30.6	5.4	50.6	26.6
959	Live grass	47.6	29.4	4.7	50.0	26.0
1006	Dicot	48.8	34.6	6.7	52.3	29.8
1005	Dicot	50.3	30.5	3.7	52.2	27.9
915	Leaf litter	52.2	35.4	3.4	54.1	33.1
916	Leaf litter	51.0	36.0	5.6	54.0	32.2
987	Dead grass	47.7	25.1	4.2	49.8	21.7
983	Dead grass	47.4	23.4	4.7	49.8	19.6
966	Live grass	47.8	27.1	4.3	49.9	23.9
963	Live grass	48.3	27.2	3.9	50.3	24.2
1002	Dicot	51.6	33.3	3.6	53.6	30.8
999	Dicot	52.9	29.6	2.6	54.2	27.7
911	Leaf litter	52.2	39.0	4.4	54.6	36.2
909	Leaf litter	53.5	39.8	3.9	55.7	37.4
990	Dead grass	47.1	25.6	5.8	50.0	21.0
992	Dead grass	47.3	23.9	5.2	49.9	19.7
968	Live grass	47.4	27.9	5.1	49.9	24.1
971	Live grass	47.0	29.0	5.4	49.6	24.9
993	Dicot	51.7	31.2	2.8	53.2	29.2
994	Dicot	53.3	30.3	2.6	54.7	28.5
864	1 hr (<0.64 cm)	51.4	27.5	1.4	52.2	26.5
861	1 hr (<0.64 cm)	51.0	30.1	1.4	51.7	29.1
881	10 hr (0.64 to 2.54 cm)	51.9	28.3	1.2	52.5	27.4
878	10 hr (0.64 to 2.54 cm)	52.5	31.3	1.2	53.1	30.4
942	Attached foliage	50.4	36.0	7.4	54.4	30.9
938	Attached foliage	47.6	43.3	11.7	53.9	35.7
930	Litter	48.7	39.8	9.9	54.0	33.2
929	Litter	49.5	41.7	9.4	54.6	35.6
950	Dicot	46.8	33.8	8.5	51.2	27.6
951	Dicot	44.3	36.0	10.5	49.4	28.5
866	1 hr (<0.64 cm)	49.3	32.2	4.4	51.5	29.1
869	1 hr (<0.64 cm)	49.7	30.7	3.5	51.5	28.1
884	10 hr (0.64 to 2.54 cm)	50.1	29.9	2.2	51.2	28.3
882	10 hr (0.64 to 2.54 cm)	49.7	25.7	1.9	50.7	24.3
886	10 hr (0.64 to 2.54 cm)	51.4	29.2	2.1	52.5	27.7
891	100 hr (2.54 to 7.62 cm)	49.2	23.8	1.1	49.8	22.9
892	100 hr (2.54 to 7.62 cm)	48.7	23.0	1.4	49.4	21.9
901	1000 hr (7.63 to 20.5 cm)	49.7	23.9	1.2	50.3	23.0
900	1000 hr (7.63 to 20.5 cm)	50.1	23.6	1.2	50.7	22.7
936	Attached foliage	50.7	37.3	5.7	53.8	33.5
935	Attached foliage	49.5	36.0	7.3	53.4	30.9
927	Litter	51.2	35.8	5.2	54.0	32.2
923	Litter	51.6	36.2	5.3	54.5	32.6
947	Dicot	48.5	35.6	7.6	52.5	30.3
946	Dicot	49.0	31.8	5.6	51.9	27.8
871	1 hr (<0.64 cm)	50.3	28.4	1.8	51.2	27.1
874	1 hr (<0.64 cm)	51.0	30.9	2.8	52.5	28.9
887	10 hr (0.64 to 2.54 cm)	50.2	27.1	2.0	51.2	25.7
896	100 hr (2.54 to 7.62 cm)	50.1	25.7	1.6	50.9	24.6
893	100 hr (2.54 to 7.62 cm)	49.8	24.3	1.4	50.4	23.2
904	1000 hr (7.63 to 20.5 cm)	50.5	23.0	0.9	51.0	22.3
907	1000 hr (7.63 to 20.5 cm)	50.2	26.2	1.4	51.0	25.1

Table 24.A5 (continued)

Sample number	Fuel type	Ash-included basis			Ash-free basis	
		C%	TG char %	TG ash %	C%	TG char %
Brazil 1991						
100	Litter	42.0	54.4	29.4	59.5	35.4
31	Dicot	48.6	34.0	5.8	51.6	29.9
163	100 hr (2.54 to 7.62 cm)	48.2	25.7	3.1	49.7	23.3
188	1000 hr (7.63 to 20.5 cm)	47.9	32.8	5.5	50.7	28.9
21	Bush	55.4	40.8	2.0	56.5	39.6
102	Litter	54.3	38.9	3.8	56.4	36.5
146	10 hr (0.64 to 2.54 cm)	53.2	34.2	2.2	54.4	32.7
178	1000 hr (7.63 to 20.5 cm)	52.0	32.5	2.6	53.4	30.8
80	Red slash	50.7	36.4	6.8	54.4	31.8
164	100 hr (2.54 to 7.62 cm)	49.9	23.3	1.9	50.9	21.8
Brazil 1992						
201	Dicot	54.1	38.5	2.2	55.3	37.1
131	Litter	55.0	36.3	3.9	57.2	33.7
121	Litter	53.6	39.9	6.7	57.4	35.6
22	1000 hr (7.63 to 20.5 cm)	54.0	33.3	1.2	54.6	32.5
148	Litter	44.4	34.3	12.3	50.6	25.1
142	Litter	45.0	40.4	13.6	52.1	31.1
182	Grass	45.7	32.7	8.9	50.2	26.1
166	Root matter	35.2	63.3	40.2	58.8	38.6
156	Red slash	50.1	33.5	2.9	51.6	31.5
58	10 hr (0.64 to 2.54 cm)	49.8	29.1	2.7	51.2	27.1
Africa 1992						
13	Leaf	55.9	39.1	5.6	59.2	35.5
221	Leaf	54.2	38.5	6.1	57.7	34.4
87	Leaf	53.8	39.5	6.5	57.6	35.3
157	Leaf	53.2	36.1	4.3	55.6	33.5
16	Dung	40.5	44.3	23.4	52.8	27.3
43	Grass	44.4	28.8	8.5	48.5	22.1
185	Grass	45.5	28.6	7.0	48.9	23.3
76	Litter	46.8	30.3	8.6	51.3	23.7
133	1 hr (<0.64 cm)	49.7	38.1	6.0	52.9	34.2
201	10 hr (0.64 to 2.54 cm)	50.0	32.9	4.9	52.5	29.5

sample due to pyrolysis or oxidation, leaving an inorganic residue. Table 24.A5 lists carbon measured by CHN analysis along with char and ash from the thermogravimetric analysis. The ash content was used to calculate carbon content and char fraction on an ash-free basis. The relation of ash-free char to ash-free carbon is shown in Figure 24.4.

Acknowledgments

Funding in partial support of this research was provided by the National Aeronautics and Space Administration for studies in Brazil and by the Environmental Protection Agency for studies in Africa.

References

Byram, G. M. 1959. Combustion of forest fuels, in *Forest Fire Control and Use*, edited by K. P. Davis, pp. 61–89, McGraw-Hill, New York.

Clements, H. B., and C. K. McMahon. 1980. Nitrogen oxides from burning forest fuels examined by thermogravimetry and evolved gas analysis, *Thermochimica Acta*, 35, 133–139.

Coutinho, L. M. 1990. Fire in the ecology of the Brazilian Cerrado, in *Fire in the Tropical Biota: Ecosystem Processes and Global Challenges*, edited by J. G. Goldammer, Ecological Studies 84, pp. 82–105, Springer-Verlag, Berlin-Heidelberg.

Crutzen, P. J., and M. O. Andreae. 1990. Biomass burning in the tropics: impact on atmospheric chemistry and biogeochemical cycles, *Science*, 250, 1669–1678.

Deeming, J. E., R. E. Burgan, and J. D. Cohen. 1977. The national fire-danger rating system—1978, *General Technical Report. INT-39*, 36 pp., Department of Agriculture, Forest Service Intermountain Research Station, Ogden, Utah.

Dignon, J., and J. E. Penner. 1991. Biomass burning: a source of nitrogen oxides in the atmosphere, in *Global Biomass Burning: Atmospheric, Climatic, and Biospheric Implications*, edited by J. S. Levine, pp. 370–375, MIT Press, Cambridge, Mass.

Ewel, J., C. Berish, B. Brown, N. Price, and J. Raich. 1981. Slash and burn impacts on a Costa Rica wet forest site, *Ecology*, 62, 816–829.

Fearnside, P. M. 1990. Fire in the ecology of the Brazilian Cerrado, in *Fire in the Tropical Biota: Ecosystem Processes and Global Challenges*, edited by J. G. Goldammer, Ecological Studies 84, pp. 106–116, Springer-Verlag, Berlin-Heidelberg.

Gnaiger, A., and G. Bitterlich. 1984. Proximate biochemical composition and caloric content calculated from elemental CHN analysis: a stoichiometric concept, *Oecologia*, 62, 289–329.

Hao, W. M., and M.-H. Liu. 1994. Spatial and temporal distribution of tropical biomass burning, *Global Biogeochemical Cycles*, 8, 495–503.

Hao, W. M., M.-H. Liu, and P. J. Crutzen. 1990. Estimates of annual and regional releases of CO_2 and other trace gases to the atmosphere from fires in the tropics, based on the FAO statistics for period 1975–1980, in *Fire in the Tropical Biota: Ecosystem Processes and Global Challenges*, edited by J. G. Goldammer, Ecological Studies 84, pp. 440–462, Springer-Verlag, Berlin-Heidelberg.

Hao, W. M., and D. E. Ward. 1993. Methane production from global biomass burning, *J. Geophys Res.*, 98, 20 657–20 661.

Kauffman, J. B., D. L. Cummings, and D. E. Ward. 1994. Relationships of fire, biomass and nutrient dynamics along a vegetation gradient in the Brazilian Cerrado, *J. Ecology* 82, 519–531.

Levine, J. S., Ed. 1991. *Global Biomass Burning: Atmospheric, Climatic, and Biospheric Implications*, MIT Press, Cambridge, Mass.

Lobert, J. M., D. H. Scharffe, W. M. Hao, T. A. Kuhlbusch, R. Seuwen, P. Warneck, and P. J. Cnutzon. 1991. Experimental evaluation of biomass burning emissions: nitrogen and carbon containing compounds, in *Global Biomass Burning: Atmospheric, Climatic, and Biospheric Implications*, edited by J. S. Levine, pp. 289–304, MIT Press, Cambridge, Mass.

Nelson, R. M., Jr. 1982. An evaluation of the carbon mass balance technique for estimating emission factors and fuel consumption in forest fires, *Research. Paper SE-231*, U.S. Department of Agriculture, Forest Service, Southeastern Forest Experiment Station, Asheville, N.C.

Olson, J. S., J. A. Watts, and L. J. Allison. 1983. Carbon in live vegetation of major world ecosystems, *Report TR004, DOE/NBB-0037*, U.S. Department of Energy, Washington, D.C.

Patel, S. N., and L. E. Erickson. 1981. Estimation of heats of combustion of biomass from elemental analysis using available electron concepts, *Biotechnology and Bioengineering*, 23, 2051–2067.

Pettersen, R. C. 1984. The chemical composition of wood, in *The Chemistry of Solid Wood*, edited by R. Rowell, Advances in Chemistry Series 207, 57–126, American Chemical Society, Washington, D.C.

Philpot, C. W. 1970. Influence of mineral content on the pyrolysis of plant materials, *Forest Science*, 16, 461–471.

Sarkanen, K. V., and H. L. Hergert. 1971. In *Lignins: Occurrence, Formation, Structure and Reactions*, edited by K. V. Sarkanen and C. H. Ludwig, pp. 43–94, Wiley-Interscience, New York.

Seiler, W., and P. J. Crutzen. 1980. Estimates of grassland net fluxes of carbon between the biosphere and the atmosphere from biomass burning, *Clim. Change*, 2, 207–247.

Shafizadeh, F. 1984. The chemistry of pyrolysis and combustion, in *The Chemistry of Solid Wood*, edited by R. Rowell, Advances in Chemistry Series 207, pp. 489–529, American Chemical Society, Washington, D.C.

Shafizadeh, F., P. S. Chin, and W. F. DeGroot. 1977. Effective heat content of green forest fuels, *Forest Science*, 23, 81–91.

Shea, R. W., B. W. Shea, J. B. Kauffman, D. E. Ward, C. I. Haskins, and M. C. Scholes, accepted for publication. 1996. Fuel biomass and combustion factors in savanna ecosystems of South Africa and Zambia, *J. Geophys. Res.*

Susott, R. A., W. F. DeGroot, and F. Shafizadeh. 1975. Heat content of natural fuels, *J. Fire and Flammability*, 6, 311–324.

Susott, R. A., F. Shafizadeh, and T. W. Aanerud. 1979. A quantitative thermal analysis technique for combustible gas detection, *J. Fire and Flammability*, 10, 94–104.

Susott, R. A., D. E. Ward, R. E. Babbitt, and D. J. Latham. 1991. The measurement of trace emissions and combustion characteristics for a mass fire, in *Global Biomass Burning: Atmospheric, Climatic, and Biospheric Implications*, edited by J. S. Levine, pp. 245–257, MIT Press, Cambridge, Mass.

Ward, D. E. 1980. Particulate matter production from cylindrical laminar diffusion flames, in *Proceedings Fall Meeting of the Western States Section of the Combustion Institute*, Los Angeles, 18 pp.

Ward D. E. 1990. Factors influencing the emissions of gases and particulate matter from biomass burning, in *Fire in the Tropical Biota: Ecosystem Proceees and Global Challenges*, edited by J. G. Goldammer, Ecological Studies 84, pp. 418–435, Springer-Verlag, Berlin-Heidelberg.

Ward, D. E., A. W. Setzer, Y. J. Kaufman, and R. A. Rasmussen. 1991. Characteristics of smoke emissions from biomass fires for the amazon region—BASE-A experiment, in *Global Biomass Burning: Atmospheric, Climatic, and Biospheric Implications*, edited by J. S. Levine, pp. 394–402, MIT Press, Cambridge, Mass.

Ward, D. E., R. A. Susott, J. B. Kauffman, R. E. Babbitt, D. L. Cummings, B. Dias, B. N. Holben, Y. J. Kaufman, R. A. Rasmussen, and A. W. Setzer. 1992. Smoke and fire characteristics for Cerrado and deforestation burns in Brazil: BASE-B experiment, *J. Geophys. Res.*, 97, 14 601–14 619.

Vogt, K., 1991. Carbon budgets of temperate forest ecosystems, *Tree Physiol.* 9, 69–86.

Biomass Burning in the Savannas of Southern Africa with Particular Reference to the Kruger National Park in South Africa

Winston S. W. Trollope

Africa is referred to as the "Fire Continent" (Komarek 1971) as a result of the widespread occurrence of biomass burning, particularly in the savanna biome. This description is equally applicable to southern Africa, where savanna is a major plant community, and the early Portuguese explorers who rounded the Cape of Good Hope in the fifteenth century recorded in their ships' logs that the interior of South Africa was "Terra dos fumos"—the land of smoke and fire (Scott 1970). This capacity of Africa to support fire stems from the fact that climatic factors are the driving force of fire ecology, and the main requirement for fire to occur anywhere on earth is to have lightning as the primary ignition source and climatic conditions that will permit the burning of vegetation and the spread of fires caused by lightning strikes. Africa is one of the continents that is highly prone to lightning storms and has a fire climate comprising dry and wet periods during which fires can burn the plant fuels during the dry period that have been produced and accumulated during the wet rainy period (Komarek 1971).

Considering lightning as a primary ignitions source for natural vegetation fires, Komarek (1971) concluded that lightning is but an expression of the earth's electrical field and a visual display of the universal force, electricity. The ingredients of the weather that are necessary for the development of thunderstorms and lightning are hot, cold, and humid masses of air, and whenever they meet and mix, thunderstorms and lightning occur. Most of these storms originate from the masses of air that circulate the globe from a westerly direction as a result of the earth's rotation. Africa is one of the continents that is most prone to thunderstorms and lightning, and considerable evidence is available on the high frequency of thunderstorms and lightning in western, central, eastern, and southern Africa (Komarek 1971; West 1965). This general circulation of air from the west occurs as a result of the heavier cold air at the two polar regions flowing to the equator and in turn the hot air at the equator rising and flowing to the poles. In each case, the direction of flow is deflected by the earth's rotation, causing the air masses to move in an easterly direction from the west,

and the intermixing of the resultant types of air masses gives rise to thunderstorms and lightning, thus providing an ignition source for fires to occur (Pyne 1982). Pyne (1982) states that lightning fires occurred on earth as soon as the atmosphere evolved and vegetation appeared. The role of lightning is to balance the electrical equilibrium of the earth. As a result of the atmosphere being able to conduct electricity to a certain extent, there is a constant leakage of electricity from the earth to the atmosphere, creating an electrical potential. When the potential is great enough, electricity discharges back to earth in the form of lightning. It has been estimated that the earth would lose its electrical charge in less than an hour (48 minutes) unless it is replenished through lightning. Thunderstorms are therefore both a thermodynamic and electromagnetic necessity. It is estimated that thunderstorms produce more than 8 million lightning strokes per day globally, which is equivalent to more than 2 billion kilowatt hours of electricity, i.e., approximately 4.9 times the amount of electricity produced in South Africa per year (Anonymous 1992). While the primary ignition role of lightning in causing vegetation fires in the savanna areas of Africa is recognized, the stage has now been reached that in most regions of the world humans have become more important than lightning as sources of ignition (Goldammer and Crutzen 1993). This is well illustrated in the savanna areas comprising the Kruger National Park in South Africa, where anthropogenic fires have now become the dominant ignition source of fires in that type of savanna community (Trollope 1993).

The other requirement for natural fires is a fire climate, which comprises dry and wet periods so that fires can burn the plant fuels during the dry period that have been produced and accumulated during the wet rainy period (Komarek 1971). These climatic conditions are characteristic of the savannas of southern Africa, which receive a strictly summer rainfall distribution followed by an extended dry winter period. Lightning is further enhanced as an ignition source by the climatic characteristic in Africa of having dry lightning storms at the end of the dry winter period,

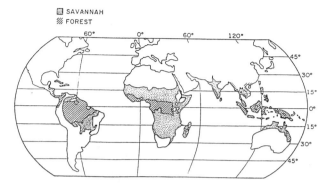

Figure 25.1 The distribution of savanna and forest types of vegetation in the world (adapted from Bartholomew 1987)

during which time the plant fuels have a very low moisture content and are highly inflammable. Finally, Africa has the most extensive area of tropical savanna in the world, which is characterized by a grassy understory that becomes extremely inflammable during the dry season (figure 25.1). As a result of Africa's having an abundant and effective ignition source in the form of lightning, an ideal fire climate, and highly inflammable vegetation during the dry season, it has come to be known as the "Fire Continent," as mentioned earlier (Komarek 1971).

As a result of African savannas being very prone to fire, burning is recognized as an important ecological factor in the savanna ecosystems of southern Africa, and research investigating the effects of the fire regime on the biotic and abiotic components of the ecosystem has been conducted in this region since the early period of the twentieth century. This has led to a general understanding of the effects of the type and intensity of fire and season and frequency of burning on the grass and tree components of the vegetation. A major portion of this research has been conducted in the savannas of the Kruger National Park, which are representative of a significant proportion of this type of vegetation on the subcontinent. Attention will be given to the known effects of the fire regime on the vegetation of the savanna areas of southern Africa, and this will be followed by a description of the current fire regime in the Kruger National Park, which is an area representative of a significant proportion of the savanna areas of southern Africa.

Fire Ecology of the Savannas of Southern Africa

Fire ecology refers to the response of the biotic and abiotic components of the ecosystem to the fire regime, that is, type and intensity of fire and the season and

frequency of burning (Trollope, Trollope, and Bosch 1990). Fire research was initiated in the savannas of southern Africa in 1916 when the first burning plots were established at Groenkloof, Pretoria in South Africa and later at the Matopos in Zimbabwe in 1947 (West 1965). An interesting feature about these early investigations and subsequent research up until 1971 was that they focused on addressing the two key questions related to the fire regime of what are the effects of season and frequency of burning on the forage production potential of the grass sward and the ratio of bush to grass in savanna areas (West 1965; Rose-Innes 1971; Scott 1971). This was undoubtedly in response to enquiries from mainly range scientists and livestock farmers who needed to know the correct season and frequency of burning for maintaining the forage production potential of rangeland and for controlling bush encroachment. Thus, until recently, fire research in southern Africa was conducted with an agricultural objective in mind rather than with the ecological objective of determining the effect of fire on all the biotic and abiotic components of the ecosystem. This was in contrast to fire research in other fire-prone habitats like the United States and Australia, where the emphasis was on studying fire behavior as a means of controlling wild fires. However, in 1971 a conference was convened by the Tall Timbers Research Station in Tallahassee, Florida on the theme of "Fire in Africa." This congress was attended by fire ecologists from throughout Africa, including southern Africa. The major benefit that accrued from this conference was the realization that in Africa the study of fire behavior and its effects on the ecosystem, as described by type and intensity of fire, had been largely ignored in all the fire research that had been conducted up until that time. In contrast, detailed knowledge on and models for predicting fire behavior had been developed by the United States Forest Service (Byram 1959; Rothermel 1972; Brown and Davis 1973) as a means for controlling wildfires in the extensive forested areas of the country. A similar situation existed in Australia where McArthur (1966), a forest fire researcher in New South Wales, had developed procedures based on fire behavior for decreasing the fire hazard in highly flammable eucalyptus forests by reducing fuel loads through controlled burning. The outcome of this congress proved to be a turning point in fire research in the savanna and grassland areas in South Africa, and a research program was initiated to determine the effect of all the components of the fire regime on the vegetation, that is, effects of type and intensity of fire and season and frequency of burn. Unfortunately, a similar research program was not initiated elsewhere in Africa as far as is known, but nevertheless the aforementioned program has gone a long way in describing the effects of the *entire* fire regime on the vegetation in the grassland and savanna areas of the continent. To follow will be an overview of the known effects of the fire regime on grass and bush vegetation in the savanna areas of southern Africa based on fire research conducted since 1971.

Effect of Type of Fire

The most common types of fire in savanna areas are surface fires burning either as head or back fires. Crown fires do occur in savanna but only under extreme fire conditions. Trollope (1978) investigated the effects of surface fires, occurring as either head or back fires, on the grass sward in the arid savannas of the Eastern Cape of South Africa. The results showed that back fires significantly (p ≤ .01) depressed the regrowth of grass in comparison to head fires because a critical threshold temperature of approximately 95°C was maintained for 20 seconds longer during back fires than during head fires.

Bush is very sensitive to various types of fires because of differences in the vertical distribution of the release of heat energy. Field observations in the Kruger National Park and in the Eastern Cape indicate that crown and surface head fires cause the highest topkill of stems and branches as compared with back fires. Unfortunately, there are only limited quantitative data to support these observations. Nevertheless, data are available from a recently completed burning trial at the University of Fort Hare in the False Thornveld of the Eastern Cape (arid savanna), where a field scale burn was applied to an area of 62 ha to control bush encroachment. The effect of surface head and back fires on the topkill of stems and branches of bush is presented in table 25.1. The data were collected in 2-m wide belt transects laid out in the areas burnt as head and back fires. The majority of the trial area was burnt as a head fire, and the results in table 25.1 indicate that the phytomass of bush was reduced by 75% in the area burnt as a head fire in comparison to 42% in the area

burnt as a back fire. This clearly illustrates the effects different types of fire have on tree and shrub vegetation.

Effect of Fire Intensity

The effect of fire intensity on the recovery of the grass sward was investigated in the arid savannas of the Eastern Cape in South Africa. After a series of fires ranging in intensity from 925 to 3326 kJ s^{-1} m^{-1}, there were no significant differences in the recovery of the grass sward at the end of the first or second growing seasons after the burns (Trollope and Tainton 1986). It was concluded that fire intensity had had no significant effect on the recovery of the grass sward after the burns, and this was considered to be a logical effect, as otherwise intense fires would not favor the development and maintenance of grassland or open savanna.

The effect of fire intensity on bush has been studied in the arid savannas of the Eastern Cape (Trollope and Tainton 1986) and the Kruger National Park (Trollope, Potgieter, and Zambatis 1990) in South Africa. This comprised determining the mortality of plants and secondly the total topkill of stems and branches of bush of different heights. The results indicated that bush is very resistant to fire alone, and in the Eastern Cape, the mortality of bush after a high-intensity fire of 3875 kJ s^{-1} m^{-1} was only 9.3%. In the Kruger National Park, the average mortality of 14 of the most common bush species subjected to 43 fires ranging in fire intensity from 110 to 6704 kJ s^{-1} m^{-1} was only 1.3%. In both areas, the majority of the trees suffered a topkill of stems and branches and coppiced from the collar region of the stem. From these research results it can be concluded that, generally, the main effect of fire on bush in the savanna areas is to cause a topkill of stems and branches, forcing the plants to coppice from the collar region of the stem.

The effect of fire intensity on the topkill of bush was investigated in the savanna areas of the Kruger National Park (Trollope, Potgieter, and Zambatis 1990). This involved determining the proportion of trees and shrubs that suffered a total topkill of stems and branches and that coppiced from the collar region at the base of the stem after a wide range of fire

Table 25.1 The effect of surface head and back fires on the topkill of bush in the False Thornveld of the Eastern Cape expressed as the reduction in the number of tree equivalents per hectare (TE ha^{-1})

| Type of fire | Transect | | Bush phytomass (TE ha^{-1}) | | Reduction (%) |
	Length (m)	Width (m)	Before	After	
Head fire	940	2	3525	888	75
Back fire	560	2	3407	1991	42

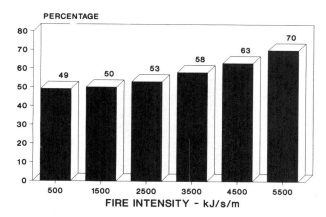

Figure 25.2 Effect of fire intensity on the topkill of bush 2 m high in the Kruger National Park in South Africa

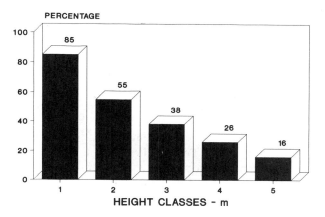

Figure 25.3 Effect of height on the topkill of bush subjected to a fire intensity of 3000 kJ/s/m in the Kruger National Park

intensities had been applied. The results of this study are illustrated in figure 25.2.

The results in figure 25.2 show that there was a significantly greater topkill of bush with increasing fire intensities. However, the research also showed that the bush became more resistant to fire as the height of the trees and shrubs increased, and this is illustrated in figure 25.3. Similar responses were obtained in the arid savannas of the Eastern Cape in South Africa (Trollope and Tainton 1986).

Effect of Season of Burning

Season of burning is one of the most controversial questions concerning the use of fire in range management. Nevertheless, very little quantitative information is available on the effect of season of burning on the grass sward. West (1965) stressed the importance of burning when the grass is dormant and advocated

burning just prior to the spring rains at the end of winter in order to obtain a high-intensity fire necessary for controlling bush encroachment. Conversely, Scott (1971) stated that burning in winter damages the grass sward and recommended burning after the first spring rains for all forms of controlled burning. However, more recent research has led to the conclusion that, for all practical purposes, burning when the grass sward in dormant in late winter or immediately after the first spring rains has very little difference in effect on the grass sward (Tainton, Groves, and Nash 1977; Dillon 1980; Trollope 1987).

It is difficult to ascertain the effect of season of burning on bush because generally it is confounded with fire intensity. This is because when the trees are dormant in winter the grass is dry and supports intense fires, whereas when the trees are actively growing during summer the grass is green and the fires are much cooler. Suffice it to say that West (1965) postulated that trees and shrubs are probably more susceptible to fire at the end of the dry season when the plant reserves are depleted due to the new spring growth. However, the results of Trollope, Potgieter, and Zambatis (1990) showed that the mortality of bush in the Kruger National Park was only 1.3% after fires that had been applied to bush ranging from dormant to actively growing plants. Therefore it would appear that bush is not sensitive to season of burn in the savanna areas of southern Africa.

Effect of Frequency of Burning

The effect of frequency of burning on the grass sward has been investigated in the False Thornveld of the Eastern Cape in South Africa, where it was found that frequent burning favors an increase in *Themeda triandra* and a decrease in *Cymbopogon plurinodis* (Robinson, Gibbs-Russel, Trollope, and Downing 1979; Forbes and Trollope 1990). The effect of frequency of burning on *T. triandra* is illustrated in figure 25.4. Similar results were obtained by Scott (1970) and Dillon (1980) in the Tall Grassveld of Natal in South Africa, where annual burning in late winter and spring favored the dominance of *T. triandra*. Conversely *Tristachya leucothrix* became dominant with complete protection from fire and with less frequent burning.

Conflicting results have been obtained on the effect of frequency of burning on bush. Kennan (1971) in Zimbabwe and van Wyk (1971) in the Kruger National Park in South Africa both found that there were no biologically meaningful changes in bush density in response to different burning frequencies. In the False Thornveld of the Eastern Cape in South Africa,

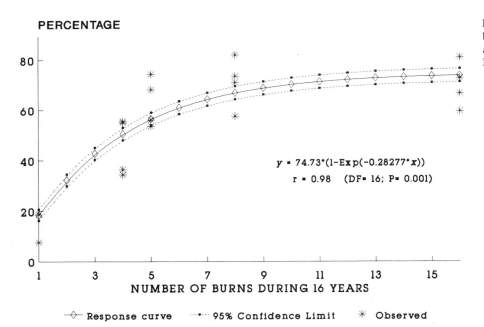

PERCENTAGE

$$y = 74.73 \cdot (1 - \text{Exp}(-0.28277 \cdot x))$$
$$r = 0.98 \quad (DF = 16; \ P = 0.001)$$

NUMBER OF BURNS DURING 16 YEARS

◇— Response curve ·····95% Confidence Limit ✳ Observed

Figure 25.4 Effect of frequency of burning on the abundance of *Themeda triandra* in the False Thornveld of the Eastern Cape in South Africa

Trollope (1983) found that after 10 years of annual burning, the density of bush increased by 41%, the majority of which were in the form of short coppicing plants. Conversely Sweet (1982) in Botswana and Boultwood and Rodel (1981) in Zimbabwe found that annual burning caused a significantly greater reduction in the density of bush than less frequent burning. It is difficult to draw any general conclusions from these contradictory results except to note that in all cases significant numbers of bushes were present even in the areas burnt annually, irrespective of whether they had decreased or increased after burning.

The effect of frequency of burning on forage production has not been intensively studied in the savannas of southern Africa, and only limited quantitative data are available. The general conclusion is that the immediate effect of burning on the grass sward is to significantly reduce the yield of grass during the first growing season after burning, but the depressive effect disappears during the second season (Tainton and Mentis 1984; Trollope 1984).

The effect of frequency of burning on the quality of forage is that generally frequent fires improve and maintain the nutritional quality of the grass sward, particularly in moist savannas (West 1965; Tainton, Groves, and Nash 1977).

There is apparently no information available on the effect of frequency of burning on the production and quality of browse by bush in the savanna areas.

This more comprehensive account of the fire ecology of the savanna areas of Africa has enabled the formulation of more ecologically acceptable and economically viable guidelines for using fire as a tool in the management of rangeland for livestock production and wildlife management in southern Africa (Tainton 1981; Trollope 1989; Trollope 1990). It has also thrown more light on the undesirable effects of fire on the environment when burning is incorrectly used in range management and has brought greater clarity to the conclusion drawn by Phillips (1965) that "fire is a bad master but a good servant."

Fire Regime of the Kruger National Park

The Kruger National Park is situated in the northeastern region of South Africa, bordering on Mozambique in the east and Zimbabwe in the north. The park comprises 1.9 million ha of mainly arid savanna, receiving 350 to 500 mm of rain per annum, but also has limited examples of moist savanna in the south and northwest regions, receiving a mean annual rainfall of approximately 700 mm. A wide variety of different soil types occur in the park, with heavier soils derived from basalt dominating in the eastern half and sandy soils derived from granite in the western half (van Wyk 1971). These variations in the climatic and edaphic environment have resulted in a wide variety of different types of savanna communities, making the study of fire ecology in the Kruger National Park applicable to a major portion of the savanna areas of southern Africa.

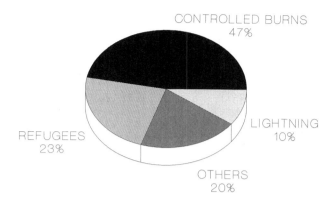

Figure 25.5 The proportion of the Kruger National Park that has been burnt by different ignition sources during the period 5 May 1980 to 30 September 1992

Ignition Sources of Fires

In describing the fire regime of the Kruger National Park, it is important to recognize that it varies according to the source of ignition of the fires, and therefore this aspect must be considered. The data presented are for the period 5 May 1980 to 30 September 1992 and comprise information on 1137 fires, which burnt a total area of 3 041 249 ha, which is equivalent to an area 1.6 times the size of the park (Trollope 1993). The different sources of ignition of fires in the Kruger National Park for this period are presented in figure 25.5.

The most important ignition source in the park was controlled burning (47%), which is applied on a rotational basis to remove moribund and/or unpalatable grass material and to maintain an optimum balance between grass and tree vegetation. The next most important ignition source was fires caused by refugees (23%) fleeing from Mozambique via the park into South Africa. These homeless people lit fires for protection, cooking, and warmth, which then spread either by accident or when left unattended. The next most important causes of fires have been grouped under a general heading called others (20%). This includes wild fires caused by poachers, tourists, arsonists, accidents, and reasons unknown. Surprisingly, lightning was not a very significant ignition source (10%), and the probable reason for this is that ignition from other sources currently has a greater probability of causing fires under the present form of management in the park and political conditions in Mozambique (Trollope 1993). This concurs with what is occurring on a global scale where, with the significant increase in the human population, the majority of the tropical savannas are being shaped and maintained by anthro-

pogenic fires, and the stage has now been reached that in most regions of the world humans have become more important than lightning as sources of ignition (Goldammer and Crutzen 1993). The incidence of lightning fires has also probably decreased, as is indicated by the fact that data collected since 1989 during the annual assessment of the condition of the rangeland in the park showed that the majority of fires caused by lightning (66%) occurred under conditions where the grass fuel load was low (< 2000 kg ha^{-1}). Furthermore, because the overall standing crop of grass in the park was also generally low, there was less of an opportunity for these fires to spread and burn a greater area. It is therefore possible that under pristine conditions that may have existed in the past, lightning was a more important source of ignition of fires in the savanna areas currently occupied by the Kruger National Park than is the situation at present (Trollope 1993).

Type of Fire

The most commonly occurring types of fires in the park are *surface fires* burning in the grass sward either as *head fires* with the wind or *back fires* against the wind. *Crown fires* do occur when the aerial portions of trees and shrubs occasionally ignite during fierce high-intensity head fires, but these are the exception rather than the rule. No data are available on what proportion of the burns caused by the different ignition sources occur as the aforementioned types of fires. Suffice it to say that the rotational burns applied up until 1990 occurred mainly as head fires because the burning program comprised dividing the park up into 456 burning blocks that were burnt on a rotational system according to certain prescribed criteria. As a result of the burning blocks being relatively small (mean = 4800 ha), the procedure of igniting the perimeter of the block tended to cause the development of a central coalescing smoke plume that set up convection currents that drew the flame fronts inward from around the perimeter, resulting in the burns being predominantly head fires burning with the wind. This was subsequently concluded to be ecologically undesirable because having the area burnt mainly by one type of fire was probably causing a relatively uniform effect on the vegetation, therefore reducing biodiversity and the range of different habitat types. In an effort to introduce a more natural type of fire regime, the 456 burning blocks were combined into 88 larger burning units in 1991. These larger areas were also being ignited around the perimeter, but because the burning units were significantly greater, the areas burnt for much longer periods, that is, days rather than hours, during which time the fire front fragmented into separate fires and was driven in different directions in response to changes in the wind direction. The diurnal variations in the atmosphere during the extended burning period also resulted in different fire behavior, causing the controlled burn to have a greater range of effects on the vegetation. It was hypothesized and anticipated that the rotational burns ignited in this way would comprise a greater range of types of fires than was previously the case (Trollope 1993).

No quantitative data are available on the relative proportions of the different types of fires caused by the other ignition sources. It is hypothesized though that because these ignition sources result in fires developing from a single point, the burns will comprise a greater range of types of fires because with the fire front spreading and increasing outward from the ignition point, it will be continually influenced by changes in the wind direction (Trollope 1993).

Fire Intensity

Research on fire behavior in the Kruger National Park by Trollope and Potgieter (1985) led to the conclusion that fire intensity as defined by Byram (1959) is the parameter best suited for describing the general behavior of fires in the park and provides a convenient means for formulating guidelines for controlled burning. Based on these data and research conducted in the Eastern Cape by Trollope (1983), a fire intensity model was developed based on the effects of fuel load, fuel moisture, relative humidity, and wind speed on fire intensity. Recognizing that fuel load and fuel moisture have the greatest effect on fire intensity and using this fire intensity model, estimates of the potential of the rangeland in the Kruger National Park to support fires of different intensities were made. Assuming that fires burn mostly during the dry season when the fuel moisture does not exceed 20%, the model indicates that intense fires (>3000 kJ s^{-1} m^{-1}) occur when the fuel load of grass is >4000 kg ha^{-1} and cool to moderately intense fires (<3000 kJ s^{-1} m^{-1}) when the fuel load is <4000 kg ha^{-1}. Field experience also indicates that when the grass fuel load is <2000 kg ha^{-1}, fires will generally not spread because of the discontinuity of the fuel bed. Using the estimates of the standing crop of grass measured with a disc meter that have been recorded annually in the park since 1989, estimates were made of proportion of the rangeland that would support fires of different intensities (figure 25.6). The data in figure 25.6 illustrate the important influence fuel load has on the potential fire intensity of the

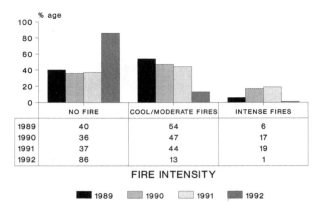

Figure 25.6 The potential intensity of head fires in the Kruger National Park as determined by the fuel load of grass during the period 1989 to 1992

Table 25.2 The mean, maximum, and minimum intensities of head and back fire recorded during the SAFARI-92 project in the Kruger National Park during August and September 1992

Type of fire	Fire intensity (kJ s^{-1} m^{-1})		
	Mean	Maximum	Minimum
Head fire	2810 ± 893	8845	22
Back fire	77 ± 29	160	20

rangeland in the park, and the effect of the devastating drought experienced during 1992 is clearly shown by the high proportion of the park (85%) that could not have supported a fire.

Obviously it is not logistically possible to monitor the intensities of the fires that occur in the Kruger National Park annually. However, the intensities of the fires that were recorded during August and September 1992, as part of the SAFARI-92 (Southern African Fire-Atmosphere Research Initiative—1992) project, give a representative indication of the range of fire intensities that occur in the park under field conditions (table 25.2). The data in table 25.2 indicate that head fires are generally far more intense than back fires but are more variable. These results corroborate the behavior of head and back fires that has been observed during the application of controlled burns on a field scale in the Kruger National Park (Trollope 1993).

Season of Burning

Data are available on the monthly occurrence of fires caused by controlled rotational burning, refugees, lightning, and other causes in the Kruger National Park since 1980. The areas burnt by these fires, ex-

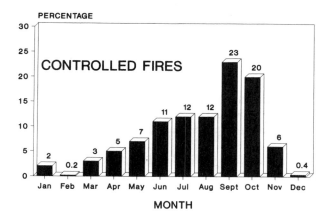

Figure 25.7 The percentage of area burnt by controlled burns during the different months of the year in the Kruger National Park for the period 1980 to 1992

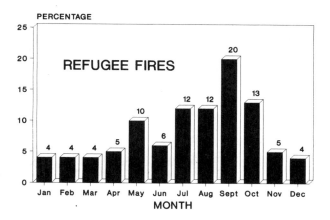

Figure 25.8 The percentage of area burnt by fires caused by refugees during the different months of the year in the Kruger National Park for the period 1985 to 1992

pressed as a percentage, are presented in figures 25.7, 25.8, 25.9, and 25.10. The data in these figures show that, except for lightning, the majority of the fires occurred during the dry, dormant winter period. Of course fires caused by lightning also happen when the grass sward is dry, but the data would suggest that these fires are associated with dry lightning storms that are prevalent from October through January (Trollope 1993).

Frequency of Burning

The frequency of burning in the Kruger National Park is highly variable and is influenced by the annual rainfall and the grazing pressure exerted by the wildlife populations that fluctuate from year to year. For example, in 1981, 37.3% of the park burned as a result of high fuel loads accumulating after a period of above average rainfall, whereas only 1.5% burned in 1992 as a result of the devastating drought severely depressing grass production. A further analysis of the data showed that the frequency of burning was greater in sourveld areas than in sweetveld areas, that is, high and low rainfall areas respectively. The results showed that on average the sourveld areas were triennially burned and the sweetveld areas octennially burned. This is a credible result, as the quality of the grass forage in the sourveld areas declines significantly over time due to the high rainfall leaching out the soluble chemical fractions of the plant material. Conversely in sweetveld areas, which receive a low rainfall, the quality of the grass forage does not decline to the extent that the herbage becomes unpalatable to grazing animals, even when the grass is dry and dormant during winter.

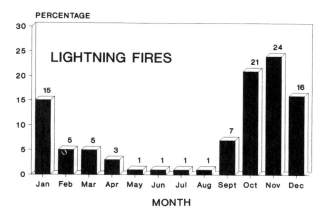

Figure 25.9 The percentage of area burnt by fires caused by lightning during the different months of the year in the Kruger National Park for the period 1980 to 1992

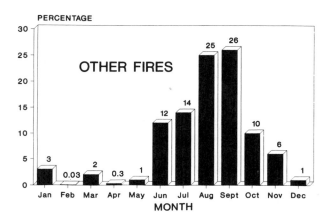

Figure 25.10 The percentage of area burnt by fires ignited by "other" causes during the different months of the year in the Kruger National Park for the period 1980 to 1992

Therefore, the level of utilization of grass forage is much higher in these areas, and the accumulation of adequate fuel loads of grass only occurs during above average rainfall years, when the grazing capacity of the rangeland far exceeds the stocking rate of grazing animals (Trollope 1993).

General Discussion and Conclusions

The aforegoing review of biomass burning in the savannas of southern Africa with particular reference to the Kruger National Park indicates that the fire regime in the savanna areas is highly variable as a result of being affected by variations in climate, soils, and anthropogenic factors. The influence of different patterns of land use is another source of variation that has not been considered in this review but obviously has a major influence on the type and intensity of fires, the season and frequency of burning, and the respective effects of these elements of the fire regime on the biotic and abiotic components of the savanna ecosystems. A positive aspect of this resultant highly variable fire regime is that its effect on the different savanna communities is also highly variable, thus ensuring the maintenance of a wide range of habitat types and biodiversity. A common theme that emerges from this review is that fire is an integral and essential part of the environment affecting the savannas of southern Africa and is highly necessary for their optimum functioning. For example, in the moist savannas the periodic removal of unpalatable and moribund grass material by burning is absolutely necessary for maintaining the palatability and vigor of the grass sward for use by either domestic or wild grazing animals. The exclusion of fire in these ecosystems is both economically and ecologically undesirable when considered in relation to the most common form of land use in the savanna areas, namely, livestock production. Fire is also necessary in the arid savannas, where it plays an important role in maintaining the optimum balance between grass and bush vegetation. Its prescribed or unintended exclusion in recent times from these savanna communities has been one of the main causes of bush encroachment, with disastrous negative effects on the carrying capacity of the rangeland.

One of the major developments in fire ecology in Africa occurred in 1992 when the *Southern African Fire-Atmosphere Research Initiative (SAFARI)* was launched to investigate the effect of biomass burning on the chemical content of the atmosphere (Andreae, Goldammer, and Lindesay 1992). The discovery of significantly elevated concentrations of ozone over large parts of the tropics, particularly over the southern tropical Atlantic Ocean between Africa and South America, during the period August to October every year provided the original stimulus for launching a major research investigation. The preliminary results of the project indicate that biomass burning does contribute to an increase of trace gases in the atmosphere that may have an effect on global climatic change. However, care must be taken to differentiate between biomass burning that results in a net release of gases to the atmosphere and biomass burning that results in a cyclic release and reabsorption of gases to and from the atmosphere. An example of the net release of trace gases is the wholesale destruction of rain forest through felling and burning in the Congo Basin in Zaire. In contrast, there is the cyclic release of gases occurring during the burning of savanna or grassland as a range management practice in South Africa where gases like carbon dioxide are reabsorbed during photosynthesis when the burnt plants regrow during the following growing season. Clearly concern must be expressed about biomass burning that results in massive net releases of gases into the atmosphere that could contribute to global climatic change such as the current widespread destruction of rain forest in the world. However, it would be both ecologically and economically unwise to suggest the conscious withdrawal of controlled burning in grassland and savanna ecosystems when used as a recommended range management practice in extensive livestock production systems. This type of biomass burning does not result in a net release of gases into the atmosphere and is often an ecologically essential treatment for the sustainable existence of most grassland and savanna ecosystems. Therefore, when assessing the impact of biomass burning on possible global climatic change, care must be taken to differentiate between the different types of biomass burning that occur in different situations in the world.

References

Andreae, M. O. Goldammer, J. G., and Lindesay, J. A., 1992. *Southern African Fire-Atmosphere Research Initiative: A subprogramme of the International Geosphere-Biosphere Program/ International Global Atmospheric Chemistry Program.* Final prospectus. University Witwatersrand, South Africa.

Anonymous. 1992. *ESCOM Annual Report.* Electricity Supply Commission, South Africa, 1–25.

Bartholomew, J. C. 1987. Plate 4—World Vegetation. In *The Times Atlas of the World.* 7th edition. John Bartholomew, Edinburgh, Great Britain.

Boultwood, J. N., and Rodel, M. G. W. 1981. Effects of stocking rate and burning frequency on *Brachystegia/Julbernadia* veld in Zimbabwe. *Proc. Grassl. Soc. S. Afr.* 16, 111–115.

Brown, A. A., and Davis, K. P. 1973. *Forest Fire: Control and Use.* McGraw-Hill, New York.

Byram, G. M. 1959. Combustion of forest fuels. In *Forest Fire: Control and Use.* Ed. K. P. Davis. McGraw-Hill, New York.

Dillon, R. F. 1980. Some effects of fire in the Tall Grassveld of Natal. M. Sc. (Agric.) Thesis, Univ. Natal. Pietermaritzburg.

Forbes, R. G., and Trollope, W. S. W. 1990. Effect of burning on sweet grassveld in Ciskei and the Eastern Cape. Unpublished paper. University Fort Hare, Alice, Ciskei.

Goldammer, J. G., and Crutzen, P. J., 1993. Fire in the environment: Scientific rationale and summary of results of the Dahlem Workshop. In Crutzen, P. J., and Goldammer, J. G. (eds.). *Fire in the Environment: The Ecological, Atmospheric and Climatic Importance of Vegetation Fires.* John Wiley, Toronto, 1–14.

Kennan, T. C. D. 1971. The effects of fire on two vegetation types of Matopos. *Proc. Tall Timbers Fire Ecology Conf.* 11, 53–98.

Komarek, E. V. 1971. Lightning and fire ecology in Africa. *Proc. Tall Timbers Fire Ecology Conf.* 11, 473–511.

McArthur, A. G. 1966. Weather and grassland fire behaviour. *Commonw. Aust. For. Timb. Bur. Leafl.* 100.

Phillips, J. F. V., 1965. Fire—as a master and servant: its influence in the bioclimatic regions of Trans-Saharan Africa. *Proc. Tall Timbers Fire Ecology Conf.* 4, 7–10.

Pyne, S. J., 1982. *Fire in America: A Cultural History of Wildland and Rural Fire.* Princeton University Press, Princeton, N.J.

Robinson, E. T., Gibbs-Russel, G. E., Trollope, W. S. W., and Downing, B. H., 1979. Assessment of short term burning treatments on the herb layer of False Thornveld of the Eastern Cape. *Proc. Grassl. Soc. S. Afr.* 14, 79–83.

Rose-Innes, R., 1971. Fire in west African vegetation. *Proc. Tall Timbers Fire Ecology Conf.* 11, 147–173.

Rothermel, R. C., 1972. A mathematical model for predicting fire speed in wildland fuels. *U.S.D.A. For. Serv. Res. Paper I.N.T.* 115.

Scott, J. D., 1970. Pros and cons of eliminating veld burning. *Proc. Grassld. Soc. S. Afr.* 5, 23–26.

Scott, J. D. 1971. Veld burning in Natal. *Proc. Tall Timbers Fire Ecology Conf.* 11, 33–51.

Sweet, R. J., 1982. Bush control with fire in *Acacia nigrescens/Combretum apiculatum* savanna in Botswana. *Proc. Grassl. Soc. S. Afr.* 17, 25–28.

Tainton, N. M. 1981. *Veld and Pasture Management in South Africa.* Shuter and Shooter, Pietermaritzburg.

Tainton, N. M., Groves, R. H. and Nash, R., 1977. Time of mowing and burning veld: short term effects on production and tiller development. *Proc. Grassld. Soc. S. Afr.* 12, 59–64.

Tainton, N. M., and Mentis, M. T., 1984. Fire in grassland. In Booysen, P. de V., and Tainton, N. M. (eds.). *Ecological Effects of Fire in South African Ecosystems. Ecological Studies* No. 48, 115–197.

Trollope, W. S. W., 1978. Fire behaviour—a preliminary study. *Proc. Grassl. Soc. S. Afr.* 13, 123–128.

Trollope, W. S. W., 1983. Control of bush encroachment with fire in the arid savannas of southeastern Africa. PhD thesis, University Natal, Pietermaritzburg.

Trollope, W. S. W., 1984. Fire in savanna. In Booysen, P. de V., and Tainton, N. M. (eds.). *Ecological Effects of Fire in South African Ecosystems. Ecological Studies* No. 48, 149–175.

Trollope, W. S. W., 1987. Effect of season of burning on grass recovery in the False Thornveld of the Eastern Cape. *J. Grassl. Soc. S. Afr.* 4, 2, 74–77.

Trollope, W. S. W., 1989. Veld burning as a management practice in livestock production. In Danckwerts, J. E., and Teague, W. R. (eds.). *Veld Management in the Eastern Cape.* Government Printer, Pretoria, 67–73.

Trollope, W. S. W., 1990. Veld management with specific reference to game ranching in the grassland and savanna areas of South Africa. *Koedoe* 33, 2, 77–86.

Trollope, W. S. W., 1993. Fire regime of the Kruger National Park for the period 1980–1992. *Koedoe,* 36, 2, 45–52.

Trollope, W. S. W., and Potgieter, A. L. F., 1985. Fire behaviour in the Kruger National Park. *J. Grassl. Soc. S. Afr.* 2, 2, 17–22.

Trollope, W. S. W., Potgieter, A. L. F., and Zambatis, N., 1990. Characterization of fire behaviour in the Kruger National Park. Final report. National Parks Board, Kruger National Park, Skukuza.

Trollope, W. S. W., and Tainton, N. M., 1986. Effect of fire intensity on the grass and bush components of the Eastern Cape Thornveld. *J. Grassl. Soc. S. Afr.* 2, 27–42.

Trollope, W. S. W., Trollope, L., and Bosch. O. J. H. 1990. Veld and pasture terminology in southern Africa. *J. Grassl. Soc. S. Afr.* 7(1), 52–61.

West, O., 1965. Fire in vegetation and its use in pasture management with special reference to tropical and sub-tropical Africa. *Mem. Pub. Commonw. Agric. Bur.*, Farnham Royal, Bucks., England. No. 1.

van Wyk, P., 1971. Veld burning in the Kruger National Park, an interim report of some aspects of research. *Proc. Tall Timbers Fire Ecology Conf.* 11, 9–13.

Contribution of the Spread-Fire Model in the Study of Savanna Fires

Micheline Moula, John-Michel Brustet, Hugh Douglas Eva, Jean-Pierre Lacaux,
Jean-Marie Grégoire, and Jacques Fontan

On a global scale, biomass burning constitutes an important source of numerous trace compounds including carbon dioxide, methane, nitrous oxide, and aerosols that are known to exert a notable influence on the radiative budget of the earth. The principal observable consequence of their rejection into the atmosphere is an increase in the amount of gases involved in the greenhouse effect and of those having an oxidizing and acidifying power. According to recent estimates, 60 to 80% of the biomass of African savannas is burnt throughout the dry season (Andreae 1990).

However, for more thorough investigations, it becomes necessary to evaluate the amount of burnt biomass over a reduced time scale, of the order of one day to one week, corresponding to the lifetime of the chemical species emitted through the combustion process. In situ satellite imagery constitutes one of the most advance technologies for the remote sensing of biomass burning. The Advanced Very High Resolution Radiometer (AVHRR) embarked on (NOAA) satellites provides both large-scale (GAC) images (5 km) and small-scale (LAC) images (1 km) of different quality allowing the seasonal evolution of biomass burning over a given region or continent to be traced (Dozier 1981; Malingreau et al. 1985; Malingreau and Tucker 1987; Matson et al. 1984, 1985, 1986, 1987; Brustet et al. 1990). However, the spatial resolution of LAC images is notably insufficient to obtain precise information on a given individual fire (e.g., fire-front length, rate of spread).

A more accurate numbering of the fires existing along the overpassed area can be achieved by exploitation of LAC images and by an appropriate treatment aimed to suppress false signals. With such improvements, AVHRR imagery allows a fine description of the fires (Vickos 1991), revealing in particular a south-bound shift of the fire zones located in the Northern Hemisphere throughout the dry season. Such a displacement appears to be dependent on climatiologic conditions and on the state of the vegetation (Franca 1994; Kennedy, Belward, and Grégoire 1994).

Current evaluations of biomass burning hardly take into account the amount of biomass burnt daily, a value that would be necessary for an accurate description of spatiotemporal scales adapted to atmospheric chemistry.

Our present objective is to propose a conceptual scheme for evaluating the daily temporal evolution of burnt biomass. In our approach, a fire simulation model (BEHAVE) is used to estimate the rate of spread and the ignition probability, whereas satellite imagery allows determination of the density of fires during the period of maximum activity (between 2 P.M. and 4 P.M.).

The model used in this chapter was validated during the course of a series of airborne and surface measurements campaigns carried out in Central African Republic. The probability of observing the fires was calculated from the model and compared with experimental data obtained from AVHRR imagery. In this chapter, these informations were used to calculate the fluctuations of the amount of biomass burnt every day.

Conceptual Approach Used for Studying the Daily Variation of Biomass Burning

Several methods aimed at determining the amount of burnt biomass have been developed. In particular, the method proposed by Seiler and Crutzen (1980) is based on the determination of burnt surfaces, whereas the one reported by Kaufman and al. (1990) rests on the data obtained from the detection of smoke by remote sensing. Vickos (1991) has developed a calculation of the amount of burnt biomass based on the measurement of the fire-front rate of spread.

Our own approach is an extension of the latter method. We first calculate the temporal evolution of the number of fires and then estimate the amount of biomass burnt within a given zone over a period of one day. The different steps required to obtain the latter value are shown in figure 26.1. The data we use for this purpose are obtained from surface measurements, air-

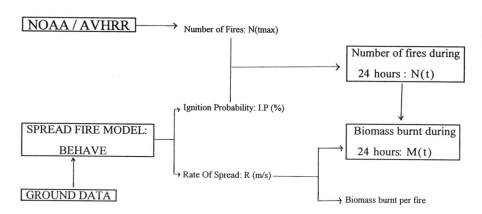

Figure 26.1 Concept to approach the daily variation of biomass burning

borne measurements, satellite imagery, and utilization of the BEHAVE model (Rothermel 1972).

The amount of burnt biomass (M) is expressed by equation (26.1), where N is the number of active fires, q is the amount of biomass per surface unit, k is the efficiency coefficient for combustion, L is the length of the fire front, and R is the rate of spread.

$$M(kg/s) = N(t) * R(t) * [q * k * L] \qquad (26.1)$$

Surface measurements provide the amount of biomass per surface unit (q), the efficiency coefficient for combustion (k), and the micrometeorological parameters (wind, relative humidity, and air temperature). Airborne measurements (by plane) allow obtainment of the average fire-front length (L). The data relevant to the biomass characteristics (total amount of biomass, percentage of dead and living matter, average height of the combustion bed, surface-volume ratio of living and dead fuel as well as their calorific capacity) are introduced as parameters of the BEHAVE model, which forecasts the behavior of the fire from meteorological and topographical conditions. The exit parameters pertinent to our investigation are the rate of spread (R) and the ignition probability (IP), calculated over a daily cycle.

With the satellites of the NOAA series under normal data acquisition conditions, we obtain the number of fires (N) at a given time (generally corresponding to the satellite overpass in the afternoon). Upon combining the number of fires thus obtained and the ignition probability, we obtain the temporal evolution of the number of fires N(t) and then the temporal evolution of burnt biomass.

Experimental Strategy

As part of the DECAFE (dynamique et chimie de l'atmosphère en forêt équatoriale) program and the FIRE (Fire in Global Resources and Environmental Monitoring) project, a measurement campaign was undertaken in the Central African Republic during the dry season, in January 1994. During the course of that campaign, a NOAA-AVHRR mobile receiving station was installed in Bangui, whereas additional measurements at ground level or by plane were carried out on different sites with the aim of testing the validity of the model on that type of vegetation and to highlight the more important parameters.

The experimental data for the present study were collected in January and February 1994 during the dry season. The principal experimental sites were located in the neighborhood of Bangui (4°N, 18°E). One of them was located 500 km north from Bangui. The dimensions of the sites ranged between 1 and 6 ha. Three types of measurements were made on six selected experimental fires.

Surface Measurements

Biomass Before and After the Fire (t/ha) An estimate of the amount of biomass was made on standard surfaces of 9 m², both before and after the fire. The collected material was sorted out as dead and living matter.

Meteorological Conditions The temperature (°C), the relative humidity (%) of the air, and the wind speed and direction were recorded. These parameters were periodically measured every 2 minutes, starting 30 minutes before the fire until the end of the fire. The instruments were placed 50 m away from the fire and at a height of 2 m to avoid any perturbation due to the fire.

Water Content of the Fuel (%) Vegetation samples were selected before each fire. Each sample (dead or living vegetation) was weighed, and the water content of the fuel was determined after drying the herbaceous material at 106°C over a period of 24 hours. The humidity of the fuel is expressed in percentage relative to the amount of dry matter.

Rate of Spread (m/s) The rate of spread of the fire was estimated by measuring the displacement time between two points separated by 5 to 20 m.

Airborne Measurements
The flights over experimental fires were made at an altitude of about 500 m. The plane was equipped with two cameras exploring respectively near-infrared (IR) and visible domains.

The focus of the IR camera was 42° and that of the visible camera was 29°.

Rate of Spread of the Fire The rate of spread was evaluated by overpassing the fire at intervals of 3 to 4 minutes.

Burnt Surfaces and Fire-Front Length The IR and visible cameras recorded images of burnt surfaces during the flights to evaluate the fire-front lengths. These data were used to calculate the average length of a fire and to evaluate the amount of biomass burnt in the investigated region.

Satellite Imagery
A receiving station collecting the data recorded by NOAA was installed in Bangui from December 1993 to February 1994. A total of 138 images were recorded during that period. After elimination of those truncated by the presence of clouds, 95 exploitable images were selected and classified as diurnal and nocturnal images. These images were processed at the MTV (Monitoring Tropical Vegetation) unit of the European Union Joint Research Center in Italy.

The tests used for the extraction of the pixels identified as fires were the following:

1) $T3 \geq 311$ K
2) $T4 > 293$ K
3) $T3 - T4 \geq 14$ K
4) 0 K $\leq T4 - T5 \leq 4$ K
5) $T5 > 291$ K

The Ti values represent the apparent temperature in channel i. The values are expressed in Kelvin degrees (K).

The three first criteria were developed by Kaufman et al. (1990), the fourth one by Franca (1993), and the fifth one was included under our experimental conditions. The values of thresholds given here were obtained by statistical methods.

As a whole, the data defining the biomass as well as those relevant to the climatic environment constitute the entry parameters for the BEHAVE model.

Forecasting the Fire Behavior: BEHAVE Model

The model forecasting the behavior of the fire is based on the propagation model established by Rothermel (1972) and was originally developed in the United States. It was subsequently modified and completed by several addenda (Albini 1976; Rothermel 1983; Andrews 1986). The model BEHAVE includes two subroutines, namely FUEL and BURN.

The subprogram FUEL, described by Burgan and Rothermel (1984) uses the vegetation characteristics to set up and test a fuel model. Several fuel model tests were reported in the literature (Everson, Van Wilgen, and Everson 1988; Sneuwjagt and Frandsen 1977; Van Wilgen, le Maitre, and Kruger 1985).

The subprogram BURN, described by Andrews (1986) and Andrews and Chase (1989), includes the fire propagation model along with other files, the role of which is to estimate some additional parameters such as the humidity of the fuel and its ignition probability.

Rate of Spread of the Fire
The rate of spread of the fire is calculated and expressed as the ratio between the adsorbed heat flux and the flux necessary to bring the fuel to the ignition temperature (Frandsen 1971).

Ignition Probability (IP)
The ignition probability is defined as the probability for a branch to produce a heat superior or equal to the preignition heat. Thus, this is the probability for an existing fire to continue to burn. The preignition heat is defined as the net heat (gain − loss) necessary to bring the fuel from its initial temperature to the ignition temperature. The equation used to calculate the ignition probability was developed by Schroeder (1969). It is based on the temperature of the fuel (depending on the cloud cover, the wind, the temperature, and the relative humidity of the air) and its water content.

Results and Discussion

Biomass Characteristics
The constitution of biomass before and after the fire is reported in table 26.1. Despite the fact that the sites were selected upon consideration of their apparent homogeneity, we note important variations of the total

Table 26.1 Dry mass of vegetation (t/ha) before and after burning

	Site					
	1	2	3	4	5	6
Total biomass before burning ǫ(t/ha)	17	13	23	14	9	5
Dead material (t/ha)	5	7	17	11	3	5
Total biomass after burning (t/ha)	4.6	3.5	5.3	4.6	2.7	0
% Biomass consumed (k)	73	73	77	67	70	100

Table 26.2 Weather and fuel conditions for the experimental fires

	Site					
	1	2	3	4	5	6
Date of burn	01/25	01/27	01/28	01/29	02/02	01/31
Time of burn (h:min)	15:10	15:44	16:22	16:13	15:14	12:20
Air temperature (°C)	29	32.5	33.6	32	32.6	35
Relative humidity (%)	46	48	48	56	20.6	13.4
Wind speed (m/s)	0.36	0.47	1.25	0.28	0.25	0.6
Dead fuel moisture (% dry weight)	5	5	6	5	5	3
Live fuel moisture (% dry weight)	81	96	106	83	92	

amount of biomass and the ratio between dead and living matter. The reported values are estimated with a precision of $\pm 20\%$. The first five sites correspond to a Guinean savanna, whereas the sixth one corresponds to a Sudanian savanna. The fraction of consumed biomass is practically constant.

We have also carried out measurements of the height of the fuel bed and of the diameter of stems, assumed to be cylindric and representing 70% of the total biomass. These parameters were estimated on all samples with the aim to define accurately the combustible and the parameters that will enable the model to forecast the rate of spread of the fire.

Environmental Conditions

The parameterization of the fuel and the meteorological conditions encountered during the experiments are shown in table 26.2. Though the temperature of the air was not subject to important variations, the fires were seen to burn under rather different conditions, since the wind speed was found to vary from 0.25 to 1.25 m/s, whereas the relative humidity of the air was varying between 13 and 56%.

Comparison Between Measured and Forecasted Rate of Spread

The propagation rate predicted by Rothermel's model was compared with the observed rate of spread. The results obtained for the predicted rate of spread (Vp) show a dispersion of the points along the diagonal line, which corresponds to a perfect agreement between forecasted and observed rates. In order to adjust the model, we have used the regression straight line between observed and forested rates. The equation found for the regression straight line was $R = 0.026 + 1.207^* Vp$ with a correlation coefficient of 0.84, where R is the corrected rate of spread. The plots showing observed versus corrected rate of spread are reported in figure 26.2.

The above result will be used to adjust the model parameters with the aim to improve the accuracy of the forecast for that type of fuel.

Burnt Surfaces and Fire-Front Length

Figures 26.3 and 26.4 represent the burnt surfaces on the 39 observed sites and the fire-front lengths based on 64 observations. The burnt surfaces considered here exhibit areas smaller than 15 ha, with an average of 4.5 ha. The fire-front lengths ranged between 30 and 800 m, with an average of 190 m. The "average" notion is used here, though the distributions are not of Gaussian type. Considering the above distributions, one may reasonably suggest that the dimensions of fire fronts and burnt surfaces are limited by the presence of the gallery forest and by the cutout of parcels.

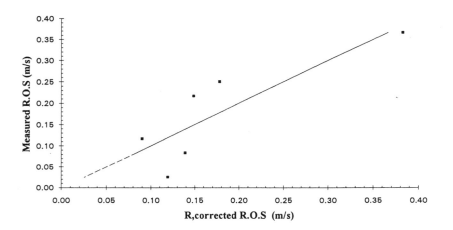

Figure 26.2 Measured rate of spread (R. O. S) against corrected rate of spread

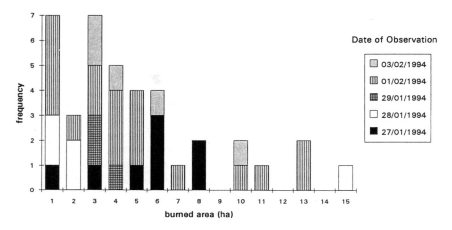

Figure 26.3 Distribution of burnt surface areas derived from airborne video

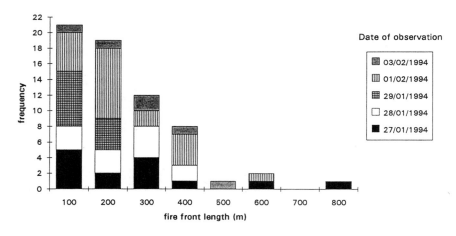

Figure 26.4 Distribution of fire-front lengths derived from airborne video

Table 26.3 Estimate of burnt biomass per fire unit

	Site					
	1	2	3	4	5	6
Corrected rate of spread: R (m/s)	0.12	0.39	0.33	0.19	0.04	0.24
Burnt biomass per fire event: M (kg/s)	28	29	117	18	2	72

Estimation of Burnt Biomass

Taking into account an average fire-front length of 190 m, we can estimate the amount of burnt biomass per fire and per time unit. The results obtained for the six sites are displayed in table 26.3.

After estimating the amount of burnt biomass for one fire, we use satellite data to calculate it for the number of fires existing in the considered zone over a period of one day.

Combined Use of Remote Sensing and Modelling

We have selected the series of images corresponding to a maximum data acquisition, namely, those recorded during 4, 5, and 6 January 1994. Independent thresholds were calculated for diurnal and nocturnal images, respectively, using a statistical method. This allowed maps to be drawn giving the distribution of pixels detecting a fire. The saturation of the sensor in channel 3 is indicative of the presence of an active fire front.

Figure 26.5 displays the number of fire pixels detected for each image during the three days over an area of 184 000 km². The plot reveals the occurrence of a daily cycle, with maximum values in the afternoon around 3 P.M. and a total extinction at night. The maximum corresponds to a saturation of 0.6% of the pixels.

During these three days, we have also calculated the ignition probability of the fire by using the model BEHAVE. The calculation requires entry parameters such as the temperature, the relative air humidity, the cloud cover, the wind speed, and the characteristics of the fuel. The temporal variation of the ignition probability over the three days is also displayed in figure 26.5. It also shows a daily cycle with a minimum value of 0% around 6 A.M. and a maximum value of 90% around 3 P.M.

Thus, under normal conditions where one image is acquired per day, teledetection and modelization can be combined for evaluating the actual number of fires: whereas satellite imagery provides the density of fires at a given time, the model calculates the temporal variation of the ignition probability. From these data, and considering that the maximum activity is reached between 2 and 4 P.M. during satellite overpasses, we can calculate the number of fires per time unit (t) via the following approximation:

$$N(t) = N(t_{max}) * \frac{PI(t)}{PI(t_{max})} \qquad (26.2)$$

Upon applying the above relationship, we can estimate the temporal evolution of burnt biomass for the considered region, assuming that the biomass, the fire-front length, and the efficiency coefficient are constant in equation (26.1).

We have calculated the variation of the rate of spread throughout the day of 5 January 1994 for a mean wind speed of 0.5 m/s. The total biomass weight is estimated to be of 15 t/ha of dry matter (average on the first five sites) containing 58% of living matter. The efficiency coefficient for combustion is 0.75 and the fire-front length is 190 m.

The results giving the variation of the rate of spread are displayed in figure 26.6. The plot shows a significant variation of the rate of spread during the day

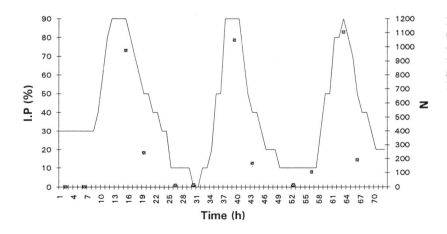

Figure 26.5 Variation of fire number (N) detected by satellite and variation of ignition probability (I. P) over a three-day period for a region located around Bangui (Central African Republic)

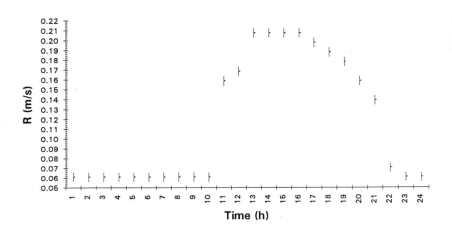

Figure 26.6 Variation of the rate of spread on 5 January 1994 for the same region around Bangui as in figure 26.5

(variation coefficient of 53%), but the variation coefficient is only 13% over the period of active fires (11 A.M. to 9 P.M.).

From equation (26.2), we obtain the temporal evolution of the number of fires N(t). From equation (26.1), we will next evaluate the temporal evolution of the amount of burnt biomass.

Conclusion

Whereas earlier investigations in this topic have dealt primarily with the amount of biomass burnt over a period of one year, the present work has focused on the daily amount of burnt biomass, which appears to be more adapted to modern requirements of atmospheric chemistry.

During the present measurement campaign, we have adapted the propagation model developed by Rothermel to the biomass characteristics encountered in Central African Republic. The rate of spread was found to vary from 0.05 to 0.4 m/s, depending essentially on the wind speed. Airborne measurements allowed us to obtain an average value of the fire-front length, which is of the order of 190 m for the region considered here. Then, we were able to estimate the amount of burnt biomass per fire unit, shown to vary between 2 and 117 kg/s, with an average efficiency coefficient for combustion equal to 0.72.

The use of satellite images revealed that 0.6% of the pixels were saturated, for a surface of 184 000 km^2 during the satellite overpass, in the afternoon. To trace the evolution of the number of fires during the day, we have combined the instantaneous data obtained from satellite images with the information regarding the ignition probability provided by the model BEHAVE. Such a combined use of remote sensing and modelling allows estimation of the number of fires at any time during the day and will lead to more precise evalua-

tions of the fluxes of the compounds emitted by these fires.

Such a method has been designed and developed on a surface of 184 000 km^2, which is representative of a humid savanna having a strong biomass density. It will be next extrapolated to study the whole Central African Republic and also the region considered in the program EXPRESSO (Experiment for Regional Sources and Sinks of Oxidants). This will require the development of surface measurements to determine the biomass characteristics in the different ecological zones (north-south), the creation of a data bank for the storage of various meteorological measurements, and the evaluation of an average fire-front length for each zone by exploitation of LANDSAT TM images.

Thus, it will soon be possible to calculate the temporal evolution of the production of chemical compounds emitted by fires in the EXPRESSO region from the temporal evolution of the amount of burnt biomass, taking into account the emission coefficients proposed by Lacaux (1995) for the wet savanna.

Acknowledgments

The authors gratefully acknowledge the scientific and logistical collaboration of the ORSTOM and ASECNA of Central African Republic (Bangui), with special thanks to J. B. Vickos. This research was supported by the Program DECAFE, the French Ministry of Cooperation (programme Campus) and the Joint Research Center of European Union.

References

Andreae, M. O. 1991. Its history, use, and distribution and its impact on environmental quality and global climate, in *Global Biomass Burning: Atmospheric, Climatic, and Biospheric Implications*, edited by J. S. Levine, pp. 3–21, MIT Press, Cambridge, Mass.

Andrews, P. L. 1986, Behave: Fire behavior prediction and fuel modeling system—Burn subsystem, Part 1. U.S.D.A. Forest Service, General Technical Report INT-194, 130 pp.

Andrews, P. L., and C. H. Chase. 1989, Behave: Fire behavior prediction and fuel modeling system—Burn subsystem, Part 2. U.S.D.A. Forest Service, General Technical Report INT-260, 93 pp.

Albini, F. A. 1976, Estimating wildfire behavior and effects. U.S.D.A. Forest Service, Gen. Tech. Rep. INT-30, 92 pp.

Brustet, J. M. et al. 1990, Remote sensing of biomass burning in West Africa with NOAA-AVHRR, in *Global Biomass Burning*, edited by J. S. Levine, MIT Press, Cambridge, Mass. pp. 47–52.

Brustet, J. M. et al. 1990, Detection and characterization of active biomass fires in West Africa with Landsat Thematic Mapper and NOAA-AVHRR. 7th International Symposium of the Commission on Atmospheric Chemistry and Global Pollution, 5–11 September Chamrousse, France.

Burgan, R. E., and Rothermel, R. C. 1984, BEHAVE: Fire behavior prediction and fuel modelling system—Fuel subsystem. U.S.D.A. Forest Service, General Technical Report INT-167, 126 pp.

Dozier, J. 1981, A method for satellite identification of surface temperature fields of subpixel resolution. *Remote Sensing of Environment*, 11, 221–229.

Everson, T. M., Van Wilgen, and Everson, 1988, Adaptation of a model for rating fire danger in the Natal Drakensberg. *S. Afr. J. Sci.*, 84, 44–49.

Franca, J. R. 1994, Télédétection satellitaire des feux de végétation en Afrique intertropicale—Application à l'estimation des flux des composés en trace émis dans l'atmosphère, Thèse Université Paul Sabatier, 215 pp., Laboratoire d'Aérologie, Toulouse, France.

Frandsen, W. H. 1971, Fire spread through porous fuels from the conservation of energy. *Combustion and Flame*, 16, 9–16.

Kaufman, Y. J., et al. 1990, Remote sensing of biomass burning in tropics. *J. Geophysical Res.* 95, 9927–9939, 20 June.

Kennedy, P. J., Belward, A. S., and Grégoire, J. M. 1994, An improved approach to fire monitoring in West Africa using AVHRR data, *Int. J. Remote Sensing*, 15, 11, 2235–2255.

Lacaux, J. P., et al. 1995, Biomass burning in the tropical savannas of Ivory Coast: An overview of the field experiment fire of savannas (FOS/DECAFE). *J. Atmos. Chem.*, in press.

Malingreau, J. P., et al. 1985, Remote sensing of forest fires: Kalimantan and North Borneo in 1982–1983. *Ambio* 4, 6, 314–321.

Malingreau, J. P., and Tucker, C. J. 1987, The contribution of AVHRR data for measuring and understanding global processes: Large-scale Deforestation in the Amazon Basin. International Geoscience and Remote Sensing Symposium. IGARSS, Proceedings, Ann Arbor, Michigan, 18–21 May, pp. 443–448.

Matson, M., et al. 1984, Fire detection using the National Oceanic and Atmospheric Administration Series Satellites. National Environment Satellite, Data and Information Service, Washington, D.C., 41 pp.

Matson, M., et al. 1985, Fire detection using NOAA polar-orbiting satellite data. Proceedings of the William T. Pecora Memorial Symposium, 10th, Remote Sensing in Forest and Range Resource Management, Fort Collins, Colorado, 20–22 August, pp. 445–446.

Matson, M., et al. 1986, Regional and global fire detection using NOAA satellite data. Proceedings of Australian AVHRR Conference, 1st, Perth, Australia, 22–24 October, pp. 201–209, CSIRO, Division of Groundwater Research, Perth, Australia 6014.

Matson, M., et al. 1987, Fire detection using data from NOAA-N satellite. *Int. J. Remote Sensing*, 7, 961–970.

Rothermel, R. C. 1972, A mathematical model for predicting fire spread in wildland fuels. U.S.D.A. Forest Service, Research Paper INT-115, 40 pp.

Rothermel, R. C. 1983, How to predict the spread and intensity of forest and range fires, Intermountain Forest and Range Experiment Station, Ogden, Utah, U.S.D.A. Forest Service, Gen. Tech. Rep. INT-143, 161 pp.

Schroeder, M. J. 1969, Ignition probability. Fort Collins, Colo., U.S.D.A. Forest Service, Rocky Mountain Forest and Range Experiment Station. Unpublished report, 27 pp.

Seiler, W., and Crutzen, P. J. 1980, Estimates of gross and net fluxes of carbon between the biosphere and the atmosphere from biomass burning, *Climate Change*, 2, 207–247.

Sneeuwjagt, R. J., and Frandsen, W. H. 1977, Behaviour of experimental grass fires vs. predictions based on Rothermel's fire model. *Can. J. For. Res.*, 7, 357–367.

Van Wilgen, B. W., le Maitre, and Kruger, 1985, Fire behaviour in South African fynbos (macchia) vegetation and predictions from Rothermel's fire model. *J. Appl. Ecol.* 22, 207–216.

Vickos, J. B. 1991, Télédétection des feux de végétation en Afrique intertropicale et estimation des émissions de constituants ayant un intérêt atmosphérique, Thèse Université Paul Sabatier, 148 pp., Laboratoire d'Aérologie, Toulouse, France.

Trace Gas and Aerosol Emissions from Savanna Fires

Meinrat O. Andreae, Elliot Atlas, Hélène Cachier,
Wesley R. Cofer III, Geoffrey W. Harris, Günter Helas,
Ralf Koppmann, Jean-Pierre Lacaux, and Darold E. Ward

Savanna fires are the single largest source of biomass burning emissions worldwide, ranking well in front of forest fires, agricultural fires, or domestic burning (Crutzen and Andreae 1990; Hao, Liu, and Crutzen 1990; Andreae 1991; Andreae 1993b; Hao and Liu 1994). The fundamental reason for the high incidence of savanna fires is the seasonal oscillation between a wet season, during which ample biomass is produced, and a prolonged dry season, during which this biomass is turned into a highly flammable material. Even in the absence of humans, savanna fires must be a common event, especially when lightning from the first thunderstorms at the end of the dry season strikes the still dry vegetation.

Africa contains about two thirds of the world's savanna regions, and the fires in African savannas alone account for about 30% of the biomass burned in the tropics worldwide (Hao and Liu 1994). African agricultural and pastoral practices typically include annual or biennial burning of the savanna vegetation. There are numerous reasons for savanna burning: clearing of dead or unwanted vegetation, fertilization of the soil by rapid pyromineralization of the nutrient content of the biomass, improvement of access for hunting and for collection of food and other vegetation products, and the elimination of pests and parasites. As a result of these practices, an area of some 820 million ha is burned in savannas annually, resulting in the combustion of about 3400 to 3700 Tg dm yr^{-1} (1 Tg $=$ 10^{12} g; dm: dry matter) of biomass (Andreae 1993b; Hao et al. 1993). The biomass burned annually in African savannas has been estimated at approximately 2000 Tg dm yr^{-1}, and the area exposed to fire in African savannas covers approximately 440 million ha (Hao et al. 1996).

In spite of the obvious importance of savanna fires to atmospheric chemistry and climate on regional and even global scales, there are still only a relatively small number of studies that have investigated the chemical and physical characteristics of the gaseous and particulate emissions from savanna fires. Following the pioneering, but also very preliminary studies in the Brazilian "cerrado" savannas by Crutzen et al. (1985), detailed investigations were made during the BASE-A and BASE-B experiments in Brazil (Ward et al. 1991; Ward et al. 1992). In Australia, where large savanna areas are burned every year in the tropical northern region, emission studies were conducted in recent years (Ayers and Gillett 1988; Hurst et al. 1994; Hurst, Griffith, and Cook 1994).

The first measurements on emissions from African savanna fires were made in West Africa by Delmas (1982). Studies of regional atmospheric chemistry over the Congo forest suggested that biomass burning in the sub-Sahelian savannas have far-reaching impact on the atmospheric environment (Cachier and Ducret 1991; Cachier et al. 1991; Andreae et al. 1992; Bingemer et al. 1992; Cros et al. 1992a; Helas and Andreae 1992; Lacaux et al. 1992). At about the same time, the evidence grew that the fires in the African savannas may be the main cause for a conspicuous tropical ozone maximum spanning from Africa across the Atlantic to South America (Cros et al. 1988; Fishman et al. 1991; Andreae et al. 1992; Cros et al. 1992b; Andreae et al. 1994a), in addition to being the source of considerable regional air pollution. As a result, a first multidisciplinary, multinational campaign to study pyrogenic emissions was organized under the title FOS/DECAFE-91 (Fire of Savannas/Dynamique et Chimie Atmosphérique en Forêt Equatoriale). This experiment took place in the humid savanna of Côte d'Ivoire, West Africa (Bonsang et al. 1995; Cachier et al. 1995; Gaudichet et al. 1995; Helas et al. 1995b; Lacaux et al. 1995; Rudolph et al. 1995). The dearth of information on the emission characteristics of fires in dry savannas was one of the main reasons why, under the auspices of the International Geosphere-Biosphere Programme (IGBP) and its Core Project International Global Atmospheric Chemistry (IGAC), a field experiment was organized in southern Africa under the title SAFARI-92 (Southern African Fire-Atmosphere Research Initiative) (Andreae et al. 1994b). This experiment took place during the 1992 dry season (August–October) and involved the effort of some 150 scientists from 14 nations. It was complemented by the TRACE-A (Transport and Atmospheric Chemistry near the Equator—Atlantic) experiment, which emphasized the study of the large-scale impact of pyrogenic emissions (Fishman et al. 1996).

During SAFARI-92, emission studies were made at a number of sites, using several aircraft platforms as well as ground-based techniques for sampling. A large array of measurements were performed in the field and in home laboratories in Africa, Europe, and North America. The detailed results are presented in a series of original papers, many of which are published in a special issue on SAFARI-92 of the *Journal of Geophysical Research* (Andreae et al. 1996b; Cofer et al. 1996; Hao et al. 1996; Lacaux et al. 1996; Le Canut et al. 1996; Maenhaut et al. 1996b; Scholes, Ward, and Justice 1996; Ward et al. 1996). Other papers are still in preparation at this time (Andreae et al. 1996a; Atlas, Pollock, and De Kock 1996; Harris et al. 1996a; Harris et al. 1996b; Helas et al. 1996). Here, we will attempt to summarize these data, compare the results obtained by the different groups with one another and with previous work, and attempt to extrapolate to savanna burning worldwide.

Methods

Study Sites and Sampling Platforms

Emission information was derived from two types of samples: smoke collected directly over experimental fires or over fires encountered during regional survey flights, and samples of ambient air from altitudes between 20 m (above ground level) and 4000 m (above sea level) over the study area in southern Africa, which extended from Swaziland via South Africa, Zimbabwe, and Zambia to southeastern Angola (figure 27.1). Most of the emission characterization is based on measurements on fresh smoke, but occasionally it was possible to derive emission parameters from plume layers embedded in the regional background. Experimental fires for SAFARI-92 were conducted in Kruger National Park (KNP), South Africa, at two savanna sites in Zambia (Kasanka National Park and near Choma), and in sugar cane fields near Big Bend, Swaziland.

The KNP study sites were in a low-fertility, moist "sourveld" savanna at Pretoriuskop, near the southern end of the Park. This vegetation is relatively nutrient poor and had not been intensively grazed in spite of the severe drought, which had prevailed in southern Africa for two years when SAFARI-92 began. Information

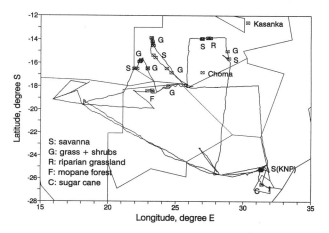

Figure 27.1 Map of the SAFARI-92 study region in southern Africa, showing the flight tracks of the DC-3 aircraft and the sampling sites for emission measurements

on the vegetation characteristics and fire behavior can be found in the papers by Trollope et al. (1996), Shea et al. (1996), and Stocks et al. (1996). Eight of the KNP fires were conducted in small plots with an area of about 6 to 7 ha each that form part of a long-term fire ecology research study conducted by the park. Two fires were set on large management blocks in KNP (No. 55 and 56, with areas of 2043 and 2333 ha, respectively). The fires in KNP were sampled from the ground using automated Fire Atmosphere Sampling Systems (FASS) (Ward et al. 1992; Ward et al. 1996), with handheld sampling devices supplied by tubing held on long poles over the fires (Lacaux et al. 1996), or by opening sampling canisters close to the fires (Atlas, Pollock, and De Kock 1996). Airborne sampling was conducted from two fixed-wing aircraft, a Cessna 310 and a DC-3, and from a helicopter. All aircraft had been fitted with instrumentation and sampling devices for atmospheric measurements (Cofer et al. 1996; Harris et al. 1996b; Helas et al. 1996).

The study sites in Zambia represent a moist miombo woodland savanna, a "dambo" grassland, a disturbed miombo savanna, and a semiarid treed savanna. The ecological conditions and fuel characteristics of these sites were described in detail by Shea et al. (1996). Measurements were conducted with the FASS systems only (Ward et al. 1996).

The sugar cane fires sampled by the DC-3 represent burns that are typical of regional agricultural practices. Such fires are set immediately before harvesting to facilitate the cutting and processing of the sugar cane. This practice results in substantial levels of air pollution in many sugar growing areas (Kirchhoff et al. 1991). During SAFARI-92, emissions from fires on two sugar cane fields of 12- and 43.7-ha area were investigated. These fields were located near Big Bend, Swaziland, and had been scheduled to be burned on the day of the study based on standard agricultural criteria. The fires burned very intensively with large flames and consumed about 25 tons of dry mass per hectare.

On its survey and transit flights, the DC-3 sampled numerous "fires of opportunity." We studied two groups of wildfires in the Drakensberg Escarpment of South Africa, where mixtures of native forest, shrub, and grass were burning. A large number of fires were observed and sampled in the savannas of Zambia, Botswana, and Angola. These fires represent native burning practices in grasslands and savannas, which typically consist of relatively small and well-controlled fires. They frequently burn in the form of backfires, that is, with the fire front moving against the prevailing wind direction. Details are given in the article by Le Canut et al. (1996).

Analytical Techniques

A wide variety of trace gases and particulate species were determined during SAFARI-92 (table 27.1). The analytical methods used include the continuous determination of some chemical species in the field by various instruments, for example, CO_2 by nondispersive infrared absorption; CO, CH_4, NO_2, HCHO, and

Table 27.1 Chemical species determined in smoke during SAFARI-92

	Gases	Particulates
Carbon species	Nitrogen species	
CO_2	NO	
CO	NO_2	Total mass
CH_4	NO_y	Number size distribution
HCHO	HNO_3	Volume size distribution
Alkanes	NH_3	
Alkenes	PAN	Soluble ions
Alkynes	PPN	Trace elements
Aromatics	Halogen species	
Alcohols	CH_3Cl	Total carbon
Aldehydes	CH_3Br	Black carbon
Carbonyls	CH_3I	
Organic acids	Various halocarbons	Carbon isotopes
H_2	SO_2	Nitrogen isotopes
H_2O_2	COS	
H_2O	O_3	

N_2O by tunable diode laser absorption spectrometry; NO and NO_y by chemiluminescence; black carbon by light absorption; and the particle number/size distribution by a laser-optical particle counter. For most other compounds, discrete samples were taken by various methods, such as filling of stainless steel flasks for the determination of various organic compounds, adsorption in liquids for measurements of organic acids, and filter sampling for aerosol investigations. The details are discussed in the original publications (Harris et al. 1993; Cachier et al. 1995; Andreae et al. 1996a; Andreae et al. 1996b; Atlas, Pollock, and De Kock 1996; Cofer et al. 1996; Hao et al. 1996; Harris et al. 1996b; Helas et al. 1996; the Koppmann et al. chapter 29, this volume; Lacaux et al. 1996; Le Canut et al. 1996; Maenhaut et al. 1996b; Scholes, Ward, and Justice 1996; Ward et al. 1996).

Results

To characterize the emissions from biomass fires, we use two related parameters, emission ratios and emission factors, rather than actual concentrations measured in the plumes. This is appropriate because the absolute concentration of trace gases in samples from a smoke plume has little meaning: it simply represents varying degrees of dilution of flame gases with ambient air, an effect that is particularly pronounced in aircraft sampling, where the transit time through the plume is comparable to the sampling time, and thus any given sample represents a mixture of smoke from different parts of the plume. The data are therefore normalized by dividing the trace species concentrations by the concentration of a simultaneously measured reference gas, such as CO_2 or CO. For example, the emission ratio of methyl chloride (CH_3Cl) relative to CO is

$$ER_{CH_3Cl/CO} = \frac{\Delta CH_3Cl}{\Delta CO} = \frac{(CH_3Cl)_{Smoke} - (CH_3Cl)_{Ambient}}{(CO)_{Smoke} - (CO)_{Ambient}}$$

Emission ratios are usually obtained by calculating the slope of the regression line between the trace species of interest and CO or CO_2 for a given data set from a fire or plume. Where long integration times are necessary, for example, in the case of aerosol samples, a different procedure is used. First, appropriate background concentrations have to be subtracted from both trace and reference species concentration data. Then the concentration of the reference gas anomaly (i.e., the concentration after background subtraction) is integrated over the same period over which the trace species sample was collected. The emission ratio is then calculated as the ratio between the integrated trace and reference species anomalies. The various techniques for these calculations and the associated errors are discussed in Le Canut et al. (1996). For gases, the results will be expressed in terms of molar ratios. For aerosols, the emission ratios were obtained initially in units of mass aerosol per kilogram of carbon in the form of CO_2 (g/kg C[CO_2]), but in this paper the results will be presented after conversion to emission factors (see below).

The choice of CO or CO_2 as reference gas is determined by the final objective of the analysis. It has been shown previously (e.g., Lobert et al. [1991]; Manø [1995]) that many species are emitted preferentially in the smoldering stage of fires. CO is a suitable reference species for smoldering combustion, as it is also emitted predominantly during this stage. The emission ratios relative to CO are largely dependent on fuel characteristics, such as the nitrogen or halogen element content of the fuel. The various smoldering-derived gases are usually highly correlated among one another and with CO, a fact that allows accurate estimation of trace gas emissions from fires for which the CO emission is known. Correlations of smoldering-derived gases with CO_2, on the other hand, tend to be relatively poor, since the relative proportion of flaming versus smoldering combustion in different fires or even different parts of the same fire results in variable trace gas to CO_2 ratios. In such cases, the best way to estimate the emission ratio of a smoldering-derived species, X, to CO_2 is by first calculating an emission ratio to CO (ER [X/CO]) and then multiplying with an independently obtained, whole-fire ER(CO/CO_2). CH_4 can be used as a reference gas in the same way as CO in these calculations. The ER(X/CO_2) permits the estimation of trace gas emission from fires for which only the amount of biomass burned, but not the amount of CO emitted, is known. Therefore, this ratio is the most suitable parameter for regional or global estimations.

Another parameter frequently used to characterize emissions from fires is the emission factor, which is defined as the amount of a compound released per amount of fuel *consumed* (g kg^{-1} dm; dm: dry matter). Calculation of this value requires knowledge of the carbon content of the biomass burned and the carbon budget of the fire (usually expressed as combustion efficiency; see Ward et al. [1996]); both parameters are difficult to determine in the field, as opposed to laboratory experiments where they are readily determined. During SAFARI-92, the composition of the biomass and the amounts of fuel consumed were determined at the burn plots in Kruger National Park, at the four sites in Zambia where ground studies were conducted,

and in the sugar cane fields in Swaziland. The mean carbon content of the African savanna fuels (consisting mostly of grass and litter) was found to be 47% (Susott et al. chapter 24 this volume). Trace species emission factors were then calculated based on a mass balance of carbon released and the various emission ratios relative to CO_2 determined in the smoke. Where fuel and residue data at the ground were not available, typically for the "fires of opportunity" sampled by the DC-3 aircraft, a fuel carbon content of 47% was assumed to derive emission factors from the emission ratios obtained by airborne measurements.

CO and CO_2

Since CO_2 is always the dominant carbon species released from the fires, it is generally used as a reference species, and the concept of emission ratios is not applicable to CO_2. Therefore, only emission factors are presented for this species. The results regarding the emission of CO and CO_2 obtained by the various groups are summarized in table 27.2 and figure 27.2. The data set collected at the ground (using the FASS system) from the fires in KNP give the highest ER(CO/CO_2); the mean ER(CO/CO_2) from this group is 7.1 $\pm$ 1.6% (Ward et al. 1996). Two other ground-based studies found significantly lower ER(CO/CO_2): Lacaux et al. (1996) gives an average of 4.7% (range 3.0–5.8), and Atlas, Pollock, and De Kock (1996) obtained a ratio of 4.9 $\pm$ 2.1% (using predominantly ground-based samples, but also some aircraft data). Where ground-based data were obtained on the same fires by both Lacaux et al. (1996) and Ward et al. (1996), the results from the latter group are typically 50 to 100% higher. The differences between these studies may be related to a different weighting between the contributions from flaming and smoldering stages to the overall mean emission ratio, where the FASS system appears to somewhat favor the smoldering stage. This view is supported by the airborne measurements from the KNP fires, which are consistent with lower ER(CO/CO_2) values: the samples collected by helicopter gave a ratio of 4.8 $\pm$ 0.9% (Cofer et al. 1996), the measurements from the Cessna 310 yielded an average of 4.0 $\pm$ 2.0% (Helas et al. 1996, and the can samples collected on board the DC-3 showed a mean of 5.1 $\pm$ 0.2% (Koppmann et al. chapter 29 this volume. 1996). Due to instrument problems, ER(CO/CO_2) values from the KNP fires are not available from the continuous analyzers on board the DC-3. Similar differences exist in the measurements made previously in the Brazilian cerrado, where airborne measurements gave an ER(CO/CO_2) of 2.2 (Ward et al. 1991),

Table 27.2 CO_2 emission factors (EF[CO_2]) and CO/CO_2 emission ratios (ER[CO/CO_2]) from SAFARI-92 and from previous studies

Ecosystem, location	EF(CO_2) (g CO_2/kg dm)	ER(CO/CO_2) (mol%)	Source
Results from SAFARI-92			
Moist, infertile savanna, Kruger National Park	1597 ± 29[a]	7.1 ± 1.6	USFS-FASS
	—	4.9 ± 2.1	NCAR
	—	4.7 (3.0–5.8)	UPS
	1598 ± 397[a]	4.8 ± 0.9	NASA
	—	5.1 ± 0.2	KFA-DC3
	—	4.0 ± 2.0	MPIC-Cessna
Moist, infertile savanna, Zambia	—	5.0 ± 1.5	MPIC-DC3
Moist and semiarid savanna, Zambia	1603 ± 40[a]	6.8 ± 2.5	USFS-FASS
Arid, fertile savanna, Zimbabwe	—	3.5 ± 1.0	MPIC-DC3
Grasslands, Zambia, Angola	—	6.4 ± 0.9	MPIC-DC3
Sugar cane fields, Swaziland	—	3.3 ± 1.1	KFA-DC3
Forest and shrub, Drakensberg, S. Africa	—	9.4 ± 1.3	KFA-DC3
All vegetation fires, southern Africa	—	7.4	Scholes, Ward, & Justice (1996)
Results from other regions			
Moist, fertile savanna, West Africa	—	5.3	Helas et al. (1995a)
	1679	6.3	Bonsang et al. (1995)
Cerrado, Brazil	1722 ± 23[a]	5.3 ± 1.0	Ward et al. (1992)
Eucalypt savanna, Australia	1595 ± 55	7.8 ± 2.3	Hurst, Griffith, & Cook (1994)
Grassland, chaparral, and savanna fires, global average	1640	6.2 ± 1.0	

a. Emission factors calculated relative to the total carbon released to the atmosphere during combustion, without accounting for fuel carbon remaining in ash.

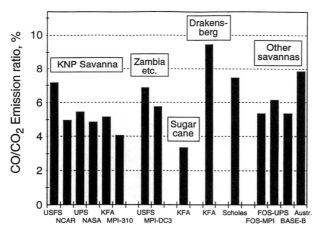

Figure 27.2 Emission ratios for CO relative to CO_2 for the experimental fires at Kruger National Park, the native and experimental fires in southern African savannas and grasslands, the sugar cane fires in Swaziland, and the forest fires in the Drakensberg. For comparison, a regional average for southern Africa (Scholes, Ward, and Justice 1996) and results from savanna fires in West Africa, Brazil, and Australia are indicated.

while ground-based FASS measurements showed an ER(CO/CO_2) of 5.3 (Ward et al. 1992).

The ER(CO/CO_2) obtained by Ward et al. (1996) from experimental fires in the Zambian savannas did not show any systematic differences from the results collected by the same group at KNP. The measurements made by the DC-3 in plumes from numerous native fires in the moist, infertile savannas and grasslands of southern Africa are slightly lower, but still in reasonable agreement with the results obtained by ground-based sampling from the experimental fires (Harris et al. 1996b; Le Canut et al. 1996). Overall, the ER(CO/CO_2) values obtained during SAFARI-92 agree well with the range of values obtained in other grassland and savanna regions and support the proposed global mean ER(CO/CO_2) value of 6.2 for savanna fires (see table 27.2). Scholes, Ward, and Justice (1996) derived a regional estimate for CO and CO_2 emissions from biomass burning in southern Africa from which a mean CO emission factor of 7.4% can be derived. Since their computation is based on the FASS-based emission factors of Ward et al. (1996), this relatively high value may be subject to the same systematic differences discussed previously.

The widespread burning of sugar cane fields before harvest and the frequent accidental fires in the forests on the Drakensberg escarpment are of considerable importance as regional sources of smoke pollution. For this reason, measurements were also made in these two environments during SAFARI-92. The sugar cane fires burn with very intensive, large flame fronts, and produce a black, soot-rich smoke. Their $ER(CO/CO_2)$ is quite low (see table 27.2), consistent with the dominance of the flaming combustion mode. Comparably low $ER(CO/CO_2)$ values were observed in the flaming phase of laboratory burns of savanna grass (Manö 1995). In strong contrast to the relatively "clean" combustion characteristic of the sugar cane fires is the high $ER(CO/CO_2)$ ($9.4 \pm 1.3\%$) from the Drakensberg fires that were burning in native forest and shrub. Such high ratios are typical of forest fires in other regions as well and are related to the considerable amount of smoldering combustion present in these fires.

To place the preceding discussion into perspective, we want to emphasize at this point that while the differences between the results from the different groups participating in SAFARI make for interesting argument, the similarities are really much greater than the discrepancies. We are now at a point where the mean $ER(CO/CO_2)$ from savanna fires is constrained in the narrow range of $6 \pm 1\%$, an uncertainty of less than 20%. Slightly larger uncertainties may prevail for other types of fires. These error ranges are trivial, however, compared to the range of uncertainty in the annual amount of biomass burned, where differences of almost a factor of five exist between regional estimates. Using remote sensing and ecological maps, Scholes, Ward, and Justice (1996) estimate a total CO_2 emission of about 330 Tg CO_2 yr^{-1} from biomass fires in Africa south of the equator, whereas from the gridded data of Hao and Liu (1994) a value of about 1630 Tg CO_2 yr^{-1} can be derived for the same region!

Methane (CH_4) and Nonmethane Hydrocarbons (NMHCs)

The first step in the combustion of vegetation is the pyrolytic decomposition of plant matter immediately ahead of the flames. It releases a wide array of reduced and partially oxidized organic compounds, with CH_4 being the most abundant species. Further, such species are produced by the recombination of radicals in the cooling zone of flames. During SAFARI-92, emission ratios for CH_4 and NMHCs were determined from canister samples by six laboratories: the Max Planck Institute for Chemistry in field laboratories located at KNP (MPIC-K) and in Pretoria (MPIC-P) (Helas et al. 1996), the National Center for Atmospheric Research (NCAR) (Atlas, Pollock, and de Kock 1996), the Forschungszentrum Jülich (KFA) (Koppmann et

al. chapter 29 1996), the National Aeronautics and Space Administration (NASA) Langley Research Center (Cofer et al. 1996), and the U.S. Forest Service Intermountain Fire Sciences Laboratory (USFS) (Hao et al. 1996; Ward et al. 1996). On the DC-3 aircraft, CH_4 was additionally measured by tunable diode laser spectrometry (Harris et al. 1996b). The results are shown in table 27.3; the NMHC data in this table were obtained either by a single measurement of total NMHC (NASA) or by summing individual measurements (all other groups). The NMHC results from KFA and NASA include aromatic and partially oxidized compounds; however, for some of these compounds, the exact calibration constants are not known. An average calibration constant representative of short-chain hydrocarbons has been applied in these cases.

The hydrocarbon emission results from SAFARI-92 are in good agreement with previous CH_4 and NMHC emission measurements from fires in similar ecosystems (figure 27.3). Previous studies on grassland, savanna, and chaparral fires showed that emissions of hydrocarbons from these flaming-dominated fires are relatively low, with CH_4 emission ratios typically in the range 2 to 4×10^{-3} (Hao and Ward 1993; Andreae and Warneck 1994). The results from the KNP fires fall within this range; there are no statistically significant differences between the results obtained by the various groups. The CH_4 measurements made by MPIC-K and KFA on independently analyzed samples collected with the Cessna 310 agree extremely well but both lie near the lower end of the range of results. This may be due to a tendency to oversample flaming-phase emissions.

The CH_4 emission ratios from the experimental savanna fires in Zambia are very similar to those from KNP, while their NMHC emission ratios are slightly lower. In contrast, very low CH_4 and NMHC emission ratios were obtained from the sugar field fires, especially the second one, consistent with the low CO emission ratios from these flaming-dominated fires. The opposite extreme is represented by the Drakensberg fire, which had very high hydrocarbon emissions, reflecting the substantial contribution from smoldering typical of forest fires. The NMHC emissions follow the same trend as the CH_4 emissions, and Hao et al. (1996) found close correlations between CH_4 and individual hydrocarbon emission ratios. The largest fraction of the NMHC is present in the form of the C_2–C_4 hydrocarbons (see table 27.3), but the contribution from higher-molecular species is by no means negligible

Table 27.3 CH_4/CO_2 emission ratios

Ecosystem, location	Emission ratio (10^{-3})				Source
	CH_4	$C_2–C_4$	$\geq C_5$	Total	
Results from SAFARI-92					
Moist, infertile savanna, KNP	4.1 ± 1.9	4.8 ± 0.7	1.2 ± 0.2	6.1 ± 1.0	USFS
	3.8 ± 1.7	≥ 2.2[b]	—	—	NCAR
	3.4 ± 1.0[a]	—	—	5.0 ± 2.5[a]	NASA
	3.9 ± 1.0	—	—	—	MPIC-TDLAS
	2.8 ± 0.7	4.6 ± 0.6	2.8 ± 1.0	7.4 ± 1.6	KFA
	2.7 ± 1.5	1.3[b]	—	—	MPIC-Cessna
Moist and semiarid savanna, Zambia	3.6 ± 1.5	—	—	4.3 ± 2.0	USFS
Sugar cane field #1, Swaziland	2.6 ± 2.2	2.6 ± 2.0	1.3 ± 0.9	3.9 ± 2.9	KFA
Sugar cane field #2, Swaziland	1.1 ± 0.9	1.2 ± 0.5	0.5 ± 0.2	1.7 ± 0.7	KFA
Forest and shrub, Drakensberg, S. Africa	10.3 ± 1.7	9.3 ± 0.4	4.4 ± 0.5	13.7 ± 0.9	KFA
All vegetation fires, southern Africa	4.3	—	—	—	Scholes, Ward, & Justice (1996)
Results from other regions					
Moist, fertile savanna, West Africa	4.2	—	—	—	Helas et al. (1995a)
	4.4	—	—	6.9	Bonsang et al. (1995)
Cerrado, Brazil	2.1 ± 0.4	—	—	—	Ward et al. (1992)
Eucalypt savanna, Australia	4.0 ± 1.4	—	—	10.5 ± 6.0	Hurst Griffith, & Cook (1994)
Grassland, Florida	3.0 ± 1.1	—	—	4.1 ± 2.1	Cofer et al. (1990)
Grassland, chaparral, and savanna fires, global average	4.0	—	—	—	Andreae and Warneck (1994)

a. Data originally reported separately for flaming and smoldering combustion. Average formed by weighting the two phases using the proportions reported by Ward et al. (1996).
b. C_2 and C_3 hydrocarbons only.

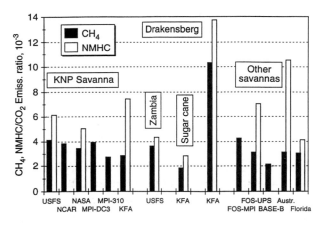

Figure 27.3 Emission ratios for methane and nonmethane hydrocarbons relative to CO_2 for the experimental fires at Kruger National Park and in the Zambian savannas, the sugar cane fires in Swaziland, and the forest fires in the Drakensberg. For comparison, results from savanna and grassland fires in West Africa, Brazil, Australia, and Florida are indicated.

(Atlas, Pollock, and De Kock 1996; Koppmann et al., chapter 29 this volume).

The results from SAFARI-92 are consistent with the value of 4×10^{-3} proposed by Andreae and Warneck (1994) as global average for the CH_4 emission ratio from savanna and grassland burning. For the NMHCs, a range of 5 to 7×10^{-3} (on a carbon atom basis) appears to be appropriate. These values are about a factor of three lower than the emission ratios characteristic of forest fires in both temperate and tropical regions. As a result, forest fires emit considerably larger amounts of hydrocarbons worldwide than savanna fires, in spite of the greater amount of biomass combusted annually in savannas.

Molecular hydrogen (H_2) is emitted from biomass fires with surprisingly high emission ratios. Previous work had suggested a global mean $ER(H_2/CO_2)$ of 3.3% (Andreae 1993b), with savanna fires at the low end of the range of emission ratios observed. For the cerrado fires in Brazil, Ward et al. (1992) gives a value of $1.0 \pm 0.3\%$, and for savanna fires in West Africa, Helas et al. (1995b) present a mean of 1.5%. The results obtained by airborne sampling over the KNP fires ($1.0 \pm 0.2\%$) (Cofer et al. 1996) are in close agreement with these previous studies and confirm the relatively low H_2 yield from the flaming-dominated savanna fires.

Table 27.4 Emission ratios for partially oxidized and aromatic hydrocarbons relative to CO_2 (all values multiplied by 10^3)

	Laboratory[a] (var. fuels)	Drakensberg forest[b]	Sugar cane[b]	Kruger National Park savanna[b]	Kruger National Park savanna[c]
Alcohols					
Methanol	0.26	—	—	—	—
Ethanol	0.086	—	—	—	—
1-Propanol	0.16	—	—	—	0.011
2-Propanol	0.12	—	—	—	—
1-Butanol	—	0.006	0.004	0.006	—
Aldehydes					
Formaldehyde[d]	—	0.19 ± 0.07	0.11 ± 0.03	0.32 ± 0.03	—
Acetaldehyde	0.64	—	—	—	0.27
Propanal	0.46	—	—	—	—
Butanal	—	0.084	0.029	0.067	—
Ketones					
Acetone	0.83	—	—	—	0.26
2-Butanon	0.31	—	—	—	—
2,3-Butanedione	0.88	—	—	—	—
Acrolein	0.68	—	—	—	—
Furans					
Furan	0.48	—	—	—	0.22
2-Methylfuran	0.60	0.29	0.020	0.080	0.11
3-Methylfuran	0.07	0.05	0.005	0.018	—
2,5-Dimethylfuran	0.28	—	—	—	—
Furfural	1.44	—	—	—	—
Esters					
Methyl formate	0.036	—	—	—	—
Methyl acetate	0.15	—	—	—	—
Total oxygenates	7.5	—	—	—	—
Aromatics					
Benzene	0.49	0.88	0.31	0.49	0.32
Toluene	0.48	0.46	0.057	0.22	0.21
Acetonitrile	0.39	—	—	—	—

a. Manö 1995.

b. Koppmann et al., chapter 29 this volume.

c. Atlas, Pollock, and De Kock 1996.

d. Harris et al. 1996b.

Partially Oxidized Hydrocarbon Species (e.g., Aldehydes, Ketones, Acids) and Aromatic Hydrocarbons

In addition to the hydrocarbons proper, a large variety of partially oxidized hydrocarbons (oxygenates) are emitted, especially during the smoldering phase of fires. These compounds include alcohols, aldehydes, ketones, carboxylic acids, and esters, among which species with up to four carbon atoms are the most abundant. Mostly due to the difficulties associated with the quantitative determination of the oxygenates, the emission of this compound class from biomass fires has not yet received much scientific attention, in spite of its obvious atmospheric chemical importance. Only gaseous species are discussed in this section; a considerable amount of high-molecular-weight organic compounds are also emitted from fires and rapidly condense into the particulate phase.

Table 27.4 gives a list of some of the most important oxygenate species detected in the smoke of experimental fires in the laboratory (Manö 1995) and in savanna fires studied during SAFARI-92 (Atlas and de Kock 1996; Koppmann et al. 1996). The abundance of oxygenates in the smoke, when expressed either on a mass or on a carbon atom basis, rivals or exceeds that of the hydrocarbons. For example, the emission ratio for the total identified oxygenates (not including the carboxylic acids) in the laboratory measurements of Manö (1995) is 7.5×10^{-3} (mol C/mol CO_2), while in similar experiments Lobert et al. (1991) found a hydrocarbon emission ratio of 9.8×10^{-3}. Atlas, Pollock, and de Kock (1996) identified a considerable number of additional oxidized species in the smoke samples, which they were unable to quantify due to the lack of appropriate standards. Adding the carboxylic acids and the numerous unidentified species present in the

Table 27.5 Emission ratios for nitrogen species determined during SAFARI-92 and from other savanna fires

Ecosystem, location	Emission ratio to CO_2 (10^3)						Source
	NO	NO_2	NO_x	NO_y	N_2O	NH_3	
Results from SAFARI-92							
Moist, infertile savanna, KNP	—	—	3.8 ± 0.5	—	—	—	UPS
	—	—	—	—	0.09	—	NASA
	1.6 ± 0.2	1.1 ± 0.1	2.7 ± 0.3	3.5 ± 0.4	—	0.7 ± 0.4^a	MPIC
Moist, infertile savanna, Zambia, Angola	2.1 ± 1.1	1.2 ± 0.9	3.3 ± 2.0	8.3 ± 1.2	—	—	MPIC
Grasslands, Zambia, Angola	1.7 ± 0.3	0.74 ± 0.37	2.4 ± 0.7	7.0 ± 0.8	—	—	MPIC
Forest and shrub, Drakensberg, S. Africa	0.19 ± 0.06	0.28 ± 0.12	0.47 ± 0.18	2.2 ± 1.1	—	—	MPIC
Sugar cane fields, Swaziland	0.68 ± 0.35	0.47 ± 0.06	1.2 ± 0.4	5.1 ± 2.5	—	—	MPIC
Results from other regions							
Moist, fertile savanna, West Africa	—	0.18 ± 0.04	—	—	—	—	Helas et al. (1995a)
	1.3	0.14	1.4 ± 0.2	1.9	0.14	0.06	Lacaux et al. (1995)
Eucalypt savanna, Australia	—	—	2.1 ± 0.4	—	0.04 ± 0.02	2.9 ± 1.9	Hurst Griffith, and Cook (1994)

a. Emission ratio for particulate ammonium only.

chromatograms would lead to a significantly higher emission ratio for the oxygenates than for the hydrocarbons.

Given the paucity of previous work on oxygenate emission from biomass burning, there is little to which our results can be compared. One exception are the measurements of formaldehyde and acetaldehyde from savanna fires in Australia made by matrix isolation FTIR (Fourier Transform Infrared Spectroscopy), a technique completely independent of ours (Hurst, Griffith, and Cook 1994). They found emission ratios of $HCHO/CO_2 = (0.21 \pm 0.14) \times 10^{-3}$ and $CH_3CHO/CO_2 = (0.30 \pm 0.22) \times 10^{-3}$. These values agree closely with our results, which were obtained by onboard TDLAS (Tunable Diode Laser Spectroscopy) in the case of formaldehyde and GC/MS (Gas Chromatography/Mass Spectroscopy) from canister samples in the case of acetaldehyde.

Also shown in table 27.4 are two aromatic species, benzene and toluene. They are the most abundant members of this substance class, which also includes the polycyclic aromatic hydrocarbons. The latter compounds are present in both aerosol and gas phase and include a number of potential marker substances for tracing pyrogenic emissions.

Nitrogen Species (NO, NO_2, NO_y, N_2O, NH_3)

Since the carbon content of biomass fuels is relatively constant, the emission ratios of carbon species are almost uniquely a function of the combustion characteristics. In contrast, the N content of biomass fuels is quite variable, and therefore the N species emissions from biomass fires are also strongly dependent on the

N content of the fuel. There is usually a distinct difference between the average N content of the aboveground biomass and the fuel N content, due to the higher fraction of dead biomass with low N content in the fuel. The biomass exposed to fire in savannas contains some 0.3 to 1.1% N by weight (Lacaux, Cachier, and Delmas 1993; Hurst, Griffith, and Cook 1994; Cook 1995; Kuhlbusch et al. 1996); for the fuels burned in the KNP fires, Kuhlbusch (1996) determined an average N content of 0.57%. This agrees well with the results of Susott et al. (chapter 24 this volume), who determined a mean N content of 0.9% in the standing biomass of savannas in Zambia and South Africa, while the actual fuel at these sites had only 0.5% N.

Most of the N in the fuel becomes volatilized during the fire and is emitted as a mixture of molecular N (N_2), NO, NO_2, N_2O, NH_3, and some other trace N species (Kuhlbusch et al. 1991; Lobert et al. 1991). Emission ratios from SAFARI-92 and from other studies on savanna fires are shown in table 27.5. On the DC-3, NO and NO_2 were measured separately with independent instruments (NO by chemiluminescence and NO_2 by TDLAS; Harris et al. 1996b), and the emission ratio given in table 27.5 for NO_x is obtained by addition. In contrast, the ground-based measurements by Lacaux et al. (1996) determined total NO_x by chemiluminescence after conversion of NO to NO_2. The determination of NO_y on the DC-3 was made by conversion of NO_y to NO using a hot gold converter, followed by chemiluminescence detection of NO (Harris et al. 1996b). The results obtained by these two groups show reasonable agreement ($2.7 \pm 0.3 \times 10^{-3}$

on the DC-3 versus $3.8 \pm 0.5 \times 10^{-3}$ on the ground), particularly if we consider that some of the NO_x measured at the ground may have already been converted to species detected as NO_y by the time the plume was probed by the aircraft. The NO_y emission ratio obtained for the KNP fires by the DC-3 ($3.5 \pm 0.5 \times 10^{-3}$) is almost identical to the NO_x emission ratio determined at the ground. The NO, NO_2, and NO_x emission ratios from the savanna and grassland fires sampled by the DC-3 in Zambia and Angola were comparable to the values obtained at KNP. Note that in all cases NO_2 represents a significant fraction of NO_x, suggesting that either some NO_2 is actually released from the fires or that rapid conversion of NO to NO_2 has taken place in the short interval while the plume was rising from the fire to an altitude of a few hundred meters. Significantly lower emission ratios were obtained from the Drakensberg forest fires and the sugar cane fires. In the case of the forest fires, this may be related to a high fraction of smoldering combustion, which would tend to favor the release of reduced N species.

Analyses of the pre- and postburn biomass from the KNP study sites by Kuhlbusch et al. (1996) showed that the N/C mole ratio in the KNP savanna fuels was 0.011 and that on average 74% of the fuel N and 88% of the fuel C were volatilized. Thus, the ratio of the sum of the N species in the smoke to the sum of the C species should be 9.4×10^{-3}. The observed amount of NO_x accounts for 24 to 35% of the fuel N and 29 to 40% of the volatilized N, depending on which NO_x emission ratio is used ($2.7 \pm 0.3 \times 10^{-3}$ or $3.8 \pm 0.5 \times 10^{-3}$). NO_y includes, besides NO and NO_2, a number of N species, such as organic nitrates and HNO_3. Together, these species account for 32% of the fuel N, or 37% of volatilized N. Unfortunately, no reliable data on NH_3 emissions from savanna fires could be obtained during SAFARI-92. However, the emission ratios for particulate NH_4^+ in the smoke aerosol can at least provide a lower limit for the total amount of NH_3 emitted (0.7×10^{-3}; see table 27.5), assuming that the particulate NH_4^+ results from the recombination of NH_3 with acidic gases in the smoke. The NH_3 emission factor of 2.9×10^{-3} from the Australian savanna (Hurst, Griffith, and Cook 1994) is the highest value reported for this species so far and probably represents an upper limit. Adding these upper and lower limits for the NH_3 emission ratio to the NO_y emission ratio brings the total determined N species to 48 to 71%. Thus, about 30 to 50% of the fuel N that has volatilized could not be chemically identified in the smoke. This fraction is comparable to the "un-detected" fraction of N emitted from savanna fires in Australia (Hurst, Griffith, and Cook 1994). Previous studies have shown that most of this unaccounted N is released in the form of molecular N_2 (Kuhlbusch et al. 1991).

The total amount of N lost per hectare can be estimated based on the biomass densities in the savannas investigated in southern Africa (3200 to 7400 kg dm ha^{-1} (Shea et al. 1996), the average N content of the fuel (0.6%) (Susott et al., chapter 24 this volume), and the average combustion factor (81%) and N release fraction (74%) (Shea et al. 1996). Based on these values, we find that 14 to 33 kg N ha^{-1}, corresponding to 60% of the aboveground N pool, is released during our savanna fires. This compares closely to the pyrogenic release rates obtained by Cook (1995) in three types of Australian savanna (17 to 28 kg N ha^{-1}), by Hurst et al. (1994) in the savanna at Kakadu National Park (24 ± 7 kg ha^{-1}), and by Isichei and Sanford (1980) in Nigerian grasslands (12 to 15 kg N ha^{-1}). The N release rate due to savanna fires is of the same order as the estimates of annual N inputs by deposition (3 to 7 kg N ha^{-1} yr^{-1}) and N fixation (7 to 30 kg N ha^{-1} yr^{-1}) in African savannas (Robertson and Rosswall 1986). Since additional N losses occur by microbial production and emission of NO and N_2O from the soils, annual or biennial burning places a severe load on the long-term N balance of savanna ecosystems.

Halogen and Sulfur Species

Plant biomass contains significant amounts of the halogen elements chlorine, bromine, and iodine, mostly in ionic form as solutes in the cytoplasm. In dry biomass, these elements may be present as crystallized salt particles within the tissue. Sulfur occurs in plants as sulfate, as well as a variety of sulfate esters and other organic sulfur compounds, particularly in proteins and other metabolites. During combustion, a large fraction of the halogen and sulfur content of the biomass is released to the atmosphere. The results of Kuhlbusch et al. (1996) from the KNP fires suggest a sulfur volatilization of 86%, which agrees closely with the observations of Hurst et al. (1994) in Australian savanna fires. The emission of halogens was studied for the two large KNP fires. The elemental composition of fuel and ash were determined by Instrumental Neutron Activation Analysis (Andreae et al. 1996b). From the halogen loadings in the fuel minus the halogen amounts remaining in the unburned residue and ash, divided by the amount of halogen in the *prefire* fuel, release fractions of 88%, 88%, and 76% were obtained

Table 27.6 Emission of sulfur and halogen species

Element	Element content in fuel (mg/kg)	Fraction emitted (%)	X_{total}/CO_2 in smoke (mol/mol [10^{-6}])	$\Delta X_{inorg}/\Delta CO_2$ (mol/mol [10^{-6}])	X_{inorg}/X_{total} (%)	$\Delta CH_3X/\Delta CO_2$ in smoke[a] (mol/mol [10^{-6}])	CH_3X/X_{total} (%)
Carbon	$480\,000 \pm 14\,000$	88%	—	—	—	—	—
Sulfur	2100 ± 400	86%	1740 ± 620	130 ± 50	7%	—	—
Chlorine	1260 ± 310	88%	960 ± 370	820 ± 300	85%	27 ± 14	2.8%
Bromine	7.0 ± 1.6	88%	2.4 ± 0.9	1.7^b	71%	0.11 ± 0.04	4.6%
Iodine	1.3 ± 0.3	76%	0.24 ± 0.09	0.17^b	71%	0.09 ± 0.05	38%

a. Mean $\Delta CH_3X/\Delta CO_2$ emission ratios for the KNP fires (Andreae et al. 1996b).
b. Based on smoke aerosol composition data in Maenhaut et al. (1996a).

for chlorine, bromine, and iodine, respectively (Andreae et al. 1996b). The emission fraction relative to the fuel *burned* can be obtained from these numbers by dividing by 0.88, the fraction of the prefire fuel consumed in the fire. This shows that some 90% of the halogen elements present in the fuel burned were actually released to the atmosphere during combustion.

Using the fuel carbon, sulfur, and halogen contents and the release fractions for these elements, we can calculate the corresponding "total element" emission ratios (table 27.6) from the mass balance of fuel and residue, independent of the atmospheric measurements. These ratios can be compared to the emission ratios of halogen and methyl halide species measured in the smoke. Table 27.6 shows that the emission ratios for the halide ions determined in the KNP fires are in reasonable agreement with the ratios predicted from the fuel-residue budget. The chloride emission ratio [$(820 \pm 300) \times 10^{-6}$] given in table 27.6 is from Andreae et al. (1996a) and approaches the amount of chlorine released by combustion [$(960 \pm 370) \times 10^{-6}$]; however, with the aerosol composition data of Maenhaut et al. (1996a), a chloride emission ratio of 510×10^{-6} and a fraction Cl_{inorg}/Cl_{total} of 56% would be obtained. These differences are probably mostly due to sampling variability, as the high standard deviation of the emission ratios suggests. Thus, our results suggest that most of the fuel halogen content is released either directly as halide particles or as hydrogen halides, which rapidly combine with alkaline substances in the aerosol or with gaseous ammonia to produce halide particles.

The emission of methyl halides from the KNP and other savanna fires was studied by four groups during SAFARI-92: the MPIC-K and MPIC-P (Helas et al. 1996), the NCAR (Atlas, Pollock, and de Kock 1996), the KFA (Koppmann et al., chapter 29 this volume). The results are summarized in a paper (Andreae et al. 1996b), from which the emission ratios in table 27.6

are obtained. The methyl chloride emission ratio from SAFARI-92 [$(27 \pm 14) \times 10^{-6}$] is slightly lower than the value of $(43 \pm 8) \times 10^{-6}$ obtained in the Guinean savanna of West Africa (Rudolph et al. 1995). Comparison of the methyl halide emission ratios with the total halogen element emissions shows that, in the case of chlorine, the methylated species (CH_3Cl) accounts for $2.8 \pm 1.9\%$ of the total element emitted. This fraction increases to $4.6 \pm 2.6\%$ for bromine and to $38 \pm 25\%$ for iodine. In the case of iodine, this fraction appears very large and may reflect measurement uncertainties and sampling variability. Nevertheless, our results show that a significant fraction of the fuel halogen content is methylated in the combustion process and that this fraction increases in the sequence $Cl < Br < I$.

Carbonyl sulfide (COS) and aerosol sulfate were the only sulfur species for which emission ratios were determined during SAFARI-92. Atlas, Pollock, and de Kock (1996) found a COS emission ratio of 0.74×10^{-6} from the KNP fires. This value is very close to the emission ratio from West African savanna fires (1.1×10^{-6}) (Nguyen et al. 1995) and in good agreement with the range of previous results (Andreae 1993b). The sulfate emission data will be discussed in the following section on aerosols.

Aerosols

The emission of aerosol particles from biomass fires makes the resulting smoke plumes visible to the unaided eye and leads to a regional reduction of visibility. Smoke from savanna and forest fires is frequently seen even from orbiting spacecraft some 400 km above the earth (Andreae 1993a). Recent estimates suggest that the aerosol emissions from biomass burning may be the second largest source of anthropogenic aerosols after the production of sulfates from SO_2 (for reviews, see Andreae [1995] and Houghton et al. [1995]).

Table 27.7 Emission factors (EF) for total particulate matter (TPM), particulate matter smaller than 2.5 μm (PM 2.5), total particulate carbon, and black carbon aerosol

Ecosystem, location	EF(TPM) (g kg^{-1} dm)	EF(PM2.5) (g kg^{-1} dm)	EF(Total C) (g kg^{-1} dm)	EF(Black C) (g kg^{-1} dm)	Source
Results from SAFARI-92					
Moist, infertile savanna, KNP	—	7.2 ± 3.8	—	—	USFS
	> 6	3.9 ± 2.0[a]	—	0.41 ± 0.20	MPIC-DC3
	10.0 ± 8.5	—	6.4 ± 5.1	0.54 ± 0.36	CFR
Arid, fertile savanna, Zimbabwe	> 10	6.4 ± 3.2	—	—	MPIC-DC3
Moist and semiarid savanna, Zambia	> 11	5.4 ± 2.2	—	—	USFS
Moist, infertile savanna, Zambia	> 6	3.7 ± 1.6	—	0.80 ± 0.20	MPIC-DC3
Grasslands, Zambia, Angola	> 3	2.1 ± 0.4	—	0.68 ± 0.20	MPIC-DC3
Forest and shrub, Drakensberg, S. Africa	—	—	—	0.28 ± 0.12	MPIC-DC3
Sugar cane field, Swaziland	—	3.7 ± 0.5	—	0.61 ± 0.18	MPIC-DC3
All vegetation fires, southern Africa	—	6.1	—	—	Scholes, Ward, & Justice (1996)
Results from other regions					
Moist, fertile savanna, West Africa	7.9 ± 3.2	—	3.0 ± 1.3	0.41 ± 0.16	Cachier et al. (1995)
Cerrado, Brazil	—	3.6 ± 1.2	—	—	Ward et al. (1992)
Grassland and savanna fires, average	10	5	4.7	0.6	

a. Emission factor for aerosols up to 3.0 μm diameter.

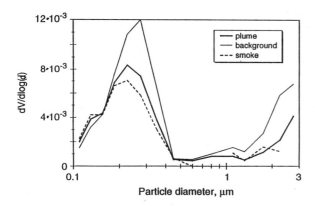

Figure 27.4 Mean volume versus size distribution of the aerosol in the smoke plumes, the regional background, and the smoke component (i.e., the difference between plume and background) as determined by the laser-optical probe on the DC-3 aircraft

Since the man-made emission of aerosols has been implicated in global climate change (Charlson et al. 1987; Charlson et al. 1992; Houghton et al. 1995), research on their origins, properties, and long-range transport has intensified over the last few years. The climatic effect of aerosols depends on their abundance, their size distribution, their optical properties (especially the light absorption capacity), their solubility, and their spatial distribution. During SAFARI-92, the aerosols emitted from various fires were analyzed for several of these properties. The size distribution was investigated with a laser-optical aerosol probe (PCASP-100X) mounted on the DC-3 (Le Canut et al. 1996), the mass concentration of aerosols was deter-

mined using an optical system at the ground (Cachier et al. 1995) and by weighing filters (Maenhaut et al. 1996a; Ward et al. 1996), and the organic and black carbon contents of the aerosols were determined by optical (Cachier et al. 1995; Andreae et al. 1996a; Maenhaut et al. 1996a) and thermochemical (Cachier et al. 1995) methods. The chemical composition of the aerosol was investigated by Proton-Induced X-ray Emission (PIXE) (Maenhaut et al. 1996a), and the soluble fraction of the aerosol was analyzed by capillary electrophoresis (Andreae et al. 1996a).

Emission factors for the aerosol components determined during SAFARI-92 are summarized in table 27.7. EF(TPM) is the emission factor for total particulate matter (TPM) without size discrimination. For the KNP fires, this value was determined using an optical method by Cachier et al. (1995). The lower limit estimates given for the fires investigated by the DC-3 are based on the data in the overscale channel of the PCASP-100X laser-optical probe. Conservative analysis of these data suggests that particles with diameters greater than 3 μm represent a mass at least as large as that present in the calibrated size channels of the instrument (Le Canut et al. 1996). EF(PM2.5) represents the emission factor for aerosols in the size range up to 2.5 μm or, in the case of the data from the DC-3, the size range up to 3.0 μm. The difference between these values is minor, as the range of 2.5 to 3.0 μm is still well below the maximum of the mass versus size distribution of the coarse particle mode (figure 27.4). For total carbon, emission factor data are only available from the KNP fires, while for black carbon the ground-based data from KNP are complemented by aircraft data from various fire types and regions. Comparison between the black carbon results obtained on the KNP fires by ground-based measurements with thermochemical analysis and the airborne measurements using optical detection with an aethalometer shows satisfactory agreement (0.54 ± 0.36 and 0.41 ± 0.20 g kg^{-1} dm, respectively).

The lowest aerosol emission factors were measured over the small, scattered fires in the savannas and grasslands and over the sugar cane fires. At the same time, these fires had the highest black carbon emission factors. This is probably related to the fact that many of the small fires burned as backfires, which are characterized by a high fraction of flaming combustion and tend to burn with a relatively clean flame. In the case of the sugar cane fires, the near absence of a smoldering phase would have a similar effect. Overall, the aerosol mass emission factors measured during SAFARI-92 agree well with measurements from other savannas, for example, in West Africa (Cachier et al. 1995) and in Brazil (Ward et al. 1992). The aerosol emissions from savanna fires are substantially lower than values from forest fires in tropical and extratropical regions, which typically fall in the range of 10 to 30 g kg^{-1} dm (Einfeld, Ward, and Hardy 1991; Radke et al. 1991; Ward and Hardy 1991; Ward et al. 1991). On the other hand, the black carbon emission ratios from savanna fires are similar to those from forest fires, resulting in a relatively high ratio of black carbon to total carbon and to submicron aerosol mass (10 to 30%). Consequently, the single scatter albedo of savanna fire smokes is well below that of most forest fires.

The chemical composition of smoke aerosols is dominated by organic material from the incomplete combustion, distillation, and recondensation of tarry substances. Taking into account that the organic matter in these aerosols has a carbon content of about 60 to 70%, the rest being hydrogen and oxygen, the measured particulate organic carbon concentrations can account for practically the total observed TPM concentrations (Andreae et al. 1988). However, the smoke aerosols contain also several percent of inorganic substances, prominent among them the cations potassium and ammonium and the anions chloride, nitrate, and sulfate (figure 27.5). The presence of potassium in submicron aerosols can be used as a diagnostic tracer for particles of pyrogenic origin, as there are few other sources for this element in fine aerosol particles (Andreae 1983; Gaudichet et al. 1995).

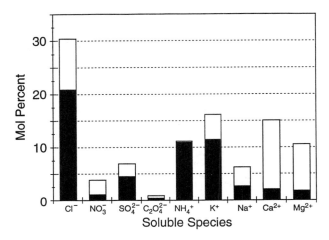

Figure 27.5 Mean composition (in mol% of the soluble fraction of the smoke aerosol from the biomass fires studied during SAFARI-92. The submicron fraction is indicated as the dark part of the bars.

Annual Emissions from Savanna Fires and Other Forms of Biomass Burning

From the data in tables 27.2 to 27.7 and published emission data on savanna and grassland fires, we derived "best guess" emission ratios and used these values to estimate the emission of trace gases and aerosols from savanna fires in Africa and worldwide (table 27.8). We calculated emission factors from the emission ratios using a mean CO_2 emission factor of 1640 g CO_2 kg^{-1} dry fuel. For the methyl halides, where the emission ratios relative to CO_2 were subject to substantial sampling uncertainty, the measured emission ratios relative to CO were converted to emission factors using the mean CO emission factor. The estimates of biomass burned are based on the assessments of Hao, Liu, and Grutzen (1990) and Andreae (1993b). As noted previously, a recent investigation based on fire detection by remote sensing and on an ecosystem database (Scholes, Ward, and Justice 1996) yielded estivates of biomass burned that were a factor of five lower for southern Africa than Hao and Liu's (1994) value. The reasons for this large discrepancy need to be resolved urgently. It should be noted, however, that when pyrogenic CO emissions as low as those proposed by Scholes, Ward, and Justice (1996) are used as boundary condition in atmospheric chemistry/transport models, unrealistically low CO concentrations are obtained for stations downwind of Africa (J. Logan, personal communication 1995).

Based on the estimates of Andreae (1993b), savanna fires in Africa and worldwide account for about 22% and 42%, respectively, of the biomass burned glob-

Table 27.8 "Best guess" emission factors and emission ratios for savanna fires and estimates for emissions from African and global savanna fires, all biomass burning, and all anthropogenic sources (including biomass burning)

Species	Molar emission ratio (10^{-3})	Emisson factor (g *species*/kg dry matter)	African savanna	Global savanna	Biomass burning	All anthrop. sources
				Tg *species*/yr		
Biomass burned			2000	3700	8910	—
Carbon burned			1000	1660	4100	—
CO_2	1000	1640	3280	6070	13 500	33 700[a]
CO	62	65	130	240	680	1600[a]
CH_4	4	2.4	5	9	43	275[a]
NMHC	6	3.1	6	11	42	100[b]
H_2	10	0.75	1.5	2.8	16	40[c]
NO_x[d]	2.8	3.1	6	11	21	70[a]
NO_y[d]	3.5	3.9	8	14	—	—
N_2O	0.09	0.15	0.30	0.56	1.3	5.5[a]
NH_3	1.5	1	2	3.7	6.7	57[e]
SO_2	0.25	0.6	1.2	2.2	4.8	160[f]
COS	0.01	0.02	0.04	0.07	0.21	0.38[g]
CH_3Cl[h]	0.95	0.11	0.22	0.41	1.1	1.1?
CH_3Br[h]	0.0083	0.002	0.004	0.007	0.019	0.11[i]
CH_3I[h]	0.0026	0.001	0.002	0.004	0.008	?
Aerosols						
TPM		10	20	37	90	390[f]
PM2.5		5	10	19	—	240[f]
Carbon		7	14	26	60	90[f]
BC		0.8	1.6	3	9	20[f]
K		0.33	0.7	1.2	1.4	—
CCN[j]		$1.2 \cdot 10^{15}$	$2.4 \cdot 10^{27}$	$4.5 \cdot 10^{27}$	$35 \cdot 10^{27}$	?

a. Houghton et al. (1995).
b. Ehhalt, Rudolph, and Schmidt (1986).
c. Warneck (1988).
d. As NO.
e. Schlesinger and Hartley (1992).
f. Andreae (1995).
g. Chin and Davis (1993).
h. Emission ratios relative to CO.
i. WMO/UNEP (1995).
j. In units of CCN particles per kilogram of dry matter.

ally. However, due to the relatively clean, flaming-dominated combustion characteristic of savanna fires, the emission of reduced trace gases (e.g., CO, CH_4, NMHC) is less than would be expected from the fraction of global biomass burning that takes place in savannas. In the case of CO, for example, savanna fires produce only 35% of the global pyrogenic emission, and in the case of CH_4, the savanna contribution is only 21%. In contrast, species that are produced by flaming combustion are favored in savanna fires, which therefore are responsible for some 52% of pyrogenic NO_x emissions. One consequence of this flaming-

dominated emission profile of savanna fires is that the ratio NO_x/NMHC in the smoke plumes is relatively high. Since the ozone production in biomass smoke plumes tends to be NO_x-limited, higher specific ozone production ($\Delta O_3/\Delta CO$, the ratio of excess ozone to excess CO relative to the plume background) can take place in savanna fire plumes than in forest fire emissions (Andreae et al. 1994a; Helas et al. 1995a).

Savanna fires contribute to methyl halide emissions roughly proportional to their fraction of biomass burned, since the lower methylation fraction resulting from the small fraction of smoldering combustion is compensated by the high halogen element concentration in the savanna biomass as compared to forest biomass. Since smoldering fires release more aerosol than flaming fires (in terms of mass, but not in terms of specific components like inorganic species or black carbon), a somewhat lower relative contribution from savannas to the total pyrogenic source strength would have been expected. The surprisingly high proportion of savanna emissions in table 27.8 may be due to the large uncertainties still present in the emission factor estimates for aerosols from most types of fires. This applies even more to the production of cloud condensation nuclei (CCN) from fires, so that the CCN emission values given in table 27.8 must be considered as very preliminary estimates. Even so, these results indicate that fires may be the largest source of CCN in the tropics.

In conclusion, we find that savanna fires in Africa and around the world are a highly significant source of many trace gases and aerosol species. Since the occurrence of these fires is centered on the dry tropics and mostly limited to the dry season, their regional and seasonal impact is even more conspicuous. The low incidence of rain during the fire season enhances the atmospheric lifetime of the pyrogenic pollutants, increasing the potential for pollutant buildup during periods of recirculation and stagnation and for long-range transport.

To accurately assess the influence of savanna fires on regional and global atmospheric chemistry, we require spatially and temporally resolved estimates of emissions, which can serve as input to regional and global models of atmospheric chemistry and transport. To construct such emission fields, knowledge of both the amounts of biomass burned as a function of space and time and of the emission characteristics (especially the emission factors) is required. While reasonably accurate emission factors have now been determined for many important species, great uncertainty persists with regard to the amount of biomass burned for many regions.

Acknowledgments

We thank the Kruger National Park team (particularly A. Potgieter, W. Trollope, L. Trollope, and J. G. Goldammer) for their untiring cooperation during the experimental fires in the park, the team at the sugar plantation at Ubombo Ranches, Big Bend, Swaziland (especially J. Gosnell and D. Shipley) for their generous help during our experiment, and the pilots and crews of the SAFARI-92 research aircraft for their assistance during the research flights. We appreciate the logistical, technical, and scientific help of a large number of colleagues, whose names are given in the authors' lists and acknowledgements of the primary publications. We acknowledge the permission of various government agencies in Angola, Botswana, Namibia, South Africa, Swaziland, Zambia, and Zimbabwe for permission to conduct research in these countries or to use the airspace over these countries.

References

Andreae, M. O. 1983. Soot carbon and excess fine potassium: Long-range transport of combustion-derived aerosols, *Science*, 220, 1148–1151.

Andreae, M. O. 1991. Biomass burning: Its history, use and distribution and its impact on environmental quality and global climate, in *Global Biomass Burning: Atmospheric, Climatic and Biospheric Implications*, edited by J. S. Levine, pp. 3–21, MIT Press, Cambridge, Mass.

Andreae, M. O. 1993a. Global distribution of fires seen from space, *Eos Trans. AGU*, 74, 129–135.

Andreae, M. O. 1993b. The influence of tropical biomass burning on climate and the atmospheric environment, in *Biogeochemistry of Global Change: Radiatively Active Trace Gases*, edited by R. S. Oremland, pp. 113–150, Chapman & Hall, New York.

Andreae, M. O. 1995. Climatic effects of changing atmospheric aerosol levels, in *World Survey of Climatology. Vol. 16: Future Climates of the World*, edited by A. Henderson-Sellers, pp. 341–392, Elsevier, Amsterdam.

Andreae, M. O., B. E. Anderson, D. R. Blake, J. D. Bradshaw, J. E. Collins, G. L. Gregory, G. W. Sachse, and M. C. Shipham. 1994a. Influence of plumes from biomass burning on atmospheric chemistry over the equatorial Atlantic during CITE-3, *J. Geophys. Res.*, 99, 12 793–12 808.

Andreae, M. O., T. W. Andreae, H. Annegarn, F. Beer, H. Cachier, W. Elbert, G. W. Harris, W. Maenhaut, I. Salma, R. Swap, F. G. Wienhold, and T. Zenker. 1996a, Airborne studies of aerosol emissions from savanna fires in southern Africa: 2. Aerosol chemical composition, *J. Geophys. Res.*, in preparation.

Andreae, M. O., E. Atlas, G. W. Harris, G. Helas, A. de Kock, R. Koppmann, W. Maenhaut, S. Manö, W. H. Pollock, J. Rudolph, D. Scharffe, G. Schebeske, and M. Welling. 1996b. Methyl halide emissions from savanna fires in southern Africa, *J. Geophys. Res.*, in press.

Andreae, M. O., E. V. Browell, M. Garstang, G. L. Gregory, R. C. Harriss, G. F. Hill, D. J. Jacob, M. C. Pereira, G. W. Sachse, A. W. Setzer, P. L. S. Dias, R. W. Talbot, A. L. Torres, S. C. Wofsy, 1988. Biomass-burning emissions and associated haze layers over Amazonia. *J. Geophys. Res.*, 93, 1509–1527.

Andreae, M. O., A. Chapuis, B. Cros, J. Fontan, G. Helas, C. Justice, Y. J. Kaufman, A. Minga, and D. Nganga. 1992. Ozone and Aitken nuclei over equatorial Africa: Airborne observations during DECAFE 88, *J. Geophys. Res.*, 97, 6137–6148.

Andreae, M. O., J. Fishman, M. Garstang, J. G. Goldammer, C. O. Justice, J. S. Levine, R. J. Scholes, B. J. Stocks, A. M. Thompson, B. van Wilgen, and the STARE/TRACE-A/SAFARI Science Team 1994b. Biomass burning in the global environment: First results from the IGAC/BIBEX field campaign STARE/TRACE-A/SA-FARI-92, in *Global Atmospheric-Biospheric Chemistry*, edited by R. G. Prinn, pp. 83–101, Plenum, New York.

Andreae, M. O., and P. Warneck. 1994. Global methane emissions from biomass burning and comparison with other sources, *Pure and Applied Chemistry*, 66, 162–169.

Atlas, E., W. Pollock, and A. de Kock. Organic compound emissions from biomass burning during the SAFARI experiment, *J. Geophys. Res.*, in preparation.

Ayers, G. P., and R. W. Gillett. 1988. Isoprene emissions from vegetation and hydrocarbon emissions from bushfires in tropical Australia, *J. Atmos. Chem.*, 7, 177–190.

Bingemer, H. G., M. O. Andreae, T. W. Andreae, P. Artaxo, G. Helas, D. J. Jacob, N. Mihalopoulos, and B. C. Nguyen. 1992. Sulfur gases and aerosols in and above the equatorial African rainforest, *J. Geophys. Res.*, 97, 6207–6217.

Bonsang, B., C. Boissard, M. F. Le Cloarec, J. Rudolph, and J. P. Lacaux. 1995. Methane, carbon monoxide and light non methane hydrocarbon emissions from African savanna burnings during the FOS/DECAFE experiment, *J. Atmos. Chem.*, 22, 149–162.

Cachier, H., and J. Ducret. 1991. Influence of biomass burning on equatorial African rains, *Nature*, 352, 228–230.

Cachier. H., J. Ducret, M.-P. Brémond, V. Yoboué, J.-P. Lacaux, A. Gaudichet, and J. Baudet. 1991. Biomass burning in a savanna region of the Ivory Coast, in *Global Biomass Burning: Atmospheric, Climatic and Biospheric Implications*, edited by J. S. Levine, pp. 174–180, MIT Press, Cambridge, Mass.

Cachier, H., C. Liousse, P. Buat-Menard, and A. Gaudichet. 1995. Particulate content of savanna fire emissions, *J. Atmos. Chem.*, 22, 123–148.

Charlson, R. J., J. E. Lovelock, M. O. Andreae, and S. G. Warren. 1987. Oceanic phytoplankton, atmospheric sulphur, cloud albedo, and climate, *Nature*, 326, 655–661.

Charlson, R. J., S. E. Schwartz, J. M. Hales, R. D. Cess, J. A. Coakley, J. E. Hansen, and D. J. Hofmann. 1992. Climate forcing by anthropogenic aerosols, *Science*, 255, 423–430.

Chin, M., and D. D. Davis. 1993. Global sources and sinks of OCS and CS_2 and their distributions, *Global Biogeochem. Cycles*, 7, 321–337.

Cofer, W. R., J. S. Levine, E. L. Winstead, P. J. Le Bel, A. M. Koller, C. R. Hinkle. 1990. Trace gas emissions from burning Florida wetlands, *J. Geophys. Res.*, 95, 1865–1870.

Cofer, W. R., III, J. S. Levine, E. L. Winstead, D. R. Cahoon, D. I. Sebacher, J. P. Pinto, and B. J. Stocks. 1996. Source composition of trace gases released during African savanna fires, *J. Geophys. Res.*, in press.

Cook, G. D. 1996. The fate of nutrients during fires in a tropical savanna, *Aust. J. Ecology.* in press.

Cros, B., R. Delmas, D. Nganga, B. Clairac, and J. Fontan. 1988. Seasonal trends of ozone in equatorial Africa: Experimental evidence of photochemical formation, *J. Geophys. Res.*, 93, 8355–8366.

Cros, B., J. Fontan, A. Minga, G. Helas, D. Nganga, R. Delmas, A. Chapuis, B. Benech, and M. O. Andreae. 1992a. Vertical profiles of ozone between 0–400 meters in and above the African Equatorial Forest, *J. Geophys. Res.*, 97, 12877–12887.

Cros, B., D. Nganga, A. Minga, J. Fishman, and V. Brackett. 1992b. Distribution of tropospheric ozone at Brazzaville, Congo, deter-mined from ozonesonde measurements, *J. Geophys. Res.*, 97, 12869–12875.

Crutzen, P. J., and M. O. Andreae. 1990. Biomass burning in the tropics: Impact on atmospheric chemistry and biogeochemical cycles, *Science*, 250, 1669–1678.

Crutzen, P. J., A. C. Delany, J. P. Greenberg, P. Haagenson, L. Heidt, R. Lueb, W. Pollock, W. Seiler, A. F. Wartburg, and P. R. Zimmerman. 1985. Tropospheric chemical composition measurements in Brazil during the dry season, *J. Atmos, Chem.*, 2, 233–256.

Delmas, R. 1982. On the emission of carbon, nitrogen and sulfur in the atmosphere during bushfires in intertropical savannah zones, *Geophys. Res. Lett.*, 9, 761–764.

Ehhalt, D. H., J. Rudolph, and U. Schmidt. 1986. On the importance of light hydrocarbons in multiphase atmospheric systems, in *Chemistry of Multiphase Atmospheric Systems* edited by W. Jaeschke, pp. 321–350, Springer-Verlag, Berlin Heidelberg.

Einfeld, W., D. E. Ward, and C. C. Hardy. 1991. Effects of fire behavior on prescribed fire smoke characteristics: A case study, in *Global Biomass Burning: Atmospheric, Climatic, and Biospheric Implications*, edited by J. S. Levine, pp. 412–419, MIT Press, Cambridge, Mass.

Fishman, J., K. Fakhruzzaman, B. Cros, and D. Nganga. 1991. Identification of widespread pollution in the southern hemisphere deduced from satellite analyses, *Science*, 252, 1693–1696.

Fishman, J., J. M. Hoell, R. J. Bendura, V. W. J. H. Kirchhoff, and R. J. McNeal. 1996. The NASA GTE TRACE-A experiment, *J. Geophys. Res.*, in press.

Gaudichet, A., F. Echalar, B. Chatenet, J. P. Quisefit, G. Malingre, H. Cachier, P. Buat-Ménard, P. Artaxo, and W. Maenhaut. 1995. Trace elements in tropical African savanna biomass burning aerosols, *J. Atmos. Chem.*, 22, 19–39.

Hao, W.-M., and M.-H. Liu. 1994. Spatial and temporal distribution of tropical biomass burning, *Global Biogeochem. Cycles*, 8(4), 495–503.

Hao, W.-M., M.-H. Liu, and P. J. Crutzen. 1990. Estimates of annual and regional releases of CO_2 and other trace gases to the atmosphere from fires in the tropics, based on the FAO statistics for the period 1975–1980, in *Fire in the Tropical Biota: Ecosystem Processes and Global Challenges*, edited by J. G. Goldammer, pp. 440–462, Springer-Verlag, Berlin.

Hao, W.-M., M.-H. Liu, D. E. Ward, M. Lorenzini, and K. D. Sing. 1993. The trend of tropical biomass burning and its impact on emission of trace gases. Paper presented at the Symposium on Challenges in Atmospheric Chemistry and Global Change, Boulder, Col.

Hao, W.-M., and D. E. Ward. 1993. Methane production from global biomass burning, *J. Geophys. Res.*, 98, 20657–20661.

Hao, W.-M., D. E. Ward, G. Olbu, and S. P. Baker. 1996. Emissions of CO_2, CO and hydrocarbons from fires in diverse African savanna ecosystems, *J. Geophys. Res.*, in press.

Harris, G. W., T. Zenker, M. O. Andreae, F. G. Wienhold, M. Welling, and U. Parchatka. 1996a. Regional distribution of trace gases over southern Africa measured during SAFARI-92, *J. Geophys. Res.*, in preparation.

Harris, G. W., T. Zenker, F. G. Wienhold, M. Welling, U. Parchatka, and M. O. Andreae. 1993. Airborne measurements of trace gas emission ratios from southern African veld fires, *Eos Trans. AGU*, 74, 104.

Harris, G. W., T. Zenker, F. G. Wienhold, M. Welling, U. Parchatka, and M. O. Andreae. 1996b. Airborne measurements of trace gas emission ratios from southern African veld fires, *J. Geophys. Res.*, in preparation.

Helas, G., H. Bingemer, and M. O. Andreae. 1992. Organic acids over equatorial Africa: Results from DECAFE 88, *J. Geophys. Res.*, 97, 6187–6193.

Helas, G., J. Lobert, D. Scharffe, L. Schäfer, J. Goldammer, J. Baudet, B. Ahoua, A.-L. Ajavon, J.-P. Lacaux, R. Delmas, and M. O. Andreae. 1995a. Ozone production due to emissions from vegetation burning, *J. Atmos. Chem.*, 22, 163–174.

Helas, G., J. Lobert, D. Scharffe, L. Schäfer, J. Goldammer, J. Baudet, A.-L. Ajavon, B. Ahoua, J.-P. Lacaux, R. Delmas, and M. O. Andreae. 1995b. Airborne measurements of savanna fire emissions and the regional distribution of pyrogenic pollutants over western Africa, *J. Atmos. Chem.*, 22, 217–239.

Helas, G., G. Schebeske, D. Scharffe, S. Manö, M. O. Andreae, and R. Koppmann. 1996. Light hydrocarbon measurements over savanna fires in South Africa, *J. Geophys. Res.*, in preparation.

Houghton, J. T., L. G. Meira Filho, J. Bruce, H. Lee, B. A. Callander, E. Haites, N. Harris, and K. Maskell. 1995. *Climate Change 1994: Radiative Forcing of Climate Change.* Cambridge University Press, Cambridge, England, 339 pp.

Hurst, D. F., D. W. T. Griffith, J. N. Carras, D. J. Williams, and P. J. Fraser. 1994a. Measurements of trace gases emitted by Australian savanna fires during the 1990 dry season, *J. Atmos. Chem.*, 18, 33–56.

Hurst, D. F., D. W. T. Griffith, and G. D. Cook. 1994b. Trace gas emissions from biomass burning in tropical Australian savannas, *J. Geophys. Res.*, 99, 16441–16456.

Isichei, A. O., and W. W. Sanford. 1980. Nitrogen loss by burning from Nigerian grassland ecosystems, in *Nitrogen Cycling in West African Ecosystems*, edited by T. Rosswall, pp. 325–331, SCOPE/UNEP, Uppsala.

Kirchhoff, V. W. J. H., E. V. A. Marinho, P. L. S. Dias, E. B. Pereira, R. Calheiros, R. André, and C. Volpe. 1991. Enhancements of CO and O_3 from burnings in sugar cane fields, *J. Atmos. Chem.*, 12, 87–102.

Kuhlbusch, T. A., J. M. Lobert, P. J. Crutzen, and P. Warneck. 1991. Molecular nitrogen emissions from denitrfication during biomass burning, *Nature*, 351, 135–137.

Kuhlbusch, T. A. J., M. O. Andreae, H. Cachier, J. G. Goldammer, J.-P. Lacaux, R. Shea, and P. J. Crutzen. 1996. Black carbon formation by savanna fires: Measurements and implications for the global carbon cycle, *J. Geophys. Res.*, in press.

Lacaux, J. P., J. M. Brustet, R. Delmas, J. C. Menaut, L. Abbadie, B. Bonsang, H. Cachier, J. Baudet, M. O. Andreae, and G. Helas. 1995. Biomass burning in the tropical savannas of Ivory Coast: An overview of the field experiment Fire Of Savannas (FOS/DECAFE '91), *J. Atmos. Chem.*, 22. 195–216.

Lacaux, J. P., R. Delmas, C. Jambert, and T. Kuhlbusch. 1996. NO_x emissions from African savanna fire, *J. Geophys. Res.*, in press.

Lacaux, J. P., R. Delmas, G. Kouadio, B. Cros, and M. O. Andreae. 1992. Precipitation chemistry in the Mayombe Forest of equatorial Africa, *J. Geophys. Res.*, 97, 6195–6206.

Lacaux, J.-P., H. Cachier, and R. Delmas. 1993. Biomass burning in Africa: An overview of its impact on atmospheric chemistry, in *Fire in the Environment: The Ecological, Atmospheric, and Climatic Im-*

portance of Vegetation Fires, edited by P. J. Crutzen and J. G. Goldammer, pp. 159–191, Wiley, Chichester, England.

Le Canut, P., M. O. Andreae, G. W. Harris, F. G. Wienhold, and T. Zenker. 1996. Airborne studies of emissions from savanna fires in southern Africa: I. Aerosol emissions measured with a laser-optical particle counter, *J. Geophys. Res.*, in press.

Lobert, J. M., D. H. Scharffe, W.-M. Hao, T. A. Kuhlbusch, R. Seuwen, P. Warneck, and P. J. Crutzen. 1991. Experimental evaluation of biomass burning emissions: Nitrogen and carbon containing compounds, in *Global Biomass Burning: Atmospheric, Climatic and Biospheric Implications*, edited by J. S. Levine, pp. 289–304, MIT Press, Cambridge, Mass.

Maenhaut, W., I. Salma, J. Cafmeyer, H. J. Annegarn, and M. O. Andreae. 1996a. Regional atmospheric aerosol composition and sources in the Eastern Transvaal, South Africa, and impact of biomass burning, *J. Geophys. Res.*, in press.

Maenhaut, W., I. Salma, M. Garstang, and F. Meixner. 1996b. Composition and origin of the regional atmospheric aerosol at Etosha, Namibia, and Victoria Falls, Zimbabwe, *J. Geophys. Res.*, in preparation.

Manö, S., 1995. Messung von partiell oxidierten Kohlenwasserstoffen in Emissionen von Biomasseverbrennung. Ph.D. thesis, Goethe-Universität, Frankfurt, 120 pp.

Nguyen, B. C., N. Mihalopoulos, J. P. Putaud, and B. Bonsang. 1995. Carbonyl sulfide emissions from biomass burning in the tropics, *J. Atmos. Chem.*, 22, 55–65.

Radke, L. F., D. A. Hegg, P. V. Hobbs, J. D. Nance, J. H. Lyons, K. K. Laursen, R. E. Weiss, P. J. Riggan, and D. E. Ward. 1991. Particulate and trace gas emissions from large biomass fires in North America, in *Global Biomass Burning: Atmospheric, Climatic and Biospheric Implications*, edited by J. S. Levine, pp. 209–224, MIT Press, Cambridge, Mass.

Robertson, G. P., and T. Rosswall. 1986. Nitrogen in West Africa: The Regional cycle, *Ecological Monographs*, 56(1), 43–72.

Rudolph, J., A. Khedim, R. Koppmann, and B. Bonsang. 1995. Field study of the emissions of methyl chloride and other halocarbons from biomass burning in western Africa, *J. Atmos. Chem.*, 22, 67–80.

Schlesinger, W. H., and A. E. Hartley. 1992. A global budget for atmospheric NH_3, *Biogeochemistry*, 15, 191–211.

Scholes, R. J., D. Ward, and C. O. Justice. 1996. Emissions of trace gases and aerosol particles due to vegetation burning in southern-hemisphere Africa, *J. Geophys. Res.*, in press.

Shea, R. W., B. W. Shea, J. B. Kauffman, D. E. Ward, C. I. Haskins, and M. C. Scholes. 1996. Fuel biomass and combustion factors associated with fires in savanna ecosystems of South Africa and Zambia, *J. Geophys. Res.*, in press.

Stocks, B. J., B. W. van Wilgen, W. S. W. Trollope, D. J. McRae, J. A. Mason, F. Weirich, and A. L. F. Potgieter. 1996. Fuels and fire behavior dynamics on large-scale savanna fires in Kruger National Park, South Africa, *J. Geophys. Res.*, in press.

Trollope, W. S. W., L. A. Trollope, A. L. F. Potgieter, and N. Zambatis. 1996. SAFARI'92—characterization of biomass and fire behavior in the small experimental burns in the Kruger National Park, *J. Geophys. Res.*, in press.

Ward, D. E., W.-M. Hao, R. A. Susott, R. A. Babbitt, R. W. Shea, J. B. Kauffman, and C. O. Justice. 1996. Effect of fuel composition on combustion efficiency and emission factors for African savanna ecosystems, *J. Geophys. Res.*, in press.

Ward, D. E., and C. C. Hardy. 1991. Smoke emissions from wild-land fires, *Environment Int.*, 17, 117–134.

Ward, D. E., A. W. Setzer, Y. J. Kaufman, and R. A. Rasmussen. 1991. Characteristics of smoke emissions from biomass fires of the Amazon region—BASE-A experiment, in *Global Biomass Burning: Atmospheric, Climatic, and Biospheric Implications*, edited by J. S. Levine, pp. 394–402, MIT Press, Cambridge, Mass.

Ward, D. E., R. A. Susott, J. B. Kauffman, R. E. Babbitt, D. L. Cummings, B. Dias, B. N. Holben, Y. J. Kaufman, R. A. Rasmussen, and A. W. Setzer. 1992. Smoke and fire characteristics for cerrado and deforestation burns in Brazil: BASE-B experiment, *J. Geophys. Res.*, 97, 14 601–14 619.

Warneck, P. 1988. *Chemistry of the Natural Atmosphere*, Academic Press, San Diego, 757 pp.

WMO/UNEP. 1995. *Scientific Assessment of Ozone Depletion: 1994*, World Meteorological Organization, Geneva.

Regional Trace Gas Distribution and Air Mass Characteristics in the Haze Layer over Southern Africa during the Biomass Burning Season (September/October 1992): Observations and Modeling from the STARE/SAFARI-92/DC-3

Thomas Zenker, Anne M. Thompson, Donna P. McNamara, Tom L. Kucsera, Geoffrey W. Harris, Frank G. Wienhold, Philippe Le Canut, Meinrat O. Andreae, and Ralf Koppmann

The IGAC/STARE/SAFARI-92/TRACE-A campaigns (International Global Atmospheric Chemistry/ South Tropical Atlantic Regional Experiment/Southern African Fire Atmospheric Research Initiative/ Transport and Atmospheric Chemistry near the Equator—Atlantic) supplied comprehensive data sets of biomass burning studies in southern Africa and Amazonia in 1992. The observations provided insight into the understanding of how the outflow of biomass burning emissions from both southern Africa and Amazonia influences the high level of tropospheric ozone over the South Atlantic Ocean (Andreae et al. 1994; Lindesay et al. 1996; Fishman et al. 1996) that has been previously found by satellite observations (Fishman et al. 1990).

This chapter contributes to the knowledge about the chemical composition and the ozone-forming potential of the continental mixed layer over southern Africa in which the biomass burning emissions are likely confined prior to their outflow over the oceans. We report on observations from SAFARI-92 made during two regional surveys from 24–28 September and 1–4, 6 October 1992 aboard a chartered DC-3 turboprop aircraft, which we outfitted for trace gas measurements over southern Africa during the end of the biomass burning season in the haze layer at cruising altitudes between 150 feet above ground level and 4 km above sea level. The flight track coverage of the TRACE-A and SAFARI-92 airborne measurements is shown in figure 28.1, and an overview of measured species aboard the DC-3 is given in table 28.1. We discuss general observations from measurements taken in the haze layer but outside of fresh smoke plumes. Also, a range of air mass characteristics and history will be discussed, as well as the first photochemical "point" modeling results based on the DC-3 data.

General Observations

Figure 28.2 shows normalized frequency distributions for NO_x, CO, and O_3 mixing ratios as well as for the NO_x/NO_y ratio that were observed during the two flight surveys (figure 28.3) in 24–28 September and 1–4, 6 October. In the NO_x distribution, more than 90% of the data show mixing ratios greater than 200 pptv, implying that in most cases there are enough active nitrogen oxides available to catalyze photochemical ozone production. The NO_x/NO_y ratio ranges over the entire scale from 0 to 1, with a median frequency at 0.4 for both periods.

The CO and O_3 distributions show that the mixing ratios were significantly more elevated during the first survey flights in September; this suggests a higher input of combustion emissions to the atmosphere, coinciding with more fire activity in September and relatively few fires in October (Kendall et al. 1996; Justice et al. 1996).

The vertical and spatial CO and O_3 distributions are illustrated in figures 28.4 and 28.5, which contain an overlay of all CO and O_3 soundings measured during the two flight surveys. For both survey periods, an overall NE–SW gradient of trace gas mixing ratios is evident, with higher values in the northeastern parts closer to more concentrated fire areas in Mozambique, north Zambia, and east Angola (Kendall et al. 1996; Justice et al. 1996). Back-trajectory calculations based on observation points of high trace gas mixing ratios on 24–26 September in the E sector (figures 28.4a and 28.5a) point to intense Mozambique fires on 21–22 September.

In various profiles, an air mass layer boundary is apparent at about 800-hPa altitude (figures 28.4 and 28.5). We identify this as the top of the (new) mixed layer; this is also based on potential temperature (θ_e) and specific humidity data (see example in figure 28.6); most of our measurements were carried out before noon local time. The residual layer, that is, the residual mixed layer from the day before, which is atop of the new mixed layer, can generally be characterized for our data by turbulent temperature conditions, followed by a temperature inversion that is described by Garstang et al. (1996) as the "3 km absolute stable layer."

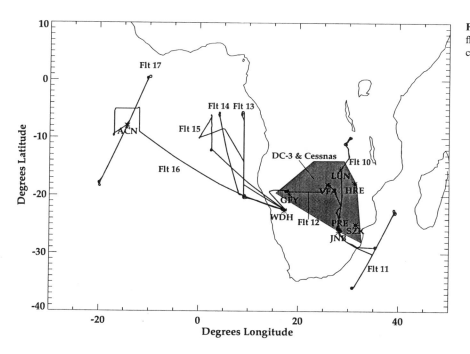

Figure 28.1 TRACE-A (numbered flights) and SAFARI-92 DC-3 flight coverage

Table 28.1 DC-3 measurement characteristics

Species	Method/Instrument	Typical time resolution
NO_2, N_2O, HCHO, CH_4	TDLAS,[a] 4-channel	60 sec
CO	TDLAS-TTFM,[b] 1-channel	1 sec
CO_2	NDIR[c]/Li-cor	10 sec
O_3	UV[d]/Thermo Environment, Model 49; SCL[e]/Unisearch Ozone Analyser	5 sec
NO, NO_y	CLD[f], Gold-CO-CLD/Tecan Instrument, NO inlet and external mounted gold converter inlet alternating	10 sec
NMHC	Grab samples, subsequent GC analysis	1-min samples every half hour
Aerosols, size distribution	Optical probe/PMS PCASP-100X, 0.1–3 μm	10 sec
Aerosols, composition	Filter samples, subsequent PIXE[g]	Hours
Black carbon	Filter sampling	Hours
Position	Global Positioning System	3 sec
P, T, rH	MKS Baratron, Vaisala sensors	1 sec

a. Tunable Diode Laser Absorption Spectroscopy.
b. Two Tone Frequency Modulation.
c. Non Dispersive Infrared.
d. Ultra Violet Absorption.
e. Surface Chemi-luminescence.
f. Chemiluminescence Detector.
g. Proton Induced X-ray Emission.

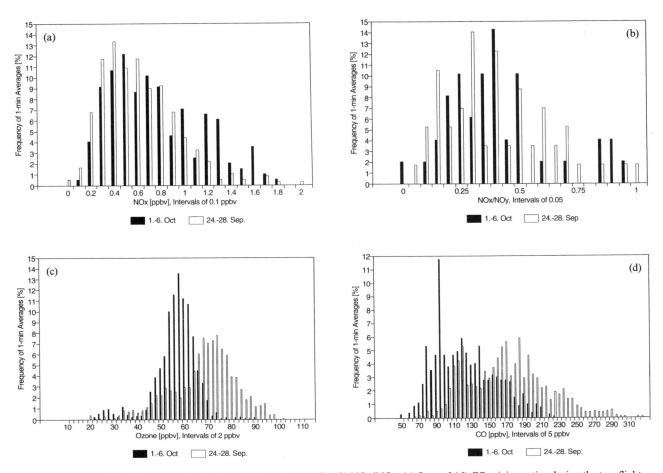

Figure 28.2 Normalized frequency distributions of observed (*a*) NO$_x$, (*b*) NO$_x$/NO$_y$, (*c*) O$_3$, and (*d*) CO mixing ratios during the two flight surveys in September and October

In general, over the continent one expects a mixed layer that starts at sunrise out of the nocturnal boundary inversion and develops during the day up to the first inversion, then mixes with the residual layer. Occasionally we observed a θ_e minimum (see example in figure 28.6) at altitudes of 3 km (700 hPa) or higher, which we barely reach due to our maximum cruise altitudes of about 630 hPa. Depending on the synoptic weather situation, the altitude of the virtual potential temperature minimum varies between 3 and 4 km for anticyclonic and westerly trough circulations, respectively, as discussed from observations at Okaukuejo (Diab et al. 1996).

A common feature of all profiles is that O$_3$ mixing ratios are elevated above the 800-hPa boundary (see figure 28.4), in the 2- to 4-km altitude band. This suggests an accumulation of aged air carrying biomass burning emissions that already instigated a first pulse of photochemical ozone formation. In the air masses

from ground level through the developing mixed layer, ozone destruction after sunset of the day before is manifested as lower mixing ratios than in the residual layer above; that is, in many O$_3$ profiles, a positive O$_3$-altitude gradient is apparent, mostly for measurements carried out before noon. Ozone removal may result from dry deposition or through NO$_x$ catalyzed destruction during the night if biogenic NO$_x$ sources are available, as has occasionally been observed (Harris, Wienhold, and Zenker 1996; Jacob and Wofsy 1988). Different air mass origins for the two layers play a role in the 1 October data.

Highly structured CO vertical profiles with distinct plumes visible as mixing ratios up to 300 ppbv from 24–26 September indicate intensive impacts from biomass burning not yet well mixed in the 2 to 4 km band. These coincide with elevated O$_3$ mixing ratios and suggest air mass origins from the Mozambique fires of 22 September. In contrast, the vertical CO profiles, still

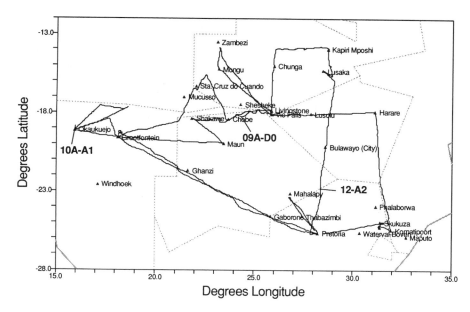

Figure 28.3 SAFARI-92/DC-3 flight track (24 September–6 October 1992). September survey flight track: Pretoria (Wonderboom)–Skukuza–Harare–Victoria Falls–Zambesi–Victoria Falls–Maun–Okaukuejo–Pretoria. October survey flight track: Pretoria–Lusaka–Livingstone–Sta. Cruz do Cuando–Grootfontain–Pretoria–Mahalapy–Pretoria. Locations of soundings 09A-D0 (26 September), 10A-A1 (27 September), and 12-A2 (1 October) containing the input data sets are indicated.

elevated above 80 ppbv CO during the October survey flights (as well as in the September southwestern flight region), are well mixed over the entire vertical profile. This suggests more aged air masses than the four days from the Mozambique fires (from back-trajectories) for the 24–26 September observations. This is supported by back-trajectory calculations for 27 September and 1 October that point back to Mozambique fires with travel times between 6 and 8 days. On 1–3 October, very recent combustion impacts are indicated by elevated CO mixing ratios captured in the developing mixed layer and not mixed with the residual layer from the day before.

Air Mass Characteristics and Modeling of Examples

Model calculations were undertaken by first restricting the data set to those measurement periods where nonmethane hydrocarbon (NMHC) grab samples are taken outside of fresh smoke plumes in "background" air. With these restrictions, we obtain a data set suitable for instantaneous photochemical model calculations representing southern African continental conditions at the end of the dry (biomass burning) season. The first results are based on a data subset including 13 NMHC grab samples.

"Point" Model Description

The GSFC (Goddard Space Flight Center) "point" model (Thompson et al. 1996) performs instantaneous photochemical equilibrium calculations using aircraft data to constrain key species. The point model includes 36 species, 7 transients, and 126 reactions and photodissociations. The kinetics scheme used in the model is nearly identical to that given in Thompson et al. (1993). Photodissociation rates (J-values) are based on the TOMS (Total Ozone Mapping Spectrometer) tropospheric ozone column typically observed over southern Africa during SAFARI-92, assuming a solar angle for 1 October.

The assimilated data from the DC-3 1-minute data set are O_3, CO, $C_{2,3}$-alkanes/alkenes, toluene, NO, temperature (T), relative humidity (rH), and pressure (P). The model computes instantaneous steady-state values for the remaining species. Among these are NO_2, HCHO, PAN, NO_y, OH, $O(^1D)$, HO_2, and R_iO_2; oxygenated organic radicals R_iO_2 include CH_3O_2, $C_2H_5O_2$, $C_3H_6OHO_2$, $C_2H_4OHO_2$, $C_3H_7OHO_2$ (both n- and i-), and CH_3CO_3. Furthermore, production and loss terms for O_3 ($P[O_3]$, $L[O_3]$) are used to calculate an ozone production potential (OPP), defined as the net ozone formation rate from

$$OPP = P - L(O_3)$$
$$= +k_1[NO][HO_2] + \Sigma k_i[NO][R_iO_2] - \{k_3[OH] + k_4[HO_2]\}[O_3] - k_5[O(^1D)][H_2O]$$

OPP, as 24-hour averaged rates in ppbv O_3/day (see table 28.2), are computed based on instantaneous rates at the series of observation points assuming a sinusoidal diurnal OPP pattern between 6 A.M. and 6 P.M. local time.

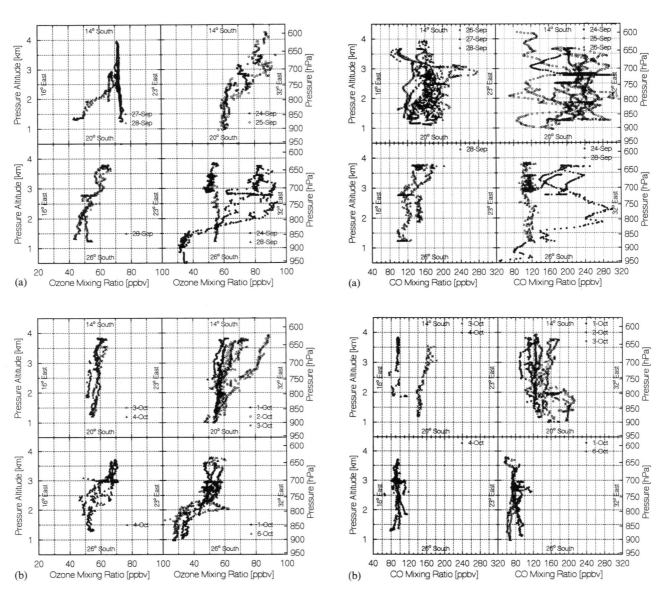

Figure 28.4 Soundings of O_3 from the STARE/SAFARI DC-3 aircraft: (*a*) 24–28 September 1992 and (*b*) 1–6 October 1992 flights. The individual data points represent 10-second averages of "background" data, that is, data from fire smoke plume transits were rejected from the overall data set. Different symbols represent individual days. Soundings are split and overlaid in four spatial sectors; boundaries between the sectors were chosen at 23°E longitude and 20°S latitude so that the SAFARI sites, Pretoria and Skukuza, Victoria Falls, and Okaukuejo, fall into SE, NE, and NW quadrants, respectively.

Figure 28.5 Soundings of CO from the STARE/SAFARI DC-3 aircraft: (*a*) 24–28 September 1992 and (*b*) 1–6 October 1992 flights (same locations as figure 28.4)

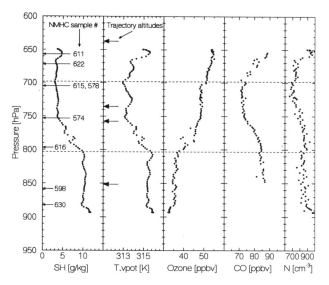

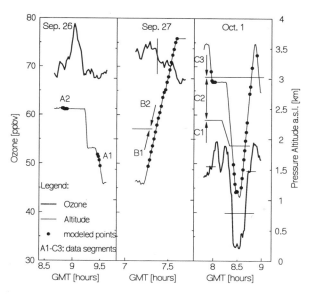

Figure 28.6 Sounding 12-A2 (SAFARI) on 1 October 1992; 10:42 local time, 22.9°S, 28.4°E. Specific humidity (SH), virtual potential temperature (T.vpot), ozone, carbon monoxide (CO), and particle density (N) versus pressure. Altitude positions and numbers of nonmethane hydrocarbon (NMHC) grab samples are marked, as well as start altitudes for calculated back-trajectories. Air mass layer boundaries (see text) are indicated by horizontal dashed lines.

Data Set Used in Model

The simulated data subset is grouped according to three different soundings on 26 and 27 September and 1 October. The locations of these soundings are marked in figure 28.3 using the code 09A-D0, 10A-A1, and 12-A2, where, for example, the code 10A-A1 means flight 10A, ascent 1.

As an example, using sounding 12-A2 (1 October), some tracer species that were used to identify different air mass layers are shown in figure 28.6. Fairly constant values of all five species below the 800 hPa level describe a well-developed mixed layer. Above 800 hPa, decreasing values for specific humidity and virtual potential temperature give strong evidence of an air mass boundary. This middle residual layer reaches to 700 hPa where the unstable and turbulent conditions of falling potential temperatures change toward stable conditions at higher altitudes. This stable layer is expected to extend above the maximum altitude of our measurements (Garstang et al. 1996; Diab et al. 1996). The air mass boundary at 700 hPa is also supported by the correlation between O_3 and CO, which suddenly switches from a negative sign below 700 hPa to a positive one above. The starting altitudes for back-trajectory calculations, which cover all three layers, are marked in the diagram. The data are grouped into segments sublabeled 1, 2, and 3, (e.g., C1, C2, and C3)

Figure 28.7 Sampling of DC-3 O_3 data used in the photochemical point model (SAFARI-92). Flight altitude and locations of data sets modeled are indicated. Indicated segments A1–C3 reflect groups of data within different air mass layers identified by using DC-3 data, including potential temperature, specific humidity, CO, and O_3.

according to the three identified layers: mixed layer (ML), residual layer (RL), and first stable layer (SL), respectively.

Similarly, in sounding 10A-A1 (27 September), the two lower air mass layers, ML and RL, could be identified, separated at 810 hPa, with corresponding data segments B1 and B2, respectively. In sounding 09A-D0 (26 September), NMHC samples were collected only at two altitudes, 825 and 747 hPa, corresponding to ML and RL and data segments A1 and A2, respectively. Trajectories were started in all of these layers, with at least one trajectory for each data segment.

The sampling time for the pressurized filling of a NMHC flask was about 1 minute. We expanded the data set of the original 13 one-minute sample points by averaging or interpolating NMHC or other missing data over a maximum five-minute period centered about a single NMHC grab sample. Precautions were taken in constructing these interpolations by noting the layer structure of the data within the soundings and noting analogous changes to similar species, such as NO, NO_2, and NO_y. The resulting data set, which we used for our modeling study, consists of 55 one-minute data points. The location of the individual data points along the altitude profile is illustrated in figure 28.7, along with the corresponding O_3 time series and back-trajectory starting altitudes.

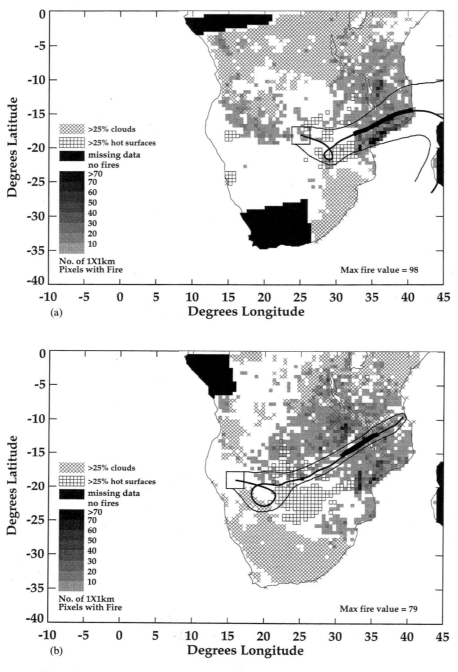

Figure 28.8 Fire count data and back-trajectories. The daily fire counts are on a 0.5°-based grid using data from the NOAA-AVHRR (National Oceanic and Atmospheric Administration-Advanced Very High Resolution Radiometer) satellite instrument (Kendall et al. 1996; Justice et al. 1996). Clusters of 11 × 11 eight-day back-trajectories cover a starting area of 2.5° × 2.5° centered at the locations and observation times of the selected soundings. The starting areas are indicated by a box, and the boundaries of the trajectory clusters are indicated by its envelop curve. The calculated center back-trajectories are displayed, and the temporal overlap between the displayed fire counts and the corresponding trajectory sections is shown as a thick line. (*a*) The back-trajectory cluster starts on 26 September at the location of sounding 09A-D0 (RL, 723 hPa, $\theta_e = 314.5$ K). Fire count data from 22 September are overlaid. A second back-trajectory calculation in the ML at 800 hPa, $\theta_e = 316.1$ K shows a nearly identical result. (*b*) The back-trajectories start on 27 September at the location of sounding 10A-A1 (RL, 771 hPa, $\theta_e = 315.3$ K). Fire

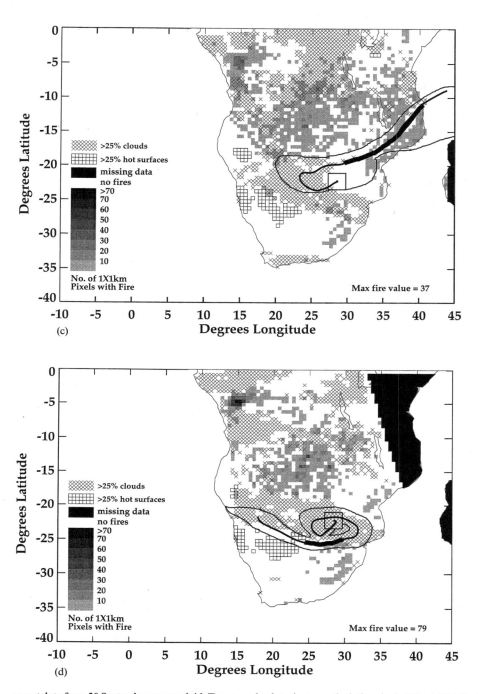

count data from 20 September are overlaid. Two more back-trajectory calculations in the RL at 671 hPa, $\theta_e = 316.5$ K and in the ML at 848 hPa, $\theta_e = 314.1$ K show very similar results with varying travel times of seven to eight days. (c) The back-trajectories start on 1 October at the location of sounding 12-A2 (RL, 738 hPa, $\theta_e = 314.6$ K). Fire count data from the average of 24 and 26 September are overlaid (25 September fire count data are only partially available due to missing satellite observations). A second back-trajectory calculation in the SL at 638 hPa, $\theta_e = 315.4$ K shows a nearly identical result. (d) The back-trajectories start on 1 October at the location of sounding 12-A2 (ML, 852 hPa, $\theta_e = 312.6$ K). Fire count data from 25 September are overlaid. A second back-trajectory calculation at 756 hPa, $\theta_e = 313.0$ K shows a nearly identical result.

Back-Trajectory Calculations

The eight-day back-trajectory calculations were carried out with the National Aeronautics and Space Administration (NASA)/Goddard isentropic trajectory model (Schoeberl et al. 1992) using 12-hour ECMWF (European Centre for Medium-range Weather Forecasts) data fields (Pickering et al. 1994). The altitude starting points of the trajectories were selected from NMHC sampling altitudes or from other unique air mass layers. The shapes of the vertical θ_e profiles in the ECMWF field were in agreement with the measured high-frequency data. However, the absolute θ_e values differed, resulting in unacceptable differences in altitude between the selected starting point and the ECMWF field at the same θ_e value, particulary under unstable conditions. This is likely due to the poor altitude resolution of the ECMWF data in the 0–4 km band. Therefore, the trajectories were computed at the surfaces of potential temperature belonging to selected pressures. Air mass layer boundaries in the ECMWF field, which might become obvious by having different calculated origins of the back-trajectories, may appear at different altitudes than those evident in the data structure of the soundings. For example, the boundary between ML and RL in sounding 12-A2 (see figure 28.6) is suggested to be about 800 hPa by the measured data, whereas different origins of the back-trajectories suggest a higher boundary at about 750 hPa (see discussion below, figure 28.8c and d). In each case, a cluster of 121 trajectories was computed, located in a $2.5° \times 2.5°$ area on a single θ_e surface, as recommended by Pickering et al. (1996). In figure 28.8, examples of back-trajectories are overlaid on the corresponding fire count data from Justice et al. (1996).

Except for ML segment C1 (1 October, figure 28.8d), all the displayed back-trajectories suggest air mass origins over the intensive Mozambique fire region with travel times of four, six, and seven to eight days for segments A (26 September, figure 28.8a), C2–C3 (1 October, figure 28.8c), and B (27 September, figure 28.8b), and corresponding fire periods on 22, 25, and 19–20 September, respectively. All trajectories point back to altitude levels around 600 hPa, which is at the top of a fully developed mixed layer for which daily mixing all the way to ground level is expected. In contrast, the air in ML segment C1 seems to have originated from western parts of southern Africa, having passed the highly industrialized Johannesburg-Pretoria area and the region to the east and south before reaching the sampling location. Along this trajectory path, no fire activities were detected by Advanced Very High Resolution Radiometer (AVHRR).

In summary, the back-trajectories suggest two types of air mass origins: All data segments except C1 are likely influenced by Mozambique biomass burning emissions that have been photochemically aged between 4 and 8 days, whereas the air mass of ML segment C1 appears to be influenced by recent industrial and urban emission and not by biomass burning.

Model Results and Discussion

The 55 single data points that have been used as model input were averaged within the data segments A1–C3 (see figure 28.7) and are presented in table 28.2 along with computed OH densities and the assimilated data of the major species, including O_3, CO, NO, and NMHC. Standard deviations given for the OPP values

Table 28.2 Resulting net O_3 formation rates modeled from DC-3 data

Segment (layer)[a]	O_3 (ppbv)	CO (ppbv)	$C_{2,3}$-Alkane (ppbv)	$C_{2,3}$-Alkene (ppbv)	Toluene (pptv)	C_3H_3/C_2H_6 (%)	C_2H_2/CO (‰)	NO_y (ppbv)	NO_x/NO_y	NO (pptv)	Particle (cm^{-3})	OH[b] (10^6/cm^3)	OPP[c] (ppbv O_3/day)
26 Sept., sounding 09A-D0, near Victoria Falls													
A2 (RL)	70	198	1.78	0.34	34	13.0	1.53			45	2135	7.8	0.9 ± 4.9
A1 (ML)	68	240	1.80	0.30	19	12.7	1.26	6.5	0.08	93	1611	8.7	11.9 ± 2.7
27 Sept., sounding 10A-A1, Ovamboland/Namibia-Angola border													
B2 (RL)	71	145	1.16	0.12	9	9.0	1.39			29	682	5.4	-0.2 ± 4.2
B1 (ML)	74	146	1.24	0.11	34	8.5	1.42			97	771	12.6	13.6 ± 0.7
1 Oct., sounding 12-A2, flight: Pretoria to Lusaka													
C3 (SL)	51	83	0.77	0.14	7	9.7	1.18	2.0	0.35	67	1020	9.6	2.6 ± 4.9
C2 (RL)	48	85	0.80	0.25	8	9.9	1.23	2.1	0.33	80	1055	10.5	9.0 ± 5.8
C1 (ML)	34	89	0.93	0.26	10	12.3	1.38			155	1052	13.2	2.1 ± 4.6

a. A1–C3: data segments (see figure 28.7). SL = stable layer; RL = residual layer; ML = mixed layer (see text).
b. 12-hour mean. Assuming sinusoidal pattern between 6 A.M. and 6 P.M. local time.
c. 12-hour mean. Assuming sinusoidal pattern between 6 A.M. and 6 P.M. local time; OPP = O_3 production potential; instantaneous net O_3 formation rate (see text).

are given primarily to reflect scatter of NO data within the data segments.

Air Mass Characteristics

A common feature of the ML data segments A1, B1, and C1 of the three DC-3 flight is high O_3 formation rate encountered: 17.5 ppbv O_3/day (average of ML single data points). These high formation rates mainly derive from high NO mixing ratios, which are more elevated than in upper layers.

The highest OPP values occur in the ML air mass of segment C1, which is not caused by biomass burning. Instead, the eight-day back-trajectories (see figure 28.8d) suggest urban and industrial impacts including domestic fuel burning from the Johannesburg area, which account for the relatively high NO and slightly elevated NMHC. For all three segments, local, very recent emissions captured in developing mixed layers that started out in the nocturnal boundary inversion (Zunckel et al. 1996) appear likely. These imply additional impacts from source regions passed during the last about 18 hours since the previous sunset.

For segment B1 (Etosha National Park [ENP]), we cannot clearly verify a possible nighttime source. Biogenic NO_x would seem unlikely because of increasing CO mixing ratios toward the ground. (During the same flight, biogenic NO_x emissions were observed only 200 km northeast of ENP [Harris, Wienhold, and Zenker 1996], and the characteristic signature was a negative NO/CO correlation in the lowest layer.) As an additional NO source, hidden under stronger sources with positive NO/CO characteristics, biogenic NO_x remains a possibility; note the back-trajectory history from the east (see figure 28.8b). Fires were not observed during this flight north and east of ENP, and an influence from south Angolan fires (200 to 300 km away) seems unlikely, since air mass origins were directly from the east.

For elevated NO and CO mixing ratios in the lowest layer A1 (see figure 28.8a; west of Victoria Falls), we suggest fresh, overnight impact from local biomass burning. However, this cannot be proved by elevated ratios of alkenes to alkanes, for example (see table 28.2), which tend to be higher near fresh biomass burning. Nevertheless, we frequently observed fires during morning and afternoon flights in this region (south and east Zambia [Le Canut et al. 1996]; see figure 28.2), mostly small fires typical of regional land use practice.

In the upper RL and SL segments (see table 28.2), ozone formation rates are much lower, in most cases

not statistically different from zero. A significant positive formation rate in segment C2 may arise from upward mixing from a ML with elevated O_3 formation rates, and not primarily from long-range transport from the Mozambique fires. Note that O_3 formation rates of the individual data points of segment C2 decrease with altitude.

For the DC-3 flights, on average, the upper air segments apparently transported from Mozambique fire region show an instantaneous O_3 formation rate of +2.8 ppbv O_3/day. An interpretation of varying photochemical ages of these air masses as suggested by four- to eight-day travel times from Mozambique is also supported by chemical signatures. The C_3H_8/C_2H_6 ratios confirm shortest photochemical age for the 26 September data (segment A) and highest for the 27 September data (segment B). In contrast, the C_2H_2/CO ratio implies the greatest photochemical age for 1 October upper layers (segment C2–C3). In addition, the long-lived CO tracer and particle density counts, which were generally found to be the most sensitive indicators of plumes, implied the highest combustion loading for the 26 September (A) air mass. (Another indicator, the NO_x/NO_y ratio, could not be used consistently due to the infrequency of NO_y data.) Following the hydrocarbon ratio interpretation scheme of Gregory et al. (1996) for approximates age of air from combustion sources, we find that both the C_2H_2/CO ratio of ≈ 0.0013 and the C_3H_8/C_2H_6 ratio of ≈ 0.1 suggest a mean photochemical age to be three to five days for all data segments. The C_3H_8/C_2H_6 ratios are very similar to those found in the entire mixed layer (0 to 4 km) on TRACE-A profile 10.4 (northern Zambia, 6 October) (Gregory et al. 1996), whereas their C_2H_2/CO ratio of ≈ 0.004 suggests fresher impacts on this flight, approximately one to two days old.

The resulting OPP values are clearly determined by the NO mixing ratios. A linear regression between OPP and NO shows a strong positive correlation of 0.167 ± 0.005 ppbv O_3/day per pptv NO ($r^2 = .96$, DF = 53). Poorer correlations but a clear trend still exists between OPP and CO as well as OPP and C_2H_6. If ML and upper layer data segments are analyzed separately, the linear regression slopes are -0.06 ± 0.02 and -0.07 ± 0.02 ppbv O_3/day per ppbv CO ($r^2 = .40$ and .17, DF = 17 and 34) with different intercepts, and for C_2H_6 -13 ± 7 and -13 ± 5 ppbv O_3/day per ppbv C_2H_6 ($r^2 = .19$ and .15), respectively. Negative correlation between OPP and ozone (-0.42 ± 0.06 ppbv O_3/day per ppbv O_3, $r^2 = .47$) can be expected. In air masses with the same NO_x, for example, air with lower ozone values is further from its steady-state composition, and the instantaneous production potential is higher. No correlations between OPP and other species or ratios of species are evident—linear correlation coefficients are all $r^2 < .1$.

Averaging over all the DC-3 flight segments shown, the rate of O_3 formations is $+7.9$ ppbv O_3/day. The O_3 formation characteristics derived from point model calculations of TRACE-A DC-8 data over southern Africa are similar, although we note that mixed layer sampling of southern Africa by the DC-8 was not as extensive as for the DC-3. Mixed layer (0 to 4 km) O_3 formation from DC-8 sampling primarily over Zambia-Botswana-Namibia was 10 to 15 ppbv/day, decreasing to lower and even net negative O_3 formation for 4 to 8 km. Above 8 to 12 km, O_3 formation was positive, at 1 to 2 ppbv/day (Jacob et al. 1996; Thompson et al. 1996).

Another way to look at the photochemical environment above the mixed layer is to note that comparing the medians for O_3 during the two surveys suggests an initial O_3 production of 10 ppbv in the September period above that in the October period. With a typical one- to two-week recirculation time for low-lying air masses over southern Africa during SAFARI-92 (Garstang et al. 1996; Tyson et al., chapter 39 this volume), this gives a formation rate of >1 ppbv O_3/day for the September period. These are similar to the rates in our RL or SL (table 28.2). Thus a first-order estimate from the DC-3 surveys and transport times is consistent with detailed modeling of the chemical data.

Model-Data Comparisons
In the model calculations, mixing ratios for HCHO and NO_2 were computed and compared to the measured ones; NO_y could not be investigated consistently due to the infrequency of the NO_y data. The simulated HCHO values agree well with the measured values at lower mixing ratios but drop to approximately 50% at higher mixing ratios. The disagreement for NO_2 is that computed NO_2 mixing ratios are on average a factor of ~ 3 lower. J-values in the model might be somewhat high, but not enough to account for the discrepancy. The sensitivity of the results to HNO_3 heterogeneous losses (to aerosols, rainout, or dry deposition) has been investigated; the changes of the OPP values were within 15%. OPP is greater when HNO_3 removal is simulated with scavenging and more closely duplicates the range of HNO_3 mixing ratios measured in the ML on the DC-8 (Talbot et al. 1996). Thus, the OPP in table 28.2 probably represent lower limits.

Conclusions

Air mass characteristics in terms of vertical layer structure, chemical gradients, and species correlations have been identified in the first analysis of a subset of STARE/SAFARI-92/DC-3 aircraft measurements carried out in the haze layer over the southern African continent. Based on these signatures and related back-trajectory history and ozone production potential (instantaneous ozone formation rate), we can draw the following conclusions:

1. In general, higher concentrations of O_3, CO, and NO_x were encountered during 24–28 September flights than during 1–6 October 1992, but chemical tracers on both sets of flights were frequently elevated above background levels due to biomass burning.

2. In many flights (before noon), a two-layer structure is evident from the variation of O_3 and CO, as well as from temperature and humidity data. A surface 2-km band (mixed layer [ML]), a 2- to 4-km band (residual layer [RL]), and occasionally the beginning of a first stable layer (SL) above 3- to 4-km altitude were observed.

3. From the fact that the September observations were 10 ppbv higher in the O_3 mixing ratio than the October data, and one- to two-week residence times for recirculating air masses over southern Africa were typical, we estimate an ozone formation rate of ≥ 1 ppbv O_3/day for continental haze layers for late September 1992, which is consistent with the results from detailed calculations.

4. In the upper layers, RL and SL, a mean ozone formation rate $+2.8$ ppbv O_3/day is calculated. Back-trajectories suggest approximately four to eight days of aging from the loading of air masses with strong biomass burning injections over intensive Mozambique fire regions. In general agreement with this observation are the chemical composition measurements, such as NMHC ratios, which suggest mean photochemical ages of approximately three to five days.

5. In the ML, for all three soundings examined here, a mean ozone formation rate of 17.5 ppbv O_3/day is evident; this tends to be photochemically driven by enhanced NO mixing ratios. Various sources for enhanced mixing ratios of various species in the ML are suggested: recent, local biomass burnings, and urban-industrial sources. Biogenic NO_x emissions are also possible, but there is insufficient data to verify this.

6. If one assumes nocturnal boundary inversions (Zunckel et al. 1996), then in all three examples, fresh

overnight emissions from different sources cause enhanced ozone formation rates in the developing ML. This apparently mixes with the RL during the late day throughout the entire haze 0- to 4-km layer, refreshing the accumulated photochemically aged upper air with fresh inputs.

The point model is useful for interpreting high-frequency DC-3 aircraft data in terms of estimating a potential O_3 formation in the haze layer over the southern African continent before the air masses flow out over either the Atlantic or Indian Ocean. Although the sample of SAFARI-92 flight data discussed here is only a small fraction of the DC-3 $O_3/CO/NO$ data set, it covers a range of conditions representative of the background chemical environment encountered on the full set of DC-3 flights. The frequency of details identified in this study will be determined by further analysis of the complete DC-3 data set, especially with a wider sampling of sounding studies in the haze layer created by biomass burning over southern Africa in 1992.

Acknowledgments

We would like to thank the crews of the DC-3 aircraft for their professionalism and expertise, particularly in navigating the aircraft just atop the canopy as well as through smoke plumes from savanna fires. We also thank those responsible for the tremendous organizational efforts required for the STARE/SAFARI-92 campaign, especially the South African groups of Janette Lindesay, Harold Annegarn, and colleagues. Without their efforts, the measurement campaign never would have taken place. We thank M. R. Schoeberl, L. R. Lait, and P. A. Newman for making the Goddard trajectory model available to us. Thanks to J. D. Kendall and C. O. Justice for providing fire count data. AMT acknowledges support from the USEPA and from NASA Programs in Tropospheric Chemistry, Atmospheric Chemistry Modeling and Data Analysis, and EOS.

References

Andreae, M. O., J. Fishman, M. Garstang, J. G. Goldammer, C. O. Justice, J. S. Levine, R. J. Scholes, B. J. Stocks, A. M. Thompson, B. van Wilgen, and the STARE/TRACE-A SAFARI-92 Science Team. 1994. Biomass Burning in the Global Environment: First Results from the IGAC/BIBEX Field Campaign STARE/TRACE-A/SAFARI-92. In: Prinn R. G. (ed.) *Global Atmospheric-Biospheric Chemistry*. Plenum Press (Environmental science research; v.48), New York, pp. 83–101.

Diab, R. D., M. R. Jury, J. M. Combrink, and F. Sokolic. 1996. A comparison of anticyclone and trough influences on the vertical distribution of ozone and meteorological conditions during SAFARI-92, *J. Geophys. Res.*, in press.

Fishman, J., J. M. Hoell, Jr., R. D. Bendura, R. J. McNeal, Jr., V. W. J. H. Kirchhoff. 1996. NASA GTE TRACE-A Experiment (September–October 1992), *J. Geophys. Res.*, in press.

Fishman, J., C. E. Watson, J. R. Larson, and J. A. Logan. 1990. Distribution of tropospheric ozone determined from satellite data, *J. Geophys. Res.*, 95, 3599–3617.

Garstang, M., P. D. Tyson, R. Swap, M. Edwards, P. Kållberg, and J. A. Lindesay. 1996. Horizontal and vertical transport of air over southern Africa, *J. Geophys Res.*, in press.

Gregory, G. L., H. E. Fuelberg, S. F. Longmore, B. E. Anderson, J. E. Collins, and D. R. Blake. 1996. Chemical characteristics of tropospheric air over the tropical South Atlantic Ocean: Relationship to trajectory history, *J. Geophys. Res.*, in press.

Harris, G. W., F. G. Wienhold and T. Zenker. 1996. Airborne observations of strong biogenic NO_x emissions from the Namibian savanna at the end of the dry season, *J. Geophys. Res.*, in press.

Jacob, D. J., and S. C. Wofsy. 1988. Photochemistry of biogenic emissions over the Amazon forest, *J. Geophys. Res.*, 93, 1477–1488.

Jacob, D. J., et al. 1996. The origin of ozone and NO_x in the tropical troposphere: Photochemical analysis of aircraft observations over the South Atlantic Basin, *J. Geophys. Res.*, in press.

Justice, C. O., J. D. Kendall, P. R. Dowty, and R. J. Scholes. 1996. Satellite remote sensing of fires during the SAFARI campaign using NOAA-AVHRR data, *J. Geophys. Res.*, in press.

Kendall, J. D., C. O. Justice, P. R. Dowty, D. D. Elvidge, and J. Goldammer. 1996. Remote sensing of fires in southern Africa during the SAFARI 1992 campaign, in *Fire in Southern African Savanna: Ecological and Atmospheric Perspectives*, edited by J. A. Lindesay, M. O. Andreae, P. D. Tyson, and B. van Wilgen, Univ. of Witwatersrand Press, Johannesburg, in press.

Le Canut, P., M. O. Andreae, G. W. Harris, F. G. Wienhold, and T. Zenker. 1996. Airborne studies of emissions from savanna fires in southern Africa: I. Aerosol emissions measured with a laser-optical particle counter, *J. Geophys. Res.*, in press.

Lindesay, J. A., M. O. Andreae, J. G. Goldammer, G. W. Harris, H. J. Annegarn, M. Garstang, R. J. Scholes, and B. W. van Wilgen. 1996. The IGBP/IGAC SAFARI-92 field experiment: Background and overview, *J. Geophys. Res.*, in press.

Pickering, K. E., A. M. Thompson, D. P. McNamara, and M. R. Schoeberl. 1994. An intercomparison of isentropic trajectories over the South Atlantic, *Mon. Wea. Rev.*, 122, 864–879.

Pickering, K. E., A. M. Thompson, D. P. McNamara, M. R. Schoeberl, H. E. Fuelberg, R. O. Loring, Jr., M. V. Waston, K. Fakhruzzaman, and A. S. Bachmeier. 1996. TRACE-A trajectory intercomparison: Effects of different input analyses, *J. Geophys. Res.*, in press.

Schoeberl, M. R., L. R. Lait, P. A. Newman, and J. E. Rosenfield. 1992. The structure of the polar vortex, *J. Geophys. Res.* 97, 7859–7882.

Talbot, R. W., et al. Submitted, 1996. Chemical characteristics of continental outflow over the tropical south Atlantic Ocean from Brazil and Africa, *J. Geophys. Res.*, in press.

Thompson, A. M., et al. 1993. Ozone observations and a model of marine boundary layer photochemistry during SAGA 3, *J. Geophys. Res.*, 98, 16 955–16 968.

Thompson, A. M., K. E. Pickering, D. P. McNamara, M. R. Schoeberl, R. D. Hudson, J. H. Kim, E. V. Browell, V. W. J. H. Kirchhoff, and D. Nganga. 1996. Where did tropospheric ozone over southern Africa and the tropical Atlantic come from? Insights from TOMS, GTE TRACE-A and SAFARI-92, *J. Geophys. Res.*, in press.

Zunckel, M., Y. Hong, K. M. Brassel, and S. O'Beirne. 1996. Characteristics of the nocturnal boundary layer: Okaukuejo Namibia during SAFARI-92, *J. Geophys. Res.*, in press.

Airborne Measurements of Organic Trace Gases from Savanna Fires in Southern Africa during SAFARI-92

Ralf Koppmann, Ahmed Khedim, Jochen Rudolph, Günter Helas, Michael Welling, and Thomas Zenker

Emissions from biomass burning are known to be a considerable source of atmospheric hydrocarbons, especially in tropical and subtropical regions (Greenberg et al. 1984; Greenberg and Zimmerman 1984; Crutzen et al. 1985; Zimmerman, Greenberg, and Westberg 1988; Rasmussen and Khalil 1988; Bonsang, Lambert, and Boissard 1991; Rudolph, Khedim, and Bonsang 1992; Rudolph et al. 1995). Due to transport, these emissions also have a significant impact on the budget of organic trace gases in the tropical marine atmosphere (Koppmann et al. 1992). Moreover, owing to their impact on the photochemical ozone formation, hydrocarbons influence the budget of tropospheric ozone (Chatfield and Delany 1990; Fishman et al. 1990). The biomass burnt annually on a global scale is about 8.6×10^{15} g of dry material, 45% of which is estimated to be savanna grassland (Levine 1991). Savanna fires are thus the largest single soruce of biomass burning emissions. In the last years, a growing number of studies were published dealing with field measurements of biomass burning emissions in Africa, which contains about two thirds of the world's savanna regions (Delmas 1982; Bonsang, Lambert, and Boissard 1991; Rudolph et al. 1995; Lacaux et al. 1991). These data were obtained from both ground-based and airborne measurements in the vicinity and over large-scale fires. The results showed that during the biomass burning season the concentrations of CO and organic trace gases in the planetary boundary layer were considerably enhanced compared to the wet season (Rudolph, Khedim, and Bonsang 1992). Meanwhile, laboratory experiments supplied more information about the factors determining the emission ratios such as different burning stages and fire conditions as well as fuel types (Lobert et al. 1991; Manoe 1995). The results showed that the burning of organic material in oxygen-deficient fires leads to the emission of methane, nonmethane hydrocarbons (NMHCs), and a variety of partially oxidized organic compounds. In some cases, the emissios of medium- and higher-molecular-weight organic compounds exceed those of light NMHCs. These emissions can contribute significantly to the budgets of organic trace gases in the regions of biomass burning and may have a considerable impact on the formation of ozone in these areas. Very recent studies showed the importance of higher-molecular-weight organic compounds in biomass burning plumes. However, the results were given unspeciated as the sum over all organic compounds on a ppb carbon basis (Hurst et al. 1994). The TRACE-A/SAFARI campaigns (Transport and Atmospheric Chemistry near the Equator—Atlantic/Southern African Fire Atmosphere Research Initiative) supplied a large data set concerning biomass burning emissions in southern Africa. The background and summary findings of these campaigns are given by Andreae et al. (1994) and Fishman et al. (1995). In this chapter, gas chromatography (GC) and gas chromatography/mass spectrometry (GC/MS) measurements of plume samples collected during SAFARI-92 from different types of fire were used to investigate the composition of biomass burning emissions with respect to organic molecules. Part of these measurements are used to estimate the impact of the various compounds on the photochemical ozone formation.

Experiment

During SAFARI-92, prescribed savanna fires in the Kruger National Park (South Africa) and sugar cane fires as well as uncontrolled fires were investigated. Before the prescribed fires were ignited, the fuel load, fuel type, and distribution were measured within the burning areas (Stocks et al. 1995). During the fires, a large set of trace compounds and aerosols was investigated using ground-based and airborne measurements (Andreae et al. 1995; Le Canut et al. 1995). The fire stages and burning conditions were controlled and monitored during the campaign. After the fires, type and amount of the remaining biomass as well as the ash were analyzed.

The measurements discussed here are based on samples collected on three flight legs. During the flight SAF04 on 19 September 1992, the plumes of two sugar cane fires on the Ubombo Ranches plantation near Big

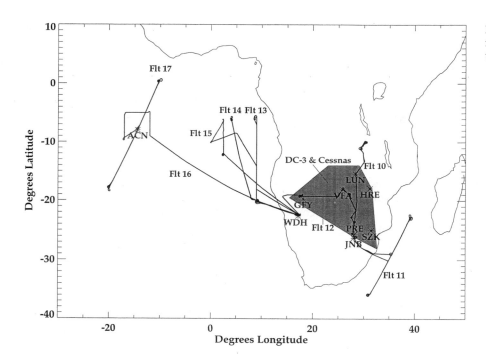

Figure 29.1 TRACE-A/SAFARI flight coverage. The shaded area marks the region of the DC-3 activities.

Bend, Swaziland, were investigated. The two fields had an area of 12 and 43.7 ha. These types of fire are typical of regional agricultural practices. They are set immediately before harvesting to facilitate the cutting and processing of the sugar cane. During flight SAF06 on 24 September 1992, a controlled burn on Block 55 (2333 ha) in the Pretoriuskop section of the Kruger National Park was investigated. Details on fuel and fire behavior are given by Stocks et al. (1995), and details on particle emissions are given by Le Canut et al. (1995). During flight SAF05 on 20 September 1992, an uncontrolled "fire of opportunity" in the Drakensberg region was encountered. The burnt biomass was a mixture of forest, shrub, and grass.

Sampling was done aboard a DC-3 aircraft operating between 100 and 4000 m altitude. Figure 29.1 gives an overview over the flight tracks of the STARE/SAFARI project. The shaded area indicates the region where the DC-3 operated.

Sampling

About 120 whole-air samples were collected in evacuated stainless steel canisters of 2-l volume. Figure 29.2 shows a schematic drawing of the air sampling system. The air intake was a stainless steel tube (6 mm ID) mounted on top of the aircraft extending beyond the boundary layer of the DC-3. The exhaust line was mounted in the back of the aircraft. Once the DC-3 was airborne, the inlet line was continuously flushed with ambient air. The air samples were pressurized

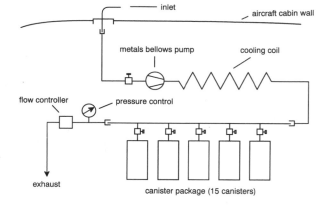

Figure 29.2 Schematic drawing of the sampling system installed on board the DC-3

with a metal bellows pump. With a valve, the pressure in the stainless steel line with the canisters was adjusted to about 300 kPa. Typical sampling times were between one and two minutes depending on the altitude of the aircraft. This time is comparable to the transit time through the plumes.

Analysis

The samples were analyzed in the laboratory in Jülich by gas chromatography for CO, CO_2, CH_4, and NMHCs. CO, CO_2, and CH_4 were determined by a gas chromatographic method very similar to the one described by Heidt (1978). The precision is 3% for CO, 0.5% for CO_2, and 1% for CH_4. NHMCs were analyzed using gas chromatography with flame ionization (FID) and electron capture (ECD) detectors. The hydrocarbons were preconcentrated cryogenically from volumes between several hundred milliliters and about 2 l depending on the type of sample and separated on a combination of a 7-m micropacked Porapack QS column and a 105-m RTX-1 fused silica capillary column. The detection limits were between 10 ppt for the measurements on the packed column (C_2–C_4 hydrocarbons) and below 1 ppt for the measurements on the capillary column (C_5–C_{10} organic trace gases). The reproducibility ranged between 10 and 30% depending on the compound. Details of the analytical techniques are given by Koppmann et al. (1992, 1993), Rudolph, Johnen, and Khedim (1986), and Rudolph et al. (1995).

For a number of plume samples, the medium-molecular-weight organic trace gases (C_5 and higher) were additionally analyzed with a mass spectrometer (Fisons, MD 800). For this purpose, the effluent of the capillary column was split with a ratio of 3:1 between the FID detector and the MS. The FID was used to quantify the compounds in the sample, the MS for identification purposes. With this technique, it was possible to record mass spectra even of compounds with mixing ratios in the ppt range. It should be noted that this technique, including the sampling in stainless steel canisters, is not an adequate technique for all compounds that can be observed in the samples. It is well known that a number of compounds are not stable in canisters or in stainless steel lines; others cannot be preconcentrated and desorbed quantitatively (Rudolph, Müller, and Koppmann 1990). Significant losses from reactions with ozone in the canisters can be excluded due to the very limited stability of ozone in stainless steel canisters (Greenberg et al. 1992) as well as during the preconcentration and desorption step for the operating conditions of our preconcentration system (Koppmann et al. 1995).

Results and Discussion

Organic Composition of Biomass Burning Emissions

Vegetation material is made up of cellulose, hemicelluloses, lignin, proteins, nucleic acids, amino acids, and volatile substances (Franke 1989). During the first phase of a fire, the later compounds such as aldehydes, alcohols, and terpenes are released. In the following phase, temperature-caused cracking of the fuel mole-

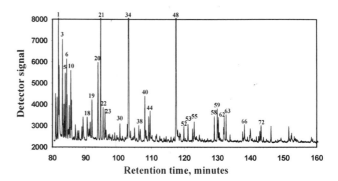

Figure 29.3 Chromatogram of sample no. 3602. The major peaks are identified by numbers. The numbers in the chromatogram correspond to the numbers given in table 29.2.

cules occurs. Higher-molecular-weight compounds are decomposed into a variety of volatile organic compounds (Lobert and Warnatz 1993). In addition to the light NMHCs, which were the only organic compounds besides methane investigated in previous field studies, a large number of higher-molecular-weight organic trace gases are expected in the plumes of biomass burning. They include alcohols, aldehydes, ketones, carboxylic acids, esters, ethers, furanes, etc. The partly oxidized compounds are believed to be predominantly emitted during the smoldering phase of fires. One task of this study was to provide a quantitative knowledge of the organic composition of biomass burning emissions. Figure 29.3 shows a sample chromatogram of the capillary column of an air sample collected inside the plume of the Drakensberg fire during flight SAF05 (sample no. 3602). Details of the flights through the plume of this fire are given in table 29.1. The corresponding mixing ratios of CO_2 and CO in this sample were 399.5 ppm and 4416.8 ppb, respectively. The average $\Delta(CO)/\Delta(CO_2)$ ratio (calculated from all passages through the plume) was $9.36 \pm 1.30\%$, which indicates that the smoldering phase dominated in this fire. The mixing ratios of the sum of C_2 and C_3 hydrocarbons were 364 ppb carbon. For the analysis of this sample, the mass spectrometer was used in parallel to the FID and ECD. More than 140 major peaks are found in this chromatogram, about 70 of which could be positively identified. Table 29.2 gives the retention times of the peaks and the identification of the compounds. In a first step, the identification of these compounds was done by analyzing the mass spectra and comparing them with the National Institute of Standards and Technology (NIST) library. In some cases, two or more compounds overlapped in one peak. These compounds are included here only in

Table 29.1 List of plume samples at the Drakensberg fire during flight SAF05 on 20 September 1992

Sample no.	Sampling time (UTC)	Latitude	Longitude	Altitude (m)	Passage no.	CH$_4$ (ppb)	CO (ppb)	CO$_2$ (ppm)
3593	1106	−25.30	31.09	1039.00	A	2022	2906	384
3592	1110	−25.28	31.07	1160.00	B	1831	1100	367
3594	1114	−25.29	31.09	1110.00	C	2039	3099	382
3595	1117	−25.28	31.11	1069.00	D	1986	2763	382
3596	1120	−25.28	31.10	1013.00	E	1902	1719	374
3598	1130	−25.47	30.69	1656.00	F	1825	975	368
3599	1133	−25.45	30.72	1701.00	G	1833	1149	371
3601	1142	−25.44	30.70	1303.00	H	2103	3413	388
3602	1145	−25.45	30.70	1463.00	I	2236	4416	399

A total of nine passages through this plume were flown during flight SAF05. Plume crossings are identified by broad maxima in the continuous CO and CO$_2$ measurements. The time for one plume passage was about two minutes; the sampling time was about one minute.

cases where a positive identification of at least one of the compounds was possible. The identification due to the analysis of the mass spectra was then verified by comparing the relative retention times and the boiling point of the compounds. As a result of this analysis, a large number of organic trace gases could be identified that were previously not known to be emitted by biomass burning. In general, the mixing ratios of the different members of the various compound groups decrease with increasing number of carbon atoms.

The chromatograms of all other samples collected inside the plumes of savanna fires were analyzed based on the identification with the mass spectrometer. From these data, emission ratios are calculated on a ppb carbon basis, and the contribution of these compounds to the total emission of organic trace gases and relative to CO and CO$_2$ can be derived.

Emission Ratios of Organic Trace Gases

CO$_2$ is the dominant carbon species emitted from fires, followed by CO as the second important compound. The emission ratio of the excess CO to excess CO$_2$ allows differentiation between burning stages. The $\Delta CO/\Delta CO_2$ ratio varies between the different burning processes. During the flaming phase, the ratio is relatively low, in general 5 to 10%. During the smoldering phase of a fire, due to low oxygen supply more CO is emitted, and the ratio of $\Delta CO/\Delta CO_2$ increases. Typical values observed in this phase of fires are in the order of 10 to 15%. The emissions of organic compounds also vary as a function of the burning stages. From correlations of organic trace gases with CO$_2$, it is possible to calculate the emission ratios of trace gases for the different types of fire and the different burning processes or burning stages. The quantitative analysis

of the organic compounds was based on the FID response. The response of the FID is proportional to the molar mass of the compounds, that is, within one group of organic trace gases the response increases linearly with the mass of the molecule. The response, however, is slightly different for groups of compounds containing hetero atoms such as oxygen, nitrogen, and sulfur or for compounds containing different functional groups. Generally, in these cases the response is lower than for "pure" hydrocarbons. Since for this analysis we treated all compounds as "pure" hydrocarbons and due to the previously mentioned problems of a quantitative determination of some of these compounds, our emission ratios can be seen as lower limits, that is, the "real" emission ratios may be higher than the values given here.

Table 29.3 summarizes the emission ratios of the two sugar cane fires, the Kruger National Park fire #2 and the Drakensberg fire for CO, CH$_4$, the sum of C$_2$–C$_4$ hydrocarbons, and the sum of organic trace gases with five or more carbon atoms. The total amount of organic trace gases with five or more carbon atoms is based on all FID-responsive compounds found in the chromatograms. Table 29.4 gives the specific emission ratios for the identified organic compounds. In all cases the highest emission ratios were observed for the Drakensberg fire, while the sugar cane fires showed the lowest emission ratios. From the emission ratio of CO relative to CO$_2$, the burning conditions of the different fires can be estimated. For the fire at Drakensberg, smoldering was observed to be the dominant burning process, while the sugar cane fires were large flaming fires. The fire of Block 55 in the Kruger National Park was a mixture of both. The flaming phase of the fire was followed by an intense smoldering phase. Since

Table 29.2 Compounds positively identified in sample no. 3602

Peak no.	Retention time (min)	Identification
1	81.61	Acetone
2	81.81	i-Pentane
3	82.82	1-Pentene
4	83.30	2-Methyl-1-butene
5	83.66	n-Pentane
6	84.08	Isoprene
7	84.21	(E) 2-Pentene
8	84.60	2-Methyl-2-propanol
9	84.89	Methylacetate
10	85.34	2-Methyl-2-butene
11	85.57	(E) 1,3-Pentadiene
13	86.36	$CS_2 + ?$
14	86.74	(Z) 1,3-Pentadiene
15	87.77	Propanenitrile
16	88.68	Tetrahydrofuran
17	89.08	Cyclopentane
18	90.33	Methylvinylketone + 2-methyl-1-pentane
19	91.75	Butanal
20	93.60	1-Hexene
21	94.42	2-Methylfuran
22	95.04	Ethylacetate
23	95.21	n-Hexane
24	95.81	3-Methylfurane
25	96.15	4-Methyl-2-pentene
26	97.26	2-methyl-1-propanol + ?
27	98.13	4-Methyl-2-pentene
28	98.77	2,3-Dihydrofuran
29	99.47	Methylcyclopentane
30	100.37	3-Methylbutanal
31	101.15	3-Methylbutanone
32	101.91	1-Butanol
33	102.59	2,4-Hexadiene
34	102.95	Benzene
35	103.59	Thiophene
36	104.51	1-Penten-3-one
37	104.82	2-Pentanone
38	106.09	Cyclopentanal
39	106.57	Cyclohexane
40	107.94	1,2-Dimethylcyclopentane
41	108.27	Trichlorethene
42	108.50	2-Ethylfuran
43	109.07	2,4-Dimethylalfuran +3-hexen-1-ol-acetate
44	109.51	n-Heptane

Table 29.2 (continued)

Peak no.	Retention time (min)	Identification
45	111.26	Phenol
46	112.47	2-Hexanone + ?
47	113.74	Dimethylsulfide
48	117.41	Toluene
49	117.96	2-Methylheptane
50	118.20	2,2-Dimethylpetanale
51	118.56	Dimethylcyclohexane
52	119.82	Hexanale
53	121.05	1-Octene
54	122.45	Acetic acid 2-methylpropylester
55	122.91	n-Octane
56	123.38	2-Octene
57	123.62	2-Furaldehyde
58	128.98	Ethylbenzene
59	129.89	p,m-Xylene +3-heptanone
60	130.55	2-Heptanone
61	131.81	Styrene
62	132.00	7-Methyl-1-octene
63	132.61	o-Xylene
64	133.81	n-Nonane
65	137.68	6-Methyl-2-heptanone
66	138.24	Benzaldehyde
67	139.18	Propylbenzene
68	139.84	1-Ethyl-3-methylbenzene
69	140.01	Benzonitrile + ?
70	140.62	1,2,4-Trimethylbenzene
71	141.85	1-Ethyl-4-methylbenzene
72	143.28	Benzofuran

this fire covered a large savanna area, in the smoke plumes emission patterns of both phases superimpose, leading to the observed emission ratios. It can be seen that the emission ratios of the sum of the light NMHCs are comparable or even higher than the emission ratio of methane. This has already been observed during previous field studies of large savanna fires with similar $\Delta CO/\Delta CO_2$ ratios in Africa and Australia (Bonsang et al. 1991; Hurst et al. 1994). Our results show further that the emission ratios of organic compounds with a chain length of five or more carbon atoms add up to at least half of the contribution of light NMHCs. The emission ratios relative to CO_2 for the total measured organic compounds were $(13.7 \pm 0.9) \times 10^{-3}$ for the Drakensberg fire, $(7.4 \pm 1.6) \times 10^{-3}$ for the Kruger

Table 29.3 Emission ratios of methane, carbon monoxide, light NMHCs, and higher-molecular-weight organic trace compounds calculated from the plume samples

Flight no.	Fire	Passage no.	$\Delta(CO)/\Delta(CO_2)$ (%)	$\Delta(CH_4)/\Delta(CO_2)$ (%)	$\Delta(\Sigma C_2-C_4)/\Delta(CO_2)$ (%)	$\Delta(\Sigma C_5-C_9)/\Delta(CO_2)$ (%)
SAF04	Sugar cane, field #1	1	6.97	0.57	0.22	0.16
SAF04	Sugar cane, field #1	2	5.90	0.35	0.59	0.27
SAF04	Sugar cane, field #1	4	2.25	0.07	0.13	0.06
SAF04	Sugar cane, field #1	5	2.41	0.05	0.10	0.04
Mean			4.38	0.26	0.26	0.13
Standard deviation			2.09	0.22	0.20	0.09
SAF04	Sugar cane, field #2	10	2.52	0.05	0.11	0.05
SAF04	Sugar cane, field #2	11	1.83	0.03	0.07	0.02
SAF04	Sugar cane, field #2	18	2.21	0.23	0.18	0.08
Mean			2.19	0.11	0.12	0.05
Standard deviation			0.28	0.09	0.05	0.02
SAF05	Drakensberg	A	9.81	1.04	0.40	0.41
SAF05	Drakensberg	B	8.77	0.93	1.92	0.54
SAF05	Drakensberg	C	11.53	1.27	0.99	0.48
SAF05	Drakensberg	D	10.46	1.05	0.87	0.43
SAF05	Drakensberg	E	9.22	1.07	0.77	0.40
SAF05	Drakensberg	F	7.32	0.78	0.68	0.37
SAF05	Drakensberg	G	7.36	0.72	0.72	0.38
SAF05	Drakensberg	H	9.86	1.17	1.00	0.49
SAF05	Drakensberg	I	9.86	1.20	1.02	0.48
Mean			9.36	1.03	0.93	0.44
Standard deviation			1.30	0.17	0.40	0.05
SAF06	KNP #2[a]	3	5.23	0.32	0.48	0.21
SAF06	KNP #2	4	5.51	0.28	0.45	0.21
SAF06	KNP #2	5	5.68	0.32	0.51	0.48
SAF06	KNP #2	6	5.86	0.35	0.50	0.20
SAF06	KNP #2	7	4.15	0.10	0.38	0.38
SAF06	KNP #2	8	4.27	0.23	0.35	0.16
SAF06	KNP #2	9	6.37	0.34	0.54	0.35
SAF06	KNP #2	10	5.26	0.28	0.44	0.23
Mean			5.29	0.28	0.46	0.28
Standard deviation			0.71	0.07	0.06	0.10

a. KNP = Kruger National Park.

Table 29.4 Mean and standard deviation of emission ratios of organic trace gases relative to the excess CO_2 (in ppb carbon) for the different types of fires

Compound	Drakensberg fire	KNP fire #2	Sugar cane fires
1-Pentene	$(1.12 \pm 0.26) \times 10^{-4}$	$(5.83 \pm 1.27) \times 10^{-5}$	$(1.65 \pm 2.47) \times 10^{-5}$
2-Methyl-1-butene	$(5.89 \pm 1.92) \times 10^{-5}$	$(0.53 \pm 1.00) \times 10^{-5}$	$(0.70 \pm 1.18) \times 10^{-5}$
2-Methyl-2-butene	$(8.67 \pm 2.03) \times 10^{-5}$	$(1.79 \pm 1.19) \times 10^{-5}$	$(5.65 \pm 9.18) \times 10^{-6}$
4-Methyl-1-pentene	$(7.89 \pm 3.47) \times 10^{-5}$	$(7.78 \pm 5.22) \times 10^{-5}$	$(2.74 \pm 3.53) \times 10^{-5}$
1-Hexene	$(1.26 \pm 0.33) \times 10^{-4}$	$(8.40 \pm 1.05) \times 10^{-5}$	$(2.59 \pm 3.99) \times 10^{-5}$
1-Octene	$(2.23 \pm 0.76) \times 10^{-5}$	$(1.56 \pm 0.79) \times 10^{-5}$	$(0.68 \pm 1.15) \times 10^{-5}$
2-Octene	$(1.98 \pm 1.78) \times 10^{-6}$	—	$(2.80 \pm 3.09) \times 10^{-7}$
2-Methyl-2-propanol	$(6.65 \pm 6.93) \times 10^{-6}$	$(6.14 \pm 7.05) \times 10^{-6}$	$(1.36 \pm 1.02) \times 10^{-5}$
1-Butanol	$(5.97 \pm 1.77) \times 10^{-6}$	$(5.93 \pm 4.64) \times 10^{-6}$	$(3.79 \pm 5.68) \times 10^{-6}$
Cyclopentanol	$(4.95 \pm 2.50) \times 10^{-5}$	$(5.07 \pm 3.84) \times 10^{-5}$	$(2.79 \pm 3.00) \times 10^{-5}$
2-Methyl-propanal	$(1.77 \pm 0.75) \times 10^{-5}$	$(1.31 \pm 0.86) \times 10^{-5}$	$(3.11 \pm 2.34) \times 10^{-6}$
Butanal	$(8.39 \pm 1.88) \times 10^{-5}$	$(6.71 \pm 6.14) \times 10^{-5}$	$(2.91 \pm 3.53) \times 10^{-5}$
3-Methyl-butanal	$(2.35 \pm 0.69) \times 10^{-5}$	$(1.61 \pm 1.45) \times 10^{-5}$	$(7.36 \pm 9.34) \times 10^{-6}$
2,2-Dimethyl-pentanal	$(5.70 \pm 4.49) \times 10^{-6}$	$(1.24 \pm 2.53) \times 10^{-6}$	$(1.39 \pm 2.05) \times 10^{-6}$
Hexanal	$(4.92 \pm 2.79) \times 10^{-5}$	$(3.89 \pm 2.16) \times 10^{-5}$	$(1.86 \pm 1.31) \times 10^{-5}$
Benzaldehyde	$(4.23 \pm 1.93) \times 10^{-5}$	$(4.45 \pm 2.99) \times 10^{-5}$	$(1.43 \pm 1.31) \times 10^{-5}$
Acetic acid-2-methylpropylester	$(1.51 \pm 0.48) \times 10^{-5}$	$(8.15 \pm 0.45) \times 10^{-5}$	$(1.70 \pm 2.44) \times 10^{-6}$
3-Methylbutanon	$(8.22 \pm 1.17) \times 10^{-6}$	$(7.45 \pm 5.16) \times 10^{-6}$	$(2.47 \pm 3.06) \times 10^{-6}$
1-Penten-3-on	$(6.59 \pm 2.70) \times 10^{-6}$	$(4.37 \pm 2.46) \times 10^{-6}$	$(2.11 \pm 3.02) \times 10^{-6}$
2-Pentanon	$(3.71 \pm 1.83) \times 10^{-5}$	$(2.53 \pm 1.80) \times 10^{-5}$	$(0.94 \pm 1.10) \times 10^{-5}$
2-Heptanon	$(2.92 \pm 1.90) \times 10^{-6}$	$(9.52 \pm 7.64) \times 10^{-6}$	$(3.92 \pm 2.19) \times 10^{-6}$
6-Methyl-2-heptanon	$(3.28 \pm 1.56) \times 10^{-5}$	$(2.46 \pm 1.77) \times 10^{-5}$	$(1.27 \pm 0.64) \times 10^{-7}$
2-Methylfuran	$(2.94 \pm 0.76) \times 10^{-4}$	$(7.99 \pm 3.44) \times 10^{-5}$	$(1.96 \pm 3.29) \times 10^{-5}$
3-Methylfuran	$(4.97 \pm 1.00) \times 10^{-5}$	$(1.81 \pm 0.20) \times 10^{-5}$	$(5.22 \pm 7.99) \times 10^{-6}$
Tetrahydrofuran	$(2.48 \pm 0.58) \times 10^{-5}$	$(2.38 \pm 1.66) \times 10^{-5}$	$(0.91 \pm 1.08) \times 10^{-5}$
2,3-Dihydrofuran	$(2.08 \pm 0.65) \times 10^{-5}$	$(1.94 \pm 1.06) \times 10^{-5}$	$(0.82 \pm 1.01) \times 10^{-5}$
2-Ethylfuran	$(5.97 \pm 3.45) \times 10^{-6}$	$(2.06 \pm 2.47) \times 10^{-6}$	$(0.86 \pm 1.28) \times 10^{-6}$
2,4-Dimethylfuran	$(4.07 \pm 1.12) \times 10^{-5}$	$(1.40 \pm 0.31) \times 10^{-5}$	$(4.16 \pm 6.53) \times 10^{-6}$
Benzofuran	$(2.94 \pm 1.22) \times 10^{-5}$	$(2.53 \pm 1.47) \times 10^{-5}$	$(7.83 \pm 9.22) \times 10^{-6}$
1-Ethyl-2-methylbenzene	$(5.21 \pm 2.48) \times 10^{-6}$	$(3.84 \pm 2.92) \times 10^{-6}$	$(1.31 \pm 1.23) \times 10^{-6}$
1,2,4-Trimethylbenzene	$(3.22 \pm 2.67) \times 10^{-6}$	$(0.86 \pm 1.30) \times 10^{-6}$	$(1.32 \pm 1.51) \times 10^{-6}$
1-Ethyl-3-methylbenzene	$(7.78 \pm 1.05) \times 10^{-6}$	$(4.86 \pm 1.95) \times 10^{-6}$	$(2.61 \pm 3.74) \times 10^{-7}$
Propylbenzene	$(3.20 \pm 2.73) \times 10^{-6}$	$(2.10 \pm 2.11) \times 10^{-6}$	$(7.01 \pm 6.67) \times 10^{-7}$
o-Xylene	$(2.89 \pm 0.76) \times 10^{-5}$	$(1.72 \pm 1.01) \times 10^{-5}$	$(6.82 \pm 9.41) \times 10^{-6}$
Ethylbenzene	$(2.82 \pm 1.08) \times 10^{-5}$	$(1.83 \pm 0.99) \times 10^{-5}$	$(6.63 \pm 1.08) \times 10^{-5}$
Toluene	$(4.58 \pm 2.31) \times 10^{-4}$	$(2.15 \pm 0.64) \times 10^{-4}$	$(5.74 \pm 8.69) \times 10^{-5}$
Benzene	$(8.77 \pm 1.26) \times 10^{-4}$	$(4.93 \pm 0.70) \times 10^{-4}$	$(3.11 \pm 1.83) \times 10^{-4}$
Phenol	$(1.02 \pm 0.62) \times 10^{-5}$	$(5.18 \pm 2.40) \times 10^{-6}$	$(1.76 \pm 2.83) \times 10^{-6}$

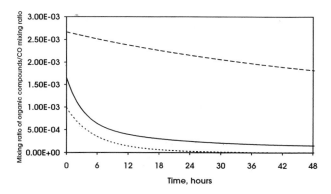

Figure 29.4 Plot of the calculated change in the mixing ratios of organic compounds in the plume of the Drakensberg fire as a function of time due to the reaction with OH. Solid line: oxygenated hydrocarbons; dotted line: alkenes; dashed line: aromatics. The mixing ratios of the organic compounds (OC) was calculated assuming an OH radical concentration of 1×10^{-6} cm^{-3} with $OC = (OC)_0 \times \exp - (t/\tau)$. The kinetic data were taken from Atkinson (1989).

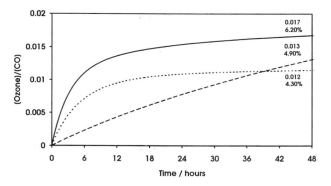

Figure 29.5 Plot of the ratio of produced ozone and excess CO as a function of time. Solid line: oxygenated hydrocarbons; dotted line: alkenes; dashed line: aromatics. The contribution of CO and light NMHCs to the ozone formation within 48 hours is 40% and 35%, respectively.

National Park fire, and $(3.9 \pm 2.9) \times 10^{-3}$ and $(1.7 \pm 0.7) \times 10^{-3}$ for the sugar cane fires #1 and #2, respectively. These ratios can be compared with laboratory measurements of Mano (1995), who gives an average emission ratio of the sum of organic compounds of 8.4×10^{-3} for his experiments and Lobert et al. (1991), who give an average of 11.9×10^{-3} based on 10 different experiments. The ratio found for the Kruger National Park fire is in good agreement with the laboratory measurements, while the ratio found for the Drakensberg fire exceeds the laboratory values. The emission ratios of the predominantly flaming sugar cane fires are considerably lower than the results of the laboratory studies. This may be due to the fact that large flaming combustion cannot be simulated realistically in confined fires, which are controlled by their artificial boundaries. Especially the response of fires to wind, which affects mainly the flaming phases, differs significantly between free and confined fires (Albini 1993).

Impact on Regional Tropospheric Chemistry

Once the organic composition of biomass burning emissions as well as the emission ratios of the different types of fires is known, their contribution to the budget of organic trace gases and their impact on the photochemical formation of ozone can been estimated. The uncontrolled fire in the Drakensberg region described previously showed the highest emission ratios found during this campaign. We used the emission data from this fire to estimate the impact of biomass burning emissions on regional photochemistry of the tropo-

sphere and the potential ozone formation. Figure 29.4 shows the mixing ratios of organic trace gases relative to CO as a function of time due to the removal by OH radicals. Plotted are the sum of oxygenated hydrocarbons, the alkenes, and the aromatics. For the calculation, an OH concentration of 1×10^6 cm^{-3} was assumed. As the initial concentration, the average concentration measured during the passages through the plume of the Drakensberg fire was used. This plot shows the loss of the photochemically reactive compounds in the plumes. The highly reactive alkenes and oxygenated hydrocarbons are removed from the plumes within the first 10 hours after emission. Within 48 hours, almost all reactive compounds are removed. The slower removal of the sum of aromatics is due to the high proportion of benzene and toluene in that group. Especially benzene has an atmopheric lifetime that is long compared to the other aromatic compounds. It is evident that most of the photochemistry and thus most of the formation of ozone occur within the first 48 hours after emission. The turnover of the organic trace gases and the subsequent formation of ozone was calculated assuming that, in an NO$_x$-rich regime, on the average two ozone molecules are produced per carbon atom of each compound as a lower limit. In figure 29.5, the ratio of the produced O$_3$ and excess CO (above background) is plotted as a function of time for a period of 48 hours. A typical value for $\Delta(CO_3)/\Delta(CO)$ in a photochemically aged biomass burning plume is about 0.3. Within 48 hours, the production of ozone due to the turnover of CO contributes about 40% to that value, while the contribution of light NMHCs (Σ C$_2$–C$_4$) is another 35%. The contribution of oxygenated hydrocarbons, aromatics, and alkenes is

6.2%, 4.9%, and 4.3%, respectively. It should be noted that the situation changes if the calculation is continued over longer time intervals, since the removal of ozone is not included in this estimation. Since this calculation was carried out for an uncontrolled fire, the plumes of which showed the highest load of organic trace gases, this can be used as an upper limit of ozone formation in biomass burning plumes. In contrast to that, the plumes of the sugar cane fires showed the lowest content of organic trace gases, which is due to the absence of a smoldering phase in these fires. This observation is supported by the measurements of aerosol and black carbon emission (Andreae 1995; Le Canut 1995). This simple estimate of the photochemical ozone formation in biomass burning plumes nevertheless indicates that higher-molecular-weight organic trace gases contribute significantly to the ozone production and cannot be neglected.

The samples discussed here were collected on a very limited scale, which represents only a fraction of the areas burnt globally each year. Therefore, any extrapolations of our emission ratios to global scales are highly uncertain. However, to get an idea about the impact of biomass burning emissions to atmospheric chemistry, it is worthwhile to compare the emissions of organic trace gases with the global budgets. From our emission ratios relative to CO_2 and the estimated range of the global biomass burning source for CO_2 of 1600 to 4100×10^{12} gC/yr (Crutzen and Andreae 1990), we are able to estimate the global emission of higher-molecular-weight organic compounds from biomass burning. Our estimates of the global emissions for the sum of organic compounds range between 2.7×10^{12} g/yr and 56×10^{12} g/yr. The lower value is based on the data of sugar cane fire #2 with the lowest emission ratios and the lower limit of the estimated CO_2 emissions, while the higher value is based on the Drakensberg fire with the highest emission ratios and the upper limit of the CO_2 emissions. The higher-molecular-weight organic compounds contribute up to 50% to that value. The source strength of organic compounds based on our measurements is in good agreement with the results of previous estimates by Bonsang et al. (1991), who derived 12×10^{12} g/yr from measurements over large savanna fires during the DECAFE project in the Ivory Coast. Laboratory studies lead to higher estimates. Lobert et al. (1991) calculated 42×10^{12} g/yr from their experiments. These results indicate that the emissions of biomass burning contribute significantly to the local and regional budget of organic compounds in the troposphere. It is also evident that the influence of higher-molecular-weight

organic compounds cannot be neglected in estimating the potential formation of ozone in the plumes of savanna fires.

Conclusions

In all samples collected in the plumes of the savanna and agricultural fires during SAFARI, a large variety of organic trace gases were found. They include alcohols, aldehydes, ketones, carboxylic acids, esters, ethers, furanes, and others. The observed emission ratios show a considerable variation depending on the fuel type and the burning stages of the fires. The emission ratio of the sum of light NMHCs was in the same order of magnitude or even higher than that of methane. From a number of recent studies, it is known that biomass burning is also a significant source for higher-molecular-weight organic trace compounds in the tropics. Our results show that these compounds, most of which are highly reactive, contribute significantly to the load of volatile organic compounds in biomass burning emissions. Presently, about 70 of about 140 major compounds could be positively identified in the samples, a number of which were previously not known to be emitted by biomass burning and thus were not included in estimates of the influence on the regional atmospheric chemistry. The uncontrolled fire in the Drakensberg forest showed emission ratios of the sum of organic compounds relative to CO_2 of $(13.7 \pm 0.9) \times 10^{-3}$, which are the highest values found during this campaign. Simple estimates of the removal due to the reaction with OH radicals in an NO_x-rich plume show that the contribution of organic compounds to the formation of ozone within the first 48 hours after emission may be up to 50% of that of CO alone. Our results show further that the emission ratio of organic compounds with a chain length of five or more carbon atoms adds up to at least half of the contribution of light NMHCs. However, due to their high reactivity, their influence on the photochemistry may be much more important than that of the lighter hydrocarbons.

The quantification of the impact of biomass burning emissions on the photochemistry of the troposphere on local, regional, and even global scales requires models investigating the atmospheric dynamics and the chemistry inside and in the vicinity of fire plumes. Model calculations, however, need reliable input data. Despite the increasing number of investigations in the last years, the data sets concerning the trace gas composition of biomass burning emissions are still very limited. Our results show that higher-molecular-weight organic trace gases contribute significantly to these emissions and play an important role for the photochemistry in biomass burning plumes. Especially their impact on the formation of tropospheric ozone is not negligible.

Acknowledgments

We thank the pilot and crews of the SAFARI-92 DC-3 research aircraft for their assistance during the research flights, the Kruger Park team (particularly A. Potgieter, W. Trollope, L. Trollope, and J. G. Goldammer) for their cooperation during the experimental fires in the park, and the team at the sugar plantation at Ubombo Ranches, Big Bend, Swaziland (especially J. Gosnell and D. Shipley) for their generous help during our experiment. We acknowledge the permission of various government agencies in Angola, Botswana, South Africa, Namibia, Swaziland, Zambia, and Zimbabwe for permission to conduct research in these countries or to use the airspace over these countries. We appreciate the help of T. W. Andreae (logistical support) and U. Parchatka (technical support) during the airborne sampling missions on the DC-3 aircraft. Partial support of the DC-3 aircraft program by the National Science Foundation is gratefully acknowledged. This research was part of IGBP/IGAC and was supported by the Max Planck Society.

References

Albini, F. A. 1993. Dynamics and Modeling of Vegetation Fires: Observations, in: P. J. Crutzen and J. G. Goldammer (eds.), *Fire in the Environment, The Ecological, Atmospheric, and Climatic Importance of Vegetation Fires, Report of the Dahlem Workshop, Berlin, 15–20 March, 1992*, Wiley, New York.

Andreae, M. O., J. Fishman, M. Garstang, J. G. Goldammer, C. O. Justice, J. S. Levine, R. J. Scholes, B. J. Stocks, A. M. Thompson, B. van Wilgen, and the STARE/TRACE-A/SAFARI Science Team. 1994. Biomass burning in the global environment: First results from the IGAC/BIBEX field campaign STARE/TRACE-A/SAFARI-92, in: R. G. Prinn (ed.), *Global Atmospheric-Biospheric Chemistry*, Plenum Press (Environmental Science Research, 48), New York, pp. 83–101.

Andreae, M. O., E. Atlas, G. W. Harris, G. Helas, A. de Kock, R. Koppmann, W. Maenhaut, S. Mano, W. H. Pllock, J. Rudolph, D. Scharffe, G. Schebeske, and M. Welling. 1995. Methyl halide emissions from savanna fires in southern Africa, *J. Geophys. Res.*, in press.

Atkinson, R. 1989. Kinetics and mechanisms of the gas-phase reactions of the hydroxyl radical with organic compounds, *J. Phys. Chem. Ref. Data, Monograph* 1.

Bonsang, B., G. Lambert, and C. Boissard. 1991. Light hydrocarbon emissions from African Savanna burnings, in J. S. Levine (ed.), *Global Biomass Burning: Atmospheric, Climatic, and Biospheric Implications*, MIT Press, Cambridge, Mass., pp. 155–166.

Chatfield, R. B., and A. C. Delany. 1990. Convection links biomass burning to increased tropical ozone; however, models, will tend to overpredict O_3, *J. Geophys. Res.*, 95, 18 473–18 488.

Crutzen, P. J., and M. O. Andreae. 1990. Biomass burning in the tropics: Impact on atmospheric chemistry and biogeochemical cycles, *Science*, 250, 1669–1678.

Crutzen, P. J., A. C. Delany, J. Greenberg, P. Haagenson, L. Heidt, R. Lueb, W. Pollock, W. Seiler, A. F. Wartburg, and P. Zimmerman. 1985. Tropospheric chemical composition measurements in Brazil during the dry season, *J. Atmos. Chem.*, 2, 233–256.

Delmas, R. 1982. On the emission of carbon, nitrogen, and sulfur in the atmosphere during bush fires in intertropical savanna zones, *Geophys. Res. Lett.*, 9, 761–764.

Fishman, J., M. Hoell, Jr., R. D. Bendura, V. W. J. H. Kirchhoff, R. J. McNeal, Jr., The NASA GTE TRACE-A Experiment (September–October 1992), *J. Geophys. Res.*, submitted, 1995.

Fishman, J., C. E. Watson, J. L. Larson, and J. A. Logan. 1990. Distribution of tropospheric ozone determined from satellite data, *J. Geophys. Res.*, 95, 3599–3617.

Franke, W. 1989. *Nutzpflanzenkunde*, Thieme-Verlag, Stuttgart, Germany.

Greenberg, J. P., and P. R. Zimmerman. 1984. Nonmethane hydrocarbons in remote tropical, continental and marine atmospheres, *J. Geophys. Res.*, 89, 4767–4778.

Greenberg, J. P., P. R. Zimmerman, L. Heidt, and W. Pollock. 1984. Hydrocarbon and carbon monoxide emissions from biomass burning in Brazil, *J. Geophys. Res.*, 89, 1350–1354.

Greenberg, J. P., and P. R. Zimmerman, W. F. Pollock, R. A. Lueb, and L. E. Heidt. 1992. Diurnal variability of atmospheric methane, nonmethane hydrocarbons, and carbon monoxide at Mauna Loa, *J. Geophys. Res.*, 97, 10 395–10 413.

Heidt, L. 1978. Whole air collection and analysis, *Atmos. Technol.*, 9, 3–7.

Hurst, D. F., D. W. T. Griffith, J. N. Carras, D. J. Williams, and P. J. Frase. 1994. Measurements of trace gases emitted by Australian savanna fires during the 1990 dry season, *J. Atmos. Chem.*, 18, 33–56.

Koppmann, R., R. Bauer, F. J. Johnen, C. Plass, amd J. Rudolph. 1992. The distribution of light nonmethane hydrocarbons over the Mid-Atlantic: Results of the Polarstern cruise ANT VII/1, *J. Atmos. Chem.*, 15, 215–234.

Koppmann, R., F. J. Johnen, A. Khedim, J. Rudolph, A. Wedel, and B. Wiards. 1995. The influence of ozone on light nonmethane hydrocarbons during cryogenic preconcentration, *J. Geophys. Res.* in press.

Koppmann, R., F. J. Johnen, C. Plass-Dülmer, and J. Rudolph. 1993. Distribution of methylchloride, dichloromethane, trichloroethene and tetrachloroethene over the North and South Atlantic, *J. Geophys. Res.* 100, 11 383–11 391.

Lacaux, J. P., J. M. Brustet, R. Delmas, J. C. Menaut, J. Baudet, M. O. Andreae, and G. Helas. 1991. DECAFE 91: Biomass burning in the tropical savannas of Ivory Coast: A general overview of the field experiment Fire of Savannas (FOS), in G. Angeletti, S. Beilke, and J. Slanina (eds.), *Air Pollution Research Report 39*, Commmission of the European Communities, Brussels, Belgium, pp. 81–86.

Le Canut, P., M. O. Andreae, G. W. Harris, F. G. Wienhold, and T. Zenker. 1995. Airborne studies of emissions from savanna fires in Southern Africa: I. Aerosol emissions measure with a laser-optical particle counter, *J. Geophys. Res.*, in press.

Levine, J. S. 1991. Global biomass burning: Atmospheric, climatic, and biospheric implications, in J. S. Levine (ed.), *Global Biomass Burning: Atmospheric, Climatic, and Biospheric Implications*, MIT Press, Cambridge, Mass., pp. XXV–XXX.

Lobert, J. M., D. H. Scharffe, W. M. Hao, T. A. Kuhlbusch, R. Seuwen, P. Warneck, and P. Crutzen. 1991. Experimental evaluation of biomass burning emissions: Nitrogen and carbon containing compounds, in J. S. Levine (ed.), *Global Biomass Burning: Atmospheric, Climatic, and Biospheric Implications*, MIT Press, Cambridge, Mass., pp. 289–304.

Lobert, J. M., and J. Warnatz. 1993. Emissions from the combustion process in vegetation, in: P. J. Crutzen and J. G. Goldammer (eds.), *Fire in the Environment, The Ecological, Atmospheric, and Climatic Importance of Vegetation Fires, Report of the Dahlem Workshop, Berlin, 15–20 March, 1992*, Wiley, New York.

Manoe, S. 1995. Messung von partiell oxidierten Kohlenwasserstoffen in Emissionen von Biomasseverbrennung, PhD thesis, Johann-Wolfgang-von-Goethe-Universität Frankfurt.

Rasmussen, P. R., and M. A. Khalil. 1988. Isoprene over the Amazon basin, *J. Geophys. Res.*, 93, 1417–1421.

Rudolph, J., F. J. Johnen, and A. Khedim. 1986. Problems connected with the analysis of hydrocarbons and halocarbons in the non-urban atmosphere, *Int. J. Environ. Anal. Chem.* 27, 97–122.

Rudolph, J., A. Khedim, and B. Bonsang. 1992. Light hydrocarbons in the tropical boundary layer over tropical Africa, *J. Geophys. Res.*, 97, 6181–6186.

Rudolph, J., A. Khedim, R. Koppmann, and B. Bonsang. 1995. Field study of the emission of methyl chloride and other halocarbons from biomass burning in western Africa, *J. Atmos. Chem.*, 22, 67–80.

Rudolph, J., K. P. Müller, and R. Koppmann. 1990. Sampling of organic volatiles in the atmosphere at moderate and low pollution levels, *Analytica Chimica Acta*, 236, 197–211.

Stocks, B. J., B. W. van Wilgen, W. S. W. Trollope, D. J. McRae, F. Weirich, and A. L. F. Potgieter. 1995. Fuels and fire behavior on large-scale savanna fires in Kruger National Park, South Africa, *J. Geophys. Res.*, in press.

Zimmerman, P. R., J. P. Greenberg, and C. E. Westberg. 1988. Measurements of atmospheric hydrocarbons and biogenic emission fluxes in the Amazon boundary layer, *J. Geophys. Res.*, 93, 1407–1416.

Regional Scale Impacts of Biomass Burning Emissions Over Southern Africa

Stuart J. Piketh, Harold J. Annegarn, and Melanie A. Kneen

Emissions from biomass burning have recently been shown to contribute significantly to atmospheric chemistry on a global scale (Crutzen and Andreae 1990). Africa is the continent where biomass burning occurs most frequently. A number of types of burning have been identified; namely, savanna fires, fires relating to deforestation, agricultural burning, industrial burning, and domestic fuel usage. Currently about 4950 million ton of biomass is burned globally each year in the tropical regions, with fires in savanna regions being the dominant phenomena, contributing 75% of all biomass burning (Lacaux, Cachier, and Delmas 1993). Savanna fires on the African continent account for 57% of biomass burned in these regions (Hao, Liu, and Crutzen 1990; Andreae 1991; Delmas et al. 1991). Fires play an important role in managing the natural biome of savanna ecosystems. Fire activity has strong seasonality, with the maximum occurring at the end of the dry season, before the first rainfall (Cahoon et al. 1992; Andreae 1991).

Particulate emissions from biomass burning are dominated by submicron, accumulation mode particles. In Venezuela, it has been shown that concentrations of particles smaller than 0.49 μm are five times higher during the burn season (Sanhueza 1991). The dramatic increase in particles in this size range could result from increased primary particles produced during the combustion process or through oxidation (secondary particles) of gaseous emissions from the fires. Particles in this size range are known to be efficient at scattering incoming of solar radiation, thus inducing climatic forcing at a regional scale. A good understanding of the atmospheric composition and behavior of particles generated by biomass burning is imperative to be able to predict their impact on the climate over southern Africa. During the Southern African Fire-Atmosphere Research Initiative (SAFARI-92) much insight was gained into the characteristics of fires and the resultant emissions in savanna ecosystems of southern Africa. SAFARI-92 only encompassed the time period coinciding with the highest frequency of burns, namely austral spring. Little is known about the impacts that

biomass burning emissions exercise on the atmosphere for the remainder of the year. The purpose of this chapter is to present results from a sampling program undertaken between January and December 1993. The program was designed to quantify the impacts of industrial and natural sources of particulate matter in the atmosphere at three remote sites in the Eastern Transvaal of South Africa. Time-resolved samples were collected at Elandsfontein, Misty Mountain, and Ulusaba Game Lodge (Ulusaba). The sites were arranged in an approximately linear pattern across the Eastern Transvaal to detect the advection of pollutants across the Escarpment region into the Lowveld. Samples for three seasons of the year, summer, winter, and spring, were selected for analysis for the inorganic elemental composition by Proton-Induced X-ray Emissions (PIXE). The elemental results were apportioned through receptor type modeling to the contributing sources.

Biomass burning-related emissions, which constituted an important component of all the samples analyzed formed the focus of this paper. Various aspects of the contribution of biomass burning at the three sites are considered in detail. The method of extracting the biomass burning component from the elemental data is outlined. Temporal and spatial characteristics of biomass burning episodes are described. Seasonal variations in both average and maximum concentrations are given, and an attempt is made to explain the transport of the burn emissions from wind data collected at Elandsfontein and Ulusaba. Biomass burning as a regional scale source during the month of September is evaluated.

Sampling and Analysis

Sampling Sites

From 4 January 1993, samples were continuously collected at three ground-based sites in the Eastern Transvaal, South Africa (figure 30.1). The sites included Elandsfontein, a site on the Highveld (1790 m

Figure 30.1 Map showing the location of the sampling sites

above sea level (asl)); Misty Mountain, located in the central Escarpment region (1990 m asl); and Ulusaba, a Lowveld site (525 m asl). Elandsfontein was established to characterize the air masses that passed over the industrial regions of the Highveld. Elandsfontein is situated 30 km southeast of Witbank and is surrounded by six coal-fired power stations that are on average 35 km away. The purpose of Misty Mountain was to detect any transportation of pollutants and other atmospheric constituents between the Highveld and the Lowveld. Ulusaba, located on a hill that is prominent above the otherwise very flat surrounding terrain, provided an ideal opportunity to characterize the remote air of the Lowveld. Ulusaba is situated 25 km northwest of Skukuza, the focus of sampling conducted during SAFARI-92.

Sampling Procedure

Ambient air samples were collected at the three sites with single-stage PIXE International Inc., circular streaker units. Samplers were equipped with 10-micron aerodynamic (μmad) size selective inlet-impactors. The streaker samples consisted of 0.4-μm pore size Nucleopore filter membrane mounted on a circular frame. The frame rotates at one revolution per month. The filter membrane once mounted in the streaker unit passes over a 1×7 mm sucking orifice, leaving behind a sample or 'streak' of particulate matter in the less than 10-μmad size range. Streaker frames were exposed for 672 hours, or 28 days. The average volume of flow for each sample was estimated from the elapsed number of hours and the total volume of air for the 28-day period. The sampling procedure and equipment have been described elsewhere (Annegarn et al. 1983, Annegarn 1987; Annegarn et al. 1992).

Filter Analysis

The streaker samples were analyzed by PIXE. Analyses were performed with a Van de Graaff Tandem

Accelerator at the University of the Witwatersrand. PIXE has in the past shown its usefulness for characterizing the elemental composition of atmospheric samples of low concentrations. The streak on each sample was analysed in 144 sequential steps of 1-mm width. Each step was equivalent to 4.5 hours of sampling. Elements analyzed and routinely detected above the detection limits included Al, Si, S, Cl, K, Ca, Ti, Cr, Mn, Fe, Cu, Zn, Br, and Pb (Annegarn 1987 and Annegarn et al. 1992). Detection limits were calculated for every element in each spectrum as part of the data validation procedure. Only values found to be above the detection limits were retained for data interpretation. Unexposed sections of the filters were analyzed as field blanks and concentrations were corrected where necessary.

Identification of the Biomass Burning Component

Concentrations of each of the detected elements were calculated from the average volume of air that passed through the filter for each 4.5 hour period. The range of detected elements were apportioned to six source categories that were identified during SAFARI-92 (Salma et al. 1992; Maenhaut et al. 1995). The source categories identified were anthropogenic-produced sulfur, soil or crustal material, marine aerosols, biomass burning particles, smelting industry heavy metal emissions, and lastly a general group of unspecified sources. Each source category was characterized by a unique set of elements or elemental ratios that constituted their identification. Where one element was found in more than one source, it was apportioned by its unique ratio to a reference element. Due to the limited number of detected elements, sources were identified and apportioned from their major elemental emissions.

According to Maenhaut et al. (1995) savanna fires are a major source of P, K, Ca, Mn, Zn, Sr, and I in the coarse fraction (larger than 2.5 μmad and smaller than 10 μmad) and Cl, K, Cu, Zn, Br, Rb, Sb, I, Cs, and Pb in the fine fraction (smaller than 2.5 μmad). K has been found to be consistently measured in elevated concentrations related to biomass burning samples (Andreae 1983; Sanhuez 1992; Salma et al. 1993; Maenhaut et al. 1995). Due to the low sample loads of material and the low background concentrations of Rb, Sr, Sb, I, and Cs, these elements were not included for analysis. Cl, Mn, and Zn were detected at the three sites, however, time series data for these elements did not coincide with peak K episodes. Ca and P were also detected in the samples but it was not possible to resolve these elements to their sources. Br and Pb were

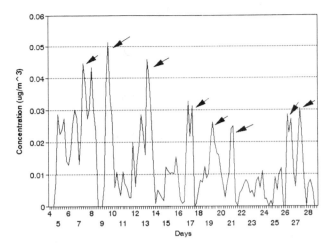

Figure 30.2 Biomass burning component at Ulusaba during February 1993. Arrows indicate characteristic short episodes.

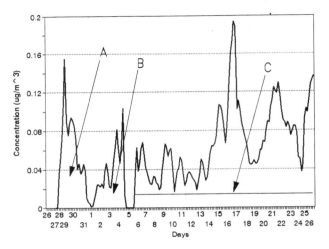

Figure 30.3 Biomass burning component at Misty Mountain during September 1993. A, B and C indicate period of extended elevated excess K.

rarely observed in the samples above the calculated detection limits and were therefore not included in the biomass burning apportionment calculations. K was therefore used as the sole indicator of biomass burning impacts in the streaker samples.

The biomass burning component was extracted using the following formula:

$$[K] - ([Cl] \cdot (K/Cl)) - ([Si] \cdot (K/Si)) \tag{30.1}$$

where $[Cl] \cdot (K/Cl)$ was K calculated from marine emissions and $[Si] \cdot (K/Si)$ referred to K calculated from crustal emissions. Justification for simplifying the biomass burning component to excess K can be found in the fact that the element constitutes between 3.1 and 3.8% of the total emissions from biomass burnt. The remaining elements listed account for only about 3% of biomass burning emissions together. In an atmosphere in which pollutants from numerous sources are well mixed, as is the case over southern Africa, it is difficult to extract the contributions from these lower concentration elements. This limitation needs to be set against the time resolution achieved from the samples, an invaluable tool in atmospheric constituent interpretation.

Characteristics of Biomass Burning Episodes

Two modes of biomass burning episodes were identified at the three sites. The first mode was characterized by periods of high concentrations that lasted between 4 and 6 hours (figure 30.2). Peak concentrations of excess K varied significantly between the sites. At Elandsfontein, these short episodes reached maximum

concentrations of 0.5 $\mu g/m^3$ during September. During June, biomass concentrations were much lower with a maximum peak of 0.1 $\mu g/m^3$. At Ulusaba and Misty Mountain samples for three seasons were analyzed. Maximum peak concentrations of 0.04 $\mu g/m^3$ were recorded at Ulusaba. Peak concentrations during the first part of June at Ulusaba were as high as 0.1 $\mu g/m^3$. During spring the regional background values dominate the biomass burning component and local peaks are of the order of 0.15 $\mu g/m^3$. During June peaks of local origin regularly reached values of 0.1 $\mu g/m^3$ at Misty Mountain.

These short episodes of elevated biomass-burning related concentrations characterizing the first mode were probably related to burns in the vicinity of the samplers (up to 120 km). Similar patterns were detected in contributions from anthropogenic sulfur at Ulusaba, the closest major source of which was about 120 km to the southwest at Ngodwana. Peaks representing the smelting component were the result of the transport of emissions from a greater distance. The peaks only occurred, however, under plain mountain and mountain plain wind conditions during June. No evidence of such longer-range transport related to the short episodes could be detected for the biomass burning component. The majority of short peak episodes were transported from within a radius of 120 km.

The second mode of biomass burning impacts detected at the three sites was characterized by extended periods of elevated biomass burning concentrations (figure 30.3). The concentrations again varied both between sites and seasonally, increasing from summer

Table 30.1 Average concentrations (μg/m^3) of biomass burning particles (excess potassium) for February, June, and September 1993

	February	June	September
Elandsfontein	—	0.048	0.129
Misty Mountain	0.016	0.036	0.114
Ulusaba	0.011	0.012	0.059

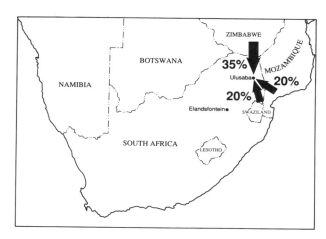

Figure 30.4 Major transport directions of biomass burning episodes to Elandsfontein during September 1993

to spring. The secondary episodes typically lasted for between 3 and 8 days, varying in concentration between 0.03 and 0.1 μg/m^3. Such excess K has been ascribed to more distant, and probably more extensive, sources of biomass burning. The lower concentrations would have resulted from dilution of the biomass burning emissions with increasing distance from the source.

Seasonal Variation of Biomass Burning Concentrations

The seasonal pattern that emerged from the data was both expected and enlightening. Firstly, both average and maximum concentrations increased steadily through the year, peaking in September (austral spring). The highest average and maximum concentrations of biomass burning related aerosols (0.59 μg/m^3) were detected at Elandsfontein during the second and third weeks of September (table 30.1). These higher values during spring were expected as it is at this time of the year that the majority of prescribed burning of the natural vegetation occurs in southern Africa (Cahoon et al. 1992).

Several incidents of biomass burning were measured at Ulusaba and Misty Mountain during February. The average concentrations recorded in February were, however, significantly lower than those observed in September, by a factor of at least two at Misty Mountain and a factor of five at Ulusaba. Biomass burning emissions during February could be ascribed to domestic wood burning as an energy source. During June, concentrations at Ulusaba had not changed significantly from February. At Misty Mountain, June concentrations of excess potassium were greatly increased from February, with maximum concentrations being equal to those recorded later on in the year during September. Similar concentrations were found to occur at Elandsfontein. Possible explanations of these elevated burn concentrations during June at Misty Mountain and Elandsfontein could be the burning of slash and fire breaks in the Escarpment region and farmers burning crops on the Highveld near the site. Additional domestic wood burning for

space heating due to colder ambient temperatures could also be partly responsible for these increased K levels.

Contributions of excess potassium were found throughout the year. Concentrations detected during the summer months were not expected to be a significant source of atmospheric particulate matter. This, however, was not found to be the case. Contributions from biomass burning to the inorganic elemental aerosol only dropped below 1% once through the year, during June at Ulusaba (0.66%). This is significant particularly since potassium represents only a small fraction of the total emissions from biomass burning.

Transportation of Biomass Burning Emissions

Wind data were collected simultaneously with streaker samples at Elandsfontein and Ulusaba during the period of sampling. Wind speed and direction data were compared with the biomass burning concentrations detected at the same times. From an analysis of local scale wind data, it was established that peak episodes of the biomass burning aerosols had two major source directions at Elandsfontein (figure 30.4) and three major transport directions at Ulusaba (figure 30.5). At Elandsfontein, the majority of elevated potassium episodes were transported from the northwest ($\sim$50%). A smaller component originated from the northeast of the sampling site ($\sim$10%). At Ulusaba, a major component of the biomass burning emissions were transported from the north (35%). A south to southeasterly component also constituted a large portion of the biomass burning particulate matter (40%). The pattern of biomass burning emission transportation is consistent throughout the year.

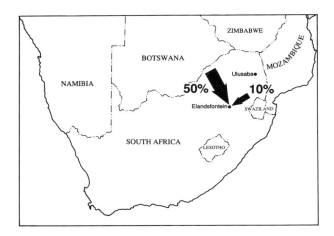

Figure 30.5 Major transport directions of biomass burning episodes to Ulusaba during September 1993

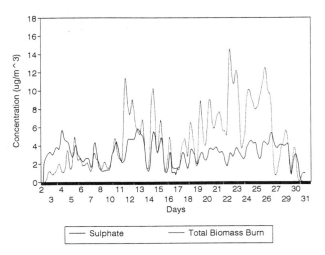

Figure 30.6 Estimated total atmospheric sulfate and biomass burning particulate load at Elandsfontein for the last three weeks of September 1993

At both sites, the component of biomass burning related emissions transported from a north to north-westerly component, particularly at Elandsfontein, was related to continental anticyclonic flow over the subcontinent. From studies of the local wind patterns over the Eastern Transvaal Highveld, this relationship between northwesterly to northeasterly winds and continental anticyclones is well established (Pretorius et al. 1986; Tyson, Kruger, and Louw 1988). Episodes of biomass burning emissions from the north are most likely to be indicative of large-scale circulation from regions to the north of South Africa. It can be expected that if the anticyclonic circulation persisted for several days (up to 10 days), recirculation of material over the sites would occur (Garstang et al. 1995). Recirculation would usually be accompanied by ever increasing loads of suspended matter in the atmosphere, depending on the strength of the source. The transport of biomass burning emissions from the northerly component was most prominent in September.

Surface circulation from the southeast at Ulusaba could usually be related to ridging anticyclonic flow that follows westerly disturbances. The source of this biomass burning is most likely to be the extensive burning of sugar cane fields that occurs in Kwazulu/Natal during spring.

Biomass Burning as a Regional Source of Particulate Matter

The wind data were not the only evidence of large scale circulation of biomass burning related material over South Africa. Anthropogenic sulfur was detected throughout the year at relatively constant concentra-

tions. This was not unexpected, as there is an uninterrupted source of this species. During February and June, there was no apparent relationship between the detected sulfur and excess potassium. During September, however, the patterns of the sulfur and biomass burning components of the aerosol were remarkably similar. As the sources of these two components are distinctly different, this could only be ascribed to common transport mechanisms.

The large increases in the atmospheric load of biomass burning particles, observed in the detected concentrations during September, make this source important at the regional scale. If the concentrations of excess potassium are calculated in terms of their total contribution from biomass burning emissions (3.8%), the biomass burning component is at least, if not more, extensive than a calculated sulfate component. The sulfate component was estimated from the detected elemental anthropogenic sulfur where the ratio of sulfur to sulfate was calculated to be 1:3. Figure 30.6 shows how the regional background of the two components are closely related, with biomass burning often exceeding the sulfate concentrations. The combined effects of sulfate and elevated biomass burning constituents are postulated as being responsible for the increase in atmospheric turbidity over the Lowveld and Highveld regions.

Conclusions

Three sampling sites were established in the Eastern Transvaal, South Africa, to investigate the regional scale impacts and transport of atmospheric constituents in the region. Samples were collected using a streaker sampler, with a time resolution of approximately four hours. Samples representing three seasons of the year, summer, winter, and spring, were analyzed using PIXE. Fourteen elements were routinely detected above detection limits. The detected elements were apportioned to the contributing sources which included, anthropogenic sulfur, crustal material, marine, biomass burning, smelting and a general group.

Concentrations of biomass burning-related emissions were detected in the ambient atmosphere throughout the year. Two modes of biomass burning episodes were found to occur. The first lasted between 4 and 8 hours and could be ascribed to local sources within a 120-km radius. The short, sharp episodes occurred regularly at Misty Mountain and Ulusaba during summer and winter, indicating impacts from local domestic wood burning, the primary source of energy in the remote areas of the Eastern Transvaal Escarpment and Lowveld regions. The second mode was representative of the regional air mass and typically lasted for 36 to 60 hours. These impacts are thought to be from more distant sources. The highest concentrations of biomass burning were found, as expected, during September (austral spring). Average concentrations during this period increased significantly (by at least a factor of two) from the previous two months at all three sites. Local wind fields recorded at Elandsfontein and Ulusaba showed that most of the biomass burning related emissions during September were transported from the northeast to northwest. This wind field is representative of continental anticyclonic circulation, which results in large-scale circulation of material over the subcontinent. It is likely that large portions of the detected biomass burning material were transported from across the South African borders. At Ulusaba secondary transport routes were from the southeast. The southeasterly wind component was normally coincident with ridging anticyclonic flow. The material during these episodes would originate from the burning of sugar cane fields in Kwazulu/Natal which reaches a maximum intensity during the spring months.

Similar patterns of anthropogenic sulfur and biomass burning components impacting at the sites during September presented further evidence of large-scale transport of biomass burning emissions. Biomass burning has thus been shown to be an important source of particulate matter at a regional scale, probably contributing at least as much as particulate sulfate to atmospheric haziness over the Lowveld during September. The simultaneous impact of biomass burning emissions and anthropogenic sulfur, two independent sources, is indicative of large-scale burning resulting in the emissions being transported over long distances and thus becoming well mixed with sulfur emissions.

Acknowledgments

This work was funded by ESKOM (TRI) as part of the ETHAR project. The authors would like to acknowledge the financial and logististical support provided for this project by ESKOM (TRI), Schonland Research Centre, and Ulusaba Game Lodge. The hospitality of the Ulusaba Game Lodge staff during site visits is gratefully acknowledged.

References

Andreae, M. O. 1983. Soot carbon and excess fine potassium: long-range transport of combustion-derived aerosols, *Science*, 220, pp. 1148–1151.

Andreae, M. O. 1991. Biomass burning: Its history, use, and distributions and its impacts on environmental quality and global climate, in Levine, J. S., (ed.) *Global Biomass Burning, Atmospheric, Climatic, and Biospheric Implications*, The MIT Press, Cambridge, pp. 240–244.

Annegarn, H. J. 1987. Time series analysis of PIXE aerosol measurements, *Nuclear Instruments and Methods in Physics Research*, B22, pp. 270–274.

Annegarn, H. J., G. M. Braga Marcazzan, E. Cereda, M. Marchionni, and A. Zucchiatti. 1992. Source profiles by unique ratios (SPUR) analysis: determination of source profiles from receptor-site streaker samples, *Atmospheric Environment*, 26A, 2, pp. 333–343.

Annegarn, H. J., A. C. D. Lesile, J. W Winchester, J. P. F Sellschop. 1983. Particle size and temporal characteristics of aerosol composition near coal-fired electric power plants of the Eastern Transvaal, *Aerosol Science and Technology*, 2, pp. 489–498.

Cahoon, R. C., B. J. Stocks, J. S. Levine, W. R. Cofer, and K. P. O'Neil. 1992. Seasonal distribution of African savanna fires, *Nature*, 359, pp. 812–815.

Crutzen, P. J., and M. O. Andreae. 1990. Biomass burning in the tropics: Impact on atmospheric chemistry and biogeochemical cycles, *Science*, 250, pp. 1669–1677.

Delmas, R., P. Loudjani, A. Podaire, and J. C. Menhaut. 1991. Biomass burning in Africa: an assessment of annually burnt biomass, in J. S. Levine (ed.), *Global Biomass Burning: Atmospheric, Climatic, and Biospheric Implications*, MIT Press, Cambridge, Mass., pp. 126–133.

Garstang, M., P. D. Tyson, R. Swap, M. Edwards, P. Kallberg, and J. A. Lindesay. 1996. Horizontal and vertical transport of aerosols over southern Africa, *Journal of Geophysical Research*, in press.

Hao, W. M., M. H. Liu and P. J. Crutzen. 1990. Estimates of annual and regional releases of CO_2 and other trace gases to the atmosphere from fires in the tropics, in J. G. Goldammer (eds.), *Fire in the Tropical Biota*, Springer Verlag, Berlin, pp. 440–462.

Lacaux, J. P., H. Cachier and R. Delmas. 1993. Biomass burning in Africa: An overview of its impacts on atmospheric chemistry, in P. J. Crutzen and J. G. Goldammer (eds.), *Fire in the Environment*, Wiley, New York.

Maenhaut, W., I. Salma, J. Cafmeyer, H. J. Annegarn and M. O. Andreae. 1996. Regional atmospheric aerosol composition and sources in the Eastern Transvaal, South Africa, and impacts of biomass burning, *Journal of Geophysical Research*, in press.

Pretorius, R. W., I. Auret, G. Held, K. M. Brassel, I. R. Danford, and D. D. Waldie. 1986. *The Climatology of the Boundary Layer Over the Eastern Transvaal Highveld and its Impacts on Sulfur Dioxide Concentrations at Ground Level*, Report to the Foundation for Research Development, CSIR, ATMOS/86/16, Atmospheric Science Division, NPRL, CSIR, Pretoria.

Salma, I., W. Maenhaut, J. Cafmeyer, H. J. Annegarn, and M. O. Andreae. 1993. PIXE analysis of cascade impactor samples collected at the Kruger National Park, South Africa, *Nuclear Instruments and Methods in Physics Research*, B85, pp. 849–855.

Sanhuez, E. 1991. Effects of vegetation burning on the atmospheric chemistry of the Venezuela Savanna, in J. S. Levine (ed.), *Global Biomass Burrning: Atmospheric, Climatic, and Biospheric Implications*, MIT Press, Cambridge, Mass., pp. 122–125.

Tyson, P. D., F. J. Kruger, and C. W. Louw. 1988. *Atmospheric pollution and its implications in the Eastern Transvaal Highveld*, South African National Scientific Programmes, 150.

Tropospheric Ozone on Both Sides of the Equator in Africa

Bernard Cros, Brou Ahoua, Didier Orange, Michel Dimbele,
and Jean-Pierre Lacaux

Biomass burning increases tropospheric ozone during the dry season in Africa (Cros et al. 1987; Cros et al. 1988; Andreae et al. 1992). This enhancement occurs from November to March in the northern hemisphere (NH) and from June to October in the southern hemisphere (SH). The main biomass burning in Africa is savanna fires, traditional charcoal making, charcoal combustion, wood fires for cooking, etc. Estimates of Andreae (1991) show that the African continent contributes about 40% of biomass burned in the world and savanna fires represent 75% of the biomass burned in Africa.

A rough estimate of seasonal biomass burned in intertropical Africa shows that 49% of the burnings occur during the dry season of the NH (from November to February) and 51% during the dry season of the SH (from June to October) (Lefeivre 1993). However, the seasonal behavior of tropospheric ozone in the intertropical belt is more emphasized in the SH than in the NH.

In the SH, Cros et al. (1987) have reported ground-level ozone of more than 50 ppbv at Brazzaville (4°S) during the dry season. Surface ozone concentration in Panama (9°N), in Venezuela (9°N), and in Ndjamena (12°N) is less than 25 ppbv during the dry season (Cros et al. 1991). In Lamto station (6°N) in the Ivory Coast, surface ozone measured from October 1990 to August 1992 has shown a high variability: the monthly mean of daily maximal is about 25 ppbv from December 1990 to February 1991 and about 45 ppbv from December 1991 to February 1992 (Ahoua 1993). This high difference comes from regional meteorological conditions and particularly the degree of dryness of the dry season. This behavior is in good agreement with the cold cloud occurrences detected over the Ivory Coast during this period (Lahuec 1992). High values of occurrence are the signature of vigorous deep convection and the potential of great amounts of precipitation. Moreover, the total carbon and the black carbon measured in the air at Lamto (Cachier, personal communication) showed the same variation as the surface ozone. Although the correlation coefficient between surface concentrations and the amount of

ozone in the troposphere can strongly vary with the site of measurements, it is only .52 at Brazzaville (Cros et al. 1992). We think that ozone temporal variability and local meteorological conditions play an important role in the seasonal variation of tropospheric ozone.

The deviation of tropospheric ozone in both hemispheres of intertropical Africa is also shown by the satellite measurements of Fishman et al. (1991). High concentrations are revealed during September and October, corresponding with the SH dry season. On the other hand, no important increase of tropospheric ozone is shown during January and February, the dry season of the NH (Fishman et al. 1991).

Ozonesonde measurements at Brazzaville were initiated on June 1990. The first year's data allowed Cros et al. (1992) to demonstrate that the seasonal cycle of ozone derived from the ozonesonde measurements is in good agreement with the climatological seasonal cycle inferred from the use of satellite data.

In this chapter, we present the first ozone soundings of the African northern tropics; they were performed at Bangui (4°N, 18°E) during a preliminary phase of the Experiment for Regional Sources and Sinks of Oxidants (EXPRESSO) from December 1993 to April 1994. These first five months' results are discussed and compared with measurements obtained at Brazzaville during pre-TRACE and TRACE-A (Transport and Atmospheric Chemistry near the Equator—Atlantic) experiments from June 1990 to May 1993. Brazzaville is exactly situated on the other side of the equator at (4°S, 15°E).

Electrochemical cell (ECC) sondes have been used at Brazzaville and at Bangui. Although we have 99 soundings at Brazzaville and 16 soundings at Bangui, we can compare the monthly mean integrated ozone in these two sites because 3 soundings were realized per month on the average.

Results

Ozonesonde Data

The monthly mean of integrated ozone amounts are given at Brazzaville (table 31.1) and at Bangui

Table 31.1 Monthly mean of integrated ozone amounts (in Dobson units) at Brazzaville from June 1990 to May 1993

Month	Number of soundings	1000–100 hPa (0–17 km)	850–600 hPa (1.5–4 km)	600–130 hPa (4–15 km)	130–100 hPa (15–17 km)
1	5	36.4	8.9	23.2	2.5
2	8	35.0	7.8	22.4	3.6
3	5	35.2	7.5	23.1	3.0
4	4	30.9	5.2	22.1	3.0
5	6	35.6	8.2	23.4	2.6
6	8	46.2	11.8	29.8	2.7
7	10	48.4	14.7	28.0	2.9
8	11	47.9	13.8	28.7	3.1
9	10	49.5	11.4	32.8	3.0
10	16	42.3	8.3	28.9	3.4
11	9	37.2	6.4	36.0	3.2
12	7	38.1	8.1	23.5	3.0

Table 31.2 Monthly mean of integrated ozone amounts (in Dobson units) at Bangui from December 1993 to April 1994

Month	Number of soundings	1000–100 hPa (0–17 km)	850–600 hPa (1.5–4 km)	600–130 hPa (4–15 km)	130–100 hPa (15–17 km)
12	2	43.73	14.52	23.73	0.34
1	4	37.95	11.21	21.54	1.98
2	3	30.17	9.14	17.10	1.26
3	4	34.63	8.48	21.54	0.99
4	3	33.77	6.75	21.45	2.69

(table 31.2). These values are calculated at different layers of troposphere. We have considered the entire troposphere betwen 1000 and 100 hPa, the trade wind layer between 850 and 600 hPa, the upper troposphere between 130 and 100 hPa, and an intermediate layer from 600 to 130 hPa, which is characterized by a decrease of ozone partial pressure when altitude increases. For each layer, the approximate altitude is indicated.

Examples of Ozonesonde Profiles during the Dry Season and the Wet Season at Bangui and at Brazzaville

We present examples of ozonesonde profiles at Bangui for the dry season (figure 31.1a) and the wet season (figure 31.1b). The same profiles are represented for the site of Brazzaville (figures 31.1c and 31.1d). During the dry season, we note in both sites an increase of the ozone partial pressure in the lower troposphere. This enhancement is localized between 1.5 and 4 km, corresponding to about 850 and 600 hPa.

The vertical distribution of air fluxes in this part of the tropical troposphere is described by Cros et al.

(1988). The dry season at Bangui (in the NH) is characterized by the Harmattan flux, which is a northeast continental trade wind. This air flux is loaded by the effluents of the savanna fires of the northern part of the tropical belt. In the SH, the dry season at Brazzaville is characterized by the southeast trade wind. During this period, the savannas of the southern part of the tropical belt are burning. We also note in the two cases the permanent monsoon flux from the southwest, which flows in the first 1000 to 1500 m. According to this air distribution, we have reported the vertical profile of ozone in the lower troposphere at Bangui and at Brazzaville for the dry and the wet seasons (figures 31.1e–h). In the two sites, the increase of the ozone partial pressure coincides with the trade wind layer.

Thus the high ozone partial pressures observed are the consequence of the intense photochemical activity in this part of the troposphere (Crutzen et al. 1985; Cros et al. 1988). For these examples (figures 31.1e and 31.1g), the ozone mixing ratios calculated from the highest ozone partial pressures are 70 ppbv at Bangui and 95 ppbv at Brazzaville. The profiles of the wet season (figures 31.1f and 31.1h) indicate that the

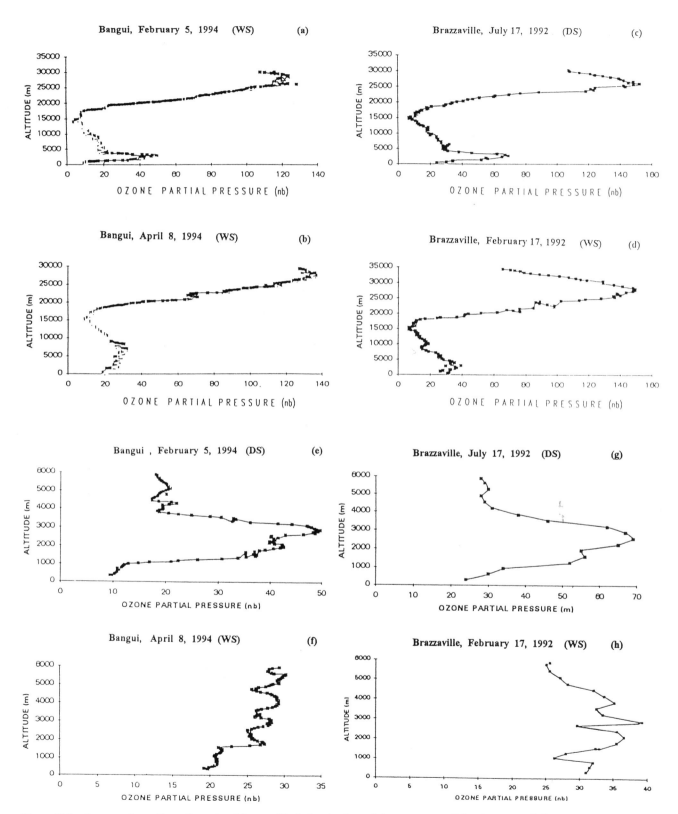

Figure 31.1 Ozonesonde profiles at Bangui and Brazzaville during the dry and the wet seasons. DS = dry season; WS = wet season.

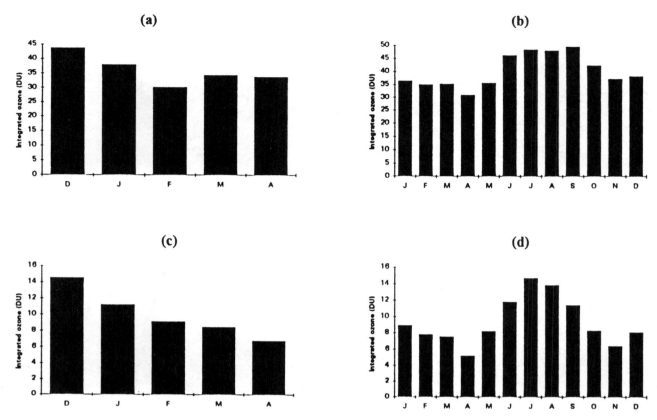

Figure 31.2 Monthly mean integrated ozone variation between 1000 and 100 hPa at Bangui (*a*) and Brazzaville (*b*) and between 850 and 600 hPa at Bangui (*c*) and Brazzaville (*d*)

background of ozone partial pressure is about 27 nb (about 37 ppbv) at Bangui and 30 nb (about 40 ppbv) at Brazzaville in the trade wind layers.

Monthly Mean Integrated Ozone at Bangui and Brazzaville

For each site, figure 31.2 illustrates the amount of integrated ozone in the entire troposphere (between 1000 and 100 hPa) and in the trade wind layer (between 850 and 600 hPa). The data from Bangui are not yet complete, so the monthly mean values of the dry season and the early wet season are presented. Nevertheless, the seasonal variation of ozone clearly appears in the two sites and in the two layers. The highest integrated values at Brazzaville occur from July to September (figure 31.2*b*), during the dry season of the SH. This variation is more emphasized in the trade wind layer (figure 31.2*d*). At Bangui, the highest integrated values correspond with December and January. Moreover, we note an increase of the monthly mean integrated ozone in March (figure 31.2*a*), which is not in agreement with the beginning of the wet season. This anomaly is also observed for the total carbon (Cachier,

personal communication) and for the fire distribution (Koffi, personal communication), so it is probably due to the fire stir up that precedes the raining season and is well known in savanna region.

Contribution of Biomass Burning to Tropospheric Ozone Enhancement during the Dry Season in Intertropical Africa

We have estimated the increase of tropospheric ozone due to the biomass burning at Bangui and Brazzaville. We report in table 31.3 the monthly mean integrated ozone (in Dobson units) for each season and for the two layers considered. The base of this estimate is that the wet season values depict roughly the background, and we assume that the ozone enhancement observed during the dry season is due to the biomass burning.

The integrated ozone amounts during the wet season are comparable on the two sites in the trade wind layer (7.6 DU at Bangui and 7.4 DU at Brazzaville) and in the entire troposphere (34.1 DU at Bangui and 35.5 DU at Brazzaville). For the dry season, we have comparable amount in the trade wind layer (11.6 DU at Bangui and 12 DU at Brazzaville), while values in

Table 31.3 Comparison of the contribution of biomass burning to the increase of tropospheric ozone in the trade wind layer (850–600 hPa) and in the entire troposphere (1000–100 hPa) at Bangui and at Brazzaville

		Monthly mean (DU)	
		Bangui (4 °N)	Brazzaville (4 °S)
Dry season	850–600 hPa (1.5–4 km)	11.6	12
	1000–100 hPa (0–17 km)	37.3	46.9
Wet season	850–600 hPa	7.6	7.4
	1000–100 hPa	34.1	35.5
Contribution of biomass burning	850–600 hPa	35%	38%
	1000–100 hPa	9%	24%

the entire troposphere are quite different (37.3 DU at Bangui and 46.9 DU at Brazzaville). The consequence is that the contribution of biomass burning to increased tropospheric ozone is almost the same in the trade wind layer (35% at Bangui and 38% at Brazzaville), but in the entire troposphere, this contribution is 24% at Brazzaville versus only 9% at Bangui.

Discussion and Conclusions

Tropospheric ozone increases during the dry season on both sides of the equator in Africa. This observation is in good agreement with the comparable amount of biomass burned in each hemisphere. The trade wind layer, more sensitive to the photochemical activity, shows comparable amounts of integrated ozone enhancement, with 35% in the NH and 38% in the SH. Therefore, we think that this increase of tropospheric ozone on each side of the equator is due to biomass burning. But, in the entire troposphere, the contribution of biomass burning to increased tropospheric ozone is more important in the SH (24%) than in the NH (9%). One of the reasons for such difference could be the transport of gases and particles emitted by biomass burning in the intertropical region. In general, the upward motion around the equator is stronger in the NH winter than in the SH winter (Shiotani and Hasebe 1994). Moreover, the upward mass flux from the troposphere to the stratosphere in the tropics has been estimated by Holton (1990) and Rosenlof and Holton (1993). They found that the upward mass flux in the tropics in maximum during the NH winter and minimum during the SH winter. Taking into account these general considerations, we calculated the integrated ozone in the upper troposphere in the two sites studied. The integrated amount between 12 and 17 km altitude is 4 DU at Bangui and 8 DU at Brazzaville. In a more thin layer, between 15 and 17 km altitude, we found 1 DU at Bangui and 3 DU at Brazzaville.

Therefore, it appears that the vertical transport of gases and particles is not the same at 4°N and 4°S. The horizontal transport of ozone and its precursors also plays an important role in the difference observed between the tropospheric integrated ozone on both sides of the equator. Krishnamurti et al. (1993) explain the maximum tropospheric ozone observed over the South Atlantic Ocean by two kinds of transport. By using three-dimensional trajectories calculated from both observed and forecast data, they found that gases from biomass burning over central Brazil could be transported south and east around an anticyclone located over northern South America. Likewise, burning by-product from central and southern Africa could be transported westward in the lower troposphere, also toward the ozone maximum (Krishnamurti et al. 1993). Therefore it appears necessary to have a comparable study of the meteorological environment of the tropospheric ozone on both the two sides of the equator in order to have a better comprehension of the differences observed.

Our results also show that the difference between integrated ozone during the dry season at Bangui (37.3 DU) and that of the wet season at Brazzaville (35.5 DU) is not very important. It could be one of the reasons why the integrated tropospheric ozone residual from Fishman et al. (1991) does not show an important variation in the intertropical Africa during the dry season of NH.

In conclusion, biomass burning increases tropospheric ozone in the lower troposphere both in the southern and the northern hemispheres of intertropical Africa. However, the dynamic is not the same on the two sides of the equator. Therefore, the distribution of the ozone yielded from the combustions is different in the two hemispheres. Consequently, the integrated ozone in the entire troposphere presents an important difference between the southern and the northern hemisphere. More ozonesonde data from Bangui and comparative meteorological studies can allow a better comprehension.

Acknowledgments

We thank Philippe Rousselot and the ASECNA people for helping during the ozone soundings at Bangui. Part of this project was funded under grant of the Campus Programme (Ministère de la Coopération— France).

References

Ahoua, B., 1993. L'ozone troposphérique dans l'hémisphère nord de l'Afrique intertropicale. DEA de l'Institut National Polytechnique et de l'Université Paul Sabatier, Toulouse, France, 38 pp.

Andreae, M. O. 1991. Biomass burning: its history, use, and distribution and its impacts on environmental quality and global climate, in *Global Biomass Burning: Atmospheric, Climatic and Biospheric Implications*, edited by J. S. Levine, MIT Press, Cambridge, Mass., pp. 3–21.

Andreae, M. O., A. Chapuis, B. Cros, J. Fontan, G. Heals, C. Justice, Y. J. Kaufman, A. Minga, and D. Nganga. 1992. Ozone and Aitken nuclei over equatorial Africa: air-borne observations during DECAFE 88. *J. Geophy. Res.*, 97, 6137–6148.

Cros, B., R Delmas, B. Clairac, J. Loemba-Ndembi, and J. Fontan. 1987. Survey of ozone concentration in an equatorial region during the rainy season. *J. Geophys. Res.*, 92, 9772–9778.

Cros, B., R. Delmas, D. Nganga, and B. Clairac. 1988. Seasonal trends of ozone in equatorial Africa: experimental evidence of photochemical formation. *J. Geophys. Res.*, 93, 8355–8366.

Cros, B., D. Nganga, R. Delmas, and J. Fontan. 1991. Tropospheric ozone and biomass burning in intertropical Africa, in *Global Biomass Burning: Atmospheric, Climatic and Biospheric Implications*, edited by J. S. Levine, pp. 143–146, MIT Press, Cambridge, Mass.

Cros, B., D. Nganga, A. Minga, J. Fishman, and V. Brakett. 1992. Distribution of tropospheric ozone at Brazzaville, Congo, determined from ozonesonde measurements. *J. Geophys. Res.*, 97, 12 869–12 875.

Crutzen, P. J., A. C. Delany, J. Greenberg, P. Haagenson, L. Heidt, R. Lueb, W. Polock, W. Seiler, A. Wartburg, and P. R. Zimmerman. 1985. Tropospheric chemical composition measurements in Brazil during the dry season. *J. Atmos. Chem.*, 2, 233–256.

Fishman, J., K. Fakhruzzaman, B. Cros, and D. Nganga. 1991. Identification of widespread pollution in the southern hemisphere deduced from satellite analysis. *Science*, 252, 1693–1696.

Holton, J. R. 1990. On the global exchange of mass between the stratosphere and troposphere, *J. Atmos. Sci.*, 47, 392–395.

Krishnamurti, T. N., H. E. Fuelberg, M. C. Sinha, D. Oosterhof, E. L. Bensman, and V. B. Kumar. 1993. The meteorological environment of the tropospheric ozone maximum over the tropical South Atlantic Ocean, *J. Geophys. Res.*, 98, 10 621–10 641.

Lahuec, J. P. 1992. Convergence intertropicale: intensité de la convection de janvier à avril 199. *Veille Climatique Satellitaire*, 41, 14–36.

Lefeivre, B. 1993. Etude expérimentale et par modelisation des caractéristiques physiques et chimiques des précipiations collectées en forêt équatoriale africaine. Thèse de Doctarat, Université Paul Sabatier, Toulouse, France, 206 pp.

Rosenlof, K. H., and J. R. Holton. 1993. Estimates of the stratospheric residual circulation using the downward control principle, *J. Geophys. Res.*, 98, 10 465–10 479.

Shiotani, M. and F. Hasebe. 1994. Stratospheric ozone variations in the equatorial region as seen in Stratospheric Aerosol and Gas Experiment data, *J. Geophys. Res.*, 99, 14 575–14 584.

Evaluation of Biomass Burning Effects on Ozone during SAFARI/TRACE-A: Examples from Process Models

Anne M. Thompson

Models of varying types have been used to interpret SAFARI/TRACE-A (Southern African Fire Atmospheric Research Initiative/Transport and Atmospheric Chemistry near the Equator—Atlantic) observations (Andreae et al. 1994; Fishman 1994). Primary emphases are in two areas: (1) photochemistry and dynamics related to the seasonal ozone maximum of focus in TRACE-A, and (2) fluxes, transport, and mapping of biomass burning emissions in SAFARI.

Examples of models that have been used in the analysis of SAFARI and TRACE-A data are given in table 32.1, along with references where details are given. In this chapter, we review the key science questions motivating the SAFARI and TRACE-A missions and show how process models have been used to answer them. These campaigns emphasized local and regional objectives. With a large, comprehensive data set collected over a limited area and time, it is possible to extract answers on major science issues with extensive use of relatively simple models—trajectory models and zero-dimensional photochemical models (point and box). Case studies form a basis for extrapolation to regional results over the September–October 1992 period.

One two-dimensional model has been applied to TRACE-A data (Chatfield et al. 1996), and two mesoscale models have been used in a tracer mode. A model for southern African fire counts, emissions factors, and ecology has been used to develop a grid of biomass burning trace gas emissions (CO, NO_x, CH_4) (Scholes, Ward, and Justice 1996). Validation of this model in a global chemical-transport model has not been carried out, although Lelieveld et al. (1995) present a generic evaluation of the biomass burning role in tropical ozone formation with the Max Planck MOGUNTIA model.

In the following sections, the motivating questions of SAFARI and TRACE-A are presented, and answers are given in a way that illustrates the application of various models.

What Are Flow Patterns over the South Atlantic Basin during SAFARI/TRACE-A?

Trajectory models have been used to determine the climatology of transport over the SAFARI/TRACE-A (Garstang et al. 1996; Garstang and Tyson 1996; Thompson et al. 1996b; Thompson et al. 1995c) as well as the fate and origin of specific air masses (see chapter 39). Gridded global wind data sets are used to track air parcels in these models. Forward trajectories give the fate of air masses and are useful for predicting where air masses enriched in biomass burning will be transported. Origins of specific air masses are determined by running backward trajectories, with the sampling location as the arrival point.

Two types of trajectory models have been used for SAFARI/TRACE-A analysis: kinematic and isentropic. In the latter, vertical motions are neglected because transit along constant potential temperature surfaces is assumed. Kinematic models use three-dimensional winds fields, although these may exaggerate vertical transport. A problem in the SAFARI/TRACE-A region is a scarcity of data over the tropical south Atlantic for verifying the winds used. For example, Pickering et al. (1996a) compared ECMWF (European Centre for Medium-Range Forecasting) and NMC (National Meteorological Center) horizontal wind components to DC-8 dropsondes from TRACE-A and found only small differences between them. Fuelberg et al. (1996) compared two model types over a sample of days on TRACE-A. In cases of disagreement, comparisons with observations were not clear-cut enough to decide in favor of one or the other model.

Flow Characteristics over Southern Africa

Garstang, Tyson, and coworkers have used a series of trajectory calculations to determine the climatology of mid- and low-level southern African transport during SAFARI and TRACE-A (Tyson et al. 1996; Garstang et al. 1996) (see also chapters 31 and 39). Trajectory calculations were performed with arrival (backward trajectories) and departure points (forward trajectories)

Table 32.1 Models used to analyze atmospheric effects of biomass burning (references are for application of model to SAFARI-92 and TRACE-A data.)

Type	Scale	Data and application, reference
Chemical		
Instantaneous	Point	Assimilate high-frequency trace gas species date—test photochemical theory, compute O_3 formation rate (Jacob et al. 1996; Thompson et al. 1996a)
Box	Homogeneous region, air mass	Simulate mean conditions, test consistency of concentrations and fluxes, track air parcel (Pickering et al. 1996a)
One-dimensional	Homogeneous region, air mass	Similar to box, but gradient and flux details can be studied
Dynamic		
Trajectory	Regional	Determine air mass origin and fate of biomass burning emissions (Garstang et al. 1996; Tyson et al. 1996; Swap et al. 1996) (see also chapters 37–39)
Mesoscale	Regional	Synoptic systems, simulation of mesoscale convective systems (Wang et al. 1996; Chatfield 1996)
Coupled Chemical-Transport (CTMs)	Multi	Coupled chemistry-dynamics; micro to global used, depending on problem (Lelieveld et al. 1996; Chatfield et al. 1996)

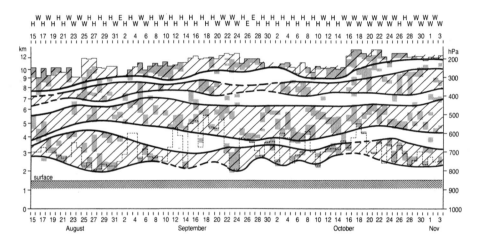

Figure 32.1 Structure of stable layers deduced from radiosondes over Pretoria, South Africa, during SAFARI/TRACE-A. Different circulation regimes are H = high, W = westerly, E = easterly. (From Garstang et al. 1996.)

located at major SAFARI sampling sites: Irene (Pretoria; 25°52'S, 28°11'E), Kruger National Park (24°58'S, 31°35'E), Okaukuejo (Etosha Park, Namibia; 19°11'S, 15°55'E), and Victoria Falls (18°10'S, 25°80'E). These locations all showed the presence of surprisingly stable layers, some of which remain intact for weeks at a time (figure 32.1). This means that vertical motions over southern Africa (south of 15°S) are suppressed, in contrast to the band from the equator south to 15°S, where convective activity leads to extensive vertical mixing and redistribution (R. Chatfield, personal communication, 1995).

Tyson et al. (1996) performed a statistical analysis of forward kinematic trajectories from the 875-, 850-, 800-, 700-, and 500-hPa levels (see also chapter 39). These were initialized daily from the four SAFARI sites mentioned previously over the SAFARI/TRACE-A period (15 September–3 November 1992). Figure

32.2 shows the predominant flow patterns. Anticyclonic flow prevails within all regimes, causing extensive recirculation of air masses at all SAFARI-92 stations. For calculating net flows, the fraction of trajectories crossing planes at 10°E (in the easterlies, i.e., exit toward Atlantic) and 35°E (westerlies, heading to Indian Ocean) are taken (table 32.2).

The flows computed by Tyson et al. (1996) have been applied to estimates of ozone and aerosol transfer (see chapter 39). The uv-DIAL aircraft instrument gives a curtain plot of ozone, showing local maxima, as do ozone soundings at four ozone sounding sites: Ascension, Brazzaville, Okaukuejo, and Irene. These excess ozone amounts, multiplied by flow rates, can be used to estimate ozone transport over the Atlantic and Indian Oceans. In the band from 15 to 35°S, excess ozone in September–October 1992 averages 10 to 15 DU, corresponding to 6 to 9 Tg of ozone exported to

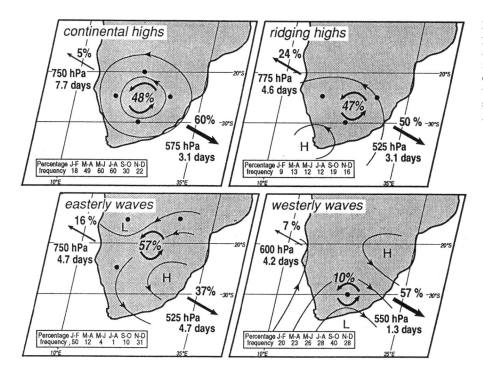

Figure 32.2 Predominant flow patterns south of 15°S over southern Africa determined from trajectories at 500 and 700 mb. Mass volume fraction exiting from continent is predominantly toward the Indian Ocean. All regimes show a high degree of recirculation over Africa. (From Tyson et al. 1996.)

Table 32.2 Southern circulation types observed during SAFARI-92

Flow type	% Air mass → Atlantic[a]	% Air mass → Indian	% Recirculation	Sept./Oct. occurs
Continental high	5	60	48	30%
Ridging high	24	47	50	20%
Easterly wave	16	57	37	10%
Westerly wave	7	10	57	40%

a. For trajectories initially at 500 to 700 mb. North-south extent of planes is from 15 to 35 °S. Sum of fractions is not unity because some air masses never leave continent in 10-day period of calculation; other air masses exit study region without crossing 10 or 35 °E.

the Atlantic Ocean and 9 to 12 Tg to the Indian Ocean. The amount of ozone recirculating over southern Africa without export, based on the statistics of Garstang et al. (1996), was ~15 Tg during SAFARI/TRACE-A.

Flows over the South Atlantic during SAFARI/TRACE-A: Where Do Emissions from Biomass Burning in South America and Africa Go?

In covering the south Atlantic Basin, TRACE-A included a larger region than SAFARI. The aim of TRACE-A was to characterize mid and upper tropospheric flows as well as chemistry in the boundary layer. TRACE-A objectives were guided by wanting to know origins of ozone sampled extensively on the DC-8 as well as the fate of emissions from the range of biomass burning regions of southern Africa and

Brazil. More specific questions such as the following are formulated: What were flows from the biomass burning regions of Africa to the south tropical Atlantic? What were the origins of ozone maxima sensed by satellite and by instrumentation on the TRACE-A DC-8 and the SAFARI DC-3? What were the origins of ozone from soundings?

Figure 32.3a shows fire counts for 1–15 October 1992 over southern Africa detected by Advanced Very High Resolution Radiometer (AVHRR) (Justice et al. 1996). Forward trajectories (8 days at ~700 mb level with the GSFC (Goddard Spau Flight Center) isentropic model) were initialized for each day within the period according to fire location and scaled to the amount of burning. Figure 32.3b is a composite of hourly locations along the trajectories. Similarities with flow patterns discussed by Tyson et al. (1996) and

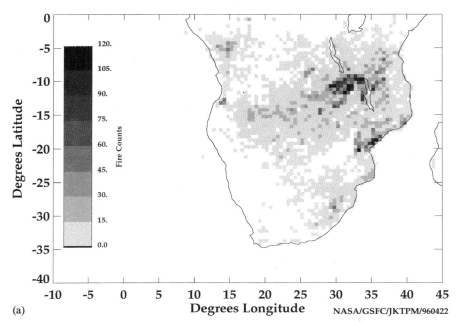

(a)

NASA/GSFC/JKTPM/960422

Figure 32.3 (*a*) Total AVHRR-derived fire counts for 1–15 October 1992. Not every day has complete coverage due to missing data, clouds, and hot surfaces. Data loss was greatest in 0–10°S band, and fires maximize between 10 and 20°S (Justice et al. 1996). (*b*) Forward trajectories (8 days) run with GSFC isentropic model using ECMWF-analyzed winds from fires in *a*. Origins at ∼ 700 mb. Trajectories are initialized each day for which there are fire counts (four missing days). Distributions of parcels at hourly locations show extensive recirculation of air masses leaving continent. Burning products south of 20°S recirculate as in Figure 32.2 or head toward the Indian Ocean. (After Thompson et al. 1996b.)

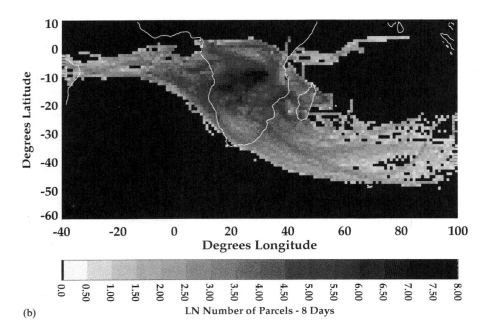

(b)

Garstang et al. (1996) are obvious (see also chapters 31 and 39). The eastern Atlantic or subtropical African anticyclone suppresses flow toward the Atlantic, and most air masses originating over fires recirculate before exiting the continent. The Atlantic exit ("Angolan plume" [Garstang et al. 1996]) is north of 15°S and is fed primarily by fires in northern Zambia and Angola. Mozambique fires, which were prominent in the SAFARI/TRACE-A period, are caught in air masses that go primarily toward the Indian Ocean ("Natal plume"). We show below that a combination of photochemical ozone formation rates determined from sampling near active fires, along with the residence times derived from trajectories, accounts for the amount of ozone seen over Africa by Total Ozone Mapping Spectrometer (TOMS) (Kim et al. 1996) and the TRACE-A uv-DIAL (Browell et al. 1995).

Figure 32.4a shows late September 1992 biomass fire regions of Brazil. These areas also coincide with the deep convection that is highly effective at injecting biomass burning emissions, with ozone-producing potential, into the middle and upper troposphere. One of the Brazilian TRACE-A flights (Flight 6, on 27 September 1992) sampled convective outflow directly (see p. 344), and enhanced ozone in soundings downwind from this convection confirmed the convection-ozone connection. Forward trajectories in Figure 32.4b show that South American biomass burning emissions pumped upward by deep convection are carried toward the Atlantic by westerlies. Ozone sounding sites downwind of this flow are Natal (Brazil), Ascension Island, and southern African SAFARI/TRACE-A sites at Okaukuejo and Irene. These soundings, as well as DC-8 ozone instrumentation and the TOMS satellite, detected tropospheric ozone over the south Atlantic that was twice the column depth observed in the nonburning season.

Ozone Origins-Trajectory Links: Where Does the Ozone in the South Atlantic Basin Come From?

Complementary to the question of biomass burning emissions transport is to ask where south Atlantic air masses elevated in tropospheric ozone originate. Figure 32.5 shows a composite of back-trajectories from a region that the TOMS-derived tropospheric ozone column shows to be elevated in ozone during early October 1992 (average column depth = 47 DU) (Kim et al. 1996). The origins of the high ozone are consistent with the forward trajectories from African and South American fires (see figures 32.3 and 32.4). The

8-day back-trajectories at lower level ($\sim$700 mb [figure 32.5a]) originate from Africa or show extensive recirculation over the Atlantic due to the anticyclone that persists over Angola and Namibia or just off the coast (see chapter 37). Upper-level origins (figure 32.5b) are from South America. These patterns imply that both continents contributed to enriched ozone over the Atlantic during SAFARI/TRACE-A. This was evident from DC-8 sampling on TRACE-A; several flights encountered high ozone, which trajectory analysis ascribes to South American upper-level outflow (above 10 km) or African mid- to lower-level outflow (usually below 10 km).

Confirmation of dual continental flow toward the Atlantic is seen in SAFARI/TRACE-A ozone soundings at Ascension Island. An example of a sounding with layers of enhanced ozone appears in Figure 32.6a. Air mass origins for this sounding were determined by taking back-trajectories at lower ($\sim$4.5 km), middle ($\sim$7.5 km), and upper (11.5 km) tropospheric locations and running the GSFC isentropic model backwards for 8 days. In Figure 32.6b, the transport patterns of air masses sampled in the 27 September 1992 sounding at Ascension Island show typical flows over the tropical Atlantic. Lower- and mid-level origins are African/Atlantic, and upper troposphere excess ozone originates from South America. When we performed this analysis for all the Ascension soundings, mid-level origins were found to be predominantly but not exclusively African/Atlantic; more than 85% of upper-level flows were from South America (Thompson et al. 1996c). Although there are no independent tracers coincident with the ozone soundings, TRACE-A DC-8 flights over the eastern central Atlantic, including three arriving or departing at Ascension Island, reveal the upper tropospheric air masses enriched in CO, hydrocarbons, and by-products of NO_x oxidation reactions (see p. 341).

Figure 32.6c gives the statistical origins for Ascension Island. This pattern of upper-level South American origin (or beyond) was followed for trajectories from Okaukuejo, Irene and a couple of Réunion Island soundings during SAFARI/TRACE-A. Only at Brazzaville were all the trajectories from Africa. Figure 32.6a shows "background" ozone, computed from a mean of March–May profiles, the nonburning season for Southern Hemisphere savannas. The region between the two curves in Figure 32.6a illustrates "excess ozone," a parameter that has been computed for every SAFARI/TRACE-A sounding. Using back-trajectories to apportion excess ozone for each sound-

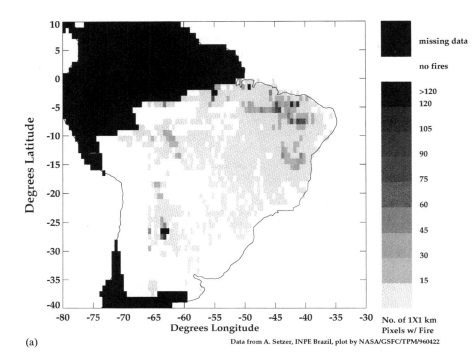

(a)

Data from A. Setzer, INPE Brazil, plot by NASA/GSFC/TPM/960422

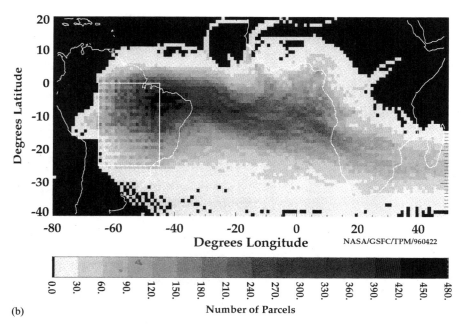

(b)

Figure 32.4 (*a*) Total AVHRR-derived fire counts for the last week of September 1992 over savanna burning areas of Brazil, subject to same caveats as in Figure 32.3*a*. (*b*) Forward trajectories (8 days) run with GSFC isentropic model using ECMWF-analyzed winds from grid in which deep convective activity over South America was prominent during that period (25 September to 20 October 1992). Origins at ~150 mb. (After Thompson et al. 1996b.)

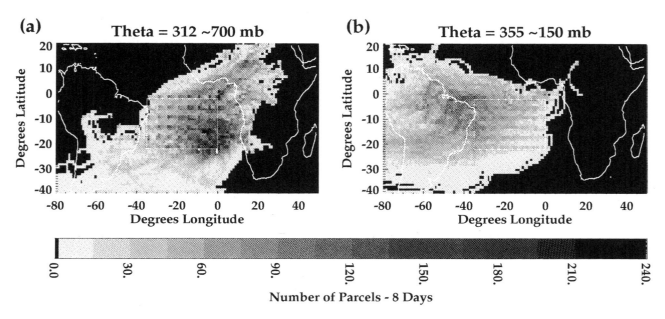

(a) Theta = 312 ~700 mb

(b) Theta = 355 ~150 mb

Number of Parcels - 8 Days

Figure 32.5 Eight-day backward trajectories (6–21 October 1992, hourly points) from region of maximum tropospheric ozone column (box) as identified from TOMS satellite in Kim et al. (1996). Two levels shown are approximately (a) 700 mb and (b) 150 mb. (After Thompson et al. 1996a; Thompson et al. 1996b.)

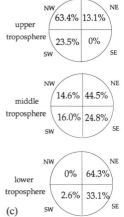

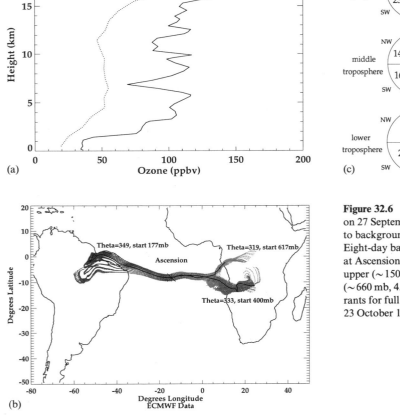

Figure 32.6 (a) Ozone profile from Ascension Island sounding on 27 September 1992, showing layers of excess ozone compared to background profile obtained from March–May average. (b) Eight-day back-trajectories, initialized as a cluster of 121 points at Ascension Island on 27 September, show air mass origins from upper (~150 mb, 11.5 km), middle (~400mb, 7.5 km), and lower (~660 mb, 4.5 km) locations. (c) Statistical sampling of origin quadrants for full set of Ascension soundings between 15 September and 23 October 1992. (After Thompson et al. 1996b)

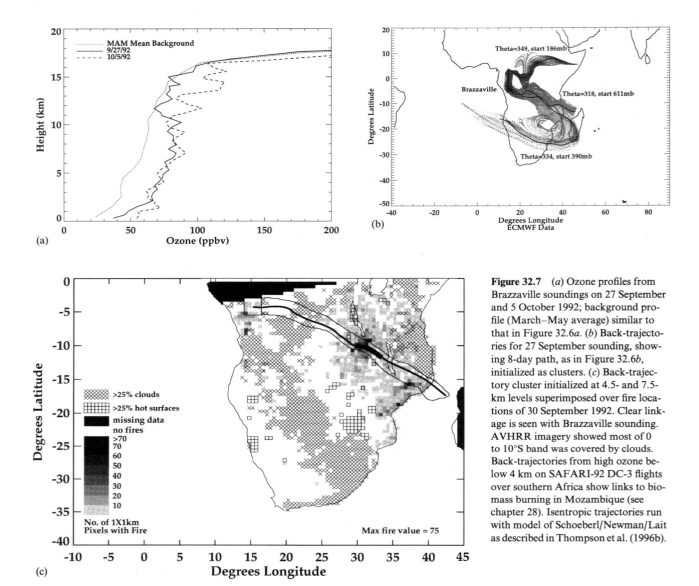

Figure 32.7 (*a*) Ozone profiles from Brazzaville soundings on 27 September and 5 October 1992; background profile (March–May average) similar to that in Figure 32.6*a*. (*b*) Back-trajectories for 27 September sounding, showing 8-day path, as in Figure 32.6*b*, initialized as clusters. (*c*) Back-trajectory cluster initialized at 4.5- and 7.5-km levels superimposed over fire locations of 30 September 1992. Clear linkage is seen with Brazzaville sounding. AVHRR imagery showed most of 0 to 10°S band was covered by clouds. Back-trajectories from high ozone below 4 km on SAFARI-92 DC-3 flights over southern Africa show links to biomass burning in Mozambique (see chapter 28). Isentropic trajectories run with model of Schoeberl/Newman/Lait as described in Thompson et al. (1996b).

ing gives the following for the origins of Ascension Island biomass burning–produced ozone: 70% from Africa, 20% from South America, with remaining ozone from air masses that recirculated over the Atlantic more than 8 days.

Figure 32.7*a* shows two typical Brazzaville soundings taken during TRACE-A. This first one was taken on the same day as the sample Ascension Island sounding (Figure 32.6*a*) with air mass origins determined from back-trajectories. As at Ascension Island, upper tropospheric ozone was enriched above background (nonburning season) at this site. Trajectories (Figure 32.7*b*) and the location of deep convection indicate that recirculation of ozone and ozone pre-

cursors supplied by deep convection probably act in this region. Trajectory analysis of excess ozone in the Brazzaville soundings showed 100% origins from Africa, frequently following recirculation over the Atlantic. In Figure 32.7*c*, back-trajectories initialized at 4.5- and 7.5-km levels from a Brazzaville sounding pass over fire locations from five days earlier.

Were Ozone Amounts over the South Atlantic Basin in SAFARI/TRACE-A Consistent with Photochemistry?

Table 32.1 shows that photochemical models range from point models, instantaneous steady-state photochemical models used to assimilate high-frequency

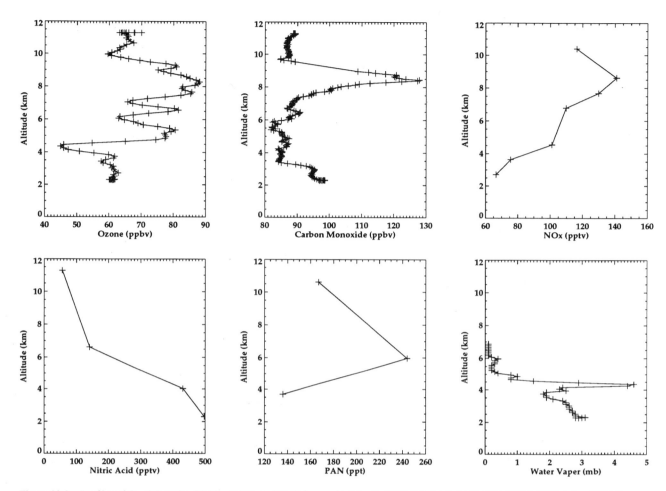

Figure 32.8 Profiles of O₃, CO, NOx, HNO₃, PAN, and H₂O from mid-Atlantic TRACE-A Flight 16 (6°S, 17°W, 20 October 1992) from Windhoek to Ascension Island. Correlation between CO and O₃ is variable. Peak at 9 km is elevated in both NOₓ and CO ozone precursors. The 4-km O₃ peak corresponds to CO levels expected for aged air mass but NOₓ remains sufficiently high to permit net ozone formation.

aircraft data to three-dimensional coupled chemistry-transport models. Point, box, two-dimensional, and a mesoscale model with tracers have been used to answer the question, Can photochemical formation of ozone be explained by the trace gas observations of SAFARI/TRACE-A?

First, it is important to note that ozone and ozone precursors (CO, NOₓ, and hydrocarbons) were ubiquitous on SAFARI and TRACE-A aircraft sampling. Examples of profiles taken by the DC-8 over the Atlantic appear in Figure 32.8. Aged haze layers over southern Africa were also enriched in these constituents (see chapter 28). Continental sources were detected in aerosol samples at Ascension Island (Swap et al. 1996). In general, stratospheric incursions or large-scale dynamics did not bring elevated ozone into the troposphere in the south Atlantic Basin. However,

Krishnamurti et al. (1996) performed tracer studies with ozone in the TRACE-A region for mid-October 1992. Sporadic communication with the lower stratosphere occurred in isolated regions over South America and east central Africa, but not in areas where observations were made. Krishnamurti et al. (1996) showed that high ozone levels observed in the lower troposphere on TRACE-A flights could not be reproduced without assuming an in situ ozone source. However, subsidence over the tropical south Atlantic causes ozone to accumulate in the lower troposphere whether photochemical formation is active or not (Krishnamurti et al. 1996).

TRACE-A Flight 10 (6 October 1992) characterized sources in fire-rich regions of northern Zambia. Sampling in the mixed layer shows elevated levels of NO, CO, and hydrocarbons, as well as ozone (Figure 32.9).

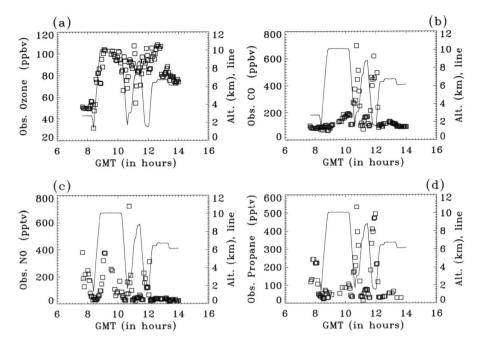

Figure 32.9 TRACE-A DC-8 data on Flight 10 (6 October 1992) used in GSFC photochemical point model: (*a*) O₃; (*b*) CO; (*c*) NO; (*d*) propane. (From merged data set prepared by G. Gardner and D. Jacob [Jacob et al. 1996]). The flight originated in Johannesburg; two later spirals to the mixed layer took place over northern Zambia. Instantaneous net O₃ production is equivalent to 14 ppbv O₃/day in 0- to 4-km layers; +0.4 ppbv O₃/day at 4 to 8 km, and 1.6 ppbv O₃/day at 8 to 12 km. Note elevated NO/CO/C₃H₈ at 10 km at 0900 GMT. At this altitude, convective injection may have increased concentrations above background levels.

Ozone maximizes above the mixed layer, whereas air masses near active fires are highest in the ozone precursors. The ozone, though elevated in the mixed layer, was not as high as it was above 6 km. The reason is that ozone formation rates are highest in the mixed layer, but so are ozone destruction rates. Just above the mixed layer, and again above 6 km, aged air masses continue to form ozone as long as mixing ratios of NO$_x$ remain above 80 to 100 pptv. Just above the mixed layer, stable air masses (see Garstang et al. [1996] and chapter 28) recirculate with the same levels of ozone found in the mixed layer, that is, 80 ppbv or greater. At higher altitudes, ozone formed slowly but steadily, as NO$_x$ levels remained at several hundred parts per trillion by volume (pptv). It is not clear whether this NO$_x$ was from recycled biomass burning, biogenic organic-nitrogen products, or lightning. In any case, the recycled products would be borne aloft by convection.

When the aircraft data are used as input into a photochemical point model, the instantaneous rate of ozone formation is 10 to 15 ppbv/day in the boundary layer, reaching a minimum in the middle troposphere, and increasing again in the upper troposphere (Jacob et al. 1995; Thompson et al. 1996). A point model assimilates the mixing ratios shown in Figure 32.9 along with PAN (peroxyacetyl nitrate), HNO$_3$, other hydrocarbons, H$_2$O, observed temperature, pressure,

ultraviolet radiation to compute OH, HO$_2$, H$_2$O$_2$, and 25 or more additional constituents. A typical reaction set for the troposphere (GSFC point model [Thompson et al. 1996b]) appears in Table 32.3. This model uses computed transient species concentrations (O[^{1}D], HO$_2$, R$_i$O$_2$) to evaluate net ozone formation as follows:

$$P(O_3) = k_1(NO)(HO_2) + \Sigma k_i(NO)(R_iO_2) - (O_3)[k_3(OH) + k_4(HO_2)] - k_5[O(^1D)](H_2O) \quad (32.1)$$

where R$_i$O$_2$, organic peroxy radicals, include CH$_3$O$_2$, C$_2$H$_5$O$_2$, C$_3$H$_6$OHO$_2$, C$_2$H$_4$OHO$_2$, C$_3$H$_7$OHO$_2$ (both *n*- and *i*-), CH$_3$CO$_3$, and the by-product of toluene oxidation. The pattern of net ozone formation above 8 km = 1 to 5 ppbv/day was ubiquitous on TRACE-A flights and somewhat unexpected. Likewise the mid-troposphere tended to be net ozone-destroying, although some of the highest ozone mixing ratios were found in layers at 6 to 8 km (Browell et al. 1996).

A closer look at ozone mechanisms in the boundary layer is given by the SAFARI DC-3 data. This aircraft made a survey set of flights around southern Africa in late September and early October 1992, with frequent profiling during ascents and descents below 4 km. Mixed layer ozone and ozone precursor concentrations were similar to those recorded on the DC-8 Flight 10, as were rates of ozone formation (see ozone

Table 32.3 Reaction scheme in GSFC one-dimensional tropospheric model (1995 version)

N #	Reaction	N #	Reaction
	Photodissociations	64	HO2 + NO2 + MMM = HNO4 + M
1	O3 + hv = O2 + O	65	HNO4 = HO2 + NO2
2	O3 + hv = O2 + O(1D)	66	HNO4 + OH = H2O + O2 + NO2
3	NO2 + hv = NO + O	67	HCO + O2 = CO + HO2
4	HNO3 + hv = OH + NO2	68	C2H6 + OH = C2H5O2 + H2O
5	H2O2 + hv = OH + OH	69	C2H5O2 + NO = C2H5O + NO2
6	NO3 + hv = NO + O2	70	C2H5O + O2 = CH3CHO + HO2
7	NO3 + hv = NO2 + O	71	C2H5O2 + HO2 = C2H5OOH + O2
8	H2CO + hv = H + HCO	72	CH3CHO + OH = CH3CO3 + H2O
9	CH3OOH + hv = OH + CH3O	73	CH3CO3 + NO = CH3 + CO2 + NO2
10	HNO4 + hv = HO2 + NO2	74	CH3CO3 + NO2 + MMM = PAN + M
11	CH3CHO + hv = CH3 + HCO	75	PAN = CH3CO3 + NO2
12	N2O5 + hv = NO2 + NO3	76	CH3OOH + OH = CH2OOH + H2O
13	H2CO + hv = H2 + CO	77	C2H5OOH + OH = C2H5O2 + H2O
14	C2H5OOH + hv = C2H5O + OH	78	C2H5OOH + OH = C2H4OOH + H2O
15	PAN + hv = CH3CO3 + NO2	79	N2O5 + H2O = HNO3 + HNO3
16	C2H5ONO2 + hv = C2H5O + NO2	80	CH3O2 + CH3O2 = CH3OH + H2CO + O2
17	PNITRATE + hv = NC3H7O + NO2	81	CH3O2 + CH3O2 = CH3O + H2CO + O2
18	INITRATE + hv = IC3H7O + NO2	82	CH3O2 + CH3O2 = CH3OOCH3 + O2
	Free-radical reactions	83	CH3CHO + NO3 = CH3CO3 + HNO3
19	O + O2 + MMM = O3 + M	84	H2CO + NO3 = HCO3 + HNO3
20	O + O3 = O2 + O2	85	C3H6 + OH + MMM = C3H6OHO2 + H2O + MMM
21	O + O + MMM = O2 + M	86	C3H6OHO2 + NO = C3H6OHO + NO2
22	O(1D) + N2 = O + N2	87	C3H6OHO + MMM = H2CO + CH3CHO
23	O(1D) + O2 = O + O2	88	C3H6OHO2 + HO2 = C3H7O3H + O2
24	NO + O3 = NO2 + O2	89	C3H6OHO2 + C3H6OHO2 = C3H6OHO + C3H6OHO + O2
25	NO2 + O = NO + O2	90	C3H6OHO2 + CH3O2 = CH3O + C3H6OH + O2
26	NO2 + O3 = NO3 + O2	91	C3H6 + O3 = CH2O2E + CH3CHO
27	NO + NO3 = NO2 + NO2	92	CH2O2E + MMM = CH2O2
28	NO + O + MMM = NO2 + M	93	CH2O2E + MMM = H2 + CO2
29	NO2 + NO3 + MMM = N2O5 + M	94	CH2O2E + MMM = H2O + CO
30	N2O5 = NO2 + NO3	95	CH2O2E + MMM = HCOOH
31	O(1D) + H2O = OH + OH	96	CH2O2 + H2O = HCOOH + H2O
32	O(1D) + CH4 = OH + CH3	97	C3H6 + O3 = C2H4O2E + H2CO
33	O(1D) + CH4 = H2 + H2CO	98	C2H4O2E + MMM = C2H4O2
34	O(1D) + H2 = OH + H	99	C2H4O2E + MMM = CH4 + CO2
35	H + O3 = OH + O2	100	C2H4O2E + MMM = CH3OH + CO
36	H + O2 + MMM = HO2 + M	101	C2H4O2E + MMM = CH3COOH
37	OH + O3 = HO2 + OH	102	C2H4O2 + H2O = CH3COOH
38	HO2 + O3 = OH + O2 + O2	103	C3H8 + OH = NC3H7O2 + H2O
39	OH + O = H + O2	104	NC3H7O2 + NO = NC3H7O + NO2
40	HO2 + O = OH + O2	105	NC3H7O + O2 = PALD + HO2
41	H2O2 + O = OH + HO2	106	NC3H7O2 + HO2 = NC3H7OOH + O2
42	OH + CH4 = H2O + CH3	107	NC3H7O2 + CH3O2 = NC3H7O + H2CO + O2
43	HO2 + NO = OH + NO2	108	NC3H7O2 + NC3H7O2 = PALD + PALD + O2
44	OH + CO = CO2 + H	109	C3H8 + OH = IC3H7O2 + H2O
45	OH + H2 = H2O + H	110	IC3H7O2 + NO = IC3H7O + NO2
46	OH + NO2 + MMM = HNO3 + M	111	IC3H7O2 + NO = INITRATE + O2
47	HNO3 + OH = NO3 + H2O	112	IC3H7O + O2 = CH3COCH3
48	OH + H2O2 = H2O + HO2	113	IC3H7O2 + HO2 = IC3H7OOH + O2
49	OH + HO2 = H2O + O2	114	IC3H7O2 + CH3O2 = IC3H7O + H2CO + O2
50	OH + OH = H2O + O	115	C2H5O2 + NO = C2H5ONO2
51	OH + H2CO = H2O + HCO	116	NC3H7O2 + NO = PNITRATE
52	HO2 + HO2 = H2O2 + O2	117	OH + C2H5ONO2 = PROD
53	OH + OH + MMM = H2O2 + M	118	OH + PNITRATE = PROD
54	H + HO2 = H2 + O2	119	OH + INITRATE = PROD
55	H + HO2 = H2O + O	120	C2H4 + OH + MMM = C2H4OHO2 + H2O + M
56	H + HO2 = OH + OH	121	C2H4OHO2 + NO = C2H4OHO + O2
57	H2CO + O = OH + HCO	122	C2H4OHO2 + HO2 = C2H4O3H + O2
58	CH3 + O2 + MMM = CH3O2 + M	123	C2H4OHO2 + CH3O2 = CH3O + C2H4OH + O2
59	CH3O2 + NO = CH3O + NO2	124	TOL + OH = HCP + HCP
60	CH3O2 + HO2 = CH3OOH + O2	125	HCP + NO = CH3CHO + HO2 + NO2
61	CH3OOH + OH = CH3O2 + H2O	126	HCP + HO2 = C3H7O2H
62	CH3O + O2 = H2CO + HO2	127	CH3O2 + NO2 + MMM = MPN + MMM
63	H2 + O = OH + H	128	MPN = CH3O2 + NO2

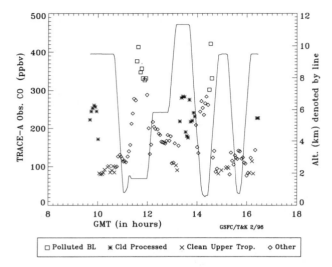

Figure 32.10 CO mixing ratios showing mixture of background-level (BL) and postconvective cloud (Cld) outflow, leading to enhanced CO in the free troposphere (Trop.). From TRACE-A Flight 6 out of Brasilia, 27 September 1992.

and CO profiles in chapter 28). The same models (GSFC isentropic trajectory model and photochemical point model) were used in this analysis as in TRACE-A analyses reported by Thompson et al. (1996a). Detailed trajectories for air masses sampled on the DC-3 show generally a 3- to 4-day removal from active fires of Zambia and Mozambique.

The enhancement of ozone formation by deep convection was studied closely on a TRACE-A flight dedicated to cloud outflow (Flight 6, 27 September 1992, from Brasilia). Figure 32.10 shows that CO mixing ratios registered at cloud outflow level (* symbol) represent a mixture of air from above (low CO, background ∼80 ppbv) and from boundary-layer fire-exposed air masses (500 ppbv or higher CO). Similarly, NO is elevated above background. The CO/NO ratios are fairly constant within the cloud outflow layer, except when NO shoots up by a factor of 2 or greater, presumably from lightning injection. On Flight 6, the CO/NO ratio variability suggested that 30 to 40% of upper tropospheric NO_x derived from lightning (Pickering et al. 1996b). These levels of NO

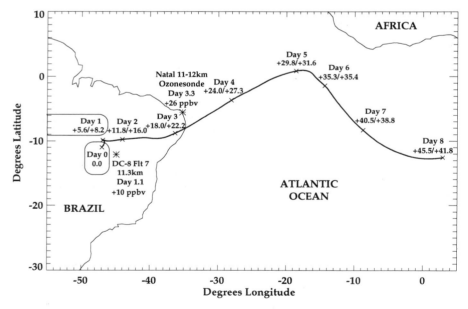

Figure 32.11 Cumulative daily ozone formed (in ppbv) at points along typical 8-day trajectory initialized at the location and altitude of TRACE-A Flight 6 convective outflow sampling (27 September 1992). GSFC tropospheric box model used. The first number is based on highest 9-minute average cloud outflow mixing ratios sampled by DC-8 at 11.3 km, and the second value is based on highest mixing ratios at 9.5 km. Ozone formation was NO_x limited, and tracer correlations suggest that 30 to 40% of upper tropospheric NO in cloud outflow layers was due to lightning (Pickering et al. 1996b). Note increase in upper tropospheric ozone recorded between 28 and 30 September 1992 with Natal soundings agrees well with calculation. TRACE-A Flight 7 (28 September 1992) shows 10-ppbv O_3 increase, consistent with calculations.

were sufficient to maintain positive ozone formation in cloud outflow layers. This ozone formation was simulated in the 8- to 12-km and 12- to 16-km layers, using the GSFC box model, which uses the same chemical reactions as in the point model. The box model starts with a composition derived from observed mixing ratios at cloud outflow level on the DC-8 in the 8- to 12-km layer. Above 12 km, profiles of ozone precursors, CO, NO, and hydrocarbons are taken from tracer runs with the Goddard Cumulus Ensemble cloud-resolving model (Tao and Simpson 1993). Figure 32.11 shows the amount of ozone formed each day at points along a typical 8-day trajectory initialized at the location and level of the Flight 6 sampling. At upper levels, the Natal ozone sounding on 30 September 1992, which was 2 to 4 days downwind from the Flight 6 sampling, recorded a 25- to 30-ppbv ozone increase in selected layers beyond the 28 September sounding. Conversely, running back-trajectories from Natal on 30 September 1992 at lower, mid, and upper levels show that upper-level air came from cloud outflow regions sampled three days earlier on 27 September. Lower-level air mass origins for the 28 and 30 September 1992 Natal ozone soundings are from Africa (Thompson et al. 1996b).

Convective venting of boundary layer emissions has been shown to be an effective route for keeping boundary-layer CO from accumulating despite steady emissions (Pickering et al. 1992; Thompson et al. 1994). Convection appeared to keep the boundary layer in Brazil and southern Africa at moderate levels as well (Pickering et al. 1996b). Stable layers aloft were frequently higher in CO than near the surface (see figures 28.5a and 28.6), and actual cloud outflow layers, as sampled during Brazilian sampling, were clearly elevated in CO. Figure 32.12 is a tracer simulation of CO using the GSFC version of the Penn State/NCAR MM5 (mesoscale) model. Cross-sections of CO mixing ratio at cloud outflow levels sampled at 9.5 km and 11.3 km are consistent with DC-8 measurements at these altitudes. The large regional extent of the feature shown in figure 32.12 confirms the importance of deep convection in transporting ozone precursors to the free troposphere.

Regional Budgets and CO/NO/HC Sources

Another set of SAFARI/TRACE-A objectives at southern Africa was directed at budgets of ozone precursors. What fraction of CO, NO, and NMHC emissions in savannas is natural, that is, from soils, from plants, from oxidation of plant emissions (in the case of CO)? What part of anthropogenic emissions are industrial, from wood fuel, from savanna burning? These questions are harder to answer because models of scales from micro to global are required. During September 1992, biogenic and pyrogenic fluxes were measured only for limited periods, areas, and types of ecosystems.

Episodically, at least, SAFARI-92 showed substantial biogenic NO and CO emissions in southern Africa. Wetting of soils increased NO fluxes greatly, and NO postprecipitation was detected by aircraft (Harris et al. 1996). Soil NO emissions postburning increased by a factor of 4 to 7 (Levine et al. 1995). However, uncertainties in the soil NO flux–biomass burning NO flux ratio span a factor of 20. Comparisons are shown in table 32.4.

For wet soil, Swap et al. (see chapter 38) extrapolated using biomass burned area, precipitation frequency, area, and duration to conclude that boundary-layer biogenic NO sources greatly outweighed biomass burning for typical southern African savannas. However, other NO sources—wood fuel burning, land clearing, fossil fuel combustion—were not evaluated. Mid- and upper-tropospheric NO sources appeared to be dominated by recycled organic nitrogen species and lightning (Talbot et al. 1996; Smyth et al. 1996).

To extrapolate fluxes measured at isolated sites to a region requires many assumptions. In the case of biogenic sources, vegetation and soil types must be assumed. In the case of biomass burning, using fluxes from controlled burns (the strategy for SAFARI-92) required assumptions about fuel type, carbon, and nitrogen content of vegetation, mass, and area burned. In SAFARI-92, CO emissions from biomass burning were estimated by Scholes, Ward, and Justice (1996) with a process model using the relation

$$E = \sum A_i B_i \alpha_i \beta_i \tag{32.2}$$

where E = the total emissions over the region in question (southern Africa), i = fuel type, A = area, B = biomass density, α = burning efficiency, and β = emission factor. Overall uncertainty is $\sim$77% rms, two thirds contributed by the first two terms. AVHRR-derived fire counts are incorporated in A.

An attempt was made to construct a regional boundary-layer CO budget for southern Africa during SAFARI-92 using the methods of Thompson et al. (1994). Figure 32.13 shows terms needed in a regional CO evaluation. Comparison of the Scholes, Ward, and Justice (1996) SAFARI-92 CO biomass burning

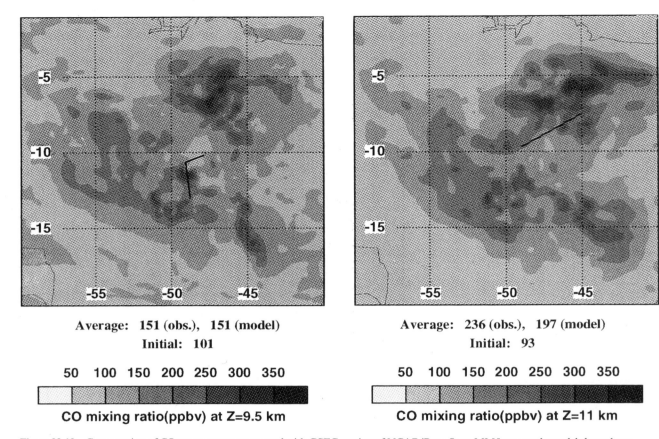

Average: 151 (obs.), 151 (model)
Initial: 101

50 100 150 200 250 300 350

CO mixing ratio(ppbv) at Z=9.5 km

Average: 236 (obs.), 197 (model)
Initial: 93

50 100 150 200 250 300 350

CO mixing ratio(ppbv) at Z=11 km

Figure 32.12 Cross-section of CO tracer output computed with GSFC version of NCAR/Penn State MM5 mesoscale model shows large regional impact of deep convection sampled on 27 September 1992. Initialization of model made with profiles from DC-8 and INPE Bandeirante. (After Pickering et al. 1996b.)

Table 32.4 Southern Africa flux measurements during SAFARI-92[a]

Trace gas	Savanna dry	Savanna wet	Biomass burn EF[b]
CH$_4$	Negligible	-0.2 mg C m^{-2} s^{-1}/day	0.24–0.36
CO	84–112 $\times$ 10^9 cm^{-2} s^{-1}	—	4.5–5.0
NO	1.3 ng N m^{-2} s^{-1}	20–70 ng N m^{-2} s^{-1}	Not given
N$_2$O	Negligible ($<$2ng N m^{-2} s^{-1})	—	0.007–0.009

a. Harris, Wienhold, and Zenker (1996) for NO wet; Levine et al. (1996) for others.

b. EF= emissions factors, volume gas/volume CO$_2$ produced (Cofer et al. 1996). Flaming phase factor given, which is approximately the same for smoldering for N$_2$O and CO, but twice the flaming amount for CH$_4$.

source to soil emissions from savannas (Zepp et al. 1996) and forest (Kirchhoff and Marinho, 1990) suggests that biogenic sources are greater than the biomass burning input. This seems inconsistent with observed CO during biomass burning season being elevated a factor of 2 above nonburning season. (The regional impact of the Southern Hemisphere savanna burning is evident in background CO monitored at Ascension Island [Novelli et al. 1992].) Other terms that need to be included in the boundary-layer CO budget are industrial and wood fuel combustion, advection from surrounding regions, photochemical formation from methane, and natural and anthropogenic nonmethane hydrocarbons. Losses of CO include OH oxidation, advection to adjoining regions, and convective venting. Accurate evaluation of the CO budget will require more extensive observations, along with small (microscale), regional (mesoscale and finer), and global scale models to disaggregate all the sources and sinks.

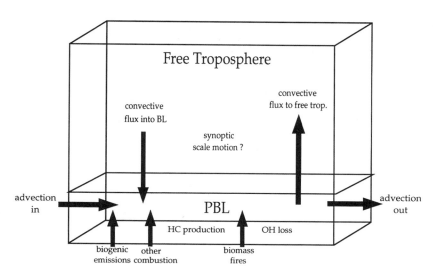

Figure 32.13 Terms in regional boundary-layer CO budget for region affected by savanna burning, wood fuel and industrial sources, and soil and vegetative emissions. Boundaries receive advective inflows from adjacent areas. Deep and shallow convection both vent and supply, via downdrafts, boundary-layer CO in exchange with free troposphere. (BL = boundary layer; PBL = planetary boundary layer).

Conclusions

The application of models to the main questions of SAFARI/TRACE-A has been reviewed. Trajectory models and simple photochemical models (comprehensive chemistry but zero- or one-dimensional dynamics) applied to the south Atlantic Basin showed the following:

1. High ozone levels throughout the troposphere can indeed be explained by photochemical processes with biomass burning–derived precursors (CO, NO, nonmethane hydrocarbons) at concentration levels observed in the aircraft experiments.

2. The buildup of ozone to 15 to 25 DU greater than background comes from (a) long ozone lifetime in the middle and upper troposphere, and (b) stable layers subject to weeks-long recirculation over southern Africa. The latter process has eventual exit of ozone toward the Atlantic (via the "Angolan" plume) and the Indian Ocean ("Natal plume").

3. The upper troposphere over the south Atlantic Basin is ozone producing because NO_x levels—either through lightning enhancement or recycling of surface-derived nitrates—are sufficient for net ozone production. Trajectories show that excess ozone above 9 km traveled primarily in the westerlies leaving South America, which was unusually active in deep convection in September 1992. Ozone at lower and mid-tropospheric levels over the south tropical Atlantic was transported via easterlies from Africa.

4. Trajectory calculations can be used to apportion excess ozone over the south tropical Atlantic during the biomass burning season: 70 to 75% from Africa, 20 to 30% from South America.

These results on ozone sources were possible because SAFARI/TRACE-A supplied an excellent database from ground-based, balloon-borne, satellite and aircraft measurements. Comprehensive meteorological and constituent measurements enabled simple (zero-dimensional, one-dimensional, trajectory) models to answer key questions fairly conclusively.

When we look at sources and sinks of reactive carbon and nitrogen over the region sampled in SAFARI/TRACE-A and ask questions about natural versus anthropogenic, biogenic versus pyrogenic, uncertainties are much greater. Extrapolating from the limited flux database is risky, and yet a better grid of fluxes is required for evaluation in regional or global transport-chemical models. At the same time, in situ observations of CO, NO_x, and hydrocarbons were too sparse to allow validation of model outputs, even if the flux information had been available.

Acknowledgments

This research was sponsored by National Aeronautics and Space Administration programs in EOS, Tropospheric Chemistry, and ACMAP (Atmospheric Chemistry Modeling and Analysis Program) and by the U.S. Environmental Protection Agency. D. McNamara (Applied Research Corporation at Goddard) performed trajectory calculations with the Goddard isentropic model of Schoeberl, Lait, and Newman. T. Kucsera (Applied Research Corporation at Goddard) carried out calculations with the Goddard chemical point model of Thompson. Thanks to P. Tyson, R. Chatfield, M. Garstang, and R. Zepp for communicating work in press.

References

Andreae M. O., J. Fishman, M. Garstang, J. G. Goldammer, C. O. Justice, J. S. Levine, R. J. Scholes, B. J. Stocks, A. M. Thompson, B. van Wilgen, the STARE/TRACE-A/SAFARI-92 Science Team. 1994. Biomass Burning in the Global Environment: First Results from the IGAC/BIBEX Field Campaign STARE/ TRACE-A/SA-FARI-92, in *Global Atmospheric-Biospheric Chemistry: The First IGAC Scientific Conference*, ed. R. Prinn, Plenum Press, New York, pp. 83–101.

Andreae, M. O. 1996. Emissions of trace gases and aerosols from southern African savanna fires, in *Fire in Southern African Savanna: Ecological and Atmospheric Perspectives*, ed. B. van Wilgen, M. O. Andreae, J. G. Goldammer, and J. A. Lindesay, Univ. Witwatersrand Press, Johannesburg, pp 161–184.

Browell, E. V., M. A. Fenn, C. F. Butler, W. B. Grant, M. Clayton, J. Fishman, S. Bachmeier, B. E. Anderson, G. L. Gregory, H. E. Fuelberg, J. D. Bradshaw, S. T. Sandholm, D. R. Blake, B. G. Heikes, G. W. Sachse, H. B. Singh and R. W. Talbot. 1996. Ozone and aerosol distributions and air mass characteristics observed over the South Atlantic Basin during the burning season, *J. Geophys. Res.*, in press.

Chatfield, R. B., J. A. Vastano, H. B. Singh, and G. W. Sachse. 1996. A general model of how fire emissions and chemistry produce African/oceanic plumes (O3, CO, PAN, smoke) seen in TRACE-A, *J. Geophys. Res.*, in press.

Cofer, W. R., III, J. S. Levine, E. L. Winstead, D. R. Cahoon, Jr., D. I. Sebacher, J. P. Pinto, and B. J. Stocks. 1996. Source compositions of trace gases released during African savanna fires, *J. Geophys. Res.*, in press.

Fishman, J. 1994. Experiment probes elevated ozone levels over the tropical south Atlantic Ocean, *Eos*, 75, 380.

Fishman J., K. Fakhruzzaman, B. Cros, and D. Nganga. 1991. Identification of widespread pollution in the Southern Hemisphere deduced from satellite analyses, *Science*, 252, 1693–1696.

Fuelberg, H. E., R. O. Loring, Jr., M. V. Watson, M. C. Sinha, K. E. Pickering, A. M. Thompson, D. R. Blake, G. W. Sachse and M. R. Schoeberl. 1996. TRACE-A trajectory intercomparison. 2. Isentropic and kinematic methods, *J. Geophys. Res.*, in press.

Garstang, M. and P. D. Tyson. 1996. Atmospheric circulation, vertical structure and transport over southern Africa during the SAFARI 1992 campaign, in *Fire in Southern African Savanna: Ecological and Atmospheric Perspectives*, ed. B. van Wilgen, M. O. Andreae, J. G. Goldammer, and J. A. Lindesay, Univ. Witwatersrand Press, Johannesburg, pp. 57–88.

Garstang M., P. D. Tyson, R. J. Swap, M. Edwards, P. Kållberg, and J. A. Lindesay. 1996. Horizontal and vertical transport of air over southern Africa, *J. Geophys. Res.*, in press.

Harris, G. W., F. G. Wienhold and T. Zenker. 1996. Airborne observations of strong biogenic NOx emissions from the Namibian savanna at the end of the dry season, *J. Geophys. Res.*, in press.

Jacob, D. J., B. G. Heikes, S-M. Fan, J. A. Logan, D. L. Mauzerall, J. D. Bradshaw, H. B. Singh, G. L. Gregory, R. W. Talbot, D. R. Blake, and G. W. Sachse. 1996. The origin of ozone and NOx in the tropical troposphere: A photochemical analysis of aircraft observations over the South Atlantic Basin, *J. Geophys. Res.*, in press.

Justice C. O., J. D. Kendall, P. R. Dowty, and R. J. Scholes. 1996. Satellite remote sensing of fires during the SAFARI campaign using NOAA-AVHRR data, *J. Geophys. Res.*, in press.

Kim J. H., R. D. Hudson, and A. M. Thompson. 1996. Derivation of time-averaged tropospheric column ozone from radiances measured by the Total Ozone Mapping Spectrometer: Intercomparison and analysis, *J. Geophys. Res.*, in press.

Kirchhoff, V. W. J. H., and E. V. A. Marinho. 1990. Surface carbon monoxide measurements in Amazonia, *J. Geophys. Res.*, 95, 16 933–16 944.

Krishnamurti, T. N., M. C. Sinha, M. Kanamitsu, D. Oosterhof, H. Fuelberg, R. Chatfield, D. J. Jacob, and J. Logan. 1996. Passive tracer transports relevant to the TRACE-A experiment, *J. Geophys. Res.*, in press.

Lelieveld J., P. J. Crutzen, D. J. Jacob, and A. M. Thompson. 1996. Modeling of biomass burning influences on ozone, in *Fire in Southern African Savanna: Ecological and Atmospheric Perspectives*, ed. B. van Wilgen, M. O. Andreae, J. G. Goldammer, and J. A. Lindesay, Univ. Witwatersrand Press, Johannesburg, pp. 217–238.

Levine J. S., D. A. B. Parsons, R. G. Zepp, R. A. Burke, D. R. Cahoon, Jr., W. R. Cofer III, W. L. Miller, M. C. Scholes, R. J. Scholes, D. I. Sebacher, S. Sebacher, and E. L. Winstead. 1996. Southern African savanna as a source of atmospheric gases, in *Fire in Southern African Savanna: Ecological and Atmospheric Perspectives*, ed. B. van Wilgen, M. O. Andreae, J. G. Goldammer, and J. A. Lindesay, Univ. Witwatersrand Press, Johannesburg, pp 135–160.

Novelli, P. C., L. P. Steele, and P. P. Tans. 1992. Mixing ratios of carbon monoxide in the tropopshere, *J. Geophys. Res.*, 97, 20 731–20 750.

Olson, J. R., J. Fishman, V. W. J. H. Kirchhoff, D. Nganga, and B. Cros, 1996. An analysis of the distribution of O3 over the southern Atlantic region, *J. Geophys. Res.*, in press.

Pickering, K. E., A. M. Thompson, J. R. Scala, W-K Tao, and J. Simpson. 1992. A regional estimate of convective transport of CO from biomass burning, *Geophys. Res. Lett.*, 19, 289–292.

Pickering K. E., A. M. Thompson, Y. Wang, W-L. Tao, D. P. McNamara, V. W. J. H. Kirchhoff, B. G. Heikes, G. W. Sachse, J. D. Bradshaw, G. L. Gregory, and D. R. Blake. 1996a. Convective transport of biomass burning emissions over Brazil during TRACE-A, *J. Geophys. Res.*, in press.

Pickering K. E., A. M. Thompson, D. P. McNamara, M. R. Schoeberl, H. E. Fuelberg, R. O. Loring, Jr., M. V. Watson, K. Fakhruzzaman, and A. S. Bachmeier. 1996b. TRACE-A trajectory intercomparison: 1. Effects of different input analyses, *J. Geophys. Res.*, in press.

Scholes, R. J., D. Ward, and C. O. Justice. 1996. Emissions of trace gases and aerosol particles due to vegetation burning in southern-hemisphere Africa, *J. Geophys. Res.*, in press.

Smyth, S., S. Sandholm, J. Bradshaw, R. Talbot, D. Blake, N. Blake, S. Rowland, H. Singh, G. Gregory, B. Anderson, G. Sachse, J. Collins, and S. Bachmeier. 1996. Factors influencing the upper free tropospheric distributions of reactive nitrogen over the South Atlantic during the TRACE-A experiment, *J. Geophys. Res.*, in press.

Swap, R. J., M. Garstang, S. A. Macko, P. D. Tyson, W. Maenhaut, P. Artaxo, P. Kållberg and R. Talbot. 1996. The long-range transport of southern African aerosols to the tropical south Atlantic, *J. Geophys. Res.*, in press.

Talbot, R., J. D. Bradshaw, S. T. Sandholm, S. Smtyh, D. R. Blake, G. W. Sachse, J. Collins, B. G. Heikes, B. E. Anderson, G. L. Gregory, H. B. Singh, B. L. Lefer, and A. S. Bachmeier. 1996. Chemical characteristics of continental outflow over the tropical south Atlantic Ocean from Brazil and southern Africa, *J. Geophys. Res.*, in press.

Tao, W. K. and J. Simpson. 1993 The Goddard Cumulus Ensemble Model: Part 1. Model description, *Terr. Atmos. Oceanic Sci.* 4, 35–72.

Thompson, A. M., K. E. Pickering, R. R. Dickerson, W. G. Eillis, Jr., D. J. Jacob, J. R. Scala, W-K. Tao, D. P. McNamara, and J. Simpson. 1994. Convective transport over the central United States and its role in the regional CO and ozone budgets, *J. Geophys. Res.*, 99, 18 703–18 711.

Thompson A. M., R. D. Diab, G. E. Bodeker, M. Zunckel, G. J. R. Coetzee, C. B. Archer, D. P. McNamara, K. E. Pickering, J. B. Combrink, J. Fishman, and D. Nganga. 1996a. Ozone over southern Africa during SAFARI-92/TRACE-A, *J. Geophys. Res.*, in press.

Thompson, A. M., K. E. Pickering, D. P. McNamara, M. R. Schoeberl, R. D. Hudson, J. H. Kim, E. V. Browell, V. W. J. H. Kirchhoff, and D. Nganga. 1996b. Where did tropospheric ozone over southern Africa and the tropical Atlantic come from in October 1992? Insights from TOMS, GTE/TRACE-A and SAFARI-92, *J. Geophys. Res.*, in press.

Thompson, A. M., T. Zenker, G. E. Bodeker, and D. P. McNamara. 1996c. Ozone over southern Africa: Patterns and influences, in *Fire in Southern African Savanna: Ecological and Atmospheric Perspectives*, ed. B. van Wilgen, M. O. Andreae, J. G. Goldammer, and J. A. Lindesay, Univ. Witwatersrand Press, Johannesburg, pp. 185–216.

Tyson P. D., M. Garstang, R. Swap, P. Kållberg, and M. Edwards. 1996. An air transport climatology for subtropical southern Africa, *Intl. J. Climatol.*, 16, pp. 265–291.

Wang, Y., W-K. Tao, K. E. Pickering, A. M. Thompson, R. Adler, J. Simpson, P. Keehn, and G. Lai. 1996. Mesoscale (MM5) simulations of TRACE-A and PRE-STORM convective events, *J. Geophys. Res.*, in press.

Zepp, R. G., W. L. Miller, R. A. Burke, D. A. B. Parsons, and M. C. Scholes. 1996. Effects of moisture and burning on soil-atmosphere exchange of trace carbon gases in a southern African savanna, *J. Geophys. Res.*, in press.

Emissions from the Combustion of Biofuels in Western Africa

Delphine Brocard, Corinne Lacaux, Jean-Pierre Lacaux, Georges Kouadio,
and Véronique Yoboué

Biomass fuels are used by more than 50% of the world's population as the major source of domestic energy (Boleij et al. 1989). In all Africa, fuelwood in the form of wood and charcoal provides over 60% of the total energy needs and more than 90% of the domestic energy needs (Vergnet 1986). These traditional fuels are used mainly for household cooking or heating. In general, biomass energy use is predominant in the rural areas of developing countries, where it often supplies over 90% of total energy requirements (Hall 1991). However, it provides an important fuel source for the urban areas. Wood is mainly used as firewood or charcoal. In rural areas, firewood is preferred in most households, due to its easy collection. In urban areas, due to an increasing demand and to the depletion of local wood resources, wood is more and more transformed into charcoal. The latter is easier to carry from the market, due to its greater energy density (29 MJ/kg) as compared with that of wood (7 to 15 MJ/kg depending on the climate) (Leach and Gowen 1987). It is also easier to handle and to store and less smoky than firewood.

Firewood is generally burnt under inefficient conditions in poorly ventilated areas, which causes an increase in the amount of health-damaging air pollutants. In the 1980s, several studies on the energy and health implications of firewood stoves in developing countries were reported (Smith, Aggarwal, and Dave 1983; Pandey 1987; Boleij et al. 1989). However, only a limited number of such studies dealt with the contribution of this source to the chemical composition of the atmosphere.

As part of the DECAFE (dynamique et chimie de l'atmosphere en foret equatoriale) program, two field campaigns were devoted to the evaluation of the atmospheric pollution caused by the use of wood as a fuel. These studies were conducted in Côte d'Ivoire. They dealt with firewood and charcoal burning as well as wood carbonization for charcoal making.

In this chapter, we present both the CHARCOAL/DECAFE-92 experiment, focusing on the emission of chemical compounds by traditional charcoal making, and the FIREWOOD/DECAFE-94, dealing with the emissions form firewood and charcoal burning. For these three types of combustion, the concentrations of carbon compounds (CO_2, CO, CH_4, light hydrocarbons, organic acids) and nitrogen compounds (NO_x, NH_3, N_2O) were measured under experimental conditions where all parameters of the combustion process were controlled. To normalize these measurements, emission ratios were calculated and subsequently used for determining the emission factors, defined as the amount of a given gas X per kilogram of dry matter. Using the available data on the amount of firewood and charcoal in West Africa, we were then able to evaluate the magnitude of the emissions relevant to biofuel combustion in West Africa. The results indicate that biofuel burning is a significant source of pollutants, compared with savanna fires.

Experimental Methods

Experimental Strategy

The aim of these experiments was to quantify the production of gases and particles emitted by traditional fuelwood fires. The concentration of each species of emitted gas was measured in the smoke plume of a fire. Since the concentration represents a proportional measurement, the data may be normalized by dividing the concentration of a given gas X by the concentration of a species being characteristic of the combustion (generally CO_2 or CO for biomass burning).

The experimental strategy consisted of simultaneous measurements, in the same volume of fuelwood smoke, of the concentration of a given compound X and of the concentration of CO_2, relative to background atmospheric concentrations. This allowed a determination of the emissions ratios ($\Delta X/\Delta CO_2$), where ΔX represents $[X]_{smoke} - [X]_{ambient}$ and ΔCO_2 represents $[CO_2]_{smoke} - [CO_2]_{ambient}$.

Description of the Experiments

CHARCOAL/DECAFE-92 The CHARCOAL experiment (Lacaux et al. 1994) took place in Lamto, in

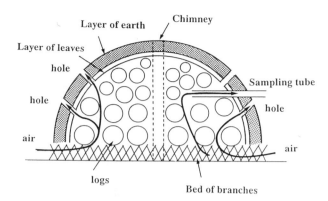

Figure 33.1 Scheme of the charcoal-making kiln. The samplings of gases and particles were performed at the exits of the metal tube and/ or of the successive holes made by the carbonizator to promote an air circulation and the propagation of the pyrolysis in the kiln.

Côte d'Ivoire, over a two-week period in April 1992. The main objective was to study the amount of gases and particles released by a traditional charcoal kiln. Four trees were felled, cut into 1.5-m-long logs, and weighed. In the clearing formed, a team of local charcoal makers built a traditional kiln, shown in figure 33.1. It consisted of a circular branch bed on which the logs were set along the circle radiuses, in successive layers. The kiln was covered with a layer of leaves and a layer of earth about 10-cm thick. At the center, a chimney allowed the fire to be set with live charcoal. The draught necessary to the propagation was facilitated by an air circulation between the basis of the kiln and a row of holes made in the horizontal plane. In such a kiln, the layers of leaves and earth tend to reduce the oxygen supply to the fire, thereby creating a pyrolytic process. Due to the lack of oxygen, three phases are produced through the combustion process, namely, a solid phase (the charcoal), a liquid phase (water, pyroligneous liquid and tars), and a gaseous phase composed of reduced or incompletely oxidized compounds. A total of 9264 kg of wood, containing 30% water and 46% carbon compared to dry matter, gave 1789 kg of charcoal with 71% carbon. The 19.3% carbonization rate obtained (27.6% compared to dry wood) is comparable to the 16% measured in Burundi (Eimer and Ndama 1987) and to the 17% measured in Senegal (Amous 1992). Twenty-eight smoke samples were collected at the exit of a metal tube or at the natural holes made in the kiln, using Teflon tubing handled with a pole.

FIREWOOD/DECAFE-94 In July 1994, several experiments were conducted in Côte d'Ivoire to study the emissions from firewood and charcoal burning.

Wood was burnt on a traditional three-stone open fire, whereas charcoal was burnt in a traditional stove. All fires were conducted by Ivorian women, who agreed to simulate the different stages of a fire representative of their cooking practices. From a discussion with these women, we were able to identify three stages, namely, (1) the cooking stage, producing high heat, due to the presence of flame, (2) the end cooking stage, where the wood is separated to produce a lot of live charcoal and small flames convenient for gentle cooking, and finally, (3) the fire end stage, where a lot of ash is produced.

When the fires were representative of any of these stages, two types of sampling were made through Teflon tubing in the smoke plume over a 10-minute period. In the first sampling procedure, the smoke was drawn up in a tedlar bag by the internal pump of an infrared absorption analyzer of CO and CO_2 (BINOS 100). In the second one, the smoke was sampled with a mist scrubber from the same volume and over the same period.

All along the sampling period, the temperature was measured with a thermocouple in several locations of the fire. Fire systems were placed on a balance to measure the fuelwood consumption by weight. All experiments were videotaped to verify if the sampling periods were representative of a homogeneous combustion process. Forty-three fires were ignited, using either only one type of local wood or a mixture of different wood, and 91 samples were collected. The water content and the nitrogen and carbon contents were determined for each type of wood.

Experimental techniques
CO and CO_2 were measured using a nondispersive infrared photometer analyzer (Binos 100, Rosemount). The concentration range of this instrument is 0 to 3000 ppm for CO_2 and 0 to 500 ppm for CO. To eliminate water, a dryer cartridge was placed at the air inlet. A zero-level adjustment was performed before each measurement period, by applying a soda lime cartridge to the air inlet. A span adjustment was performed before and after each campaign, using a standard gas (200 ppm CO and 2000 ppm CO_2). Smoke samples were either taken as close as possible to the fire using Teflon tubing handled with a pole or taken from 20-l Teflon bag samples. The latter method gave an integrated value of CO and CO_2 over the sampling period into the bag (generally 10 minutes).

NO_x was measured with a Scintrex LMA3 chemiluminescence instrument. This instrument detects NO_2 in a concentration range of 0 to 1 ppm by a chemiluminescence reaction with a 1-ppm luminol solution.

The air stream can be connected with a bypass to a CrO_3 chemical converter, which oxidizes NO to NO_2. Then, the analyzer detects NO_x. Ammonia and nitrogen oxides were also measured by chemiluminescence with an Environment SA NH30M. This instrument has two thermal converters. One, made of molybdenum, reduces NO_2 into NO, at 350°C, and then gives a value of nitrogen oxides (NO_x). NO is detected by a chemiluminescence reaction with ozone produced by an internal generator. The second converter, made of stainless steel, oxidizes NH_3 into NO, subsequently detected by chemiluminescence in the same way. The concentration of ammonia is automatically calculated from the difference between the two channels. These instruments were calibrated during field experiments, using an Environment SA VE3M with certified permeation tubes of NH_3 and NO_x (VICI Metronics).

Gas Chromatography Air samples were taken with 50-ml syringes, either directly in the smoke plume or from Teflon bag samples. In the Teflon bag samples, air was drawn up through polyethylene tubing by the internal pump of the infrared absorption analyzer. Air samples were then injected in preevacuated glass tubes closed by Teflon septums and crimped with aluminum caps. Analyses were performed by gas chromatography during the next two to three weeks.

CH_4, CO, and CO_2 were measured, using a PoraplotQ-filled column (Chrompack GC 9000) and a flame ionization detector (FID). CO_2 and CO were converted into methane at 650 K through a reduced Ni catalytic converter. Samples were introduced into a 400-μl sampling loop.

N_2O was measured with a ^{63}Ni electron capture detector (ECD) and a PoraplotQ-filled column. A standard gas (CO 1.3 ppm, CH_4 1.9 ppm, CO_2 497 ppm, N_2O 1 ppm) was used for calibration.

Other samples were collected in preevacuated electropolished stainless steel in order to measure light hydrocarbons. Analyses were performed in the "Centre des faibles Radioactivités, France," using a capillary column (AL_2O_2/KCl) and a FID detector according to a technique described by Bonsang and Lambert (1985) and Kanakidou, Bonsang, and Lambert (1989).

Liquid Chromatography Scrubber sampling was made for gaseous organic acids. In a vessel containing approximately 15 ml of deionized water, air was drawn up through Teflon tubing in the mist scrubber instrument, which produces a highly dispersed mist. All through the sampling period, the mist scrubber was refrigerated to prevent the loss of water vapor and the mist droplets. Organic acids were then absorbed in

the solution. Teflon prefilters were applied to extract the particle phase of organic acid. The samples were preserved by freezing to avoid degradation as a result of microbial activity. CH_3COO^- and $HCOO^-$ were measured within two to three weeks in our laboratory using a Dionex 4000i liquid chromatograph with a Dionex AS5A 5-μ column (Anion Separation), a Dionex ATC column (Anion Trap Column), a gradient of NaOH solutions as eluant, and H_2SO_4 as regenerant in a Dionex AMMS (Anion Micro Membrane Suppressor). The lower detection limits were 0.03 ppm (in water) for CH_3COO^- and 0.02 ppm for $HCOO^-$, corresponding to 0.14 ppbv and 0.16 ppbv, assuming a sample volume of 1 m^3 of air.

Aerosol Measurements Aerosol measurements were made in the "Centre des faibles Radioactivités" according to a technique described by Cachier et al. (1996).

Chemical Composition of Wood Wood was oven dried at 105°C to determine its water content. The calcium and nitrogen contents of the wood and of the ash were measured at the Service Central d'Analyse du CNRS in Vernaison, France. The carbon and nitrogen analyses were performed on samples of finely ground dried material through a total combustion process. The CO_2 generated by the complete combustion of carbon was then measured by infrared absorption, thus giving the carbon content. The NO_x generated by the complete combustion of nitrogen was reduced in molecular nitrogen, which was then measured by using a catharometric method.

Results

Emission Ratios

The emission ratio $\Delta CO/\Delta CO_2$ and the mean temperature in the furnace, which represent the main parameters characterizing the burning process, are reported in table 33.1. These parameters indicate that charcoal making is a typical smoldering combustion. Indeed, the corresponding low combustion temperature (380°C) produces high amounts of reduced or incompletely oxidized compounds ($\Delta CO/\Delta CO_2 = 24\%$ [with $\delta = 3$]). During charcoal burning, a lot of carbon monoxide is produced ($\Delta CO/\Delta CO_2 = 15.5\%$ [with $\delta = 3$]) at high temperatures (850°C). Hence, charcoal combustion can be classified as a kind of glowing combustion. According to Lobert and Warnatz (1993), glowing combustion follows the pyrolytic stage of a fire, at about 850 K if oxygen is present, resulting in

Table 33.1 Combustion efficiency (± standard deviation) and mean temperature, measured in the kiln for charcoal making and in the furnace for charcoal burning and firewood combustion

	Charcoal making	Charcoal burning	Firewood combustion			
			Ignition	Cooking	End cooking	End fire
$\Delta CO/\Delta CO_2$(%)	24.0 ± 3.0	15.5 ± 3.0	26.1 ± 4.8	5.7 ± 1.1	15.0 ± 2.8	21.0 ± 2.7
Temperature (°C)	380	850		800	690	530

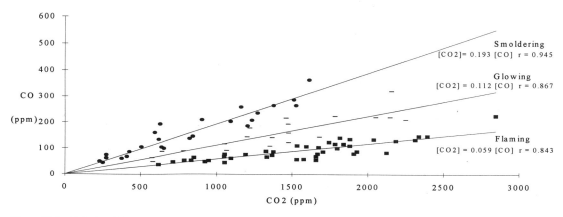

Figure 33.2 Relationship between the excess CO and CO_2 in firewood plumes. The flaming process is involved during the cooking stage and represents 80% of the initial wood mass. The glowing process is involved during the end-cooking fire and consumes 15% of the burnt material. At the end of the fire, the combustion process is a smoldering where 5% of the initial mass is burned.

char oxidation to CO, which is further oxidized to CO_2. For firewood combustion, the process depends on the stage of cooking. For the cooking stage, $\Delta CO/\Delta CO_2$ is 5.7% (with $\delta = 1.1$) with a temperature of 800°C. The process, which is highly exothermic and emits oxidized compounds, is of flaming type (Lobert and Warnatz 1993). The end cooking stage has the same characteristics as the charcoal burning process. The end of fire has the same characteristic as the carbonization process. The marked relationship between the excess of CO and CO_2 in smoke plumes at different stages of firewood combustion is shown in figure 33.2.

In order to adopt an average $\Delta CO/\Delta CO_2$ value being representative of the whole firewood combustion process, the contribution of each stage was evaluated by taking into account the corresponding weight of burnt wood. After discussions with the women and as a result of our experimental study, it was found that the cooking stage consumes 80% of the initial wood mass, whereas the end-cooking stage represents 15% of the weight burned and the end fire stage consumes 5% of burnt material. From these estimates, a mean $\Delta CO/\Delta CO_2$ of 7.9% is obtained by weighted mass.

The emission ratios for the three domestic combustion processes are presented table 33.2. For firewood

burning, a weighted average of the emission ratios respectively obtained for the three fire stages has been calculated, taking into account the proportion of wood burnt during each phase throughout the day. These emission ratios were compared with those obtained from savanna burning during the FOS (Fire of Savanna)/DECAFE-91 experiment. The emissions of firewood were found to be of the same magnitude as those of savannas fire, thereby confirming that firewood combustion involves mainly a flaming process. Charcoal burning releases high amounts of CO, but very little methane and light hydrocarbons, whereas charcoal making produces important quantities of reduced compounds (CH_4, nonmethane hydrocarbons, NH_3) or incompletely oxidized compounds (CO), compared to savanna fire. This confirms that the respective processes involved in these combustion are different: charcoal burning involves mainly a glowing combustion, whereas charcoal making is a smoldering process.

Emission Factors

The emissions factors are defined as the amount of a compound X emitted per unit of burnt dry matter. During both experiments, the amount of consumed dry matter was controlled.

Table 33.2 Emission ratios (in %) of carbon and nitrogen compounds from the combustion of biofuels compared with emission ratios from savanna fires (FOS/DECAFE-91 experiment [Lacaux et al. 1993, 1995; Bonsang et al. 1995])

	Combustion of biofuels (%)			Fire of savannas (%)
	Charcoal making	Charcoal burning	Firewood	
$\Delta CO/\Delta CO_2$	24.0 ± 3.0	15.5 ± 3.0	7.9 ± 1.5	6.1
$\Delta CH_4/\Delta CO_2$	6.8 ± 0.6	0.25 ± 0.20	0.38 ± 0.11	0.31
$\Delta NMHC^a/\Delta CO_2$	1.3 ± 0.3^b	0.060 ± 0.007	0.57 ± 0.24	0.70
$\Delta OA^c/\Delta CO_2$	$HCOOH = 0.010 \pm 0.010$		0.008 ± 0.004	
	$CH_3COOH = 0.160 \pm 0.140$		0.065 ± 0.034	
$\Delta Aer^d/\Delta CO_2$	3.3 ± 0.7^e		1.17 ± 0.63	0.72^f
$\Delta NO_x/\Delta CO_2$	0.02 ± 0.01		0.15 ± 0.04	0.14
$\Delta NH_3/\Delta CO_2$	0.05 ± 0.06			0.005
$\Delta N_2O/\Delta CO_2$	0.0075 ± 0.0120			0.015

a. Nonmethane hydrocarbons (C_2–C_{10}).
b. Bonsang, Kanakidou, and Boissard 1994.
c. Organic acids (formic and acetic).
d. Aerosols.
e. Hery and Cachier 1993.
f. Cachier et al. 1996.

To evaluate these emission factors, it is first necessary to evaluate the relative contribution of each compound to the carbonaceous emission of fire (Delmas, Lacaux, and Brocard 1995). For that purpose, the total amount of atmospheric carbon is assumed to be emitted only under the form of CO_2, CO, CH_4, light hydrocarbons, and organic acids and by the carbonaceous aerosols:

$$C(atmospheric) = C(CO_2) + C(CO) + C(CH_4)$$
$$+ C(NMHC) + C(OA) + C(aerosols)$$
$$(33.1)$$

where NMHC is nonmethane hydrocarbons, and OA is organic acids.

The relative contribution of a given compound X to the carbonaceous emission is then given by the expression C(X):

$$C(X) = \frac{\frac{\Delta X}{\Delta CO_2}}{1 + \frac{\Delta CO}{\Delta CO_2} + \frac{\Delta CH_4}{\Delta CO_2} + \frac{\Delta NMHC}{\Delta CO_2} + \frac{\Delta OA}{\Delta CO_2} + \frac{\Delta aerosols}{\Delta CO_2}} \quad (33.2)$$

where C(X) is the percentage of carbon released in the form of X and $\Delta X/\Delta CO_2$ is the molar emission ratio of X. The amount of carbon emitted into the atmosphere is then estimated for each combustion type. For charcoal making, three phases are produced. The carbon released in the atmophere is

$$C(atmospheric) = C(wood) - C(charcoal)$$
$$- C(pyroligneous\ liquid) \quad (33.3)$$

The carbon present in initial wood and in charcoal has been measured. The carbon present in pyroligneous liquid, mainly composed of polyphenols, furfural, and tars, has been estimated from laboratory measurements by Dumont and Gelus (1982) at 20.6% of the initial carbon. Figure 33.3 represents the global carbon budget for charcoal making, expressed in percent of carbon in the dry matter of the initial wood, and the percentage of atmospheric carbon released in the form of each species in percent of atmospheric carbon.

For firewood and charcoal burning, the atmospheric carbon content is

$$C(atmospheric) = C(wood) - C(ash) \quad (33.4)$$

Due to the high efficiency of these combustions, ash is produced in negligible quantities. Five percent of ashes (compared to initial mass), containing 11% of carbon, is produced during firewood combustion and only 1%, containing 26% of carbon, during charcoal burning. Figures 33.4 and 33.5 represent the global carbon budget for firewood burning and charcoal burning, respectively, expressed in percent of carbon in the dry matter of the initial wood, and the percentage of atmospheric carbon released in the form of each species, expressed in percent of atmospheric carbon. For firewood, the percentage of carbon in the form of particles has been estimated with the emission ratios obtained during FOS/DECAFE-91.

Taking into account the fractions of emitted carbon and the quantity of atmospheric carbon released per

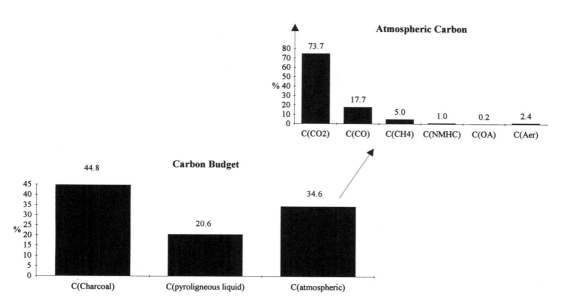

Figure 33.3 CHARCOAL/DECAFE-92—charcoal making. Global budget of carbon in the dry matter of the initial wood and budget of atmospheric carbonaceous compounds in percent of atmospheric carbon: C(atmospheric) = $C(CO_2) + C(CO) + C(CH_4) + C(NMHC) + C(OA) + C(aerosols)$. NMHC = nonmethane hydrocarbons; OA = organic acids; Aer = aerosols.

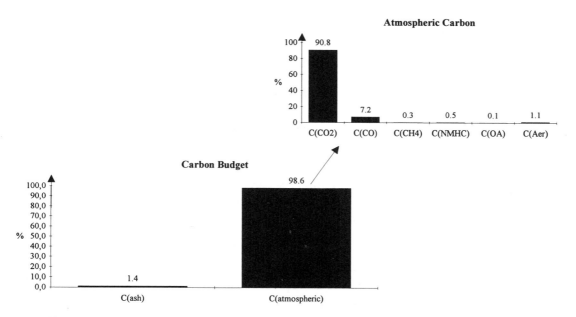

Figure 33.4 FIREWOOD/DECAFE-94—firewood burning. Global budget of carbon in the dry matter of the initial wood and budget of atmospheric carbonaceous compounds in percent of atmospheric carbon: C(atmospheric) = $C(CO_2) + C(CO) + C(CH_4) + C(NMHC) + C(OA) + C(aerosols)$. NMHC = nonmethane hydrocarbons; OA = organic acids; Aer = aerosols.

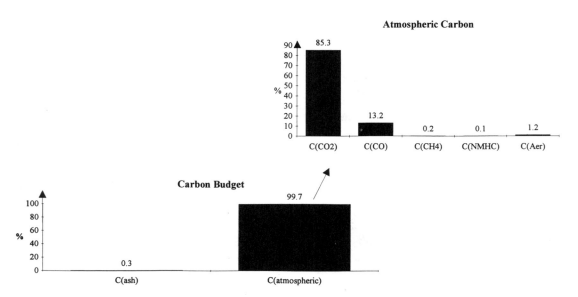

Figure 33.5 FIREWOOD/DECAFE-94—charcoal burning. Global budget of carbon in the dry matter of the initial wood and budget of atmospheric carbonaceous compounds in percent of atmospheric carbon: C(atmospheric) = $C(CO_2) + C(CO) + C(CH_4) + C(NMHC) + C(Aer)$. Carbon in the form of aerosol is calculated with estimated emission ratio. NMHC = nonmethane hydrocarbons; OA = organic acids; Aer = aerosols.

kilogram of dry matter burned, the emission factors are calculated according to the following expression:

$$EF(g\ C\ in\ the\ form\ of\ X/kg\ dm) = C\ (atmospheric) \times C(X)$$
(33.5)

where EF is the emission factor, dm is dry matter, C(atmospheric) is the atmospheric carbon, and $C(X)$ is the percentage of carbon released in the form of X.

Taking into account the molar emission ratios of nitrogen compounds (compared to CO_2), the nitrogen emission is calculated relative to the emission of carbon in the form of CO_2. With the emission factor of CO_2, the emission factors of the nitrogen compounds (in grams of nitrogen in the form of $X/kg\ dm$) are calculated. Table 33.3 presents the emission factors for the three domestic combustions in grams of calcium per kilogram of dry wood and in grams of nitrogen per kilogram of dry wood. During charcoal making, atmospheric carbon is mainly emitted in the form of CO_2, CO, and CH_4. For firewood, atmospheric carbon is mainly emitted in the form of CO_2, whereas during charcoal burning atmospheric emission are mainly in the form of CO_2 and CO.

Estimate of the Emission from the Combustion of Biofuels in Western Africa

Fuelwood Consumption in Western Africa The total amount of burnt fuelwood is difficult to estimate due

to the high percentage of gathered wood compared to the marketed wood. We are actually working on a new database, based on population and per capita use statistics. With national energy balances or with local survey, a daily per capita consumption of wood and charcoal is proposed for each country of Western Africa (table 33.4).

For wood, the daily per capita consumption is minimum in Mauritania, where 0.53 kg is used, and maximum in Guinea, where 2.46 kg of wood is used. For charcoal, the minimum used is 0.007 kg daily per capita in Niger, and the maximum is 0.17 kg in Senegal. As a matter of fact, these data are closely related to the fuelwood resources of each country. Indeed, in Guinea, Côte d'Ivoire, and most of the west coast countries, vegetation is mainly forest, thereby supplying all the fuelwood needs. On the other hand, Mauritania and Niger are mostly desert, and so there people tend to save their resources by decreasing their consumption. If data are missing, a mean consumption is calculated with the consumption of the bordering countries. The yearly fuelwood consumption in West Africa is estimated to be 100 Tg for firewood and 4.5 Tg for charcoal.

Emissions from Domestic Combustion in Western Africa
Table 33.5 represents both fuelwood emissions and savanna burning emissions in western Africa. The yearly emission from savanna burning in western

Table 33.3 Emission factors of the atmospheric carbon and nitrogen compounds for domestic combustion

	Charcoal making (g C/kg dry wood)	Charcoal burning (g C/kg dry wood)	Firewood (g C/kg dry wood)
$C(CO_2)$	120 ± 25	170 ± 40	400 ± 70
$C(CO)$	30 ± 8	25 ± 10	30 ± 10
$C(CH_4)$	8 ± 2	0.5 ± 0.4	1.5 ± 0.6
$C(NMHC^a)$	2.0 ± 0.6	0.10 ± 0.03	2.5 ± 1.3
$C(OA^b)$	0.2 ± 0.2		0.3 ± 0.2
$C(Aer^c)$	4.0 ± 1.5		5 ± 3
	(g N/kg dry wood)		(g N/kg dry wood)
$N(NO_x)$	0.02 ± 0.02		0.69 ± 0.18
$N(NH_3)$	0.07 ± 0.10		
$N(N_2O)$	0.02 ± 0.04		

a. Nonmethane hydrocarbons (C_2–C_{10}).
b. Organic acids (formic and acetic).
c. Aerosols.

Table 33.4 Daily per capita consumption of wood and charcoal in western Africa

	Source	Population[a] (1995)	Wood consumption (kg/capita/day)	Charcoal consumption (kg/capita/day)	Yearly wood consumption (T)	Yearly charcoal consumption (T)
Senegal	Sow 1989[b]	8 468 000	1.04	0.173	3 214 453	534 712
Gambia	Estimate[c]	1 071 000	1.04	0.173	406 552	67 628
Guinea-Bissau	Diombera 1993[d]	1 084 000	1.32	0.139	522 271	54 997
Guinea	Girod 1993[e]	6 618 000	2.46	0.037	5 942 302	89 376
Sierra-Leone	Estimate[c]	4 707 000	1.79	0.099	3 075 318	170 087
Liberia	Estimate[c]	2 379 000	1.79	0.099	1 554 320	85 965
Côte d'Ivoire	Plan national energie, 1991	14 342 000	1.12	0.160	5 863 010	837 573
Ghana	Estimate[c]	17 236 000	1.01	0.156	6 354 051	1 006 582
Togo	Alioune Tamchir Thiam 1991	4 266 000	0.91	0.152	1 418 442	237 315
Benin	Girod 1993[d]	5 470 000	1.25	0.008	2 495 688	15 972
Mali	ENDA 1995[f]	9 833 000	1.67	0.017	5 993 705	61 014
Burkina-Faso	Sow 1989[b]	10 439 000	1.48	0.033	5 639 148	125 738
Mauritania	ENDA 1995[f]	2 255 000	0.53	0.086	434 584	70 784
Niger	Sow 1989[b]	9 037 000	0.73	0.007	2 407 909	23 090
Nigeria	Obioh et al. 1994[g]	111 273 000	1.32	0.025	53 678 710	1 009 151
Western Africa population		208 478 000			99 E + 06	4.4 E + 06

a. Estimates and projections with related demographic statistics (World Bank 1994).
b. Data for 1987.
c. Estimate with the consumption of the bordering countries.
d. Data for 1992.
e. Data for 1990.
f. Data for 1989.
g. Data for 1988.

Table 33.5 Emissions from biofuel and savannas burning in western Africa

	Annual burned biomass				
	Savanna[a]	Charcoal production	Firewood		
	290 Tg dry matter	4.5 Tg (16 Tg dry wood)	100 Tg (82 Tg dry wood)		
	Emissions[b] in (Tg)				
	FOS	Charcoal making	Charcoal burning	Firewood	Fuelwood/Savanna (%)
$C(CO_2)$	90.6	1.9	2.7	33	42
$C(CO)$	5.6	0.5	0.4	2.5	63
$C(CH_4)$	0.29	0.15	0.01	0.12	97
$C(NMHC)$	0.66	0.03	2×10^{-3}	0.21	37
$C(OA)$	—	0.01		0.03	
$C(Aer)$	2	0.06		0.4	
$N(NO_x)$	0.14	0.5×10^{-3}		0.06	
$N(NH_3)$	5×10^{-3}	1.5×10^{-3}			
$N(N_2O)$	0.03	0.3×10^{-3}			

a. The yearly emission by the burning of savanna in Western Africa is calculated with an estimation of burnt biomass by Menaut et al. (1991) and emissions factors of the FOS/DECAFE-91 Experiment (Lacaux et al. 1995).
b. Emissions are given as Tg C/year for CO_2, CO, CH_4, nonmethane hydrocarbons (NMHC), aerosols (Aer), and organic acids (OA) and as Tg N/year for NO_2, NH_3, and N_2O.

Africa is calculated using an estimation of burnt biomass by Menaut et al. (1991) and emissions factors of FOS experiment by Lacaux et al. (1993, 1995). Global fuelwood emissions represent 50% of savanna fire emissions for CO_2 and CO and are equal for CH_4, confirming that fuelwood emissions are nonnegligible compared to savanna emissions.

Conclusions

During two field experiments within the DECAFE program, the emissions from fuelwood combustion were studied in Côte d'Ivoire. Carbon compounds (CO_2, CO, CH_4, light hydrocarbons, organic acids) and nitrogen compounds (NO_x, NH_3, N_2O) were measured in the smoke plume released by firewood and charcoal burning as well as by wood carbonization for charcoal making. All experiments were conducted by local people, in order to ascertain their representativity of traditional burnings.

The emission ratios, calculated from these measurements, show that the processes involved in these three combustions are different: Firewood combustion involves, like savanna combustion, mostly a flaming process, whereas charcoal making is basically a smoldering combustion. Charcoal burning is a kind of glowing combustion (Lobert and Warnatz 1993). Depending on the process involved, oxidized compounds (emitted mainly during flaming process) or reduced and

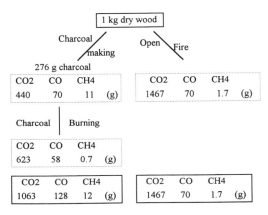

Figure 33.6 Comparison of the emission resulting from the combustion of 1 kg of dry wood, when it burns in an open fire and when it is transformed into charcoal

incompletely oxidized compounds (produced mainly during smoldering combustion) are produced. These compounds are either chemical reactive species or nonreactive species and have then a different impact on atmospheric chemistry. From these emission ratios, a set of emission factors was proposed. These emission factors allow comparison of the emission from 1 kg of dry wood when it burns in an open fire with the emission given out when it is transformed into charcoal (figure 33.6). If 1 kg of dry wood is burned, 1470 g of CO_2 is released, 70 g of CO, and 2 g of CH_4. If 1 kg of dry wood is transformed into charcoal and burned, at the end of the process, 1060 g of CO_2 is emitted, 130 g of CO, and 12 g of CH_4. Charcoal fuel emits less CO_2 but much more reduced compounds (CO and CH_4) than wood.

Combined with a data bank of burned biomass, based on national energy balance, these emission factors give an estimate of the emissions from fuelwood combustion. Global fuelwood emissions represent 50% of savanna fire emissions for CO_2 and CO and are equal for CH_4. These first original campaigns have shown that fuelwood emissions are an important source compared to savannas emissions in western Africa. They will have to be taken into account in future global estimates of biomass burning, especially if we consider that the African population will increase along with their corresponding energy needs. However, this work is still in progress. With the aim to confirm these first results, we hope to be able to develop our study on "per capita" use statistics, especially on cooking practices in Africa. Indeed, the results on firewood emission are based on the assessment that 80% of mass is burnt during the flaming process. Sample survey in households of the main cooking and heating practices is needed to quantify the atmospheric fuelwood emissions in the different regions of Africa.

Acknowledgments

The CHARCOAL/DECAFE-92 and the FIRE-WOOD/DECAFE-94 experiments were supported by the Environment Program of the CNRS, the French Ministry of Environment, and the French Ministry of Cooperation (Campus program). We would like to thank B. Ahoua, M. Assa-Achy, G. Conan, B. Coulibali, M. Koffi, X. Leroux, and J. Dessens for their assistance in the field experiments. We wish also to thank J. L. Tireford and R. Vuattoux from the station of Lamto for their help in the organization of this experiment. Special thanks to the women of N'Zéré and to Elisabeth, who patiently made fire for us.

References

Amous, S. 1992. Systèmes energétiques et changements globaux au Sénégal. Conf. IGBP Africa and Global Change, ENDA Energy, Niamey, Niger, 24 pp.

Boleij, J. S. M., P. Ruigewaard, F. Hoek, H. Thairu, E. Wafula, F. Onyango, and H. De Koning. 1989. Domestic air pollution from biomass burning in Kenya, *Atmos. Environ.*, 23, (8), 1677–1681.

Bonsang, B., C. Boissard, M. F. Le Cloarec, J. Rudolph, and J. P. Lacaux. 1995. Methane, carbon monoxide and light non methane hydrocarbons emissions from African Savanna burnings during the FOS/DECAFE experiment, in an oceanic atmosphere, *J. Atmos. Chem.*, Special issue FOS/DECAFE experiment, Vol. 22, Nos. 1 & 2, 149–162.

Bonsang, B., and G. Lambert. 1985. Non methane hydrocarbons in an oceanic atmosphere, *J. Atmos. Chem.*, 2, 257–271.

Bonsang, B., M. Kanakidou, and C. Boissard. 1994. Contribution of tropical biomass burning to the global budget of hydrocarbons, carbon monoxide and tropospheric ozone, in J. Van Ham et al. (eds.), Non CO_2 greenhouse gases, 261–270, Kluwer, The Netherlands.

Bos, E., M. Y. T. Vu, E. Massiah, and R. A. Bulatao. 1994. *World Population Projections 1994–1995: Estimates and Projections with Related Demographic Statistics*, Johns Hopkins University Press, Baltimore.

Cachier, H., C. Liousse, A. Gaudichet, F. Echalar, T. Kuhlbusch and J. P. Lacaux. 1996. Particulate emission during savannas fires in Kruger National Park, *J. Geophys. Res*, submitted.

Delmas, R., J. P. Lacaux, and D. Brocard. 1995. Determination of biomass burning emission factors: methods and results, *Environmental Monitoring and Assessment*, 38, 181–204.

Diombera, K. 1993. Recolte, transformation et consommation de bois en Guinée Bissau: Filières bois d'oeuvre et bois energie, Economie et sociologie rurale, ENGREF-INRA, Nancy, France.

Dumont, H., and M. Gelus. 1982. *La Valorisation Chimique du Bois*, Masson, Paris, 4–49.

Eimer, P., and C. Ndmana. 1987. Carbonisation: Les ratios de transformation au Burundi, Rapp. Mission CRETEZAIRENIL, Ministère de l'agriculture de Burundi, 36 pp.

ENDA. 1995. L'energie en Afrique, situation énergétique de 34 pays, enda-éditions, Dakar, Sénégal, 467 pp.

Girod, J. 1993. Bénin, Côte d'Ivoire, Guinée, Guinée-Bissau, Guinée Equatoriale, Togo: Aperçu de la situation énergétique, *Liaison Energie-Francophonie*, 19, 4–7.

Hall, D. O. 1991. Biomass energy, *Energy Policy*, 711–737.

Hery, J. L., and H. Cachier. 1993. Caractérisation des émissions particulaires par combustion de biomasse végétale: Exemple d'une savane sèche, Fabrication artisanale de charbon de bois. *Rapport de DEA Chimie de la pollution atmosphérique et physique de l'environnement*, Universite de Paris 7, Paris 12, Chambery, Grenoble 1, Orléans, France.

Kanakidou, M., B. Bonsang, and G. Lambert. 1989. Light hydrocarbons vertical profiles and fluxes in a French rural area, *Atmos. Env.*, 921–927.

Lacaux, J. P., D. Brocard, C. Lacaux, R. Delmas, B. Ahoua, V. Yoboué, and M. Koffi. 1994. Traditional charcoal making: An important source of atmospheric pollution in the African tropics, *Atmos. Res.*, 35, 71–76.

Lacaux, J. P., J. M. Brustet, R. Delmas, J. C. Menaut, L. Abbadie, B. Bonsang, H. Cachier, J. Baudet, M. O. Andreae, and G. Helas. 1995. Biomass burning in the tropical savannas of Ivory Coast: An overview of the field experiment Fire of Savannas (FOS/DECAFE91), *J. Atmos. Chem*, Special issue FOS/DECAFE experiment, Vol. 22, Nos. 1 & 2, 195–216.

Lacaux, J. P., H. Cachier, and R. Delmas. 1993. Biomass Burning in Africa: an overview of its impact on atmospheric chemistry, in *Fire in the Environment: Ecological Climatic and Atmospheric Chemical Importance*, edited by P. J. Crutzen and J. G. Goldammer, Wiley, New York, 159–191.

Leach, G., and M. Gowen. 1987. *Household Energy Handbook : An Interim Guide and Reference Manual*, World Bank Technical Paper 67, Washington, D.C.

Lobert, J. M., and J. Warnatz. 1993. Emissions from the combustion process in vegetation, in *Fire in the Environment: The Ecological, Atmospheric, and Climatic importance of Vegetation Fires*, edited by P. J. Crutzen and J. Goldammer, Wiley, New York, 15–37.

Menaut, J. C., F. Lavenu, P. Loudjani, and A. Podaire. 1991. Biomass burning in West African savannas, in *Global Biomass Burning*, edited by J. Levine, MIT Press, Cambridge, Mass. 133–143.

Obioh, I. B., A. F. Oluwole, F. A. Akeredolu, and O. I. Asuboijo. 1994. *National Inventory of Air Pollutants in Nigeria: Emissions for 1988*, SEEMS Ltd., Lagos, Nigeria.

Pandey, M. R., R. P. Neupane, and A. Gautam. 1987. Domestic smoke pollution and acute respiratory infection in Nepal, in (edited by B. Seifert et al.) vol 3, pp 25–30, Institute for Water, Soil and Air Hygiene, Berlin.

Smith, K. R. 1991. Biomass cookstoves in global perspectives: Energy, health, and global warming, *Report to EWC/ESMAP/UNDP evaluation of improved cookstove programs*, Energy Sector Management Assistance Program, World Bank, Washington, D.C.

Smith, K. R., A. L. Aggarwal, and R. M. Dave. 1983. Air pollution and rural biomass fuels in developing countries: A pilot village study in India and implications for research and policy, *Atmos. Environ*, 17, (11), 2343–2362.

Sow, H., Le bois energie au Sahel, *ACCT-CTA*, 1989.

Vergnet, L. F. 1986. La place du bois dans les preoccupations énergétiques en zone inter-tropicale Africaine, Séminaire sur la politique énergétique, Lome, Togo, 17–21 février 1986, Centre International de formation en politique énergétique (C.I.F.O.P.E), Paris, France, 1986.

Emissions of Trace Gases from Fallow Forests and Woodland Savannas in Zambia

Wei Min Hao, Darold E. Ward, Gerald Olbu, Stephen P. Baker, and James R. Plummer

Biomass burning is an important source of many trace gases in the atmosphere (Crutzen and Andreae 1990). These gases may play important roles in the chemistry of troposphere and stratosphere and in the warming of global climate. Most of biomass burning occurs in the tropics for deforestation, shifting cultivation, savanna fires, fuelwood use, and clearing of agricultural residue (Hao and Liu 1994). Considerable research has been done to quantify the emissions of trace gases from biomass burning in tropical forests and savannas (Greenberg et al. 1984; Greenberg, Zimmerman, and Chatfield 1985; Andreae et al. 1988; Bonsang, Lambert, and Boissard 1991; Delmas et al. 1991; Hao et al. 1995). However, there is little knowledge on emissions from fires used for shifting cultivation. In shifting cultivation, forests are often cleared and cultivated for several years. When nutrients in the soil are depleted, the land is abandoned for about 5 to 15 years to allow regrowth to secondary forests (fallow period). The amount of biomass burned in shifting cultivation is about 2.5 times greater than that in deforestation (Hao and Liu 1994). Approximately two-thirds of the biomass burned due to shifting cultivation occurs in closed forests and one-third in open forests. About 40% of the biomass burned in shifting cultivation taken place in tropical Africa. Shifting cultivation is also called chitemene agriculture in Africa.

We present the results of trace gas emissions from biomass fires in Zambia in August, September, and October 1993. The field experiments were conducted in three different ecosystems: open forests in various fallow periods, open fallow forests being actively cultivated, and semiarid miombo woodland savannas. Concentrations of CO_2, CO, CH_4, C_2–C_6 alkanes, alkenes, and alkynes, and aromatic compounds emitted from these fires are measured. The emission of CH_4 as a function of combustion efficiency is determined. Combustion efficiency is the molar ratio of emitted CO_2 to CO and CO_2. Correlations between emitted compounds and CH_4 are examined. The results are compared with those in other African ecosystems.

Experiment

Field Sites

The meteorological conditions, the composition of biomass, the moisture content in different components of biomass, and the species of trees or understory shrub layer at each site have been described in detail by Shea, Shea, and Kauffman (1995). Table 34.1 summarizes the number of sites, location, elevation, annual precipitation and temperature, and aboveground fuel biomass in the three ecosystems.

The open fallow forests are moist miombo woodland. The trees of 8- to 10-m high are dominated by *Brachystegia longifolia, B. floribunda, Julbernadia paniculata*, and *Pterocarpus angolensis*. The primary species in the shrub layer are coppice *Brachyssegia* spp., *Uapaca sansibarica*, and *U. kirkiana*. Most grass is *Rhynchelytrum repens* in fallow areas and is *Hyparrhenia* spp. and *Pteridium aquilium* in adjacent areas. The aboveground fuel biomass ranges from 2.8 to 15.3 tons ha^{-1} at five sites with fallow periods of 1, 7, 11, 17, and 29 years. In cultivated fallow forests, tree branches were cut, collected, and stacked for several months before the rainy season. The aboveground fuel biomass at these sites is substantially higher than that in open forest during fallow periods because of the large amount of wood debris.

These were two sites at the Chisamba Forest Reserve, which are semiarid miombo woodland savannas. One site has not been disturbed, and the other site has been deforested as a result of intensive charcoal production. In the undisturbed site, the major trees are *Brachystegia boehmii, B. spiciformis*, and *Combretum collinum*, and the sparse shrub layer is dominated by *B. spiciormis, Diploryhyus condylocarpon*, and *Julbernardia globiflora*. Grass is predominantly *Hyparrhenia* spp., *Shizachyrium jeffreysii*, and *Panicum maximum*. In the disturbed site, there are no trees, and the dense scrub layer is mostly *J. globiflora, Hymenocardia acida, Ochna pulchra*, and *B. spiciformis. Hyparrhenia* spp. dominates the grass layer.

Table 34.1 Description of field sites

Type of ecosystem	Open fallow forest		Miombo woodland savannas
	In fallow period	In cultivation	
No. of sites	5	2	2
Location	Kasama, North Province (10°02′S–10°16′S, 31°19′E–31°21′E)	Kasama, North Province (10°24′S, 30°53′E)	Chisama Forest Reserve, Central Province (15°02′16″S, 28°17′22″E; 15°04′19″S, 28°21′15″E)
Elevation	~1200–1400 m	~1200–1400 m	1170 m
Annual precipitation	~1277 mm	~1277 mm	~700–900 mm
Annual temperature (min., max.)	15°C, 30°C	15°C, 30°C	15°C, 26°C
Aboveground fuel biomass (tons ha^{-1}) & fallow period	6.9; 1 year 15.3; 7 years 5.5; 11 years 2.8; 17 years 6.3; 29 years	75.8; 95 years	9.1 years undisturbed 11.5 years disturbed

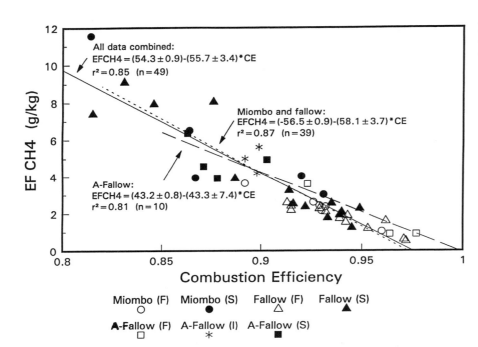

Figure 34.1 Relationship between emission factor of CH_4 (EFCH4) and combustion efficiency. Fallow: open forests during fallow periods; A-Fallow: actively cultivated fallow forests; F: flaming combustion; S: smoldering combustion.

Sampling and Analytical Methods

The sampling and analytical methods have been described in detail by Hao et al. (1995). In summary, smoke samples were drawn from the top of a 5- to 10-m tower through a 1/4-inch Teflon line to electropolished stainless steel canisters that were enclosed in an aluminum box and buried below ground. Carbon dioxide, CO, CH_4, C_2–C_6 alkanes, alkenes and alkynes, and aromatic compounds were analyzed by two Hewlett Packard 5890 Series II gas chromatographs equipped with flame ionization detectors (FID). Carbon dioxide and CO were separated by a packed Carbosphere column (1/8 inch O.D. × 6 feet long) with helium as the carrier gas and were reduced to CH_4 by a methanizer before reaching the FID detector. Methane, C_2–C_3 alkanes, alkenes, and alkynes were separated by a HP-1 precolumn (0.53 mm ID × 4 m long) and a J&W Scientific GS-Q megabore column (0.53 mm ID × 30 m long) with helium as the carrier gas. A DB-1 capillary column (0.25 mm ID × 30 m long) with N_2 carrier gas was used to separate hydrocarbons greater than C_3 after 100 ml of the sample has been cryogenically concentrated at −196°C. The peaks of the chromatograms were integrated by the HP Chemstation

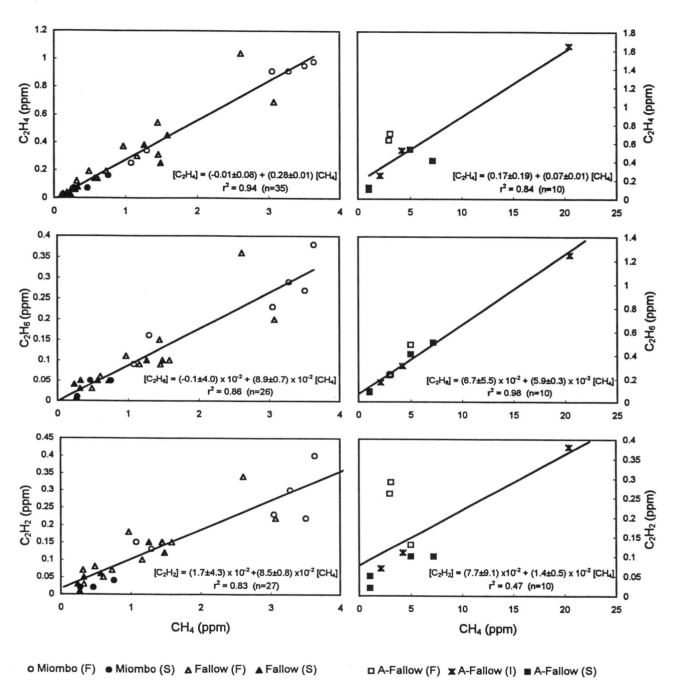

Figure 34.2 Relationship between emitted C_2 compounds and CH_4. Fallow: open forests during fallow periods; A-Fallow: actively cultivated fallow forests; F: flaming combustion; S: smoldering combustion.

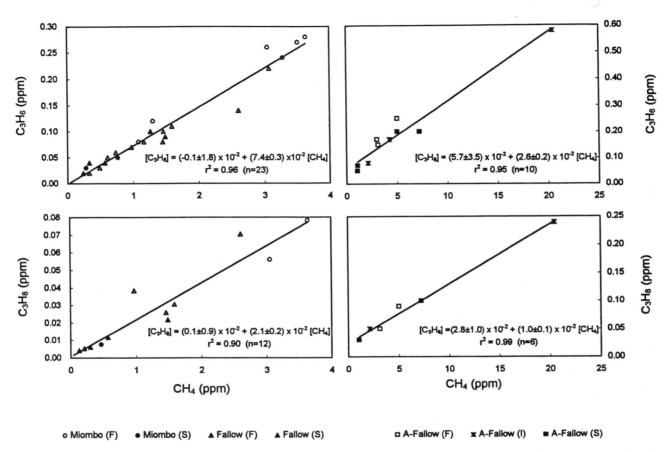

Figure 34.3   Relationship between emitted C_3 compounds and CH_4. Fallow: open forests during fallow periods; A-Fallow: actively cultivated fallow forests; F: flaming combustion; S: smoldering combustion.

software. A standard gas of 2-ppm propane and the response factors of different compounds to the FID were used to calculate the concentrations of different compounds.

Results and Discussion

Emission Factor of CH_4 Versus Combustion Efficiency
The amount of CH_4 emitted per unit weight of dry biomass burned, or the emission factor of CH_4, is strongly and linearly correlated with combustion efficiency (figure 34.1). The less efficient the combustion is, the larger the emission of CH_4 is. There is no difference statistically in the linear relationship for flaming and smoldering combustion. There is also no difference in the slope of the linear correlation for open forests during different fallow periods and semiarid miombo woodland savannas (-58.1 ± 3.7 g kg^{-1}, $r^2 = .87$, $n = 39$) because the aboveground fuel bio-

mass is very similar in both ecosystems. The biomass is predominantly grass and litter. The slope of the linear relationship between the emission of CH_4 and combustion efficiency is -43.3 ± 7.4 g kg^{-1} ($r^2 = .81$, $n = 10$) in cultivated open fallow forests. Most fuel biomass is stacked wood debris at these sites. It is difficult to ascertain whether there is clear difference in the linear relationship between ecosystems dominated by grass and litter and ecosystems mostly wood debris, since there are only limited data available in cultivated fallow forests. The slope is -55.7 ± 3.4 g kg^{-1} ($r^2 = .85$, $n = 49$) for all the measurements in the three ecosystems.

The linear relationships between the emission of CH_4 and combustion efficiency derived from these measurements are considerably different from the previous results (Lobert et al. 1991; Ward et al. 1992; Hao and Ward 1993; Hao et al. 1995). For a given combustion efficiency, the amount of CH_4 emitted from fires in open forests during fallow periods and semiarid

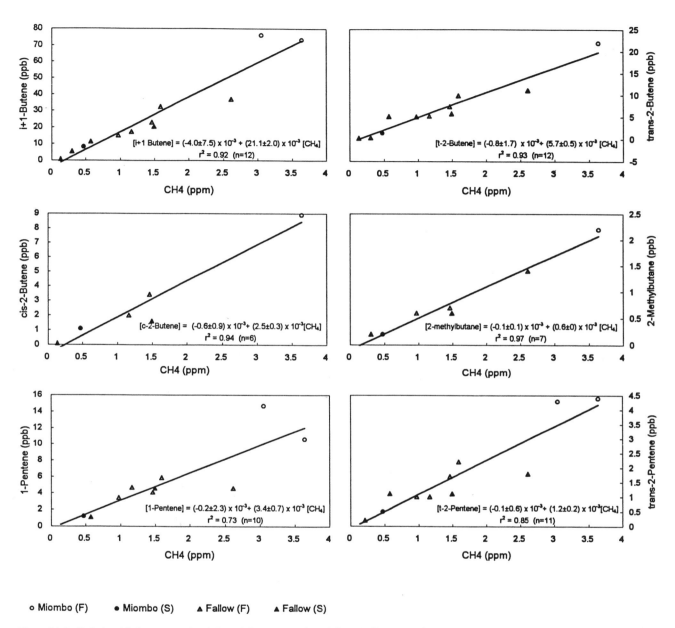

Figure 34.4 Relationship between emitted C_4 and C_5 compounds and CH_4. Fallow: open forests during fallow periods; F: flaming combustion; S: smoldering combustion.

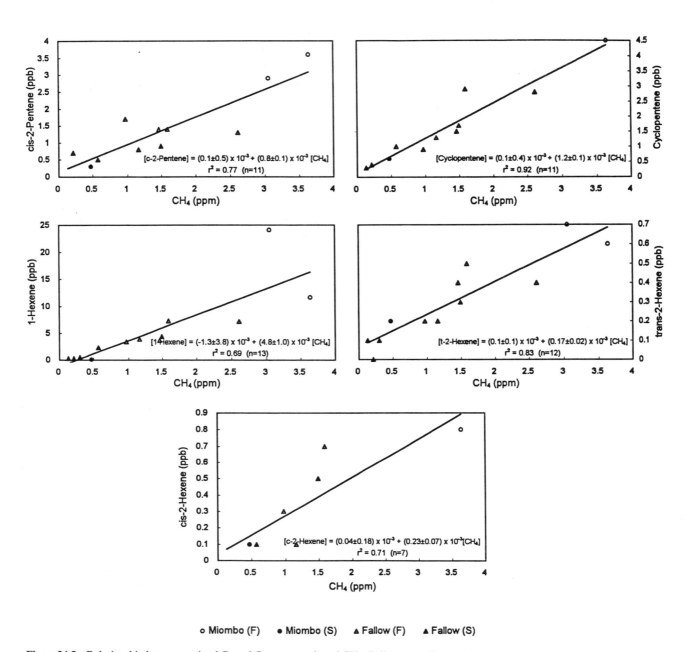

Figure 34.5 Relationship between emitted C_5 and C_6 compounds and CH_4. Fallow: open forests during fallow periods; F: flaming combustion; S: smoldering combustion.

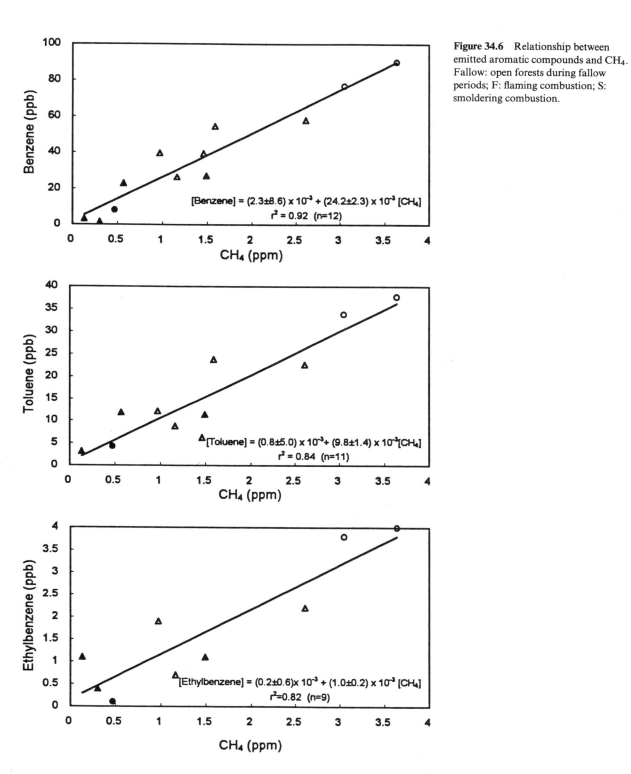

Figure 34.6 Relationship between emitted aromatic compounds and CH₄. Fallow: open forests during fallow periods; F: flaming combustion; S: smoldering combustion.

o Miombo (F) ● Miombo (S) ▵ Fallow (F) ▲ Fallow (S)

Table 34.2 CO and hydrocarbon emissions from African savanna fires

| Compound | This study | | | Hao et al. (1995) |
	Emission ratio to CH_4 (10^{-3})	r^2 relative to CH_4	Number of measurements (n)	Emission ratio to CH_4 (10^{-3})
CO	$15\,370 \pm 550$	94	49	$19\,190 \pm 560$
Methane	1000	1		
Ethene	280 ± 10	94	35	230 ± 10
Ethane	89 ± 7	86	26	74 ± 4
Ethyne	85 ± 8	83	27	73 ± 5
Propene	74 ± 3	96	23	68 ± 3
Propane	21 ± 2	90	12	19 ± 2
i- & 1-Butene	21 ± 2	92	12	17 ± 1
trans-2-Butene	5.7 ± 0.5	93	12	3.4 ± 0.3
cis-2-Butene	2.5 ± 0.3	94	6	2.1 ± 0.1
1-Pentene	3.4 ± 0.7	73	10	2.6 ± 0.2
trans-2-Pentene	1.2 ± 0.2	85	11	
cis-2-Pentene	0.8 ± 0.1	77	11	
2-Methylbutane	0.59 ± 0.05	97	7	0.9 ± 0.1
Cyclopentene	1.2 ± 0.1	92	11	2.2 ± 0.4
1-Hexene	4.8 ± 1.0	69	13	3.5 ± 0.2
trans-2-Hexene	0.17 ± 0.02	83	12	
cis-2-Hexene	0.23 ± 0.07	71	7	
Benzene	24.2 ± 2.3	92	12	20.6 ± 1.3
Toluene	9.8 ± 1.4	84	11	10.2 ± 0.3
Ethylbenzene	1.0 ± 0.2	82	9	

miombo woodland savannas in Zambia is about 1.5 times larger than the CH_4 emissions from fires in other African savanna ecosystems or burning grasslike materials. It is difficult to explain the differences with the available information. Further research is clearly needed, especially in open forests during fallow or cultivation periods.

Approximately 1310×10^{12} g of biomass was burned per year in closed and open fallow forests in the late 1970s (Hao and Liu 1994). Assuming that our measurements from burning stacked wood debris in cultivated fallow forests are representative for fires in fallow forests and that the average combustion efficiency is 0.9, we estimate that about 1.5×10^{14} g of CO is emitted per year. Hence, shifting cultivation in the tropics contributes about 6% of the total source of atmospheric CO. In a similar way, we estimate that 5.5×10^{12} g of CH_4 was emitted per year from fires in shifting cultivation, which contributes about 1% of the atmospheric CH_4.

Nonmethane Hydrocarbons Versus CH_4

Concentrations of most of the emitted C_2–C_6 alkanes, alkenes and alkynes, and aromatic compounds are linearly correlated with emitted CH_4 concentrations for flaming and smoldering combustion in semiarid woodland savannas and in open forests during various fallow periods (figures 34.2 through 34.6). Nonmethane hydrocarbon data for cultivated fallow forest are not available. The coefficients of determination (r^2) are greater than .7 for these compounds. Ethene is the dominant C_2 compound, and propene is the dominant C_3 compound. The major C_4 compounds are i-butene and 1-butene, which cannot be separated by the DB-1 column. The most concentrated aromatic compounds is benzene.

The ratios of emitted nonmethane hydrocarbons to CH_4, r^2, and the number of measurements (n) in figures 34.2 through 34.6 are summarized in table 34.2. The aboveground fuel biomass in both ecosystems is mostly litter and grass. The results are compared with our previous results in moist grassland and woodland savannas in Zambia and in semiarid woodland savannas in South Africa and Zambia in 1992 (Hao et al. 1995). There is no significant difference in the emission ratios of nonmethane hydrocarbons to CH_4 between the present and previous measurements. They strongly indicate that the emission ratios are the same regardless of the phase of combustion, the climatic zone, the meteorological condition, the herbaceous species, and the amount of aboveground fuel biomass. A similar conclusion has been made on the basis of previous measurements (Hao et al. 1995). These results further strengthen our previous conclusion that savanna fires in Africa are an important source of atmospheric ethyne, propene, and benzene.

Conclusions

Emissions of CO_2, CO, and hydrocarbons from fires were quantified in open forests during fallow and cultivation periods and in semiarid woodland savannas in Zambia. The linear relationship between the emission factor of CH_4 and combustion efficiency is the same for flaming and smoldering combustion in ecosystems that are mostly grass and litter. Concentrations of emitted nonmethane hydrocarbons are linearly correlated with emitted CH_4 concentrations for flaming and smoldering combustion in these ecosystems. These results are consistent with our previous measurements in other African savanna ecosystems. African savanna fires are an important source of atmospheric CO, ethyne, propene, and benzene. The amount of CO produced from fires due to shifting cultivation in the tropics could contribute about 6% of the atmospheric CO.

References

Andreae, M. O. et al. 1988. Biomass-burning emissions and associated haze layers over Amazonia, *J. Geophys. Res.*, 93, 1509–1527.

Bonsang, B., G. Lambert, and C. C. Boissard. 1991. Light hydrocarbons emissions from African savanna burnings, in *Global Biomass Burning: Atmospheric, Climatic, and Biospheric Implications*, edited by J. S. Levine, pp. 155–161, MIT Press, Cambridge, Mass.

Crutzen, P. J., and M. O. Andreae. 1990. Biomass burning in the tropics: Impact on atmospheric chemistry and biogeochemical cycles, *Science*, 250, 1669–1678.

Delmas, R. A., A. Marenco, J. P. Tathy, B. Cros, and J. G. R. Baudet. 1991. Sources and sinks of methane in the African savanna: CH_4 emissions from biomass burning, *J. Geophys. Res.*, 96, 7287–7299.

Greenberg, J. P., P. R. Zimmerman, and R. B. Chatfield. 1985. Hydrocarbons and carbon monoxide in African savanna air, *Geophys. Res. Lett.*, 12, 113–116.

Greenberg, J. P., P. R. Zimmerman, L. Heidt, and W. Pollock. 1984. Hydrocarbon and carbon monoxide emissions from biomass burning in Brazil, *J. Geophys. Res.*, 89, 1350–1354.

Hao, W. M., and M.-H. Liu. 1994. Spatial distribution of tropical biomass burning in 1980 with 5° × 5° resolution, *Global Biogeochem. Cycles*, 8, 495–503.

Hao, W. M., and D. E. Ward. 1993. Methane production from global biomass burning, *J. Geophys. Res.*, 98, 20 657–20 661.

Hao, W. M., D. E. Ward, G. Olbu, and S. P. Baker. 1995. Emissions of CO_2, CO, and hydrocarbons from fires in diverse African savanna ecosystems, *J. Geophys. Res.*, in press.

Lobert, J. M., D. H. Scharffe, W. M. Hao, T. A. Kuhlbusch, R. Seuwen, P. Warneck, and P. J. Crutzen. 1991. Experimental evaluation of biomass burning emissions: Nitrogen and carbon containing compounds, in *Global Biomass Burning: Atmospheric, Climatic, and Biospheric Implications*, edited by J. S. Levine, 289–304, MIT Press, Cambridge, Mass.

Shea, R. W., B. W. Shea, and J. B. Kauffman. 28 April 1995. Fire effects on African savanna ecosystems, *Final Report to USDA Forest Forest*.

Ward, D. E., R. A. Susott, J. B. Kauffman, R. E. Babbitt, D. L. Cummings, B. Dias, B. N. Holben, Y. J. Kaufman, R. A. Rasmussen, and A. W. Setzer. 1992. Smoke and fire characteristics for cerrado and deforestation burns in Brazil: BASE-B experiment, *J. Geophys. Res.*, 97, 14 601–14 619.

Biomass Burning, Biogenic Soil Emissions, and the Global Nitrogen Budget

Joel S. Levine, Wesley R. Cofer III, Donald R. Cahoon, Jr., Edward L. Winstead,
Daniel I. Sebacher, Mary C. Scholes, Dirk A. B. Parsons, and Robert J. Scholes

At more than 78% by volume, molecular nitrogen (N_2) is the overwhelming constituent of the atmosphere. The amount of molecular nitrogen in the atmosphere is controlled by processes that "fix" or remove nitrogen from the atmosphere, primarily in the forms of nitrate (NO_3^-) and ammonium (NH_4^+), and by processes that return nitrogen to the atmosphere in the form of N_2, nitrous oxide (N_2O), nitric oxide (NO), and ammonia (NH_3). Human activities are perturbing the global reservoirs of nitrogen. The reservoirs of nitrogen compounds include the atmosphere, the biosphere, the soil, and the oceans. The amount of nitrogen in these reservoirs is summarized in table 35.1. As already noted, molecular nitrogen is the overwhelming form of nitrogen in the atmosphere, with smaller amounts of nitrous oxide, nitric oxide, nitrogen dioxide (NO_2), and ammonia. Compared to the atmosphere, the terrestrial biosphere, the soil, and the oceans contain relatively small quantities of nitrogen.

The fixation of nitrogen results from biological processes, from the industrial production of nitrogen fertilizer (forming NH_4^+), and from atmospheric lightning and high temperature combustion (forming NO and eventually NO_3^- through atmospheric chemical reactions). The magnitudes of the various nitrogen fixation processes are summarized in table 35.2. The two largest nitrogen fixation processes, biological fixation on land and the production of nitrogen fertilizer, are both impacted by human activities. As shown in table 35.3, biological fixation on land is dominated by the forests, grasslands, and agricultural lands, which are all perturbed to some extent by human activity. Humans destroy forests and grasslands by burning for land clearing and land use change, as well as control the amount and type of agricultural production, which impacts nitrogen fixation. Human fixation of atmospheric nitrogen in the production of fertilizer has increased by more than a factor of 30 over the last 54 years from 3 Tg N yr^{-1} in 1940 to over 90 Tg N yr^{-1} in 1994 (table 35.4) (1 Tg $= 10^{12}$ g $= 10^6$ metric tons). The projections are for increased growth of fertilizer production to feed the ever increasing population of our planet. The return of nitrogen from the biosphere to the atmosphere results from fossil fuel combustion (producing NO), biomass burning (producing NO and smaller amounts of N_2O), and the microbial metabolic processes of denitrification (producing N_2 and some N_2O) and nitrification (producing NO). The magnitudes of the microbial processes of nitrogen production are summarized in table 35.5.

Two sources not included in global budgets of nitrogen are both related to fire: (1) nitrogen compounds (N_2, NO, and N_2O) produced by the combustion of biomass burning and instantaneously returned to the atmosphere and (2) fire-enhanced biogenic soil emissions of N_2, N_2O, and NO emitted to the atmosphere over time periods that range from days to weeks following burning. The instantaneous combustion of biomass material returns significant quantities of N_2, N_2O, NO, and NH_3 to the atmosphere. Over the last few years, biomass burning field experiments have quantified both the type and amount of the nitrogen gases returned to the atmosphere during burning. Estimates indicate that biomass burning may return 20 to 36 Tg N yr^{-1} to the atmosphere, but it may be as high as 27 to 63 Tg N yr^{-1} (table 35.6).

This chapter deals with the second impact of burning, the slower, longer-term emission into the atmosphere of NO and N_2O from the soil following burning, which is less precisely understood. The measurements reported in this chapter were obtained in the savannas of southern Africa during the Southern African Fire-Atmosphere Research Initiative (SAFARI-92). The savannas of southern Africa are burned several times each year and, in terms of total mass consumed by fire, are the fire center of the planet. Burning in the tropical savannas consumes about 3690 Tg dm yr^{-1} (dm:dry matter), compared to the total mass consumed by all burning, estimated to be about 8700 Tg dm yr^{-1} (Andreae 1991).

The oxides of nitrogen, NO and N_2O, are key gases involved in the chemistry of the troposphere and stratosphere and in global climate change. NO is an important species involved in the photochemical pro-

Table 35.1 Global nitrogen reservoirs (Tg N) (Soderlund and Rosswall 1982; Schlesinger 1991)

N_2 in atmosphere	3.9 E(9)
Fixed N (NO, NO_2, NH_3)	1.4 E(3)
N_2O	1.4 E(3)
Terrestrial biosphere	3.5 E(3)
Soil	9.5 E(4)
N_2 in ocean	2.2 E(7)
N_2O in ocean	2.0 E(4)

Table 35.2 Nitrogen fixation and other nitrogen transfers (Tg N yr^{-1}) (Soderlund and Rosswall 1982; Schlesinger 1991)

Nitrogen fixation	
Biological	
Land	140 (44–200)
Oceans	30
Lightning	10–30
Combustion	20
Fertilizer production (1994)	94
Nitrogen transfer	
Washout over oceans (NO_3, NH_4)	50
River input to oceans	36

Table 35.3 Nitrogen fixation/land (Tg N yr^{-1}) (Soderlund and Rosswall 1982)

Agriculture	
Legumes	35
Rice	4
Other crops	5
Grasslands	45
Forests	40
Other	10
Total	139

Table 35.4 Estimated world use of industrially fixed nitrogen fertilizer (Tg N yr^{-1}) (Delwiche 1981; Levine 1992)

1940	3
1950	5
1960	10
1970	26
1980	55
1990	75
1994	94

Table 35.5 Surface sources of nitrogen to atmosphere (Tg N yr^{-1}) (Schlesinger 1991)

Denitrification/land	130 (13–233)
Denitrification/oceans	110
Biogenic NO_x production/soil	11

Components not included in global nitrogen budget:
1. Biomass burning N combustion products.
2. Fire-enhanced biogenic soil emissions of nitrogen species (N_2, N_2O, NO).

Table 35.6 Biomass burning nitrogen contribution to atmosphere (Tg N yr^{-1})

1. Combustion of biomass matter
   N_2O emissions = 0.8(Andreae 1991)
   NO_x emissions = 8.5 (Andreae 1991)
   NH_3 emissions = 5.3 (Andreae 1991)
   N_2 emissions = 5.6–21.3 (if 22% of fuel N is released in form of N_2 [Lobert et al. 1991])
   = 12.7–48.4 (if 50% of fuel N is released in form of N_2 [Lobert et al. 1991])

   Total biomass burning combustion nitrogen emissions
   = 20–36 (27.3–63.0)

2. Fire-enhanced biogenic soil emission of nitrogen gases (see text)

duction of ozone (O_3) in the troposphere; in the chemical production of nitric acid (HNO_3), the fastest growing component of acidic precipitation; and in the chemistry of the hydroxyl radical (OH), the major chemical scavenger in the troposphere (Williams, Hutchinson, and Fehsenfeld 1992). NO_2 leads to the chemical destruction of ozone in the stratosphere and is an important greenhouse gas with a global warming potential more than 200 times that of carbon dioxide on a per molecule basis (Williams, Hutchinson, and Fehsenfeld 1992). Both NO and N_2O appear to be increasing with time (Houghton, Jenkins, and Ephraums 1990).

At the present time, there are major uncertainties concerning the global sources of N_2O (Houghton, Jenkins, and Ephraums 1990; Levine 1992). Recent studies indicate that two previously believed significant sources of N_2O are only minor sources at most—the combustion of fossil fuels (Muzio and Kramlich 1988) and the burning of biomass, that is, trees, vegetation, grass, and agricultural waste (Cofer et al. 1991; Hao et al. 1991). These two discoveries indicate that the source for more than 30% of the global production of N_2O is as yet not known (Houghton, Jenkins, and Ephraums 1990).

It is generally agreed that microbial soil emissions are significant global sources for both N_2O and NO

(Williams, Hutchinson, and Fehsenfeld 1992). It has been found that biogenic soil emissions of NO and N_2O are enhanced following natural rainfall or artificial irrigation to simulate rainfall (Anderson et al. 1988; Levine et al. 1988; Johansson, Rodhe, and Sanhueza 1988; Hao et al. 1988; Firestone and Davidson 1989; Davidson 1992). Recently, it was discovered that burning also enhances the biogenic soil emissions of NO and N_2O (Anderson et al. 1988; Levine et al. 1988; Johansson, Rodhe, and Sanhueza 1988; Levine et al. 1990). It has been hypothesized that increased soil concentrations of ammonium, the substrate for the microbiological process of nitrification, enhance biogenic soil emissions of NO and N_2O (Levine et al. 1988). Ammonium is a major component of the burn ash (Levine et al. 1988). Recently, the postfire enhancement of biogenic soil emissions of NO has been included in an empirical global model of biogenic soil emissions of NO (Yienger and Levy 1995).

This chapter deals with microbial soil emissions of NO and N_2O from the savannas of South Africa and the effect of natural rainfall, artificial wetting, and fire on these emissions. The vast and extensive savannas of Africa may be an important site for biogenic soil emissions of NO and N_2O. However, there are no measurements of biogenic soil emissions of NO and N_2O from the African savanna in the literature. Fires are a regular feature of the African savanna. A study based on nighttime satellite measurements obtained with the Defense Meteorological Satellite Program (DMSP) satellites indicates that burning in the African savannas is very frequent and very widespread (Cahoon et al. 1992). Studies indicate that the burning of tropical savannas consumes about 3690 Tg dm yr^{-1} (Hao, Liu, and Crutzen 1990; Andreae 1991). For comparison, burning consumes the following mass of dry matter per year in the indicated processes and/or ecosystem: agricultural waste, 2020 Tg dm yr^{-1}; tropical forests, 1260 Tg dm yr^{-1}; and temperate and boreal forests, 280 Tg dm yr^{-1} (Andreae 1991). Two-thirds of the world's savanna burning occurs in Africa (2430 Tg dm yr^{-1} out of a total of 3690 Tg dm yr) (Hao, Liu, and Crutzen 1990; Andreae 1991). The SAFARI experiment provided a unique opportunity to obtain the first measurements of biogenic soil emissions of NO and N_2O from the African savanna and to assess the response of these emissions to both wetting and burning. While there are no measurements of biogenic soil emissions of NO and N_2O from the African savanna in the literature, measurements have been reported for the savannas in Venezuela. These studies have considered biogenic NO emissions (Johansson

and Sanhueza 1988; Johansson, Rodhe, and Sanhueza 1988; Sanhueza et al. 1990) and biogenic N_2O emissions (Hao et al. 1988; Sanhueza et al. 1990) from savannas in Venezuela.

Measurement Site

The microbial soil emissions of NO and N_2O reported in this chapter were obtained in the Kruger National Park, South Africa, in September 1992, as part of SAFARI-92, a component of the Southern Tropical Atlantic Research Experiment (STARE). STARE is a major research activity of the International Global Atmospheric Chemistry (IGAC) Project, International Geosphere-Biosphere Program (IGBP). The Kruger National Park lies in the northeastern section of the Transvaal Lowveld (30°50′–32°00′S; 22°25′–25°32′E). The study area is located in the southwestern corner of the park. The climate is characteristic of a semiarid savanna—a hot, wet season lasting from October to April and the remainder of the year warm and dry with an absence of frost. Daytime temperatures around 35°C are common during the summer. Winter temperatures are moderate. The annual rainfall varies from 600 to 1000 mm with an average of 740 mm (Venter and Gertenbach 1986). However, southern Africa was experiencing a period of extreme drought during the SAFARI-92 experiment.

The burned and unburned control sites (both Shabeni and Kambeni) are situated on catenal ridges in the Lowveld Sour Bushveld of Pretoriuskop (Figure 35.1). The soil is red to yellow-brown in color and varies from a sand to a sandy loam (6 to 15% clay) and is deeply leached. The soils are classified as Hutton and Clovelly forms with the Portsmouth/Moriah and Paleisheuvel/Denhere, respectively, as the dominant soil series. This soil nomenclature follows the South African Soil Classification of Macvicar et al. (1977). In the Food and Agricultural Organization (FAO) system, these soils would be classified as luvisols; in the U.S. Department of Agriculture (USDA) Soil Taxonomy classification, these soils would be alfisols. Soil characteristics, including soil composition, bulk density, pH, base saturation, cation exchange capacity (CEC), total nitrogen, and organic carbon were measured by Parsons et al. (1996) and are summarized in Table 35.7.

The vegetation of these sites is characteristic of open tree savanna, with the woody component dominated by *Dichrostachys cinerea* and *Termimalia sericea*. The herb layer is dominated by sourveld grass species such as *Hyperthelia dissolute* and *Elionurus muticus*.

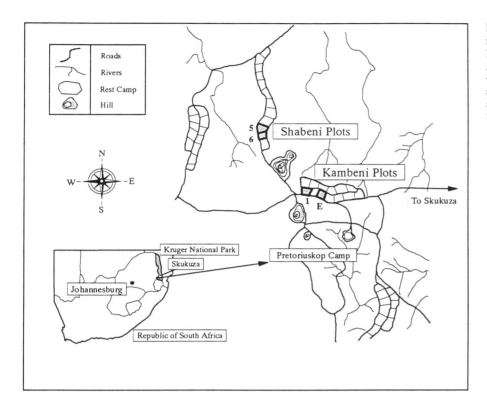

Figure 35.1 Experimental fire plots in the Pretorioskop area of the Kruger National Park, Republic of South Africa. Measurement sites are located within the Shabeni fire plots (#5 and #6) and the Kambeni fire plots (#1 and #E). (Prepared by William L. Miller, USEPA.)

Table 35.7 Soil characteristics for the SAFARI Shabeni and Kambeni sites[a]

	Shabeni sites	Kambeni sites
Sand (%) (Diameter = 0.074 to 2.0 mm)	84.8[b] (1.5)[c]	82.5 (0.7)
Silt (%) (Diameter = 0.005 to 0.074 mm)	5.0 (1.4)	6.5 (0.7)
Clay (%) (Diameter < 0.05 mm)	10.2 (1.7)	11.0 (1.4)
Bulk density (kg m^{-3})	1.6	1.6
pH (H$_2$O)	5.3 (1.5)	6.1 (0.5)
Base saturation (%)	63.0 (11.9)	82.5 (43.1)
Cation exchange capacity (CEC) (c mol(+) kg^{-1})	33.9 (7.5)	50.2 (4.5)
Total N (μg N g dry soil^{-1})	450	900
Organic carbon (%)	1.9	2.5

a. Based on measurements described in detail in Parsons et al. (1996).
b. Mean value.
c. Standard deviation.

The measurements took place right before the beginning of the South African rainy season, although it did rain lightly (0.8 mm) on 17 September in the middle of the measurement program. Measurements were obtained at two locations—Shabeni (last burned 2 years age) and Kambeni (last burned 35 years ago) sites (see Figure 35.1). Shabeni is an "open" or grass savanna site; Khambeni is a "closed" or wooded savanna site. During SAFARI-92, Shabeni #6 and Kambeni #1 were the unburned control sites. Shabeni #5 was burned on 15 September and Kambeni Extra was burned on 22 September. Measurements were obtained at both dry and wet sites at Shabeni and Kambeni. The dry and wet sites were only several feet away from each other. The close proximity of the dry and wet sites was to minimize possible spatial variations in soil parameters that may impact emissions of NO and N$_2$O. The wet sites were artificially irrigated by the addition of 600 ml of deionized water (equivalent to 1.5 mm of rainfall) prior to each measurement day. As already indicated, the first natural rainfall occurred on 17 September. During this rainfall event, Kambeni Extra was shielded from the rain using a plastic ground cover and remained dry throughout the entire experiment. However, Shabeni #5 and the control sites on both Shabeni and Kambeni were naturally wetted by

Table 35.8 Summary of NO flux measurements: uncorrected fluxes, soil temperature, and temperature-corrected fluxes

	Dry site			Wetted site[b]		
Date[a]	Uncorrected NO flux (ng N m^{-2} s^{-1})	Soil temp (°C)	Corrected NO flux (ng N m^{-2} s^{-1})	Uncorrected NO flux (ng N m^{-2} sec^{-1})	Soil temp (°C)	Corrected NO flux (ng N m^{-2} s^{-1})
Shabeni #6 (control site) (Figure 35.2/uncorrected NO flux)						
11	**0.4** (0.1)[c]	34.3	0.1	**34.0** (0.8)	29.7	17.5
13	**2.6** (0.2)	27.9	1.5	**13.3** (0.8)	29.5	7.0
14	**3.6** (0.3)	34.4	1.4	**5.7** (0.9)	31.9	2.5
16	**1.4** (0.3)	40.5	0.4	**14.1** (0.9)	31.5	6.3
			First rain			
17	**39.1** (0.6)	23.2	31.2	**8.2** (0.3)	22.4	7.0
18	**2.4** (0.2)	32.5	1.0	**5.7** (0.7)	27.7	3.3
21	BDL[d]	39.0	BDL	**10.3** (0.9)	32.0	4.4
22	**0.6** (0.1)	23.0	0.4	—	—	—
25	**2.4** (0.4)	39.4	0.6	**28.1** (0.9)	31.8	12.4
26	**1.3** (0.3)	35.7	0.4	**24.3** (0.8)	30.0	12.2
Shabeni #5 (burn site) (Figure 35.3/uncorrected NO flux)						
11	**0.6** (0.1)	35.3	0.2	**21.1** (0.6)	35.3	7.4
13	**2.4** (0.3)	23.5	1.9	**7.0** (0.3)	25.4	4.8
14	**3.2** (0.4)	28.6	1.8	**4.7** (0.3)	25.1	3.3
15			Fire			
16	**13.3** (0.7)	33.0	5.5	**61.8** (1.0)	27.7	35.9
17			First rain			
17	**52.1** (8.0)	19.7	53.6	**26.7** (3.7)	19.2	29.6
18	**9.8** (0.4)	25.3	6.8	**6.8** (0.2)	23.4	5.3
21	**0.6** (0.2)	42.5	0.1	**29.3** (1.3)	34.0	11.1
25	**4.6** (0.4)	39.9	1.2	**20.1** (1.0)	30.3	9.7
26	**4.4** (0.3)	37.2	1.3	**19.1** (1.0)	30.5	9.1
Kambeni #1 (control site) (Figure 35.4/uncorrected NO flux)						
13	**1.6** (0.3)	24.6	1.1	**9.5** (0.5)	22.8	7.8
17			First rain			
19	**3.9** (0.6)	25.4	2.6	**9.8** (0.5)	27.0	6.0
21	**3.9** (0.5)	33.0	1.6	**7.4** (0.8)	30.3	3.5
23	**1.4** (0.2)	19.6	1.4	**16.1** (0.4)	18.0	18.8
24	**2.9** (0.4)	31.0	1.4	—	—	—
26	**2.2** (0.4)	29.7	1.1	**9.8** (0.6)	26.9	6.0
Kambeni Extra (burn site) (Figure 35.5/uncorrected NO flux)						
13	**1.3** (0.2)	26.8	0.8	**23.4** (0.9)	25.3	16.2
19	**2.0** (0.3)	31.5	0.9	**16.8** (1.0)	27.7	9.7
21	**6.2** (0.5)	26.5	3.9	**12.7** (0.8)	23.5	9.9
22			Fire			
23	**15.2** (1.5)	17.7	18.2	**61.5** (9.2)	17.2	76.9
24	**15.0** (0.7)	24.6	11.0	**20.1** (1.9)	22.0	17.6
26	**2.4** (0.4)	28.5	1.3	**17.9** (0.8)	25.0	12.8

Note: Flux values represent the mean of three flux measurements obtained within a 45-minute time interval; values of uncorrected (uncorrected for soil temperature) NO flux are plotted in Figures 35.2 through 35.5.
a. September 1992.
b. Wetted site measurements were artificially irrigated with 600 ml of distilled water prior to each measurement period. The area of the flux chamber to which the water was applied was 0.4 m^{-2}.
c. The number in parentheses following the mean NO flux values represents the standard error of the mean $(s/\sqrt{n})$, where s is the standard deviation and n is the number of measurements (Arkin and Colton 1953). For all values, a minimum of three flux measurements were obtained.
d. BDL = below detection limit.

the 17 September rainfall. The 17 September rainfall event was a unique opportunity to assess how unburned and burned, dry and artificially wetted sites reacted to natural rainfall.

At each of the four measurement sites (Shabeni #6, Shabeni #5, Kambeni #1, and Kambeni Extra), three collars were placed in the soil. The collars were removed at the burn sites prior to burning and then replaced in their original location after the fire. At each of the four sites, collar 1 was for dry or natural soil conditions, collar 2 was for wetted or naturally irrigated conditions (600 ml of distilled water, which is equivalent to about 1.5 mm of rainfall, a representative value), and collar 3 was for supersaturated wetting (10 l of distilled water was applied to this site). NO flux measurements obtained at collar 3 are not reported in this chapter. Hence, at each of the 12 measurement sites, NO flux measurements were obtained at a single collar site. Three NO flux measurements obtained over about a 40-minute time period were averaged for the mean value and the standard error of the mean given in table 35.8.

During the SAFARI-92 experiment, biogenic soil emissions of carbon dioxide, carbon monoxide, and methane were made at the same sites and times as the NO and N_2O measurements (Zepp et al. 1996). This provided simultaneous measurements of biogenic soil emissions of the nitrogen and carbon gases from the same sites. Following SAFARI-92, measurements of biogenic soil emissions of NO continued from October to December at both the Shabeni and Kambeni sites, as well as at a third site. These post-SAFARI measurements are reported in Parsons et al. (1996).

Instrumentation

Nitric oxide and nitrous oxide fluxes were determined using a closed chamber flux technique (Anderson and Levine 1987). At each of the plots selected for this study, rectangular aluminum collars were inserted into the soil to a depth of at least 3 cm. The collars covered an area measuring about 0.4 m^2. The top edges of the collar formed a V-shaped groove into which the flux box could be set. Water in the groove provided a seal for the flux box. The inner surfaces of both the collar and the flux box were coated with Teflon. The flux box was insulated with a closed-cell foam covered with highly reflective aluminized Mylar. The combined volume of the flux box plus collar varied from about 170 to 190 l, depending on the depth to which the collar was inserted into the soil. A muffin fan inside the box stirred the air at the rate of 3 m^3 min^{-1} at zero

static back pressure. A 1.5-cm vent at the top of the box prevented development of a pressure differential when air was pumped out of the box for analysis of NO. Teflon tubing mounted in a bulkhead fitting and extending 20 cm into the box was used for sampling air for NO and NO_x analysis. Fittings with silicone rubber septa were used for removal of chamber air by syringe for N_2O analysis. Another fitting allowed insertion of probes used to measure the temperatures of both soil and air within the box. Care was taken to minimize disturbance to soil and vegetation when collars were inserted. Collars were installed at least 1 week prior to beginning the series of measurements and remained undisturbed for the duration of the study. The flux box was left in place for a maximum of 15 minutes for determination of the NO flux and 30 minutes for the N_2O flux.

The measurements of NO, nitrogen dioxide (NO_2), and $NO_x(NO + NO_2)$ were made with a modified LMA-3 Luminox NO_2 monitor manufactured by Scintrex/Unisearch of Concord, Ontario, Canada. The Luminox monitor operates by detecting the chemiluminescence when NO_2 interacts with a surface wetted with a specially formulated luminol solution. The luminol solution is oxidized, producing chemiluminescence at a wavelength of 425 nm, which is measured by a photomultiplier tube. The signal from the photomultiplier tube is directly proportional to the mixing ratio of NO_2. Since the NO_2/luminol reaction is temperature sensitive, the NO_2 monitor is equipped for temperature compensation. The signal for a given NO_2 mixing ratio is constant over the temperature range of 10 to 40°C. For the measurement of NO and NO_x, a chromium trioxide converter system was developed for the conversion of NO to NO_2 prior to introduction into the Luminox detector. The NO-to-NO_2 converter consists of a Teflon-lined stainless steel tube (6.0 cm × 1.2 cm) packed with a Chromosorb-support material coated with chromium trioxide. The converter material is prepared by soaking Chromosorb P, 30 to 60 mesh (manufactured by Johns Manville Corporation), in a 17% chromium trioxide aqueous solution, decanting off excess solution, drying in an oven at 40°C, and then exposing the material to ambient air conditions for 24 hours as described by Levaggi et al. (1974). The conversion of NO to NO_2 is nearly 100% efficient provided that the relative humidity of the sample airstream is less than 25%. This is accomplished by pumping the sample airstream at approximately 2.2 l min^{-1} through a Teflon filter and then through a 1-m-long Nafion tube (Type 815, Dupont perfluorinated polymer, 1.0 mm OD × 0.875 mm ID,

Perma Pure, Incorporated), packed in Drierite. The Nafion dryer lowers the water content of the air to an acceptable level without the loss of NO or NO_2. The dried air is directed either through the chromium trioxide converter, where NO is converted to NO_2 for the measurement of NO_x, or through an unpacked column for the measurement of NO_2. Sample air from the NO-to-NO_2 converter system is pumped to the Luminox detector at a flow rate of approximately 1.5 l min^{-1}, Excess air is vented to prevent pressurization of the LMA-3 detector. NO is calculated as the difference between the converted and unconverted signals. The detector was calibrated for NO_2 using a permeation tube system approved by the National Bureau of Standards (NBS) (now the National Institute of Standards and Technology [NIST]). The detector was calibrated for NO_2 using a field calibration master gravimetric standard certified by Scott Specialty Gases (Plumsteadville, Pennsylvania) at the $\pm 1\%$ level. A calibration curve was obtained by mass flow dilution of the standard with ultra zero ambient monitoring air. The field calibration source was checked frequently against a NBS standard reference material (SRM). The detector has a detection limit of at least 20 parts per trillion by volume (pptv = 10^{-12}) at a signal-to-noise ratio of 2. The minimum detectable flux with this instrument is 13 pg (10^{-12} g) N m^{-2} s^{-1} of NO over a 12-minute interval at 298°K.

In a ground-based intercomparison, Fehsenfeld et al. (1990) tested the photolysis/chemiluminescence and luminol techniques for NO_2 detection. The results indicated that peroxyacetyl nitrate (PAN) and ozone interfered with the NO_2 measurements made using the luminol technique. However, measurements that were obtained in our flux chambers indicate that neither PAN nor ozone interference is a problem in our measurements (Anderson and Levine 1987).

The N_2O measurements were made using a Shimadzu model GC Mini 2 gas chromatograph/electron capture detector with a 1-ml sample loop and stainless steel Poropack Q column (4.7 m × 0.3 cm). The detector temperature was 340°C; oven temperature was 80°C. The carrier gas, 5% methane in argon, was supplied at a flow rate of 30 ml min^{-1}. The precision of this instrument was 1% (standard deviation/mean). The minimum flux of N_2O that could be detected with this system over a 30-minute period at 298°K was 2 ng (10^{-9} g) N m^{-2} s^{-1} of N_2O.

NO Measurements

The measurements of NO are summarized in table 35.8 and in figures 35.2 through 35.5: figure 35.2 for the Shabeni #6 unburned control site, figure 35.3 for the Shabeni #5 burned site, figure 35.4 for the Kambeni #1 unburned control site, and figure 35.5 for the Kambeni Extra burned site. All flux measurements are given in units of nanograms (1 ng = 10^{-9} g) of nitrogen (N) meter^{-1} second^{-1} (ng N m^{-2} s^{-1}). Table 35.8 lists the following information: date of measurement, NO flux not corrected for soil temperature effects, the soil temperature at the time of the measurement, and the temperature-corrected NO flux. The temperature-corrected NO fluxes have been corrected to a temperature of 20°C based on the Q-10 temperature correction, that is, for a soil temperature of 30°C, which is 50% greater than 20°C, the flux is reduced by 50% with respect to the flux at 20°C (Anderson and Levine 1986; Levine et al. 1988; Anderson et al. 1988; Williams, Hutchinson, and Fehsenfeld 1992; Yienger and Levy 1995). However, the values of NO fluxes in figures 35.2 to 35.5 are for the NO fluxes not corrected for soil temperature, which are given in table 35.8 in bold type.

During the course of the measurements, it became apparent that the amount of water artificially added or the amount of natural precipitation falling on the measurement sites had a very significant impact on the NO flux from the soil. Unfortunately, we were not equipped to measure the soil moisture or water pore space before, during, and after the application of water to the measurement sites simultaneously with the NO flux measurements. Our only information on the amount of water applied to the sites is the amount of precipitable water applied or the precipitable rainfall. While this not an index of soil moisture, it is an index of the amount of water added to the measurement sites. Future measurements of biogenic soil emissions of NO from the southern African savanna should include simultaneous measurements of soil moisture.

In the Shabeni #6 dry control site (figure 35.2), the NO flux increased from below 3.6 to 39.1 ng N m^{-2} s^{-1} following the first rain on 17 September. The day after the rain, the NO flux dropped to the prerain level of below 3.6 ng N m^{-2} s^{-1}. In contrast, the wetted site consistently exhibited higher NO fluxes, except the day of the first rain. Before and after the rain, the NO fluxes from the wetted site ranged from 5.7 to 34 ng N m^{-2} s^{-1} and exceeded the NO flux from the dry site by factors ranging from 1.5 to 85. However, the day that it rained (17 September), the NO flux from the dry site exceeded the flux from the wet site by a factor of 5.

In the Shabeni #5 dry burn site (figure 35.3), the NO flux increased from below 3.2 ng N m^{-2} s^{-1} prior to the 15 September fire, to 13.3 ng N m^{-2} s^{-1} the day following the fire, and then increased to

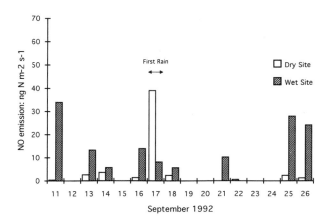

Figure 35.2 NO emissions from dry and wetted sites, Shabeni #6 (unburned control site), before and after first rain. (All NO measurements have not been corrected for soil temperature. Based on data in table 35.8.)

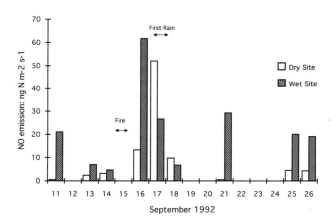

Figure 35.3 NO emissions from dry and wetted sites, Shabeni #5 (burned site), before and after first rain. (All NO measurements have not been corrected for soil temperature. Based on data in table 35.8.)

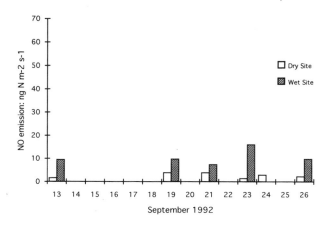

Figure 35.4 NO emissions from dry and wetted sites, Kambeni #1 (unburned control site), before and after first rain. (All NO measurements have not been corrected for soil temperature. Based on data in table 35.8.)

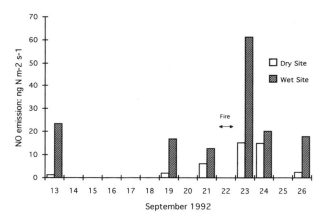

Figure 35.5 NO emissions from dry and wetted sites, Kambeni Extra (burned site), before and after first rain. (All NO measurements have not been corrected for soil temperature. Based on data in table 35.8.)

52.1 ng N m^{-2} s^{-1} following the rain on 17 September. Four days after the rain, the NO flux returned to the prefire/prerain range of 4.6 ng N m^{-2} s^{-1} or less. As in the Shabeni #6 site, Shabeni #5 wetted site consistently exhibited higher NO fluxes, except the day of the rain and the day following the rain. Before and after the rain, the NO fluxes from the wetted site ranged from 4.7 to 29.3 ng N m^{-2} s^{-1} and exceeded the NO flux from the dry site by factors ranging from 1.5 to 48. However, the day it rained and the day following the rain, the NO fluxes from the dry site exceeded the NO flux from the wet site by factors of 2 and 1.4, respectively.

In the Kambeni #1 dry control site (figure 35.4), the NO flux increased from 1.6 to 3.9 ng N m^{-2} s^{-1}

following the 17 September rainfall. In the wet site, the NO flux ranged from 7.4 to 16.1 ng N m^{-2} s^{-1}. The NO flux from the wet site always exceeded the NO flux from the dry site by factors that ranged from 2 to 11.

In the Kambeni Extra dry burn site (figure 35.5), the NO flux ranged from 1.3 to 6.2 ng N m^{-2} s^{-1} prior to the 22 September fire and increased to 15 ng N m^{-2} s^{-1} for two days following the fire (as already noted, the Kambeni Extra site was protected from the 17 September rainfall). Four days after the fire, the NO flux was back to its prefire level of 2.4 ng N m^{-2} s^{-1}. The NO flux from the wet site was consistently higher than the dry site, with fluxes that ranged from 12.7 to 23.4 ng N m^{-2} s^{-1} before the fire. A flux of

61.5 ng N m^{-2} s^{-1} was measured the day after the fire. Two and four days after the fire, the NO flux was back to its prefire level in the range of 17.9 to 20.1 ng N m^{-2} s^{-1}.

N$_2$O Measurements

The IPCC Scientific Assessment on Climate Change (Houghton, Jenkins, and Ephraums 1990) stated that the role of the world's savannas on the global budget of N$_2$O is uncertain due to the paucity of measurements. The SAFARI experiment provided a unique opportunity to add to the sparse database of measurements of N$_2$O emissions from the African savannas. However, soil emissions of N$_2$O were not detected from any site—Shabeni or Kambeni, dry or wet, before or after burning. The lack of measurable N$_2$O soil emissions indicates that the emissions were below 2 ng N m^{-2} s^{-1}, the minimum detection limit of the instrumentation. Previous measurements of microbial soil N$_2$O emissions in temperate ecosystems showed a very significant enhancement following wetting and burning (Levine et al. 1988). In tropical savanna soils of Venezuela, burning did not lead to increased biogenic soil emissions of N$_2$O, but wetting resulted in significant increases after both unburned and burned soils were watered (Hao et al. 1988). The detection of the system used in the Hao et al. (1988) study was about a factor of five more sensitive than the system used in the present study.

Discussion

Before burning or from unburned control sites, NO emissions from the dry sites ranged from a low of 0.4 to a high of 6.2 ng N m^{-2} s^{-1}, and emissions from the wet sites ranged from a low of 5.7 to a high of 34.0 ng N m^{-2} s^{-1}. Several days following burning, after the fire-induced emission enhancement, NO emissions from the dry sites ranged from a low of 0.6 to a high of 4.6 ng N m^{-2} s^{-1} and, from the wet sites, from 6.8 to 29.3 ng N m^{-2} s^{-1}. Hence, it appears that typical natural, "undisturbed" background NO emissions from the South African savanna are in the range of 0.4 to 6.2 ng N m^{-2} s^{-1} from the dry sites and are in the range of 5.7 to 34 ng N m^{-2} s^{-1} from the wet sites. The biogenic soil emissions of NO reported in this chapter represent the first measurements obtained from the savannas in southern Africa. However, biogenic soil emissions of NO have been obtained from savannas in Venezuela both during the dry season (Johansson, Rodhe, and Sanhueza 1988) and during

the rainy season (Johansson and Sanhueza 1988). During the dry season, Johansson, Rodhe, and Sanhueza (1988) reported NO emissions in the range of 3 to 15 ng N m^{-2} s^{-1}, with a mean value of 8 ng N m^{-2} s^{-1}. They also reported NO emissions in the range of 2 to 250 ng N m^{-2} s^{-1}, with a mean value of 56 ng N m^{-2} s^{-1}, following artificial irrigation and increases ranging from 100 to 300 times the background emissions following the addition of nitrate to the soil. During the rainy season, Johansson and Sanhueza (1988) reported NO emissions in the range of 9.5 to 38 ng N m^{-2} s^{-1} for soil moisture of 7% and NO emissions in the range of 16 to 117 ng N m^{-2} s^{-1} for soil moisture greater than 15%. Following a heavy rainfall (15.2 mm of rain), they reported NO fluxes in the range of 150 to 250 ng N m^{-2} s^{-1}.

Fire had a significant impact on NO emissions from both dry and wet sites. Within a day after burning, NO emissions from the dry sites ranged from 13.3 ng N m^{-2} s^{-1} (Shabeni) to 15.2 ng N m^{-2} s^{-1} (Kambeni) and, from the wet sites, from 61.5 ng N m^{-2} s^{-1} (Kambeni) to 61.8 ng N m^{-2} s^{-1} (Shabeni). The postfire NO emissions from the wet Kambeni site of 76.9 ng N m^{-2} s^{-1} are among the very highest NO emissions reported in the literature (Williams, Hutchinson, and Fehsenfeld 1992). Johansson, Rodhe, and Sanhueza (1988) observed about a tenfold increase in soil emissions of NO following burning. The NO emissions were found to decrease slowly over a four-day interval. The Kambeni Extra burn site also exhibited about a tenfold increase in NO emissions following burning.

The NO emissions from both the dry and wet Kambeni sites were consistently higher than the emissions from the dry and wet Shabeni sites. The higher emissions from the Kambeni sites, which were last burned 35 years ago, are probably related to the fact that these sites contained more standing and dead biomass than the Shabeni sites, which were last burned only two years ago. Total soil nitrogen values were twice as high at the Kambeni site than at the Shabeni site (see table 35.1) (Parsons et al. 1996). The Kambeni sites also contained more combustible matter, and hence, burning produced more ammonium in the ash. Ammonium is the substrate utilized by nitrifying bacteria in the microbial production of NO. The ratio of the postburn fluxes of NO the N$_2$O measured in a temperate chaparral ecosystem was found to be similar to the ratio of these fluxes obtained in controlled laboratory experiments using nitrifying bacteria (Anderson and Levine 1986). In addition, postfire emissions of NO and N$_2$O in this temperate chaparral ecosystem have been correlated to increased postfire levels of ammonium in the soil (Anderson and Levine 1987; Anderson et al. 1988; Levine et al. 1984, 1988, 1990, 1991). Johansson, Rodhe, and Sanhueza (1988) concluded that for dry soil conditions, the NO emission from the Venezuelan savanna may also be due to nitrification. However, Johansson and Sanhueza (1988) reported that the NO emission increased dramatically in response to the addition of nitrate, the substrate in the microbiological process of denitrification. Johansson and Sanhueza suggested that perhaps two different mechanisms (nitrification and denitrification) are responsible for the emission of NO from dry and wet savanna soil. Additional research is needed on the question of the mechanism or mechanisms responsible for NO production in savanna soil.

During the SAFARI experiment, aircraft measurements indicated substantial levels of NO_x from a several-hundred-square-kilometer savanna area region in northern Namibia (Harris et al. 1996). These very high levels of NO_x were not associated with savanna burning. Harris et al. (1996) calculated that biogenic soil emissions of NO in the range of 20 to 40 ng N m^{-2} s^{-1} would be needed to produce the measured airborne levels. Our measurements indicate biogenic soil emissions of NO in the range of 40 to 60 ng N m^{-2} s^{-1} from both unburned and burned sites following the first rain on 17 September, 1994. Harris et al. (1996) concluded that biogenic soil emissions of NO following rainfall during and at the end of the dry season may be an important regional source of NO. The biogenic soil source of NO may be comparable to or greater than the source due to biomass burning of the savanna. Harris et al. (1996) also concluded that biogenic soil emissions of NO may have important consequences for regional production of tropospheric ozone.

The SAFARI-92 experiment provided a unique opportunity to investigate biogenic soil emissions of NO and N_2O from the South African savanna. The measurements reported here contain two interesting findings—the high biogenic soil emissions of NO following artificial wetting, natural rainfall, and burning and the lack of detectable biogenic soil emissions of N_2O even after artificial wetting, natural rainfall, and burning. Biogenic soil emissions of NO may be comparable to NO production resulting from the combustion of biomass. The lack of measurable emissions of N_2O from the South African savanna indicates that we must continue the search for additional global sources of this environmentally important gas, which impacts global climate via its greenhouse effect and leads to the chemical destruction of stratospheric ozone.

References

Anderson, I. C., and J. S. Levine. 1986. Relative rates of nitric oxide and nitrous oxide production by nitrifiers, denitrifiers, and nitrate respirers, *Appl. Environ. Microbiol.*, 51, 938–945.

Anderson, I. C., and J. S. Levine. 1987. Simultaneous field measurements of biogenic emissions of nitric oxide and nitrous oxide, *J. Geophys. Res.*, 92, 964–976.

Anderson, I. C., J. S. Levine, M. A. Poth, and P. J. Riggan. 1988. Enhanced biogenic emissions of nitric oxide and nitrous oxide following surface biomass burning, *J. Geophys. Res.*, 93, 3893–3898.

Andreae, M. O. 1991. Biomass burning: Its history, use, and distribution and its impact on environmental quality and global climate, in *Global Biomass Burning: Atmospheric, Climatic, and Biospheric Implications*, edited by J. S. Levine, pp. 3–21, MIT Press, Cambridge, Mass.

Arkin, H., and R. R. Colton. 1953. *Statistical Methods*, pp. 126–127, Barnes and Noble, New York.

Blackmer, A. M., S. G. Robbins, and J. M. Bremner. 1982. Diurnal variability in rate of emission of nitrous oxide from soils, *Soil Sci. Soc. Am. J.*, 46, 937–942.

Cahoon, D. R., B. J. Stocks, J. S. Levine, W. R. Cofer, and K. P. O'Neill. 1992. Seasonal distribution of African savanna fires, *Nature*, 359, 812–815.

Cofer, W. R., J. S. Levine, E. L. Winstead, and B. J. Stocks. 1991. New estimates of nitrous oxide emissions from biomass burning, *Nature*, 349, 689–691.

Davidson, E. A. 1992. Sources of nitric oxide and nitrous oxide following wetting of dry soil, *Soil Sci. Soc. Am. J.* 56, 95–102.

Delwiche, C. C. 1981. The nitrogen cycle and nitrous oxide, *Denitrification, Nitrification, and Atmospheric Nitrous Oxide*, edited by C. C. Delwiche, pp. 1–15, Wiley, New York.

Fehsenfeld, F. C., J. W. Drummond, U. K. Roychowdhury, P. J. Galvin, E. J. Williams, M. P. Buhr, D. D. Parrish, G. Hubler, A. O. Langford, J. G. Calvert, B. A. Ridley, F. Grahek, B. G. Heikes, G. L. Kok, J. D. Shetter, J. G. Walega, C. M. Elsworth, R. B. Norton, D. W. Fahey, P. C. Murphy, C. Hovermale, V. A. Mohnen, K. L. Demerjian, G. I. Mackay, and H. I. Schiff. 1990. Intercomparison of NO_2 measurement techniques, *J. Geophys. Res.*, 95, 3579–3597.

Firestone, M. K., and E. A. Davidson. 1989. Microbiological basis of NO and N_2O production and consumption. *Exchange of Trace Gases Between Terrestrial Ecosystems and the Atmosphere*, edited by M. O. Andreae and D. S. Schimel, pp. 7–21, Wiley, New York.

Focht, D. D., and W. Verstraete. 1977. Biochemical ecology of nitrification and denitrification, *Adv. Microbiol. Ecol.*, 1, 135–214.

Galbally, I. E. 1989. Factors controlling NO_x emissions from soils, in *Exchange of Trace Gases Between Terrestrial Ecosystems and the Atmosphere*, edited by M. O. Andreae and D. S. Schimel, pp. 189–207, Wiley, New York.

Hao, W. M., M. H. Liu, and P. J. Crutzen. 1990. Estimates of annual and regional release of CO_2 and other trace gases to the atmosphere from fires in the tropics, based on FAO statistics for the period 1975–1980, in *Fire in the Tropical Biota: Ecosystem Processes and Global Challenges*, edited by J. O. Goldammer, pp. 440–462, Ecological Studies 84, Springer-Verlag, Berlin-Heidelberg.

Hao, W. M., D. Scharffe, P. J. Crutzen, and E. Sanhueza. 1988. Production of N_2O, CH_4, and CO_2 from soils in the tropical savanna during the dry season, *J. Atmos. Chem.*, 7, 93–105.

Hao, W. M., D. H. Scharffe, J. M. Lobert, and P. J. Crutzen. 1991. Emissions of nitrous oxide from the burning of biomass in an experimental system, *Geophys. Res. Lett.*, 18, 999–1002.

Harris, G. W., T. Zenker, F. G. Weinhold, and V. Parchatka. 1996. Airborne observations of strong biogenic NO_x emissions from the Namibian savanna at the end of the dry season, *J. Geophys. Res.*, 101, in press.

Houghton, J. T., G. J. Jenkins, and J. J. Ephraums. 1990. *Climate Change: The IPCC Scientific Assessment*, Cambridge University Press, Cambridge, England.

Johansson, C., H. Rodhe, and E. Sanhueza. 1988. Emission of NO in tropical savanna and a cloud forest during the dry season, *J. Geophys. Res.*, 93, 7180–7192.

Johansson, C., and E. Sanhueza. 1988. Emission of NO from savanna soils during rainy season, *J. Geophys. Res.*, 93, 14193–14198.

Levaggi, D., E. L. Kothny, T. Belsky, E. de Vera, and P. K. Mueller. 1974. Quantitative analysis of nitric oxide in presence of nitrogen dioxide at atmospheric concentrations, *Environmental Science and Technology*, 8, 348–350.

Levine, J. S. 1992. The global budget of nitrous oxide, *Proceedings of the 5th International Workshop on Nitrous Oxide Emissions*, edited by H. Moritomi, pp. 1–9, National Institute for Resources and Development, Tsukuba, Japan.

Levine, J. S., T. R. Augustsson, I. C. Anderson, J. M. Hoell, and D. A. Brewer. 1984. Tropospheric sources of NO_x: Lightning and biology, *Atmospheric Environment*, 18, 1797–1804.

Levine, J. S., W. R. Cofer, D. I. Sebacher, R. P. Rhinehart, E. L. Winstead, S. Sebacher, C. R. Hinkle, P. A. Schmalzer, and A. M. Koller, Jr. 1990. The effects of fire on biogenic emissions of methane and nitric oxide from wetlands, *J. Geophys. Res.*, 95, 1853–1864.

Levine, J. S., W. R. Cofer, D. I. Sebacher, E. L. Winstead, S. Sebacher, and P. J. Boston. 1988. The effects of fire on biogenic soil emissions of nitric oxide and nitrous oxide, *Global Biogeochem. Cycles*, 2, 445–449.

Levine, J. S., W. R. Cofer, E. L. Winstead, R. P. Rhinehart, D. R. Cahoon, D. I. Sebacher, S. Sebacher, and B. J. Stocks. 1991. Biomass burning: Combustion emissions, satellite imagery, and biogenic emissions, in *Global Biomass Burning: Atmospheric, Climatic, and Biospheric Implications*, edited by J. S. Levine, pp. 264–271, MIT Press, Cambridge, Mass.

Lobert, J. M., D. H. Scharffe, W. M. Hao, T. A. Kuhlbusch, R. Seuwen, P. Warneck, and P. J. Crutzen. 1991. Experimental evaluation of biomass burning emissions: Nitrogen and carbon compounds, in *Global Biomass Burning: Atmospheric, Climatic, and Biospheric Implications*, edited by J. S. Levine, pp. 289–304, MIT Press, Cambridge, Mass.

Macvicar, C. N., R. F. Loxton, J. J. N. Lambrechts, J. Le Roux, H. J. von Harmes, J. M. De Villiers, E. Verster, F. R. Merryweather, and T. H. Van Rooyen. 1977. *Soil Classification: A Binomial System for South Africa*, The Department of Agricultural Technical Services, South Africa.

Muzio, L. J., and J. C. Kramlich. 1988. An artifact in the measurement of N_2O from combustion sources, *Geophys. Res. Lett.*, 15, 1369–1372.

Parsons, D. A. B., M. C. Scholes, R. J. Scholes, and J. S. Levine. 1996. Biogenic NO emissions from savanna soils as a function of soil nitrogen and water status, *J. Geophys. Res.*, 101, in press.

Sanhueza, E., W. M. Hao, D. Scharffe, L. Donoso, and P. J. Crutzen. 1990. N_2O and NO emissions from soils of the northern part of the Guayana shield, Venezuela, *J. Geophys. Res.*, 95, 22481–22488.

Schlesinger, W. H. 1991. *Biogeochemistry: An Analysis of Global Change*, Academic Press, San Diego.

Shepherd, M. F., S. Barzetti, and D. R. Hastie. 1991. The production of atmospheric NO_x and N_2O from a fertilized agricultural soil, *Atmospheric Environment*, 25A, 1961–1969.

Soderlund, R., and T. Rosswall. 1982. The nitrogen cycles, *The Handbook of Environmental Chemistry: Volume 1, Part B. The Natural Environment and the Biogeochemical Cycles*, pp. 61–81, edited by O. Hutzinger, Springer-Verlag, Berlin.

Venter, F. J., and W. P. D. Gertenbach. 1986. A cursory review of the climate and vegetation of the Kruger National Park, *Koedoe*, 29, 139–148.

Williams, E. J., G. L. Hutchinson, and F. C. Fehsenfeld. 1992. NO_x and N_2O emissions from soils, *Global Biogeochem. Cycles*, 6, 351–388.

Yienger, J. J., and H. Levy II. 1995. Empirical model of global soil-biogenic NO_x emissions, *J. Geophys. Res.*, 100, 11447–11464.

Zepp, R. G., W. L. Miller, R. A. Burke, D. A. B. Parsons, and M. C. Scholes. 1996. Effects of moisture and burning on soil-atmosphere exchange of trace gases in a southern African savanna, *J. Geophys. Res.*, 101, in press.

Dynamics of Carbon Monoxide Emissions from Soil and Vegetation in a Southern African Savanna

Richard G. Zepp, William L. Miller, Matthew A. Tarr, Roger A. Burke, Dirk A. B. Parsons, and Mary C. Scholes

Biomass burning is known to be a major source of atmospheric carbon monoxide (CO), accounting for 300 to 700 Tg yr^{-1} (Logan 1994), with much of the emission coming from savannas. In addition to the direct emissions from burning, postburn alterations in CO emissions may occur as a consequence of changes in soil and vegetation biogeochemistry. Moreover, it is likely that land use changes that often accompany fire, such as conversion of savanna to pasture (Keller and Matson 1994; Johansson, Rodhe, and Sanhueza 1988; Sanhueza 1991; Menaut, Abbadie, and Vitousek 1994; Sanhueza et al. 1994a, 1994b; Sanhueza and Santana 1994) or intense agriculture (Woomer 1993), may have important effects on soil and vegetation emissions. There are very few studies of such effects in tropical savannas (Sanhueza et al. 1994a, 1994b; Hao and Liu 1994), although most fire activity occurs in these regions.

Soils have also been proposed as sinks for atmospheric CO on a global basis (Watson et al. 1992). However, the data are few and conflicting with regard to both the magnitude and direction of CO fluxes from tropical savanna soils (Sanhueza et al. 1994a; Conrad and Seiler 1982; Scharffe et al. 1990; Conrad and Seiler 1985a; Seiler and Conrad 1987). The net CO flux results from a competition between chemical oxidation of soil organic matter to produce CO (Conrad and Seiler 1985b), which is favored by dry soil conditions (Conrad and Seiler 1982), and biological oxidation of CO by soil microorganisms (Seiler and Conrad 1987). Data concerning the effects of fires on soil CO fluxes are not available.

Vegetation also has been recognized as a source of atmospheric CO. It has long been known that hydrocarbons emitted by vegetation are oxidized to CO in the troposphere. Recent estimates of this source have been scaled back to about 200 Tg yr^{-1} (Logan 1994), considerably smaller than earlier estimates that ranged up to 1400 Tg yr^{-1} (Watson et al. 1992). Older estimates of direct emissions of CO from plants fall in the range of 50 to 200 Tg yr^{-1} (Watson et al. 1992). Field studies in the Amazon basin have shown that this ecosystem produces a significant amount of CO that is not accounted for by oxidation of hydrocarbons. Several authors have attributed this source to direct emissions of CO from vegetation (Jacob and Wofsy 1990; Harriss et al. 1990; Kirchhoff and Marinho 1990). These investigations also indicate that higher wet-season CO levels exist than were previously estimated.

Little is known about the effects of human perturbations of savannas on the biosphere-atmosphere exchange of CO, although this biome covers about 11% of the global land surface (≈ 16 million km^2). Because the ecology of savannas is strongly influenced by periodic burning, evaluation of the interconnection between fire and the soil and vegetation CO emissions in these ecosystems is especially pertinent. In this chapter, we report field and laboratory studies of CO emissions from savanna soils and plant matter obtained from sites at the Kruger National Park (KNP) in South Africa with well-defined burning histories. Experimental manipulations of soil moisture in the field also were conducted to compare the relative impact of fire and rain. We also attempted to identify the major abiotic components controlling field fluxes of CO by measuring CO fluxes in the laboratory from collected ground litter, standing dry grass, and soil.

Experiment

Field Sites

Field studies were carried out during the dry season in the KNP, Republic of South Africa, near the Pretoriuskop camp in a broad-leafed savanna. The dominant tree and grass species have been described in Trollope et al. (1995). Average annual rainfall at Pretoriuskop is 740 mm (Venter and Gertenbach 1986), occurring almost entirely in the months of October through April. Daytime air temperatures measured during our field study (11–26 September 1992) ranged from 17 to 43°C. Specific flux chamber locations were established within four plots (≈ 300 m $\times$ 200 m) that have been maintained in a controlled experimental burning program carried out by the National Park Service of South Africa continuously since 1954 (van Wyk 1971). All sites that were investigated were in close proximity and had the same climate. Two plots from the Kambeni series (Kambeni Extra and Kambeni 1) were protected from fire for more than 35 years prior to this study. Two plots from the Shabeni series (Shabeni 5 and Shabeni 6) had been burnt biennially since 1954. The sandy to sandy loam soils found in these plots were deeply leached with a total carbon content of 5.2 mg/g soil (0.5%) (Otter 1992).

Two pairs of sites were investigated. All sites that were investigated had the same granitic (low nutrient) soil type and the same climate. Each of the sites of a given pair had the same vegetation type and density and the same burning history. One of each pair of sites was burned, and the other site was used as a control. The Shabeni sites were burned biennially and thus were predominantly grassland. The Kambeni sites, which had not been burned for more than three decades, were partially wooded and had a much higher aboveground biomass density and more litter at the sites investigated.

Flux Measurements and Instrumentation

A closed-chamber technique (Sebacher and Harriss 1982; Sebacher, Harriss, and Bartlett 1983; Levine et al. 1988; Levine et al. 1990) was used to determine CO fluxes in the field. Details that pertain to the Southern African Fire-Atmosphere Research Initiative—1992 (SAFARI-92) experiments are provided in Zepp et al. (1996).

Two standard gas cylinders were used for calibration in the field. A Scott-Marrin standard contained CO at a level of 47 ppb $\pm$ 10%. CO concentrations also were determined using a 9.50 ± 0.14 ppm CO National Institute of Standards and Technology (NIST) cylinder (Cyl. No. FF-33955, Samp. No. 23-24-E). A NIST standard curve for CO was generated each day using a series of calibrated sample loops.

Laboratory Studies of CO Production

Biomass and soil samples were brought back to the laboratory in Athens, Georgia, for thermal and photochemical CO production studies. Samples of standing dry grass, leaf litter composite, and unsieved soil were weighed and placed in ashed glass bottles (about 37 cm^3 volume) with crimp-capped silicone septa. Each bottle was overpressured by 10 cm^3 with CO free air and submerged in a water bath at temperatures from 36 to 66°C. After a minimum of 15 minutes equilibra-

tion, a 1-ml headspace sample was analyzed for CO as a function of time using the instrumentation described. Activation energies for each component were calculated using the Arrhenius equation:

$$p = Ae^{-(E_a/RT)} \qquad (36.1)$$

where p is CO production rate at a temperature of T (°K), R is the gas constant, and E_a is the activation energy in units of kJ mol^{-1}.

Studies of the photoproduction of CO from dead plant matter collected at the KNP were performed using simulated sunlight provided by a Spectral Energy (Westwood, New Jersey) Series II illuminator. Filters with various lower wavelength cutoffs (Schott Glass, Inc., Duryea, Pennsylvania) were employed to vary the amount of ultraviolet (UV) radiation reaching the sample. The filters had the following cutoffs, represented by the wavelength at which transmittance was lowered to 50%: 305 nm, 320 nm, 345 nm, 360 nm, and 425 nm. Wavelength studies were achieved by sequentially exposing a sample using filters of increasingly longer cutoff wavelengths. The total irradiance was held constant in these experiments, and only a small percentage of the total light was removed by changing to a filter of higher cutoff. By subtracting the CO flux with the various UV filters in place from the full spectrum flux (305-nm cutoff filter), the contribution of each wavelength region, as defined by the available filters, could be assessed. In addition to the aforementioned filters, an air mass 1 filter (AM1, Spectral Energy) was also used in some exposures. This filter is intended to simulate solar irradiance at the earth's surface with a solar zenith angle of zero.

Light intensities were monitored using an International Light (Newburyport, Massachusetts) IL 1700 radiometer. Full-spectrum irradiance was measured with an SED623 probe, UV-A irradiance was estimated using an SED033 probe (maximum response at 360 nm), and UV-B irradiance was estimated using an SED240 probe (maximum response at 290 nm).

Results and Discussion

As in the case of carbon dioxide, both soils and vegetation contribute to the biosphere-atmosphere exchange of CO. Here we present data obtained under controlled conditions in the laboratory concerning CO emissions from soil and plant matter that were obtained from an African savanna that was investigated during SAFARI-92. Results of these studies are then used in a simple empirical model to compute CO emission fluxes for SAFARI-92 sites. Results are compared to observed fluxes and to CO emissions that were directly produced by fires at the sites.

Climatic Conditions

Sample collection and flux measurements were conducted near the end of the dry season in the KNP in South Africa during SAFARI-92. Most vegetation at the sites was either dead or in a state of dormancy. The soils were extremely dry and remained so during most of the field campaign with the exception of a light shower midway through the campaign. Weather conditions were extremely variable during SAFARI-92 (Lindesay et al. 1996), and the soil temperatures at a depth of 2 cm ranged from a low of 17°C to a high of 42°C. These wide variations provided ample opportunity for investigation of the effects of temperature on soil emissions through field experiments. Distilled water was deliberately added to some of the chamber locations to permit studies of the effects of varying soil moisture content on the CO fluxes.

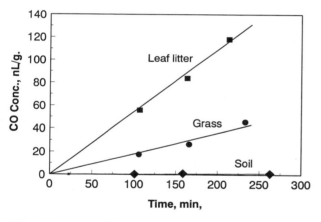

Figure 36.1 Dark production of CO from leaf litter, dead grass, and soils from the Kruger National Park (35°C)

Temperature Effects

Experiments in the laboratory showed that CO was produced by dark (thermal or biotic) decomposition of leaf litter, dry grasses, and soil collected from the SAFARI sites (figure 36.1). Dark production of CO from all the samples was linear with time at temperatures in the 20 to 66°C range. Using equation (36.1), activation parameters for production of CO were computed (table 36.1). The leaf litter and standing dry grass, with their much higher organic content per unit weight, produced CO in the dark much more rapidly than the soil at the same temperature. The studies also showed that the activation energies for CO production were similar for soil and grass, but much lower for the leaf litter composite. The activation parameters for the soil are close to those previously determined for a similar soil from the Transvaal (Conrad and Seiler 1985b).

These results indicate that the temperature dependence for CO emissions from a wooded savanna may be quite different from those for an open savanna. For example, the leaf litter emissions increased only 2.5-fold when temperature increased from 20°C to 40°C, but emissions from the dead grass, which dominates in open savanna, increased more than an order of magnitude in the same temperature range. Furthermore, because CO sources have variable temperature responses, the response of CO fluxes in the field to temperature changes must depend on soil organic matter and the relative abundance of standing grass and litter.

Effects of Solar Radiation

For some time it has been known that synthetic and natural organic substances produce CO when exposed to light (Wilks 1963). In this study, exposure of the leaf litter and standing dead grass collected at the KNP to simulated solar radiation resulted in large increases in CO production (figure 36.2). Controls showed that the

Table 36.1 Kinetic parameters for dark production of carbon monoxide from various components of a South African savanna

CO Source	Organic C (%)	E_a (kJ mol⁻¹)	A (nl h⁻¹ g⁻¹ d.w.)	Rate, 20°C (nl h⁻¹ g⁻¹ d.w.)	Rate, 30°C (nl h⁻¹ g⁻¹ d.w.)	Rate, 40°C (nl h⁻¹ g⁻¹ d.w.)
Leaf litter	45	35	2.4×10^7	16.3	26.0	40.3
Grass	43	88	1.0×10^{16}	1.9	6.3	19.2
Soil (KNP)	0.52	77	2.9×10^{12}	0.067	0.19	0.50
Soil (Transvaal)[a]	1.2	80	2×10^{13}	0.11	0.33	0.92

a. Conrad and Seiler 1985b.

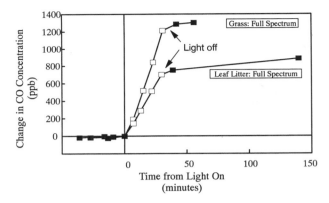

Figure 36.2 Effects of simulated solar radiation on the production of CO from dead standing grass collected in an open savanna, Kruger National Park, R.S.A.

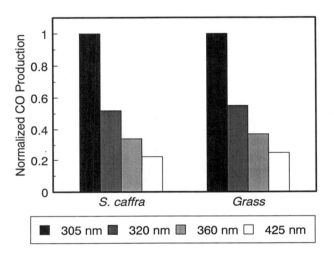

Figure 36.3 Wavelength effects on the photoproduction of CO from leaf litter (marula tree, *Sclerocarya caffra*) and dead standing grass from Kruger National Park, R.S.A.

CO production was not due to enhanced thermal or biological processes caused by warming by the light, but instead was mainly attributable to photodegradation. Observed fluxes were higher for the dead grass (3.2×10^{11} molecules cm^{-2} s^{-1} at total irradiance of 650 W m^{-2}) than for leaf litter from the marula tree (*Sclerocarya caffra*) (1.2×10^{11} molecules cm^{-2} s^{-1}). Removal of selected regions of the UV radiation by light filters caused large reductions in the photoproduction rate of CO (figure 36.3). These results indicate that previous measurements of CO exchange using dark chambers underestimated production rates. Other results have shown that dead plant matter

photoproduces CO much more efficiently than living plants (Tarr, Miller, and Zepp 1995). Thus, it is likely that CO emissions via this pathway are highest during the dry season when grasses are dead and sunlight exposures are highest.

Field and Modeling Studies

Combined CO fluxes from soils, leaf litter, and standing dead grass were measured at sites in the KNP using closed environmental chambers. All measurements of CO fluxes at the KNP were strongly positive, with values from the untreated sites ranging up to 3.6×10^{11} molecules cm^{-2} s^{-1} (figures 36.4 and 36.5). It has been previously established that microorganisms consume CO (Jones and Morita 1983; Jones, Morita, and Giffiths 1984) and that moist soils can provide a sink for atmospheric CO (Seiler and Conrad 1987). In the dry KNP soils, however, microbial activity was minimal. For example, the buildup of CO in the chambers did not plateau in the fashion that was previously observed in the Transvaal (Conrad and Seiler 1985a), indicating that CO sink activity was close to nil. After burning, the fluxes rose dramatically but dropped back to preburn levels within a few days (see figures 36.4 and 36.5). Comparisons with data obtained at other savanna sites indicated that these CO emissions were considerably higher than those previously reported for the Transvaal during nondrought conditions (table 36.2). Arrhenius plots of the field data illustrate the strong dependence of the fluxes on temperature (figure 36.6). The data do not exhibit the tight correlation that was observed in previous field studies in the Transvaal (Conrad and Seiler 1985a), possibly because we included all leaf litter and standing dried grass within the chambers. Previous studies have eliminated standing grass from the chambers in order to measure soil fluxes with minimum ambiguity. A system composed of multiple components with differing activation energies for thermal CO production would not necessarily produce a simple correlation between CO flux and measured field temperature.

Based on these results, previous studies of thermal production (Conrad and Seiler 1985a; Sanhueza et al. 1994a), photochemical production (Tarr, Miller, and Zepp 1995), and microbial consumption (Seiler and Conrad 1987), we propose a conceptual model in which the net exchange of CO between the terrestrial biosphere and the atmosphere is described by three major processes: thermal (abiotic) production from biomass on the surface and upper layer of the soil, photoproduction from plant matter (living and non-

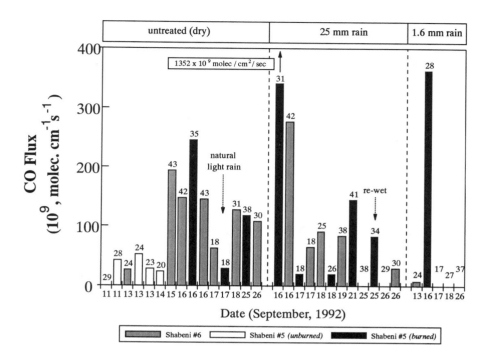

Figure 36.4 Carbon monoxide fluxes at the Shabeni plots (open savanna). Treatments are described in the text. Numbers with no bars are soil temperatures for measurements where the flux was below the detection limit. (From Zepp et al. 1996. Copyright by the American Geophysical Union.)

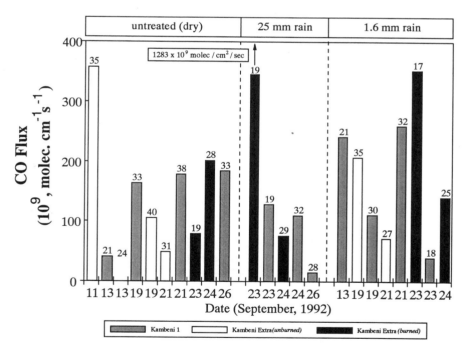

Figure 36.5 Carbon monoxide fluxes at the Kambeni plots (wooded savanna). Treatments are described in the text. Numbers with no bars are soil temperatures for measurements where the flux was below the detection limit. (From Zepp et al. 1996. Copyright by the American Geophysical Union.)

Table 36.2 Soil CO fluxes (10^9 molecules cm^{-2} s^{-1}) for tropical and subtropical savannas and grasslands (KNP averages are for untreated, unburned plots $\pm$ 1 SD)

Soil CO fluxes	Location	Season	References
≤ 60.5	Karoo[a], RSA	Wet	Conrad and Seiler 1982
29	Transvaal, RSA	Wet	Conrad and Seiler 1985a
33	Transvaal, RSA	Dry	Conrad and Seiler 1985a
12	Karoo[a], RSA	Wet	Conrad and Seiler 1985a
5.6 ± 0.7[b] 100[c] (Low 0, high 200)	Venezuela	Wet	Scharffe et al. 1990
0.8[b] 30[c] (Low 0, high 53)	Venezuela	Wet	Sanhueza et al. 1994a
34[b] 56[c] (Low 15, high 103)	Venezuela	Wet	Sanhueza et al. 1994b
78 ± 59 (Low 0, high 192) ($n = 12$)	South Africa, KNP Shabeni (*open*)	Dry (*untreated*)	Zepp et al. 1996
134 ± 113 (Low 0, high 356) ($n = 8$)	Kambeni (*wooded*)	Dry (*untreated*)	Zepp et al. 1996

a. Semidesert.
b. Average 24-hour net flux.
c. Approximate daylight average.
Source: Modified from Zepp et al., 1996. Copyright by the American Geophysical Union.

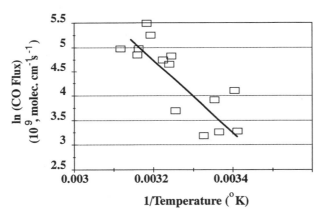

Figure 36.6 Arrhenius plots of CO fluxes observed at Kruger National Park during SAFARI-92

living), and microbial uptake by soil microorganisms. A general empirical equation can be expressed for the average rate of change in the mass of CO, dM_a/dt, within a UV-transparent closed chamber over the soil and vegetation:

$$\frac{dM_{CO}}{dt} = F_t + F_p - F_b \qquad (36.2)$$

where M_{CO} is the mass of CO in the chamber and the other terms are fluxes (e.g., molecules s^{-1}): F_t is the flux of CO produced by thermal decomposition, F_p is the flux produced by photodecomposition of plant

matter, and F_b is the flux consumed by soil microbiota. To partially evaluate the validity of this conceptual model, we computed values of the flux derived from thermal decomposition (F_t) using the activation energies for individual components (see table 36.1) and biomass density estimates from field data obtained at the KNP sites as part of the SAFARI campaign for comparison with fluxes of CO that were measured during our SAFARI-92 field campaign. No photochemical contribution to CO production was included in the model, since our field measurements were done in dark chambers. Using equation (36.1), kinetic parameters listed in table 36.1, field temperatures, and average biomass estimates for dry grass and leaf litter at the KNP sites (T. Kuhlbusch, personal communication), we calculated expected CO production rates. The best model results were obtained when the contribution of CO by soils was calculated using a two-box approach with surface-box temperatures (0 to 5 cm) based on direct measurements and deep-box production calculated for a temperature 80% of the measured 2-cm temperature. This temperature structure appears reasonable based on more sophisticated measurements and models of heat transfer in dry savanna soils (B. Scholes, personal communication). The computed initial fluxes are compared with field observations for the unburned savanna site (Shabeni 6) in figure 36.7. It is clear that much of the variability in the CO flux data

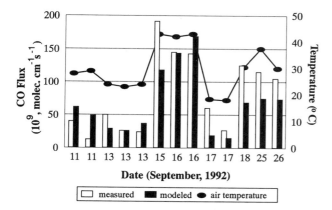

Figure 36.7 Comparison of CO production calculated by model with observed CO fluxes at Shabeni 6 site (open savanna). Studies were conducted in dark chambers and thus do not include photo-production.

reflects a strong response to the large changes in air temperature that occurred during our study period (see figure 36.7).

The fluxes for photoproduction and biotic consumption of CO have been discussed in other recent publications. Photoproduction rates are dependent on factors such as solar irradiance and leaf area indices (Tarr, Miller, and Zepp 1995). An exact estimate of the CO photoproduction is hampered by lack of data on leaf area indices for the sites. Available data for another well-studied grass-dominated savanna at Nylsvley, South Africa, indicate that leaf area indices range from 0.5 to 1.5 (Scholes and Walker 1993). Using this range, our data suggest that photoproduction fluxes are at least as large as those derived from thermal and biotic production. Taken together, these calculations indicate that leaf litter and dead standing grass were the primary sources of CO during the dry season at KNP. Thus, reductions in the amounts of above-ground biomass, for example, via land conversions or drought, are likely to significantly reduce net CO emissions from similar savanna sites.

Microbial consumption fluxes are proportional to CO deposition velocities and atmospheric CO concentrations (Seiler and Conrad 1987). During SAFARI-92 we observed no microbial uptake of CO in unwetted soils, presumably because a severe drought had reduced the soil microbial population. To investigate the effects of changing soil moisture levels, distilled water was added to some of the chamber locations. Moisture was added either as a repetitive light rain simulation (1.6 mm) similar to that experienced during the wet season or as a heavy rain simulation (25 mm) comparable to that experienced during initial rains at the

onset of the wet season. The heavy rain simulation was followed by flux measurements as the soil dried out over a two-week period. In contrast to large increases that were observed for CO_2 and nitric oxide (NO) fluxes, addition of moisture to the soil had no major effect on the CO fluxes, which remained virtually unchanged for at least two weeks of wetting. The generally higher CO fluxes in the dry chambers compared to those repeatedly wet to simulate 1.6 mm of rain (see figures 36.4 and 36.5) suggest the induction of a biological sink by the addition of water. This difference, however, could also result from changes in activation energies in samples with varied moisture content (Conrad and Seiler 1985b).

Conclusions

Laboratory experiments showed that an equivalent mass of surface litter and nonliving grass produced CO more efficiently in the dark than soil by up to two orders of magnitude. Other studies showed CO production was stimulated by the exposure of leaf litter and dead grass to solar radiation. Field studies during the dry season in South Africa showed that soil CO fluxes were positive under all conditions and generally increased with increasing soil temperature. The fluxes were strongly enhanced by burning, but the effect was short-lived. Added moisture has little effect on the fluxes over the length of these experiments.

Acknowledgments

We express our gratitude to T. Kuhlbusch for providing data on the biomass distribution at our KNP sites. We also thank R. Shea for providing samples of leaf litter, standing grass, and soils for the determinations of activation energies for CO production. Finally, we thank J. Levine for use of the collars and chambers for the SAFARI-92 experiments. W. Miller and M. Tarr acknowledge financial support of the National Research Council for these studies.

References

Andreae, M. O. 1991. Biomass burning: Its history, use, and distribution and its impact on environmental quality and global climate, in *Global Biomass Burning*, pp. 3–21, edited by J. S. Levine, MIT Press, Cambridge, Mass.

Conrad, R., and W. Seiler. 1992. Arid soils as a source of atmospheric carbon monoxide, *Geophys. Res. Lett.*, 12, 1353–1356.

Conrad, R., and W. Seiler. 1985a. Influence of temperature, moisture, and organic carbon on the flux of H_2, and CO between soil and atmosphere: Field studies in subtropical regions, *J. Geophys. Res.*, 90, 5699–5709.

Conrad, R., and W. Seiler. 1985b. Characteristics of abiological carbon monoxide formation from soil organic matter, humic acids, and phenolic compounds, *Environ. Sci. Technol.*, 19, 1165–1169.

Crutzen, P. J., and M. O. Andreae. 1990. Biomass burning in the tropics: Impact on atmospheric chemistry and biogeochemical cycles, *Science*, 250, 1669–1678.

Hao, W. M., and M. H. Liu. 1994. Spatial and temporal distribution of tropical biomass burning, *Global Biogeochem. Cycles*, 8, 495–503.

Harriss, R. C., G. W. Sachse, G. F. Hill, L. O. Wade, and G. L. Gregory. 1990. Carbon monoxide over the Amazon basin during the wet season, *J. Geophys. Res.*, 95(D10), 16 927–16 932.

Johansson, C., H. Rodhe, and E. Sanhueza. 1988. Emission of NO in a tropical savanna and cloud forest during dry season, *J. Geophys. Res.*, 93D, 7180–7192.

Jacob, D. J., and S. C. Wofsy. 1990. Budgets of reactive nitrogen, hydrocarbons, and ozone over the Amazon forest during the wet season, *J. Geophys. Res.*, 95(D10), 16 737–16 754.

Jones, R. D., and R. Y. Morita. 1983. Carbon monoxide oxidation by chemolithotrophic ammonium oxidizers, *Can. J. Microbiol.*, 29, 1545–1551.

Jones, R. D., R. Y. Morita, and R. P. Griffiths. 1984. Method for estimating *in situ* chemolithotrophic ammonium oxidation using carbon monoxide, *Mar. Ecol. Prog. Ser.*, 17, 259–269.

Keller, M., W. A. Kaplan, and S. C. Wofsy. 1986. Emissions of N₂O, CH₄, and CO₂ from tropical forest soils, *J. Geophys. Res.*, 91, 11 791–11 802.

Keller, M., and P. Matson. 1994. Biosphere-atmosphere exchange of trace gases in the tropics: Evaluating the effects of land use changes, in *Global Atmospheric-Biospheric Chemistry*, edited by R. G. Prinn, pp. 103–118, Plenum, New York.

Kirchhoff, V. W. J. H., and E. V. A. Marinho. 1990. Surface carbon monoxide measurements in Amazonia, *J. Geophys. Res.*, 95(D10), 16 933–16 943.

Levine, J. S., W. R. Cofer, D. I. Sebacher, R. P. Rhinehart, E. L. Winstead, S. Sebacher, C. R. Hinckle, P. A. Schmalzer, and A. M. Koller. 1990. The effects of fire on biogenic emissions of methane and nitric oxide from wetlands, *J. Geophys. Res.*, 95, 1853–1864.

Levine, J. S., W. R. Cofer, D. I. Sebacher, E. L. Winstead, S. Sebacher, and P. J. Boston. 1988. The effects of fire on biogenic soil emissions of nitric oxide and nitrous oxide, *Global Biogeochem. Cycles*, 3, 445–449.

Lindesay, J. A., M. O. Andreae, J. G. Goldammer, G. Harris, H. J. Annegarn, M. Garstang, R. J. Scholes, and B. W. van Wilgen. 1996. The IGBP/IGAC SAFARI-92 field experiment: Background and overview, *J. Geophys. Res.*, in press.

Logan, J. A. 1994. The global budget of CO, in Report of the WMO Meeting of Experts on Global Carbon Monoxide Measurements, *WMO/GAW Report* 98, 37.

Menaut, J.-C., L. Abbadie, and P. M. Vitousek. 1994. Nutrient and organic matter dynamics in tropical ecosystems, in *Fire in the Environment: The Ecological, Atmospheric, and Climatic Importance of Vegetation Fires*, edited by P. J. Crutzen and J. G. Goldammer, pp. 222–231, Wiley, Chichester, UK.

Otter, L. B. 1992. *Soil Carbon Fractionation of Sand and Clay Soils Under Different Burning Regimes*, Dissertation, University of Witwatersrand, Johannesburg, South Africa.

Sanhueza, E. 1991. Effects of vegetative burning on the atmospheric chemistry of the Venezuelan savanna, in *Global Biomass Burning*, edited by J. S. Levine, pp. 122–125, MIT Press, Cambridge, Mass.

Sanhueza, E., L. Cárdenas, L. Donoso, and M. Santana. 1994b. Effect of plowing on CO₂, CO, CH₄, N₂O, and NO fluxes from tropical savannah soils, *J. Geophys. Res.* 99(D8), 16 429–16 434.

Sanhueza, E., L. Donoso, D. Scharffe, and P. J. Crutzen. 1994a. Carbon monoxide fluxes from natural, managed or cultivated savannah grasslands, *J. Geophys. Res.*, 99(D8), 16 421–16 427.

Sanhueza, E., and M. Santana. 1994. CO₂ emissions from savannah soil under first year cultivation, *Interciencia*, 19, 20–23.

Scharffe, D., W. M. Hao, L. Donoso, P. J. Crutzen, and E. Sanhueza. 1990. Soil fluxes and atmospheric concentration of CO and CH₄ in the northern part of the Guayana Shield, Venezuela, *J. Geophys. Res.*, 95D, 22 475–22 480.

Scholes, R. J., and B. H. Walker. 1993. *An African Savanna: Synthesis of the Nylsvley Study*, pp. 164–167, Cambridge University Press, Cambridge, England.

Sebacher, D. I., and R. C. Harriss. 1982. A system for measuring methane fluxes from inland and coastal wetland environments, *J. Environ. Qual.*, 11, 34–37.

Sebacher, D. I., R. C. Harriss, and K. B. Bartlett. 1983. Methane flux across the air-water interface: Air velocity effects, *Tellus*, 35B, 103–109.

Seiler, W., and R. Conrad. 1987. Contribution of tropical ecosystems to the global budgets of trace gases, especially CH₄, H₂, CO, and N₂O, in *The Geophysiology of Amazonia: Vegetation and Climate Interactions*, edited by R. E. Dickinson, pp. 133–160, Wiley, New York.

Tarr, M. A., W. L. Miller, and R. G. Zepp. 1995. Direct carbon monoxide photoproduction from plant matter, *J. Geophys. Res.*, 100, 11 403–11 413.

Trollope, W. S. W., L. A. Trollope, A. L. F. Potgieter, and N. Zambatis. 1996. SAFARI'92—characterization of biomass and fire behavior in the small experimental burns in the Kruger National Park, *J. Geophys. Res.*, in press.

van Wyk, P. 1971. Veld burning in the Kruger National Park: An interim report of some aspects of research, *Ann. Proc. Tall Timbers Fire Ecology Conference*, No. 11, 9–31.

Venter, F. J., and W. P. D. Gertenbach. 1986. A cursory review of the climate and vegetation of the Kruger National Park, *Koedoe*, 29, 139–148.

Watson, R., L. G. Meira-Filho, E. Sanhueza, and A. Janetos. 1992. Greenhouse gases: Sources and sinks, in *1992 IPCC Supplement to Climate Change: The IPCC Scientific Assessment*, edited by J. T. Houghton, B. A. Callander, and S. K. Varney, pp. 27–46, Cambridge University Press, Cambridge, England.

Wilks, S. S. 1963. Toxic photooxidation products in closed environments, *Aerospace Med.*, 838–841.

Woomer, P. L. 1993. The Impact of cultivation on carbon fluxes in woody savannas of Southern Africa, in *Terrestrial Biospheric Carbon Fluxes: Quantification of Sinks and Soures of CO₂*, edited by J. Wisniewski and R. N. Sampson, pp. 403–412, Kluwer, Dordrecht. The Netherlands.

Zepp, R. G., W. L. Miller, R. A. Burke, D. A. B. Parsons, and M. C. Scholes. 1996. Effects of moisture and burning on soil-atmosphere exchange of trace carbon gases in a southern African savanna, *J. Geophys. Res.*, in press.

Large-Scale Transports of Biogenic and Biomass Burning Products

Michael Garstang, Peter D. Tyson, Edward Browell, and Robert Swap

Trajectories have been widely used to describe atmospheric transports. Typically, transports have been implied by superimposing many individual trajectories on a single diagram. Although these spaghetti diagrams provide a good visual impression of transport, the method does not lend itself to quantitative determination of transports. Cluster analysis of a field of trajectories provides statistical measures of the mean transport envelope and estimates of dispersion (Moody et al. 1995). The cluster analysis approach as currently used is two-dimensional in form and still does not provide quantitative estimates of transport volume or mass flux. An alternative ensemble approach to trajectory analysis of large-scale transports is presented in this chapter (Garstang et al. 1996; Tyson et al. 1996). The approach is based on prior classification of circulation types allowing studies to be placed in a climatological context. Inferences can be drawn about transports in terms of the circulation climatology and not as a function of individual trajectories. The method discussed incorporates atmospheric stability and relates three-dimensional atmospheric transport to stable and inversion layers present in the atmospheric column.

By dealing with trajectories as ensembles and not as individual Lagrangian parcels, the two-dimensional (x,z and y,z) frequency distributions of the trajectories can be determined and statistical limits placed on these distributions. Statistical measures of dispersions in the transport plume can be simultaneously displayed. The two-dimensional frequency distributions on vertical planes within the atmosphere can be converted to three-dimensional volume fluxes once the transport velocity of the plume has been determined. These transport pathways can then be compared to observed vertical sections of the atmosphere as derived from the ultraviolet differential absorption lidar (UV-DIAL) (Browell et al. 1995) and lidar in-space technology experiment (LITE) (McCormick et al. 1993) systems. Knowledge of concentrations of aerosols or trace gases then permits the computation of mass fluxes of those quantities. This chapter details and establishes the methodology described above. These methods will subsequently be used and expanded upon in the two succeeding chapters. (See Swap et al. chapter 38 this volume; Tyson et al. chapter 39 this volume.)

Methodology

Classification of Dominant Circulation Types

The circulation classification example chosen is drawn from southern Africa (Garstang et al. 1996). Figure 37.1 shows three dominant circulation types occurring over southern Africa. Frequency distributions can be determined for each type and are shown in figure 37.2. Three transport modes are found over southern Africa (figure 37.3): direct transports, indirect transports, and recirculation. The direct and indirect transports are to the east and west with exit into the atmosphere over the respective oceans occurring quickly for the direct case and more slowly in the indirect case. Recirculation is predominantly in the form of anticyclonic gyres over the subcontinent. Moody et al. (1995) and Dorling et al. (1992) have used a similar but more limited approach that uses synoptic climatology to place the transport findings in the context of larger scale and longer term processes.

Determination of Transport Pathways

A large number of trajectories are calculated from a horizontal grid of points for each circulation type. Trajectories at the points of origin are calculated at five levels: 875, 850, 800, 700 and 500 hPa from the European Center for Medium range Weather Forecasting (ECMWF) operational analyses. The ECMWF operational analysis of the three-dimensional wind field are available globally every 6 h at 31 levels, 13 of which are below 500 hPa. Bengtsson (1985) gives a general description of the ECMWF data assimilation system and Kållberg (1984) a description of the trajectory program. Each forward trajectory is calculated over a 10-day period using the principle of Lagrangian advection. The horizontal (u,v) and vertical (w) wind

Figure 37.1 Major circulation types controlling tropospheric transports over and beyond southern Africa. The heavy lines give the circulation at the 500-hPa level; the light lines give surface conditions as sea level isobars over the oceans and contours of the 850 hPa surface over the subcontinent (modified after Tyson 1986).

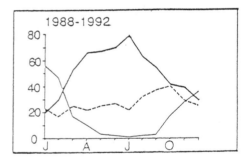

Figure 37.2 The monthly frequency of occurrence over the period 1988–1992 of the combined continental and ridging anticyclone classes (heavy solid line), the combined westerly wave/troughs and cut-off lows (light solid line), and the easterly wave disturbances (dashed line) over southern Africa (modified after Tyson 1986).

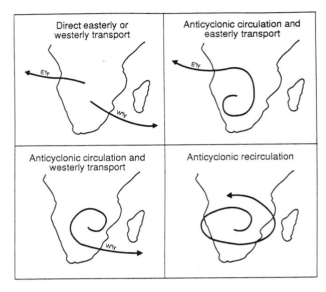

Figure 37.3 Schematic representation of major low-level transport trajectory modes likely to result from the circulation classes shown in figure 37.1 (after Garstang et al. 1996).

components at the starting point are used to compute a new downstream location 15 min later. This position is then iteratively redetermined by using the components at the new position to calculate the next location. The vertical velocities are calculated from diabatic non-linear normal-mode initialization which permits consideration of diabatic as well as adiabatic processes that influence vertical motion. Trajectories from all locations and levels are calculated for each day falling into each circulation class.

Meridional and latitudinal planes, as illustrated in figure 37.4 (a meridional plane along longitude 20°E), are then erected from the surface to 100 hPa at 5° intervals of latitude and longitude over the continent and at 10° intervals over the adjacent oceans (only heights to 500 hPa are shown in figure 37.4). The meridional and latitudinal planes are gridded at 5° and 100 hPa intervals and the location and number of trajectory strikes in each grid cell are recorded. Trajectory strikes originating from the point of origin in the direction of the large-scale flow are recorded separately from strikes that may occur due to reversal (recirculation) of the flow. The time taken from the point of origin to the grid cell or from one grid cell on a given plane to another grid cell on an adjacent plane is also recorded.

The percentage of number of hits per grid cell of the total number of direct or recirculated trajectories is recorded at the center of each grid cell such that the envelope including 98% of strikes on a given plane can

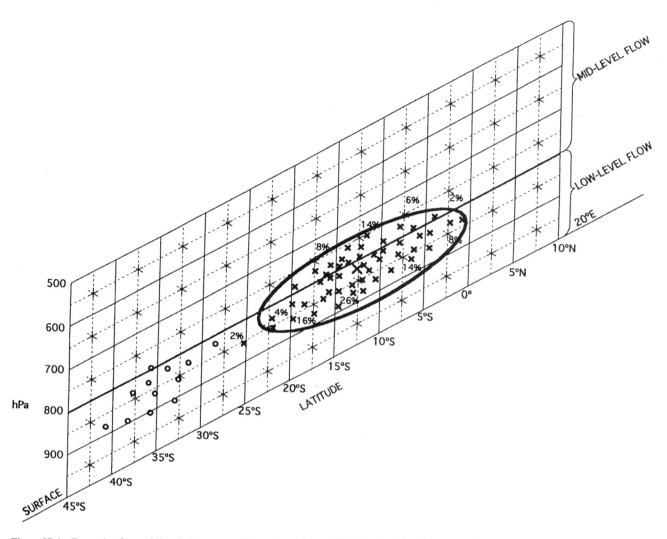

Figure 37.4 Example of a meridional plane erected from the surface to 500 hPa along the 20th east meridian from 45°S to 10°N latitude. The crosses indicate individual direct trajectory strikes from an upstream point of location, the circles indicate indirect or recirculated trajectory strikes by trajectories embedded in return flow. The percentage number of strikes for the direct circulation is shown at the center of each cell and are enclosed by the 98% frequency contour (after Tyson et al. 1996).

be constructed as shown in figure 37.4. Connecting envelopes on successive planes constitute the transport pathway as shown in figure 37.5. The central core or maximum frequency transport pathway, is shown by the heavy dark line and arrowheads in figure 37.5 and the recirculated core transport is shown by the heavy dashed lines and associated arrowheads.

Calculation of Volume and Mass Fluxes

Once the 98% transport area has been identified on the x,z and y,z planes, a volume transport can be calculated by multiplying the area on the plane by the transport velocity. This volume flux may be obtained in the zonal or meridional directions for direct and re-

circulated transport.

Tyson et al. (1996) have estimated the mean daily zonal volume fluxes associated with each circulation type. The annual frequency of occurrence of each circulation type established over a 5-year period allows an estimate of the annual volume fluxes of air across 10°E and 35°E longitude off southern Africa to be determined. These results are shown in table 37.1.

Volume fluxes can be converted to mass fluxes if information on concentrations is available. Tyson et al. (1996) use estimated mean surface concentrations of aerosols over southern Africa to obtain annual mean mass fluxes of aerosols into the atmosphere over the Atlantic (29 Mtons) and Indian Oceans (45 Mtons).

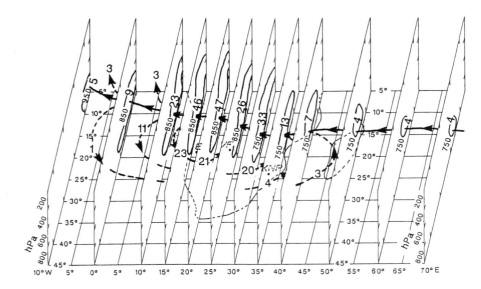

Figure 37.5 An example of the loci of maximum transport in trajectories embedded in easterly zonal flow between 875–500 hPa. The results shown are derived from 960 trajectories originating at five levels and three locations over southern Africa for an 80-day period between 16 August and 31 October 1992. On each meridional plane the percentage of the initial total number of trajectories passing through the plane is indicated in bold type; the height of the locus of highest frequency is shown in hPa in light type. The boundary line enclosing 98% of the trajectory passages on each meridional plane together with the recirculating flow is also shown. Solid lines given direct transport; dashed lines indicate recirculation (after Garstang et al. 1996)

Table 37.1 Mean annual zonal volume flux as a function of four circulation classes through 10°E (upper) and 35°E (lower) in m³ y⁻¹ × 10¹⁶ and in percentage in columns three, four, and five

Circulation type	Mean annual frequency of occurrence (%)	Annual zonal volume flux (m³ y⁻¹ × 10¹⁶)	Percent volume flux eastwards and westwards	Percent of total zonal volume flux
Continental	40	31.9	67.7	26.6
anticyclone		46.2	63.5	38.5
Ridging	13.6	2.7	5.7	2.2
anticyclone		8.5	11.7	7.1
Westerly	27.6	6.1	13.0	5.1
wave		7.7	10.6	6.4
Easterly	18	6.4	13.6	5.3
wave		10.4	14.3	8.7
TOTALS		47.1		39.2
		72.8		60.7

They further estimate that some 60 Mtons of material is deposited over the subcontinent in the course of a single year.

Comparisons with UV-DIAL Measurements

Ultraviolet differential absorption lidar measurements were made over and off southern Africa during the Transport and Atmospheric Research Chemistry near the Equator-Atlantic (TRACE-A) (Fishman 1994; Browell et al. 1995). Vertical sections from altitudes above 500 hPa were obtained for aerosol concentrations. Maximum frequency pathways calculated by the methods described above can be compared with the UV-DIAL aerosol sections. Two examples are shown in figure 37.6. Absolutely stable layers where the lapse rate of the atmosphere is less than the moist or wet adiabatic lapse rate are superimposed on the UV-DIAL sections. Both aerosols and the transport pathways are constrained below the stable layers shown in figure 37.6. Near-surface (900 hPa), low- (700 hPa) and mid-level (500 hPa) transports are shown in figure 37.6a. Each trajectory passes through the vertical section at an aerosol maximum (>50 × 10³ relative aerosol scattering). The low-level (700 hPa) transport shows explicit continental origin, while the surface (900 hPa) and mid-level (500 hPa) trajectories enter the UV-DIAL section from over the Atlantic Ocean, but with clear anticyclonic recurvature. All three transports reflect the circulation class of ridging anticyclones, one of which prevailed on 3 October 1992 (Garstang and Tyson 1996).

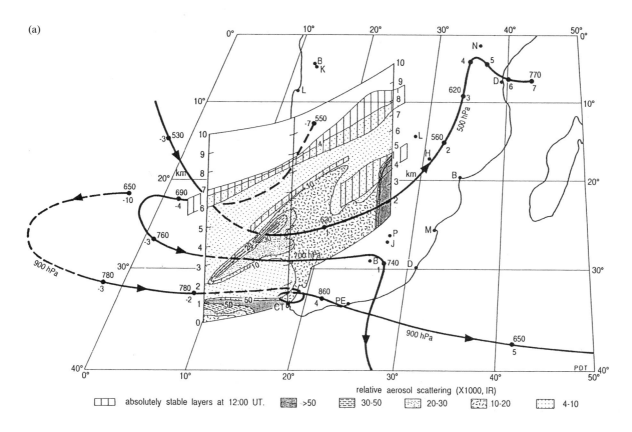

(a)

relative aerosol scattering (X1000, IR)

absolutely stable layers at 12:00 UT.   >50   30-50   20-30   10-20   4-10

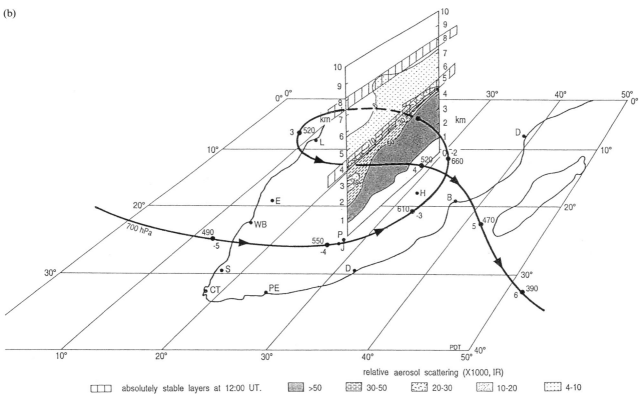

(b)

relative aerosol scattering (X1000, IR)

absolutely stable layers at 12:00 UT.   >50   30-50   20-30   10-20   4-10

Figure 37.6 UV-DIAL sections of relative aerosol backscatter below flight levels of the track of the NASA DC-8 aircraft during TRACE-A. Absolutely stable layers are indicated on each UV-DIAL section. (*a*) 3 October 1992 section showing trajectories at three heights: 900, 700, and 500 hPa. The locations of towns are shown by upper case letters. Successive 24-h locations and heights in hPa are shown along each transport pathway. (*b*) 6 October 1992 as for (*a*) showing transport at 700 hPa.

On 6 October 1992, figure 37.6*b* shows that the 700-hPa trajectory originated over the Atlantic Ocean more than 5 days earlier. Over the subcontinent the transport is subject to recirculation in an anticyclone gyre which penetrates the highly stratified UV-DIAL section in the region of highest relative aerosol backscatter between 1 and 4 km. On return through the UV-DIAL section, the trajectory, after a brief excursion over the western regions, is elevated to above 500 hPa and penetrates the UV-DIAL section at a higher level and lower concentration (relative backscatter 30–50 × 103). Both penetrations of the UV-DIAL section are, however, below the lower level stable layer between 4 and 5 km above sea level (ASL).

Conclusions

Trajectories, when determined from multiple locations and levels under uniform atmospheric circulation conditions, can be treated in a collective sense to yield transport pathways. These transport pathways, dealt with in terms of zonal and meridional flow, can be treated statistically to provide the location of the maximum frequency core transport in three-dimensional space, the envelope containing a specific frequency of transport (such as the 98% level), and an associated representation of the spatial dispersion of the transport envelope containing the specific frequency of individual trajectories.

The vertical thermodynamic structure of the atmosphere typical of the circulation prevailing on a given day can be shown in direct relationship to the meridional and zonal transport pathways. All cases examined in the spring of 1992 over southern Africa showed transport pathways constrained in altitude by pervasive stable layers in the atmospheric column. Concentrations of materials, such as aerosols, were seen to be similarly confined to discrete layers, conforming to the inversion or stable layer heights. Correspondence was thus seen between the UV-DIAL relative aerosol backscatter and the locations of the major transport cores as determined from the trajectory analysis.

Determination of an envelope of transport defines an area on a vertical plane that can be transformed to a volume flux by determining the transport velocity. The transport velocity is known from the time and distance of individual trajectories from their points of origin. Mean travel, times, and distances for each grouping of trajectories yield a mean transport velocity for that group of trajectories. Once volume fluxes are known, mass fluxes can be determined given knowledge of the mean concentrations of the quantity under consid-

eration. The volume fluxes determined as a function of the circulation type can be extended to longer time scales (months, seasons, or years) based on the climatology of the circulation types. Similarly, mass fluxes can be extended over time if time-dependent concentrations are known.

Presentation of transport in terms of circulation classes also provides insight into source regions, source types, and transport characteristics such as direction, recirculation, and deposition. Knowledge of transport times as well as pathways is particularly important for photochemically active elements. The extent to which recirculation takes place over southern Africa is a direct result of partitioning the transport into circulation classes. The methodology rests heavily upon the large-scale model-generated data base. As such, the method best describes large-scale transports and must be modified or treated with caution when applied in regions or at times when smaller scale motions in the atmosphere are playing a significant role. Parameterization of the smaller scale processes is possible and is subject to the ultimate goal of determining net transport.

Acknowledgments

This work has been supported under grant ATM92-07924 from the National Science Foundation to the University of Virginia and by the South African Foundation for Research and Development.

References

Bengtsson, L. 1985. Medium-range forecasting at ECMWF, *Adv. Geophys.*, 238, 3–56.

Browell, E. V., M. P. McCormick, C. F. Butler, M. A. Fenn, G. D. Nowicki, W. B. Grant, and S. Ismail. 1995. Airborne and spaceborne lidar observations of biomass burning plumes over Africa and South America. Paper presented at the AGU Chapman Conf. Biomass Burning and Global Change, March 13–17, Williamsburg, Va.

Dorling, S. R., T. D. Davies, and C. E. Pierce. 1992. Cluster analysis: a technique for estimating the synoptic meteorological controls on air and precipitation chemistry—results from Eskdalemuir, South Scotland, *Atmos. Environ.*, 26A, 2583–2602.

Fishman, J. 1994. Experiment probes elevated ozone levels over the tropical south Atlantic Ocean, *EOS*, 75, 380.

Garstang, M. and P. D. Tyson. 1996. Atmospheric circulation, vertical structure and transport. Chapter 5 in *Fire in Southern African Savannas: Ecological and Atmospheric Perspectives*, Witwatersrand University Press, Johannesburg, South Africa.

Garstang, M., P. D. Tyson, R. Swap, M. Edwards, P. Kållerg, and J. A. Lindesay. 1996. Horizontal and vertical transport of air over southern Africa, *J. Geophys. Res.*, 101, 23,721–723, 736.

Kållberg, P. 1984. Air parcel trajectories from analyzed or forecast windfields, Swedish Meteorological and Hydrological Institute, Research and Development Note No. 37. Norrköping, Sweden.

McCormick, M. P., D. M. Winker, E. V. Browell, J. A. Coakley, C. S. Gardner, R. M. Hoff, G. S. Kent, S. H. Melfi, R. T. Menzies, C. M. R. Platt, D. A. Randall, and J. A. Reagan. 1993. Scientific Investigations Planned for the Lidar In-Space Technology Experiment (LITE), *Bull. Amer. Meteor. Soc.*, 74, 205–214.

Moody, J. L., S. J. Oltmans, H. Levy II, and J. T. Merrill. 1995. Transport climatology of tropospheric ozone: Bermuda, 1988–1991, *J. Geophys. Res.*, 100, 7179–7194.

Tyson, P. D. 1986. Climatic Changes and Variability in Southern Africa, Oxford University Press, Cape Town, South Africa, 220 pp.

Tyson, P. D., M. Garstang, R. Swap, P. Kållberg, and M. Edwards. 1996. An air transport climatology for subtropical southern Africa, *Internat. J. Clim.*, 16, 265–291.

Comparison of Biomass Burning Emissions and Biogenic Emissions to the Tropical South Atlantic

Robert J. Swap, Michael Garstang, Stephen A. Macko, Peter D. Tyson, and Per Kållberg

Biomass burning of African savannas is known to produce large amounts of photochemically active aerosols and trace gases that are necessary precursors of tropospheric ozone (Crutzen and Andreae 1990). Biomass burning emissions from both the African and South American continents are hypothesized to be the major source of oxidative precursors for an area of elevated concentrations of mid-tropospheric ozone in the tropical south Atlantic Ocean (Fishman et al. 1991; Andreae et al. 1994). The atmospheric feature is highly seasonal and most pronounced during the austral winter-spring transition (August to October) (Fishman et al. 1991).

The near coincidence of the maximum in Southern Hemisphere biomass burning and the ozone maximum over the south Atlantic gave rise to the hypothesis that the transport of biomass burning emissions from southern Africa to the tropical central south Atlantic are responsible for the formation and maintenance of this feature (Fishman et al. 1991; Cahoon et al. 1992). The excess in mid-tropospheric ozone maximum is most pronounced primarily between 1 and 5 km above the ocean surface (Fishman et al. 1991; Thompson et al. 1996).

Krishnamurti et al. (1993), Fakruzzaman et al. (1993) and Pickering et al. (1994) examined the meteorological environment of this region through the use of model-generated large-scale atmospheric circulation fields and trajectory analyses. They concluded that the maximum in ozone is supported by products derived from biomass burning over both southern Africa and Brazil. Krishnamurti et al. (1993) have also noted the possibility of contributions of stratospheric intrusions of ozone to the maximum.

Two large field campaigns were conducted during the austral winter-spring transition of 1992. These experiments, the Southern African Fire-Atmosphere Research Initiative (SAFARI) and Transport and Atmospheric Chemistry near the Equator-Atlantic (TRACE-A) set out to characterize the regional atmospheric chemistry and aerosol production processes of the south Atlantic and surrounding land masses (Andreae et al. 1994; Fishman 1994). SAFARI focused more on African biomass burning whereas TRACE-A concentrated on the formation and maintenance of the mid-tropospheric ozone maximum.

Although the meteorological environment and modeled transports in this region have been described, few studies have actually observed direct aerosol transport from either South America or Africa to the tropical south Atlantic. Spatial and vertical distributions of aerosols correlate well with the distributions of more reactive, shorter-lived trace gases that are often difficult to measure (Cros et al. 1988). Determination of aerosol distributions and transports may yield an improved understanding of the transports of photochemically active trace gases.

Swap et al. (1996) document two discrete episodes of low-level (900–800 hPa), long-range transport of southern African aerosols to the region of Ascension Island during SAFARI-92. The types of aerosols produced and transported from southern Africa are from a combination of biomass burning, mineral dust and biogenic sources and reflect this marked seasonality of influences (Swap et al. 1996). The pronounced wet and dry seasons that dominate this region create this strong annual cycle (Tyson 1986). Aerosols produced from arid soils and savanna vegetation precede the beginning of the rainy season and are temporally and spatially in phase with aerosols and trace gases produced from biomass burning. The addition of spring rains to seasonally arid soil surfaces also produces nitrogenous trace gas emissions and hydrocarbons from otherwise dormant vegetation (Williams et al. 1992; Fehsenfeld et al. 1992; Levine et al. 1996). The presence of these aerosols from different sources suggest the possibility of multiple origins for photochemically active trace gases.

This chapter furthers the findings of earlier research of Swap et al. (1996), Garstang et al. (1996) and Tyson et al. (1996) and couples information about the meteorological environment, air parcel transports, and surface trace gas and aerosol production processes in southern Africa to compare and contrast the con-

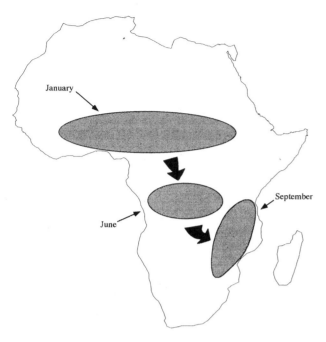

Figure 38.1 Seasonal progression of maximum African biomass burning from January to September-October. The shaded areas encompass the areas of maximum fire frequency in each of three months. (Data from Cahoon et al. 1992.)

tributions of oxidative precursor emissions due to biomass burning and biogenic processes. The importance of biomass burning and rainfall seasonality in African aerosol and trace gas production and emission is examined. Air parcel transports and source regions determined from three-dimensional trajectory analyses are related to African fire and rain seasons. A case study presents the relative contributions to NO_x production in southern Africa by microbial emissions due to pyrogenic and biogenic causes during SAFARI.

Seasonality of Fire and Rain

Two major African aerosol and trace gas production processes are fire and rain. These follow distinct temporal and spatial progressions across the continent and give rise to the marked seasonality in aerosol production over Africa (Swap et al. 1996). Southern Africa exhibits a strong seasonal pattern in biomass burning and rainfall (Cahoon et al. 1992; Nicholson 1986; Tyson 1986; Preston-Whyte and Tyson 1989; Swap et al. 1996). Biomass burning progresses down the subcontinent from the equator southwards from the month of January to October (figure 38.1).

Maximum burning occurs at the end of the dry season and minimal burning is evident during the height

of the rainy season (Cahoon et al. 1992). Seasonal maxima in southern African biomass burning related to anthropogenic activities precedes the onset of the southern African rainy season anywhere from a week to a number of weeks (Cahoon et al. 1992). In contrast, natural fires produced by lightning strikes coincide with the onset of the rainy season storms (Siegfried 1981).

The onset of the rainy season for southern-hemisphere Africa follows a spatial progression that lags behind the biomass burning season at a given location by time periods from weeks to a month (Swap et al. 1996). Spring rains begin in July-August near the equator and in October-November around 20°–25°S (figure 38.2). During SAFARI TRACE-A, the rains had progressed well into the wet season from 0° to 10°S latitude with early spring rains just beginning from 10° to 20°S.

Ascension Island Air Parcel Source Regions

Aerosol and trace gas characteristics are inferred from the determination of the transport and source regions of air parcels arriving at a given location. Ten-day, three-dimensional air parcel trajectories are utilized to describe atmospheric and aerosol transports in and around southern Africa during SAFARI-92. The three-dimensional trajectory package of the Swedish Meteorological and Hydrological Institute (SMHI) is employed for the purposes of this study. This package utilizes the European Centre for Medium range Weather Forecasting (ECMWF) operational analyses of the three-dimensional global windfields that are available every six hours. Results of intercomparative studies have found that the global windfield analyses provided by ECMWF produce the best current trajectory results of the global windfield data sets currently available (Pickering et al. 1994).

The results of the SMHI trajectory package have been verified by comparison with chemical, thermodynamic, and remotely sensed observations for this region of the world (Swap et al. 1995; Garstang et al. 1996; Tyson et al. 1996). Swap et al. (1996) find consistent results between the surface aerosol chemistry and air parcel transports from southern Africa and Ascension Island. The modeled air parcel trajectories reflect the southern African sources only when surface particulate chemistry indicates terrestrial signals at Ascension Island. When particulate chemistry shows no temporal connection at Ascension Island, trajectory analyses show no links between Ascension Island and southern Africa. Results of air parcel trajectory

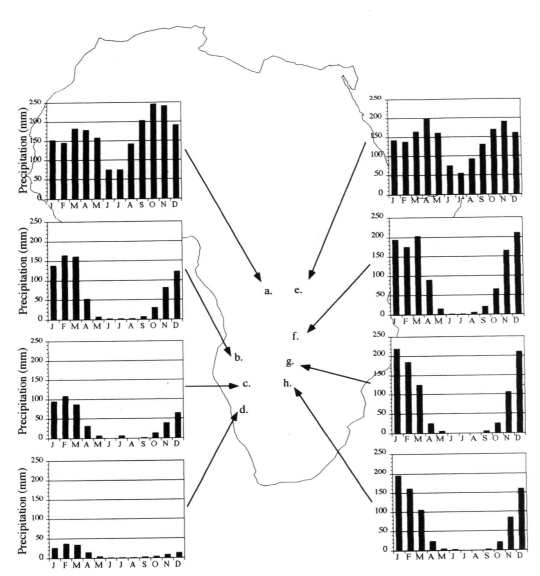

Figure 38.2 Histograms of mean monthly rainfall (mm/month) along two longitudinal transects (17°E, a-d; and 26°E, e-g) for selected southern African sites from the equator to approximately 20°S. The stations and their approximate locations are (*a*) Salonga, Zaire, 2°S, 21°E; (*b*) Bikuar/Mupa, Angola, 15°S, 15°E; (*c*) Etosha National Park, Namibia, 19°S, 16°E; (*d*) Namib/Nakluft, Namibia, 23°S, 15°E; (*e*) Maiko, Zaire, 2°S, 28°E; (*f*) Upemba, Zaire, 10°S, 27°E; (*g*) Kafue, Zambia, 15°S, 26°E; and (*h*) Chobe, Botswana, 18°S, 25°E. Rainfall data are from the Global Change Database Project, 1990, sponsored by the International Council of Scientific Unions, International Geosphere-Biosphere Program and Panel for World Data Centers, Clark University, Worcester, MA, 06610.

calculations during SAFARI also behave according to the vertical constraints placed on them by persistent and often ubiquitous layers of absolute stability present over the region (Garstang et al. 1996; Tyson et al. 1996; Swap et al. 1996). And finally, correspondence is high between the location of the core of three-dimensional trajectory transport and locations of maximum aerosol scattering as observed by airborne UV-DIAL lidar (see chapter 37 this volume by Garstang et al. and chapter 39 this volume by Tyson et al.).

Verification of southern African air parcel trajectories generates a high degree of confidence in the results of the analysis of SAFARI low- (900–800 hPa) and mid-level (700–500 hPa), three-dimensional, 10-day air parcel trajectories. Low- and mid-level trajectories are used because the multiple layers of absolute stability over this region during this season effectively prohibit upward motion of aerosols (Garstang et al. 1996). Descriptive statistics about the backward air parcel trajectories arriving in the lower and middle troposphere at Ascension Island for a 56-day period during SAFARI-92 are generated using the SMHI trajectory package. Five trajectories, each originating at a different initial pressure level for each of the 56 days are grouped and analyzed as an ensemble in the method described in chapter 37.

For air arriving at the lower troposphere (900–800 hPa), just over 70% of the trajectories have their origins over Africa with the remainder originating from the west. The percentages are similar (65% from E, 35% from W) for air parcel trajectories arriving in the middle troposphere (700–500 hPa). The African source region extends over a large, latitudinal band from 20°S to 10°N. Within this broad latitudinal extent, major contributions to the air parcel transports originate between 0° and 10°S (Swap et al. 1996; figure 38.3).

The surface cover and vegetation type are known to vary latitudinally on the African continent with the equatorial rainforest changing gradually polewards from the equator to progressively drier savanna ecosystems. The types of emissions supplied from these different geographical locations are a function of the underlying vegetation and physical processes (biomass burning and rainfall). These physical processes vary seasonally in maximum and minimum intensities. The major air parcel source (for the tropical south Atlantic during September and October) is the region of the African continent lying between 0° and 10°S. At this time this region is at the height of its rainy season and at a concurrent minimum in its fire frequency. The

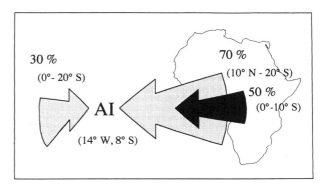

Figure 38.3 Geographical representation of source regions of low-level (900–800 hPa), three-dimensional air parcel transports arriving at Ascension Island during SAFARI-92 (after Swap et al. 1996)

period September to October corresponds with the most pronounced occurrence of the mid-tropospheric ozone maximum over the tropical south Atlantic. The timing of the rains and the lack of burning in the 0° to 10°S region suggests that biomass burning may not be the sole source of ozone and its precursor gases fueling the tropical south Atlantic maximum in ozone.

Case Study of Microbial Production of NO_x During SAFARI

Biomass burning and rainfall follow each other closely and give rise to increased microbial emissions due to pyrogenic and biogenic processes. Both processes are capable of producing large quantities of NO_x. Order of magnitude calculations of pyrogenic and biogenic production are presented for regions in southern Africa during the SAFARI observational period.

Microbial production in arid regions of photochemically important trace gases, such as NO_x, is well known and sensitive to the physical controls of fire and water (Levine et al. 1996; Poth et al. 1996; Anderson and Poth 1989; Hartley and Schlesinger 1993; table 38.1). Increases of 10 to 15 times are seen in the microbial production of NO_x from arid soils above background levels after the addition of water (Levine et al. 1996). This increase is even greater for arid soils that have been burned before the addition of water (Levine et al. 1996; Poth et al. 1996).

Relatively minor amounts of water are required to initiate the production of NO_x from arid soils at the end of the dry season. The values of NO_x production reported by Levine et al. (1996) reflect the input of less than 1 mm of precipitation. These soils are capable of the continued production of NO_x with the next several wetting and drying cycles as the region progresses into

Table 38.1 Ranges of NO$_x$ fluxes from seasonally arid lands after the addition of water to an unburned surface and after the addition of water to a burned surface. NO$_x$ fluxes have the units of NO-N ng m^{-2} s^{-1}.

Study	NO$_x$ (dry)	NO$_x$ (fire)	NO$_x$ (water)	NO$_x$ (fire and water)
Levine et al. 1995[a]	0.83–1.88	5.51–18.22	5.16–11.93	35.86–76.90
Harris et al. 1995[a]	—	—	~30	—
Poth et al. 1995[b]	3.0–6.0	6.0–18.0	18.0–27.0	87.0–126.0
Hartley & Schlesinger 1993[c]	0.34–0.42	—	3.61–36.1	—

a. Southern Africa savanna.
b. Brazilian savanna.
c. New Mexico arid grassland and shrubland.

the rainy season (Anderson, personal communication). This elevated production is known to persist for several days and even, in some cases, for months after burning (Levine et al. 1996; Poth et al. 1996). During the SAFARI campaign when biomass burning in southern Africa was taking place, significantly elevated concentrations of biogenically produced NO$_x$ (~1 ppb) were found (Harris et al. 1996).

Microbial emissions of NO$_x$ resulting from the wetting of burned areas of southern Africa during two weeks (from 20 September to 4 October 1992) of SAFARI are calculated and compared to those fluxes resulting from a precipitation event over central, southwestern Africa on 28 September 1992. The location of maximum biomass burning during these two weeks was northeastern South Africa, southern Mozambique, and eastern Zimbabwe (Kendall, personal communication). The area burned, estimated from satellite observations during these two weeks, is approximately 4.5×10^4 km^2. The area of rainfall over western Botswana and eastern and central Namibia on 28 September 1992, as determined from South African Weather Bureau daily synoptic charts and METEOSAT visible imagery, encompasses approximately 1.4×10^5 km^2. The area contributing microbial emissions of NO$_x$ is halved (to 7.0×10^4 km^2) to reflect the convective nature of the rainfall over this region on 28 September 1992. The area burned during these two weeks represents ~0.8% of the area of southern Africa from 5° to 25°S. The area wetted by the rainfall event of 28 September 1992 represents approximately 1.2%. More precipitation in this region was observed from 20 September to 4 October 1992.

Rates of NO$_x$ emissions representative of these regions are used. An average NO$_x$ flux rate of 50 ng N m^{-2} s^{-1} from Levine et al. (1996) is taken as representative for the region of the burned area during 20 September to 4 October 1992. The rate of 30 ng N m^{-2} s^{-1} is the rate taken as representative of the conditions in Namibia and Botswana on 28 September 1992. This rate is the estimated flux rate necessary to produce the concentrations of strong biogenic emissions ([NO$_x$] ~1 ppb) over central northern Namibia observed by Harris et al. (1996). These strong biogenic emissions were anticorrelated with biomass burning products and most likely the result of a day or two of local thunderstorms. These elevated NO$_x$ flux rates are assumed to persist for at least 48 h for both the area burned and the area receiving rainfall.

Total emission calculations are made for the following cases: burned area assuming the entire area is burned and then wetted, the burned area without any addition of water, the area receiving precipitation during the one-day rain event (50% of area contributing to production), and for the same area assuming no precipitation. The calculations are made as follows:

$$\text{total emission of NO}_x \text{ in g N} = \text{rate(ng N m}^{-2}\text{ s}^{-1})$$
$$\times \text{area(m}^2)$$
$$\times \text{duration}(1.728 \times 10^5 \text{ s}^{-1}).$$

The results are presented in table 38.2 and shown schematically in figure 38.4. These results show that microbial NO$_4$ emissions due to the pyrogenic (burning and wetting) and biogenic processes when dry soils are wetted by rain are of the same order of magnitude. This result is obtained notwithstanding having halved the area wetted by the precipitation.

Discussion and Conclusions

Direct, low-level (900–800 hPa) links have been documented during SAFARI between the African continent and the central tropical south Atlantic midtropospheric ozone maximum (Swap et al. 1996). The transport of southern African surface produced aerosols is governed vertically by the presence of multiple layers of absolute stability (see chapters 37 and 39 this volume). Forward and backward low-level air parcel

Table 38.2 Calculation of total microbial production of NO$_x$ over southern Africa for the following four cases: (A) due to two-week burning from 20 September to 4 October 1992 in southern Africa and subsequent wetting; (B) due to two-week burning alone; (C) due to precipitation event over western Botswana and central Namibia on 28 September 1992; and (D) for the same area as (C), but without the effects of burning or precipitation. Burning data courtesy of J. Kendall. Rainfall coverage data from South African Weather Bureau Monthly Weather Bulletin.

	A	B	C	D
Area	~4.5 × 10^4 km^{2b}	~4.5 × 10^4 km^2	~7.0 × 10^4 km^{2a}	~7.0 × 10^4 km^2
Avg. microbial NO$_x$ production rate	50 ng N m^{-2} s^{-1b}	10 ng N m^{-2} s^{-1b}	30 ng N m^{-2} s^{-1c}	1 ng N m^{-2} s^{-1b}
Duration	1.728 × 10^5 s (48 h)	1.728 × 10^5 s	1.728 × 10^5 s	1.728 × 10^5 s
Total microbial production of NO$_x$	0.39 Gg N	0.08 Gg N	0.36 Gg N	0.01 Gg N

a. This is the area assuming 50% coverage.
b. Levine et al. 1995.
c. Harris et al. 1995.

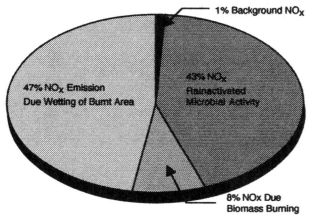

Figure 38.4 Schematic partitioning in percentage of microbial emissions of NO$_x$ during the two-week period specified in table 38.2 due to biomass burning only, wetting of the burnt area, rain-activated microbial production over the wetted area, and background production

space and time of biomass burning and rainfall across Africa that gives rise to coupled pyrogenic and biogenic aerosol and trace gas emissions. Results of the comparison of the microbial production of NO$_x$ due to pyrogenic processes (fire and rain) and due to biogenic processes (rain) show that these NO$_x$ emissions are of the same order of magnitude. This suggests that biogenic sources of NO$_x$ species may be much larger than those resulting from biomass burning for two reasons: the spatial extent of rainfall is greater than that of biomass burning in a given year and repeated rainfalls early in the spring can induce continued, elevated emission rates. Biomass burning is important as a physical process, however, because it enhances microbial NO$_x$ emission rates anywhere from 5 to 10 times that of rates due only to rainfall. Future biomass burning experiments must take into account the seasonality of both fire and rain within the region of interest, the coupled pyrogenic and biogenic emission processes that result from the seasonality, and the fact that multiple sources of photochemically important aerosols and trace gases, with similar magnitudes, are possible.

Acknowledgments

This work has been supported under grant ATM92-07924 from the National Science Foundation to the University of Virginia. The authors would also like to acknowledge contributions of constructive comments, data and useful discussion from the following persons: I. C. Anderson, J. D. Kendall, J. S. Levine, M. A. Poth, and M. E. Uhle.

transports between southern Africa and the central tropical south Atlantic suggest that Africa is the major source (~70%) for air parcel transports between 900 and 500 hPa during SAFARI. These trajectory results have been independently verified (Swap et al. 1996; Tyson et al. 1997, chapter 39 this volume). This African source region is broad (10°N–20°S), undergoes changes in the seasonality of biomass burning and rainfall, and contributes multiple types of aerosol and trace gas emissions. A significant amount (>50%) of the African transport is from the latitudinal region of 0°–10°S. This region is at the height of its climatological rainy season during September and October and experiences a minimum in biomass burning.

There is little doubt that biomass burning emissions are important sources of photochemically active emissions. However, during the austral winter to spring transition, there is a nearly synchronous progression in

References

Anderson, I. C., and M. A. Poth. 1989. Semiannual losses of nitrogen as NO and N$_2$O from unburned and burned chaparral, *Global Biogeochem. Cycles*, 3, 121–135.

Andreae, M. O., J. Fishman, M. Garstang, J. G. Goldammer, C. O. Justice, J. S. Levine, R. J. Scholes, B. J. Stocks, A. M. Thompson, B. van Wilgen, and the STARE/TRACE-A/SAFARI-92 Science Team. 1994. Biomass burning in the global environment: First results from the IGAC/BIBEX field campaign STARE/TRACE-A/SAFARI-92, *Glob. Atmos.-Biosp. Chem.*, The First IGAC Scientific Conf., Plenum Press, New York, 83–101.

Cahoon, D. R., B. J. Stocks, J. S. Levine, W. R. Cofer, and K. P. O'Neill. 1992. Seasonal distribution of African savanna fires, *Nature*, 359, 812–815.

Cros, B., R. Delmas, D. Nganga, B. Clairac, and J. Fontan. 1988. Seasonal trends of ozone in equatorial Africa: Experimental evidence of photochemical formation, *J. Geophys. Res.*, 93, 8355–8366.

Crutzen, P. J., and M. O. Andreae. 1990. Biomass burning in the tropics: Impact on atmospheric chemistry and biogeochemical cycles, *Science*, 250, 1669–1678.

Fakhruzzaman, K. M., J. Fishman, V. G. Brackett, J. O. Kendall, and C. O. Justice. 1993. Large-scale circulation patterns associated with high concentrations of tropospheric ozone in the tropical south Atlantic Ocean, *Proc. Quad. Ozone Symp.*, 158–161.

Fehsenfeld, F., J. Calvert, R. Fall, P. Goldar, A. B. Guenter, C. N. Hewitt, B. Lamb, S. Liu, M. Trainer, H. Westberg, and P. Zimmerman. 1992. Emissions of volatile organic components from vegetation and the implications for atmospheric chemistry, *Global Biogeochem. Cycles*, 6, 389–430.

Fishman, J. 1994. Experiment probes elevated ozone levels over the tropical south Atlantic Ocean, *EOS*, 75, 380.

Fishman, J., K. Fakhruzzaman, B. Cros, and D. Nganga. 1991. Identification of widespread pollution in the Southern Hemisphere deduced from satellite analyses, *Science*, 252, 1693–1696.

Fishman, J., C. E. Watson, J. C. Larsen, and J. A. Logan. 1990. The distribution of tropospheric ozone determined from satellite data, *J. Geophys. Res.*, 95, 3599–3617.

Garstang, M., P. D. Tyson, R. Swap, M. Edwards, P. Kållberg, and J. A. Lindesay. 1996. Horizontal and vertical transport of air over southern Africa, *J. Geophys. Res.*, 101, 23, 721–723, 736.

Harris, G. W., T. Zenker, F. G. Weinhold, and V. Parchatka. 1996. Airborne observations of strong biogenic NO_x emissions from the Namibian savanna at the end of the dry season, *J. Geophys. Res.*, 101, 23, 707–723, 712.

Hartley, A. E., and W. H. Schlesinger. 1993. Nitric oxide emission from arid grassland and shrubland ecosystems in southern New Mexico, *Bull. Ecol. Soc. Amer.*, 74, 268.

Krishnamurti, T. N., H. E. Fuelberg, M. C. Sinha, D. Oosterhof, E. L. Bensman, and V. B. Kumai. 1993. The meteorological environment of the tropospheric ozone maximum over the tropical south Atlantic Ocean, *J. Geophys. Res.*, 98, 10 621–10 641.

Levine, J. S., E. I. Winstead, D. A. B. Parsons, M. C. Scholes, W. R. Cofer, III, D. R. Cahoon, Jr., and D. I. Sebacher. 1996. Biogenic soil emissions of nitric oxide (NO) and nitrous oxide (N_2O) from savannas in South Africa: The impact of wetting and burning, *J. Geophys. Res.*, 101, in press.

Nicholson, S. E. 1986. The nature of rainfall variability in Africa south of the equator, *J. Clim.*, 6, 515–530.

Pickering, K. E., A. M. Thompson, D. P. McNamara, and M. R. Schoeberl. 1994. An intercomparsion of isentropic trajectories over the south Atlantic, *Mon. Wea. Rev.*, 122, 864–879.

Poth, M. A., I. C. Anderson, H. S. Miranda, A. C. Miranda, and P. J. Riggan. 1995. The magnitude and persistence of soil NO, N_2O, CH_4, and CO_2 fluxes from burned tropical savanna in Brazil, Submitted to *Glob. Biogeochem. Cycles*, 9, 503–513.

Preston-Whyte, R. A., and P. D. Tyson. 1989. *The Atmosphere and Weather of Southern Africa*, Oxford University Press, Cape Town, 374 pp.

Siegfried, W. R. 1981. The incidence of veldfire in the Etosha National Park, 1970–1979, *Madoqua*, 12, 225–230.

Swap, R. J., M. Garstang, S. A. Macko, P. D. Tyson, W. Maenhaut, P. Artaxo, P. Kållberg, and R. Talbot. 1996. The long-range transport of southern African aerosols to the tropical south Atlantic, *J. Geophys. Res.*, 101, 23, 777–783, 792.

Thompson, A. M., K. E. Pickering, D. P. McNamara, M. R. Schoeberl, R. D. Hudson, J. H. Kim, E. V. Browell, V. W. J. H. Kirchhoff, and D. Nganga. 1996. Where did tropospheric ozone over southern Africa and the tropical Atlantic come from in October 1992? Insights from TOMS, GTE/TRACE-A and SAFARI-92, *J. Geophys. Res.*, 101, 24, 251–264, 278.

Tyson, P. D. 1986. *Climatic Changes and Variability in Southern Africa*, Oxford University Press, Cape Town, 220 pp.

Tyson, P. D., M. Garstang, R. J. Swap, P. Kållberg, and M. Edwards. 1996. An air transport climatology for subtropical southern Africa, Accepted for *Internat. J. Clim.*, 16, 265–291.

Williams, E. J., G. L. Hutchinson, and F. C. Fehsenfeld. 1992. NO_x and N_2O emissions from soil, *Glob. Biogeochem. Cycles*, 16, 351–388.

Transport and Vertical Structure of Ozone and Aerosol Distributions over Southern Africa

Peter D. Tyson, Michael Garstang, Robert J. Swap, Edward V. Browell,
Roseanne D. Diab, and Anne M. Thompson

Any global change forced by greenhouse warming will be the result of integrated hemispheric effects, which in turn will be the product of the additive consequences of many regional changes. The degree to which regional effects will be contained within their localities, or will expand to encompass those of adjacent or even far-removed counterparts, will depend on production and destruction, and atmospheric mixing and transport of aerosols and trace gases. Knowledge of biogenic and anthropogenic sources is necessary if regional changes responsible for the production are to be identified. Residence time in the atmosphere will influence production and destruction of trace gases and concentrations of aerosols.

The Transport and Atmospheric Chemistry near the Equator-Atlantic (TRACE-A) experiment was conducted in 1992 over the South American-Atlantic Ocean-African region of the Southern Hemisphere and was directed at determining the existence, characteristics, and formation of tropospheric ozone in an area previously reported to be one of enhanced austral spring concentrations over the tropical South Atlantic Ocean (Fishman 1991; Fishman et al. 1992). The origin of this tropospheric maximum has been ascribed to long-range transport of biomass burning products from Africa and South America (Crutzen and Andreae 1990; Watson et al. 1990; Moody et al. 1991; Pickering et al. 1994a,b). Krishnamurti et al. (1994) and Thompson et al. (1996) have suggested possible enhancements in tropospheric ozone originating in the stratosphere. One of the objectives of TRACE-A was to determine the actual extent to which late-winter and early spring biomass burning products from southern Africa might be transported out of the continent and over the ocean to contribute to the maximum. Concurrently, the Southern African Fire-Atmosphere Research Initiative (SAFARI) was conducted and directed towards the determination of production of biomass burning products over the subcontinent and their transport vertically and horizontally in the atmosphere within and beyond the region (Andreae et al. 1994). Atmo-

spheric transport during SAFARI has been examined in detail (Garstang et al. 1996; Swap et al. 1996; Garstang and Tyson 1996) and a generalized and seasonal air transport climatology be atmospheric circulation type has been presented for southern Africa (Tyson et al. 1996). Chapter 37 of this volume synthesizes these methods specifically and addresses the transport of surface-generated species over and out of southern Africa.

In this chapter, TRACE-A and SAFARI measurements of the horizontal and vertical variation of ozone and aerosol concentrations at fixed locations and along transects over southern Africa and the Atlantic Ocean for the period August through October 1992 are considered. The basic methodology that considers a statistical ensemble of many trajectories within a climatological framework of dominant circulation types to demonstrate the existence of major transport pathways is described by Garstang et al. (see chapter 37 this volume). Centers of elevated ozone concentrations are shown to coincide with spatial sections determined from airborne UV-DIAL measurements. The origin and transport destinations of gaseous and particulate material appearing in the centers of maximum concentration on the UV-DIAL sections are examined using forward and backward trajectories from these maxima. The findings will be compared to the regional bulk transportation patterns during SAFARI, as well as against established air transport climatology for southern Africa (Tyson et al. 1996).

Background

Garstang et al. (1996) and Tyson et al. (1996) have combined three analytical elements in order to arrive at an estimate of atmospheric transport: determination of the dominant atmospheric circulation types on a climatological time scale for the region under consideration (southern Africa), incorporation of atmospheric stability both on a local and regional scale, and finally treatment of trajectories as an ensemble of many points in three-dimensional space (see chapter 37).

This approach is used to interpret independently determined airborne Ultra Violet Differential Absorption Lidar (UV-DIAL) vertical distributions of aerosol concentrations on planes comparable to those used above. The UV-DIAL measures relative backscatter from aerosols in the column below and above the aircraft. Simultaneous measurements of the vertical distribution of ozone concentrations are made (Browell et al. 1993). Absolutely stable layers determined from rawinsonde data are combined with each UV-DIAL section. The DIAL vertical averaging interval is 60 m and the horizontal averaging interval is 3.55 sec, which equates at the average cruising speed of the DC-8 to approximately 700 m. Flight paths on different days included Cape Town to Johannesburg, Johannesburg to northern Zambia, Johannesburg to Windhoek via Victoria Falls and Etosha, an offshore transect along the east coast from south of Port Elizabeth to north of Maputo, and finally, an offshore transect along the west coast from Walvis Bay to north of Luanda.

Each of the UV-DIAL sections are placed within the context of the circulation classification according to the day on which the UV-DIAL transect was complied and the transport field for that day (Garstang et al. 1996). Within this context individual forward and backward trajectories are then calculated originating from the UV-DIAL–determined maxima in ozone and aerosols.

Transports through Ozone and Aerosol Maxima in the Vertical

Ozone Transport

Figures 39.1 and 39.2 show UV-DIAL vertical sections of ozone for 6 and 9 October 1992, respectively. Absolutely stable layers as determined from the regional rawinsonde network have been superimposed upon each cross-section. Forward and backward trajectories have been calculated through origins at the UV-DIAL–determined locations of maximum concentration in ozone. The most noteworthy characteristic of the trajectories shown in figures 39.1 and 39.2 are that an absolutely stable layer at about 6 km altitude appears to act as a material barrier to penetration of trajectories below the stable layer.

On 6 October 1992 (figure 39.1), the upper-level ozone maximum over Zambia at 15°S, 30°E appears to have been associated with air that had a tropical tropospheric origin over Africa, whereas some 1000 km to the south the maximum over Zimbabwe at 20°S, 30°E

was associated with air that the 300-hPa trajectory shows was 10 days earlier at 9°S, 14°W over the tropical Atlantic Ocean off Brazil.

On 9 October 1992 (figure 39.2), the tropical stratospheric ozone in the upper troposphere over the east coast at about 25°S appears likewise to have originated in the lower stratosphere over tropical Africa just over one day previously to become trapped within stable air at around 10 km before being transported quickly eastward out of the tropics and over the Indian Ocean in the westerlies. The stratosphere is here defined conventionally as being the base of the inversion of temperature in the stratosphere. Further to the south, over the east coast south of Port Elizabeth at around 34°S, ozone at the 300-hPa level had originated more than five days earlier off the coast of southern Brazil. After passing southern Africa it was transported eastward over the ocean in a zonal westerly stream towards Australia.

Figure 39.3 shows time-height cross-sections of ozone for the SAFARI period: Ascension Island (8°S, 14°W) and Brazzaville (4°S, 15°E); figure 39.4 those for Okaukuejo (19°S, 16°E) and Irene (25°S, 28°E). The time sections of ozone show the same characteristic seen on the individual space sections from the UV-DIAL, namely a general separation in ozone concentrations above 12 km (tropopause) and below 2 or 3 km. The stable layer at 6 km over Africa would appear to maintain a nearly continuous separation in time between ozone concentrations in the upper and lower troposphere. Over Ascension Island the pervasive stable layer lies between 3 and 4 km altitude. The penetrations of stratospheric air seen on the UV-DIAL sections for 9 and 14 October 1992 are reflected in the time section at Irene shown in figure 39.4. Excess O_3 within each sounding shown in figures 39.3 and 39.4 has been computed normalizing to a mean ozone profile appropriate for the nonburning season (Cros et al. 1992). Below 5 km, excess ozone is large (11 Dobson Units) where a Dobson Unit $= 2.69 \times 10^{16}$ cm^{-2} for individual soundings at the continental sites. Th average excess ozone in the 0 to 5 km layer is 4.5 DU at Brazzaville, 4.8 DU at Etosha Park, and 3.3 DU at Irene. Aircraft sampling in stable layers up to 4 km over the Kruger-Victoria Falls-Etosha Park regions showed ozone formation increasing by as much as 15 ppbv O_3 d^{-1} (see chapter 28 this volume by Zenker et al.).

Aerosol Transport

Transport of surface-produced aerosols over southern Africa during the TRACE-A/SAFARI experiment

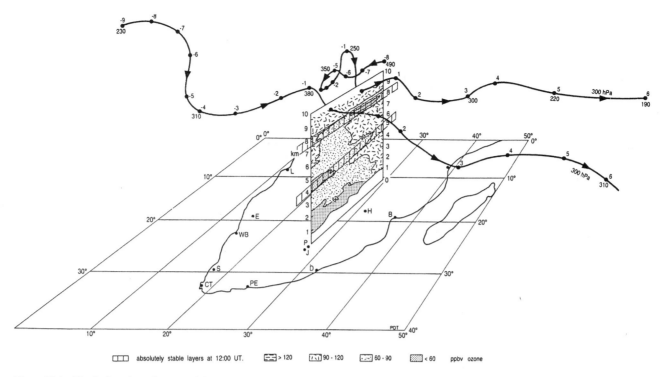

Figure 39.1 Vertical section of ozone mixing ratio, ppb, for 6 October 1992, as determined by the NASA DIAL system along a transect over central South Africa together with absolutely stable layers and 300 hPa back and forward trajectories. Days of travel and geopotential levels are given for selected points on the trajectories. CT = Cape Town, PE = Port Elizabeth, S = Springbok, B = Bloemfontein or Beira, D = Durban or Dar-Es-Salaam, J = Johannesburg, P = Pretoria, M = Maputo, W = Windhoek, E = Etosha, VF = Victoria Falls, L = Luanda or Lusaka, and H = Harare.

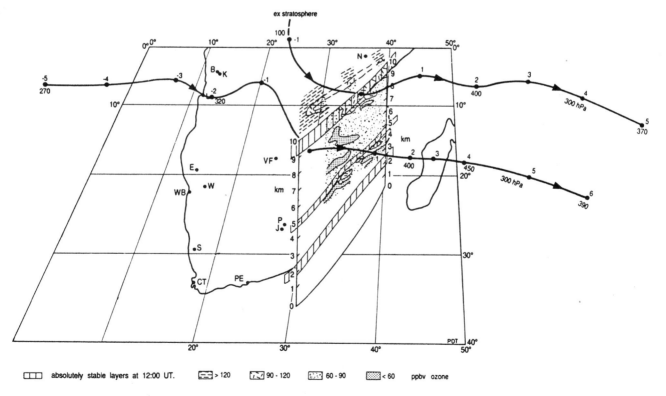

Figure 39.2 Vertical sections of ozone mixing ratio, ppb, for 9 October 1992, as determined by the NASA DIAL system along a transect off the east coast of southern Africa. Remainder as for figure 39.1.

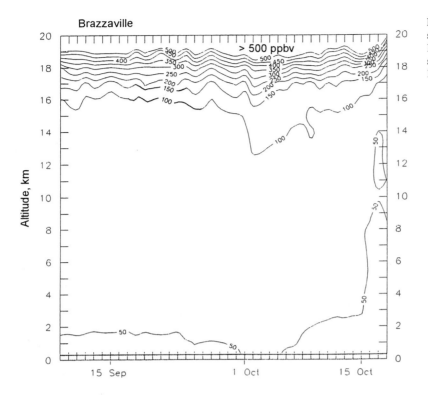

Brazzaville

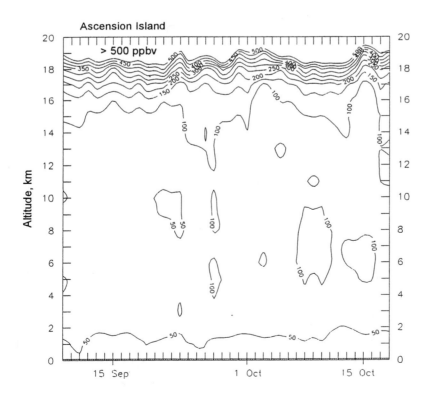

Ascension Island

Figure 39.3 Time-height cross-sections of ozone (ppbv) for the SAFARI-92 period derived from ozonesondes at Ascension Island and Brazzaville

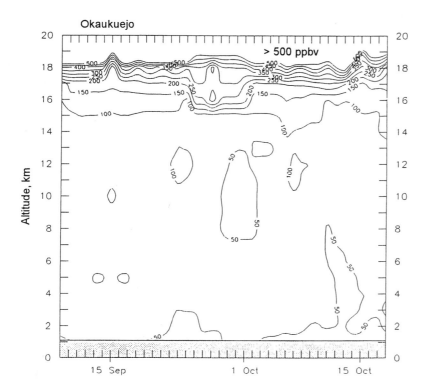

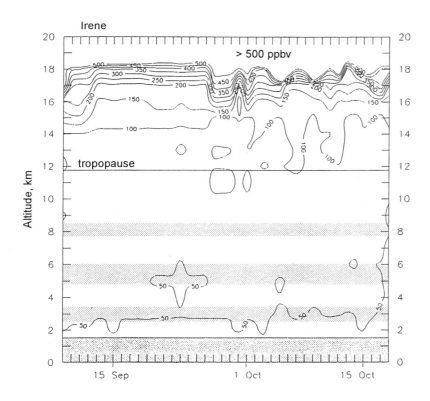

Figure 39.4 Time-height cross-sections of ozone (ppbv) for the SAFARI-92 period derived from ozonesondes at Okaukuejo and Irene; shading indicates absolutely stable layers

was controlled in the vertical by the occurrence of a persistent absolutely stable mid-level layer at about 5 km altitude, and to a lesser extent, one at about 3 km (Garstang et al. 1996). The extent to which the mid-level layer inhibits upward mixing of aerosols is evident in the examples given in figure 39.5. Unlike the quasi-zonal transport of ozone in the upper troposphere, lower tropospheric transport of both ozone and aerosols was dominated by anticyclonic recirculation. Such was the case with the transport of aerosols on 6 and 11 October 1992 (figure 39.5). On 6 October air recirculated for over 8 days over the subcontinent; on 11 October air that originated at 580 hPa just off the east of southern Africa circulated over the subcontinent for more than 14 before exiting at about the same latitude. The clear vertical structure of the anticyclonic transport vortex below the 5-km stable layer is illustrated in figure 39.6 (*upper*).

Accumulation of surface-derived species appears evident in the sequence of UV-DIAL cross-sections of aerosol concentrations shown for 3, 6, and 11 October 1992 in figures 39.5 and 39.6. Enhanced ozone seen at the three continental locations in figures 39.3 and 39.4 starting in the second week of October 1992 also suggest increased production of ozone at the surface. The vertical plume of ozone at Okaukuejo on 13 October and after 15 October at Irene and Brazzaville suggest a continent-wide production and vertical transport of ozone reflected in part in aerosol distributions shown on 11 October 1992 (figure 39.5) UV-DIAL cross-section.

Safari Transport Fields

Ground and other observations were made at three primary locations during the mid-August to end-October 1992 field observation period of SAFARI. They were Etosha National Park ($\sim 19°$S, $16°$E) in Namibia, Victoria Falls ($\sim 18°$S, $26°$E) in Zimbabwe, and Kruger National Park ($\sim 25°$S, $32°$E) in South Africa. Trajectory analyses were completed using these localities as points of origin. Nine hundred and sixty forward trajectories were analyzed, 320 per station. From these transportation fields were constructed as described in chapter 37. As an example the fields for Victoria Falls will be considered.

Of all forward trajectories emanating from Victoria Falls between the surface and 800 hPa, 83% are initially easterly (i.e., transport to the west) in the first hour (figure 39.7). Only 4% move initially to the east. The remainder either move north or south or stagnate in calm conditions. The pattern of flow with easterly transport (figure 39.7, *top*) shows that local recirculation around the origin is apparent (102% of trajectories pass through the 20°E meridional plane and this can only happen with recirculation). Only 13% of trajectories cross the Greenwich Meridian. Some 14% recirculate from the east and southeast to approach the station from the opposite direction to that from which they originated. Westerly transport out of Africa to the southeast is considerable, with the plume within which 95% of trajectories are enclosed being clearly defined and orientated (figure 39.7, *bottom*). Eighty-two percent of low-level trajectories originating from Victoria Falls exit southern Africa at 35°E at around 30°S and at a height of 600 hPa after an average travel time of 4.6 days; 58% reach 70°E at a latitude of 40°S and a height of 450 hPa after 6.2 days.

Similar analyses were completed for the layer 700–500 hPa. Knowing the horizontal and vertical dimensions of the entire transport plume and the speed with which it is moving, the total zonal volume flux of air may be determined. Coupling this quantity to mean-measured aerosol concentrations at Victoria Falls allows the mass transport of particulate material to be estimated. For the SAFARI period approximately 0.03 Mtons d^{-1} was transported westward through 10°E; 0.04 Mtons d^{-1} was transported eastwards through 35°E.

Taking all three stations, it is possible to define vertically integrated surface to 800 hPa transport along the east-west transect across the subcontinent defined by the stations. The maximum frequency pathways are similar (figure 39.8, *top*) and the plume envelopes for the different pathways overlap considerably. Transport to the west is constrained vertically by the 3-km absolutely stable layer; that to the east by the 5-km layer. Very few of the 960 trajectories analyzed could penetrate the 5-km stable layer over the continent and adjacent coastal areas. Low-level recirculation occurred on a massive scale (figure 39.8, *bottom*), with the spatial average recirculation index (defined as the sum of the percentages recirculating to the east and west together with the percentage calms/stagnation) for the three points of origin being 74%.

Similar results are obtained for vertically integrated 700–500 hPa transport. For combined surface to 500 hPa transport, a preliminary estimate of aerosol mass flux from the three stations during SAFARI is approximately 0.11 Mtons d^{-1} to the west through 10°E and 0.19 Mtons d^{-1} to the east through 35°E. The export of ozone from this region can now be estimated. The Etosha Park and Irene soundings show that ozone averages 27 ppbv above background level between 875

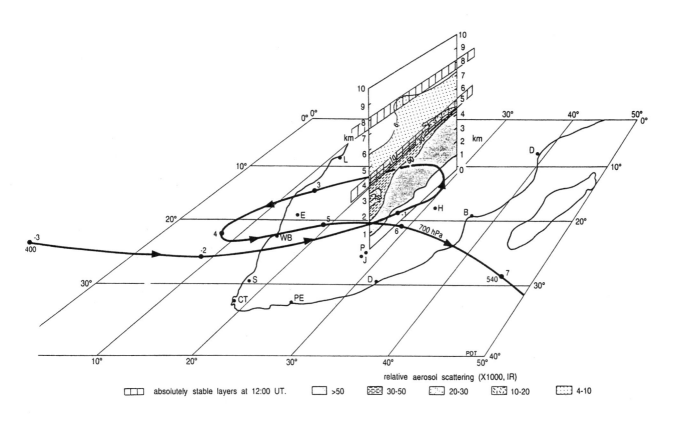

absolutely stable layers at 12:00 UT.   >50   30-50   20-30   10-20   4-10

relative aerosol scattering (X1000, IR)

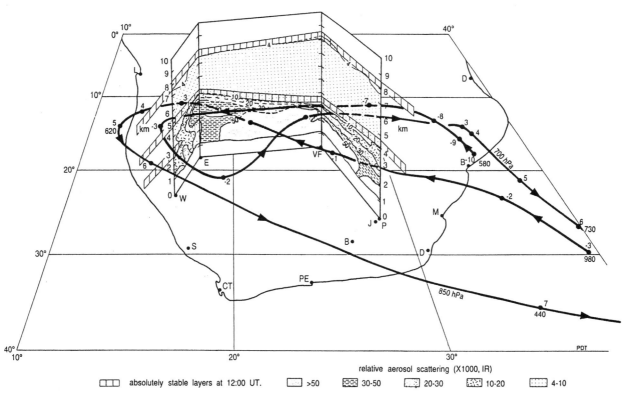

absolutely stable layers at 12:00 UT.   >50   30-50   20-30   10-20   4-10

relative aerosol scattering (X1000, IR)

Figure 39.5 Vertical sections of relative aerosol concentration as determined by the NASA/UV-DIAL system along transects from Johannesburg to northern Zambia on 6 October 1992 and from Johannesburg to Victoria Falls, Etosha, and Windhoek on 11 October 1992 and the relationship of aerosol concentrations to absolutely stable layers and 700 and 850 hPa back and forward trajectories. Days of travel and geopotential levels are given for selected points on the trajectories; places are defined as in figure 39.1.

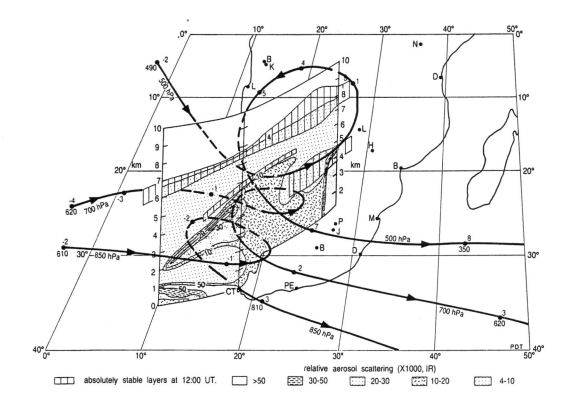

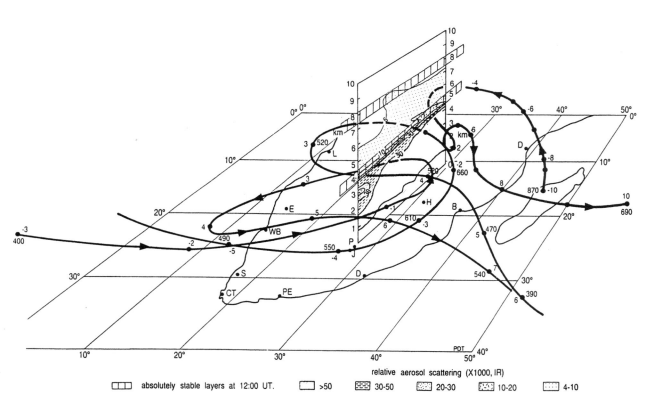

Figure 39.6 Vertical sections of relative aerosol concentration as determined by the NASA UV-DIAL system along transects from southwest of Cape Town to Johannesburg on 3 October 1992 and from Johannesburg to northern Zambia on 6 October 1992. The vertical structure of the transport vortex in the lower atmosphere is illustrated in the top section by the 850, 700, and 500 hPa back and forward trajectories. In the lower section the complicated nature of the recirculation is illustrated by 700 hPa trajectories crossing the plane of measurement in different places. Days of travel and geopotential levels are given for selected points on the trajectories; places are defined as in figure 39.1.

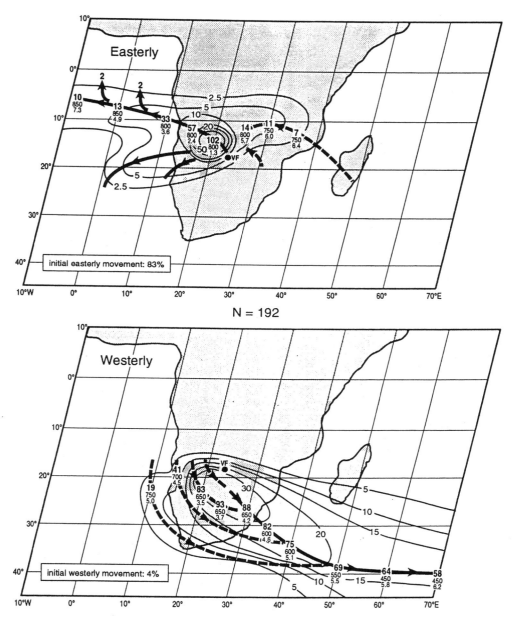

Figure 39.7 Contours of percentage of vertically integrated surface-to-800-hPa zonal transport in easterly and westerly directions from Victoria Falls during the SAFARI field observation period, August through October 1992. Number of trajectories analyzed is 192. Solid heavy lines indicate the core of direct maximum frequency transport; broken lines indicate the core of maximum frequency recirculated air transport. Bold figures give the total horizontally and vertically integrated percentage transport at a particular meridian; light figures give the height of this transport at the location of maximum frequency transport, and italic figures the mean transit time for all trajectories to reach the meridian.

maximum frequency transport pathways

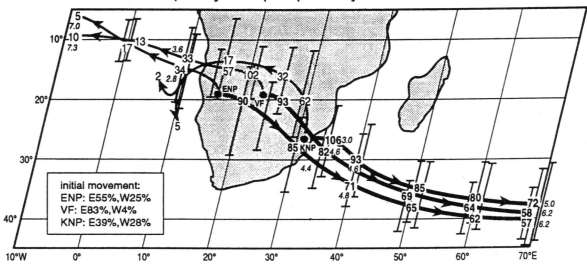

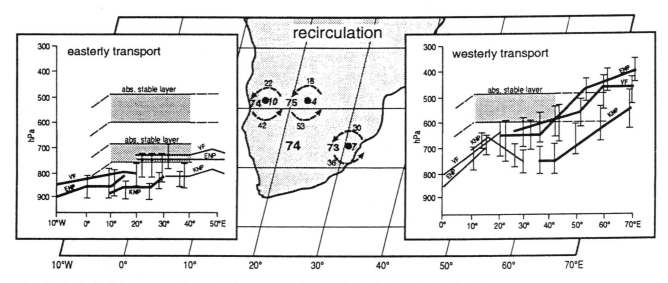

Figure 39.8 Vertically integrated surface-to-800-hPa transport from Etosha National Park, Victoria Falls, and Kruger National Park during SAFARI. Number of trajectories analyzed is 960. *Top*: cores of maximum frequency pathways. Heavy solid lines indicate direct transport. *Inset*: initial forward trajectory percentage directions within the first hour of transport are given. *Bottom*, percentage recirculated transport for different localities. The figure to the right of the station gives percentage initial stagnation (calm conditions within the first hour); those above and below the arrows give easterly and westerly percentage recirculated transport; the bold figures to the left indicate the total percentage recirculation for the station. The heavy bold figure gives the areal mean. Within the insets the variation of transport with pressure level, latitude, and absolutely stable layers (occurring when the observed lapse rate is less than the saturated adiabatic lapse rate) is given. Ninety-eight percent confidence limits are given for the horizontal and vertical dispersion of transport is given. The remaining symbol convention is as defined in figure 39.7.

and 500 mb. Taking this to be representative of southern Africa, excess ozone export during SAFARI is 0.03 Tg O_3 d^{-1} to the west and 0.08 Tg O_3 d^{-1} to the east.

Recirculation

Since recirculation is such a feature of transport fields over subtropical southern Africa, it is worth examining the phenomenon in more detail. On the basis of a study of 2050 trajectories for the 7-year period from 1986 to 1992 at nine locations in Africa south of the equator, a fine resolution ($2° \times 2°$) analysis by circulation type has been completed. The results for the semipermanent anticyclonic circulation type and trajectories originating at 25°S, 30°E (on the plateau to the west of Kruger National Park) are presented in figure 39.9. In the vertically integrated surface-to-800-hPa layer the zonal field dominates the transport patterns. Taking the zonal and meridional fields together, the general nature of the anticyclonic transport vortex (illustrated earlier in a particular case in figure 39.4) is clearly evident. In this particular case, two transport vortices within the overall anticyclonic gyral motion are evident. The larger has a radius of 600 to 800 km. Within the large vortex, a smaller one with a radius of 200 km is apparent. The total recirculation for this locality of origin, based on 189 cases for the 7-year period, is 71%. Using independent data, the spatially averaged figure across southern Africa for the 54-day period during the SAFARI was 74%.

Transport by Circulation Type

The circulation types are classified into dominant modes as described in chapter 37 to allow the month-by-month variation in circulation types to be determined. Trajectory-based transport analyses then allow the annual variation of air volume transports and mass fluxes to be estimated. Semipermanent subtropical anticyclonic circulations dominate the atmospheric circulation over southern Africa, particularly in winter when the combined frequency of occurrence of anticyclonic variants may reach 80%. Transport patterns associated with the continental anticyclonic circulation type are given in figure 39.10. The point of origin for the example given is 20°S, 25°E, that is, in central southern Africa in the vicinity of Victoria Falls. Surface-to-800-hPa zonal transport within the first hour of transport, determined over the 1986–1992 period and based on the analyses of 189 trajectories, is 90% to the west and zero to the east (the remainder is

either to the north, south, or becalmed). Almost no transport takes place into the Atlantic Ocean. A substantial amount of return recirculated transport (17%) occurs 10° to the northeast of the locality of origin (figure 39.10, *top*). Westerly recirculated transport occurs in a broad swathe to the southeast of the subcontinent. At 35°E, 83% of air transported initially to the west at 20°S exits at 30°–35°S after 4.5 days in transit. In the tropics to the northwest, 17% of air is transported back by recirculation towards its origin, having initially circulated in the opposite direction.

Four points of origin were used in the analysis of the transport associated with continental anticyclones (figure 39.11). The maximum frequency pathways of easterly and westerly transport define clear plumes exiting a little way into the Atlantic Ocean and being carried far over the Indian Ocean en route towards Australia and New Zealand. Most of the transport to the west over the Atlantic Ocean takes place in a subsiding airstream, whereas that to the east over the Indian Ocean is embedded in an airstream that rises once it moves out of the influence of the subsiding semipermanent subtropical highs and into the baroclinic westerlies (figure 39.11, *lower*). Recirculation is a major feature of transport associated with continental anticyclones. The spatial average of the recirculation index throughout the year is 64%.

Similar analyses have been completed for the vertically integrated 700–500-hPa transport associated with continental highs and other major circulation types, namely, ridging highs, westerly waves, and easterly waves. The analyses allow clear regional patterns to be discerned. In the case of surface-to-800-hPa transport with continental highs (figure 39.12), only 4% of air from the subcontinent passes west through the 10°E meridian. This it does between latitudes 15°–20°S, at a mean height of 750 hPa and with a transit time of 7.2 days. By contrast, 90% exits to the east across 35°E at 30°–35°S, at a height of 775 hPa and after 3.7 days.

More air is transported by ridging highs into the Atlantic Ocean than by continental highs and less into the Indian Ocean. Westerly waves produce little flow of air into the Atlantic Ocean. Most transport in these systems is into the Indian Ocean. Only with the occurrence of summer quasi-stationary barotropic easterly waves are transport patterns reversed. With easterly waves, 60% transport takes place into the Atlantic Ocean around 15°–20°S. It takes place at a mean height of 850 hPa with a transit time of 5.1 days. Only 18% is transported into the Indian Ocean with such systems.

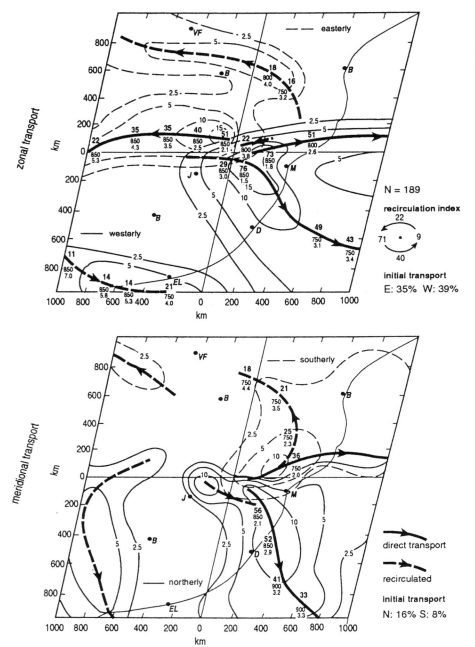

Figure 39.9 High-resolution transport fields for vertically integrated surface-to-800-hPa transport from 30°E, 25°S in semipermanent subtropical continental anticyclones. The symbol convention is as defined in figures 39.7 and 39.8.

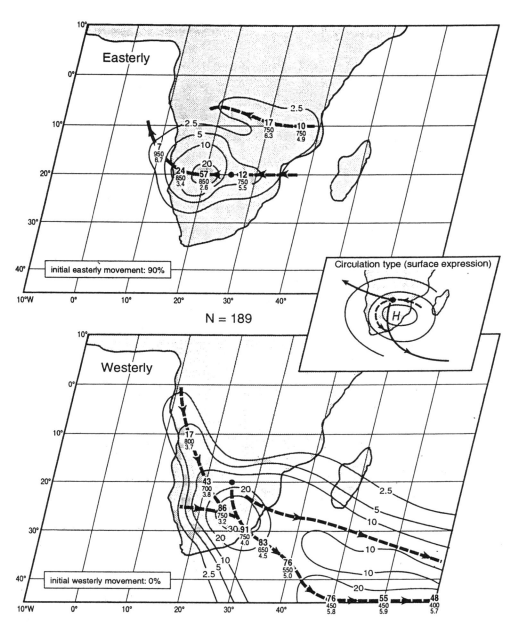

Figure 39.10 Contours of percentage of vertically integrated surface-to-800-hPa zonal transport in easterly and westerly directions in continental anticyclones. Number of trajectories analyzed is 189. Locality of origin is 25°E, 20°S. The symbol convention is as defined in figure 39.7.

total zonal transport and maximum frequency pathways (875-800hPa)

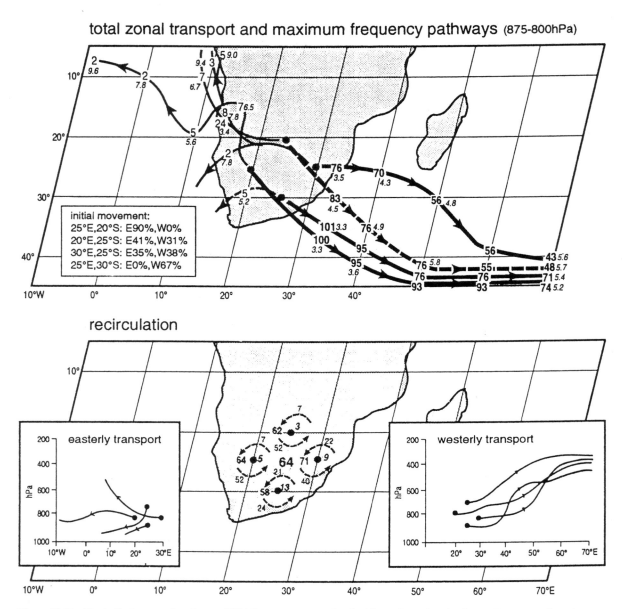

Figure 39.11 Vertically integrated surface-to-800-hPa transport associated with continental anticyclones. *Top*, cores of maximum frequency pathways. Number of trajectories analyzed is 756. *Bottom*, percentage recirculated transport for different localities. *Insets*: variation of transport with pressure level and latitude. The remaining symbol convention is as defined in figure 39.8.

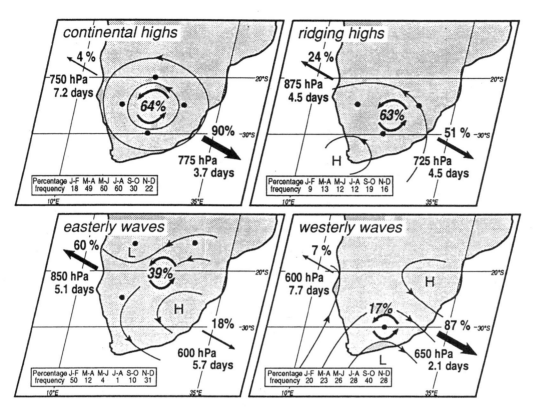

Figure 39.12 Mean percentage surface-to-800-hPa transport and recirculation by circulation type over and out of southern Africa. Mean heights of transport and transit times to reach beyond the west coast at 10°E and beyond the east coast at 35°E are given.

The patterns for transport in the 700–500-hPa layer are similar in many respects, with two important exceptions. The first is that recirculation is less in the middle atmosphere in all cases (figure 39.13). Second, the transport fields associated with tropical easterly waves are very different in surface and middle layers. Since easterly waves are largely induced by surface heating in summer and are shallow as a consequence, differences are to be expected. At the lower level the transport into the Atlantic Ocean by these systems is 60%; at middle levels it is only 16%. In addition, whereas with the other systems recirculation decreases with height, with easterly waves it increases, given the frequent anticyclonic curvature of the airstream overlying the surface waves.

From plume dimensions and exit characteristics mass fluxes and deposition rates may be estimated.

Mass Fluxes and Deposition Rates

For each transportation type, plume envelopes are determined within 98% confidence limits around the maximum transport pathway, as described in chapter 37. The horizontal advection velocity of air through each meridional plane is determined by the mean travel time of the trajectories to reach that plane and the distance from the centroid of origin of the air to the plane. The volume flux through the plane is fixed by the product of the plume area on the plane and the advection velocity. For each circulation type a greater volume flux of air occurs into the Indian Ocean by comparison to that transported into the Atlantic Ocean (figure 39.14). Only with the easterly wave type are the fluxes approximately the same. When considering total fluxes, a considerably greater volume of air is transported into the Indian than into the Atlantic Ocean.

Aerosol mass fluxes are estimated as the product of volume flux and aerosol concentration. Mean aerosol concentrations have been obtained from simultaneous filter samples taken at the three SAFARI sites for a 54-day interval during the field observation period. Metals, organic and inorganic carbon, and total nitrogen were determined from PIXIE and TRMS analyses (Maenhaut et al. 1993, 1996; Swap 1996). The method

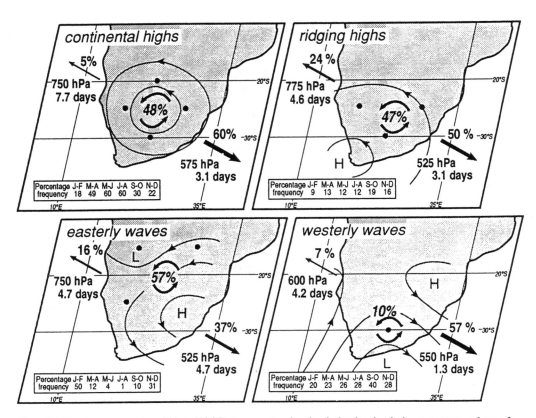

Figure 39.13 Mean percentage 700 to 500-hPa transport and recirculation by circulation type over and out of southern Africa. Mean heights of transport and transit times to reach beyond the west coast at 10°E and beyond the east coast at 35°E are given.

assumes zero settling rates and thus provides only a first and rather crude estimate of mass fluxes and bulk deposition.

The total annual zonal aerosol mass flux out of southern Africa is 74 Mtons, of which approximately 29 Mtons is transported through 10°E into the Atlantic Ocean and about 45 Mtons moves through 35°E into the Indian Ocean (figure 39.15). Of the 74 Mtons, about 26 Mtons is recirculated mass. It is estimated that approximately 60 Mtons recirculates over the continent without ever being exported. The total mass flux over the continent is thus about 135 Mtons per annum.

The total zonal mass flux is about half that reported into the North Atlantic Ocean from North Africa (D'Almedia 1987; Duce et al. 1991). Given the much greater production area and more desertlike characteristics of North Africa for dust production, this is not surprising. The breakdown of the total zonal mass flux by circulation type is given in figure 39.15. Anticyclonic circulation in continental highs is responsible for most of the flux into the Indian Ocean. Easterly

waves produce the greatest flux into the Atlantic Ocean.

Much deposition of material occurs between 10°E and the Greenwich Meridian. About 24 Mtons is deposited into the Atlantic Ocean in this longitudinal band per year. By contrast between 40° and 50°E, off the east coast, only about 4 Mtons is deposited annually into the Indian Ocean. A large amount is simply transported across the ocean in the atmosphere. Nearly 31 Mtons per year is conveyed beyond 70°E in the westerlies. By contrast, only 4 Mtons each year is transported over the central Atlantic Ocean beyond 10°W.

Conclusions

Coupled backward and forward trajectory analysis for specific points of interest on vertical profiles of ozone mixing ratios and relative aerosol concentrations measured using the NASA DIAL system during the TRACE-A experiment have demonstrated that it is possible to explain many features of the observed tro-

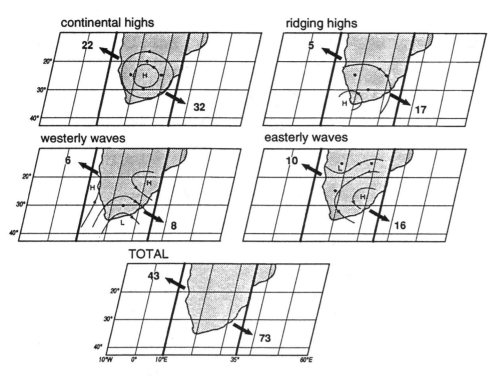

Figure 39.14 Mean annual zonal volume fluxes (m^3 d^{-1} × 10^{14}) across 10°E and 35°E

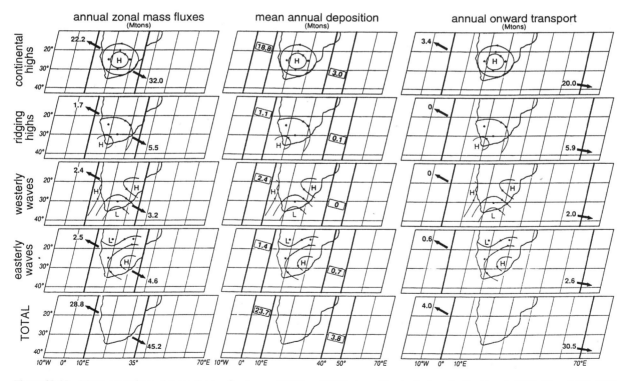

Figure 39.15 Mean annual zonal aerosol mass flux beyond 10°E and 35°E, deposition in the longitudinal bands 10°E–0° and 40°–50°E, and possible onward transport beyond the Greenwich Meridian to the west and 70°E to the east

pospheric distributions over southern Africa in a simple, graphic, three-dimensional way.

From the cases analyzed, the control exerted by absolutely stable layers on vertical mixing from the surface upwards in the case of aerosols and from the tropopause downwards in the case of stratospheric ozone is immediately apparent. The spatially prevalent and temporally persistent 5-km absolutely stable layer, which is produced by large-scale subsidence in the semipermanent subtropical anticyclones, which dominate the circulation, particularly in winter, is the most important of them. Trajectories originating in the lower and middle troposphere were observed to rise through the 5-km layer over the subcontinent on only very few occasions during SAFARI. Time sections from three continental soundings of ozone show concentrations in agreement with the UV-DIAL sections and the transport pathways.

The disadvantage of studying individual trajectories resides in the difficulty of assigning meaning or significance to collections of individual trajectories. In order to overcome this problem, analyses involving up to a thousand trajectories at a time were undertaken. They show that the individual back and forward trajectories used to explain TRACE-A observations of tropospheric aerosol and ozone concentrations over southern Africa were indeed meaningful and sensible. The analyses allowed mean transport pathways and plumes to be delimited in the horizontal and vertical for a 54-day period in August through October 1992 during SAFARI. From these volume fluxes of air and aerosol mass fluxes were determined for the austral spring. It turns out that most of the transport out of subtropical southern Africa, south of 15°S is not over the Atlantic to the west, but over the Indian Ocean to the east. Transport into the Atlantic would appear to be concentrated in the latitudinal band between the equator and 10°S (see chapter 38).

Extending the analysis to a 7-year period, for which more than two thousand trajectories were analyzed, shows that it is possible to partition transport over and out of southern Africa according to circulation types and their seasonal variations. Almost all synoptic circulation systems produce major transport throughout the year into the Indian Ocean just south of 30°S. Only with the occurrence of shallow quasi-stationary barotropic easterly waves that form over northern Botswana and adjacent areas in summer is there significant direct transport into the Atlantic Ocean in the region of 15°–20°S. This only occurs in the lower troposphere. These easterly wave systems occur at a time when little biomass burning takes place. Indirect transport, the

frequency of occurrence, and extent of anticyclonic circulation combine to effect transport into the atmosphere over the tropical south Atlantic.

Preliminary estimates reveal that over the year as a whole, 60 Mtons of aerosols recirculate over the subcontinent without ever being exported. The total mass flux out of the continent into the Atlantic and Indian oceans is 74 Mtons. Of this figure, about 29 Mtons per year is transported through 10°E into the Atlantic Ocean; about 45 Mtons moves through 35°E into the Indian Ocean. Over the west coast 24 Mtons are deposited per year between 10°E and 0°; only 4 Mtons may be transported onward over the central tropical South Atlantic Ocean beyond the Greenwich Meridian. By contrast, beyond the east coast, only about 4 Mtons are deposited between 40° and 50°, whereas as much as 31 Mtons may be conveyed to the east of 70°E towards Australia and New Zealand.

The volume fluxes of air being transported out of Africa in the major plumes are likely to be reasonably reliable and accurate. No claim is made that the aerosol mass fluxes are anything more than a first attempt to quantify particulate material being transported from the subcontinent. They are crude preliminary estimates and will need to be refined as time goes by. Since air is the carrier of trace gases and aerosols, the air transport climatology that has been presented will provide a means for such refinement to be effected. Traditionally the assessment of changing regional greenhouse gas composition is made from in situ measurements and/or from emission source inventories country by country. The framework provided in this chapter will assist in providing an alternative method for determining inter-regional transfers of greenhouse gases and regional inputs to greenhouse gas changes on the basis of vertical mixing and horizontal transport in the atmosphere. The implications for global change of transport from region to region are likely to be significant.

Acknowledgments

This work has been supported under grant ATM92-07924 form the National Science Foundation to the University of Virginia and by the South African Foundation for Research and Development.

References

Andreae, M. O., J. Fishman, M. Garstang, J. G. Goldammer, C. O. Justice, J. S. Levine, R. J. Scholes, B. J. Stocks, A. M. Thompson, B. van Wilgen, and the STARE/TRACE-A/SAFARI-92 Science Team. 1994. Biomass burning in the global environment: First results from the IGAC/BIBEX field campaign STARE/TRACE-A/ SAFARI-92, in *Global Atmospheric-Biospheric Chemistry: The First IGAC Scientific Conference*, edited by R. Prinn, Plenum, New York, 83–101.

Browell, E. V. 1993. TRACE-A airborne DIAL ozone and aerosol data, NASA/Langley Research Center, Hampton, Va.

Cros, B., D. Nganga, A. Minga, J. Fishman, and V. Brackett. 1992. Distribution of tropospheric ozone at Brazzaville, Congo, determined from ozonesonde measurements. *J. Geophys. Res.*, 97, 12 869–12 876.

Crutzen, P. J., and M. O. Andreae. 1990. Biomass burning in the tropics: Impact on atmospheric chemistry and biogeochemical cycles, *Science*, 250, 1669–1678.

D'Almeida, G. A. 1987. Desert aerosol characteristics and effects on climate, in *Paleoclimatology and Palaeometeorology: Modern and Past Patterns of Global Atmospheric Transport*, M. Leinen and M. Sarnthein (eds.), Kluwer Academic Publishers, Dordrecht, 909 pp.

Duce, R. A., P. L. Liss, J. T. Merrill, E. I. Atlas, P. Buat-Menard, B. B. Hicks, J. M. Miller, J. M. Prospero, R. Arimoto, T. M. Church, W. Ellis, J. N. Galloway, L. Hansen, T. D. Jickels, A. H. Knap, K. H. Reinhardt, B. Scheider, A. Soudine, J. J. Tokos, S. Tsunogai, R. Wollast, and M. Zhou. 1991. The atmospheric input of trace species to the world ocean, *Glob. Biogeochem. Cycles*, 5, 193–259.

Fishman, J. 1991. Probing planetary pollution from space, *Environ. Sci. Technol.*, 25, 612–621.

Fishman, J., K. Fakhruzzaman, B. Cros, and D. Mganga. 1992. Identification of widespread pollution in the Southern Hemisphere deduced from satellite analyses, *Science*, 252, 1693–1696.

Garstang, M., P. D. Tyson, R. Swap, M. Edwards, P. Kållberg, and J. A. Lindesay. 1996. Horizontal and vertical transport of air over southern Africa, *J. Geophys. Res.*, 101, 23, 721–23, 736.

Garstang, M., and P. D. Tyson. 1996. Atmospheric circulation, vertical structure and transport, Chapter 6 in *Fire in Southern African Savannas: Ecological and Atmospheric Perspectives*, B. W. van Wilgen, M. O. Andreae, J. G. Goldammer, and J. A. Lindesay (eds.), Witwatersrand University Press, Johannesburg, South Africa.

Kållberg, P. 1984. Air parcel trajectories from analyzed or forecast windfields, Swedish Meteorological and Hydrological Institute Research and Development Note No. 37.

Krishnamurti, T. N., H. F. Fuelberg, M. C. Sinha, D. Oosterhof, E. L. Bensman, and V. B. Kumai. 1994. The meteorological environment of the tropospheric ozone maximum over the tropical South Atlantic Ocean, *J. Geophys. Res.*, 98, 10 621–106.

Maenhaut, W. I., M. Salma, M. Garstang, and F. Meixner. 1993. Size-fractionated atmospheric aerosol composition and aerosol sources at Etosha, Namibia, and Victoria Falls, Zimbabwe, Proc. 1993 AGU Fall Meeting, San Francisco, CA, 6–10 December, 104.

Maenhaut, W. I., M. Salma, J. Cafmeyer, H. J. Annegarn, and M. O. Andreae. 1996. Regional atmospheric aerosol composition and sources in the Eastern Transvaal, South Africa, and the impact of biomass burning, *J. Geophys. Res.*, 101, 23, 631–23, 650.

Moody, J. L., A. A. P. Pszenny, G. Grandry, W. C. Keene, J. N. Galloway, and G. Polian. 1991. Precipitation composition and its variability in the southern Indian Ocean: Amsterdam Island, 1980–1987, *J. Geophys. Res.*, 96, 20 769–20 786.

Pickering, K. E., A. M. Thompson, D. P. McNarmara, and M. R. Schoeberl. 1994a. An intercomparison of isentropic trajectories over the south Atlantic, *Mon. Wea. Rev.*, 122, 864–879.

Pickering, K. E., A. M. Thompson, D. P. McNamara, M. R. Schoeberl, L. R. Lait, P. A. Newman, C. O. Justice, and J. D. Kendall. 1994b. A trajectory modeling investigation of the biomass burning tropical ozone relationship, *Proc. Quad. Ozone Symp.*, 101–104, NASA Conf. Publication 3266, NASA/Goddard Space Flight Center, Washington, D.C.

Swap, R. J. 1996. Transport and impact of southern African aerosols, Unpublished PhD dissertation, Dept. of Environmental Sciences, University of Virginia, 311 pp.

Swap, R. J., M. Garstang, S. A. Macko, P. D. Tyson, W. Maenhaut, P. Artaxo, P. Kållberg, and R. Talbot. 1996. The long-range transport of southern African aerosols to the tropical south Atlantic, *J. Geophys. Res.*, 101, 23, 777–23, 792.

Thompson, A. M., K. E. Pickering, D. P. McNamara, M. R. Schoeberl, R. D. Hudson, J. K. Kim, E. V. Browell, V. W. J. H. Kirchhoff, and D. Nganga. 1996. Where did tropospheric ozone over southern Africa and the tropical Atlantic come from in October 1992? Insights from TOMS, GTE/TRACE-A and SAFARI-92, *J. Geophys. Res.*, 101, 24, 251–24, 278.

Tyson, P. D., M. Garstang, R. Swap, P. Kållberg, and M. Edwards. 1996. An air transport climatology for subtropical southern Africa, accepted *Inter. J. Clim*, 16, 265–291.

Watson, C. E., H. Fishman, and H. J. Reichle. 1990. The significance of biomass burning as a source of carbon monoxide and ozone in the southern hemisphere tropics: A satellite analysis, *J. Geophys. Res.*, 95, 16 433–16 450.

Bulk and Compound-Specific Isotopic Characterization of the Products of Biomass Burning: Laboratory Studies

Vaughan C. Turekian, Stephen A. Macko, William P. Gilhooly, Donna C. Ballentine, Robert J. Swap, and Michael Garstang

The role of organic compounds released during biomass burning in regional and global atmospheric chemistry and biogeochemical cycling has received much attention (Crutzen and Andreae 1990). Stable carbon and nitrogen isotopes may provide a useful tool in the characterization of the sources of organic compounds found in the atmosphere. A fundamental question that must be addressed prior to wide use of this type of analysis is how the process of combustion affects the isotopic signature of the products of vegetation burns. This chapter investigates how pyrolytic processes affect the bulk and molecular-level isotopic fractionations of products of controlled laboratory burns.

Carbonaceous compounds of marine origin are isotopically distinct from the continentally produced compounds. These differences have been used to delineate between marine and continental sources of aerosols (Cachier et al. 1986; Chesselet et al. 1981). Isotopic identification of plant material derived from C_3 and C_4 vegetation is possible owing to the different isotopic fractionations associated with the two photosynthetic pathways (O'Leary 1988). Determination of sources of continentally produced carbonaceous aerosols on a bulk isotope level is complicated by the contributions of the variety of sources for compounds to the total isotopic signal. In addition, there is limited knowledge of how the products of vegetation burns can be related to C_3 or C_4 vegetation sources (Cachier et al. 1985; Cachier et al. 1986).

Specific organic compounds found in the atmosphere have been used as biomarkers (Simoneit et al. 1993). Sources of biogenically derived material may be identified by examining the presence and the relative distributions of particular chemical constituents. Resin acids and polycyclic aromatic hydrocarbons (PAH) are representative of thermally altered compounds, and have been used as tracers of biomass burning (Simoneit et al. 1993; Standley and Simoneit 1987). Lipids are useful tracers of biogenic sources and can be used to determine a general source of organic material. However, identifying specific sources of lipids is complicated by their ubiquity in vegetation. Gagosian et al. (1982) have investigated the sources of various lipid compounds to a remote marine location. Their conclusion was that these compounds were the product of the elevation of plant debris and were not the result of biomass burning.

Accurate source identification for individual compounds is possible with gas chromatography combustion isotope ratio mass spectrometry (GC-C-IRMS). This new technology allows for isotopic characterization at the molecular level (Freeman et al. 1990). Identification of specific organic compounds coupled with carbon isotopic analysis of these compounds provides a unique tool for determining sources of transported material.

Nitrogen isotopic distributions in vegetation are related to the aridity or wetness of the region of growth (Heaton et al. 1986), and may be used to indicate the source region of vegetation material. Studies assessing the isotopic signatures of nitrogen compounds as aerosols have focused on NH_4^+ and NO_3^- ions attached to particles or in rain (Hoering 1957; Moore 1973) and not the presence of organic nitrogenous compounds captured as aerosols. Burning results in the elevation of nitrogen containing aerosols, however, there is limited understanding of bulk nitrogen isotopic fractionations that take place during the combustion of organic material.

Methods

Plant samples of C_3 and C_4 vegetation from different regions around the globe were freeze dried and ground to a powder using a #60 Wiley mill. Controlled laboratory burns were performed by placing individual vegetation samples on pre-ashed (550°C) aluminum foil and ignited. Oxygen gas was supplied qualitatively to promote and sustain flaming combustion. Excess oxygen was not supplied for the smoldering vegetation. Burns were classified qualitatively as flaming or smoldering. Temperature profiles taken during the burns with a thermocouple indicate that the oxygen-assisted flame had temperatures higher than 500°C. The lower intensity smoldering combustion had temperatures lower than 200°C. Aerosols were captured above the combustion for the duration of the burn on a 0.4-μm pre-ashed (550°C for 1 hour) glass fiber filter (Gelman Type AE) under low-volume pumping. Ash samples represent the total residual material left on the foil following the burn. The vegetation, ash, and filter samples were stored in the freezer until isotopically analyzed.

For isotopic analysis, samples were placed in ashed quartz tubes along with copper and copper oxide. The tubes were evacuated, sealed, and combusted at 850°C for 2 hours and allowed to cool slowly using a modified Dumas technique (Macko 1981). This technique leads to the conversion of the carbonaceous compounds to CO_2 and of nitrogenous materials to N_2. A cryogenic separation of the CO_2 and the N_2 was performed by the standard laboratory technique. The $\delta^{15}N$ and $\delta^{13}C$ values, relative to the isotopic standards of atmospheric nitrogen and PDB, respectively, were obtained using a VG Prism Series II isotope ratio mass spectrometer (IRMS).

The total lipid fractions were Soxhlet extracted from the vegetation and the glass fiber filters, for 16 hours, using glass-distilled dichloromethane. The lipids were saponified under a KOH reflux for 3 hours. The saponified fatty acids were derivatized to their corresponding methyl esters with BF_3-CH_3OH (Metcalfe and Schmitz 1961). (See chapter 43 by Ballentine et al. for a flowchart for this extraction). Specific fatty acids were identified with a Hewlett Packard 5890 Series II gas chromatograph with an HP-1 column interfaced to a Hewlett Packard 5971A mass selective detector (GC/MS). Isotopic analysis on the fatty-acid methyl esters was performed by using a Hewlettt Packard 5890 Series II gas chromatograph (described above) with on-line combustion coupled to the VG Prism Series II isotope ratio mass spectrometer (GC/C/IRMS).

Results

Aerosols captured over controlled laboratory burns of C_3 vegetation (table 40.1) on average are enriched in ^{13}C compared to the source vegetation ($\Delta = 0.5‰ \pm 0.9‰$). The ^{13}C enrichment is enhanced for higher intensity burns ($\Delta = 1.3‰ \pm 0.6‰$) whereas the lower intensity burns show little difference in the carbon isotopic signal between vegetation and aerosol ($\Delta = -0.2‰ \pm 0.5‰$). Aerosols captured over the controlled pyrolysis of C_4 vegetation (table 40.1) are depleted in ^{13}C when compared to the source vegetation

($\Delta = -3.5‰ \pm 2.1‰$). This fractionation is similar for both high- and low-temperature burns. ($\Delta_{flame} = -4.0‰ \pm 2.7‰$ versus $\Delta_{smolder} = -3.0‰ \pm 1.8‰$).

Bulk nitrogen isotope data (table 40.2) for the aerosols collected over burns shows an average ^{15}N enrichment of 6.6‰ ($\pm 4.7‰$) compared to the source vegetation. There is essentially no difference in ^{15}N enrichments in the aerosols captured over the flaming and smoldering vegetation ($\Delta_{flame} = 6.9‰ \pm 5.5‰$ versus $\Delta_{smolder} = 6.3‰ \pm 4.6‰$). The ash from these samples is also enriched in ^{15}N compared to the parent vegetation with increased fractionations being associated with the higher intensity burns ($\Delta_{flame} = 3.28‰ \pm 1.3‰$ versus $\Delta_{smolder} = 1.7‰ \pm 0.9‰$).

Structurally intact fatty acids survived pyrolysis and are transported in the aerosols. GC traces (figures 40.1 and 40.2) for fatty acids extracted from Mopane vegetation and aerosols captured above the controlled smoldering burn illustrate the conservation of fatty acids through the burn. Carbon isotope data for the fatty acids extracted from the Mopane vegetation and the corresponding filter (table 40.3) indicate a possible slight ^{13}C enrichment for the individual fatty acids captured as aerosols. Results from isotope analysis fatty acids from *Zea mays* (table 40.3) and aerosols captured above the smoldering combustion of the vegetation also indicate a conservation of the carbon isotope signature for the individual compounds.

Discussion

Bulk Carbon Isotopes

The enrichment in products associated with the combustion of C_3 vegetation differ from the observed isotope effect reported by Cachier et al. (1985) (table 40.1). Controlled burns of C_4 vegetation had similar fractionations to those reported by those researchers, although the isotopic depletion was typically less than reported by Cachier et al. The data of Cachier et al. for aerosols collected over burning savanna may have been influenced by the isotopic variability of the mixture of vegetation sources. The combustion products of this present study were derived from single, known vegetation sources with known carbon isotopic values. Thus, the products of these burns could be traced back to a unique vegetation source and were not subject to the variability of a mixing of sources.

The ^{13}C enrichment observed during the heating of C_3 vegetation may result from the preferential loss of isotopically light carbon as lower molecular compounds such as CO_2, CO, CH_3, and other trace gases

Table 40.1 Bulk carbon isotope data for controlled laboratory burns

Sample	$\delta^{13}C$ (‰)			(‰)	
	Vegetation	Ash	Aerosol	ΔAsh-vegetation	ΔAerosol-vegetation
Eucalyptus[a] (F)	−29.2	27.1	−27.4	2.1	1.8
Eucalyptus[b] (F)	−27.7	−27.0	−26.2	0.6	1.4
Mopane[c] (F)	−24.4	−23.8	−23.8	0.6	0.6
Sugar cane[d] (F)	−12.9	−12.9	−16.2	0.0	−3.3
Cenchris[e] (F)	−12.9	−14.7	−19.9	−1.8	−7.0
Antephora[e] (F)	−13.3	−13.8	−15.0	−0.5	−1.7
Eucalyptus[b] (S)	−27.6	−27.0	−27.3	0.6	0.3
Mopane[c] (S)	−24.4	−24.4	−24.6	0.0	−0.2
Mopane[c] (S)	−24.4	−24.0	−25.0	0.4	−0.6
Sugar cane[d] (S)	−12.9	−12.6	−13.8	0.3	−0.9
Cenchris[e] (S)	−12.9	−12.7	−16.8	0.2	−3.9
Antephora[e] (S)	−13.3	−13.5	−17.5	−0.2	−4.2
Grasses[f]	−12.5	−14.5	−19.5	−2.0	−7.0

(F) indicates vegetation exposed to flame temperatures; (S) indicates vegetation exposed to smoldering temperatures.
Sample collected from [a]Central California, [b]Southern California, [c]Southern Africa, [d]South Africa, [e]Namibia.
[f]Data taken from Cachier et al. (1985) for a natural burn in Lamto in the Ivory Coast.

Table 40.2 Bulk nitrogen isotope data for controlled laboratory burns

Sample	$\delta^{15}N$(‰)			(‰)	
	Vegetation	Ash	Aerosol	ΔAsh-vegetation	ΔAerosol-vegetation
Eucalyptus[b] (F)	6.4	8.3	13.0	1.9	6.6
Mopane[c] (F)	3.3	8.2	2.0	4.9	−1.3
Sugar cane[d] (F)	6.9	9.2	12.8	2.3	5.9
Cenchris[e] (F)	10.0	14.3	22.7	4.3	12.7
Antephora[e] (F)	6.7	9.7	17.7	3.0	11.0
Eucalyptus[b] (S)	6.4	8.5	19.5	2.1	13.1
Mopane[c] (S)	3.3	5.3	4.7	2.0	1.4
Sugar cane[d] (S)	6.9	6.9	15.1	0.0	8.2
Cenchris[e] (S)	10.0	12.3	15.9	2.3	5.9
Antephora[e] (S)	6.7	8.7	9.8	2.0	3.1

(F) indicates vegetation exposed to flame temperatures; (S) indicates vegetation exposed to smoldering temperatures.
Sample collected from [a]Central California, [b]Southern California, [c]Southern Africa, [d]South Africa, [e]Namibia.

not captureable on the filter. Two possible mechanisms could control the formation of these isotopically light gases. The destruction of bonds and the preferential oxidation of the ^{12}C during incomplete combustion would result in a residual pool of material that was ^{13}C enriched. The $\delta^{13}C$ data (table 40.4) for carbon dioxide produced during the incomplete low-temperature combustion of eucalyptus indicates that the isotope signal is conserved from the vegetation to the CO_2. A second mechanism for ^{13}C depletion of trace gases is the preferential cleavage of ^{12}C bonds and the subsequent loss of this isotopically light carbon as lower molecular weight hydrocarbons. The latter process is believed to play an important role in the ^{13}C depletion of methane

produced during the thermal maturation of kerogen material (Peters et al. 1981; Sackett 1984; Sackett 1978).

Normal kinetic isotope effects (Kaplan 1975) can be used to predict that the bonds between the light isotopes can be ruptured with the least amount of energy. With increasing temperatures fractionations caused by kinetic isotope effect decrease, because cracking of bonds is less discriminant. It would be expected that at the higher temperature of vegetation fires the isotopic fractionation of the products should decrease.

Carbon isotopic compositions of methane from biomass burning support little evidence of isotopic fractionation from the parent vegetation (Quay et al.

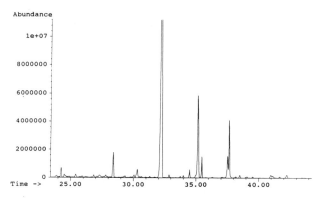

Figure 40.1 GC/MS trace in selected ion mode for fatty-acid methyl esters in Mopane vegetation

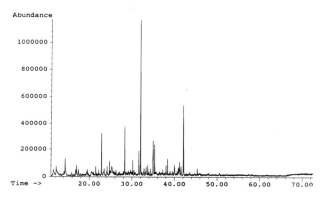

Figure 40.2 GC/MS trace in selected ion mode for fatty-acid methyl esters in aerosols collected over a controlled smoldering laboratory burn of Mopane vegetation

Table 40.3 Compound-specific carbon isotope analysis of fatty acids collected over controlled smoldering burns of *Zea mays* and Mopane (in ‰)

Fatty acid	Mopane vegetation	Mopane filter	*Zea mays* vegetation	*Zea mays* filter
C 14:0	−35.5 ± 0.4	−34.1 ± 0.4	—	—
C 16:0	−34.5 ± 0.4	−32.9 ± 0.5	−19.9	−19.7
C 18:0	−32.2 ± 0.2	−31.5 ± 0.4	−18.1	−18.2
C 18:1	—	—	−17.7	−18.0

Table 40.4 Carbon isotope data for carbon dioxide formed at different temperatures (Data for eucalyptus from Southern California)

δ^{13}C for whole leaf: -27.7‰	
Temperature (°C)	δ^{13}C (‰)
150	−27.5
350	−27.4
550	−27.6
650	−27.7
850	−27.6

1991). However, the results of this study point to an increasing fractionation for the flaming burns of C_3 vegetation than is observed for smoldering burns. Possible explanations for these observed results are (1) The flame temperatures were low enough to promote the kinetic isotope effects that caused the preferential formation and loss of isotopically light low-molecular weight gases; and (2) At the lower smoldering temperatures, compounds are distilled from the vegetation (Simoneit et al. 1993) and conserve their isotopic signature as they are captured as aerosols. These different fractionation effects may be able to be used to differentiate the products derived from flaming and smoldering burns.

Aerosols captured above controlled burns of C_4 vegetation are depleted compared to the source vegetation implying that the compounds not captured as aerosols on the filter are ^{13}C enriched compared to the source vegetation. The ^{13}C depletion observed for

aerosols and ash from C_4 vegetation burns would imply that there is a different process of bond rupture or carbon oxidation chemistry occurring during the combustion of C_4 vegetation than occurs during C_3 pyrolysis. Carbon-13 depletion of the aerosols produced during the combustion of C_4 vegetation coupled with the observed ^{13}C enrichment for aerosols produced during the higher temperature burns of C_3 vegetation imply that there may be some difficulty differentiating between C_3 or C_4 sourced products of biomass burning using bulk isotopic compositions.

Compound Specific Isotope Analysis

Fatty acids are biogenic compounds that can be used as tracers of natural materials to remote locations. The advantage of the isotopic characterization of specific organic compounds over bulk isotope techniques is that the isotopic signals from these fatty acids are diluted only by the addition of the same compound from another source. For this reason, organic compounds provide better source resolution than is possible with the use of bulk-stable isotopes alone. Distributions of specific fatty acids extracted from filters collected over the controlled smolder of Mopane and of *Zea mays* when combined with the isotope data for these fatty acids, indicate that the fatty acids are both structurally and isotopically conservative during low-temperature smolders. The fatty acids from these two vegetation sources themselves reflect the differences in the carbon isotopic signatures that exist between C_3 and C_4 vegetation. The conservation of the isotopic signal throughout the burn indicates compound-specific isotope analysis (CSIA) may be used to separate the sources of similar biogenic compounds that are derived from C_3 and C_4 vegetation.

Nitrogen Isotopes

The nitrogen isotope data for ash and aerosol sampling collected from the controlled burns of the vegetation indicates that there is enrichment of these product. By mass balance,

$$\delta^{15}N \text{ vegetation} = f_1 \delta^{15}N \text{ ash} + f_2 \delta^{15}N \text{ aerosols}(> 0.4 \ \mu m)$$
$$+ f_3 \delta^{15}N \text{ aerosols}(< 0.4 \ \mu m),$$

where f_1, f_2, and f_3 are the fractional contribution of each product. An enrichment in the aerosol $> 0.4 \ \mu m$ and the ash must be balanced by ^{15}N depletion in the aerosols $< 0.4 \ \mu m$. The degree of this depletion depends on both the total enrichment of the other products and their fractional contributions. Nitrogenous gases such as NH_3, NO_x, and N_2O, which are produced during combustion of vegetation (Ward 1993; Logan 1983; Crutzen et al. 1979), would be representative of the isotopically depleted fraction. The isotopic signal of these gases would also depend on whether they are formed from nitrogen with a vegetation source or if they are derived from the oxidation of atmospheric N_2. The proposed mechanism for the ^{15}N depletion of the nitrogenous gases, and the subsequent heavy isotope enrichment of the captured aerosols and ash, is the preferential destruction of the ^{14}N bonds and the release of nitrogenous compounds during such processes as the deamination of amino acids and other nitrogen-based compounds within the vegetation during heating.

Conclusions

Stable carbon and nitrogen isotopes and compound-specific isotope analysis (CSIA) may allow for better characterization of aerosols produced during natural vegetation. Factors influencing the isotopic fractionation of the products of laboratory burns appear to be both the intensity of the burn and the type of vegetation exposed to the combustion (grass or leaf). Kinetic isotope effects that influence the bulk carbon and nitrogen isotopic fractionations at the temperature of these laboratory combustions may be less significant at the higher temperatures of natural burns. The isotope effects reported for smoldering vegetation during controlled laboratory burns may approximate the fractionations that occur during the smoldering stages of open burns. With greater understanding of the processes controlling the isotopic fractionations of both the bulk and chemical components of the products, insights will be gained into the applicability of isotopes to source identification of the products of vegetation burns. Ultimately, the influence of these products on atmospheric chemistry and biogeochemical cycles may be better understood.

References

Cachier, H., P. Buat-Menard, and M. Fontugne. 1985. Source Terms and Source Strengths of the Carbonaceous Aerosol in the Tropics, *Journal of Atmospheric Chemistry*, 3, 469–489.

Cachier, H., P. Buat-Menard, M. Fontugne, and R. Chesselet. 1986. Long-range Transport of Continentally Derived Particulate Carbon in the Marine Atmosphere: Evidence from Stable Isotope Studies, *Tellus*, 38B, 161–177.

Chesselet, R., M. Fontugne, P. Buat-Menard, U. Ezat, and C. E. Lambert. 1981. The Origin of Particulate Organic Carbon in the Marine Atmosphere as Indicated by Its Stable Carbon Isotopic Composition, *Geophysical Research Letters*, 8, 345–348.

Crutzen, P. J., and M. O. Andreae. 1990. Biomass Burning in the Tropics: Impact on Atmospheric Chemistry and Biogeochemical Cycles, *Science*, 250, 1669–1678.

Crutzen, P. J., L. E. Heidt, J. P. Krasnec, W. H. Pollock, and W. Seiler. 1979. Biomass Burning as a Source of Atmospheric Gases CO, H_2, N_2O, NO, CH_3Cl, and COS, *Nature*, 282, 253–256.

Freeman, K. H., J. M. Hayes, J. M. Trendel, and P. Albrecht. 1990. Evidence from Carbon Isotope Measurements for Diverse Origins of Sedimentary Hydrocarbons, *Nature*, 343, 254–256.

Gagosian, R. B., O. C. Zafiriou, E. T. Peltzer, and J. B. Alford. 1982. Lipids in Aerosols from the Tropical North Pacific: Temporal Variability, *Journal of Geophysical Research*, 87, 11 133–11 144.

Heaton, T. H. E., J. C. Vogel, G. von la Chevallerie, and G. Collet. 1986. Climatic Influence on the Isotopic Compositions of Bone Nitrogen, *Nature*, 322, 822–823.

Hoering, T. C. 1957. The Isotopic Composition of the Ammonia and the Nitrate Ion in Rain, *Geochimica, Cosmochimica acta*, 12, 97–102.

Kaplan, I. R. 1975. Stable Isotopes as a Guide to Biogeochemical Processes, *Proceedings of the Royal Society of London*, Series B, 189, 183–211.

Logan, J. A. 1983. Nitrogen Oxides in the Troposphere: Global and Regional Budgets, *Journal of Geophysical Research*, 88, 10 785–10 807.

Macko, S. A. 1981. Stable Nitrogen Isotope Ratios as Tracers of Organic Geochemical Processes, Ph. D. Thesis, University of Texas, Austin.

Metcalfe, L. D., and A. A. Scmitz. 1961. The Rapid Preparation of Fatty Acid Methyl Esters for Gas Chromatographic Analysis, *Analytical Chemistry*, 33, 363–364.

Moore, H. 1973. Isotopic Measurement of Atmospheric Nitrogen Compounds, *Tellus*, 26, 169–174.

O'Leary, M. H. 1988. Carbon Isotopes in Photosynthesis, *BioScience*, 38, 328–336.

Peters, K. E., B. G. Rohrback, and I. R. Kaplan. 1981. Carbon and Hydrogen Stable Isotope Variations in Kerogen During Laboratory-Simulated Thermal Maturation, *The American Association of Petroleum Geologists Bulletin*, 65, 501–508.

Quay, P. D., S. L. King, J. Stutsman, D. O. Wilbur, L. P. Steele, I. Fung, R. H. Gammon, T. A. Brown, G. W. Farwell, P. M. Grootes, and F. H. Schmidt. 1991. Carbon Isotopic Composition of Atmospheric CH_4: Fossil and Biomass Burning Source Strengths, *Global Biogeochemical Cycles*, 5, 25–47.

Sackett, W. M. 1978. Carbon and Hydrogen Isotope Effects during the Thermocatalytic Production of Hydrocarbons in Laboratory Simulation Experiments, *Geochimica, Cosmochimica Acta*, 42, 571–580.

Sackett, W. M. 1984. Determination of Kerogen Maturity by the Pyrolisis-Carbon Isotope Method, *Organic Geochemistry*, 6, 359–363.

Simoneit, B. R. T., W. F. Rogge, M. A. Mazurek, L. J. Standley, L. M. Hildermann, and G. R. Cass. 1993. Lignin Pyrolisis Products, Lignans and Resin Acids as Specific Tracers of Plant Classes in Emissions from Biomass Combustion, *Environmental Science and Technology*, 27, 2533–2541.

Standley, L. J., and B. R. T. Simoneit. 1987. Characterization of Extractable Plant Wax, Resin and Thermally Matured Components in Smoke Particles from Prescribed Burns, *Environmental Science and Technology*, 21, 164–169.

Ward, D. E. 1993. Trace Gasses and Particulate Matter from Fires—A Review. From: Background Paper for Proceedings of the Victoria Falls Workshop, 2 to 6 June.

African Fire Particulate Emissions and Atmospheric Influence

Hélène Cachier, Catherine Liousse, Marie-Hélène Pertuisot,
Annie Gaudichet, Francisco Echalar, and Jean-Pierre Lacaux

Combustion of vegetation is now recognized as a major source of atmospheric pollution. Indeed, 90% of the biomass that burns is due to anthropogenic activities and the total amount of biomass carbon that is yearly submitted to fire rivals the amount of fossil fuel used each year (of the order of 3.10^{15} gC/yr for biomass carbon, chapter 47 this volume by Liousse et al.; and $4.8.10^{15}$ gC/yr for fossil fuels Sepauski et al. 1990).

Biomass burning activities are primarily (80%) situated in the intertropical belt. Due to severe increase of population and needs in developing countries, study of emissions are getting urgent and mandatory to assess present contribution of tropical biomass burning and to construct realistic scenarios for near future. Causes that lead to the burning of biomass in the tropics encompass an important range of activities that may be rougly classified in forest-clearing practices, savanna burning, agricultural burns, vegetation fuel burning (wood and charcoal).

Some burnings such as savanna or forest fires and, to a lesser extent, agricultural burns display a high seasonal dependence; others may prevail all year long. The resulting atmospheric pollution over continents and at more background locations will follow these seasonal patterns especially for species with a residence time of a few days.

On field, during the burning of vegetation, large amounts of living and dead material and any particles previously deposited on leaves are mobilized, chemically transformed, and injected into the atmosphere. Major emissions of carbon dioxide (CO_2) are accompanied by that of numerous gaseous and particulate species (Levine 1990; Crutzen and Andreae 1990; Cachier 1992) most of which are chemically and radiatively active. The fate of vegetation carbon consumed during combustion is to be transformed into either CO_2 or other minor constituents.

A major part (70–95%) will evolve as CO_2. The burning will constitute a net flux of CO_2 for forest fires or related combustions (domestic fires using charcoal or wood) only. For grass fires, the CO_2 flux is actually counteracted by vegetation regrowth a few months later.

Carbon may be emitted as minor or trace constituents such as CO, CH_4, volatile organic carbon (VOC) or aerosols. The abundance of these species relative to CO_2 is closely related to the mode of combustion (flaming, smoldering, or glowing), itself linked to the nature of the vegetation, its size and humidity, and last to the physical parameters of the burn (oxygen supply, temperature, primarily). As an example CO, which is the most abundant emitted species after CO_2, displays an important variability as shown by the various CO/CO_2 ratio values recorded in different fire plumes (2–30%); the smaller ratio, the better combustion conditions. An important feature is that most of the non-CO_2 carbonaceous species co-emitted during the burning of vegetation are preferentially produced when oxidative conditions are less favorable to good combustion. Their concentrations are found to be globally correlated for a given type of biomass burning, a feature which applies either in field (Bonsang et al. 1995) or in combustion chamber experiments (Lobert et al. 1991). As an example, Bonsang et al. (1995) found a satisfactory linear relationship between CH_4 and CO concentrations in savanna burning plumes. It is now commonly assumed that CO abundance may serve as an indicator of pyrolitic conditions in combustions.

Aerosols are an important by-product of biomass combustions. Recently, there has been an upsurge of interest in the potential of smoke particles to alter the radiative and chemical balance of the tropical and even global troposphere. Although these aerosols contain both black and organic particles they are, on average, primarily of organic nature (Cachier et al. 1991, 1995; Mazurek et al. 1991) and have the potential to create a cooling of the atmosphere directly by scattering incoming solar radiation. But their most important (but extremely poorly known) property is probably that they are good CCN (Rogers et al. 1991; Dinh et al. 1994) certainly due to the presence of hydrophilic organic material at surface of particles (Clain 1995). Therefore, at tropical latitudes biomass burning aerosols are likely to modify the albedo and lifetime of the cloud cover (Levine 1990; Kaufman et al. 1993), which could reinforce their cooling effect.

In light of the paucity of data in spite of the potential atmospheric importance of combustion particulates emitted in tropical zones, we studied the production in Africa of aerosols produced by different sources that are expected to gain importance in the future: savanna fires and domestic fires. Experiments were performed in the frame of the DECAFE and SAFARI programs. Our strategy was primarily to gain information on particulate emission factors and the relative abundance of organic particles in the aerosols.

Savanna burning is worldwide the major combustion source and roughly represents 50% of the tropical biomass burning. This combustion was studied during test fires in two very different ecosystems that could be representative of humid dense and dry savannas.

In several developing countries where wood is still available without shortage, domestic cooking and heating are based on the use of wood. However, due to increasing urbanization and subsequent fuel-storage problems in houses, on a global scale charcoal is now overwhelmingly used (10 times more than wood, Delmas et al. 1995). The "charcoal" source was studied for its two different steps, charcoal making and use under traditional conditions actually found in West Africa. It will be shown that this source, poorly described in most particulate carbon budgets, is a very efficient producer of atmospheric particles. The domestic use of wood was simulated in experiments actually pertaining in West African homes.

Finally, we are also investigating the regional and global influence of the biomass burning particles in atmosphere at tropical latitudes. A first step is the buildup of aerosol concentration records at representative sites. We present here results of monitoring carbonaceous aerosols from the savanna site of Lamto (Ivory Coast) and Amsterdam Island (Indian Ocean) and show the repetitive seasonal influence of tropical fires.

Savanna Test Fires

Savanna fires are surface fires affecting the grass cover and litter. They may spread either as backing or heading fires depending on orographic conditions, but they are assumed to consume more than 90% of the vegetation fuel during one of these two flaming stages. Smoldering often observed behind the firefront in Kruger National Park test burns has to be considered as an artifact due to the presence of wild animals leaving on field numerous dungs and broken trees. Observations on field of smoke plumes and ashes left on the ground suggest that atmospheric emissions are, at least qualitatively, different for these two types of combustion. Such differences could be linked to fire spread, which is much more important in heading fires and sometimes does not create very favorable combustion conditions. The conjunction of fire intensity and pertaining meteorological conditions is of major importance for the injection height and fate of smoke pollutants in the tropospheric reservoir.

Experiments

Samplings were conducted in the guinean savanna of Lamto (Ivory Coast) during the heart of the dry season 1991. The whole experiment (FOS-DECAFE) is described by Lacaux et al. (1995). Grass submitted to fires was mostly ludetia and hyparrhenia and during this special experiment very much still in a humid state. At Lamto, savanna is intentionally burnt each year, which is the primary cause of the quasi-absence of litter. At the opposite, the SAFARI-92 experiment which took place at the end of the dry season in the Kruger National Park (KNP) (South Africa) offered the very different conditions of an infertile bush savanna that had dried during an exceptional drought. Average experimental conditions are presented in table 41.1. It may be seen that there are striking

Table 41.1 Characterization of the different Lamto and KNP plots and test burns, mean values; (data from Trollope et al. [1996]; fire intensity calculated following Trollope et al. [1981] as I = H.W.R where H is heat yield in kJ/kg [taken here as 17 000], W is mass of dry fuel in kg/m², and R the rate of spread of firefront in m/s)

	Size (ha)	Temperature (°C)	Atmospheric humidity (%)	Fuel load (dm · kg/ha)	Litter (%)	Fire efficiency (%)	Moisture (%)	Head fire frontal intensity (kJ/m/s) (I)
Lamto	1	28	70	9000	4	77	83	625
KNP (plots)	6.2	28	41	5600	41	81	15	2670
KNP (block)	2100	32	34	5400	39	80	15	5140

ecological differences between the two experiments mainly due to the nature of fuel submitted to fires and its moisture content.

Test burns for small plots always began with backing fires, which were allowed to develop during 30 to 45 mins. Then the rest of the plot was ignited and could burn rapidly (about 10 minutes) under heading conditions. Due to differences in fire spread rates (about a factor 10), backing fires are overrepresented in our sampling. The pole system we used was designed to sample on field directly in the smoke plume. Conditions were to maintain mobility and to sample the same air parcel with the different gas and aerosol samplers. Different aerosol filters could be obtained in parallel with real-time data for total particulate matter (TPM) and CO-CO$_2$. Mean sampling duration was 3 to 6 mins. Back in laboratory, aerosols were weighed and further analyzed by coulometric titration for their total and black carbon content, and by PIXE for their trace element content. Some electron microscope observations were also performed. Real-time TPM (RAM-1, Mie Inc.) and CO and CO$_2$ (BINOS, Rosemount Inc.) data were used either for instantaneous data by means of peaks intensity, or in parallel with the aerosol samples by integrating record curves. TPM real-time data being obtained by optical measurements (themselves highly dependent on particle size and chemical nature) were used only for characterization of fire behavior. CO and CO$_2$ data have been used for normalization of emissions and emission factor calculations. Experimental details are given in Cachier et al. (1995, 1996), Echalar et al. (1995) and Gaudichet et al. (1995).

Real-Time Record during the Fire Stages
In figure 41.1, we show a typical record of real-time measurements of TPM and CO/CO$_2$. This record was obtained during the burn of plot Shabeni 5 in KNP. As expected, there is an apparent covariation of the intensity of aerosol emissions and the CO/CO$_2$ ratio which is inferred to relate the quality of combustion. The different stages may thus be easily differentiated in both records.

• *Backing fires* have on average less important variability than other stages, especially for aerosol emissions. TPM emissions are the lowest on average.

• *Head fires* appear as a firefront of a few meters in depth. The flame heat is hypothesized to create first dehydration and volatilization of vegetation, which could be the cause of a systematic small increase of

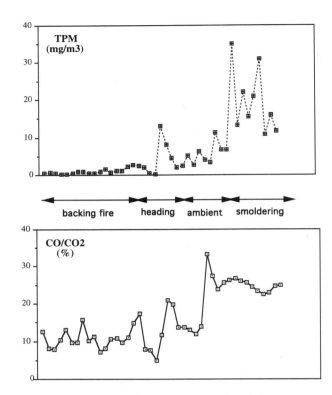

Figure 41.1 Real-time record of total aerosol (TPM) concentrations and CO/CO$_2$ ratios for the different stages of a savanna test fire (Shabeni plot 5 in KNP)

CO/CO$_2$ value. Then the best conditions are found to sustain (during a few seconds only) a high-quality combustion and the CO/CO$_2$ ratio displays for a short while its lower most value (4%). This phase is rapidly followed by an abrupt change of the CO/CO$_2$ ratio, which may reach values over 20%. This phenomenon is probably the result of oxygen or thermal quenching occurring during rapid and violent combustions (Cofer et al. 1989; Cachier et al. 1995). TPM data parallels CO/CO$_2$ data and points out the important heterogeneity of emissions during heading fire stages.

• *Ambient smoke* is found on field immediately after the flame has passed. This smoke may be the result of pyrolysis, cracking and oxidation of organic material suspended in the atmosphere. It may also be caused by the presence of organic gases escaping the smoldering vegetation. During this phase the CO/CO$_2$ ratio is of the order of 12%.

• *Smoldering samples* were obtained from branches left on field. They display the highest TPM concentrations associated with the highest (~25%) CO/CO$_2$ ratios.

Integrated Sample Data: Intra- and Interfire Variability

Integrated sample data are used for the characterization of aerosol production. This characterization is conducted in two complementary directions: comparison of aerosols from different savanna ecosystems and differences observed along the stages of the burns.

Aerosol particles in the fire plume contain primarily carbonaceous material: assessing an average C/O elemental ratio ~ 9 in the plume (Cachier et al. 1995), the abundance of carbonaceous matter is 64% with relatively small black carbon content (Cb/TPM $\sim 7\%$). Carbon amount is found to differ importantly from one sample to another, probably due to variablity in the resuspension of dust or other exogenous particles in relation with fire intensity. During the flaming phase, 75% of particle mass is found in the fine mode ($d \leq 2 \mu m$). After the firefront had passed (posthead conditions), smoking of vegetation and coagulation of freshly emitted particulate material produce a coarser aerosol.

Among the several trace elements detectable in biomass burning aerosols (Artaxo et al. 1994; Echalar et al. 1995; Gaudichet et al. 1995) the notable and quite constant enrichment of two metals, potassium and zinc, could serve as biomass burning indicators (K/Cb ~ 1.3; Zn/Cb $\sim 5.4‰$). Chlorine and sulfur are also found in significant concentrations but their variable abundance probably reflects variable inputs of predeposited dust and precludes any possibility for these elements to serve as biomass burning tracers.

Average results for the two savanna experiments are presented in table 41.2. From this table it appears, first, that there are no significant differences between the two savanna ecosystems and, second, that backing and heading fires produce quite similar aerosols. Smoldering and ambient postheading aerosols are however notably different. Our results point out the importance of the nature of the combustion stage for determining particulate emissions, ecological variability appearing of little significance.

During KNP test fires, a great interfire variability was observed for the CO/CO_2 ratios. Interestingly, it may be seen in figure 41.2 that CO/CO_2 and Cb/Ct mean ratios (obtained for integrated samples) are negatively correlated, the "better" combustion, the "greater" importance of black carbon in the aerosols. This result reinforces the potential of CO/CO_2 to be an indicator of the quality of the combustion process and the prevalence of this quality for particulate emissions.

Domestic Burnings

Experiments are described in Lacaux et al. (1994), Brocard et al. (see chapter 33 this volume) by the persons who made conception and realization of protocols. These experiments were achieved in the Ivory Coast and closely followed by Ivorian people so that they may be considered representative of local traditional domestic practices. Aerosols were collected on filters and sent back to our laboratory for subsequent analyses.

Charcoal Making

A traditional kiln was built at Lamto in April 1992, using local wood. Briefly, the kiln was a hemispheric volume stuffed with different layers of wood logs of decreasing size. The outside was constituted by a mixture of leaves and earth forming a 10-cm-thick proof cover. Air was carefully allowed to enter at the base of the kiln and ventilation was ensured by means of a central pipe. Conditions created inside the kiln are a reduced environment leading to pyrolysis processes producing carbon-rich solid and liquid material, accompanied by the evolvement of atmospheric gases and particulates.

Among others, about 15 aerosol samplings were performed for carbon analysis purpose, at different places near the kiln: at the exit of teflon pipes or holes situated at the upper half of the system or along natural vents due to some cracking of the cover. The experiment lasted 10 days and aerosols were sampled evenly during 1 week. During the whole process, 9264 kg wood was burned representing 2920 kgC (30% moisture in fuel and 45% carbon in dry fuel). At the end, 1789 kg charcoal was obtained; tar and liquid residues were estimated to be of the order of 15% of the initial amount of dry wood. Considering the carbon content of the end products (72% and 60%, respectively), it may be roughly calculated that 44% of the initial fuel carbon is transformed to charcoal, and 21% to tar and liquid residue. Consequently, it may be assumed that 35% of the initial fuel wood carbon evolved as atmospheric gaseous and particulate carbon-containing pollutants.

CO/CO_2 and Cb/Ct ratio data are available for 13 aerosol samples (figure 41.3). Both sets show a significant varibility. Mean CO/CO_2 value is $26.2 \pm 5.5\%$ attesting very reducing conditions inside the kiln. Interestingly, Cb/Ct ratios are very variable (Cb/Ct = $9.7 \pm 2.8\%$) but on average for this experiment, not so low as would have been expected by comparison

Table 41.2 Characterization of biomass burning aerosols sampled during the two savanna experiments at KNP and Lamto (from Cachier et al. 1997)

		Back fire	Head fire	Posthead fire	Smoldering fire (tree)
Fine/total TPM (%)	KNP	84 ± 6	80 ± 10	71 ± 13	—
	Lamto	74 ± 8	73 ± 12	36	—
	mean	79	77	54	
Ct/TPM (%)	KNP	46 ± 1	37 ± 12	43 ± 2	—
	lamto		mean: 70		
	mean	58	53	43	
Cb/Ct (%)	KNP	9.5 ± 1.7	10.7 ± 2	11.4 ± 2	7.0 ± 0.1
	Lamto	12.5 ± 2.6	11.7 ± 3.3	3	7.1
	mean	11	11.2	8.7	7.0
Kfine/Ktotal (%)	KNP	94 ± 2	88 ± 10	71 ± 20	64 ± 30
	Lamto	89 ± 4	70 ± 2	72	—
	mean	92	79	71	64
K/Cb	KNP	1.14 ± 0.48	0.98 ± 0.33	0.32 ± 0.17	0.011
	Lamto	1.6 ± 1.5	1.4 ± 0.9	0.40	—
	mean	1.4	1.2	0.36	0.011
Znfine/Zntotal (%)	KNP	85	89	73	—
	Lamto	81	76	55	—
	mean	83	83	64	—
Zn/Cb (‰)	KNP	6.7 ± 5.4	4.8 ± 2.4	3.35 ± 1.8	0.005
	Lamto	3.7 ± 2.1	6.2 ± 3.0	4.7	—
	mean	5.2	5.5	4.0	0.005
Clfine/Cltotal (%)	KNP	91	88	61	41
	Lamto	86	77	62	—
	mean	89	83	61	41
Cl/Cb	KNP	1.10 ± 0.57	0.84 ± 0.30	0.17 ± 0.16	0.005
	Lamto	0.43 ± 1.5	0.33 ± 0.12	0.37	—
	mean	0.77	0.59	0.27	0.005
Sfine/Stotal (%)	KNP	76	71	70	48
	Lamto	100	79	86	—
	mean	88	75	78	—
S/Cb	KNP	0.22 ± 0.7	0.19 ± 0.04	0.20 ± 0.14	0.01
	Lamto	0.036 ± 0.025	0.044	0.16	—
	mean	—	—	—	—

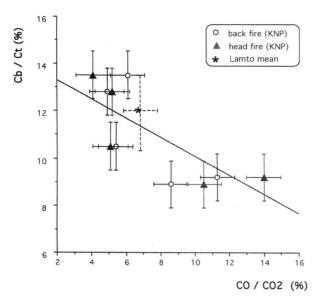

Figure 41.2 Relationship between Cb/Ct and CO/CO_2 ratio mean values obtained during the flaming phase of various savanna fires (from Cachier et al. 1996)

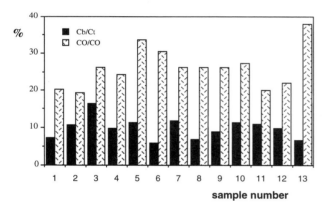

Figure 41.3 Variability of Cb/Ct and CO/CO_2 values in emissions sampled during traditional charcoal making in the Ivory Coast

with results obtained for open field fires. Contrary to savanna fires too, there is no apparent relationship between the two CO/CO_2 and Cb/Ct indicators, confirming that for charcoal making other parameters than the "reducing conditions" may also play a determinant role for particle formation.

Household Fires
The design used for the wood-burning experiments was a 3-stone open base. During numerous experiments, it has been possible to gain a clear picture of the successive stages of the combustion (Brocard et al.

1997): the initial stage, the hot cooking stage, the gentle cooking stage, and at last the end of the fire. All stages are well characterized by their temperatures and their CO/CO_2 ratios.

A few aerosol samples were obtained during the hot cooking stage of representative fires using different sorts of wood used in the Ivory Coast. The high temperature recorded (800°C) and CO/CO_2 ratio values (mean: $5.5 \pm 1.5\%$) attest to prevailing flaming conditions. From our results, Cb/Ct mean value is of the order of $10.4 \pm 1.9\%$ with no clear covariation with CO/CO_2 ratio (table 41.3).

Aerosol Emission Factors

Definitions
The emission factor (EF_x) is used to quantify emissions. It refers to the ratio of the mass of the emitted species X to either the mass of dry vegetation consumed (expressed as $gX/kg_{dry\ plant}$) or to the amount of carbon in the vegetation fuel (expressed as $gX/kgC_{dry\ plant}$). These two parameters are linked by the amount of carbon in plants, which is in the range of 0.41 to 0.47. Here, we only use the reference to carbon.

The emission factor may be experimentally obtained when key species are sampled in the same air parcel. In a given plume air parcel, the combustion produces excess concentration of X and carbon-containing species. The sum of the main carbon species excess concentrations and carbon in ashes (referring to a carbon mass) accounts for the plant carbon that has been consumed (Cachier et al. 1995), so that EF may be expressed in per mil as

$$EF_x = 1000 \cdot \Delta X / \Sigma (\Delta CO_2 + \Delta CO + \Delta CH_4 + NMHC + Ct_{(aerosol)} + VOC_{(unidentified)} + C_{(ashes)})$$

As CO_2 and CO are the main emissions during the oxidative and reducing conditions, emissions may be and are often calculated as the "apparent emission factor" EF' which is a concentration normalization by $CO_2 + CO$ only. In this work we often refer to EF'.

$$EF'_x = 1000 \cdot \Delta X / \Sigma (\Delta CO_2 + \Delta CO)$$

From previous work on savanna fires (Cachier et al. 1995, 1997) we found that relationships between EF and EF' may be roughly expressed as

$$EF_x = EF'_x \cdot 0.93 \text{ (flaming phase)}$$

$$EF_x = EF'_x \cdot 0.89 \text{ (smoldering phase)}$$

Table 41.3 Biofuel experiment (CO/CO_2, Cb/Ct results for the high-temperature cooking step)

Sample number	ΔCb ($\mu gC/m^3$)	ΔCt ($\mu gC/m^3$)	Cb/Ct %	CO/CO_2 %	$\Delta(CO+CO_2)$ ($\mu gC/m^3$)	EF'(Cb) ‰	EF'(Ct) ‰
1	632	5979	10.6	7.6	679 100	0.90	8.80
2	718	7397	9.7	8.1	444 700	1.59	16.6
3	434	5436	8.0	4.1	849 800	0.49	6.34
4	434	4376	9.9	4.5	582 200	0.70	7.42
5	642	4625	13.9	5.1	712 600	0.90	6.45
7	350	2729	12.8	5.2	313 500	1.10	8.71
8	323	3500	9.2	3.6	655 500	0.49	5.27
11	475	5603	8.5	5.5	304 800	1.60	18.4
12	363	3339	10.9	5.7	375 700	1.00	8.92

The use of EF' is substantiated by the linear correlation generally found between excess aerosol and $CO + CO_2$ during a given stage of a burn, either in instantaneous data (figure 41.4a) or in integrated samples (figure 41.4b). Apparent emission factors will be directly calculated from slopes obtained in such graphs or on a sample basis by the ratio of integrated concentration values.

Savanna Aerosols

Savanna aerosol emission in relation with fire phases is illustrated in figure 41.4a for the test fires of KNP Shabeni 5 plot. Backing fires always show a good correlation between excess concentrations and $CO + CO_2$, which may be interpreted as an indicator of a regular production at a constant rate. However, intrafire variability may be important between backing and heading stages and even during heading fires which produce very heterogenous plumes. This intrafire variability may exceed one order of magnitude and be more important that interfire variability (Cachier et al. 1995, 1997). Smoldering fires were always found to produce the highest particulate emissions.

Scatter is much less important for our aerosol samples, which have integrated variable situations during a few minutes, and surprisingly still less important when averaging for the whole experiment in a given savanna ecosystem. As expected backing fires are the most reproducible fires. A careful look at the TPM and carbon data presented in table 41.3 suggests that scatter recorded in aerosol emissions from different fires is not due to the carbon component of the aerosol, which is remarkably constant. A suggestion is that scatter could be primarily due to variation in the amount of dust resuspended during the fire, this resuspension being more important during intense heading fires.

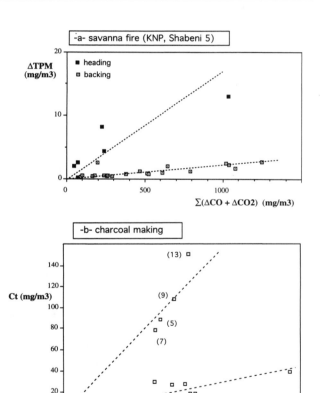

Figure 41.4 Determination of aerosol apparent emission factors (EF') as the ratio of excess concentrations to excess $CO + CO_2$: (a) Savanna fire showing two emission rates during backing and heading fires (TPM real-time data), (b) Evidence for two burning regimes in the kiln during the formation of charcoal (total particulate carbon filter data)

Table 41.4 Characterization of the mean aerosol emission for African burnings (data adapted from Cachier et al. 1995 and 1997; Brocard et al. 1997)

	EF'(Ct) ‰	EF'(Cb) ‰	EF'(TPM) ‰	Cb/Ct %	CO/CO_2	Tpre °C
KNP back fire	7.3 ± 3.7	0.6 ± 0.3	15.3 ± 6.8	9.5	7.3	600
KNP head fire	8.5 ± 4.2	0.9 ± 0.3	23.0 ± 12.8	10.7	7.8	>800
KNP (mean)	8.9	0.90	20.7	10.4	7.7	800
Lamto (mean)	8.0	1.10	12.2	11.9	6.3	600
Smoldering (mean)	52	2.9	69	7	27	350
Comb chamb	25.3	2.35	—	8.5	12	700
Charcoal making	26.0	2.5	—	9.7	26	850
Cooking	9.6	0.97	—	10.4	5.5	800

In light of this hypothesis, it may be understood that Lamto and KNP data show nearly identical carbon emission factors (about 10% discrepancy) and somewhat different TPM emission factors. Lamto fires, which are less intense fires, produce less aerosol dust. It must be recalled here that TPM data used for EF calculations are only those obtained by filter weighing.

Domestic Burning Aerosols

The same exercise was applied to charcoal making results obtained for integrated samples and gave access to the determination of emission factors. It may be seen in figure 41.4*b* that particulate production apparently follows two main distinct regimes. These two modes are not related to either absolute concentrations, day of sampling, or type of vent that have been sampled (pipe, hole, or natural cracking). The uneven production may perhaps be explained by irregular incursions of oxygen in the kiln, following collapse of wood piles and/or cracking of the kiln walls. Variation of EF' between the two modes is in a factor of 6 (5.6 ± 2.0‰ and 29.5 ± 2.7‰). Considering that our data set is representative of average conditions, weighing means obtained for Ct and Cb are 26.0 ± 4.0‰ and 2.5 ± 0.5‰, respectively. These important EF' values clearly recall those obtained for smoldering processes where reducing conditions are also observed.

Moreover, during traditional charcoal making, 35% of the initial fuel wood carbon is lost as atmospheric species and in smoldering combustions the sum CO + CO_2 represents about 89% of the carbon emitted during the process. Then, emission factors EF to be used directly for global estimates based on amount of fuel wood (expressed as gC) submitted to charcoal making process are 10.2‰ for Ct and 0.98‰ for Cb.

On the contrary, domestic cooking of wood that approaches flaming conditions is a less important pro-

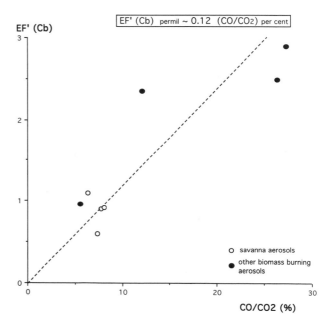

Figure 41.5 Evidence for a relationship between black carbon aerosol EF' and CO/CO_2 characteristic mean values obtained for different types of tropical burning

ducer of particles: mean apparent emission factors EF' are found to be 9.6 ± 4.6‰ for Ct and 0.97 ± 0.41‰ for Cb.

Remarks

In table 41.4, mean data have been summarized for all types of burnings. Noteworthy is the prominent role on average, of CO/CO_2. Indeed, this factor appears to be an important determinant for both the amount of aerosols produced during the burning process (as attested by the fairly good correlation displayed in figure 41.5) and the chemical composition of the carbonaceous fraction (Cb/Ct ratios also reasonably covary with CO/CO_2 ratios). For types of burning that cannot be directly studied for their aerosol emissions, it should be possible to roughly estimate the production of carbonaceous particles from the above relationships, provided access to representative CO/CO_2 ratio of the combustion process.

From a quantitative point of view, it may be considered that savanna fires, which represent a major source in the global budget of atmospheric particulate carbon, are now satisfactorily characterized. Major uncertainty to be solved in the future is the amount of fuel submitted annually to fires and in light of several field observations (see chapter 26 by Moula et al.), revisions of the global budget are needed.

Considering that the use of charcoal for cooking has the characteristics of smoldering processes (Brocard et al. 1997), it may be inferred that the accelerating shift of biofuel use from wood to charcoal in many cities of the developing world is likely to produce increasing local and regional pollution problems. Domestic burns will also gain increasing importance in global budgets.

Background Records

Lamto Background Record

Lamto (5°02W, 6°13N) is a savanna site apart from major anthropogenic activities, yet with people constantly present and available for field work. We had the unique opportunity to constitute a continuous aerosol record during about 3 years representing about 300 samples for carbon measurements and 80 for trace elements. Aerosols were collected each week during 24 hours with a flow rate of 1.5 to 2 m^3/hr to get information on the various fires prevailing in the region. Part of this data is presented here.

An expected but still striking feature is that savanna fires are clearly recorded in the background concentrations (figure 41.6) and reappear each year during the dry season. During some episodes, background concentrations may be multiplied by one order of magnitude or more during several days, attesting to the overwhelming importance of fires on a regional scale. An important year-to-year variability of the set-up and generalization of fires and their intensity is also clearly shown. Secondary peaks appear in March, which may be attributed to late agricultural burns for preparation and fertilization purposes before the beginning of the rainy season. Noteworthy is the fact that these data obtained by simple ground-based sampling are confirmed by satellite observations of the temporal distri-

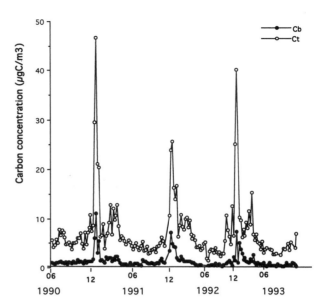

Figure 41.6 Lamto background concentrations of black and total carbon aerosols for the years 1990 to 1993 (24-hr weekly samples)

bution of fires in this region (Langaas 1992; Liousse et al. 1994). During the FOS DECAFE experiments, additional day/night samples were also obtained. In this set, a striking pattern shows diurnal concentrations always higher ($\sim$50%) than those measured at night (Liousse and Cachier 1992). This pattern has been interpreted as an indication that most savanna fires are short-lived and extinguish at night by changes to less favorable weather conditions. If so, satellite night observations (Cahoon et al. 1992) could report only the uncontrolled fires, whereas day observations would integrate both controlled burnings (mainly for agricultural purpose) and wild fires (Malingreau 1990; Langaas 1992).

When the savanna fire season is over, there remains consistent background of combustion particles in the Lamto atmosphere. This background is probably due to residual agricultural burns and domestic fires. On a global temporal scale, however, in spite of source changes there is no characteristic change in the carbon composition of the aeosols as shown by the correlation between Cb and Ct values displayed in figure 41.7. Mean Cb/Ct is 19.8 $\pm$ 4% and apparently does not show seasonal trend. This value is higher than that obtained in fresh biomass burning aerosols (table 41.4) and could point out that organic aerosols rapidly undergo chemical oxidation in the atmosphere. But this Cb/Ct value is, however characteristically smaller than that obtained for industrial aerosols (Cachier 1995).

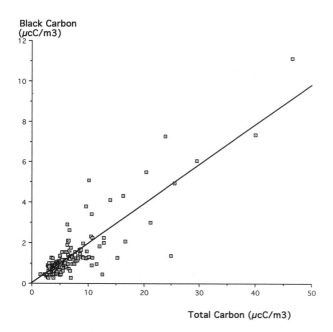

Figure 41.7 Correlation between black carbon and total particulate carbon concentrations for Lamto background aerosols (24-hr weekly samples, 1990–1993)

The background aerosol also contains significant amounts of trace elements which in fresh aerosols have proved to be biomass burning indicators. Indeed, zinc and potassium concentrations are found to covary with black carbon, an unambiguous tracer of combustions (figure 41.8), which shows that at Lamto on average for these constituents, other sources are minor. Ratios found (K/Cb $\sim$0.5; Zn/Cb $\sim$4.10^{-3}) compare with those obtained in plumes (1.2 and 5.10^{-3}, respectively).

The overwhelming atmospheric influence of biomass burning all year long is also visible in African rains. We have shown elsewhere that either in West Africa (Ducret and Cachier 1992) or in equatorial regions (Cachier and Ducret 1991), rains are highly polluted by fire emissions.

Kruger National Park Background Record Data

A few samples (7) were obtained in Kruger National Park (31°16E, 25°10S) during the SAFARI-92 experiment. To avoid contamination from tourist traffic and numerous barbecues it was decided to sample at night with batteries at the remote site of GranoKop a 200-m-high hill favorably situated off the village. Average filtered air volume was 18 m^3.

A simple look at filters, colored from very light to deep grey, indicates that at this site the atmospheric

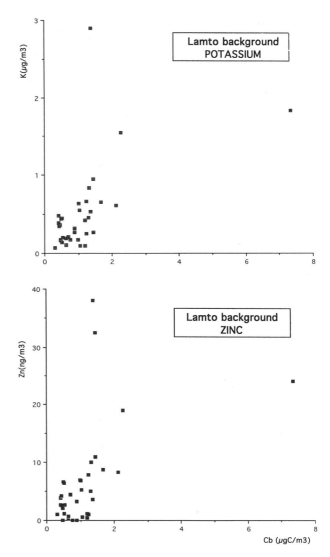

Figure 41.8 Correlation between potassium or zinc and black carbon concentrations for Lamto background aerosols

composition is highly variable. Contrary to the Lamto site that is constantly and primarily influenced by combustions especially during the dry season at KNP, fluctuation of meteorological conditions (Trollope et al. 1996) governs atmospheric loadings of species. KNP may indeed receive air masses from the east (mostly maritime), from the north (with fire influence), or from the west (with influence of the vicinal industrial basin). Transition to opposite situations may be very abrupt as attested by drastic changes in filter color and black carbon content (Cb from 0.4 to 5 μgC/m^3).

The elemental composition of aerosols is still more variable. The small influence of biomass burning is

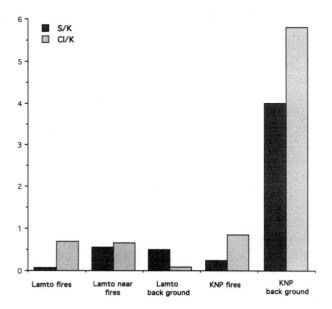

Figure 41.9 Sulfur and chlorine concentrations (normalized to potassium) in savanna burning and background aerosols at two savanna sites. The aging of biomass burning aerosols is evidenced in Lamto aerosols whereas KNP background aerosols do not hold the biomass burning signature.

pointed out when concentrations are normalized with potassium (Gaudichet et al. 1995). In figure 41.9, it may be observed that biomass burning flaming aerosols contain characteristic amounts of sulfur and chlorine. From the Lamto data it may be seen that when these aerosols undergo aging processes, they typically lose part of their chlorine content and are enriched in sulfur. However, although KNP fresh aerosol data is in accordance with our Lamto observations, KNP background data is in total disagreement, indicating that major sources other than biomass burning are prevailing. This is also confirmed by the very important range found for either S/K or Cl/K ratios (nearly two orders of magnitude), whereas the Lamto data variability is less than a factor of 1.5.

Amsterdam Island Data

In the frame of IFRTP (Institut Français de Recherche des Terres Australes) research programs, Amsterdam Island has been dedicated for long-term studies of atmospheric species. Amsterdam Island is a small (55 km^2) island located in the southern Indian Ocean (77°31E–37°47S) and exposed to westerlies.

Black carbon has been measured with a MAGEE-Scientific aethalometer installed since the beginning of 1991. At present, data for 3 years are available. The principle of the instrument is based on attenuation of

a visible light beam passing through a fibrous filter where atmospheric aerosol is continuously deposited. Black carbon is assessed to be the only significant light absorber among other aerosol particles. Due to the pertaining low concentrations, a time-base of 2 hours has been adopted. Results presented in this chapter are those obtained with the classical calibration of 19 m^2 g.

Radon (Rn) has been measured on active deposit following the procedure set up by Polian et al. (1986). Because radon is produced mainly over continental surfaces and due to its short lifetime of 3.8 days, concentrations were used to first rule out local anthropogenic influence and second to support recent continental advection. Following Gaudry et al. (1987) it is assessed that at the site, typically marine air masses will have Rn concentrations of the order of 1 pCi/m^3, whereas any concentration above 2 pCi/m^3 will denote a continental influence from South Africa and/or Madagascar.

Monthly averages of black carbon concentrations are presented in figure 41.10a. They show a persistent and reproducible background of black carbon particles throughout the year which, in this region of the world, may be primarily attributed to tropical (wild, agricultural, or domestic) biomass burning. Mean black carbon background values are low, of the order of a few nanograms per cubic meter, however one order of magnitude higher than that found at Antarctic sites (Hansen et al. 1988). The dry season of the tropical southern atmosphere is strikingly recorded each year with however some temporal variation of the peak and its intensity. It must be emphasized that these peaks are due to the conjunction of two favorable factors, the establishment of the bushfire season and the preferential transport off Africa in the eastern direction (see chapter 39 by Tyson et al.; Garstang et al. 1996). Secondary maxima are observed in spring and also attributed to transport changes at a synoptic scale (Tyson, personal communication). Rapid transport of pollutants from the South African continent is confirmed by radon measurements, showing important continental inputs and by airmass back trajectories. However, an odd aspect of our results is that concentations that are measured are ground-based measurements, whereas trajectories that may help in data interpretation are generally situated higher (700 mbar or less).

A typical and striking episode of rapid long-range transport of combustion particles is shown in figure 41.10b. This episode lasted about one day and a half and displayed an enhancement of black carbon con-

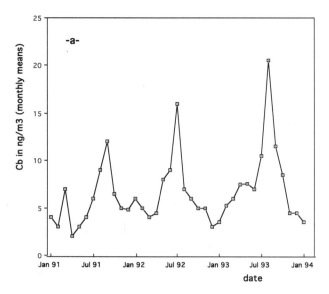

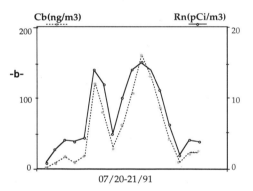

Figure 41.10 Amsterdam Island data for black carbon aerosols: (*a*) Monthly means obtained during three consecutive years (1991 to 1993), (*b*) Correlation observed between radon and black carbon concentrations during a "radon storm" indicating a rapid transport of particles from the African continent

centrations of a factor of 50 within a few hours (each concentration is obtained for a 2-hour interval). The quasi-perfect correlation between black carbon and radon concentrations shows first that black carbon particles are not produced by local pollution. It also indicates that both species originate from continents and followed similar atmospheric pathways. This study case may be considered representative of the dry season when background situation is severely altered during short and repetitive episodes, by long-range continental inputs. At this period of the year, the African southern continent is mostly under the influence of bushfires, the atmospheric emissions of which may be exported far from sources. Pollution episodes at Amsterdam Island are obviously connected with

"radon storms" and meteorological conditions. The role of source intensity which, on a regional scale, is uniform during several days is by far less important.

Finally, the prominence of transport at remote sites may be also stressed by the observation of yearly mean concentrations: the apparent variability of biomass burning inputs of black carbon (Cb means are 6.87, 8.98, and 8.42 ng/m^3) is totally buffered by radon variabilty (Rn means are 0.82, 1.18, and 1.12 pCie/m^3) and the constancy of the Cb/Rn ratio confirms that rapid transport from the African continent is the first determinant for black carbon concentration in the Amsterdam Island atmosphere. Long-term studies should confirm this result and give further indications on the role of atmospheric perturbations such as those observed during El Niño years.

Acknowledgments

Financial support from CEA, CNRS, Ministère de l'Environnement, and IFRTP in France made possible such various experiments. We wish to thank SAFARI PIs and staff in KNP, and DECAFE PIs and staff at the Lamto station for cooperation and help in the organization of the test fires. Untiring support and help from several persons during fire sampling and long-term monitoring is also gratefully acknowledged. Warm and enthusiatic exchanges with people isolated at Amsterdam Island have been a great help. This is CFR contribution no. 1864.

References

Artaxo P., F. Gerab, M. A. Yamasoe, and J. V. Martins. 1994. Fine mode aerosol composition at three long-term atmospheric monitoring sites in the Amazon basin, *J. Geophys. Res.* 99, 22857–22868.

Bonsang, B., C. Boissard, M. F. LeCloarec, J. Rudolph, and J. P. Lacaux. 1995. Methane, carbon monoxide and light nonmethane hydrocarbon emissions from African savanna burnings during the FOS DECAFE Experiment. *J. Atmos. Chem.* special DECAFE issue, 22, 149–162.

Cachier H. 1992. Biomass burning sources. In W. A. Nierenberg (ed.) *Encyclopedia of the Earth Science System*, Academic Press, San Diego, 1, 377–385.

Cachier, H. 1995. Combustion carbonaceous aerosols in the atmosphere: implications for ice-core studies. In R. Delmas (ed.) *Ice-Core Studies of Biogeochemical Cycles*, NATO-ASI series 30, 313–346, Springer Verlag, Berlin.

Cachier H. and J. Ducret. 1989. Influence of biomass burning on equatorial African rains, *Nature* 340, 371–373.

Cachier H., J. Ducret, M. P. Brémond, V. Yoboué, J. P. Lacaux, A. Gaudichet, and J. Baudet. 1991. Biomass burning aerosols in a savanna region of the Ivory Coast. In J. S. Levine (ed.) *Global Biomass Burning: Atmospheric, Climatic, and Biospheric Implications.* MIT Press, Cambridge, Mass., pp. 174–180.

Cachier H., C. Liousse, P. Buat-Ménard, and A. Gaudichet. 1995. Particulate content of savanna fire emissions, *J. Atmos. Chem.* special DECAFE issue, 22, 123–148.

Cachier H., C. Liousse, P. Buat-Ménard, A. Gaudichet, F. Echalar, T. Kuhlbush, and J. P. Lacaux. 1997. Particulate emissions during savanna fires in Kruger National Park (South Africa), *J. Geophys. Res.*, submitted.

Cahoon D. R., B. J. Stocks, J. S. Levine, W. R. Cofer, and K. P. O'Neil. 1992. Seasonal distribution of African savanna fires, *Nature* 359, 812–815.

Clain M. P. 1995. Etude de la composante organique de l'aérosol atmosphérique, Ph. D. dissertation, Université de Savoie, pp. 158 (in French).

Cofer W. R., J. S. Levine, D. I. Sebacher, E. L. Winstead, P. J. Riggan, B. J. Stocks, J. A. Brass, V. G. Ambrosia, and P. J. Boston. 1989. Trace gas emissions from chaparral and boreal forest fires, *J. Geophys. Res.* 94, 2255–2259.

Crutzen, P. J., and M. O. Andreae. 1990. Biomass burning in the tropics: Impact on atmospheric chemistry and biogeochemical cycles. *Science* 250, 1669–1778.

Delmas R., J. P. Lacaux, and D. Brocard. 1995. Determination of biomass burning emission factors: methods and results, *Environmental Monitoring and Assessment* 38, 181–204.

Dinh P. V., J. P. Lacaux, and R. Serpolay. 1994. Cloud active particles from African savanna combustion experiments, *Atmos. Res.* 31, 41–58.

Ducret J., and H. Cachier. 1992. Particulate carbon content in rains at different temperate and tropical locations, *J. Atmos. Chem.* 15, 55–67.

Echalar F., A. Gaudichet, H. Cachier, and P. Artaxo. 1995. Aerosol emissions by biomass burning in Africa and in the Amazon basin: characteristic trace elements and fluxes, *Geophys. Res. Lett.*, 22, 3039–3042.

Gaudichet A., F. Echalar, B. Chatenet, J. P. Quisefit, G. Malingre, H. Cachier, P. Buat-Ménard, P. Artaxo, and W. Maenhaut. 1995. Trace elements in Tropical African savana biomass burning aerosols, *J. Atmos. Chem.* special DECAFE issue, 22, 19–39.

Gaudry A., P. Monfray, G. Polian, and G. Lambert. 1987. The 1982–1983 El Niño episode: a 6-billion ton CO_2 release, *Tellus* 39B, 209–213.

Garstang M., P. D. Tyson, H. Cachier, J. P. Lacaux, and L. Radke. 1996. Atmospheric transport of particulate and gaseous products by fires, In J. Clark (ed.) *Sediment Records of Biomass Burning and Global Change*, NATO-ASI series, Springer Verlag, Berlin, in press.

Hansen A. D. A., B. A. Bodhaine, E. G. Dutton, and R. C. Schnell. 1988. Aerosol black carbon measurements at the Souh Ple: initial results, 1986–1987, *Geophys. Res. Lett.*, 15, 1193–1196.

Kaufman Y. J., and T. Nakajima. 1993. Effect of Amazon smoke on cloud microphysics and albedo: analysis from satellite imagery, *Applied Meteorology* 32, 729–744.

Lacaux J. P., D. Brocard, C. Lacaux, R. Delmas, A. Brou, V. Yoboué, and M. Koffi. 1994. Traditional charcoal making: an important source of atmospheric pollution in the African Tropics, *Atmos. Res.* 35, 71–76.

Lacaux J. P., J. M. Brustet, R. Delmas, J. C. Menaut, L. Abbadie, B. Bonsang, H. Cachier, J. G. Baudet, M. O. Andreae, and G. Helas. 1995. Biomass burning in the tropical savannas of Ivory Coast: an overview of the field experiment Fire of Savanna (FOS/DECAFE), *J. Atmos. Chem.* special DECAFE issue, 22, 195–216.

Langaas S. 1992. Temporal and spatial distribution of savanna fires in Senegal and The Gambia, West Africa, 1989–1990, derived from multitemporal AVHRR night images, *Int. J. Wildlands Fire* 2, 21–36.

Levine J. S. 1990. Global biomass burning: atmospheric, climatic, and biospheric implications, *EOS* 71, 37–39.

Liousse C., and H. Cachier. 1992. Measurements of black carbon aerosols in the atmosphere of two different source regions: real-time data for the Paris region and a savanna site of the Ivory Coast, *Environ. Technol.* 13, 959–967.

Liousse C., J. R. Franca, H. Cachier, and F. Dulac. 1994. Monitoring of carbonaceous aerosols at Lamto, Ivory Coast, and comparison with fire pixel numbers retrieved from AVHRR satellite data, Paper presented at the 5th International Conference on Carbonaceous Particles in the Atmosphere, Berkeley, Calif.

Lobert J. M., D. H. Scharffe, W. M. Hao, T. A. Kuhlbusch, R. Seuwen, P. Warneck, and P. J. Crutzen. 1991. Experimental evaluation of biomass burning emissions: Nitrogen and carbon containing compounds. In Levine J. S. (ed.) *Global Biomass Burning: Atmospheric, Climatic, and Biospheric Implications.* MIT Press, Cambridge, Mass., pp. 289–304.

Malingreau J. P. 1990. The contribution of remote sensing to the global monitoring of fires in tropical and subtropical ecosystems. In J. G. Goldammer ed., *Fires in the Tropical Biota*, Ecological studies 84, Springer-Verlag, Berlin, 337–370.

Mazurek M. A., W. R. Corfer, and J. S. Levine. 1991. Carbonaceous aerosols from prescribed burning of boreal forest ecosystem, In J. S. Levine (ed.) *Global Biomass Burning: Atmospheric, Climatic, and Biospheric Implications.* MIT Press, Cambridge, Mass., pp. 258–263.

Polian G., G. Lambert, B. Ardouin, and A. Jegou. 1986. Long-range transport of continental radon in subantarctic and antarctic areas, *Tellus* 38B, 178–189.

Rogers C. F., J. G. Hudson, J. Hallett, and J. E. Penner. 1991. Cloud droplet nucleation by crude oil smoke and coagulated oil/wood smoke particles, *Atmos. Environ.* 25A, 2571–2580.

Sepauski R. J., and F. W. Stoss. 1990. *Trends '90*, Boulder, Colo.

Trollope, W. S. 1981. Recommended terms, definitions and units to be used in fire ecology in Southern Africa, *Proc. Grassld Soc. Sth Afr* 16, 107–109.

Trollope W. S., A. L. Potgieter, N. Zambatis, and L. A. Trollope. 1996. SAFARI-92: Characterization of biomass and fire behaviour in controlled burns in the Kruger National Park, *J. Geophys. Research* special SAFARI issue, in press.

Aerosol Optical Properties over Southern Africa during SAFARI-92

Philippe Le Canut, Meinrat O. Andreae, Geoffrey W. Harris, Frank G. Wienhold,
and Thomas Zenker

During the past one-hundred years, the mean temperature of Earth's surface has increased by approximately 0.5°C, possibly due to the anthropogenic input of greenhouse gases (Houghton et al. 1995). This temperature increase, which is likely to continue, might have been even higher if aerosols had not also been added in large amounts to the troposphere (Hansen et al. 1993; Andreae 1995). Aerosols have direct as well as indirect effects on climate: they are believed to have a cooling effect, both by reflecting solar radiation before it reaches Earth's surface and by leading to the formation of more numerous but smaller cloud droplets and thus increasing the cloud albedo (Charlson et al. 1991, 1992; Penner et al. 1992; Hegg et al. 1993; Kiehl and Briegleb 1993; Andreae 1995). Even more than the long-lived greenhouse gases, aerosols have a spatially and temporally variable effect on climate. Their sources are unequally distributed around the world, their lifetime is short, and their optical properties depend upon their chemical composition, the ambient humidity, and their size distribution (Pilinis et al. 1995).

Recently, sulfate aerosols have been extensively studied because of their dominant contribution to anthropogenic aerosol cooling in the northern hemisphere (Charlson et al. 1991, 1992; Kiehl and Briegleb 1993). Biomass-burning particulates, on the other hand, which are most important in the tropics (Penner et al. 1992), have received much less scientific attention. Local and instantaneous measurements of the aerosol properties, at various distances from the sources, are necessary to fully understand their role and to provide data for global models. Studies have been done in remote continental and marine regions (Patterson et al. 1980), in Alaskan wildfires (Nance et al. 1993), in atmospheric dust over Tadzhikistan (Fraser 1993), over Algonquin Park in Canada (Isaac et al. 1986), and in smoke from the Kuwait oil fires (Weiss and Hobbs 1992; Pilewskie and Valero 1992). Nevertheless, more studies are needed, especially from the tropical regions. They are especially necessary in southern Africa, where few data are available on the optical properties resulting from smoke due to savanna fires. Savannas account for about half of the biomass burned worldwide (Andreae 1993) and for 10 to 20% of pyrogenic aerosols released globally (Le Canut et al. 1996). Though regionally and seasonally limited, aerosols from savanna fires may therefore have an important role in the global radiation budget.

During the international measurement campaign, SAFARI-92 (Southern African Fire/Atmosphere Research Initiative), a subcomponent of the IGAC/BIBEX (International Global Atmospheric Chemistry Program/Biomass Burning Experiment) project STARE (Southern Tropical Atlantic Region Experiment) (Andreae et al. 1994), which took place during the 1992 dry season, particle concentrations were measured aboard an instrumented DC-3 aircraft. Here, we present the aerosol loadings and size distributions measured over southern Africa, and the column burdens and optical depths derived from these measurements.

Flight Profiles

Flight profiles were either staircaselike transects with long horizontal legs, or spiral climbs and descents made for soundings. From all the flights during the airborne component of SAFARI-92, we obtained 48 vertical profiles (24 descents and 24 ascents). Although most of the ascents and descents were adjacent in time and space, some were far apart because the aircraft occasionally flew level for some time at a constant altitude. Most of the vertical profiles spanned the altitude range of 1000 to 4000 m asl (above sea level), and the average duration of an ascent or descent was 30 minutes, which corresponded to a climbing or descending rate of approximately 100 m/min. Some other ascents and descents were made during the experiment, most of which spanned an altitude of about 500 m. These are not considered in this chapter, for they are not representative of the total lower troposphere. Since data from fresh smoke plumes have been excluded from the present data set, the measurements presented here can be considered representative of the regional aerosol loading. Each profile has been assigned a code number based on flight number and flight condition (*A* for ascent and *D* for descent). These profiles are listed with their sampling date, time, and height in table 42.1, and their geographic location is marked on a map of southern Africa (figure 42.1).

Methods

Measurement Techniques

The particle counter used on board the aircraft was a Passive Cavity Aerosol Spectrometer Probe (model PCASP-100X; Particle Measuring Systems Inc., Boulder, CO). This probe classifies particles as a function of their size into 15 bins in the range of 0.10 to 3.0 μm. The aerosol concentration was recorded after averaging for 10 seconds. The instrument was calibrated by the manufacturer using spherical latex particles. The instrument calibration is known to be somewhat sensitive to differences between the optical properties of the calibration aerosol and the actual aerosol sampled in the field (Kim and Boatman 1990). However, since aerosol composition and refractive index data were not available with the temporal and particle-size resolution required, we were not able to make the corrections suggested by Kim and Boatman (1990). This limitation may have resulted in underestimates of the actual particle size, so that our aerosol mass loadings and the parameters derived from them represent lower limits of the actual values.

The CO instrument was a Tunable Diode Laser Spectrometer (TDLS), which used two-tone frequency modulation (Wienhold et al. 1993), and recorded data on a 1-second time base. The CO_2 instrument was a nondispersive infrared gas analyzer (LI-COR Corp., Lincoln, NB, USA), which operated with an internal time constant of 10 sec (Zenker et al., chapter 28 this volume).

Two O_3 instruments were on board: (1) a Thermo Electron Model 49 (Thermo Electron Instruments, Hopkinton, MA, USA) using UV absorption, and (2) a Scintrex Luminox Ozone monitor Model LOZ-4 (Scintrex/Unisearch, Concord, Ontario, Canada), using liquid chemiluminescence. The Scintrex instrument was calibrated in the field using the Thermo Electron as a reference. For the present analysis, data from the Thermo Electron were used preferentially. Only where these were not available are data from the Scintrex substituted. In a few cases, low-frequency

oscillations of unknown origin made the Thermo Electron data difficult to interpret. In these instances a 10-min. moving average of the Thermo Electron data was used to eliminate the oscillations.

Table 42.1 shows, for each profile, whether the CO, CO_2, and O_3 data are available (A), partially available (PA), or not available (NA), and lists the type of ozone data used. We averaged the CO, CO_2, and O_3 concentrations on a 10-sec basis, and eliminated differential time lags between the various instruments by using temporal markers, such as a short passage through the plume of a smokestack or through a heavily polluted layer. The particle concentrations (N) are given per cubic centimeter (cm^{-3}) and are corrected to standard temperature and pressure (STP; $T = 20°C$, $P = 1013.25$ hPa) in the section dealing with particle size distributions. For assessing the optical properties, the ambient concentrations were used.

Particle Size Distributions

The total aerosol size distribution can be approximated as the sum of several log-normal distributions called *modes*. Since the PCASP-100 X's range is 0.1 to 3.0 μm, resulting in truncation of the coarse mode, we could fit only a log-normal distribution to the accumulation mode using the method described by Horvath et al. (1990).

Two important properties of log-normal distributions are: (1) if the number size distribution is log-normal, then the volume size distribution is also log-normal, and (2) number and volume size distributions have the same geometric standard deviation.

The log-normal number distribution is described by the relation:

$$\frac{dN}{d\ln d} = \frac{N_0}{\sqrt{2\pi}\ln \sigma_g} \exp\left(-\frac{(\ln d - \ln d_g^N)^2}{2\ln^2 \sigma_g}\right) \qquad (42.1)$$

The units of $dN/d\ln d$ are in cm^{-3}. Although the distribution is really in differentials rather than in differences, we approximate d's with Δ's because of the finite size resolution of the aerosol-measuring probe.

The three characteristic parameters describing the particle population are:

1. d_g^N, the mean number geometric diameter, which corresponds to the intercept of the cumulative size distribution with the 50% abundance line.

2. σ_g, the geometric standard deviation, which is related to the slope of the cumulative size distribution by the relation:

Table 42.1 Sampling dates, times, and heights of the vertical profiles studied during SAFARI-92

Vertical profiles	Date	No.	Time, LT	Height, m	CO	CO$_2$	O$_3$
SAF07-A1	09/24/92	1	15:44:49–16:11:39	3287	A	A	A
SAF07-D1	id.	2	16:12:09–16:47:29	2730	PA	A	A
SAF07-A2	id.	3	16:47:29–17:16:39	2433	PA	A	A
SAF07-D2	id.	4	17:16:39–17:35:19	1210	A	A	A
SAF08A-D1	09/25/92	5	10:18:05–10:33:35	1537	PA	A	PA[a]
SAF08A-A2	id.	6	10:33:35–11:01:55	3295	A	A	A[a]
SAF08A-D2	id.	7	11:11:35–11:31:05	1840	PA	PA	PA[a]
SAF09A-A1	09/26/92	8	12:10:00–12:28:00	2024	A	A	A[b]
SAF09A-D1	id.	9	12:33:00–12:56:00	1975	PA	A	A[b]
SAF09B-A1	09/26/92	10	15:42:04–16:10:44	2327	A	A	A[b]
SAF09B-D1	id.	11	16:10:44–16:27:54	2327	A	A	A[b]
SAF09B-A2	id.	12	17:12:34–17:32:34	1975	PA	A	A[b]
SAF09B-D2	id.	13	17:32:34–17:52:14	2227	A	A	A[b]
SAF10A-A1	09/27/92	14	09:13:23–09:44:43	2425	A	A	A[b]
SAF10A-D1	id.	15	09:44:43–10:08:43	2314	A	A	A[b]
SAF10B-A1	09/27/92	16	16:14:23–16:41:03	2646	A	A	A[b]
SAF10B-D1	id.	17	16:41:03–17:01:43	2661	A	A	A[b]
SAF11-A1	09/28/92	18	11:32:45–11:48:45	1201	PA	PA	PA[a]
SAF11-D1	id.	19	11:48:45–12:23:05	1653	A	A	A[a]
SAF11-A2	id.	20	11:23:05–12:49:05	2502	PA	A	A[a]
SAF11-D2	id.	21	12:49:05–13:32:05	2600	A	A	A[a]
SAF12-A1	10/01/92	22	09:38:10–09:52:10	1495	NA	A	A[a]
SAF12-D1	id.	23	09:52:10–10:33:40	2557	NA	A	A[a]
SAF12-A2	id.	24	10:33:40–10:51:50	2557	NA	A	A[a]
SAF12-D2	id.	25	11:54:00–12:09:00	1721	A	A	A[a]
SAF12-A3	id.	26	12:09:00–12:29:50	2517	A	A	A[a]
SAF12-D3	id.	27	12:59:10–13:13:10	1522	A	A	A[a]
SAF14-A1	10/02/92	28	08:36:54–08:54:54	2586	A	A	A[a]
SAF14-D1	id.	29	08:58:54–09:05:24	1564	NA	A	PA[a]
SAF14-A2	id.	30	10:24:28–10:42:38	2659	A	A	A[a]
SAF14-D2	id.	31	10:42:38–10:52:38	1578	A	A	A[a]
SAF15-A1	10/03/92	32	08:40:17–08:53:37	1778	PA	A	A[a]
SAF15-D1	id.	33	08:53:37–09:07:07	1789	A	A	A[a]
SAF15-A2	id.	34	09:38:27–09:58:17	2739	A	A	A[a]
SAF15-D2	id.	35	09:58:17–10:22:47	2027	A	A	A[a]
SAF15-A3	id.	36	11:44:46–12:02:46	2361	A	PA	A[a]
SAF16-A1	10/04/92	37	09:38:31–09:50:41	1442	A	A	A[a]
SAF16-D1	id.	38	09:54:21–10:07:31	1758	A	A	PA[a]
SAF16-A2	id.	39	10:07:31–10:24:01	1883	A	PA	A[a]
SAF16-D2	id.	40	10:24:11–11:11:11	2390	A	A	A[a]
SAF16-A3	id.	41	11:11:11–11:28:11	2314	PA	PA	A[a]
SAF16-D3	id.	42	11:28:11–11:47:51	1121	PA	NA	A[a]
SAF17-A1	10/06/92	43	09:00:04–09:10:44	1473	NA	A	A
SAF17-D1	id.	44	09:10:44–09:15:44	816	PA	A	A
SAF17-A2	id.	45	09:15:44–09:51:54	1039	A	A	A
SAF17-D2	id.	46	09:51:54–10:06:34	1991	A	A	A
SAF17-A3	id.	47	10:06:34–10:26:44	2840	A	A	A
SAF17-D3	id.	48	10:26:44–10:41:54	1842	A	A	A

NA: Nonavailable.
PA: Partially available.
Ozone data: Thermo electron instrument, [a]Scintrex instrument, [b]10-mn average thermoelectron.

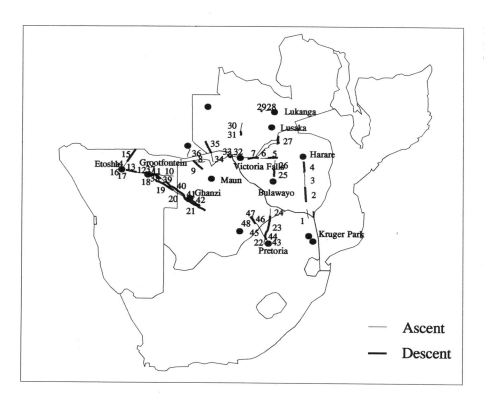

Figure 42.1 Location of the vertical profiles made by the DC-3 over southern Africa during SAFARI-92

$$\sigma_g = \exp\left(\frac{1}{\text{slope}}\right) \qquad (42.2)$$

and,

3. N_0, the total particle number concentration of the best-fitted log-normal number distribution, which is determined by a least-squares fit procedure (Horvath et al. 1990):

$$N_0 = \frac{\sum_{i=1}^{7} w_i \times \Delta n(d_i) \times \Delta N_i}{\sum_{i=1}^{7} w_i(\Delta n(d_i))^2} \quad \text{with} \quad w_i = \sqrt{\Delta n(d_i)} \qquad (42.3)$$

$\Delta n(d_i)$ is the normalized log-normal distribution and w_i is the weighting factor for the ith bin. We consider only the first seven bins because they account for about 99.5% of the particles in the range 0.1 to 3.0 μm. In the same manner, the log-normal volume distribution is characterized by three similar parameters: d_g^V, σ_g, and V_0. The units of $\Delta V/\Delta \ln d$ are in μm^{-3}/cm^{-3}.

Optical Depth

The total optical depth (δ_{total}) of the lower troposphere in the visible spectrum is the sum of: (1) the optical depth due to the scattering and adsorption of the fine aerosol particles (δ_{fine}) determined by the Mie theory, (2) the optical depth due to the large particles (δ_{coarse}),

(3) the optical depth due to Rayleigh scattering by air molecules (δ_{Rayleigh}), and (4) and optical depth due to ozone absorption (δ_{ozone}) in the Chappuis bands (400–800 nm):

$$\delta_{\text{total}} = \delta_{\text{fine}} + \delta_{\text{coarse}} + \delta_{\text{Rayleigh}} + \delta_{\text{ozone}} \qquad (42.4)$$

Since the optical depth due to the coarse mode is negligible compared to that of the accumulation mode, as we demonstrate in detail below, we do not consider it in this chapter. Except under very dusty conditions, more than 90% of the extinction is due to particles with diameters less than 2 μm (Andreae 1995). All the partial optical depths are wavelength dependent and are derived from the corresponding extinction coefficient $\sigma(\lambda, z)$ vertical distributions:

$$\delta(\lambda) = \int_{z_1}^{z_2} \sigma(\lambda, z)\, dz \qquad (42.5)$$

z_1 is the altitude at the bottom of the profile and z_2 is the altitude at the top. Since the starting and ending altitudes differ from one vertical profile to the other, the optical depths will be normalized to a standard layer thickness of 3000 m, which corresponds approximately to the maximum height of our profiles. This choice of a standard layer thickness can be justified based on the lidar observations made by Browell et al. (1996) during TRACE-A. During several flights over southern Africa between 3 October and 11 October, a sharp inversion was present at a height of 2.5 to 3.5 km above ground, with aerosol-scattering intensities below the inversion being some 10 to 20 times higher than above. As a result, the aerosol in the free troposphere above 4 km asl contributed little to the total tropospheric aerosol optical depth. This inversion persisted at approximately the same height during the entire period (Garstang et al. 1996).

Optical Depth due to the Fine Aerosol

The size of the particles measured by our probe is comparable to the wavelengths of the incident light in the 0.3 to 3.0-μm atmospheric window. The optical size parameter $\alpha = 2\pi r/\lambda$ therefore falls in the range of 0.1 to 50, justifying use of the Mie theory (Van de Hulst 1957). The extinction coefficient for the fine fraction of the aerosol is defined by:

$$\sigma(\lambda, z) = \int \pi r^2 Q(r, \lambda) n(r, z) \, dr \qquad (42.6)$$

where $n(r, z)$ is the size-number distribution at altitude z, with $n(r, z) = dn(z)/dr$, and $Q(r, \lambda)$ is the Mie efficiency factor. For each value of $\alpha = 2\pi r/\lambda$ there is a corresponding value of $Q(r, \lambda)$, which is obtained using the subroutine BHMIE from Bohren and Huffman (1983).

To perform the Mie calculations, we need to choose a complex refraction index representative of the fine particles in the lower troposphere. The complex index of refraction for organic particles is 1.43-0.0035i; it is much higher for graphitic carbon at 2.00-0.66i (Holben et al. 1991). Since the black-carbon content of the aerosol was very high during the experiment (Andreae et al. 1996), we have to consider its influence on the optical depth. The regional aerosol consists predominantly of diluted smoke, containing organic particles and black carbon with indices of refraction of 1.43-0.0035i and 2.0-0.66i, respectively. We therefore estimate separately the optical depths resulting from these two components and then add them together. We obtain identical results by estimating a mean index of refraction from the black-carbon percentage in the aerosol.

Since the probe partitions the particle size range of 0.1 to 3.0 μm into 15 bins, we approximate the extinction coefficient for the actual measured distribution by the sum:

$$\sigma(\lambda, z) \cong \sum_{i=1}^{i=15} \pi r_i^2 Q(r_i, \lambda) \Delta N_i(z) \qquad (42.7)$$

where $\Delta N_i(z)$ is the ambient particle concentration in bin i at the altitude z. When using the fitted log-normal distributions, we choose narrower bins with constant width.

The aerosol scattering and absorption coefficients at 0.55 μm ($[\sigma] = $ m^{-1}) for the accumulation mode can also be approximated at each altitude z from the mass scattering and absorption coefficients of diluted smoke ($[\alpha] = $ m^2 g^{-1}) at 0.55 μm and the mass concentration of the size distribution at the altitude z ($[m] = $ g m^{-3}):

$$\sigma(z) = \alpha \times m(z) \qquad (42.8)$$

For dry smoke from biomass burning, several values of the mass scattering coefficient are found in the literature: Radke et al. (1988) measured 5 m^2 g^{-1}, Ferrare et al. (1990) found 4.5 m^2 g^{-1}, and recently Nance et al. (1993) suggested values as low as 3.5 m^2 g^{-1}. Since our measurements include not only particles from biomass burning, but also sulfates and minor amounts of other components, we need to use a mass scattering coefficient appropriate for this kind of mixed aerosol. The mass scattering coefficient for dry sulfate aerosols is typically about 5 m^2 g^{-1} (Andreae 1995, and references therein). The mass *absorption* coefficient of smoke varies between 0.4 m^2 g^{-1} (Radke et al. 1991) and 0.7 m^2 g^{-1} (Radke et al. 1988). Considering these previous estimates for scattering and absorption coefficients, we chose a mass extinction coefficient of 4.5 m^2 g^{-1} as appropriate for our study.

To calculate the total particle mass in the size range 0.1 to 3.0 μm, we obtain first the volume distribution by multiplying each size-class number concentration with the average volume of a particle in the corresponding size class. Then the mass distribution is obtained by multiplying the volume distribution with the average density of a particle. Finally, summing the mass distribution over the total size range gives us the total aerosol mass (m). The average density of a particle emitted by biomass burning with a mean geometric mass diameter of 0.3 μm has been estimated to be 1.0 g cm^{-3} (Stith et al. 1981; Radke et al. 1991). This would be an underestimate for the total aerosol in our region, because its dust and sulfate components are denser. Patterson et al. (1980) and Isaac et al. (1986) assumed, for a similar study, a uniform density of 2.0 g cm^{-3}, which we think is an overestimate because the density of sulfates is about 1.7 g/cm^3 (Hegg et al. 1993; Kiehl and Briegleb 1993). Therefore, a value between 1.0 and 1.7 would be appropriate for our study. Given the preponderance of pyrogenic particles in our aerosol, we chose an average particle density of 1.3 g cm^{-3}.

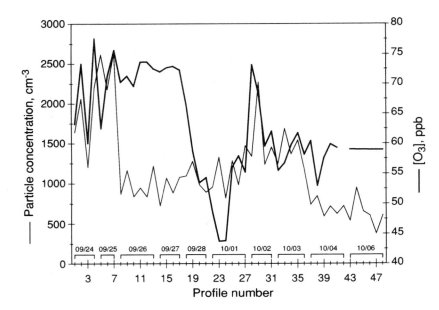

Figure 42.2 Column-averaged ozone (O_3) and aerosol-particle concentrations in the planetary boundary layer for each vertical profile

Ozone Absorption

The optical depth due to absorption by ozone is not discussed in detail because this term turns out to be negligible compared to the optical depths resulting from Mie and Rayleigh scattering. For comparison, we estimate this quantity at 0.55 μm where the ozone absorption cross-section (Chappuis bands) at 273 K and at 1013 mb is about 3.4×10^{-21} cm^2 molecule^{-1} (Griggs 1968). The absorption cross-section for the selected wavelength at any altitude is corrected for temperature and pressure. Using measured ozone concentrations, we estimate the absorption extinction coefficients, which are then vertically integrated to obtain the ozone-absorption optical depths for the various profiles.

Results and Discussion

Average Trace-Gas and Particle Concentrations

We column-averaged the trace-gas and particle concentrations over the total height for each profile. The resulting time sequence of the average concentrations of O_3, CO_2, CO, and particle number are shown in figures 42.2 and 42.3. For CO and CO_2, the concentrations decrease steadily, over the time of the experiment, from 200 to 70 ppb and from 360 to 354 ppm, respectively. This decrease corresponds to the end of the fire season and the two values at the end of the experiment are representative of relatively clean air outside the burning season over southern Africa. The time evolution of O_3 is very similar to those of CO and CO_2,

except for the steep drop between 28 September and 1 October. This drop is related to the intrusion of marine air from the Indian Ocean and is also seen in the ozone records at Victoria Falls, Zimbabwe, and Kruger Park (F. X. Meixner, personal communication 1995). At the beginning of the experimental period, high ozone concentrations (around 70 ppb) were measured at these sites as well as from the DC-3, whereas lower values (around 45 ppb) were present at the end of the experiment. The temporal evolution of the particle concentration can be divided into three periods: (1) 24–25 September, (2) 26 September, 3 October, and (3) 4–6 October. The first period is characterized by relatively high average particle concentrations (2000–2500 cm^{-3}). During the second period, moderate aerosol loadings are found (1000 cm^{-3}) with an exceptional value (2000 cm^{-3}) for SAF14-D1, in the vicinity of Lusaka, where a strong inversion had trapped polluted air masses in the lower troposphere. Finally, the third period shows low concentrations (500 cm^{-3}), representing conditions of low input of pyrogenic aerosols. The drop in trace-gas and aerosol concentrations coincides with the transition between the burning season and the nonburning season, which occurred at the end of September (Scholes et al. 1996).

A detailed study of the vertical profiles shows that the lower troposphere is not at all well mixed. During the soundings we encountered generally two to three layers with different trace-gas and particle loadings. We present an example using the profile for SAF07-A1, which was flown on 24 September around 1600 local time (LT) over the northern part of the Kruger

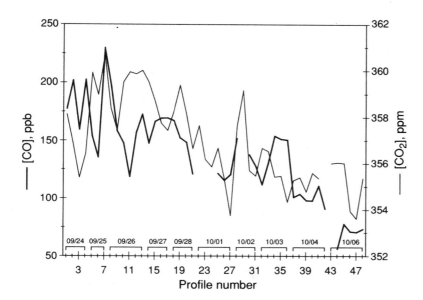

Figure 42.3 Column-averaged carbon monoxide (CO) and carbon dioxide (CO$_2$) concentrations in the planetary boundary layer for each vertical profile

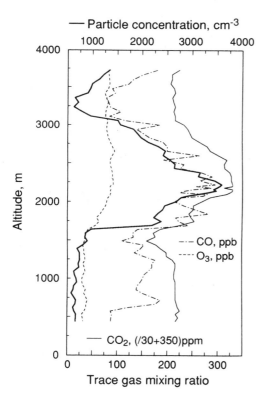

Figure 42.4 Aerosol-particle, carbon monoxide (CO), carbon dioxide (CO$_2$), and ozone (O$_3$) concentrations versus altitude for the vertical profile SAF07-Al

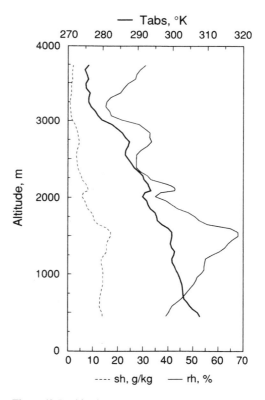

Figure 42.5 Absolute temperature (T_{abs}), relative humidity (rh), and specific humidity (sh) versus altitude for the vertical profile SAF07-Al

National Park. Figure 42.4 shows the vertical distribution of O_3, CO_2, CO, and aerosol particle concentrations and Figure 42.5 shows the temperature and humidity profiles. The ozone and specific humidity profiles present a sharp change at 1600 m asl that delimits two air masses. The lower air mass (500–1600 m) has low ozone concentrations (35 ppb) and high specific humidity (15 g/kg), whereas the upper layer has high ozone concentrations (80–90 ppb) and low specific humidity (<5 g/kg). The boundary between the two layers corresponds approximately to the altitude of the African high plateau. The highly polluted upper air mass would therefore come from the plateau region, where air masses had been recirculating for several days, becoming enriched in pyrogenic and anthropogenic emissions. The lower air mass would represent a marine inflow typical of the lowveld region.

Particle Size Distribution

For all vertical profiles and altitudes, we have fitted a log-normal distribution to both the number and volume distributions in the accumulation mode, using the data from the first seven bins. These first seven bins, which correspond to the size range 0.1 to 0.4 μm, account for about 99.5% of all particles. The total abundance in the last eight bins is about 0.5% (range 0.4–3.0 μm), and the standard deviation of the abundance is often greater than the average abundance in these bins. These two remarks apply to all profiles.

Figures 42.6a, b show the actual and log-normal number distributions at the bottom of SAF07-A1 with a log and then a normal ordinate axis. We consider the fit valid up to 0.7 μm for the number distribution, and up to 0.5 μm for the volume distribution, at which diameters the two curves intersect. Small irregularities, like the peak on the left side of the actual number distribution in figure 42.6b, cannot be reproduced by the log-normal approximation.

Figure 42.7 shows the vertical profiles of the geometric-mean number (d_g^N) and volume (d_g^V) diameters in the accumulation mode for SAF07-A1. At the bottom of the profile, d_g^N and d_g^V are estimated at 0.17 μm and 0.245 μm, whereas at the top they are estimated at 0.21 μm and 0.265 μm, respectively. The mean geometric standard variation (σ_g) varies from 1.4 at the bottom to 1.3 at the top. The number and volume distributions are therefore slightly narrower and shifted toward larger particles at the top of the profile compared to the bottom. This difference presumably is related to the presence of two air masses, one above the other, as discussed above.

The geometric-mean number and volume diameters are plotted for the bottom and top of all vertical profiles in figure 42.8. Two conclusions can be drawn from these data: (1) As the fire season ends, particles tend to be smaller at all altitudes. At the profile top, d_g^N and d_g^V decrease from 0.21 μm to 0.17 μm and from 0.27 μm to 0.25 μm respectively, whereas at the profile bottom they decrease from 0.18 μm to 0.15 μm and from 0.25 μm to 0.24 μm respectively. This relation reflects the transition from a situation in which a recirculating air mass is present with high levels of aged pyrogenic aerosols to conditions of more rapid air-mass exchange with reduced fire emissions. (2) For most of the profiles, the particles are larger and the size distributions narrower at the top of the profile than at the bottom. This variability reflects the layered structure of the lower troposphere and may be related to increasing age of the aerosol population with height.

Hygroscopic growth can be excluded as an explanation for the vertical change in aerosol size for SAF07-A1 because the relative humidity decreases with height (figure 42.5). Also, for almost all altitudes and for all the profiles studied, the relative humidity is well below the threshold of 60%, where swelling begins. The average relative humidity for all profiles is about 40%.

During the last three profiles, the aircraft succeeded in reaching the free troposphere. This altitude explains why d_g^N and d_g^V drop sharply from 0.17 μm to 0.12 μm and from 0.25 μm to 0.23 μm, respectively, at the end of the time series (figure 42.8).

Averages over the 48 vertical profiles suggest that mean geometric number diameters increase from 0.17 μm to 0.20 μm and mean geometric volume diameters from 0.24 μm to 0.26 μm with increasing altitudes (figure 42.9). Our estimates fall well within the range given in the literature. Kim et al. (1988) found average volume geometric-mean diameters of 0.22, 0.22, 0.27, and 0.21 μm with geometric deviations of 2.11, 1.77, 1.72, and 1.98 for winter, spring, summer, and fall in the lower troposphere over the central United States. Holben et al. (1991) estimated the mean mass radius in the mixing layer to be between 0.13 μm and 0.22 μm, and in the free troposphere to be between 0.20 μm and 0.27 μm. Isaac et al. (1986) found volume geometric mean diameters around 0.3 μm over Algonquin Park, Canada. Kim et al. (1988) also measured lower nighttime geometric volume diameters for low altitudes than for high altitudes: d_g^V increased from 0.21 μm at 1450 m to 0.24 μm at 2450 m in winter, from 0.27 μm to 0.29 μm in summer, and from 0.20 μm to 0.21 μm in

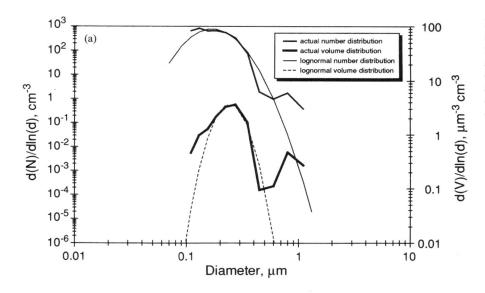

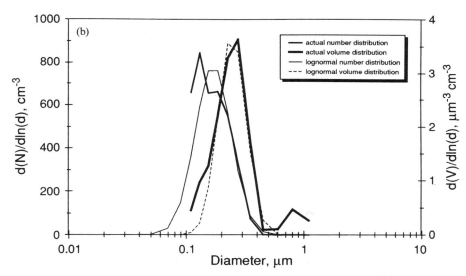

Figure 42.6 (*a*) Log-normal and actual number and volume distributions at the bottom of profile SAF07-A1, using a logarithmic ordinate axis. (*b*) Log-normal and actual number and volume distributions at the bottom of profile SAF07-A1, using a linear ordinate axis.

fall. This increase can be explained by the same phenomenon: the air in the residual layer (upper layer) is older than that in the stable nocturnal layer (lower layer) and therefore the particles are larger at higher altitudes.

From the size distribution measured by the PCASP probe, we can calculate the total particle mass concentration ($\mu g/m^{-3}$) in the size range 0.1 to 3.0 μm and the average particle mass in femtogram (1 fg = 10^{-15} g). We use a particle density of 1.3 g cm^{-3} (cf. discussion in the method section). Figure 42.10 displays the total particle mass concentration and the average particle mass for our example profile, SAF07-

A1, as a function of the altitude. The total particle mass distribution follows the particle number vertical distribution (Figure 42.4) and peaks between 1600 m and 3100 m above sea level (asl) at 32 $\mu g\ m^{-3}$, whereas the value at the bottom of the profile is around 5 $\mu g\ m^{-3}$. On the same day (24 September), Maenhaut et al. (1996) collected aerosol samples at ground level at two sites within the Kruger National Park, not far from where SAF07-A1 was flown. They found at one site, Skukuza, a total mass of about 9.2 $\mu g\ m^{-3}$ for the fine aerosol (1.3 $\mu g\ m^{-3}$ for soil aerosols, 4.6 $\mu g\ m^{-3}$ for biomass burning aerosols, 0.3 $\mu g\ m^{-3}$ for sea salt, and 3.0 $\mu g\ m^{-3}$ for sulfates), and at another site, Pretoriuskop, a

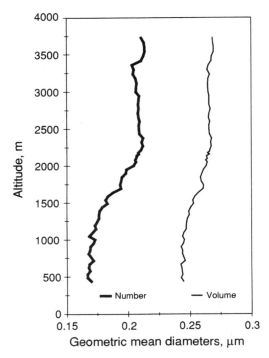

Figure 42.7 Geometric mean volume and number diameters of the accumulation mode versus altitude for profile SAF07-Al

total of 12.3 μg m^{-3}. The difference between our value measured at the bottom of the vertical profile SAF07-A1 and their value obtained at ground level can be accounted for by uncertainties in particle density, size cutoffs, instrument calibrations, and bounce-off of large particles. A factor-of-two difference may not be unexpected for this type of comparison.

For SAF07-A1, the average particle mass increases with height from 6.5 to 13 fg, reflecting the trend in particle size, and is independent of the total particle mass concentration (figure 42.10). The total column loading, determined by integrating the total vertical mass distribution over the height of the profile (3287 m), is 42 mg m^{-2}. The column burdens for all profiles, normalized to a standard layer thickness of 3000 m, are shown in figure 42.11. The three time periods previously defined for the average particle loadings in the lower troposphere (figure 42.2) also apply to the column burdens. They decrease from about 50 mg m^{-2} during the first period (24–25 September), to about 25 mg m^{-2} during the second period (26 September, 3 October), and to about 15 mg m^{-2} during the third period (4–6 October).

Optical Depth due to the Fine Aerosol

The optical depths due to the fine aerosol are given for a constant 3000-m layer thickness. As discussed above, this thickness corresponds to the height of the lowest of the persistent absolutely stable layers observed over the subcontinent during SAFARI-92. The measured size distributions over the range 0.1 to 3.0 μm are used.

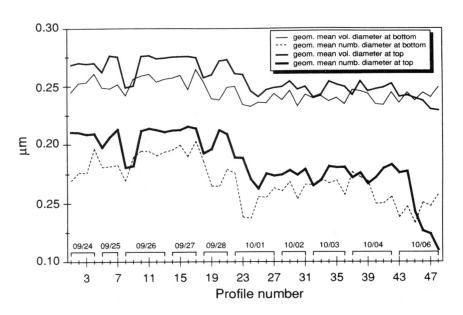

Figure 42.8 Geometric-mean volume and number diameters of the accumulation mode at the top and bottom of all vertical profiles

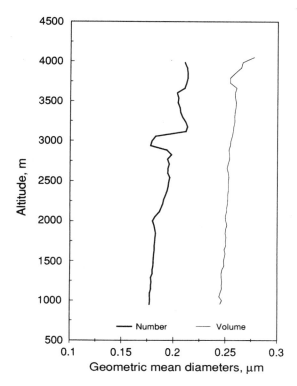

Figure 42.9 Average vertical profile (of 48 soundings) of the geometric-mean volume and number diameters

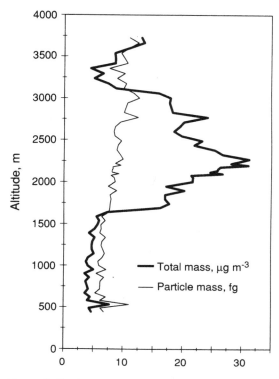

Figure 42.10 Aerosol-particle mass and aerosol total mass concentration versus altitude for profile SAF07-Al

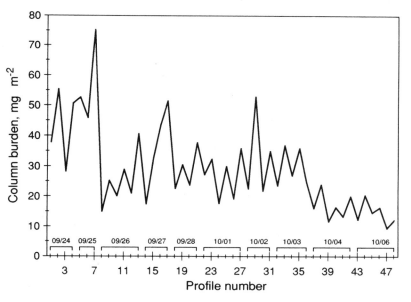

Figure 42.11 Time series of the aerosol column burden

We use a black-carbon content of 15%, representative of the values found over southern Africa (Andreae et al. 1996). The optical depths (δ_{fine}) resulting from the Mie calculations at 0.55 μm with a refractive index of 1.43-0.0035i for organic particles (85% by weight) and 2.0-0.66i for black carbon (15%) are listed for all profiles in table 42.2 and plotted in figure 42.12. We obtain the same results when using a complex index of refraction of 1.5-0.1i for the total particle population; that is, by assuming internal rather than external mixture. The three time periods previously defined for the average particle loadings in the lower troposphere are again evident in the optical depths. δ_{fine} is about 0.17 during the first period (24–25 September), about 0.08 during the second period (26 September, 3 October), and about 0.04 during the third period (4–6 October). The optical depths due to the fine aerosol are strongly correlated with the average particle loadings. The regression yields:

$$\delta_{\text{fine}} = 8.2E-5 \pm 4.8E-6 \times [\text{avg. part}] - 0.011 \pm 0.018$$
$$\text{with} \quad r^2 = 0.86, N = 48 \qquad (42.9)$$

The particle optical depths ($\delta_{\text{approximation}}$) calculated from the particle volumes, specific gravities, and extinction coefficients, using the values $\alpha = 4.5$ m^2 g^{-1} and $\rho = 1.3$ g cm^{-3}, are also plotted in figure 42.12. Again, the same three periods of time show up distinctly. $\delta_{\text{approximation}}$ is about 0.17 during the first period (24–25 September), about 0.09 during the second period (26 September, 3 October), and about 0.05 during the third period (4–6 October).

As a comparison, the vertical extinction profiles for SAF07-A1 resulting from the Mie calculations with $m = 1.5$-0.1i and from the approximation with $\alpha = 4.5$ m^2 g^{-1} and $\rho = 1.3$ g cm^{-3} are displayed in figure 42.13. The two profiles are very close, with an average difference of about 7%.

Since the fine-particle optical depths were estimated using two independent methods (direct Mie calculations and approximation with the mass extinction coefficient), but at the same wavelength (0.55 μm), their comparison is appropriate. The fine-particle optical depths resulting from the approximation are equal to those obtained with the Mie code during the first period (24–25 September), whereas they are slightly higher during the third period (4–6 October) (figure 42.12). These discrepancies could have several causes: (1) the global index of refraction that we used is too low because there is actually more than 15% of black carbon, (2) the mass extinction coefficient is too high and is not constant throughout the period of the

experiment, (3) the particle density is too high, and (4) the total mass (m) of the fine particulates is slightly overestimated because some particles belonging to the coarse mode appear at the far right of the size range 0.1 to 3.0 μm. Their influence upon the mass, and therefore upon the approximated extinction coefficient, varies with the cube of the radius, whereas their influence upon the Mie calculations is proportional only to the particle number. All these reasons can potentially apply at the same time or just one may account for the difference. Nevertheless, because the difference between the optical depths calculated by the two methods is around 3% during the 15 first profiles, we consider that the values chosen ($\alpha = 4.5$ m^2 g^{-1}, $\rho = 1.3$ g cm^{-3}, and $m = 1.5$-0.1i) are representative of the regional air over southern Africa during the fire season.

A shift in aerosol optical properties occurs between 26 and 27 September. This change closely coincides with that already observed for the geometric-mean diameters. Until 26 September (the first 15 vertical profiles), the particle optical depths resulting from the Mie calculations and from the approximation with the mass extinction coefficient are highly correlated (figure 42.14). The regression gives:

$$\delta_{\text{approximation}} = 0.96(\pm 0.02) \times \delta_{\text{fine}} + 0.006(\pm 0.005)$$
$$\text{with} \quad r^2 = 0.99, \quad N = 15 \qquad (42.10)$$

After 27 September (the last 33 vertical profiles), the correlation remains high, but the parameters are different from those of the first period (figure 42.14). The regression gives:

$$\delta_{\text{approximation}} = 1.21(\pm 0.06) \times \delta_{\text{fine}} + 0.03(\pm 0.01)$$
$$\text{with} \quad r^2 = 0.93, \quad N = 33 \qquad (42.11)$$

This shift reflects the change in the size distribution of the aerosols, as discussed in detail above. The difference in the two methods for estimating the fine-aerosol optical depth is about 25% for the last 33 profiles, whereas it was only 3% for the first 15.

Wavelength Dependence of the Fine-Mode Optical Depth

The preceding discussion dealt with aerosol optical depths at a wavelength of 0.55 μm. We now discuss the wavelength dependence of the optical depth due to the accumulation-mode aerosol. The fine-particle index of refraction is again 1.5-0.1i (as discussed above). We consider the wavelength range of 0.3 to 2.8 μm for the incoming radiation, corresponding to the main atmospheric window, and use a wavelength increment of 0.1 μm for our Mie calculations. Figure 42.15 shows

Table 42.2 Optical depths due to the accumulation mode, Rayleigh scattering, and ozone absorption for a 3000-m layer at 0.05 μm

Profile name	Profile number	Opt. depth fine mode	Rayleigh scattering	Ozone adsorbption	Total opt. depth
SAF07-A1	1	0.13	4.8E-02	8.7E-04	0.18
SAF07-D1	2	0.19	4.9E-02	9.9E-04	0.24
SAF07-A2	3	0.09	4.8E-02	8.2E-04	0.14
SAF07-D2	4	0.17	5.1E-02	9.5E-04	0.22
SAF08A-D1	5	0.19	4.6E-02	9.6E-04	0.23
SAF08A-A2	6	0.16	5.0E-02	9.3E-04	0.21
SAF08A-D2	7	0.25	5.0E-02	9.8E-04	0.31
SAF09A-A1	8	0.05	4.8E-02	9.9E-04	0.10
SAF09A-D1	9	0.08	4.8E-02	9.9E-04	0.13
SAF09B-A1	10	0.07	5.0E-02	9.3E-04	0.12
SAF09B-D1	11	0.09	5.0E-02	9.7E-04	0.15
SAF09B-A2	12	0.07	5.1E-02	9.3E-04	0.12
SAF09B-D2	13	0.12	5.1E-02	9.2E-04	0.17
SAF10A-A1	14	0.06	4.9E-02	9.8E-04	0.11
SAF10A-D1	15	0.11	4.9E-02	9.7E-04	0.16
SAF10B-A1	16	0.09	5.1E-02	9.2E-04	0.14
SAF10B-D1	17	0.11	5.1E-02	9.0E-04	0.16
SAF11-A1	18	0.07	4.9E-02	8.8E-04	0.12
SAF11-D1	19	0.10	4.9E-02	8.1E-04	0.14
SAF11-A2	20	0.07	5.0E-02	6.9E-04	0.12
SAF11-D2	21	0.09	5.0E-02	7.0E-04	0.14
SAF12-A1	22	0.07	5.2E-02	5.9E-04	0.12
SAF12-D1	23	0.09	4.9E-02	5.9E-04	0.14
SAF12-A2	24	0.05	4.9E-02	6.0E-04	0.09
SAF12-D2	25	0.08	4.9E-02	7.7E-04	0.13
SAF12-A3	26	0.05	5.0E-02	7.5E-04	0.10
SAF12-D3	27	0.10	5.2E-02	6.5E-04	0.15
SAF14-A1	28	0.07	5.0E-02	9.6E-04	0.12
SAF14-D1	29	0.17	4.7E-02	9.7E-04	0.21
SAF14-A2	30	0.06	5.0E-02	7.8E-04	0.11
SAF14-D2	31	0.09	5.2E-02	7.4E-04	0.14
SAF15-A1	32	0.07	4.7E-02	8.1E-04	0.12
SAF15-D1	33	0.11	4.7E-02	8.3E-04	0.16
SAF15-A2	34	0.08	5.0E-02	8.0E-04	0.13
SAF15-D2	35	0.10	5.1E-02	7.6E-04	0.15
SAF15-A3	36	0.06	5.0E-02	7.7E-04	0.11
SAF16-A1	37	0.04	5.6E-02	6.1E-04	0.10
SAF16-D1	38	0.06	5.0E-02	6.9E-04	0.11
SAF16-A2	39	0.03	5.1E-02	7.2E-04	0.08
SAF16-D2	40	0.04	5.0E-02	7.7E-04	0.10
SAF16-A3	41	0.03	5.0E-02	7.8E-04	0.09
SAF16-D3	42	0.05	5.3E-02	7.1E-04	0.10
SAF17-A1	43	0.03	4.7E-02	5.2E-04	0.08
SAF17-D1	44	0.06	4.8E-02	6.1E-04	0.11
SAF17-A2	45	0.04	4.8E-02	6.1E-04	0.09
SAF17-D2	46	0.04	4.7E-02	5.9E-04	0.09
SAF17-A3	47	0.02	4.8E-02	6.0E-04	0.07
SAF17-D3	48	0.03	5.0E-02	6.8E-04	0.08
avg		0.09	5.0E-02	8.0E-04	0.14
std		0.05	1.8E-03	1.4E-04	0.05

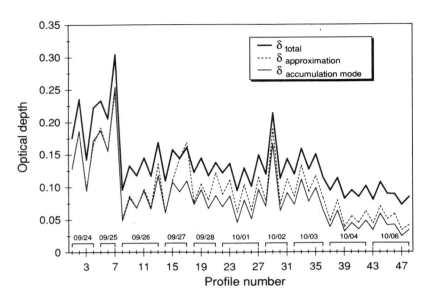

Figure 42.12 Optical depths due to the accumulation mode, estimated with the mass extinction coefficient method ($\delta_{approximation}$) and with a Mie code (δ_{fine}) for all vertical profiles. The total optical depth (δ_{total}) due to the accumulation-mode aerosol and to Rayleigh scattering is also shown.

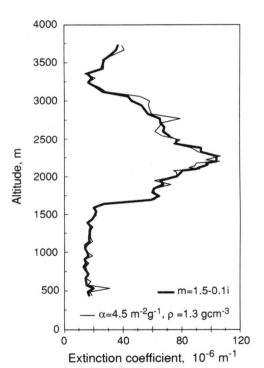

Figure 42.13 Extinction coefficient versus altitude for the vertical profile SAF07-Al estimated with the mass extinction coefficient and using the Mie calculations

the fine-aerosol optical depth as a function of wavelength for the vertical profile SAF07-A1. The data can be fitted by a decreasing exponential, defined by:

$$\delta_{Mie}(\lambda) = A \times e^{-B \times \lambda} + C \tag{42.12}$$

For this example, the parameters are estimated at: $A = 0.73$, $B = 3.3$, and $C = 0.016$ with $r^2 = 0.998$. The quality of the fit is comparable for the other vertical profiles. The parameters A and C are related to the average aerosol concentrations, whereas the parameter B is related to the mean geometric diameters of the vertical profiles. Parameter A decreases steadily toward the end of the fire season and correlates very well with the average particle concentration of the vertical profiles (figure 42.16a). A linear regression yields these results:

$$A = 4.6 \times 10^{-4}(\pm 1 \times 10^{-5}) \times [\text{average particle number}]$$
$$- 0.056(\pm 0.046) \quad \text{with} \quad r^2 = 0.97$$
$$\tag{42.13}$$

Parameter B increases slightly toward the end of the fire season and correlates very well with the geometric-mean volume diameters (figure 42.16b). A linear regression yields these results:

$$B = -20.3(\pm 1.9) \times [VGMD] + 8.6(\pm 0.1)$$
$$\text{with} \quad r^2 = 0.72 \tag{42.14}$$

For parameter C, we obtain (figure 42.16c) this linear regression:

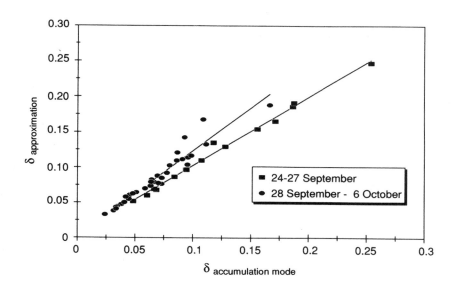

Figure 42.14 Optical depth obtained with the mass extinction coefficient approximation ($\delta_{\text{approximation}}$) versus the optical depth obtained using the Mie code ($\delta_{\text{accumulation}}$) for all vertical profiles

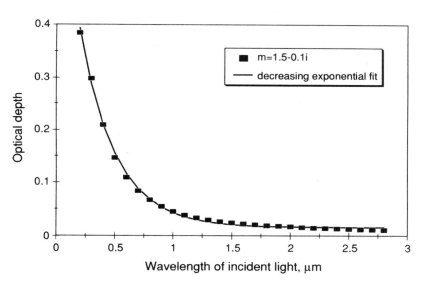

Figure 42.15 Optical depth due to the accumulation mode obtained with the Mie code versus the wavelength of the incident light for the vertical profile SAF07-Al. The data are fitted with a decreasing exponential.

$$C = 9.3 \times 10^{-6}(\pm 1 \times 10^{-6}) \times [\text{average particle number}]$$
$$+ 0.003(\pm 0.003) \quad \text{with} \quad r^2 = 0.68 \qquad (42.15)$$

Therefore, knowing only the log-normal size distributions, the aerosol optical depth can be derived at any wavelength by equation 42.12 using the relations for parameters A, B, and C.

Optical Depth due to the Coarse-Mode Aerosol

We assume the volume-size distribution of the coarse-mode aerosol to be log-normal and centered on 6 μm (d_g^V) with a geometric standard deviation of 1.7 (σ_g). The value of 6 μm is deduced on the basis of the mass median aerodynamic diameter (MMAD) data for the mineral dust that is the main component of the coarse fraction of the aerosol (W. Maenhaut, personal communication 1995). The value of 1.7 for the geometric standard deviation is based on the study by Horvath et al. (1990). We use the coarse-mode and fine-mode total particle masses measured by Maenhaut et al. (1996) at ground level in Skukuza to estimate the average ratio of the coarse-mode mass to the fine-mode mass (1.3) and assume furthermore that this ratio applies to all our vertical profiles.

With a maximum total fine-mode particulate mass of 32 μg m^{-3} for the vertical profile SAF07-Al at about 2200 m asl (see above), the corresponding total coarse-mode mass is estimated at 42 μg m^{-3}. The total

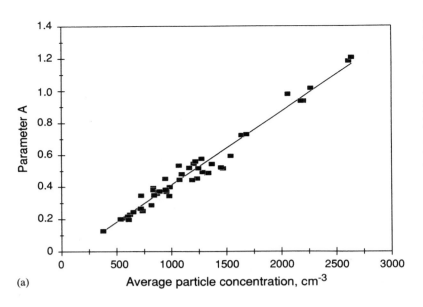

(a)

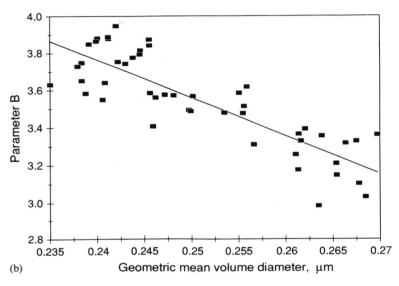

(b)

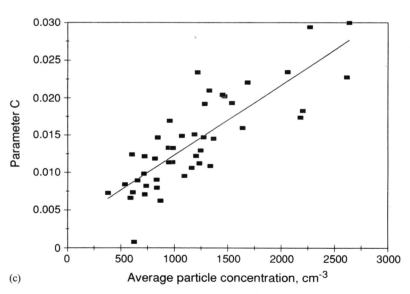

(c)

Figure 42.16 (*a*) Parameter *A* of the exponential fit [$\delta = A \cdot \exp(-B\lambda) + C$] versus the average particle concentration for all vertical profiles. (*b*) Parameter *B* of the exponential fit [$\delta = A\exp(-B\lambda) + C$] versus the geometric-mean volume diameter of the accumulation mode size distribution for all vertical profiles. (*c*) Parameter *C* of the exponential fit [$\delta = A\exp(-B\lambda) + C$] versus the average particle concentration for all vertical profiles.

particle-number concentration of the coarse-mode log-normal distribution is then calculated at about 1 cm^{-3}. Using the scattering theory applicable to large particles (the extinction is twice the geometric cross-sectional area), we obtain an extinction coefficient due to scattering by the coarse aerosol of about 8×10^{-6} m^{-1}. The extinction due to the fine-mode aerosol is about 105×10^{-6} m^{-1} at the altitude of the aerosol maximum (figure 42.13). Therefore, the coarse mode accounts for only about 8% of the optical thickness due to the fine aerosol, and, since the coarse aerosol fraction contributes relatively little to the total extinction, even relatively large errors in the assumptions made above are not critical.

Ozone-Absorption Optical Depth
The ozone-absorption optical depth at 0.55 μm varies from profile to profile, depending on the amount of ozone present in the lower troposphere. It decreases from about 0.001 at the beginning of the experiment to about 0.0005 at the end (table 42.2), with an average of 0.0008 ± 0.0001. This optical depth is nevertheless negligible compared with those due to the fine aerosol and to the Rayleigh scattering (table 42.2). Tropospheric ozone concentrations between the top of our profiles (3–4 km asl) and the tropopause were often significantly elevated (around 100 ppb) (Thompson et al. 1996). However, due to decreasing atmospheric pressure with increasing altitude, shortwave absorption by ozone in the middle and upper troposphere does not contribute significantly to the total optical depth.

Total Optical Depth
During SAFARI-92, at 0.55 μm and for the 3000-m column of air representing the planetary boundary layer, the average fine-aerosol optical depth (obtained by the Mie theory) is 0.09 ± 0.05 for the total period of the experiment and that due to the Rayleigh scattering is 0.05 ± 0.002 (table 42.2). Therefore, under the same conditions, the average total optical depth of the lower troposphere, δ_{total}, is estimated at 0.14 ± 0.05, with negligible contributions from coarse-particle scattering and absorption by ozone (table 42.2). Taking into account the calculated accumulation-mode optical depth and the Rayleigh scattering for each period, we obtain $\delta_{total} = 0.22$ for the first period (24–25 September), $\delta_{total} = 0.13$ for the second period (26 September, 3 October), and $\delta_{total} = 0.09$ for the third period (4–6 October).

Conclusions

The airborne component of SAFARI-92 took place at the end of the burning season in southern Africa and therefore provided information on both the burning and the nonburning seasons. With the fire season ending, the levels of pollution, which were quite high at the beginning of the experiment ([CO$_2$] = 360 ppm, [CO] = 200 ppb, [O$_3$] = 70 ppb, [particle] = 2000 cm^{-3}), decreased substantially toward the end of SAFARI-92 ([CO$_2$] = 354 ppm, [CO] = 70 ppb, [O$_3$] = 45 ppb, [particle] = 500 cm^{-3}).

The total optical depths (scattering and absorption of the accumulation mode and Rayleigh scattering) decreased from about 0.22 (burning season) to about 0.09 (nonburning season). The contribution of the trace gases to the total optical depth is negligible, although high concentrations were measured.

A mass extinction coefficient of 4.5 m^2 g^{-1} yields the same optical depths as the Mie calculations, with an index of refraction estimated at 1.5-0.1i. The value of 4.5 m^2 g^{-1} is therefore representative of regional air over southern Africa during the fire season. For all profiles, the measured size distributions in the accumulation mode could be fitted with good agreement to log-normal distributions. The volume and number geometric-mean diameters follow a different temporal trend than that observed for the absolute aerosol loadings. Volume and number geometric-mean diameters both decrease sharply around 28 September, suggesting a change in the air masses over the region around that time. This change is related to the synoptic conditions. Before 27 September, recirculation of air masses was the dominant phenomenon, whereas after this date, the air masses were of direct maritime origin (Garstang et al. 1996; Maenhaut et al. 1996). We can therefore conclude that during the experiment two phenomena affected the aerosol population over southern Africa. The first is a drop in the absolute loadings resulting from a seasonal decrease in the amount of burning. The second is related to a change in aerosol characteristics, in particular the geometric-mean diameter, which was caused by the transition from recirculation to inflow of oceanic air masses as the dominant synoptic condition.

Volume and number geometric-mean diameters were found to increase with increasing height. This increase is a result of the highly layered structure of the lower atmosphere over southern Africa during this time of year. Upper layers, which are likely to be older than lower layers, show narrower log-normal size and volume distributions and larger mean particle sizes than lower layers.

Acknowledgments

We thank the pilots and crews of the SAFARI-92 DC-3 aircraft for their assistance during the research flights. We acknowledge the permission granted by various government agencies in Angola, Botswana, South Africa, Namibia, Swaziland, Zambia, and Zimbabwe to conduct research in these countries or to use the airspace over these countries. We appreciate the help of T. W. Andreae (logistical support) and U. Parchatka (technical support) during the airborne sampling missions on the DC-3 aircraft. We also thank S. Bormann for helping us to obtain a Mie code, W. Maenhaut for providing total particulate matter measurements at ground level, and P. D. Tyson, W. B. Grant, and Y. Kaufman for their helpful comments on the manuscript. This research was supported by the Max Planck Society.

References

Andreae, M. O. 1993. The influence of tropical biomass burning on climate and the atmospheric environment. In *Biogeochemistry of Global Change: Radiatively Active Trace Gases*, edited by R. S. Oremland. Chapman & Hall, New York, pp. 113–150.

Andreae, M. O. 1995. Climatic effects of changing atmospheric aerosol levels. In *World Survey of Climatology*, vol. 16: *Future Climates of the World*, edited by A. Henderson-Sellers. Elsevier, Amsterdam, pp. 341–392.

Andreae, M. O., et al. 1994. Biomass burning in the global environment: First results from the IGAC/BIBEX field campaign STARE/TRACE-A/SAFARI-92. In *Global Atmospheric-Biospheric Chemistry*, edited by R. G. Prinn. Plenum, New York, pp. 83–101.

Andreae, M. O., P. le Canut, T. W. Andreae, W. Elbert, G. W. Harris, F. G. Wienhold, T. Zenker, H. Annegarn, F. Beern, H. Cachier, W. Maenhaut, I. Salma, and R. Swap. 1994. Aerosol characteristics over southern Africa during the 1992 fire season. *J. Geophys. Res.*, in preparation.

Bohren, C. F., and D. R. Huffman. 1983. *Absorption and Scattering of Light by Small Particles*. New York, Wiley, 530 pp.

Browell E. V., M. A. Fenn, C. F. Butler, W. B. Grant, M. E. Clayton, J. Fishman, A. S. Bachmeier, B. E. Anderson, G. L. Gregory, H. E. Fuelberg, J. D. Bradshaw, S. T. Sandholm, D. R. Blake, B. G. Heikes, G. W. Sachse, H. B. Singh, R. W. Albot. 1996. Ozone and aerosol distributions and air mass characteristics over the South Atlantic basin during the burning season. *J. Geophys. Res.*, in press.

Charlson, R. J., J. Langner, H. Rodhe, C. B. Leovy, and S. G. Warren. 1991. Perturbation of the northern hemisphere radiative balance by backscattering from anthropogenic sulfate aerosols. *Tellus*, 43AB, 152–163.

Charlson, R. J., S. E. Schwartz, J. M. Hales, R. D. Cess, J. A. Coakley, J. E. Hansen, and D. J. Hofmann. 1992. Climate forcing by anthropogenic aerosols. *Science*, 255, 423–430.

Ferrare, R. A., R. S. Fraser, and Y. J. Kaufman. 1990. Satellite measurements of large-scale air pollution: Measurements of forest fire smoke. *J. Geophys. Res.*, 95, 9911–9925.

Fraser, R. S. 1993. Optical thickness of atmospheric dust over Tadzhikistan. *Atmos. Environ.* 27A, 2533–2538.

Garstang, M., P. D. Tyson, R. Swap, M. Edwards, P. Kållberg, and J. A. Lindesay. 1996. Horizontal and vertical transport of air over southern Africa. *J. Geophys. Res.*, in press.

Griggs, M. 1968. Absorption coefficients of ozone in the ultraviolet and visible regions. *J. Chem. Phys.*, 49, 857.

Hansen, J., A. Lacis, R. Ruedy, M. Sato, and H. Wilson. 1993. How sensitive is the World's climate? *National Geographic Research and Exploration*, 9, 142–158.

Hegg, D. A., R. J. Ferek, and P. V. Hobbs. 1993. Light scattering and cloud condensation nucleus activity of sulfate aerosol measured over the Northeast Atlantic Ocean. *J. Geophys. Res.*, 98, 14887–14894.

Holben, B. N., Y. J. Kaufman, A. W. Setzer, D. D. Tanré, and D. E. Ward. 1991. Optical properties of aerosol emissions from biomass burning in the tropics, BASE-A. In *Global Biomass Burning: Atmospheric, Climatic, and Biospheric Implications*, edited by J. S. Levine, MIT Press, Cambridge, Mass. pp. 403–411.

Horvath, H., R. L. Gunter, and S. W. Wilkison. 1990. Determination of the coarse mode of the atmospheric aerosol using data from a forward-scattering spectrometer probe. *Aerosol Science and Technology*, 12, 964–980.

Houghton, J. T., L. G. Meira Filho, J. Bruce, H. Lee, B. A. Callander, E. Haites, N. Harris, and K. Maskell. 1995. *Climate Change 1994: Radiative Forcing of Climate Change*. Cambridge, U.K.: Cambridge University Press, 339 pp.

Isaac, G. A., W. R. Leaitch, J. W. Strapp, and K. G. Anlauf. 1986. Summer aerosol profiles over Algonquin Park, Canada. *Atmos. Environ.*, 20, 157–172.

Kiehl, J. T., and B. P. Briegleb. 1993. The relative role of sulfate aerosols and greenhouse gases in climate forcing. *Science*, 260, 311–314.

Kim, Y. J., H. Sievering, and J. F. Boatman. 1988. Airborne measurements of atmospheric aerosol particles in the lower troposphere over the central United States. *J. Geophys. Res.*, 93, 12631–12644.

Kim, Y. J., and J. F. Boatman. 1990. Size calibration corrections for the active scattering aerosol spectrometer probe (ASASP-100X). *Aerosol Science and Technology*, 12(3), 665–672.

Le Canut, P., M. O. Andreae, G. W. Harris, F. G. Wienhold, and T. Zenker. 1996. Airborne studies of aerosol emissions from savanna fires in southern Africa: I. Results from the laser-optical particle counter. *J. Geophys. Res.*, in press.

Maenhaut, W., I. Salma, J. Cafmeyer, H. J. Annegarn, and M. O. Andreae. 1996. Regional atmospheric aerosol composition and sources in the eastern Transvaal, South Africa, and impact of biomass burning. *J. Geophys. Res.*, in press.

Nance, J. D., P. V. Hobbs, L. F. Radke, and D. E. Ward. 1993. Airborne measurements of gases and particles from an Alaskan wildfire. *J. Geophys. Res.*, 98, 14873–14882.

Patterson, E. M., C. S. Kiang, A. C. Delany, A. F. Wartburg, A. C. D. Leslie, and B. J. Huebert. 1980. Global measurements of aerosols in remote continental and marine regions: Concentrations, size distributions, and optical properties. *J. Geophys. Res.*, 85, 7361–7376.

Penner, J. E., R. E. Dickinson, and C. A. O'Neill. 1992. Effects of aerosol from biomass burning on the global radiation budget. *Science*, 256, 1432–1434.

Pilewskie, P., and F. P. J. Valero. 1992. Radiative effects of the smoke clouds from the Kuwait oil fires. *J. Geophys. Res.*, 97, 14 541–14 544.

Pilinis, C., S. N. Pandis, and J. H. Seinfeld. 1995. Sensitivity of direct climate forcing by atmospheric aerosols to aerosol size and composition. *J. Geophys. Res.*, 18 739–18 754.

Radke, L. F., D. A. Hegg, P. V. Hobbs, J. D. Nance, J. H. Lyons, K. K. Laursen, R. E. Weiss, P. J. Riggan, and D. E. Ward. 1991. Particulate and trace gas emissions from large biomass fires in North America. In *Global Biomass Burning: Atmospheric, Climatic and biospheric Implications*, edited by J. S. Levien. MIT Press, Cambridge, Mass., pp. 209–224.

Radke, L. F., D. A. Hegg, J. H. Lyons, C. A. Brock, P. V. Hobbs, R. Weiss, and R. Rasmussen. 1988. Airborne measurements on smoke from biomass burning. In *Aerosols and Climate*, edited by P. V. Hobbs and M. P. McCormick. A. Deepak Publishing, Hampton, Va, pp. 441–422.

Scholes, R. J., D. Ward, and C. O. Justice. 1996. Emissions of trace gases and aerosol particles due to vegetation burning in southern-hemisphere Africa. *J. Geophys. Res.*, in press.

Stith, J. L., L. F. Radke, and P. V. Hobbs. 1981. Particle emissions and the production of ozone and nitrogen oxides from the burning of forest slash. *Atmos. Environ.*, 15, 73–82.

Thompson, A. M., K. E. Pickering, D. P. McNamara, M. R. Schoeberl, R. D. Hudson, J.-H. Kim, E. V. Browell, J. Fishman, V. W. J. H. Kirchhoff, and D. Nganga. 1996. Where did tropospheric ozone over southern Africa and the tropical Atlantic come from in October 1992? Insights from TOMS, GTE/TRACE-A and SAFARI 92. *J Geophys. Res.*, in press.

Van de Hulst, H. C. 1957. *Light Scattering by Small Particles*. New York, Wiley, 470 pp.

Weiss, R. E., and P. V. Hobbs. 1992. Optical extinction properties of smoke from the Kuwait oil fires. *J. Geophys. Res.*, 97, 14 537–14 540.

Wienhold F. G., T. Zenker, G. W. Harris. 1993. A dual channel two tone frequency modulation tunable diode laser sepectrometer for ground-based and airborne trace gas measurements. In *Proceedings of the International Symposium on Optical Sensing for Environmental Monitoring*, Atlanta, Ga., 11–15 October.

Chemical and Isotopic Characterization of Aerosols Collected during Sugar Cane Burning in South Africa

Donna C. Ballentine, Stephen A. Macko, Vaughan C. Turekian,
William P. Gilhooly, and Bice Martincigh

Emissions of organic materials during biomass burning have been suggested to influence the biogeochemical distribution of nutrients in a range of ecosystems (Crutzen and Andreae 1990). Additionally, some organic components survive pyrolytic processes and are of regional and global biogeochemical significance because they may serve as tracers for transport of biomass burning products. Two classes of compounds of interest in determining the transport of these products are polycyclic aromatic hydrocarbons (PAH) and fatty acids.

Many PAHs are known carcinogens (Dipple 1985) and their deposition in the environment via anthropogenic sources is of great concern. These compounds are typically produced during a variety of combustion processes (LaFlamme and Hites 1978), including biomass burning (Masclet 1995). PAHs have been observed in both urban and remote sediments worldwide (Hites et al. 1980) and are useful indicators of historical vegetation fires (Youngblood and Blumer 1975). The presence of PAHs in atmospheric aerosols collected over both rural and urban locations has also been confirmed (Simoneit and Mazurek 1982; Simoneit et al. 1991). Fatty acids are stable to atmospheric degradation and have been implicated as potential biomarkers for long-range atmospheric transport of raw plant material (Gagosian et al. 1982). These compounds have also been observed to survive during combustion of vegetation and are capturable as particulate aerosols (Standley and Simoneit 1987). Because the chemical distributions of both PAHs and fatty acids in aerosols are similar for a variety of combustion sources, simple chemical analysis does not allow an unambiguous determination of the origin of these species. Compound-specific stable isotope analysis (CSIA) of PAHs in sediments has been indicated as a means by which source determinations of these materials may be made (O'Malley et al. 1994). Prior to this investigation, isotopic signatures of individual PAH and fatty acids in aerosols have not been reported.

In this chapter, the chemical and isotopic data for PAHs and fatty acids emitted during controlled low- and high-temperature burns of sugar cane are presented. In order to determine if these species are suitable biomarkers for the transport of biomass burning materials, aerosols collected during sugar cane burning in South Africa have also been characterized by CSIA.

Experiments

Unburned Vegetation

Sugar cane leaves were obtained from a farm 40 kilometers from Durban and 10 kilometers from Verulam on the northern Natal coast, South Africa. The chromatographic procedures for extraction of both the plant and aerosol filters are summarized in figure 43.1. The unburned plant was Soxhlet extracted for 16 hours using 1:1 CH_2Cl_2/CH_3OH. The extracted lipids were saponified using 1N ethanolic KOH under reflux of 3 hours. The saponifiable fatty acids were derivatized to their methyl ester forms with BF_3-CH_3OH, following a modified procedure by Metcalfe and Schmitz (1961). The fatty-acid methyl esters (FAMES) were chemically characterized with a Hewlett Packard 5890 Series II gas chromatograph equipped with an HP-1 (12 m × 0.2 mm i.d.) column and interfaced to a Hewlett Packard 5971A mass selective detector (GC/MS). Bulk and compound-specific carbon isotope data were obtained using a VG Prism Series II isotope ratio mass spectrometer (IRMS) and a Hewlett Packard 5890 Series II gas chromatograph equipped with the same column as above and coupled through a combustion furnace and water trap to the VG Prism (GC/C/IRMS).

The fractionation associated with the addition of a methyl group to the fatty acids and the kinetic isotope effect resulting from esterification (Silfer et al. 1991) was determined using pure fatty-acid standards. The magnitude of the fractionation varied with chain length, and the isotopic data presented for fatty acids have been corrected accordingly.

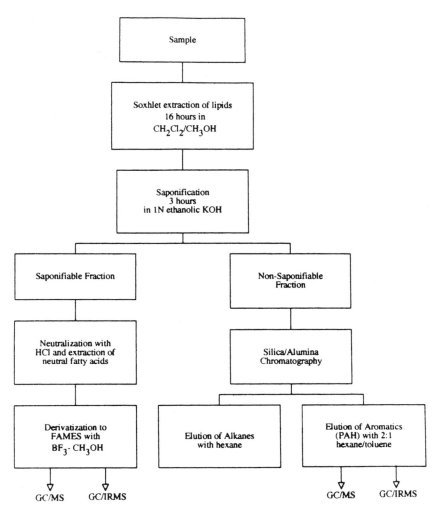

Figure 43.1 Extraction procedure for isolation and analysis of lipids in the unburned plant and aerosol filters

Laboratory Burns

Aerosol samples were collected during controlled laboratory burns of sugar cane leaves. Oxygen was added qualitatively to the flame using a gas regulator in order to increase the burn temperature. In separate combustions, particulate aerosols were collected at "low" ($\approx 150°C$) and "high" ($\approx 500°C$) temperatures on ashed (550°C, 1 hour) 0.4 μ glass fiber filters using a low-volume air sampler (60–100 L/min). During the low-temperature burn, smoldering of the vegetation was observed, whereas during the high-temperature burn a flame was always present. This procedure is described further in chapter 40 this volume by Turekian et al. Lipids captured on the filters were Soxhlet extracted with CH_2Cl_2 and the fatty acids were isolated and derivatized as before. The nonsaponifiable lipid fraction was charged to a column containing 2.5 g silica and 1.6 g alumina. Nonpolar alkanes were re-

moved from the column with 18 mL hexane, and the PAHs were eluted with 45 mL 2:1 hexane/toluene. The FAMES were chemically and isotopically characterized with GC/MS and GC/C/IRMS, respectively. The selected ion monitoring (SIM) mode of the GC/MS was used to identify trace quantities of various PAHs that were not detectable above background in the total ion chromatogram. Isotopic analysis of the PAHs in the aerosols from the laboratory burns was not possible due to an insufficient amount of sample collected.

Field Burn

The sugar cane farm was located in a rural area, so that airborne pollutants arising from vehicular and industrial emissions were not significant. Particulate aerosols were collected on glass fiber filters using a high-volume air sampler (400–850 L/min). The lipids

Table 43.1 Bulk and compound specific carbon isotopic signatures (‰) for fatty acids of unburned sugar cane plant and aerosols collected during laboratory and field sugar cane burns. Data have been corrected for derivatization

	Sugar cane plant	Low-temperature burn	High-temperature burn	Natural burn
Bulk neutral saponifiable fraction	−19.2[a]	−18.9	−20.2	−21.9
C12:0	−24.1 ± 0.4	−21.1 ± 0.6	−21.5 ± 0.4	−24.8 ± 0.7
C14:0	−24.4 ± 0.7	−23.4 ± 0.6	−25.4 ± 0.2	−24.6 ± 0.3
C16:0	−21.4 ± 0.5	−22.0 ± 0.5	−22.8 ± 0.5	−25.5 ± 0.2
C16:1	—	−21.1 ± 0.3	−23.5 ± 0.6	−24.3 ± 0.4
C18:0	−19.9 ± 0.6	−19.8 ± 0.3	−23.9 ± 0.4	−25.6 ± 0.1
C18:1	—	−20.7 ± 0.4	−23.3 ± 0.5	−26.0 ± 0.1
C20:0	−21.7 ± 0.6	−22.8 ± 1.1	−28.0[b]	−27.7 ± 0.3
C22:0	−23.6 ± 0.5	−23.8 ± 0.6	−26.7 ± 0.3	−26.6 ± 0.4

a. Instrumental precision for bulk data was less than 0.05‰.
b. Only one data point.

were Soxhlet extracted from the filters and the fatty acids and PAHs were isolated and characterized as described above. Because each of the individual samples collected did not provide enough material for isotopic analysis of PAHs, several samples having similar bulk carbon isotope signatures were combined to form one sample that was suitable for GC/C/IRMS analysis.

Results

Unburned Plant
The vegetation had a bulk carbon isotopic signature of −12.9‰, typical of C$_4$ plants, and the lipid fraction was −17.9‰, or depleted relative to the vegetation by 5‰. Fatty acids that were identified with GC/MS included even-chain–saturated species from C12 to C22. Table 43.1 shows the isotopic data for the individual fatty acids isolated from the unburned vegetation and from the laboratory and field burns. The sugar cane fatty acids were considerably lighter than the bulk lipid material. In order confirm the validity of the fatty-acid isotope signatures, the entire procedure for extraction and fatty-acid analysis of the unburned plant was repeated, and the results were consistent with the first analysis.

Laboratory Burns
Significant levels of fatty acids were found to survive the combustion process during both the low- and high-temperature burns. Quantification of fatty acids collected under different combustion conditions is presently being examined for other vegetation species

(Turekian). The isotopic signatures of the individual fatty acids were maintained during low-temperature burning, but a depletion occurred during high-temperature burning (figure 43.2).

The PAHs found both in the laboratory and field burn aerosols included biphenyl, trimethylnaphthalenes, phenanthrene, anthracene, methylphenanthrenes, fluoranthene, pyrene, methylpyrenes, chrysene, and benzanthracene. The distribution of PAHs in the aerosols varied with burn temperature as expected (Neff 1979), with a greater variety of alkylated naphthalenes, biphenyls, and phenanthrenes being produced in the laboratory burns than in the field burn. Figure 43.3 shows a GC/MS SIM mode trace of PAHs collected during the field burn, but spectra for the laboratory burns were essentially similar. The shifts in the baseline of this spectrum correspond to changes in the ions selected for scanning. Isotopic data for these species were not obtained due to insufficient amounts (<0.5 nmol) of the desired compounds in the samples.

Field Burn
The bulk carbon isotope signature for the field burn aerosol was even more depleted relative to the plant than the high-temperature laboratory burn aerosol. This trend was also observed with the individual fatty acids. In addition to the even-chain fatty acids, several odd and branched-chain species were detectable in the field burn aerosol. However, as observed from the ion counts on the GC/MS, these species were relatively much lower in concentration than the even-chain fatty acids. The PAHs produced during the field burn were

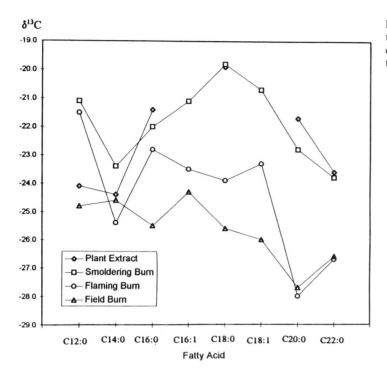

δ¹³C

Figure 43.2 Carbon isotopic distribution of fatty acids in the unburned sugar cane plant and aerosols from laboratory and field burns

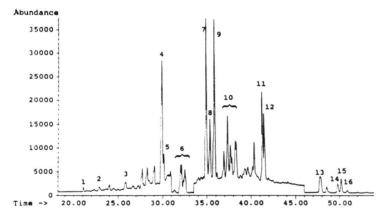

Figure 43.3 GC/MS trace in selected ion mode of PAH in aerosols collected during a field burn of sugar cane

1	Biphenyl	10	Methylpyrenes
2	Acenaphthylene	11	Chrysene
3	Fluorene	12	Benzanthracene
4	Phenanthrene	13	Benzo[b]fluoranthene
5	Anthracene		Benzo[k]fluoranthene
6	Methylphenanthrenes	14	Benzo[e]pyrene
7	Fluoranthene	15	Benzo[a]pyrene
8	Benz[e]acenaphthylene	16	Perylene
9	Pyrene		

Table 43.2 Compound specific carbon isotopic signatures (‰) for PAH collected during sugar cane field burn

Compound	$\delta^{13}C$
Phenanthrene	-24.5 ± 0.2
Fluoranthene	-25.4 ± 0.8
Pyrene	-22.9 ± 0.8

similar to those from the laboratory burn, but a combination of several samples having similar bulk carbon isotopic signatures allowed isotopic characterization of phenanthrene, fluoranthene, and pyrene (table 43.2). The PAHs were isotopically more depleted in ^{12}C than the total lipid extract and similar to the vegetation fatty acids.

Discussion

The dominance of saturated even chain fatty acids in the laboratory and field burn aerosols is consistent with a terrigenous plant-wax source (Kvenvolden 1967). The odd and branched-chain fatty acids observed in the field burn aerosol were probably bacterial in origin (Leo and Parker 1966) and may have arisen from burning of extraneous organic matter in the surface soil.

Parker (1964) showed that the fatty-acid faction of a plant is typically isotopically depleted relative to the bulk vegetation, but that the magnitude of this depletion varies with the species. In the case of sugar cane, the fatty acids were depleted relative to the plant by about 7 to 11‰. Collister et al. (1994) have shown that straight-chain alkanes from plant wax sources were considerably more depleted than the bulk lipid material. They hypothesized that fatty-acid isotopic signatures should be similar to those of straight-chain alkanes because their structural similarity indicates similar biosynthetic pathways. This hypothesis is consistent with the data presented here. The isotopic fractionation observed in the aerosol fatty acids at higher temperatures could result from several mechanisms. Simoneit et al. (1993) noted that volatilization of oxygenated compounds with low vapor pressures during a burn contributes to aerosol organic matter. Perhaps at lower temperatures preferential distillation of the isotopically lighter species would account for the observed fractionation, but it seems unlikely that this mechanism would dominate at higher temperatures. A more likely possibility is that at higher temperatures a cracking of higher molecular weight waxes and triglycerides occurs, which liberates even-chain fatty acids.

If this more tightly bound source material is isotopically lighter than the fatty acids that are readily accessible during the Soxhlet and saponification procedures, an isotopic depletion of the aerosol fatty acids would result. Still another possibility is that preferential oxidation and destruction of isotopically heavier material occurs, leaving behind isotopically depleted organic material that is captured as aerosols.

The isotopic differences observed between the high-temperature and field burns could result from an increase in temperature that facilitates one of the mechanisms described above, or from the incorporation of isotopically lighter soil organic matter into the aerosol during the burn. A correlation between temperature and the fatty-acid isotopic signatures was observed with the laboratory burns, and a field burn could clearly achieve a much higher temperature than the controlled laboratory burns. However, the presence of odd-chain fatty acids in the field burn aerosol suggests that soil matter may have been incorporated into the aerosol, so that neither explanation can be dismissed.

PAHs are derived from incomplete combustion of organic matter (Suess 1976). During pyrolysis, cracking of complex organic molecules results in the formation of free radicals that recombine rapidly to produce PAHs. Schmeltz and Hoffman (1976) have shown that highly reduced lipids, including saturated fatty acids, branched-chain isopropenoid alcohols, and sterols, are suitable precursors for pyrosynthesized PAHs. From the isotopic data obtained in this study, the delineation of exactly which lipids served as the PAH precursors is not possible at present. However, the isotopic distribution for PAHs in the sugar cane field burn aerosol can potentially be used to distinguish a C_4 vegetation combustion source from other vegetation or fossil fuel pyrolysis sources.

Conclusions

The above data emphasize the importance of pyrolytic processes in the production of organic aerosols. Although alterations of the fatty acid isotopic signatures can occur during high-temperature burning, comparison of the chemical and isotopic distributions of these compounds may allow a distinction between organic aerosols resulting from dust transport of raw plant material and those produced during biomass burning. The isolation and isotopic characterization of PAHs from aerosols provides another variable which, together with the fatty acid signatures, may allow distinction between a variety of biogenic and anthropogenic source materials. In order to strengthen the utility of this technique as a means to trace biomass burning products, further analysis of aerosols from burns of other C_4 as well as C_3 plants is planned.

References

Badger G. M., Kimber R. W. L., and Spotswood T. M. 1960. Mode of formation of BP in human environment. *Nature* 187, 663–665.

Collister J. W., Rieley G., Stern B., Eglinton G., and Fry B. 1994. Compound-specific δ^{13}C analyses of leaf lipids from plants with differing carbon dioxide metabolisms. *Org. Geochem.* 21, 619–627.

Crutzen P. J., and Andreae M. O. 1990. Biomass burning in the tropics: Impact on atmospheric chemistry and biogeochemical cycles. *Science* 250, 1669–1678.

Dipple A. 1985. PAH carcinogenesis: An introduction. In *Polycyclic Hydrocarbons and Carcinogenesis*; ACS Symposium Series 283; Washington D.C.: American Chemical Society.

Gagosian R. B., Zafiriou O. C., Oektzer E. T., and Alford J. B. 1982. Lipids in aerosols from the tropical north Pacific: Temporal variability. *J. Geophys. Res.* 87, 11 133–11 144.

Hites R. A., LaFlamme R. E., and Windsor, Jr., J. G. 1980. Polycyclic aromatic hydrocarbons in an anoxic sediment core from the Pettaquamscutt River (Rhode Island, U.S.A.). *Geochim. et Cosmochim. Acta* 44, 873–878.

Kvenvolden K. A. 1967. Normal fatty acids in sediments. *J. Amer. Oil Chem.* 44, 628–636.

LaFlamme R. E., and Hites R. A. 1978. The global distribution of polycyclic aromatic hydrocarbons in recent sediments. *Geochim. et Cosmochim. Acta* 42, 289–303.

Leo R. F., and Parker P. L. 1966. Branched-chain fatty acids in sediments. *Science* 152, 649–650.

Masclet P., Cachier H., Liousse C., and Wortham H. 1996. Emissions of polycyclic aromatic hydrocarbons by savanna fires. *Anal. Chem.*, in press.

Metcalfe L. D., and Schmitz A. A. 1961. The rapid preparation of fatty-acid esters for gas chromatographic analysis. *Anal. Chem.* 33, 363–364.

Neff J. M. 1979. *Polycyclic Aromatic Hydrocarbons in the Aquatic Environment: Sources, Fates, and Biological effects.* London: Applied Science, 266 p.

O'Malley V. P., Abrajano, Jr., T. A., and Hellou J. 1994. Determination of the ^{13}C/^{12}C ratios of individual PAH from environmental samples: Can PAH sources be apportioned? *Organ. Geochem.* 21, 809–822.

Parker P. L. 1964. The biogeochemistry of the stable isotopes of carbon in a marine bay. *Geochim. et Cosmochim. Acta* 28, 1155–1164.

Schmeltz I., and Hoffman D. 1976. Formation of polynuclear aromatic hydrocarbons from combustion of organic matter. In *Carcinogenesis—A comprehensive survey*, vol. 1. Polynuclear aromatic hydrocarbons. Chemistry, metabolism, and carcinogenesis. New York: Raven Press.

Silfer J. A., Engel M. H., Macko S. A., and Jumeau E. J. 1991. Stable carbon isotope analysis of amino acid enantiomers by conventional isotope ratio mass spectrometry and combined gas chromatography/isotope ratio mass spectrometry. *Anal. Chem.* 63, 370–374.

Simoneit B. R. T., and Mazurek M. A. 1982. Organic matter of the troposphere-II. Natural background of biogenic lipid matter in aerosols over the rural western United States. *Atmos. Environ.* 16, 2139–2159.

Simoneit B. R. T., Rogge W. F., Mazurek M. A., Standley L. J., Hildemann L. M., and Cass G. R. 1993. Lignin pyrolysis products, lignans, and resin acids as specific tracers of plant classes in emissions from biomass combustion. *Environ. Sci. Technol.* 27, 2533–2541.

Simoneit B. R. T., Sheng G., Chen X., Fu J., Zhang J. 1991. Molecular marker study of extractable organic matter in aerosols from urban areas in China. *Atmos. Environ.* 25A, 2111–2129.

Standley L. J., and Simoneit B. R. T. 1987. Characterization of extractable plant wax, resin, and thermally matured components in smoke particles from prescribed burns. *Environ. Sci. Technol.* 21, 163–169.

Suess M. J. 1976. The environmental load and cycle of polycyclic aromatic hydrocarbons. *Sci. Total Environ.* 6, 239–250.

Turekian V. C. 1996. Bulk and molecular level isotope fractionation during vegetation burns. Laboratory studies. University of Virginia, Charlottesville, Va. MSc. thesis.

Youngblood W. W., and Blumer M. 1975. Polycyclic aromatic hydrocarbons in the environment: Homologous series in soils and recent marine sediments. *Geochim. et Cosmochim. Acta* 39, 1303–1314.

Stable Carbon Isotopic Analysis of Charcoal from Single Plant Sources

William P. Gilhooly, Stephen A. Macko, Vaughan C. Turekian, Robert J. Swap,
Michael Garstang, and William F. Ruddiman

Current estimates of the production and preservation of elemental carbon produced by biomass burning suggest that the sequestration of carbon as charcoal may constitute an important sink for atmospheric carbon in addition to terrestrial plant uptake and dissolution in the oceans (Seiler and Crutzen 1980; Crutzen and Andreae 1990). Seiler and Crutzen (1980) estimate that from 0.5 to 1.7 Pg C/yr of elemental carbon is produced from the incomplete combustion of biomass to charcoal. The deposition of fossil charcoal in aquatic and marine environments can be a proxy indicator of terrestrial climate, biomass composition, and biomass burning extent and intensity. This study uses bulk stable carbon isotopes analyzed on modern charcoals from southern African savannas to determine the effect of combustion on charcoal isotopic composition. Isotopic analysis of single plant sources of charcoal can be used to better identify fossil charcoal isolated from lake or deep-sea cores.

Background

Charcoal is a refractory carbonaceous material produced during the combustion of vegetative material. Ryan and MacMahan (1976), Sandberg (1979), and Lobert and Warnatz (1993) have shown that charcoal is primarily formed during the flaming stage of combustion. Heat initially distills plant waxes and pyrolysis ruptures molecular bonds producing gases and combustion products that fuel the visible flame (Ward 1990). Charcoal is then formed by the heating of the fuel material in the absence of oxygen (Ward 1990; Seiler and Crutzen 1980). The escape of gases and the buildup of carbon on the fuel surface, also called *char*, inhibits the diffusion of oxygen. Heat buildup and fire turbulence increase the penetration of oxygen, which accelerates pyrolysis of solid fuel and produces a glowing combustion. It is at this stage that a surface reaction of oxygen with carbon forms charcoal (Trollope 1984a; Ward 1990).

Charcoal can be delivered by both fluvial and eolian transport processes to aquatic (e.g., Clark 1988; Horn 1993) and marine environments (e.g., Smith et al. 1973; Griffin and Goldberg 1975, 1979; Herring 1985). Charcoal is thought to be resistant to microbial oxidation during sedimentation and diagenesis (Cope and Chaloner 1980) and preserved in the sedimentary record. Correlating fossil charcoal with modern charcoal isotopic signatures could help identify the original vegetative character of the charcoal material.

Plant stable carbon isotopes have distinct and nonoverlapping signatures traceable to either a C_3 or C_4 photosynthetic pathway (Smith and Epstein 1971). This isotopic separation is a result of enzymatic fractionation effects during the fixation of atmospheric carbon (O'Leary 1988). Calvin cycle (C_3) plants have depleted isotopic signatures ranging from -32 to $-20‰$ with a mean of $-27‰$. Hatch-Slack (C_4) vegetation has a mean $\delta^{13}C$ of $-13‰$ and ranges from -17 to $-9‰$ (Boutton 1991). Vogel and Fulls (1978) demonstrated clearly different isotopic distributions among C_3 and C_4 grass species in South Africa. An additional photosynthetic pathway, Crassulacean acid metabolism (CAM), ranges isotopically from -28 to $-10‰$ (Boutton 1991). The apparent overlap of CAM with C_3 and C_4 does not convolute isotopic interpretation for several reasons: (1) CAM plant metabolism is not a predominant photosynthetic pathway for any South African grass species (Smith and Brown 1973) and (2) CAM plants are not of significant abundance in the grassland interior of South Africa (Werger and Ellis 1981).

Our study uses stable carbon isotopes to determine whether a fractionation effect exists between charcoal and the vegetation originally pyrolized to become charcoal. Results reported in this chapter indicate a 1 to 2 ‰ enrichment between plant and charcoal. Through quantification of the isotopic fractionation between charcoal and plant one can infer the original isotopic signature of the parent plant vegetation for comparison with the $\delta^{13}C$ of fossil charcoal in aquatic and marine sediments.

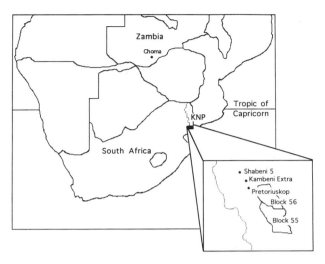

Figure 44.1 Map of southern Africa and sampling sites in Kruger National Park

Study Sites

Savanna vegetation, ash, and its charcoal constituent were collected during the Southern Africa Fire Atmospheric Research Initiative (SAFARI) in 1992 from sites within the Kruger National Park (KNP), South Africa, and Choma, Zambia (figure 44.1). Burning Blocks 55 and 56 were prescribed veld management burns in KNP, and experimental burn samples were collected from KNP in Kambeni, Kambeni Extra, and Shabeni 5, and in Choma, Zambia. The Kruger landscape is an arid to moist savanna dominated by grasses with scattered trees and shrubs. *Hyperthelia dissoluta and Diheteropogon ampletens* are the dominant C_4 grasses (R. Shea 1995 pers. com.). Kambeni Extra vegetation is different from other KNP locations because of 38 years of fire suppression; the primary grass is *Panicum maximum* (R. Shea 1995 pers. com.). Annual precipitation at KNP decreases south to north from 400 to 800 mm (van Wilgen et al. 1990). The KNP sites, around Pretoriuskop, are within a region of high annual precipitation ranging from 800 to 900 mm. Vegetation from Choma, Zambia, is markedly different from that in Kruger. The ecosystem is a more dense, less open semi-arid savanna woodland that receives an average rainfall of 780 mm/yr but has a higher wood biomass than the selected KNP sites (R. Shea 1995 pers. com.). These study sites offer a broad range of C_3 and C_4 vegetation for isotopic experimentation.

Additionally the study was supplemented with samples from a controlled field burn in Fluvanna County, Virginia, (November 1994) to investigate isotopic fractionation during combustion of a single vegetative source. Eastern red cedar (*Juniperus virginia*) wood and ash was collected for isotope analysis from a Virginia Forestry Service prescribed burn. The region is temperate deciduous woodland (mixed pine and hardwood forest) with a moist climate that receives about 1000 mm rain per year.

Methodology

Samples were homogenized in a Wiley Mill using a #60 sieve to ensure complete combustion during isotopic analysis. Charcoal was isolated from ash samples by in situ acidification with concentrated nitric acid to remove carbonates and oxidize organic carbon (Verardo, in prep). The isotopic value of the sample carbon is expressed as the "per mil" (‰) deviations of the sample relative to the PDB standard:

$$\delta^{13}C(‰) = [(R_{sample} - R_{standard})/R_{standard}] \times 1000;$$
$$R = C^{13}/C^{12}$$

Samples for carbon isotope analysis were combusted with copper and copper oxide in sealed evacuated tubes at 850°C for 1 hour. Bulk isotope determinations were made on a dual inlet, triple collector VG PRISM stable isotope mass spectrometer. Machine precision of the sample analysis was 0.05‰ or better; procedural sample replicate precision is approximately 0.1‰.

The chemical assay used to isolate charcoal was tested for fractionation effects on two refractory carbonaceous standard materials. Acidification of black carbon powder (Aesar, 14734), a graphitic carbon, and NORIT SXI (American Norit Company, A-7046), an acid-washed, steam-activated peat-based carbon, showed no significant change in the carbon isotopic signatures. Black carbon powder prior to isolation had a $\delta^{13}C$ of $-21.4 \pm 0.08‰$ (n = 5) and after $-21.5 \pm 0.05‰$ (n = 4), and NORIT SXI was $-26.3‰$ (n = 2) before and $-26.1‰$ (n = 1) after.

Results

Charcoals isolated from an ash produced during natural fire events in southern Africa (table 44.1) show a consistent enrichment, ranging from 0.6 to 2.7‰ with a mean of 1.5‰. The slope of 0.87, from the linear regression equation (figure 44.2), indicates an enrichment of 1.3‰, further illustrating this isotopic enrichment. Block 55, KNP ash samples had the smallest isotopic difference whereas Choma, Zambia, ash had the largest. These isotopic differences may reflect different degrees of burn intensity. Carbon concentration

Table 44.1 Carbon isotopic enrichment between ash and charcoal samples from burns in southern Africa

Sample name and location	Ash δ^{13}C (‰)	Charcoal δ^{13}C (‰)	Δ (‰)	% Charcoal
Blook 55, KNP ash*[a]	−14.2	−13.6	0.6	13
Blook 56, KNP grass ash	−15.0	−14.0	1.0	15
Blook 56, KNP litter ash	−17.5	−15.1	2.4	21
Kambeni, KNP ash	−15.9	−14.1	1.8	20
Kambeni, Extra, KNP ash	−15.4	−14.2	1.2	19
Shabeni 5, KNP ash	−22.5	−21.4	1.1	19
Choma, Zambia ash	−25.1	−22.4	2.7	21

a. * = mean, n = 2.

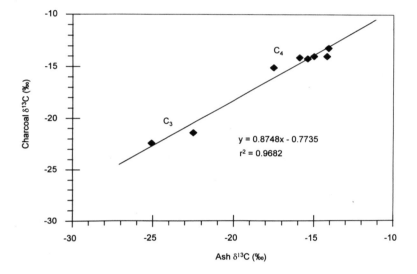

Figure 44.2 The isotopic enrichment between southern African ash and charcoal samples distributed among C_3 and C_4 photosynthetic pathways

calculations derived from manometric measurements of CO_2 gas generated during isotopic analysis indicate that charcoal carbon constitutes about 20% of the ash carbon in most southern African burn samples (table 44.1). Least squares analysis of the data indicates a strong correlation between the charcoal and the ash isotopic compositions ($r^2 = 0.97$, figure 44.2) and t-test results show the isotopic separation between C_3 and C_4 plant in charcoal ($p < 0.005$). Statistical analysis of cedar ash and cedar wood collected from a controlled field burn in Virginia (table 44.2) shows no significant difference between burned and unburned wood δ^{13}C ($p < 0.005$, t-test). Unburned and burned material from Kambeni Extra, KNP, provides an example of mixing between C_3 and C_4 isotopic end members (table 44.3).

Discussion

Few isotopic characterizations of charcoal have been published (Cachier et al. 1985; Cachier et al., 1986;

February and Van Der Merwe 1990) and a consistent fractionation effect during charcoal formation has not been reported. Results show an isotopic enrichment between ash and charcoal but no change between plant and ash. The isotopic enrichment may be a result of charcoal resistance to heat breakdown and the loss of isotopically light combustion gases. This isotopic enrichment remains the same for both C_3 and C_4 plants. High temperatures reduce fractionation between ash and plant. The isotopic enrichment between charcoal and ash and the conservation between ash and plant indicate an overall enrichment between plant and charcoal. Mixing of different plant sources must be accounted for when interpreting the bulk δ^{13}C signal. Burned and unburned samples from Kambeni Extra, KNP, provide an example of mixing between C_3 and C_4 material.

Stable carbon isotopes indicate a consistent 1 to 2‰ enrichment between ash and charcoal (table 44.1). A possible mechanism for enrichment is the resistance of charcoal to thermal degradation and the subsequent

Table 44.2 Conservation of the stable carbon isotopic signature between burned and unburned cedar wood in virginia

Unburned cedar $\delta^{13}C$ (‰)		Cedar ash $\delta^{13}C$ (‰)	
	−24.1		−23.7
	−23.6		−23.8
	−23.6		−23.7
mean	−23.8	mean	−23.7
s.d.	0.29	s.d.	0.06

Table 44.3 Relative carbon isotopic contributions of unburned grass and litter to a representative ash in Kambeni Extra, Kruger National Park

Sample	$\delta^{13}C$ (‰)	Relative contributions to isotopic mass balance
Unburned grass		
Aristida congesta barbicallis	−13.2	
Diheteropogen amplectens	−12.3	
Heteropogon contortus	12.9	
Hyperthelia dissoluta[a]	−11.8	
Melinis repens	−12.9	
Panicum maximum	−12.4	
Pogonarthrig squarrosa	−13.5	
Schizachyrium sanguincum	−12.9	
grass mean	−12.7	77%
Unburned litter		
Litter	−24.4	23%
Burned vegetation		
Representative ash	−15.4	100%

a. * = mean, n = 2.

release of isotopically light carbon as CO_2, CO, CH_4, and other trace gases. Charcoal is an assemblage of refractory carbonaceous material formed at high temperature. Beall and Eickner (1970) reported that charcoal was not completely carbonized even at 1500°C during laboratory experiments. The resistance to heat breakdown may encourage the release of isotopically light carbon during the formation of charcoal. ^{12}C-^{12}C bonds are more readily ruptured than ^{13}C-^{12}C bonds (Kaplan 1975; Hoefs 1987). Thus, cleavage of carbon-carbon bonds from the charcoal substrate are expected to produce isotopically depleted products. This is consistent with kerogen thermal maturation experiments. Peters et al. (1981) found a 5‰ enrichment for sapropelic kerogen and 1.6‰ for humic kerogen when heated to a maximum temperature of 500°C. Rayleigh fractionation calculations suggested that the sapropelic kerogen enrichment resulted from the generation of isotopically light CH_4. Humic kerogen enrichment appeared to be controlled by factors other than CH_4 generation. Similarly, Chung and Sackett (1979) reported an enrichment of 0.5‰ between pyrolytic residues and a parent coal (humic kerogen). The en-

richment was attributed to the compensatory effect of generating isotopically heavy CO_2 and isotopically light CH_4.

The isotopic enrichment between ash and charcoal is consistent for both C_3 and C_4 plants and enables isotopic partitioning between C_3 and C_4 plant biomass burning sources (e.g., Cachier et al. 1986). The savanna fire regime in southern Africa is mainly confined to the austral dry season prior to the spring rains when grasses are dormant and tree leaves are senescent (van Wilgen et al. 1990). The fire events are surface fires where typically only dried grass and leaf litter are burned whereas most trees and shrubs are fire tolerant (Vogel 1974; Trollope 1984b). Although the isotopic signature of the ash and charcoal combustion products is primarily a function of fuel type, other combustion factors that may affect the isotopic signature include ambient temperature, oxygen availability, and moisture content. Two discrete isotopic groups distributed between C_3 and C_4 sources from KNP and Choma illustrate the influence of fuel type on isotopic composition (figure 44.2). Ash and charcoal samples from Shabeni 5, KNP, and Choma, Zambia, have a depleted C_3 isotopic signal and samples from the remaining KNP sites have an enriched C_4 carbon isotopic signature. For all samples, regardless of whether they are C_3 or C_4 plants, the isotopic separation between ash and charcoal is linearly correlated ($r^2 = 0.97$).

Further isotopic analysis of cedar wood and cedar ash from Virginia (table 44.2) showed no apparent isotopic depletion. The ash isotopic conservation relative to the original plant $\delta^{13}C$ is possibly related to the temperature of the burn. Kinetic isotope effects, causing the isotopic separation between product and reactant, decrease at higher temperatures (Kaplan 1975; Hoefs 1987). South African savanna maximum fire temperatures have been reported to range from 570°C for head fires to 390°C for back fires (Trollope 1984a). At these elevated temperatures, isotopic fractionation is reduced, conserving the original plant isotopic signature in the ash.

An enrichment in ^{13}C between ash and charcoal and conservation of the isotopic signal between plant and ash imply an overall enrichment between plant and charcoal. This information provides a correlative tool with which carbon isotopic analysis of fossil charcoal isolated from sediments can be related to the parent plant material. The isotopic shift between plant and charcoal is the same for both C_3 and C_4 pathways. Unless fossil charcoal undergoes bacterial mediated fractionation, the fossil charcoal $\delta^{13}C$ should represent the original plant source (i.e., Calvin cycle or Hatch-Slack) with a 1 to 2‰ enrichment.

Mixing of heterogeneous vegetative materials complicates interpretation of the bulk $\delta^{13}C$ signal. Vogel and Fulls (1978) indicated that C_3 and C_4 grasses are geographically separated and coverage by C_4 grasses predominates most of South Africa. This distribution is ascribed to the ability of C_4 plants to photosynthesize at higher ambient air temperatures during the growing season and to a lesser degree their enhanced water use efficiency in arid environments (Vogel and Fulls 1978; Ellis et al. 1980; Cowling 1983). KNP lies within a region composed of at least 95% C_4 grasses (Huntley 1982). Most of the trees in KNP are C_3 and their fallen leaves constitute a high percentage of the litter component (R. Shea 1995 pers com). Although the major component is C_4, the actual fuel content can vary dramatically between C_3 and C_4 contributions. With exception to the Shabeni 5 location, all of the KNP $\delta^{13}C$ values typified strong C_4 isotopic influences (table 44.1). Shabeni 5 is not florally different from the rest of KNP; however, observations by R. Shea (1995 pers com) noted a large amount of litter and wood debris consumed by fire. The depleted $\delta^{13}C$ valued of -22.5 may be attributed to effects of drier conditions during which grass growth is diminished and more wood material falls from trees. Choma, Zambia, also has a depleted value of $-25.1‰$ but it is congruent with its dense woodland vegetative makeup (P. Dowty 1995 pers. com.).

An isotopic experiment with Kambeni Extra, KNP, data illustrate the use of stable isotope ratio mass spectrometry as a powerful tool to distinguish between plant combustion sources. Cachier et al. (1985) reported a 2 to 9‰ depletion between burned grass and unburned dry grass, and an 11‰ depletion between unburned grass and carbonaceous aerosols, indicating a profound isotopic separation during plant pyrolysis (for more on carbonaceous and nitrogenous aerosols see chapter 40 by Turekian et al.). Kambeni Extra ash (table 44.3) depletion relative to the unburned grass material is not a result of fractionation, rather a mixture of different plant sources. The grass vegetation samples are representative of the plant materials combusted to form the (P. Dowty 1995 pers. com.). The mean isotopic value for unburned grass was $-12.7‰$, consistent with a C_4 plant signal. Unburned litter had a depleted $C_3\delta^{13}C$ of $-24.4‰$ and the ash was $-15.7‰$. A mixture of the C_4 materials with the isotopically depleted litter explains the overall depleted ash signal. Mass balance of the ash, grass, and litter indicates that grasses contribute 77% and litter 23% of the $-15.4‰$ ash value. The partitioning of 77% grass and 23% litter is consistent with the Cachier et al. (1985) finding that

burning of C_4 plant material accounts for 75% of the carbonaceous aerosol particulate production.

Conclusions

The results show a 1 to 2‰ enrichment between ash and charcoal. There is no significant isotopic fractionation between plant and its ash product, indicating an overall 1 to 2‰ enrichment between plant and charcoal. The enrichment remains constant for both C_3 and C_4 isotopic groups. Conservation of the ash isotope values and enrichment of the charcoal values is suggested to be a function of two mechanisms: (1) ash samples are not fractionated isotopically and (2) the refractory carbon of charcoal is resistant to heat breakdown and generation of isotopically depleted gases is the source of enrichment. Mixing of different parent plant charcoal sources complicates the determination of the relative C_3 and C_4 vegetative contributions. A case study of Kambeni Extra, KNP, vegetation, ash, and charcoal shows that at the species level, isotopic partitioning among C_3 and C_4 photosynthetic pathways is a powerful and exact analytical tool.

Although there is a 1 to 2‰ isotopic enrichment between plant and charcoal, the charcoal maintains an isotopic signal similar to that of the parent plant; therefore, fossil charcoal $\delta^{13}C$ reflects the original photosynthetic pathway of either C_3 or C_4 metabolism. This modern charcoal study provides a foundation for ongoing fossil charcoal isotopic research. Carbon isotopic analysis of the fossil charcoal sedimentary record has the potential for reconstruction of terrestrial paleoclimates, biomass composition, and extent and intensity of past conflagrations. Future bulk isotope work can be complemented by current research using compound-specific isotopic analysis (see chapter 40 by Turekian et al. and chapter 43 by Ballentine et al.) to better determine the mechanisms for isotopic enrichment in charcoal and conservation in ash. Additional study is necessary to form a better consensus on the chemical composition of charcoal and to determine if isolation of elemental carbon from charcoal can be used for isotopic analysis of multiple sources (i.e., sedimentary carbon).

Acknowledgments

W. Gilhooly would like to thank the several contributing authors for advice and encouragement. R. Black and S. Pitts of the Virginia Forestry Service gave us the oppportunity to witness a controlled burn and provided insight on fire control and prevention.

P. Dowty and R. Shea collected the burned and unburned African vegetation samples from Kruger National Park and Choma, Zambia. Both offered important background information about biomass burning in southern Africa. A portion of the SAFARI-92 project was funded by NSF ATM 9207924 (M. Garstang and S. Macko). The charcoal research was conducted under NSF EAR 526948 (W. Ruddiman).

References

Beall, F. C., and H. W. Eickner. 1970. Thermal Degradation of Wood Components: A Review of the Literature, *US Department of Agriculture, Forest Service Research Paper FPL* 130, 26.

Boutton, T. W. 1991. Stable Carbon Isotope Ratios of Natural Materials: II. Atmospheric, Terrestrial, Marine, and Freshwater Environments, *Carbon Isotope Techniques*, edited by D. C. Coleman and B. Fry, Academic Press.

Cachier H., P. Buat-Ménard, and M. Fontugne. 1985. Source Terms and Source Strengths of the Carbonaceous Aerosol in the Tropics, *Journal of Atmospheric Chemistry*, 3, 469–489.

Cachier H., P. Buat-Ménard, M. Fontugne, and R. Chesselet. 1986. Long-Range Transport of Continentally-Derived Particulate Carbon in the Marine Atmosphere: Evidence from Stable Isotope Studies, *Tellus*, 38B, 161–177.

Chung, H. M., and W. M. Sackett. 1979. Use of Stable Carbon Isotope Compositions of Pyrolitically Derived Methane as Maturity Indices for Carbonaceous Materials, *Geochimica et Cosmochimica*, 43, 1979–1988.

Clark, J. S. 1988. Effect of Climate Change on Fire Regimes in Northwestern Minnesota, *Nature*, 334, 233–235.

Cope, M. J., and W. G. Chaloner. 1980. Fossil Charcoal as Evidence of Past Atmospheric Composition, *Nature*, 283, 647–649.

Cowling, R. M. 1983. The Occurrence of C_3 and C_4 Grasses in Fynbos and Allied Shrublands in the South Eastern Cape, South Africa, *Oecologia*, 58, 121–127.

Crutzen, P. J., and M. O. Andreae. 1990. Biomass Burning in the Tropics: Impact on Atmospheric Chemistry and Biogeochemical Cycles, *Science*, 250, 1669–1678.

Ellis, R. P., J. C. Vogel, and A. Fulls. 1980. Photosynthetic Pathways and the Geographical Distribution of Grasses in South West Africa/Namibia, *South African Journal of Science*, 76, 307–314.

February, E. D., and N. J. Van Der Merwe. 1992. Stable Carbon Isotope Ratios of Wood Charcoal during the Past 4000 Years: Anthropogenic and Climatic Influences, *Suid-Afrikanse Tydskrif vir Wetenskap*, 88, 291–292.

Griffin, J. J., and E. D. Goldberg. 1975. The Fluxes of Elemental Carbon in Coastal Marine Sediments, *Limnology and Oceanography*, 20, 456–463.

Griffin, J. J., and E. D. Goldberg. 1979. Morphologies and Origin of Elemental Carbon in the Environment, *Science*, 206, 563–565.

Herring, J. R. 1985. Charcoal Fluxes into Sediments of the North Pacific Ocean: The Cenozoic Record of Burning, In *The Carbon Cycle and Atmospheric CO_2: Natural Variations Archean to Present*, edited by E. T. Sundquist and W. S. Broecker, American Geophysical Union, Washington, DC.

Hoefs, J. 1987. *Stable Isotope Geochemistry*, 3rd ed., Springer-Verlag, New York.

Horn, S. P. 1993. Postglacial Vegetation and Fire History in the Chirripo Paramo of Costa Rica, *Quaternary Research*, 40, 107–116.

Huntley, B. J. 1982. Southern African Savannas, In *Ecology of Tropical Savannas*, edited by B. J. Huntley and B. H. Walker, Springer-Verlag, New York.

Kaplan, I. R. 1975. Stable Isotopes as a Guide to Biogeochemical Processes, *Proceedings of the Royal Society of London*, Series B, 189, 183–211.

Lobert, J. M., and J. Warnatz. 1993. Emissions form the Combustion Process in Vegetation, In *Fire in the Environment: The Ecological, Atmospheric, and Climatic Importance of Vegetation Fires*, edited by P. J. Crutzen and J. G. Goldammer, John Wiley, New York.

O'Leary, M. H. 1988. Carbon Isotopes in Photosynthesis, *BioScience*, 38, 328–336.

Peters, K. E., B. G. Rohrback, and I. R. Kaplan. 1981. Carbon and Hydrogen Stable Isotope Variations in Kerogen during Laboratory-Simulated Thermal Maturation, *The American Association of Petroleum Geologists Bulletin*, 65, 501–508.

Ryan, P., and W. McMahon. 1976. Some Chemical and Physical Characteristics of Emissions from Forest Fires, Paper no. 76-23 presented at 69th Annual Meeting of Air Pollution Control Association, Portland, 22 June–1 July.

Sandberg, D. V., J. M. Pierovich, D. G. Fox, and E. W. Ross. 1979. Effects of Fire on Air, US Department of Agriculture, *Forest Service, General Technical Report, WO-9*.

Seiler, W., and P. J. Crutzen. 1980. Estimates of Gross and Net Fluxes of Carbon between the Biosphere and the Atmosphre from Biomass Burning, *Climate Change*, 2, 207–247.

Smith, B. N., and W. V. Brown. 1973. The Kranz Syndrome in the Graminae as Indicated by Carbon Isotopic Ratios, *American Journal of Botany*, 60, 505–513.

Smith, B. N., and S. Epstein. 1976. Two Categories of 13C/12C Ratios of Higher Plants, *Plant Physiology*, 47, 380–384.

Smith, D. M., J. J. Griffin, and E. D. Goldberg. 1973. Elemental Carbon in Marine Sediments: A Baseline for Biomass Burning, *Nature*, 241, 268–270.

Trollope, W. S. W. 1984. Fire in Savanna, Ecological Effects of Fire, in *South African Ecosystems*, edited by P. de V. Booysen and N. M. Tainton, Springer-Verlag, New York.

Trollope, W. S. W. 1984b. Fire Behaviour, Ecological Effects of Fire, In *South African Ecosystems*, edited by P. de V. Booysen and N. M. Tainton, Springer-Verlag, New York.

van Wilgen, B. W., C. S. Everson, and W. S. W. Trollope. 1990. Fire Management in Southern Africa: Some Examples of Current Objectives, Practices, and Problems, In *Fire in the Tropical Biota. Ecosystem Processes and Global Challenges*, edited by J. G. Goldammer, Springer-Verlag, New York.

Vogel, J. C., and A. Fulls. 1978. The Geographical Distribution of Kranz Grasses in South Africa, *South African Journal of Science*, 74, 209–215.

Vogl, R. J. 1974. Effects of Fire on Grasslands, In *Fire and Ecosystems*, edited by T. T. Kozlowski and C. E. Ahlgren, Academic Press, New York.

Ward, D. E. 1990. Factors Influencing the Emissions of Gases and Particulate Matter from Biomass Burning, In *Fire in the Tropical Biota, Ecosystem*.

Laboratory Investigations on Aerosols from the Combustion of Savanna Grass and Cereal Straw

Pham-Van-Dinh, Roger Serpolay, and Jean-Pierre Lacaux

Except for some rare pioneering works (e.g., Warner and Twomey 1967; Twomey 1977), the emission of gases and particularly of aerosol particles from biomass burning, especially in tropical countries, has really drawn attention just recently, and their effects on the atmosphere require more investigation (e.g., Crutzen and Andreae (1990)).

Combustion aerosol particles, released primarily in the submicron size range, induce a dual effect upon the atmosphere: warming and cooling, by absorption and by scattering of solar radiation, respectively. The role of sulfate aerosol as a principal component of anthropogenic aerosols in (radiative) climate forcing has been the subject of numerous works in past years. As a result, forcing by sulfate aerosols in clear air has been considered capable of counteracting or significantly reducing the greenhouse-gas effect, at least in the Northern Hemisphere (Charlson et al. 1991, 1992). Nevertheless, such calculations have aroused some questions, as from Kellog (1992), who even supported the possibility that under clear-sky conditions these aerosols might further contribute to the warming that results from greenhouse gases, but Kiehl and Briegleb (1993), though they agree with the sign (negative) of radiative forcing by these sulfates, have shown that their effects were lower, from about -1 Wm^{-2} for annual direct global forcing to ~ -0.3 Wm^{-2}. In any case, the uncertainties remain important. Therefore, Penner et al. (1994), in a comprehensive article, attempt to evaluate various uncertainty factors contributing to the final uncertainty and assign a central value of -0.6 Wm^{-2} within the $(-0.26, -1.4$ $Wm^{-2})$ range for direct global mean forcing from sulfate aerosols.

Concerning the effects due to biomass burning aerosols that were neglected in the triggering article by Charlson et al. (1992), their importance is now recognized by everyone and their effects are only now being evaluated. Nevertheless, because of lacking experimental data dealing with the main characteristics of these aerosols, few results have been distributed, and uncertainty about those remains large. Penner et al.

(1992) have evaluated global-average direct forcing by such aerosols as -0.8 Wm^{-2}, with an uncertainty factor estimated at 2.7 (Penner et al. 1994), so that the range is from -0.3 to -2.2 Wm^{-2}.

If absorption by the black-carbon fraction present in these combustion aerosols is taken into account, the estimated range widens to -0.1 to -2.2 Wm^{-2}.

Another important aspect of a large fraction of the anthropogenic aerosols is their ability to form droplets that evolve to form haze and clouds. Inorganic water-soluble aerosols (sulfates) and particularly organic aerosols from combustion of vegetation can act as cloud-condensation nuclei (CCN), increase cloud albedo by increasing droplet concentration, reducing droplet size and thus coalescence rate, and so the likelihood of precipitation—that is, increasing cloud lifetime (Twomey 1977; Hudson 1993). This indirect forcing of the CCN aerosols is most likely to be significant because the global heat balance is very sensitive to cloud albedo, particularly through marine stratus clouds, which cover about a quarter of the earth. Such clouds are in turn very sensitive to change in CCN populations, as spectacularly shown by the occurrence of ships' smoky trails on a large scale that can be observed by satellite (Coagley et al. 1987; Radke et al. 1989; Hindman et al. 1994).

Indirect radiative forcing via CCN behavior has not yet been satisfactorily modeled because, among other reasons, there are very few data on actual CCN populations in remote areas potentially influenced by anthropogenic emissions, and because of our poor understanding of the mechanisms that govern production of new CCN and their lifetime in the atmosphere (Penner et al. 1994). Therefore, more data must be gathered dealing with aerosols and CCN characteristics, especially from biomass burning such as agricultural waste and savanna fires. Cereal-waste burning remains a common practice in numerous countries, including industrialized countries such as Spain and England (Kilsby 1990), and savanna burning is routinely practiced in tropical areas, especially in Africa, for various purposes such as favoring regrowth of grass for cattle, clearing ground for agriculture, and getting rid of parasites.

Several authors have evaluated particulate-matter emission from biomass burning, but their estimates were based essentially on forest-fire data and most of them on the same sources of data (Levine 1990; Crutzen and Andreae 1990; Ward and Hardy 1991; Andreae 1991; Levine et al. 1995). Furthermore, when we examine the CCN production rates, if any, given by a few authors (e.g., Radke and Hobbs 1976; Radke et al. 1991), we find that all their evaluations were based primarily on the sole known data from Warner and Twomey (1967), who had carried out combustion on a small amount (3 g) of sugarcane leaf in an electric furnace. Thus, as noted by Rogers et al. (1991), data are not yet available that would allow us to estimate the number concentration flux of CCN due to emission by tropical biomass burning. Therefore, attempting to partly fill such a gap, chamber-combustion experiments were carried out on Guinean savanna grasses (Loudetia and Hyparrhenia), from Lamto, Ivory Coast, as part of the FOS/DECAFE experiment (Lacaux et al. 1995) to quantify the emission factor jointly with other related gases and aerosols. Some (limited) combustion experiments were also performed on savanna grass from Kruger National Park, South Africa, in relation to the SAFARI-92 experiment, and also on cereal straw from the Spanish Basque Country, where cereal straw wastes are systematically gotten rid of by fire every year in autumn (Ezcurra et al. chapter 73 this volume). The essential results dealing with particulate and CCN production from three main series of combustion experiments in July 1991, June 1992, and October–November 1993 are presented here.

Experiments

Test Chamber

An aerosol dilution chamber ($10 \times 4 \times 4$ m) previously built and used for work on ice nuclei (Pham-Van 1969) was renovated and transformed into a dark test chamber allowing combustion experiments, aerosol as well as gas sampling, and monitoring from the outside. Specifically, combustion of savanna grass, when placed in the test chamber, and various devices requiring inside work such as homogenization fans and humidifiers, could be started and controlled from outside. The other gas and aerosol analyzers are around the outside walls (figure 45.1).

Each whole experiment took several hours, so that the test chamber (160 m³) had to be checked for its ability to preserve the aerosol and gas content from wall loss or leak within acceptable limits during the test time. The decay of CO_2, continuously recorded with a binos-100 (Rosemount), does not exceed 5% per hour except when the relative humidity (RH) is kept quite high (>80%) for 1 hour or more; in that case the decay exceeds 10% h⁻¹ at 89%. The variation in total particulate matter (TPM), also continuously monitored with a RAM-1 (MIE Inc.), is higher because of its sensitivity to larger size-range particles subjected to deposition as dry particles or as droplets (for details, see Pham-Van 1994).

Burning Arrangement

In the first series of experiments, about 25 g of Guinean savanna grass were put down on an electric resistance wire set on a metallic tray in the middle of the chamber. When switched on (from outside the chamber), the wire quickly heats to red and is then able to ignite the biofuel. When heated, the savanna grass emits smoke for a few seconds as a result of the pyrolysis stage; when the temperature rises enough, the flaming stage starts, lasting about 30 sec, and is followed by the smoldering stage. The three stages in combustion occur within about 1 min. That combustion is referred to hereafter as *mixed combustion*. The ash and unburned residue average in the 11 to 13% range, and the specific humidity of these species, as determined at constant temperature (110°C) in an oven for 24 h, ranges from 7.3 to 8.5% by weight. The carbon content from various samples has been determined to be quite constant: 42% by weight (Lacaux et al. 1995).

For other series of experiments, a burning system similar to a fireplace was built and placed outside the test chamber. The combustion products were introduced at a velocity of ~3 m s⁻¹ into the test chamber through a stack 1.5 m high and 15 cm in diameter. This system allowed us to separate the flaming phase from the other phases by obstructing the stack before the flames started and just after their complete extinction, thus preventing products of the pyrolysis and smoldering phases respectively from penetrating the test chamber. That combustion is referred to here as *flaming combustion*, generally characterized by a ($\Delta CO/\Delta CO_2$) ratio averaging around 5% .

To access only the smoldering process, the vegetative species were wrapped inside a metal pipe aerated with two lines of holes, thus being consumed like a cigar, smoking the which could be facilitated by a small pump blowing about 10 *l* min⁻¹. Such combustion is referred to here as *smoldering combustion*. That

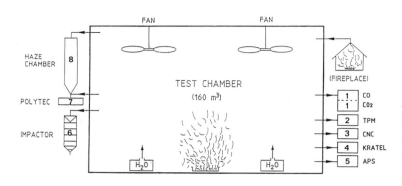

Figure 45.1 Experimental setup for investigating combustion aerosol particles: (1) test chamber, with main devices involved in analyzing CO/CO_2; (2) total particulate matter; (3) number of condensation nuclei; (4) geometric size distribution of particles; (5) aerodynamic size distribution of particles; (6) aerodynamic size mass distribution of particles; (7, 8) CCN supersaturation spectrum

Table 45.1 Main gas and aerosol devices used in this study

Parameter	Method	Range used	Resolution	Manufacturer/Model
Carbon dioxide (1)	Nondispersive infrared	0–500 ppmv	1 ppmv	Rosemount/Binos 100
Carbon monoxide (1)	Nondispersive infrared	0–500 ppmv	1 ppmv	Binos 100
Total particulate matter (2)	Light-scattering	0–2 mg/m^{-3} 0–20 mg/m^{-3}	1 μg 0.01 mg	M.I.E./RAM-1
Condensation nuclei (3)	Pulse count, Photometry	0–10^4 nuclei 10^4–10^7	1 nucleus 1%	TSI/Model 3022
Particle-size distribution (4)	Light-scattering	0.5–10 μm diam. (4 size channels)	1 particle	Kratel/Partoscope
Particle-size distribution (5)	Laser velocimetry	0.5–30 μm diam. (60 channels)	1 particle	TSI/APS 3310
Particle-mass distribution (6)	Impaction	0.5–13.6 μm D_{50} (5 stages)	—	Polytec/HC-15
Particle-size distribution (7)	Light-scattering	0.4–15 μm diam. (64 channels)	1 particle	In-house/CRA-88
Droplet-size distribution (+CCN number) (7,8)	Light-scattering	0.8–38 μm diam. (64 channels)	1 droplet (1 nucleus)	Polytec/HC-15 + In-house Isothermal Haze chamber

smoldering is characterized by a ($\Delta CO/\Delta CO_2$) ratio centered on ~30%. Both CO and CO_2 contents were sampled at the highest part of the stack with a dilution device when necessary. Combustion as flaming was much shorter than smoldering, so that to reach the steady state required more fuel, about 50 to 100 g versus ~20 g for the smoldering experiment. Such amounts allowed a duration of combustion of up to ~1 min for flaming and up to ~2 min for smoldering.

Instrumentation

Various devices for sampling and analyzing aerosols and gases emitted from different combustion processes were located outside and around the test chamber; their main features are summarized in table 45.1. Some instruments were not available for all experiments; particularly the CO content and condensation nucleus (CN) number were unfortunately not measured during the first series of experiments. Classified as mixed

combustion, these amounts were likely to be intermediate between flaming and smoldering, as discussed in the following sections.

The CCN measurements are performed with an optical spectrometer (Polytec-HC 15) associated with a cylindrical isothermal haze chamber (IHC), 2 m high and 12.5 cm in diameter (figure 45.2). To improve the temperature uniformity of its inner walls, the whole IHC is coaxially placed in a larger tube so that the annular space can be swept quickly by ambient air by means of a pump (Garmy and Serpolay 1986; Serpolay 1994).

The IHC is designed to simulate formation of haze, as in natural conditions when RH tends to reach 100%; therefore the IHC inner wall is entirely covered with a linen cloth kept steadily wet. Laktionov (1972, 1973), according to Alofs and Podzimek (1974), proposed its use to measure CCN concentrations by activating the droplet formation at critical supersaturation lower

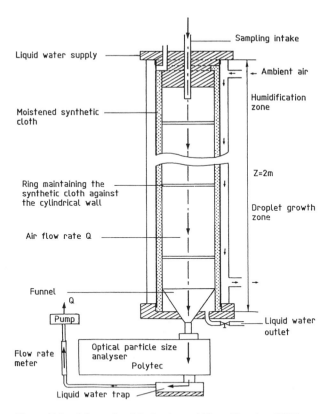

Figure 45.2 Schematic of the Isothermal Haze Chamber (IHC, laboratory model) associated with the Optical Particle Analyzer (OPC) Polytec HC-15 (adapted from Garmy and Serpolay 1986)

than 0.2%. The basic operating principle of the IHC is related to the fact that the equilibrium radius (r_{100}) of a haze droplet at 100% RH (supersaturation = 0) is solely related to the critical supersaturation (SS_c) of the nucleus by a simplified Köhler equation:

$$SS_c(\%) = \frac{0.041}{r_{100}(\mu m)} \quad \text{at ambient temperature (20°C)}$$

$$(45.1)$$

According to Hoppel and Fitzgerald (1977), cited by Hudson et al. (1981) and by Fitzgerald et al. (1981), that relationship is unique if the particle is at least 1% soluble and so weakly depends upon the nature of the nucleus. Equation (45.1) can be applied if droplets are grown to their equilibrium size in a water-saturated environment, and then they are counted and sized by the Polytec-HC 15 optical particle analyzer. The droplet size is related to the critical supersaturation, and so a curve may be drawn from a cumulative number of CCN versus SS_c (figure 45.3). The SS_c is the maximum supersaturation that can be attained by a solution droplet during its growth in such conditions.

Therefore the IHC is suitable for investigating CCN activated under very light supersaturation, as it prevails in stratus clouds. For our experimental conditions, the maximum SS_c involved in our CCN detection system is around 0.1%.

In the IHC, the equilibrium size of droplets is achieved within 15 to 20 min. The actual number of droplets for channels of each size is assumed to be equal to the difference between the number of particles that have stayed in the haze chamber and the number of particles regarded as "dry," which may not apply at high RH (e.g., at 80% RH).

Results and Discussion

The results presented hereafter are compiled from data from various experiments carried out in 1991, 1992, and 1993, some details about which can be found in Pham-Van (1994).

Particulate Emission from Savanna Species

Table 45.3 summarizes particulate emissions from various savanna species burned in different conditions, as described above on pages 473–474. The measurement uncertainties keep us from discerning the difference, if any, in emission factor among the various savanna species involved, such as Loudetia and Hyparrhenia from Ivory Coast or species from South Africa (two samples only). The combustion process is characterized by emission of CO_2 and CO, and especially by the ($\Delta CO/\Delta CO_2$) ratio (Δ denotes excess above background), averaging 5% for flaming, which is slightly lower (i.e., better flaming) than that obtained over prescribed savanna fires at Lamto, Ivory Coast, during the field experiment FOS-91/DECAFE, averaging 6.1% (± 1.7) at ground level from steel-canister sampling (Bonsang et al. 1995), or than that measured by Hurst et al. (1994) over Australian fires during the 1990 dry season (5.8 $\pm$ 2.0%). Our smoldering experiment with a ($\Delta CO/\Delta CO_2$) ratio slightly exceeding 30% may be regarded as a superior limit for such an open-field combustion process. The CO_2 release, averaging 1220 mg/g dm (mg per gram of *dry matter*) (± 180) in our flaming conditions, is very close to that from the FOS-91 experiment, averaging 313 g C/kg/dm (Lacaux et al. 1993); that is, 1146 mg CO_2/g dm. Nevertheless, this CO_2 emission factor (EF) remains lower than the EF from our mixed combustion conditions (~ 1400 mg/g dm), the latter being very close to that obtained from two experimental series by Delmas et al. (1991, 1995) in similar chamber-combustion conditions, namely 1370 and 1300 mg/g dm respectively. Because

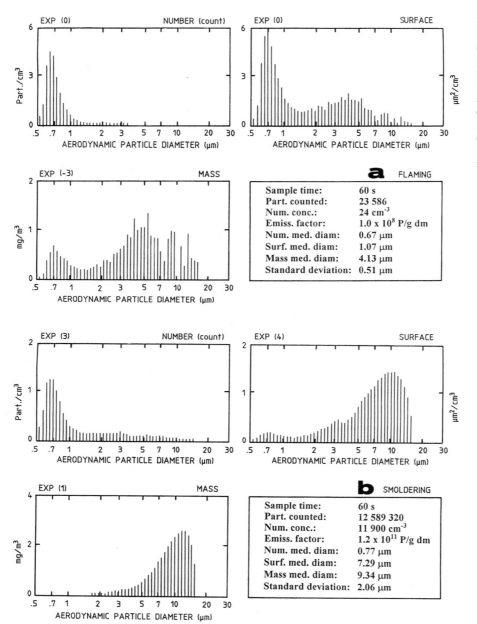

Figure 45.3 Examples of size distributions of "coarse" particles (aerodynamic diam. ≥ ∼0.5 μm) as a function of the combustion process (*a*) flaming, (*b*) smoldering, as displayed by the APS (TSI Inc., model 3310) for a same savanna species (Loudetia). The emission factor averaged $(1.1 \pm 0.1) \times 10^{11} \mathrm{g}^{-1}$ dm for the smoldering ($n = 4$) and $(1.8 \pm 3.5) \times 10^{8} \mathrm{g}^{-1}$ dm for the flaming experiments ($n = 10$).

Table 45.2 Summary of aerosol-particle emission from various *savanna species* (Ivory Coast and South Africa) in different combustion-experiment series (n = number of experiments; CN: condensation nuclei with diam. ≥ 0.01 μm; Coarse Particle: particles with diam. ≥ 0.4 μm; TPM: total particulate matter; P: particle; dm = dry matter; NA: nonavailable)

	Burned dm (g)	$\Delta CO/\Delta CO_2$ (%)	ΔCO_2 (mg/g dm)	CN (P/g dm)	Coarse Particle (P/g dm)	TPM (mg/g dm)
Flaming	48.70 ± 17.02 ($n = 12$)	5.0 ± 1.4 ($n = 12$)	1220 ± 180 ($n = 10$)	$(2.7 \pm 0.8) \times 10^{12}$ ($n = 12$)	$(2.0 \pm 1.1) \times 10^9$ ($n = 4$)	0.28 ± 0.26 ($n = 10$)
Smoldering	17.57 ± 0.93 ($n = 6$)	32 ± 7 ($n = 6$)	850 ± 330 ($n = 4$)	$(3.8 \pm 1.0) \times 10^{12}$ ($n = 6$)	$(6.4 \pm 1.0) \times 10^{11}$ ($n = 4$)	184 ± 61 ($n = 4$)
Mixed	19.85 ± 0.52 ($n = 10$)	NA	1399 ± 144 ($n = 10$)	NA	$(1.1 \pm 0.4) \times 10^{11}$ ($n = 5$)[a]	14.6 ± 5.5 ($n = 10$)

a. Particles with diam. $\geq 0.45 \mu m$.

Table 45.3 Emission of CN (diam. ≥ 0.01 μm) and "coarse" particles (diam. ≥ 0.4 μm) from cereal straw combustion as a function of the ($\Delta CO/\Delta CO_2$) ratio. Δt is age of coarse particles (CP) when their concentrations are measured; the CN measurements are carried out within the first three min. dm = dry matter

Combustion process	Burned dry matter (g)	$\Delta CO/\Delta CO_2$ %	CN (cm^{-3})	CN emission (g^{-1}dm)	Coarse particles (cm^{-3})	Δt (min.)	CP emission (g^{-1}dm)
Flaming-1	22.82	6	6.5×10^5	4.6×10^{12}	1206	58	8.4×10^9
Flaming-2	83.44	3	1×10^6	1.9×10^{12}	1666	9	3.2×10^9
Smoldering-1	14.33	30	3.5×10^5	3.9×10^{12}	62 017	35	6.9×10^{11}
Smoldering-2	19.50	35	4.1×10^5	3.4×10^{12}	81 235	2	6.7×10^{11}

the burning occurred outside the test chamber, the combustion products may not have been entirely introduced into the test chamber and may then have been underevaluated for both flaming and smoldering experiments.

The total particulate matter (TPM) reported in table 45.2 has been corrected by assuming the combustion-particle density to be 1 g cm^{-3}; that is, 2.6 times less heavy than the Arizona Road Dust used by the manufacturer in calibrating the (light-scattering) RAM-1. Such a density value, already proposed in our previous works (Pham-Van et al. 1982), results from considerations about the morphology of smoke particles (mostly as aggregates of small spherules) and about the density of bulk amorphous carbon (~ 2 g cm^{-3}) to adjust the density of combustion particles downward. No correction has been attempted for the refractive index and the size range of the combustion particles, which are certainly more light-absorbing and smaller than the dust particles, whose mean mass diameter is 1.5 μm with a geometric standard deviation of 4.1 μm (Rubow and Marple 1983).

To minimize spurious accounts owing to wall deposition, the TPM accounts apply only to measurements performed just a few minutes after the end of the burning. In such conditions the TPM ranges from 0.28 mg/g dm for the flaming stage to 184 mg/g dm for the smoldering one; that is, ~ 650 times higher for the latter, with an intermediate value for mixed combustion (14.6 mg/g dm). This emission factor is close to those determined by Radke et al. (1988, 1991) by weighing filters from airborne samples above 17 biomass fires, averaging 15.0 (± 10.6) g/kg of fuel burned for particles with aerodynamic diameter < 3.5 μm, but it is noticeably higher than those averaging 7.7 mg/g dm or than those related to the sole flaming samples (5.0 ± 2.0 mg/g dm) reported respectively by Gaudichet et al. (1995) and by Liousse (1993) for prescribed savanna fires (FOS-91 experiment).

The TPM dominated by large particles is closely related in our experiment to the larger particle fraction detected by an optical counter such as the Polytec (particles with diameter > 0.4 μm, and referred to hereafter as "coarse" particles). The coarse fraction is more than two orders of magnitude higher for smoldering than for flaming; mixed combustion provides an intermediate EF (1.1×10^{11} P/g dm), closer to the EF from smoldering than to that from flaming (2.0×10^9 P/g dm), for which any smoke from the first and the last stages of the combustion was allowed to penetrate into the test chamber. Consequently, the EF from that flaming is most likely not to be representative of the savanna flaming in the open field for both TPM and coarse particles (CP), whereas, as shown above, it is quite representative of the CO_2 release from flaming combustion in field ground conditions, because the CO_2 is emitted primarily from the most active combustion phase depicted by the flames.

Figure 45.3 shows a typical example of the number, surface, and mass (density $= 1$ g cm^{-3}) distributions of particle size as a function of the combustion stage processed by the aerodynamic particle sizer (APS), TSI Inc., model 3310, for a same savanna grass species (Loudetia). All size distributions presented are like those displayed by the APS within the aerodynamic size range 0.5 to 30 μm; that is, no adjustment has been introduced to correct the size distributions from spurious counts due to a dual effect from "phantom-particle" creation or from coincidence loss when measuring the "time of flight" of particles between two laser beams in an accelerated air flow (Heitbrink et al. 1991; Karg et al. 1991; Heitbrink and Baron 1992; Horn 1993), both of which certainly are important within the smallest size range (0.5–0.7 μm), where the combustion-particle concentrations are particularly high. Nevertheless, these number counts are very consistent with those processed by the Polytec optical counter in the size range (0.4–15 μm) as reported in table 45.2 for both flaming and smoldering experiments. In fact, all the particle numbers determined from APS measurements (aerodynamic diam. ≥ 0.5 μm) are in good agreement with those from the Polytec (diam. ≥ 0.4 μm).

Contrary to TPM and to CP, the number of smaller particles (in the Aitken particle-size range; that is, with diameter ≥ 0.01 μm, regarded as representative of the total number of aerosol particles), which is continuously detected and counted by a condensation nucleus counter (TSI, model 3022) for concentrations up to 10^7 cm^{-3}, exhibits no really noticeable dependence upon the combustion process. Smoldering combustion yields an EF of $(3.8 \pm 1.0) \times 10^{12}$ P/g dm versus $(2.7 \pm 0.8) \times 10^{12}$ P/g dm for flaming. That slight difference in favor of smoldering may not be significant because the EF dominated by the smallest size-range concentration (diameter $\sim 0.01 \mu$m) is very sensitive to coagulation effect, which is most likely to be very strong within the first instants in the test chamber, when the concentrations range from $\sim 3 \times 10^5$ cm^{-3} to more than 10^6 cm^{-3}, the highest concentrations deal with the flaming experiment when the amount of biofuel burned is 2 to 4 times higher than in the smoldering experiment, so that the related coagulation rate is much higher. Furthermore, the concentration inside the stack by means of which the CN are injected into the test chamber are about 500 times higher; that is,

possibly up to $\sim 10^9$ cm^{-3}, so that during the first seconds that concentration may be reduced by half, as a result of theorical considerations solely on the coagulation process (Twomey 1977; Reist 1984). In fact, the loss rate is more drastic due to a combination of processes, as the experimenter can see when following the temporal evolution of the CN concentration inside the test chamber. Therefore, to limit particle loss due to any cause, the EF (CN) displayed in table 45.2 results from measurements undertaken within a few minutes (namely ~ 3 min) after the end of the combustion, when homogeneous distribution of particles throughout the chamber is achieved.

Particulate Emission from Cereal Straw

Every year, throughout Spain and especially in the Basque Country area, waste-cereal fires are authorized and practiced during a short period in autumn (namely, one month); the emission of combustion products can thus be high. Therefore some (limited) experiments on cereal straw from the Spanish Basque Country were carried out in our facilities, in association with a field experiment in Spain (Ezcurra et al. chapter 73 this volume).

Table 45.3 reports, with more measurement details than for savanna species, particulate emission from a few flaming and smoldering experiments, displaying the coagulation effects discussed above: when the particles are less subject to coagulation; that is, when the amount of burned fuel is smaller, the particulate-emission factor appears to be higher, and this excess applies to any combustion process or any particle-size range, whether CN or CP. Nevertheless, for CP, the coagulation seems less effective in the test chamber, first because the number concentration is comparatively lower than that of CN, and second, the particle number loss by coagulation can be replaced by the coagulated particles from a smaller size range that has reached a size detectable by the Polytec, so that the EF deduced from CP concentrations measurements does not seem affected by the age (Δt) of the aerosols.

The emission factors from cereal straw display the same order of magnitude as those from savanna grass for both CN and CP size ranges; perhaps they are somewhat higher, particularly from flaming combustion, but these limited experiments do not allow us to draw further information.

Emission of CCN from Savanna Species

CCN measurements are performed with the isothermal haze chamber (IHC), as described on page 474 about the Polytec HC-15; the latter counts and sizes droplets through 64 channels from ~ 0.8 to ~ 38 μm (this size range may vary slightly, depending upon the high voltage applied to the detector (photomultiplier), so that after equation (45.1), the highest SS_c involved in the IHC does not exceed 0.1% (smallest droplet size detectable: 0.8 μm) and the lowest SS_c is around 0.01% (very few droplets larger than 10 μm are detected) (figure 45.4).

Table 45.4 summarizes the CCN emission from tropical savanna species (Loudetia and Hyparrhenia) as a function of combustion process. The averaged EF ranges from 0.6×10^{11} g^{-1} dm for flaming to 4.2×10^{11} g^{-1} dm at 0.1% SS_c for smoldering, with a middle value (1.6×10^{11} g^{-1} dm) for mixed combustion. Each EF accounted for corresponds to the highest concentration encountered by the Polytec during each run; the concentration reaches its maximum level after a delay depending on the droplet growth, on the combustion process, and on the initial size of particles emitted. For flaming, this lag time, depicted by the age of the aerosol, is noticeably longer than for smoldering, as a result of a smaller size range, and also it is most likely, as a result of the difference in nature and amounts of the various products emitted with different ratios at different stages in combustion.

The lesser EF of the CCN for flaming can also be expressed by a lower ratio of CCN concentration to that of the CN measured at the same aging time (CCN/CN), or by the ratio of CCN to coarse particles (CCN/CP) dramatically high (>30), which shows that most of the flaming particles are too small to be activated at 0.1% SS_c. By contrast, the (CCN/CP) ratio nears unity for smoldering; that is, most combustion particles with diameter larger than 0.4 μm may act as CCN.

Emission of CCN from Cereal Straw
For cereal-straw combustion, additional details are presented in table 45.5, but for a limited series of experiments dealing with CCN emission.

Cereal-straw combustion provides approximately the same EF from smoldering ($\sim 3.6 \times 10^{11}$ g^{-1} dm) as for savanna grass; by contrast, the EF from flaming is much lower ($\sim 0.18 \times 10^{11}$ g^{-1} dm), resulting in a smoldering-to-flaming ratio ranging from 20 for the cereal straw to ~ 7 for the savanna grass. This difference can also be seen in the very low (CCN/CP) ratio (~ 4), compared to the same ratio from savanna species (~ 34). This feature requires confirmation by further investigation to allow well-grounded discussions.

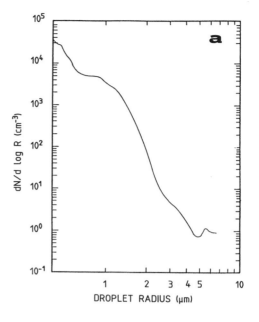

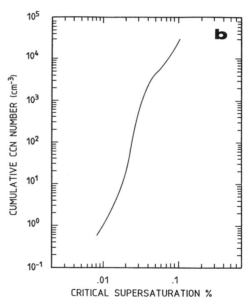

Figure 45.4 An example of number-size distribution of droplets that have grown on the combustion particles to their equilibrium radius at (*a*) 100% RH, and (*b*) the related critical supersaturation spectrum

Table 45.4 Summary of CCN emission from tropical *African savanna* as a function of combustion process as expressed by the ($\Delta CO/\Delta CO_2$) ratio; Age is the age of the aerosols when the CCN and CN concentrations are measured; CP denotes "coarse" particles with diam. $>0.4 \mu m$

Combustion process	$\Delta CO/\Delta CO_2$	Age	CCN/CN	CCN/CP	CCN emission
	%	(min)	(%)		(g^{-1}dm)
Flaming ($n = 4$)	4.0 ± 0.7	63 ± 48	19 ± 11	34 ± 18	$(0.6 \pm 0.4) \times 10^{11}$
Smoldering ($n = 4$)	33 ± 8	27 ± 9	26 ± 11	0.61 ± 0.25	$(4.2 \pm 1.5) \times 10^{11}$
Mixed ($n = 4$)	NA	57 ± 20	NA	1.5 ± 0.49	$(1.6 \pm 0.8) \times 10^{11}$

Table 45.5 CCN emission at 0.1% critical supersaturation from *cereal straw* as a function of the combustion process. The experimental conditions are indicated: Δt is the combustion duration; the ($\Delta CO/\Delta CO_2$) ratio, as an index of the combustion process, is determined in the test chamber after the combustion; Age is the age of the aerosols when the CCN and CN concentrations are measured at the ambient relative humidity (RH); CP denotes the "coarse particles with diam. $\geq 0.4 \mu m$.

Combustion process (Δt)	Burned dry matter (g)	$\Delta CO/\Delta CO_2$ (%)	Amb. RH (%)	CCN Conc. (cm^{-3}) air	Age (min)	CCN/CN (%)	CCN/CP	CCN emission (g^{-1} dm)
Flaming-1 (25 s)	22.82	6	67	2402	27	2	2.0	1.68×10^{10}
Flaming-2 (28 s)	83.44	3	64	9137	239	22	5.5	1.75×10^{10}
Smoldering-1 (49 s)	14.33	30	67	31 059	12	14	0.50	3.47×10^{11}
Smoldering-2 (60 s)	19.50	35	55	41 097	134	72	0.55	3.65×10^{11}

Hygroscopicity of Combustion Particles

In the preceding section we dealt with particles grown in an environment of 100% RH. These particles can also uptake vapor water at RH well below water saturation, as shown in the following experiment.

On some experimental days the atmospheric RH was often high (60–80%) so that the test chamber could easily be humidified up to more than 90%. Figure 45.5 illustrates a typical behavior of combustion aerosol particles larger than 0.5 μm in diameter (displayed by the Kratel OPC) in an atmosphere becoming more and more humid. The RH starts at 60% and stays roughly constant for 1 h, all cumulative numbers (N_p) of particles with diameter ≥ 0.5, 1.8, 2.5, or $\geq 3.5 \mu m$ decrease sharply, with a steeper slope for larger particles, which is very likely to be due to more rapid deposition. As RH increases to $\sim 65\%$, N_p continues to decrease. When RH exceeds $\sim 65\%$ the N_p of smaller size range begins to augment, due to the growth of particles $<0.5 \mu m$ in diameter to detectable size ($\geq 0.5 \mu m$); meanwhile, the decrease goes on for larger particles for ~ 30 min., because the number of particles growing to such a size range is still too low to make up the loss by deposition. When RH reaches $\sim 80\%$ (within ~ 1 h), the N_p of larger-size particles also begins to increase, up to a maximum ~ 90 min. later. And then the different maxima for the four size channels stay stationary, even with a quite high RH, for

several hours. This condition may be due to equilibrium between wall deposition and formation of new large particles; nevertheless, a saturation effect (coincidence count) cannot be disregarded at such particle concentrations for this device. Such a possible artifact will not harm the evidence of hygroscopicity in savanna-combustion particles for RH well below water saturation. The critical RH at which the more hygroscopic particles deliquesce seems to be around 65% and 80%.

This hygroscopicity may be explained, at least partly, by the chemical composition of savanna-combustion particles as sampled by means of a round mono-jet 5-stage impactor (plus a backup filter) and analyzed by ion chromatograph for the water-soluble inorganic components. The ionic species that may be dissociated from deliquescent salts were found at quite high concentrations, for example, several hundreds of $\mu g/g$ dm, and mostly distributed into smaller size-range stages; that is, around 0.5 μm or below (Yoboué 1991; Pham-Van 1994). In fact, KCl has been found as a dominant species in smoke plumes over savanna fire in Ivory Coast and associated with other salts such as $CaSO_4$ or K_2SO_4, present in a lesser proportion (Gaudichet et al. 1995). Furthermore, the organic species that may enhance the hygroscopicity of particles such as light carboxylic acids are found in great amounts in savanna-fire plumes, averaging 0.26 (± 0.11) mg/g dm, 0.90

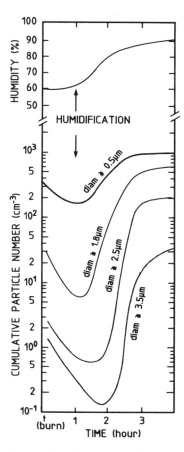

Figure 45.5 An example of hygroscopic growth of combustion particles as expressed by increase in particle numbers of different sizes when RH is increased from ~60% to ~90%

(± 0.39) mg/g dm. and 0.10 (± 0.04) mg/g dm respectively for formic, acetic, and propionic acids in the gas phase and in much smaller amounts in the particulate phase (Pham-Van et al. 1991). Our chamber combustion experiments show that the organic acids are released predominantly from the smoldering stage, with a smoldering-to-flaming emission ratio higher than ~50 (Pham-Van et al. 1993).

Table 45.6 sets out main possible CCN from "pure" inorganic species with the related relative humidity at which water vapor begins to condense at room temperature onto the particle (known as critical RH or RH_c) compiled for $RH_c \geq 50\%$ from Low (1969), Twomey (1977), James and Lord (1992), and from Tang and Munkelwitz (1994). Each salt can be characterized by more than one RH_c, due to these various sources. For mixed particles very likely to prevail in combustion smoke, the deliquescence of such a mixture of soluble components has experimentally been

shown to occur at a RH_c lower than the one corresponding to any component (Tang and Munkelwitz 1993, 1994; Koloutsou-Vakakis and Rood 1994), so that the particles released from biofuel, very likely to be mixed particles (Liousse 1993), may condense water vapor at lower RH_c. Therefore, the shifting up of N_p of different size classes, as observed in figure 45.5 at ~65% and ~80%, may be due, for example, to the deliquescence of mixtures based, respectively, on NH_4NO_3 and $(NH_4)_2SO_4$ or KCl.

Summary and Conclusions

As a response to renewed interest in biomass-combustion particles relevant to both direct and indirect radiative climate forcings, several series of laboratory investigations on particle emissions from cereal straw and savanna grass are carried out in connection with field experiments, to determine the emission factors of the total particle number, of the coarse particle fraction, and of the particle fraction able to act as CCN at low water supersaturation.

Our experimental arrangement has allowed us to classify the combustion experiments into three combustion processes: "flaming," "smoldering," and "mixed" combustion with specific characteristics; it has also permitted us to demonstrate evidence of the hygroscopicity of combustion particles.

Although our flaming experiments on savanna species characterized by ($\Delta CO/\Delta CO_2$) ~ 5% is quite close to that from field experiment over prescribed savanna fires (~6%) in Ivory Coast (Lacaux et al. 1993), the TPM, as weighted by the fraction of coarse particles, is too low (~0.3 mg/g dm) to be representative of the actual emission in free conditions. Nevertheless the total number of particles as dominated by the smaller size-range fraction (CN) does not depend noticeably upon the ($\Delta CO/\Delta CO_2$) ratio as an index of combustion process, so that the flaming EF averages ~3×10^{12} CN/g dm versus ~4×10^{12} CN/g dm for smoldering, displaying a likely undervaluation, as previously discussed.

The TPM emission from smoldering combustion, characterized by a quite high ($\Delta CO/\Delta CO_2$) ratio (~30%) may be regarded as a possible superior limit for such a process (~180 mg/g dm).

Mixed combustion yields an intermediate EF for TPM (~15 mg/g dm), not very different from the EF determined over savanna fires in field conditions (~8 mg/g dm) by Gaudichet et al. (1995). Subsequently, our mixed combustion may be regarded as close to natural savanna fires in particulate emission. Further-

Table 45.6 Main possible CCN species from soluble components as revealed by ion chromatography and their related critical relative humidity ($RH_c \geq 50\%$) for hygroscopic growth to droplets at room temperature (compiled from various authors; see text)

CCN	$Ca(NO_3)_2$	$NaHSO_4$	$Mg(NO_3)_2$	NH_4NO_3	$NaNO_2$	$NaNO_3$	$NaCl$	NH_4Cl
RH_c (%)	50	55	55–56	62–65	66	74–76	76	77–80
CCN	$(NH_4)_2SO_4$	Na_2SO_4	Na_2SO_3	KCl	$KHSO_4$	$MgSO_4$	KNO_3	K_2SO_4
RH_c (%)	80–81	84–86	85	85–86	87	91	92–93	97

more, this assertion is supported by the ratio of carbon black to total carbon (Cb/Ct) determined by Cachier et al. (1995) in the same chamber experimental conditions ($\sim 9\%$), which is closer to that of the flaming samples ($\sim 12\%$) than that of the smoldering ones ($\sim 5\%$), both taken from savanna fires in FOS-91 field experiments.

The cereal straw emits approximately the same number of CN and the same number of coarse particles per gram of dry matter as the savanna grass. In those conditions, one can expect a similar EF, for the TMP, compared to that of savanna species.

A nonnegligible part of the particles emitted are hygroscopic even at a RH far below water saturation. The total number of particles activated at low SS_c (0.1%) is quite high; an EF of 2×10^{11} CCN/g dm can be regarded as representative of savanna or straw combustion, at least within the first hour after emission. That EF falls between the one earlier determined—but not well known—by Twomey (1960) on burning wood (3×10^{10} g^{-1} at 0.5% SS) and the well-known 5×10^{12} g^{-1} (at 0.5% SS) established by Warner and Twomey (1967) on burning sugarcane leaf and widely used by various authors to estimate the flux of CCN from global biomass burning (e.g., Radke et al. 1991).

One should keep in mind that the number of CCN greatly depends on supersaturation; as shown in figure 45.3, it increases by a factor of more than 10^4 when the SS_c augments only from 0.01% to 0.1%. One should also keep in mind when modeling that water vapor condenses primarily on larger particles, so that the number of "dry" particles of size relevant to the radiative forcing decreases as the CCN number increases, and this decay is very likely to be more rapid than the growth of the CCN number. A similar effect can be seen in the (CCN/CN) ratio, which increases with the age of aerosols, as a result of the (rapid) decay of CN concentration by coagulation and other processes. The mechanism of water activation of combustion particles seems to result from complex mixing of hygroscopic matter from inorganic as well as organic species into and/or onto the particles of larger size range (internal or external mixing).

Therefore the hygroscopicity and the activation spectrum of combustion particles at very low supersaturation (<0.1%) requires more investigation because of their dramatic effect on the albedo of remote maritime areas where background CCN concentration does not exceed 100 cm^{-3}.

Such effects have recently been a subject of coverage by the international press from Southeast Asia (Jayasankaran and McBeth 1994; Pomonti 1995), where in September and October 1994, Malaysia and Singapore were dramatically plunged for many weeks into a heavy "smog" induced by vegetation fires in South Sumatra and southeastern Borneo (Indonesia); that is, at a distance of hundreds of km or much more. The smoke particles, swept away by southeast winds over the South China Sea, reached to Malaysia after uptaking water vapor and growing to a persistent haze that was perceived popularly as a "smog" by association with the well-known smog in London or Los Angeles caused by industrial sources and having similar harmful effects. This event appears to us to spectacularly demonstrate the hygroscopicity of biomass-combustion particles and their immediate effects at a regional scale.

Acknowledgments

We gratefully acknowledge Pierrette Boudinhon's CCN data processing, Anne Postal, Delphine Brocard, Eric Série, Jean-Pierre Cazes, Lucien Lahaille, and Marcel Abadie for their help with various phases of the experimentation, Régine Réchal and Eric Pique for preparing the manuscript. This work was part of the DECAFE Program, funded by CNRS, Ministère de l'Environnement and Ministère de la Coopération (programme Campus).

References

Alofs, D. J., and J. Podzimek. 1974. A review of Laktionov's isothermal cloud nucleus counter. *J. Appl. Meteor.*, 15, 511–512.

Andreae, M. O. 1991. Biomass burning: Its history, use and distribution and its impact on environmental quality and global climate. In J. S. Levine, ed., *Global biomass burning: Atmospheric, climatic and biospheric implications*. MIT Press, Cambridge, Mass., 1–21.

Bonsang, B., C. Boissard, M. F. Le Cloarec, J. Rudolph, and J. P. Lacaux. 1995. Methane, carbon monoxide and light non-methane hydrocarbon emissions from African savanna burnings during the FOS/DECAFE experiment. *J. Atmos. Chem.*, 22, 149–162.

Cachier, H., C. Liousse, P. Buat-Ménard, and A. Gaudichet. 1995. Particulate content of savanna fire emissions. *J. Atmos. Chem.*, 22, 123–148.

Charlson, R. J., J. E. Lovelock, M. O. Andreae, C. B. Loevy, and S. G. Warren. 1991. Perturbation of the Northern Hemisphere radiative balance by backscattering from anthropogenic sulfate aerosols. *Tellus*, 43B, 152–163.

Charlson, R. J., S. E. Schwartz, J. M. Hales, R. D. Cess, J. A. Coakley, J. E. Hansen, and D. J. Hofmann. 1992. Climate forcing by anthropogenic aerosols. *Science*, 255, 423–430.

Coakley J. A., Jr., R. L. Bernstein, and D. A. Durkee. 1987. Effect of ship-stack effluents on cloud reflectivity. *Science*, 237, 1020–1022.

Crutzen, P. J., and M. O. Andreae. 1990. Biomass burning in the tropics: Impacts on atmospheric chemistry and biogeochemical cycles. *Science*, 250, 1669–1678.

Delmas, R., A. Marenco, J. P. Tathy, B. Cros, and J. G. R. Baudet. 1991. Sources and sinks of methane in African savanna: CH_4 emission from biomass burning. *J. Geophys. Res.*, 96, D4, 7287–7299.

Delmas, R., J. P. Lacaux, J. C. Menaut, L. Abadie, X. Leroux, G. Helas, and J. Lobert. 1995. Nitrogen compound emission from biomass burning in tropical African savanna, FOS/DECAFE 91 experiment (Lamto, Ivory Coast). *J. Atmos. Chem.*, 22, 175–193.

Fitzgerald, J. W., C. F. Rogers, and J. C. Hudson. 1981. Review of isothermal haze chamber performance. *J. Rech. Atmos.*, 15, 333–346.

Garmy, M., and R. Serpolay. 1986. An isothermal tube for getting low supersaturation spectra of CCN by coupling with a Polytec-HC15 spectrogranulometer. *J. Aerosol. Sci.*, 17, 401–405.

Gaudichet, A., F. Echalar, B. Chatelet, J. P. Quisefit, and G. Malingre. 1995. Trace elements in tropical African savanna biomass burning aerosols. *J. Atmos. Chem.*, 22, 19–39.

Heitbrink, W. A., P. A. Baron, and K. Willeke. 1991. Coincidence in time-of-flight spectrometers: Phantom particle creation. *Aerosol Sci. Technol.*, 14, 112–126.

Heitbrink, W. A., and P. A. Baron. 1992. An approach to evaluating and correcting aerodynamic particle sizer measurements for phantom particle count creation. *Am. Ind. Hyg. Assoc. J.*, 53, 427–431.

Hindman, E. E., W. M. Porch, J. G. Hudson, and P. A. Durkee. 1994. Ship-produced cloud lines of 13 July 1991. *Atmos. Environ.*, 28, 3393–3403.

Horn, H.-G. 1993. A new technique for automatic phantom and coincidence count reduction in the aerodynamic particle sizer APS 3310. *Report Series in Aerosol Sci.*, 23, Finnish Assoc. Aerosol Res., 185–190.

Hudson, J. G., C. F. Rogers, and G. Keyser. 1981. Simultaneous operations of three CCN counters and an isothermal haze chamber at the 1980 international CCN workshop. *J. Rech. Atmos.*, 15, 271–283.

Hudson, J. G. 1993. Cloud condensation nuclei. *J. Appl. Meteor.*, 32, 596–607.

Hurst, D. R., D. W. T. Griffith, J. N. Carras, D. J. Williams, and P. J. Fraser. 1994. Measurements of trace gases emitted by Australian savanna fires during the 1990 dry season. *J. Atmos. Chem.*, 18, 33–56.

James, A. M., and M. P. Lord. 1992. *Macmillan's Chemical and Physical Data*. Macmillan Press, London, p. 16, 565 pp.

Jayasankaran, S., and J. McBeth. 1994. Hazy days: Forest fires in Indonesia irritate its neighbours. *Far Eastern Economic Review* (Hong Kong), 157, 42 (20 Oct. 1994), 66–67.

Karg, E., S. K. Dua, and J. Tschiersch. 1991. Experimental counting efficiency of the TSI Aerodynamic Particle Sizer in the submicron size range. *J. Aerosol Sci.*, 22, Suppl. 1, S351–S354.

Kellog, W. W. 1992. Aerosols and global warming, *Science*, 256, 598.

Kiehl, J. T., and J. T. Briegleb. 1993. The relative roles of sulfate aerosols and greenhouse gases in climate forcing. *Science*, 260, 311–314.

Kilsby, C. G. 1990. A study of aerosol properties and solar radiation during a straw-burning episode using aircraft and satellite measurement. *Q. J. R. Meteor. Soc.*, 116, 1173–1192.

Koloutsou-Vakakis, S., and M. J. Rood. 1994. The $(NH_4)_2SO_4$–Na_2SO_4–H_2O system: Comparison of deliquescence humidities measured in the field and estimated from laboratory measurements and thermodynamic modeling. *Tellus*, 46B, 1–15.

Lacaux, J. P., H. Cachier, and R. Delmas. 1993. Biomass burning in Africa: An overview of the impact on the atmospheric chemistry. In P. J. Crutzen and J. G. Goldammer, Editors, *Fire in the Environment: The Ecological, Atmospheric and Climatic Importance of Vegetation Fires*. Wiley, Chichester, U.K., 159–191.

Lacaux, J. P., J. M. Brustet, R. Delmas, J. C. Menaut, L. Abadie, B. Bonsang, C. Cachier, J. Baudet, M. O. Andreae, and G. Helas. 1995. Biomass burning in the tropical savannas of Ivory Coast: An overview of the field experiment Fire of Savannas (FOS/DECAFE 91). *J. Atmos. Chem.*, 22, 195–216.

Laktionov, A. G. 1972. A constant temperature method of determining the concentrations of cloud condensation nuclei. *Atmos. Ocean. Phys.*, 8, 382–385.

Laktionov, A. G. 1973. Spectra of cloud condensation nuclei in the supersaturation range 0.02–1%. VIIIth Int. Conf. on Nucleation. Leningrad, 24–29 Sept. 1973, Proc. vol., 437–444, Gidrometeoizdat, Moscow, 1975.

Levine, J. S. 1990. Global biomass burning: Atmospheric, climatic and biospheric implications. *EOS Trans. AGU*, 71, 37, 1075–1077.

Levine, J. S., W. R. Cofer III, D. R. Cahoon, and E. L. Winstead. 1995. Biomass burning: A driver for global change. *Environ. Sci. Technol.*, 29, 3, 120A–125A.

Liousse, C. 1993. Emissions carbonées particulaires des feux de savane d'Afrique: Mesures au sol et télédétection spatiale des panaches. Ph. D. diss., Univ. Paris VII, 273 pp.

Low, R. D. H. 1969. A theoretical study of nineteen condensation nuclei. *J. Rech. Atmos.*, 4, 65–78.

Penner, J. E., R. E. Dickinson, and C. A. O'Neill. 1992. Effects of aerosols from biomass burning on the global radiation budget. *Science*, 256, 1432–1434.

Penner, J. E., R. J. Charlson, J. M. Hales, N. S. Laulainen, R. Leifer, T. Novakov, J. Ogren, L. F. Radke, S. E. Schwartz, and L. Travis. 1994. Quantifying and minimizing uncertainty of climate forcing by anthropogenic aerosols. *Bull. Am. Meteor. Soc.*, 75, 375–400.

Pham-Van, D. 1969. Contribution à la technologie des modèles de dilution des noyaux d'iodure d'argent. *J. Rech. Atmos.*, 4, 187–193.

Pham-Van, D., B. Bénech, and W. Diamant. 1982. Evaluation des propriétés énergétiques et microphysiques d'une source de convection artificielle à partir d'une étude de combustion organisée de fuel-oil en milieu naturel. *Atmos. Environ.*, 16, 1219–1230.

Pham-Van, D., V. Yoboué, J. P. Lacaux, L. Schäfer, and G. Helas. 1991. Organic acids from savannah fire experiments in Ivory Coast. Presented at the Fall Meeting AGU, San Francisco. 9–13 Dec.

Pham-Van, D., V. Yoboué, and G. Helas. 1993. Organic acids from biomass combustion experiments. Presented at the XVIIIth EGS Meeting, Wiesbaden, Germany, 3–7 May 1993.

Pham-Van, D., J. P. Lacaux, and R. Serpolay. 1994. Cloud-active particles from African savanna combustion experiments. *Atmos. Res.*, 31, 41–58.

Pomonti, J. C. 1995. L'Asie défigurée. *Le Monde* (Paris), no. 15 673 (17 June 1995), 14.

Radke, L. F., and P. V. Hobbs. 1976. Cloud condensation nuclei on the Atlantic seaboard of the United States. *Science*, 193, 999–1002.

Radke, L. F., D. A. Hegg, J. H. Lyons, C. A. Brock, and P. V. Hobbs. 1988. Airborne measurements on smokes from biomass burning. In P. V. Hobbs and M. P. McCormick, Editors, *Aerosols and Climate*. A. Deepak Publishing, Hampton, Va., 411–422.

Radke, L. F., J. A. Coakley Jr., and M. D. King. 1989. Direct and remote sensing observations of the effects of ships on clouds. *Science*, 246, 1146–1149.

Radke, L. F., D. A. Hegg, P. V. Hobbs, J. D. Nance, J. H. Lyons, K. K. Laursen, R. E. Weiss, P. J. Riggan, and D. E. Ward. 1991. Particulate and trace gas emissions from large biomass fires in North America. In J. S. Levine, editor, *Global Biomass Burning: Atmospheric, Climatic and Biospheric Implications*. MIT Press, Cambridge, Mass., 209–224.

Reist, P. C. 1984. *Introduction to Aerosol Science*. Chap. 16. Macmillan, New York.

Rogers, C. F., J. G. Hudson, B. Zielinska, R. L. Tanner, J. Hallett, and J. G. Watson. 1991. Cloud condensation nuclei from biomass burning. In J. S. Levine, editor, *Global Biomass Burning: Atmospheric, Climatic and Biospheric Implications*. MIT Press, Cambridge, Mass., 431–438.

Rubow, K. L., and V. A. Marple. 1983. Instrument evaluation chamber: Calibration of commercial photometers. In V. A. Marple and B. Y. H. Liu, editors, *Aerosols in the Mining and Industrial Work Environments*, vol. 3, Instrumentation. Ann. Arbor Sci., 774–795.

Serpolay, R. 1994. Efficacy test of a haze chamber with the CCN produced by the burning of savannah grasses. *Atmos. Res.*, 31, 99–107.

Tang, I. N., and H. R. Munkelwitz. 1993. Composition and temperature dependence of the deliquescence properties of hygroscopic aerosols. *Atmos. Environ.*, 27A, 467–483.

Tang, I. N., and H. R. Munkelwitz. 1994. Aerosol phase transformation and growth in the atmosphere. *J. Appl. Meteor.*, 33, 791–796.

Twomey, S. 1960. On the nature and origin of natural cloud nuclei. *Bull. Obs. Puy-de-Dôme*, 1, 1–19.

Twomey, S. 1977. *Atmospheric Aerosols*. Elsevier, Amsterdam.

Ward, D. E., and C. C. Hardy. 1991. Smoke emissions from wildland fires. *Environ. International*, 17, 117–134.

Warner, J., and S. Twomey. 1967. The production of cloud nuclei by cane fires and the effect on cloud droplet concentration. *J. Atmos. Sci.*, 24, 704–706.

Yoboué, V. 1991. Caractéristiques physiques et chimiques des aérosols et des pluies collectés dans la savane humide de Côte d'Ivoire. Ph. D. diss., no. 914, Univ. Paul Sabatier, Toulouse, France, 146 pp. + annexes.

Biomass Burning Effects on Satellite Estimates of Shortwave Surface Radiation in Africa

Charles H. Whitlock, Donald R. Cahoon, and Thomas Konzelmann

Radiative fluxes reaching the Earth's surface are principal elements in the energy exchange between the atmosphere and surface. These fluxes also influence biomass productivity and the use of solar energy in industrial applications. The most important component of the radiation balance at the surface is the downward shortwave irradiance.

The World Climate Research Program (WCRP) Surface Radiation Budget (SRB) climatology project was initiated as a result of findings of the Workshop on Surface Radiation Budget for Climate Applications held at Columbia, Maryland, 18–21 June, 1985, under sponsorship of NASA, the WCRP, and the International Association of Meteorology and Atmospheric Physics (Suttles and Ohring 1986). Scientific justification is based on uses of SRB data to improve understanding of the four major climate system components; the oceans, the land surface, the atmosphere, and the cryosphere. Validation of both radiative parameterizations and cloud generation algorithms for Global Climate Models (GCMs) is also a key use of long-term SRB data sets, as are applications of the data to the photo-voltaic industry. Easily derived from the satellite-estimated SRB parameters are certain time-integrated maximum, minimum, and storage statistics that are required for the design of solar-cell augmented electrical equipment.

The objective of the WCRP/SRB climatology project is to produce and archive a satellite-derived global data set of shortwave (SW) and longwave (LW) surface parameters for the 12-year period July 1983 through June 1995 with a monthly uncertainty of less than 10 W m^{-2} or 5%, whichever is smaller. This level of precision is not required for many applications, however. Data with uncertainties less than 20 W m^{-2} (or 10%) are urgently needed for Global Energy and Water Cycle EXperiment (GEWEX) projects. After a long period of development and testing, the WCRP authorized processing of the first WCRP/SRB SW data set (labeled Version 1.1) for a 46-month period from March 1985 through December 1988. A user's manual (Whitlock et al. 1993) is available and limita-

tions are described in Whitlock et al. (1995). The Version 1.1 data set is available from the Earth Observation System Distributed Information System (EOSDIS) Distributed Active Archive Center (DAAC) located at the NASA Langley Research Center (telephone (804) 864-8656 or email userserv@eosdis.larc.nasa.gov) via electronic transfer methods and CD-ROM for UNIX, Macintosh, and IBM-compatible computers.

As described in Whitlock et al. (1995), the satellite-derived SRB data generally meet WCRP accuracy requirements in most mid-latitude regions. There is an issue with results over ice and snow covered surfaces because the satellite data used for input in the Version 1.1 SRB algorithms cannot distinguish between clouds and some ice/snow surfaces (Rossow et al. 1991). There are also large errors in the tropics that were initially blamed on ground truth uncertainties. The satellite-derived Version 1.1 SRB algorithms make generalized assumptions on aerosol climatology. Anthropogenic-induced changes of the atmospheric conditions may not be taken into account in the calculations. If these atmospheric disturbances are widespread, the satellite-derived downward surface irradiance is affected on a regional scale.

In equatorial regions, human-induced fires are common practice during the dry season. Extended areas are burned every year in South America (Setzer and Pereira 1991) and Africa (Cahoon et al. 1994). In South America, fire is a tool for clearing savanna and forest, and the clearing of savanna prevails in Africa. Similar fires occur in northern Australia (Penner et al. 1991). The authors also have Defense Meteorological Satellite Program (DMSP) data showing massive biomass burning in eastern India, Burma, and Indonesia. Due to variations in the fuels and weather, the combustion efficiency of the fires can vary (Ward and Hardy 1994). This leads to differences in the gaseous composition, the amount of particles, and the particle size distributions in the smoke released by the fires. The emissions cause severe atmospheric pollution and significantly increase the amount of aerosols in the atmosphere (Crutzen and Andreae 1990; Talbot et al.

1990). The prevailing wind systems disperse the smoke and haze clouds over large geographical areas. The radiative impact of such extended smoke layers depends strongly on the smoke optical depth and the single scattering albedo (Lenoble 1991).

It is the purpose of this article to synthesize differences between the WCRP satellite-derived SRB values and ground truth data during biomass burning events in central Africa. The analysis will be conducted for the 1986 and 1987 period when both satellite and ground truth measurements are available. It is believed that large errors in the satellite-derived values can be explained by anthropogenic effects that were not taken into account in the Version 1.1 SRB product.

Analysis for Central Africa

The Version 1.1 satellite-estimated SRB values were produced using two different methods (Whitlock et al. 1993). Figure 46.1 shows typical values for downward SW irradiance over the globe for both a winter and a summer month in 1987 for the Pinker SRB algorithm. The Staylor method produces similar results for most regions of the globe. From these charts, it may be observed that values typically range between 200 and 300 W m^{-2} in the tropics. Figure 46.2 shows comparisons of Pinker satellite results with ground truth from the Global Energy Balance Archive (GEBA) (Ohmura and Gilgen 1993). Only a subset of ground sites (away from mountains and coastlines) was used so as to remain consistent with limitations of the satellite data spatial resolution (280 × 280 km). Bias errors are low, but rms values (including bias) are high. It can be seen that in 1987 the satellite results are 30 to 80 W m^{-2} too large for a number of sites where the ground truth is in the 200 to 300 W m^{-2} range. Many of these sites are in the tropics. The Staylor algorithm has errors similar to those of the Pinker method in this region, suggesting that neither satellite method accounts for the effects of intensive biomass burning.

Penner et al. (1991) suggest that 58% of the global total of soot carbon emissions (5.7 × 10^{12} g/yr) may originate from central Africa. Figure 46.3 shows seasonal changes in the differences between satellite-derived and ground-site values for four sites in central Africa. Both the ground-site and satellite-derived SRB data are for 1986 when all four sites were available. Locations of the fires are taken from 1987 DMSP satellite observations. Cahoon et al. (1992) note that the spatial distribution of fires in this region was similar for both years. Figure 46.4 shows 1987 fire locations and wind patterns in the region of both the fires and

ground sites. For the fire regions, spacing between lines of the diagonal patterns is qualitatively similar to the spatial density of the fires observed by DMSP. Fire spatial density was high in January, very low in April, and medium in both July and October. Average winds (Leroux 1983) are shown for an altitude of 2 km based on haze altitude observations of Andreae et al. (1988) and Talbot et al. (1988) for a geographically similar region in South America. The synthesis of wind patterns and fire spatial density are in agreement with the magnitude of satellite-algorithm error shown in figure 46.3. In figure 46.4, smoke from northern high-density fires blows south over the ground sites in January. Low-density fires exist near the sites in April, but the smoke is blown to the west away from the sites. In July, a large area of medium density fires exists to the south of the sites. Winds carry the smoke northward directly over the sites. The largest area of fires (medium spatial density) is southeast of the sites in October. That smoke is mostly carried south of the sites. There is also a small region of very low-density fires northwest of site B, but smoke from those fires does not influence sites C and D in October 1987. Figure 46.3 suggests that there was little influence of smoke for any of the sites in October 1986.

A theoretical sensitivity study was carried out to demonstrate the effect of aerosol optical depth on downward SW irradiance in the tropics. The Streamer radiative transfer model (Key 1994) was used assuming clear skies and a surface albedo of 0.13 (Oguntoyinbo 1970). A tropical standard atmosphere was chosen, and aerosol optical depth was varied between 0.3 (undisturbed conditions) and 1.0 (Amazonas measurements) (Kaufman et al. 1992). Changing the optical depth resulted in a decrease of downward SW irradiance of 45 W m^{-2} at the surface. The introduction of additional soot would cause an additional decrease. This finding suggests that a significant portion of the satellite-product error in central Africa can be attributed to aerosol changes as a result of combustion products from extensive biomass burning.

Effects for Other Regions

Regions that are believed to experience significant biomass burning are shown in figure 46.5. The months shown represent the seasons when most fires occur. For the tropics, the seasons correlate to periods when monthly precipitation is 25 mm or less. For the boreal forests of North America and Russia, the seasons correspond to the historic period of intense electrical storms. Detailed correlations of Version 1.1 satellite-

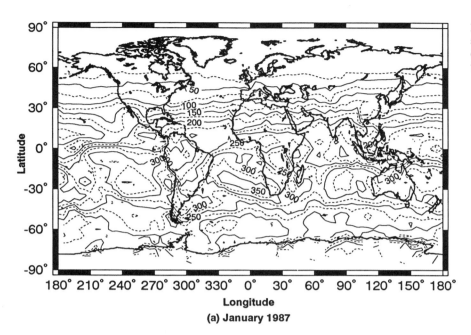

(a) January 1987

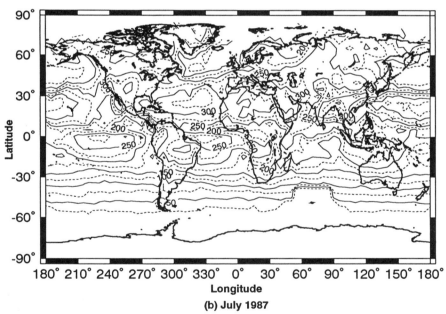

(b) July 1987

Figure 46.1 Global distribution of surface downward shortwave flux (W m^{-2}) for Pinker algorithm from the Version 1.1 WCRP/SRB product

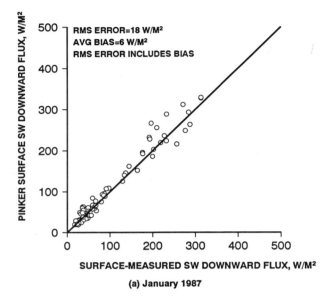

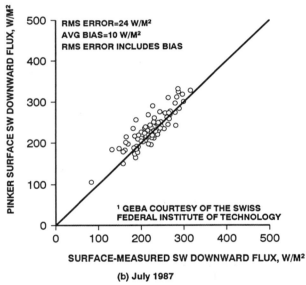

Figure 46.2 Comparison of Pinker algorithm with GEBA subset surface measurements from the Version 1.1 WCRP/SRB product. (The subset consists of surface sites away from mountains and coastlines.)

derived SRB errors relative to satellite-observed fire spatial densities are not available at this time for any region except Africa. However, an analysis of satellite SRB error relative to ground-site measurements is possible for some of the regions.

May through September is the dry season for the Boreal Forest regions of the northern United States, southern Canada, and Russia. These fires are usually created by lightning and are very intense (W. R. Cofer III et al., in chapter 79 this volume). Unlike low-altitude smoke from tropical fires, Boreal Forest smoke plumes rise rapidly to the upper troposphere or through the tropopause to spread widely and influence stratospheric chemistry. Usually, fire locations are widely dispersed spatially and the number of fires fluctuates greatly from year to year. Depending on the month, the GEBA monthly-average data contains surface measurements from approximately 20 sites in the Boreal Forest regions. Careful review of the existing 46 months of data indicates that the Version 1.1 WCRP SRB data set is usually within 10 W m^{-2} (5%) of the surface measured values in these regions during the dry season. There are a few infrequent instances in which satellite-derived values are as much as 40 W m^{-2} too large on a single-month, random-location basis. These larger-than-normal errors may be correlated with infrequent Boreal Forest fires near the particular site in question, but synthesis of wind patterns and spatial fire density had not been completed at the time this chapter was written.

May through October is the dry season for Australia. Penner et al. (1991) suggest that 9% of the total global soot from biomass burning originates from tropical northern Australia. European Center for Medium range Weather Forecasts (ECMWF) wind patterns suggest that much of this is transported to the Pacific Ocean Warm Pool region. At the time this chapter was written, we had no measurements from surface sites in this region for the 46-month period of the WCRP satellite-derived SRB data. Based on results from Africa, it is expected that the satellite-derived values for this region may not meet the WCRP accuracy goal.

December through April is the dry season for the tropical India/Southeast Asia/Central Warm Pool region. Penner et al. (1991) suggest that 7% of the total global soot originates in this region. Ten surface sites are available in this region during the period of the Version 1.1 SRB satellite data. Monthly-average errors exceed 25 W m^{-2} for 80 % of the time and 50 W m^{-2} 20% of the time. The largest errors occur in the late spring and the smallest in the early fall. Year-round average bias ranges between 35 and 50 W m^{-2}

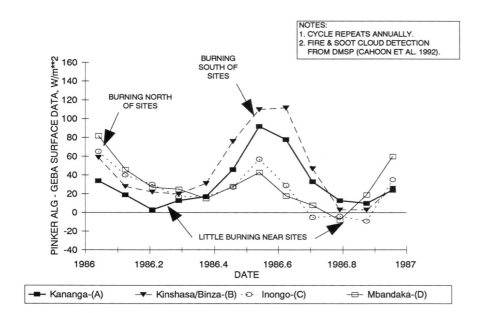

NOTES:
1. CYCLE REPEATS ANNUALLY.
2. FIRE & SOOT CLOUD DETECTION
   FROM DMSP (CAHOON ET AL. 1992).

Figure 46.3 Version 1.1 Pinker algorithm errors in central Africa

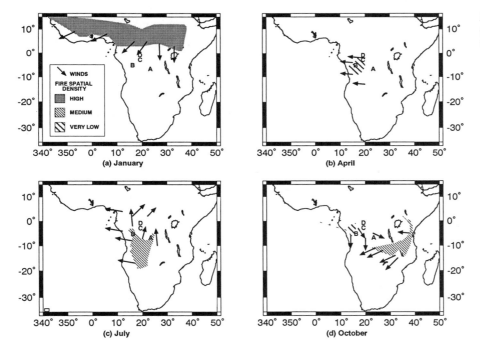

Figure 46.4 Biomass burning in 1987. (Sites A, B, C, and D are the same as those shown in figure 46.3.)

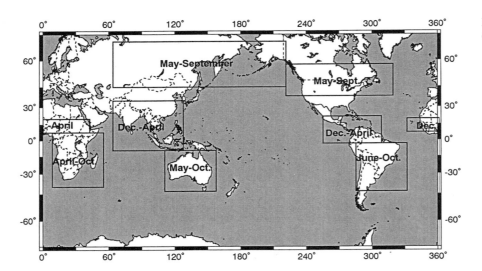

Figure 46.5 Seasonal vegetation fire cycles

along the Southeast Asia coast portion of the region (seven surface sites). Most of this error has been attributed to the fact that many of the sites are located on the coast and are not representative of the 280×280 km satellite SRB cells because of coastal clouds. It is now clear that a portion of this error may be attributed to biomass burning types of aerosols that are not accounted for in the satellite algorithms.

Twenty-eight percent of the global soot carbon emissions are believed to originate in Central and South America (Penner et al. 1991). The dry season is from December to April north of the Equator and from June through October in the southern regions. No surface site data for the interior region of South America except the Guyana Highlands (five sites) are available. The largest errors are in March (75 W m^{-2}) and the smallest are in November (25 W m^{-2}), in phase with the northern dry season shown in figure 46.5. Additional surface measurements are needed to assess errors in the southern regions, especially in the Amazonas rain forest.

Conclusion

Satellite data, surface measurements, and meteorological data from a number of different sources are consistent in suggesting that biomass burning has a significant effect on monthly-average SRB SW irradiance when fire spatial density is high. The Version 1.1 WCRP SRB satellite algorithms were computed without considering biomass burning. The error in satellite-derived values relative to actual surface measurements is one indicator of the effects of biomass burning aerosols on regional SRB SW climatology.

During dry seasons in the tropics, monthly-average errors often exceed 25 W m^{-2} and 50 to 100 W m^{-2} values sometimes occur. During wet seasons, satellite values are usually within 25 W m^{-2} of observed values. We cannot accurately assess satellite errors (hence, biomass burning effects) in many tropical regions at this time because of the lack of available surface measurement sites in our present archive of ground site data. We are also quite limited in our knowledge of fire spatial density, particularly in the climatically-important Pacific Ocean Warm Pool region. It is clear that future satellite-derived SRB data sets should consider the effects of biomass burning if long-term accuracy goals of the WCRP are to be achieved.

Acknowledgments

We are indebted to Dr. R. A. Schiffer (WCRP Radiation Projects Office, NASA Headquarters, U.S.A.) and Dr. Robert Curran (NASA Headquarters, U.S.A.) for their guidance and support during development and implementation of the WCRP/SRB activity. Dr. A. Ohmura (Swiss Federal Institute of Technology, Zurich) and the Swiss National Foundation (Grant 81EZ-039069) made it possible for Dr. T. Konzelmann to visit the NASA Langley Research Center to work on this study. W. R. Cofer III and Dr. J. S. Levine (NASA Langley Research Center, U.S.A.) provided consultation and help in understanding the current state of science relative to biomass burning issues. The authors are also grateful to R. C. DiPasquale and P. Quigley (Lockheed Martin, U.S.A.) for their computational support.

References

Andreae, M. O., E. V. Browell, M. Garstang, G. L. Gregory, R. C. Harriss, G. F. Hill, D. J. Jacob, M. C. Pereira, G. W. Sachse, A. W. Setzer, P. L. Silva Dias, R. W. Talbot, A. L. Torres, and S. C. Wofsy. 1988. Biomass-burning emissions and associated haze layers over Amazonia, *J. Geophys. Res.*, 93, 1509–1527.

Cahoon, D. R., Jr., B. J. Stocks, J. S. Levine, W. R. Cofer III, and K. P. O'Neill. 1992. The seasonal distribution of African savanna fires, *Nature*, 359, 812–815.

Cahoon, D. R., Jr., J. S. Levine, W. R. Cofer III, and B. J. Stocks. 1994. The extent of burning in African savannas, *Adv. Space. Res.*, 14, No. 11, 447–454.

Crutzen, P. J., and M. O. Andreae. 1990. Biomass burning in the tropics: Impact on the atmospheric chemistry and biogeochemical cycles, *Science*, 250, 1669–1678.

Kaufman, Y. J., A. Setzer, D. Ward, D. Tanre, B. N. Holben, P. Menzel, M. C. Pereira, and R. Rasmussen. 1992. Biomass burning airborne and spaceborne experiment in the Amazonas (BASE-A), *J. Geophys. Res.*, 97, 14 581–14 599.

Key, J. 1994. *Streamers User's Guide, Version 2.0p*, 66 pp., Cooperative Institute for Research in Environmental Science (CIRES), University of Colorado, Boulder.

Lenoble, J. 1991. The particulate matter of biomass burning: a tutorial and critical review of its radiative impact, in *Global Biomass Burning: Atmospheric, Climatic, and Biospheric Implications*, edited by J. S. Levine, pp. 381–386, MIT Press, Cambridge, Mass.

Leroux, M. 1983. *The Climate of Tropical Africa Atlas*. Editions Champion, Paris.

Oguntoyinbo, J. S. 1970. Reflection coefficient of natural vegetation, crops and urban surfaces in Nigeria, *Quart. J. Roy. Met. Soc.*, 96, 430–441.

Ohmura, A. and H. Gilgen. 1993. Re-evaluation of the global energy balance, *Geophys., Monograph 75, IUGG*, 15, 93–100.

Penner, J. E., S. J. Ghan and J. J. Walton. 1991. The role of biomass burning in the budget and cycle of carbonaceous soot aerosols and their climate impact, in *Global Biomass Burning: Atmospheric, Climatic, and Biospheric Implications*, edited by J. S. Levine, pp. 387–393, MIT Press, Cambridge, Mass.

Rossow, W. B., L. C. Garder, P.-J. Lu and A. Walker. 1991. International Satellite Cloud Climatology Project (ISCCP) documentation of cloud data, *WMO/TD-No. 266*, revised March 1991, World Meteorological Organization, Geneva, 76 pp. plus three appendices.

Setzer, A. W. and M. C. Pereira. 1991. Amazonia biomass burnings in 1987 and an estimate of their tropospheric emission, *Ambio*, 20, 19–22.

Suttles, J. T., and G. Ohring. 1986. Surface radiation budget for climate applications, *NASA RP 1169*.

Talbot, R. W., M. O. Andreae, T. W. Andreae, and R. C. Harriss. 1988. Regional aerosol chemistry of the Amazon Basin during the dry season, *J. Geophys. Res.*, 93, 1499–1508.

Talbot, R. W., M. O. Andreae, H. Berresheim, P. Artaxo, M. Garstang, R. Harris, K. M. Beecher, and S. M. Li. 1990. Aerosol chemistry during the wet season in central Amazonia: the influence of long-range transport, *J. Geophys. Res.*, 95, 16 955–16 969.

Ward, D. E., and C. C. Hardy. 1994. Advances in the characterization and control of emissions from prescribed fires, *Proc. Annu. Meet. APCA*.

Whitlock, C. H., T. P. Charlock, W. F. Staylor, R. T. Pinker, I. Laszlo, R. C. DiPasquale, and N. A. Ritchey. 1993. WCRP Surface radiation budget shortwave data product description–Version 1.1, *NASA TM 107747.*

Whitlock, C. H., T. P. Charlock, W. F. Staylor, R. T. Pinker, I. Laszlo, A. Ohmura, H. Gilgen, T. Konzelmann, R. C. DiPasquale, C. D. Moats, S. R. LeCroy, and N. A. Ritchey. 1995. First global WCRP surface radiation budget data set, *Bull. Am. Meteorol. Soc.*, 76, 905–922.

Modeling Biomass Burning Aerosols

Catherine Liousse, Joyce E. Penner, John J. Walton, Harold Eddleman,
Catherine Chuang, and Hélène Cachier

Biomass burning is the most ancient anthropogenic pollution Delmas et al. (1995). However, research on this topic only began 15 years ago with the work of Seiler and Crutzen (1980). The first focus was on emissions of major components such as gases. Interest in minor pollutants—especially in smoke particles—is recent, and this subject has been increasingly taken into consideration in recent experimental programs such as FOS/DECAFE, BASE, TRACE, and SCAR. Even if a number of questions remains unsolved, a general picture of biomass burning aerosol has been described in a few papers (Cachier et al. 1991; Ward et al. 1991; Cachier 1992; Cachier et al. 1995; Echalar et al. 1995; Gaudichet et al. 1995), particularly for aerosols emitted by tropical forests and savannas, which are the most important biomass burning sources. Only very recently have a few studies been devoted to other important sources such as fuel wood, agricultural fires, and temperate and boreal forest fires (Turn et al. 1993; Levine 1995; Brocard et al., chapter 33 this volume; Cofer et al., chapter 79 this volume; Ezcurra et al., chapter 73 this volume).

Biomass burning aerosols are mainly composed of organic particles (organic carbon with hydrogenated and oxygenated functions) and black carbon particles. Part of the organic particles, which are the predominant component of the mixture, seems to condense very rapidly—within a few seconds—on the surface of primary pollutants such as black-carbon particles (Cachier 1995; Clain 1995), which leads to the formation of heterogeneous particles. A small percentage of the mass of biomass burning aerosols is inorganic including sulfate, potassium, and nitrate particles. The relative importance of black carbon and organic particles primarily varies as a function of the type of combustion. Estimates of global aerosol fluxes have shown that biomass burning sources were the second largest source of aerosols after production of SO_2 by industrial activities (Andreae et al., chapter 27 this volume). This tendency might be reversed in the future due to both the increase of biomass burning and the decrease of SO_2 emissions as a consequence of in-dustrial anti-pollution policies. Particles emitted by biomass burning sources have been shown to be in the submicrometer-size range; therefore, they may be found far from their sources in remote areas, such as in polar atmospheres (Jaffrezo et al., 1993). Also these particles have noticeable optical properties. Black carbon particles will generate an atmospheric heating effect due to absorbing properties, whereas organic particles will produce a cooling effect due to scattering properties. Let us note that the formation of heterogeneous particles containing both black carbon and organic particles has an important impact on the optical properties of the particles, a phenomenon already stressed by Liousse et al. (1993), and Haywood and Shine (1995).

The first calculations of Penner et al. (1992) using a 1D box model suggested that a global cooling effect of $-1W/m^2$ could be expected from biomass burning particles. However, this simple model did not take into account the important spatial and temporal variability of these aerosols. A way to improve the calculation was to use a 3D transport model and then to obtain the radiative effect of biomass burning particles with a global climate model. In Penner et al. (1991), transport of biomass burning particles emitted by the two main sources (savanna and forest fires) was studied with the Grantour model (Walton et al., 1988) confirming the previous radiative results.

We reevaluated the biomass burning sources introduced in Grantour by Penner et al. (1991). Indeed, in accordance with new experimental results, it was no longer correct to assume that particle emission factors are identical for savanna and forest combustion (Cachier et al., 1995). Also, in our inventory, we took into consideration not only tropical savanna and forest burning, but also other important sources such as agricultural waste fires, fuel wood, charcoal, and dung. In developing countries, increasing amounts of fuel are needed for resources to support growing populations. In regions where there is a lack of wood, substitutes have been found with agricultural wastes or dung. For example, rice straw and sugar cane by-products are

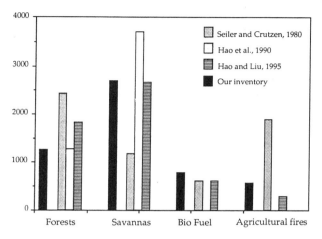

Figure 47.1 Variation of burned biomass (in Tg of dry matter) estimates from different authors

intensively burned for heating or cooking, respectively, in Southeast Asia and South America. Biofuel is the main source of energy in Africa as shown by Brocard et al. (chapter 33 this volume). The use of fuel wood in India is predicted to increase by a factor of 3 between 1985 and 2015 (*World Resources 1992–1993*). The other important source of combustion particles—the fossil fuel sources mainly occuring in the northern hemisphere—and the natural source of secondary organic particles (obtained from terpenes naturally emitted by vegetation) have been taken into account in our run (see their description in Liousse et al. 1996).

In this chapter, we describe biomass burning sources that we introduced in the Grantour model. Spatial and temporal distributions are detailed for black carbon and organic particulate emissions. Observations of aerosol concentrations in the atmosphere and in precipitations south of 20N are used to constrain the model's predicted biomass aerosols.

Description of Source Inventory: Estimate of Worldwide Burned Biomass

Determination of worldwide burned biomass is still a difficult and open-ended problem. Biomass burning sources are usually divided into four groups—savanna fires, forest fires, fuel wood and charcoal consumption, and agricultural fires. As pointed out in figure 47.1 that shows burned biomass estimates found in different studies, accurate quantitative assessments for each source are still unresolved problems.

For field fires (savanna and forest fires), the burned biomass (M) is obtained from the relationship:

$$M = A \times B \times \alpha \times \beta$$

where A is the yearly burned area, B, the biomass density, α, the fraction of aboveground biomass and β, the burning efficiency. M is highly dependent on the choice of the turnover of the burns (which determines A) and also on that of mean α and β parameters.

For the other biomass burning sources, calculations are usually based on global population, vegetation, production, or resources data. In developing countries, where these sources are important and expanding, such an estimate needs exhaustive field survey: indeed, as shown in a few other scientific papers, large differences may occur between global estimates and field data.

Savanna and Forest Fires

First estimates are found in Seiler and Crutzen (1980) (figure 47.1) including both tropical, temperate, and boreal areas. In this pioneer inventory, forest fires are notably important—more so than savanna fires. A few years later, Hao et al. (1990) assumed that the savanna was cleared each year. Consequently, considering areas yearly submitted to fires (748 million ha and 7.2 million ha for savannas and tropical forests respectively), and the average biomass burning efficiency (80 and 30%, respectively, for savannas and tropical forests), Hao et al. (1990) gave a large predominance to savanna fires. In 1995, Hao and Liu corrected these data again. Estimates of forest and savanna burned biomass were respectively increased and decreased due to a higher mean density of aboveground biomass of forest trees and to a smaller turnover of savanna fires according to Menaut et al. (1991) (55% instead of 75% of African savannas were considered to be submitted to fires yearly).

When we began this study available data were primarily those from Seiler and Crutzen (1980) and Hao et al. (1990). In 1991, Penner et al. have used biomass burning sources given by Hao et al. (1990) for tropical forests and savannas and that of Dignon (personal communication, 1990) for biomass burning inventory in Australia. We decided to reevaluate the inventory of Penner et al. (1991), taking into consideration new experimental results. We did not change the forest inventory taken from Hao et al. (1990); instead, our effort focused on the improvement of African savanna sources. According to Menaut et al. (1991) and results given by FOS/DECAFE experiments (Lacaux et al. 1995), we considered that African savannas were not only constituted of humid savanna ecosystems. Indeed, as shown by table 47.1, which presents estimates of burned biomass for African savannas, four groups of savannas may be distinguished due to different

Table 47.1 Estimate of annual burned biomass (dry matter) for different types of African savannas. These data are adapted from Menaut et al. 1991.

Savanna type	Surface area in m²	Burning frequency in %	Burned area A, in m²	Biomass density B, in g/m²	Above ground biomass α, in %	Combustion efficiency ce, in %	Burned biomass M,[a] in g
North Sudanian	0.77e + 12	25	0.19e + 12	400	71	80	44e + 12
South Sudanian	0.81e + 12	50	0.40e + 12	600	71	80	138e + 12
Guinean	0.17e + 12	80	0.14e + 12	800	71	80	62e + 12
Sahelian	0.53e + 12	15	0.08e + 12	250	71	80	11.3e + 12

a. Burned biomass is obtained from $M = A \times B \times \alpha \times ce$ following 1980 (A is the yearly burned area, B, the biomass density, α, the fraction of aboveground biomass and ce the burning efficiency).

frequencies of burning and biomass density (Menaut et al. 1991). From this new inventory, it may be seen that biomass burned in humid savannas (or Guinean savannas represents only 24% of the biomass burned in all African savannas. Hence in our new inventory, we assumed that 55% instead of 75% of African savannas are submitted to fires yearly, a new assumption also integrated in Hao and Liu (1995). Also, in our work, the mean dry biomass density of 660 g/m² chosen for savanna by Hao et al., (1990) was changed into 500 g/m². The value of 80% for burning efficiency was adopted according to Seiler and Crutzen (1980), Menaut et al. (1991) and Lacaux et al. (1995).

Agricultural Waste Fires, Fuel Wood Charcoal, and Dung Sources

Because of lack of data on these sources, we had to form many assumptions for our inventory. In developed and developing countries, fuels are neither used in the same way nor are they used for the same purposes (Seiler and Crutzen 1980, Strehler and Stutzle 1987). Consequently, a prerequisite was to do a rough classification of countries; we used FAO definitions to classify each country as either in the developed country group or the developing country group.

Agricultural Waste Combustion M, the vegetation biomass burned during agricultural waste combustion was calculated as:

$$M = P \times W/P \times Wf/W \times ce$$

where P is the crop production, W the by-product amount (W/P is defined as harvest index), Wf the by-product amount yearly submitted to fires, and ce the burning efficiency. M may be burned either on field as open fires or later as domestic fire fuel.

Table 47.2 summarizes the different steps we used to calculate M.

Crop production of various grains (rice, wheat, barley, rye, corn, and others) and of sugar cane (noted as P in table 47.2) were taken from data compiled by FAO (*FAO Yearbook* 1991). Major grain production is rice, wheat, and corn. Recently, Zhuange et al. (chapter 71 this volume) have shown that same result applies for crop production in China.

Agricultural wastes were estimated from the harvest index for each crop given by Irvine (1983), Fisher and Palmer (1983), Fisher (1983), Barnard and Kristoferson, (1985), Strehler and Stutzle (1987). In table 47.2, it may be seen that the difference between developed and developing countries is particularly important for the corn harvest index. It is important to note that the harvest index chosen in our work for sugar cane wastes is smaller than that reported in the literature (Irvine 1983). This is because sugar cane fires considered in this paragraph are open fires on field only. Use of sugar cane wastes as fuel (also called bagasse) will be taken into account in the next paragraph.

The fate of agricultural wastes varies widely with the type of crop and development of a country. Although data are scarce, we propose a set of values for the amount of wastes actually submitted to fires; the fate, however, of the unburned fraction is not clearly known.

For developed countries, according to recent observations (Jeff Amthor, personal communication, 1994; Ezcurra et al., chapter 73 this volume) on a global scale only 5% of corn, wheat, and various grain wastes and 10% of rice and sugar cane wastes could be burned. It is interesting to note that in the recent agricultural fire inventory described in Yevich and Logan (1995), nearly 7% of American agricultural wastes are considered to be submitted to fires each year, a value in the range of our assumptions. Further investigations based on local observations would be needed to improve these factors.

Table 47.2 Estimate of annual burned biomass of agricultural waste fires for developed and developing countries from crop production amount for different types of crop. Details on the use of burned biomass as open fires and fuel

	Main products P in Tg	W/P	Wf/W	ce in %	Burned biomass M^a in Tg	M as open fires in Tg	M as fuel use in Tg
Rice							
Developed	32	1.2	0.10	85	3.26	3.26	0
Undeveloped	422	1.2	0.50	85	215.2	86.1	129.1
Wheat and others							
Developed	1064	1.3	0.05	70	48.4	48.42	0
Undeveloped	526	1.3	0.30	70	143.6	71.8	71.8
Corn							
Developed	271	1	0.05	35	4.75	4.75	0
Undeveloped	187	1.85	0.30	35	36.3	10.9	25.42
Sugar Cane							
Developed	59	0.5[b]	0.10	35	1.0	1.0	0
Undeveloped	909	0.5[b]	0.70	35	111.35	77.9	33.45

a. $M = P \times W/P \times Wf/W \times ce$ where P is the main crop production amount, W the by-product amount, Wf the by-product amount submitted to fires and ce, the combustion efficiency (1996).
b. Most of sugar cane by-products considered here are burned as open fires on field. Fuel use of sugar cane is described in biofuel inventory.

In developing countries, it is still difficult to gain access to these factors, which are found to differ widely. For example, Mahtab and Islam (1984) assumed that at least 33% of rice waste is used as fuel in Bangladesh, whereas the value of 100% is given by Yevich and Logan (1995) for the same country. In addition, this parameter varies from one country to another; indeed, in Africa, 1 to 33% of agricultural wastes are burned each year (Barnard and Kristoferson 1985), whereas in Vietnam, 75% of rice straw could disappear as open fires on field (Yevich and Logan 1995). From these studies and according to Barnard and Kristoferson (1985) and to Kenneth Cassman (personal communication) we assumed, in our inventory that in developing countries 30% of wheat, corn, and other grain wastes were submitted to fires each year, and 50% and 70% of rice straw and sugar cane by-products respectively.

Calculations of M are achieved using adequate burning efficiency. Different burning efficiencies were chosen as a function of the burning property of each crop. For example, rice has a higher combustion efficiency than the other grains. Much lower values of combustion efficiency are given for fiber plants such as corn and sugar cane than for grains (Robinson 1989).

From table 47.2, it may be seen that spatial distribution of P (the main grain product) and M, the biomass which is burned as open fires and fuel do not coincide. For example, P and M for rice are, respectively, 13 and 66 times larger in developing that they are in developed countries. To point out the importance of the choice of parameters presented in table

47.2, we may also compare our data with those of Seiler and Crutzen (1980). Global data for waste production and burned biomass given by our inventory are, respectively, 3713 Tgdm and 564 Tgdm, whereas Seiler and Crutzen (1980) obtained 3600 Tgdm and 1900 Tgdm, respectively. Such a difference in estimates for burnt biomass is mainly linked to the detailed parametrizations of agricultural waste fires we have performed in this study and that were much higher in the work of Seiler and Crutzen. These authors, based on Middleton and Darley 1973, have considered that each year 50% and 80% of the global agricultural wastes were burned, respectively, in developed and developing countries. These values are much higher than ours.

As shown by previous work (Seiler and Crutzen 1980; Crutzen and Andreae 1990), agricultural wastes may be burned for two main reasons: first, to prepare or clean fields before or after harvest, and second, to use the by-products as fuel for cooking or heating. Of course, to make a clear division between these two different uses is difficult. This choice has an important impact on the spatial and temporal fire distribution. Open fires will occur before or after the harvest, whereas wastes used as fuel will be burned all year long. Open fires will also have a larger buoyancy than domestic fires, and consequently, pollutants will be injected higher. Their particulate composition, dependent on the combustion quality, will also be different. In table 47.2, we may note that the relative importance of the two different uses differs for each crop. Globally, we found that biomass burned as open fires may be 304 Tgdm, an estimate larger than that (180 Tgdm) of

Table 47.3 Estimate of burned biomass of biofuel from main products submitted to fires: biofuel including fuel wood, bagasse, charcoal, and dung

	Products exposed to fire, A in Tg	ce in %	Burned biomass M in Tg
Developed countries	—	—	—
Fuelwood and bagasse	161.7	90	145.5
Charcoal consumption	1.2	75	0.88
Developing countries	—	—	—
Fuelwood and bagasse	898.9		
Domestic use	449.4	90	404.5
Charcoal making	449.5	carbonization rate: 25	121.3[a]
Charcoal consumption	20.2	75	15
Dung	202.2	50	101

a. Burned biomass due to charcoal making may be obtained as follows: A' (products exposed to fires in TgC) $= A \times 0.45$
Charcoal produced, C in TgC $= A' \times 0.55$ (Héry 1993)
Carbonaceous emissions in TgC $= Ec = A' - C$
Carbon represents 75% of charcoal mass

Yevich and Logan (1995). This estimate is also larger than that of Hao and Liu (1995), who took into account only tropical areas, considering also that 40% of all the agricultural wastes were submitted to fires yearly.

Fuel Wood, Bagasse, Charcoal, and Dung As shown in figure 47.1, estimates for fuel wood and charcoal consumption are less scattered than for the other sources even if different procedures were used for the calculations. Seiler and Crutzen (1980) took into consideration the population distribution and consumption per capita, whereas Hao and Liu, (1995) took directly the annual data of fuel wood production from coniferous and nonconiferous forests given by the *FAO Yearbook* (1989). It must be pointed out here that "official" consumption data are frequently an underestimate of "real" consumption due to important "underground" trading and uses, especially in developing countries. We improved the inventory first by considering that a part of fuel wood was used for charcoal making and, second, by introducing a new source—the dung source—which was never taken into account previously. This source is very important in the poorest countries of the developing world such as India and Ethiopia. Our estimates are, then, much larger than previous ones.

Fuel Wood, Bagasse Global maps of fuel wood and bagasse (a fuel derived form sugar cane wastes) production were obtained from FAO data on production, importation, exportation, stock changes, and losses (*FAO Yearbook* 1991) and were gridded following the population distribution. On a global scale, 1060 Tgdm of fuel wood and bagasse are yearly submitted to fires (table 47.3). This estimate could be revised rapidly. In

Africa, for example, fuel wood and bagasse consumption is of the order of 295 Tgdm but very recent estimates based on field observations and not on FAO data (Brocard et al. chapter 33 this volume) gave a consumption of 120 Tg for West Africa only. If we assumed that this area represents 20–30% of the whole continent, Africa would burn at least 400–500 Tg of vegetation fuel for domestic purposes. This discrepancy clearly points out the need for further studies.

In developing countries, observations have shown that fuel wood products were used directly for domestic purposes such as cooking and heating, but were also for charcoal making, which may later serve as fuel. Indeed, in villages and cities, the storage of charcoal is more convenient than wood. According to Openshaw (1974) and Delmas et al. (1991), we assumed that 50% of fuel wood products are devoted to direct use, whereas the other 50% are devoted to charcoal making. Recent work of Brocard et al. (chapter 33 this volume) has stressed that the ratio is actually included in a wide range (2–50%) depending on each country's resources.

In developed countries, we have considered that all fuel wood products were directly burned for used such as residential fireplaces, wood stoves, and barbecues. A combustion efficiency of 90% was chosen for converting fuel wood products into burnt biomass according to Seiler and Crutzen (1980) and Delmas et al. (1991). This value is relative to flaming combustion and in agreement with that found by Brocard et al.

Charcoal Making Charcoal-making calculations were based on the work of Héry (1993). A rate of 25% was assumed for the production of charcoal from a certain amount of fuel wood submitted to carbon-ization and expressed in TgC/Tgdm. Biomass actually burned was obtained by substracting the amount of charcoal produced to the amount of the primary fuel wood. For this calculation, we considered that carbon represents 75% of the charcoal mass (Héry, 1993). In developing countries, charcoal produced from fuel wood may be estimated to be on the order of 148 Tg.

Charcoal Consumption To create the charcoal con-sumption inventory, we did not use the charcoal inventory we have produced from fuel wood inven-tory. Actually, we have considered the FAO data on charcoal production, importation, exportation, stock changes, and losses (*FAO Yearbook* 1991) and gridded following the population distribution. It is interesting to note that for developing countries, this procedure based on FAO data gives an amount of 20.2 Tg of charcoal annually submitted to fires, whereas charcoal products available for combustion and obtained from the fuel wood inventory were estimated to be on the order of 148 Tg. This discrepancy is certainly due to FAO underestimates described below and deserves significant further investigation.

A burning efficiency of 75% was assumed. This choice of a low combustion efficiency is in agreement with Brocard et al. (chapter 33 this volume) who have shown that domestic fires using charcoal have a lower efficiency than those with fuel wood.

Dung One of the most original parts of our inventory is to take into account a dung source. Indeed, in the poorest countries, substitutes for fuel wood and agri-cultural wastes are found in dung products (or breed-ing animal wastes). Our assumptions were based on a study by Barnard and Kristoferson (1985). First, as shown in table 47.4, a group of developing countries has been selected for their use of dung products. Ani-mal population was listed by countries. Buffalo, cattle, horses, camels, mules, asses, pigs, sheep, and goats have been taken into account. Dung production is quite variable and was thus calculated from animal population using different factors and given in dry tons of dung/animal/year by Barnard and Kristoferson (1985). For example, the factor for buffalo is on the order of 1.5, that for cattle, 1.0 whereas, that for sheep and goats is 0.15 only. Although in some countries such as India, agricultural wastes and dung are mixed together to obtain a better fuel, we did not consider such mixing. Data for dung production were mapped following population distribution. We assumed that 18.5% may represent the average fraction of dung yearly submitted to fires. Further investigation is

Table 47.4 Estimate of dung production for a subset of countries where such a fuel is used for heating or cooking. Due to lack of field data for the use of dung as fuel, a global average is assessed to be 18.5%.

	Dung production in Tg	Fraction of dung submitted to fires in %
Burkina	4.8	—
Chad	5.8	—
Egypt	10.1	15
Ethiopia	42.4	—
Kenya	16.4	—
Lesotho	1.1	34
Malawi	1.4	—
Mali	7.3	—
Niger	4.1	—
Nigeria	25.2	—
Sudan	28.5	20
Tanzania	15.1	—
Zambia	3.2	—
Zimbabwe	6.6	—
Total Africa	**172**	—
Argentina	58.7	—
Bolivia	9.1	—
Colombia	28.5	—
Peru	6.7	3
Total South America	**103**	—
Costa Rica	1.9	—
Guatemala	2.2	—
Nicaragua	2.1	—
Total Central America	**6.2**	—
Mexico	**45.1**	
Bangladesh	28.2	15
India	344.4	23
Indonesia	20.5	—
Nepal	12.1	15
Pakistan	53.3	12
Philippines	8.2	—
Sri Lanka	3.4	—
Thailand	14.5	—
Asia	**484.5**	—
China	**282.3**	15
Global	**1093**	**18.5**

Table 47.5 Global annual emissions of black carbon (BC) particles and particulate organic matter (POM) from burned biomass for all biomass burning sources

Sources	Burned biomass Tg dm	EF(BC) gC/kg dm	POM/BC %	BC TgC	POM Tg	References[a]
Forests	1259	1.53	8.6	1.93	16.6	(Radke et al. 1988; Ward et al. 1991)
Savannas	2682	0.81	7.15	2.17	15.5	(Cachier et al. 1995a,b; Le Canut et al. 1995)
Biofuel	788.2			1.02	9.38	
Fuelwood and bagasse						
Developed	145.5	1.32	5.7	0.2	1.10	(Butcher and Sorenson 1979)
Undeveloped/domestic	404.5	1.2	5.7	0.5	2.8	(Smith et al. 1993; Butcher et al. 1984)
Undeveloped/charcoal making	121.3	1.84	16.75	0.22	3.74	(Héry 1993)
Charcoal consumption	15.9	1.5	4.5	0.02	0.11	(Smith et al. 1993; Butcher et al. 1984)
Dung	101	1	16.3	0.101	1.65	(Cachier, personal communication)
Total agricultural fires	564			0.53	3.11	(Méland and Boubel 1996)
Rice						(Gerstle and Kemnitz 1967)
Developed	3.3	0.6	2.2	0.002	0.004	(Carroll et al. 1977)
Undeveloped	215.1	0.86	3.9	0.186	0.72	(Darley et al. 1979)
Wheat and others						(Seiler and Crutzen 1980)
Developed	48.4	0.9	5.2	0.04	0.23	(Jenkins et al. 1991–1993)
Undeveloped	143.6	1.2	7.3	0.17	1.26	(Turn et al. 1993)
Corn						
Developed	4.7	0.7	4.7	0.003	0.015	
Undeveloped	36.3	0.95	9.5	0.03	0.32	
Sugar cane						
Developed	1.0	0.7	4.15	0.0007	0.003	
Undeveloped	111.4	0.8	6.5	0.085	0.56	

a. Complete references may be found in Liousse et al. 1996.

needed to determine more this factor and its variability accurately. As no experimental measurements exist, burning efficiency was estimated taking into account different experimental observations. Barnard and Kristoferson (1985) indicated that dung combustion occurs with a steady flame and a low calorific intensity. Also, during SAFARI experiments Cachier et al. (1996) revealed that combustion of elephant pellets occured mainly under smoldering conditions. For these reasons, a burning efficiency of 50% was assumed, as for other smoldering processes.

Emission Inventory: Global Estimates for Black Carbon Particles and Organic Particles

Estimates on emission intensity may be obtained from emission factor values (EF), which are defined as the ratio of the mass of the emitted species to either the mass of dry vegetation consumed (expressed as g/kgdry plant) or to the amount of carbon in the vegetation fuel (expressed as g/kgCdry plant) (Cachier, 1992; Cachier et al. 1995). Here we only use the reference to dry matter. Table 47.5 presents emission factor values for black-carbon (BC) particles and organic particles (POM) adapted from the literature for each of the

sources previously described. These values were selected according to the type of combustion prevailing for the different fires sources.

For example, in our study we assumed that savanna burnings occur primarily under flaming conditions whereas, smoldering conditions have been considered to dominate forest fires. This should be revised in the future since a significant fraction of the forest biomass may burn under flaming conditions (Susott et al. 1991). We stressed the importance of forest fires as a producer of particulate organics. Indeed, it has been shown previously that smoldering conditions are accompanied by a large particulate emission factor and POM/BC ratio. Mean values for savanna fires given in table 47.5 were calculated from either ground and aircraft experiments. Those for forest fires were from aircraft studies only.

In developed countries, coniferous, synthetic, and decideous wood are the main fuel used for residential fireplaces. Consequently, a mean BC emission factor of 1.32 gC/kgdm was chosen in the middle of the range given by Butcher and Sorenson (1979) of 0.84–1.8. However, in developing countries, domestic wood in nonconiferous wood and, therefore, a different value (1.2 gC/kgdm) was adopted for this source (Smith et al. 1983; Butcher et al. 1984). It is important to note that domestic fires were considered to occur under flaming conditions. Nearly the same emission factor has been used for charcoal consumption assumed to occur primarily under flaming conditions. Considering that smoldering conditions dominate charcoal making and dung consumption, we attributed to these sources large-particulate emission factors (EF(BC)) on the order of 1.45 gC/kgdm) with large POM/BC ratios (on the order of 16%, see table 47.5) according to Cachier et al. (1996) and Héry (1993).

In developed countries, agricultural wastes were considered to burn as open fires on field exclusively. This type of burning occurs under flaming conditions. In developing countries, wastes are burned either as open fires on field or for domestic purposes. For domestic fires, combustion occurs under smoldering conditions. Such a difference has been taken into account when fixing particulate emission factors and POM/BC ratios, and it may be seen in table 47.5 that both sets of values are systematically higher in developing than in developed countries. Moreover, different emission factors were used for different crops. The most striking example is for rice burning, which has a lower emission factor than other crop burnings, because the combustion of rice straw is known to occur under very intense flaming conditions (Turn et al. 1993).

Now, calculating the total amount of BC particles and organic particles produced yearly by all biomass burning sources, we may confirm that the savanna source is more important than forest fires for BC particle emissions, but it is less important for organic particle emissions. However, the particulate production of other sources, such as agricultural waste fires and biofuel (fuel wood, charcoal, and dung sources), represent almost 33% of the global emission inventory.

Global values for BC aerosol (5.65 TgC) and POM (44.6 TgPOM) are of the same order as those found for fossil fuel sources (respectively, 6.64 and 30.4 Tg Liousse et al. 1996). Our biomass burning estimate of BC emissions compares satisfactorily with that recently proposed by Cooke and Wilson (1996). However, this agreement may be coincidental since the two inventories are based on quite different assumptions. For example, Cooke and Wilson (1996) considered only the burning of savannas, forests in temperate and tropical areas, and a few sorts of domestic fires. They did not include agricultural fires and the contributions of Australia and Russia, which are surely significant. Further investigation and joint effort will be necessary to improve carbonaceous aerosol source and emission inventories.

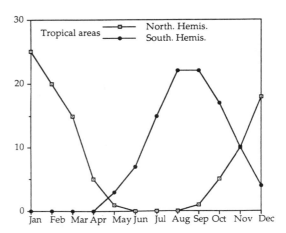

Figure 47.2 Temporal distribution (% of burned biomass) of savanna and forest fires in northern and southern hemispheres of tropical areas (adapted from Hao et al. 1990)

Temporal Distribution and Height Injection

Temporal Distribution

Most sources are not distributed regularly throughout the year, and their input in the model has to take into account their temporal distribution. The temporal distribution of tropical savanna and forest fires for the northern and southern hemispheres presented in figure 47.2 has been estimated according to Hao et al. (1990). This distribution is in global agreement with experimental data and satellite observations registered for the last 10 years. An example is given with BC concentration records, which in tropical areas may be considered as a good tracer of biomass burning. Long-term patterns of BC concentrations measured in sources areas such as Alta Floresta in Brazil (Artzxo et al. 1994) or Lamto in the Ivory Coast (Liousse et al. 1994) have peaks that can be attributed to regional-scale biomass burning occurring at the same time as that deduced from the Hao et al. (1990) temporal distribution. Moreover, this timing of the biomass burning season is corroborated by fire pixel counting on satellite images obtained for each month of the dry seasons (Cahoon et al. 1992; Franca et al. 1993). However, it must be pointed out that, recently, using ozone concentrations and rain amounts, Hao and Liu (1995) have adopted a different temporal distribution leading to important disagreements over observations. As an example, they assume that in Africa savanna areas north of the equator were burned from March to June, an assumption that is not acceptable since these regions burn at their maximum intensity between December and January.

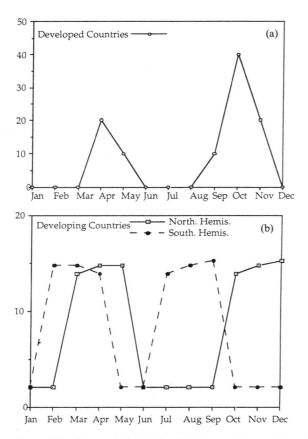

Figure 47.3 Temporal distribution (% of burned biomass) of agricultural waste fires (rice, wheat and others, corn and sugar cane by-product fires are included) in developed countries (*top*) and in developing countries (*bottom*)

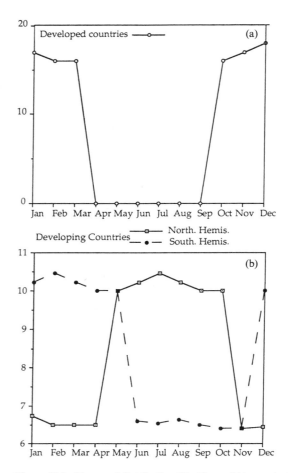

Figure 47.4 Temporal distribution (% of burned biomass) of biofuel consumption (including fuel wood, bagasse, charcoal making, charcoal consumption, and dung consumption) in developed countries (*top*) and in developing countries (*bottom*)

As mentioned above, the temporal distribution of agricultural waste fires is linked to their different uses. Figures 47.3a and 47.3b show, respectively, the case of developed countries where fires occur after the harvest period (Kenneth Cassman, personal communication, 1994) and a more complex case for the developing countries of the northern and southern hemispheres. Indeed, in such regions, agricultural wastes are burned either as open fires on field after the harvest, or as domestic fires for cooking and heating purposes. Domestic fires are considered to be used all year long for cooking purposes and during the wet season for heating purposes (almost 30% of the total).

In figure 47.4a and 47.4b, it may be seen that in developing countries, 6–7% of the fuel wood, bagasse, charcoal, and dung is used for cooking purposes. Furthermore, when heating is necessary (during the wet season) the biomass burned monthly as biofuel increases up to 10–11%. Charcoal making is assumed to

occurs during the dry season in developing countries. In developed countries, use of residential fireplaces occurs mainly during the winter season.

Height Pollutant Injection

To our knowledge, only a few works exist on the vertical distribution of biomass burning pollutants in the atmosphere. Forest fires are known to have a large buoyancy (Kaufman et al. 1992). Even if energy released in savanna fires is smaller, important atmospheric convective processes prevailing in tropical areas allow pollutants to be entrained in regional circulation (Lacaux et al. 1995; Garstang et al. 1996). We decided to inject particles emitted by savanna, forest, and open agricultural fires within a height of 2000 m. A smaller box (1000 m height) was chosen for biofuel and agricultural fires used as domestic fires. Liousse et al. (1996) stressed the importance of height injection and its important impact on aerosol concentrations recorded in remote areas.

Model Simulations

Description of the Model

Global atmospheric transport, transformation, and removal of BC particles and organic particles were simulated using the Lawrence Livermore National Laboratory chemistry/climate model called the Grantour model (Walton et al. 1988). Grantour model is a Lagrangian model with a horizontal resolution of 4.5° latitude by 7.5° longitude, and 12 vertical levels. It may be run either off-line using the wind and precipitation fields obtained from a general circulation model or interactively (on-line) with the CCM1 model, a model that allows alteration of the wind and precipitation fields by the presence of aerosols. Only the off-line mode is described here. The model was run first for a whole year with 25 000 constant mass air parcels and 12 hours as an operator split-time in order to create a steady atmosphere. Then, it was run for the second year, with 50 000 parcels and a split time of six hours, a choice which has been previously seen to have reasonable numerical accuracy (Penner et al. 1991 and 1993). Results for January and July of this second-year treatment are reported in this chapter.

Spatial and temporal distributions of BC and organic particles from biomass burning sources as described above have been implemented in the model. In addition, fossil fuel sources for BC and POM aerosol and natural sources for POM aerosol have been introduced following Liousse et al. (1996).

Previous works have shown the ubiquitous presence of BC and POM particles in rain either in source regions or at remote sites (Cachier and Ducret 1991; Ducret and Cachier 1992). There is agreement that, on average, biomass burning carbonaceous aerosols behave as hydrophilic particles, which may be easily scavenged by rain or cloud droplets (Novakov and Penner 1993; Dinh et al. 1994). In the model we used the same scavenging coefficient for both BC and POM particles of 2.1 cm^{-1} for stratiform precipitations and 0.6 cm^{-1} for convective precipitations (Ducret and Cachier 1992; Penner et al. 1993). The scavenging coefficient is defined as:

$$S(\text{in cm}^{-1}) = [W/h][\rho_w/\rho_a]$$

where W, is the scavenging ratio, h, the vertical height of the atmospheric column in which particles are re-moved by precipitation (equal here to 1000 m), and ρ_a and ρ_w, the air and water densities respectively. Model results reported below are very sensitive to the deposition velocity and precipitation scavenging coefficient chosen values. In Liousse et al. (1996), it may be seen that an increase of 15% in the mean scavenging coefficient implies a decrease of BC concentrations by a factor of 2 in remote areas. Considering that the range of uncertainty for particle scavenging coefficients is much larger than 15%, this test points out the urgent need for other measurements, especially off source regions. Following Cooke and Wilson (1996), the formulation could also be improved by considering a different removal rate of aerosol in relation to particle aging.

Finally, both BC and organic particles were assumed to have a dry deposition velocity of 0.1 cm/s.

Model Results

An overall test of the emission inventory previously described and the parameters chosen for height injection and for wet and dry deposition may be to compare the biomass burning source emission rate with the total deposition rate. Dry and wet deposition of BC aerosols were found to be 1.20 TgC/yr and 4.39 TgC/yr respectively, whereas 5.65 TgC/yr is the total amount of BC particles emitted by the global biomass burning sources. The resulting mean globally-averaged lifetime of these BC aerosols, is 4.5 days, which appears reasonable for submicron particles following Lambert et al. (1982) and Ogren et al. (1984). With their own parametrization, Cooke and Wilson (1996) obtained a value of 7.6 days.

Figure 47.5a and 47.5b present the global distribution of surface concentrations for organic particles in January and July, obtained from Grantour when all biomass burning sources and as well as natural and fossil fuel sources are included. As expected, large atmospheric concentrations situated north of 20°N are mainly governed by fossil fuel emissions, whereas those situated south of 20°N are dominated by biomass burning. For this latter area, as illustrated for Africa, in January, larger surface concentrations (of 5 μg/m^3) are situated north of the equator. In July, concentration maxima have shifted southwards and 3 concentration maxima are produced, respectively, in South America, southern Africa and Australia. Also, it may be noticed that concentrations in Antarctica are higher by a factor of 5 in July than they are in January, a feature mainly explained by seasonal variability of biomass burning sources.

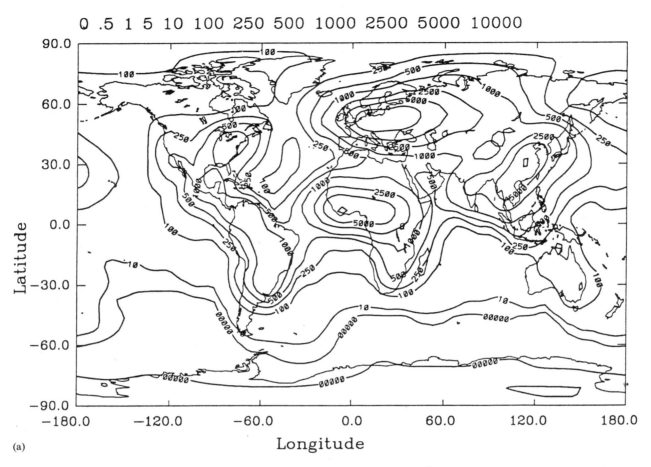

Figure 47.5 Global distribution of particulate organic matter surface concentrations in ng/m³. Natural, fossil fuel, and biomass burning sources are included ((a), January; (b), July).

In table 47.6, we present a comparison of literature data and simulated values for BC and OC surface concentrations for areas south of 20°N. Good agreement is generally found between the two sets of data for BC concentrations. For some rural locations such as Ecuador or Brazil, the correlation is better in our work than in a previous one (Penner et al. 1991), which did not take into account fossil fuel sources that may be important in some tropical countries.

A few problems linked to model or source formulation still remain. First, calculated concentration values attributed to any site, are representative of the whole grid of the model where the site is located. In some cases, there may be disagreement between this mean value and the real experimental value measured at a given site. For example, at Cape Grim and in the China Sea, our model overestimates BC concentrations. This discrepancy may be explained by the model-mapping procedure at the coastline as concentrations

from continental source regions may contribute to the concentrations at a grid point off the continent. Another example is obtained with source regions. The model is not always able to reproduce concentration records: this is observed at Lamto (Ivory Coast) and in Cuiaba, a rural location in Brazil (see figure 47.6a) for which modeled concentrations are systematically lower than for experimental ones. Underestimation by the model occurs for a subset of points in areas around 10°N, 165°E and in the southern Pacific Ocean. Modeled values are monthly means; observations are typically 6–12 hour means over the short period of an experiment.

At the South Pole, BC aerosol concentrations are also underestimated by the model. The same problem is encountered with most of BC data for antarctic snow: for example, except for Siple Point, where simulated and observed concentrations are in agreement, for Halley Bay and Amundsen Point snows, BC

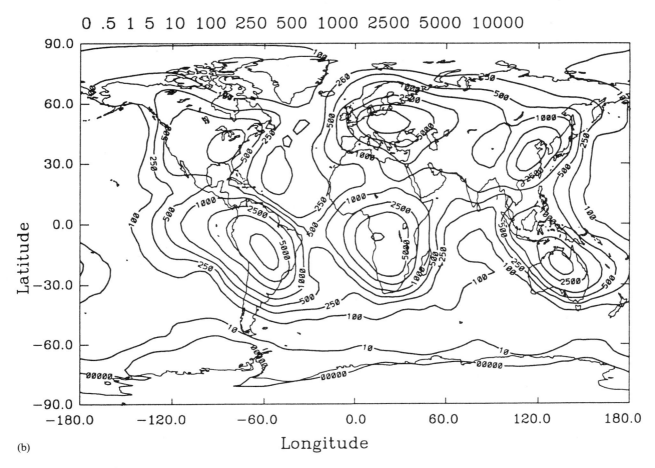

(b)

Figure 47.5 (continued)

concentrations are estimated to be on the order of 3 and 0.01 μg/l, respectively, by the model, whereas values of 18 and 0.35μg/l are recorded in experiments (Clarke and Noone 1985; Chylek et al. 1987; Warren and Clarke 1990). Among all possible causes (underestimation of biomass burning sources, overestimation of precipitation scavenging ratio, poor representation of winds near the Pole), the last one (in relation to the model formulation of the winds) seems the more plausible.

For POC too, comparison between simulated and observed surface concentrations appears globally satisfactory. The same types of problems encountered for BC particles are also found. An example is given using Enewetak Island data obtained during a short experiment (SEAREX program) that could not be correctly simulated by the model. In Enyelle (Congo) rain, it may be seen that simulated POC values are four times larger than observed data. This apparent discrepancy could be due to partial dissolving of OC, which would

decrease particle concentrations in rain water and hence alter the calculation of scavenging ratio.

Long-term studies of surface concentrations at rural and remote sites allowed us to test the capability of the model to reproduce source temporal variations. Figure 47.6a, b, c presents aerosol results for three locations influenced by biomass burning sources. The first one is Cuiaba, a Brazilian rural location where the atmosphere is highly contaminated by savanna and forest fires during dry seasons (Artaxo et al. 1995). Figure 47.6b represents monthly BC variations measured at Amsterdam Island, where recent works have shown that the atmosphere was polluted by southern African savanna fires during the dry season (Cachier et al. 1994). In figure 47.6c, we present a long-term pattern comparison for Charles Point, in Australia (Artaxo, personal communication, 1995), a site also largely influenced by savanna fires during the dry seasons, with possible contamination by both Australian and Indonesian fires at the beginning of the dry seasons.

Table 47.6 Comparison of simulated values given by Grantour and observations of surface concentrations at different locations.

Locations	Time period	Observed	Simulated	References
	Black Carbon Concentrations in ng C/m^3			
Northern Hemisphere				
10,30, -160 (Pacific Ocean)	Oct., Nov., Dec., May, June, July	7	6.6	(Andreae et al. 1984; Clarke 1989)
11.3, 162.1	April	50	20	(Cachier et al. 1990)
10, 165	July	25	9	(Parungo et al. 1994)
10, 113 (China Sea)	June	33	87	(Parungo et al. 1994)
0, 150 (Western Pacific Ocean)	July	9	10	(Parungo et al. 1994)
0, -160 (Pacific Ocean)	Annual	3	2.95	(Andreae et al. 1984; Clarke 1989)
0, -120 (Equatorial Pacific)	June	10	14	(Andreae et al. 1984)
Southern Hemisphere				
-2, -77.3 (Ecuador)	Annual	100–520	289	(Andreae et al. 1984)
-10,-30, -160 (Pacific Ocean)	Oct., Nov., Dec. May, June, July	8	4	(Andreae et al. 1984; Clarke 1989)
-10, -55 (Brazil)	Annual	200–620	536	(Andreae et al. 1984)
-10, -76 (Peru)	April	24	25.5	(Cachier et al. 1986)
-13.6, -172 (Samoa)	July	19	14	(Cachier et al. 1986)
-20, -40 (Atlantic Ocean)	October	20	74	(Andreae et al. 1984)
-37.5, 77.3 (Amsterdam Island)	Annual	5.5	4.5	(Cachier et al. 1994)
-40.7, 144.4 (Cape Grim)	Annual	3	8.9	(Heintzenberg and Bigg 1990)
-41, 174 (New Zealand)	August	22	15.2	(Cachier et al. 1986)
-75.4, -27 (Halley Bay)	Annual	0.3-3	0.34	(Cachier et al. 1986; Hansen et al. 1988)
-87, -102 (South Pole)	Annual	0.3	0.16	(Hansen et al. 1988)
-87, -102 (South Pole)	Annual	1.3	0.16	(Bodhaine et al. 1995)
	Organic Carbon Concentrations in ng/m^3			
Northern Hemisphere				
11.3, -162.1 (Enewetak)	April–May	700	70	(Chesselet et al. 1981)
18.2, -62.5 (Puerto Rico)	March	660	500	(Novakov and Penner 1993)
15, -27 (North Atlantic)	Nov.	330–1600	384	(Ketsederis et al. 1976; Andreae 1983)
1.6-5.5, -81 (Pacific Ocean)	Feb.	425	417	(Wolff et al. 1986)
Southern Hemisphere				
-37.5, 77.3 (Amsterdam Island)	Annual	22	27.4	(Cachier et al. 1994)
-41, 174 (New Zealand)	August	130	81	(Cachier et al. 1986)
-2, -77.3 (Ecuador)	Annual	510	406–5000	(Andreae et al. 1984)
-10, -76 (Peru)	March–April	160	385	(Cachier et al. 1986)
-13.55, -172 (Samoa)	August	59	52	(Cachier et al. 1986)
-40.7, 144.7 (Cape Grim)	Annual	230	50–127	(Andreae 1983)
-20, -40 (South Atlantic)	Oct.–Nov.	280	538	(Andreae et al. 1984)
-2.4-0, -81 (Pacific Ocean)	Feb.	152	242	(Wolff et al. 1986)

There is good agreement between the simulated and observed sets of data for these three locations. All year, both concentrations are of the same order. Even for the worst case, Cuiaba, where the model systematically underestimates BC concentrations, the monthly BC variability reproduced by the model is also as found for Charles Point and Amsterdam Island.

Another validation of the parameters used in our modeling work may be found in the comparison of the modeled and the observed long-term patterns of BC concentrations in wet deposition for two African sites. Figures 47.7a and 47.7b, respectively, show results for Enyelle (Congo) and Lamto (Ivory Coast) rain. Both sets of data peak simultaneously and with the same intensity. The model is able to point out at the Enyelle site the background contamination of the atmosphere by biomass smoke throughout the year, as already shown by Cachier and Ducret (1991). It is possible from figure 47.7b to reconstruct the seasonal fire pattern in Lamto region.

Conclusions

This chapter is an exhaustive description of global biomass burning sources including tropical savanna and forest fires, agricultural waste fires, and biofuel sources. Construction of a detailed inventory of burnt biomass has been performed in accordance with the

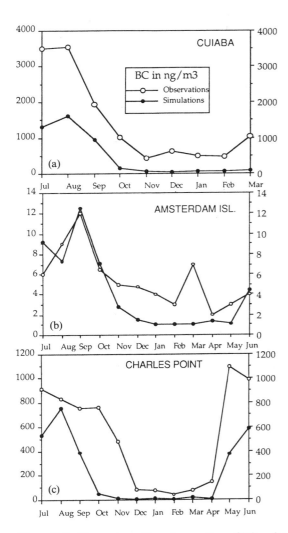

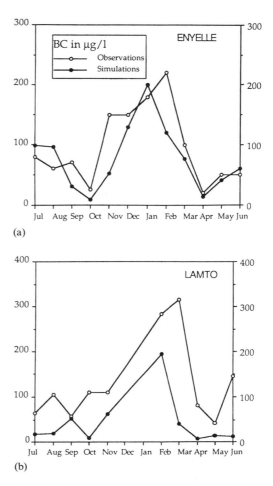

Figure 47.6 Comparison of long-term pattern of BC surface concentrations in ng/m^3 between simulated and observed data. (*a*) Cuiaba, Brazil (17.53S, 57.27W). Measurements from Artaxo et al. 1995. (*b*) Amsterdam Island (37.5S, 77.3E). Measurements from Cachier et al. 1994. (*c*) Charles Point, Australia (12.3S, 130.3E). Measurements from Artaxo (personal communication).

Figure 47.7 Comparison of long-term pattern of BC concentrations in wet depositions in µg/l between simulated and observed data. (*a*) Enyelle (2.8N, 18.06E), Congo. Measurements from Cachier and Ducret 1992. (*b*) Lamto (6.21N, 5.03W), Ivory Coast. Measurements from M. H. Pertuisot (personal communication).

most recent available literature. Black carbon and organic particle emissions have been estimated from recent experimental works.

The detailed inventory constructed here has been used in order to model the transport and deposition of black-carbon and organic carbon particles, using the 3D Lagrangian transport model of the L.L.N.L., Grantour. Comparisons of BC experimental concentrations found in the literature and obtained as output of the Grantour model have been seen to compare satisfactorily for a large subset of points, validating particulate transport and removal driven by the model.

However, this work must be considered as a first step only and efforts should be continued with this focus. First, due to scarce data in some regions such as Asia, some important sources such as agricultural waste fires could not be tested. Next, source improvement would consist of the introduction of boreal and temperate forests in our inventory, as at present, modeled concentrations during the summer in Greenland are much lower than those observed (M. H. Pertuisot, personal communication, 1995).

Sensitivity studies have shown that model results in remote areas are very sensitive to scavenging ratio and pollutant height injection. A better formulation of particle removal should be used to take into account the numerous phenomena occurring during particle aging. Experimental data on vertical distribution of

BC particles are also required to improve simulation of height injection.

The emission inventory for black-baron and organic particles that we have presented in this chapter has allowed for the first time to our knowledge, to obtain a global map of single-scattering albedo: this map points out that a global cooling effect may be expected from emissions of particulates by combustion sources. Such a preliminary result has been confirmed by coupling our inventory to that of the NCAR Community Climate Model, the CCM1 (Penner et al., 1996). However, the model radiative results were shown to be particularly sensitive to the relative amount of BC in particles. Further investigation is needed with a focus on this parameter and its variation with particle aging for all types of combustion sources.

Acknowledgments

This work was performed under the auspices of the U.S. Department of Energy by the Lawrence Livermore National Laboratory under Contract No. W-7405-Eng-48. Support from the NASA Aerosol Program and the DOE Quantitative Links program is gratefully acknowledged.

We would like to thank Paulo Artaxo and Marie Hélène Pertuisot for their helpful, unpublished black carbon data that allowed us to constrain Grantour simulations. We wish also to thank Greg Ayers and Robert Gillet for their important contribution to the Australian site. Francois Dulac is gratefully acknowledged for reviewing this chapter.

References

Andreae, M. O. 1983. Soot carbon and excess potassium: long range transport of combustion-derived aerosols. *Science* 220, 1148–1151.

Andreae, M. O., T. W. Andreae, R. J. Ferek, and H. Raemdonck. 1984. Long range transport of soot carbon in the marine atmosphere. *Sci. Total Environ.* 36, 73–80.

Artaxo, P., F. Gerab, M. A. Yamasoe, J. V. Martins, and A. Setzer. 1994. Trace elements and black carbon in aerosol particles in the Amazon basin, Paper presented at the Fifth Conference on carbonaceous aerosols, Berkeley, August 23–26.

Artaxo, P., F. Gerab, M. A. Yamasoe, J. V. Martins, A. Setzer, and A. H. Miguel. 1995. Large-scale and long-term biomass burning monitoring in the Amazon Basin, Paper presented at the Chapman Conference "Biomass Burning and Global Change", Williamsburg, Va., March 13–17.

Barnard, G., and L. Kristoferson. 1985. Agricultural residues as fuel in the third world, *Technical Report No 4*, Earthscan, 178 pp.

Bodhaine, B. A. 1995. Aerosol absorption measurements at Barrow, Mauna Loa, and the South Pole, *J. of Geophys. Res.*, 100, 8967–8975.

Butcher, S. S., and E. M. Sorenson. 1979. A study of wood stove particulate emissions, *J. Air Pollut. Contr. Assoc.*, 29, 724–728.

Butcher, S. S., U. Rao, K. R. Smith, J. Osborn, P. Azuma, and H. Fields. 1984. Emission factors and efficencies for small-scale open biomass combustion: towards standard measurement techniques, Paper presented at the annual meeting of the American Chemical Society, Division of Fuel Chemistry, Philadelphia, August 26–31.

Cachier, H., P. Buat-Ménard, M. Fontugne, and R. Chesselet. 1986. Long-range transport of continentally-derived particulate carbon in the marine atmosphere: evidence from stable carbon isotope studies. *Tellus* 38B, 161–177.

Cachier, H., M. P. Brémond, and P. Buat-Ménard. 1990. Organic and black carbon aerosols over marine regions of the Northern Hemisphere, In *Proceedings of the International Conference on Global Atmosphheric Chemistry*, Beijing, Newman and Kiange eds., Brookhaven National Laboratory Press, 241–261.

Cachier, H., and J. Ducret. 1991. Influence of biomass burning on Equatorial African rains, *Nature*, 352, 228–230.

Cachier, H., J. Ducret, M. P. Brémond, V. Yoboué, J. P. Lacaux, A. Gaudichet, and J. Baudet. 1991. Biomass burning aerosols in a savanna region of the Ivory Coast, in *Global Biomass Burning*, J. S. Levine ed., The MIT Press, Cambridge, Mass., 174–180.

Cachier, H. 1992. Biomass burning sources. In *Encyclopedia of the Earth System Science*, Vol 1, Academic Press, 377–385.

Cachier, H., C. Liousse, A. Cachier, B. Ardouin, G. Polian, V. Kazan, and A. D. A. Hansen. 1994. Black carbon aerosols at the remote site of Amsterdam Island. Paper presented at the Fifth Conference on Carbonaceous Aerosols, Berkeley, August 23–26.

Cachier, H. 1995. Combustion carbonaceous aerosols in the atmosphere: implications for ice-core studies, Nato ASI series, in "Ice Core Studies of Biogeochemical Cycles," R. Delmas ed., Springer-Verlag, (1) 30, 347–360.

Cachier, H., C. Liousse, P. Buat-Ménard, and A. Gaudichet. 1995. Particulate content of savanna fire emissions, *J. Atmos. Chem.*, 22, 123–148.

Cachier, H., C. Liousse, A. Gaudichet, F. Echalar, T. Kuhlbusce, and J. P. Lacaux. 1996. Particulate emissions during savanna fires in Kruger National Park (South Africa), *submitted to J. of Geophys. Res.*

Cahoon, D. R., B. J. Stocks, J. S. Levine, W. R. Cofer III, and K. P. O'Neil. 1992. Seasonal distribution of African savanna fires. *Nature* 359, 812–815.

Carroll, J. J., G. E. Miller, J. F. Thompson, and E. F. Darley. 1977. The dependance of open field burning emissions and plume concentrations on meteorology, field conditions and ignition technique, *Atmos. Env.*, 11, 1037–1050.

Chesseley, R., M. Fontugne, P. Buat-Menard, U. Ezat, and C. E. Lambert. 1981. The origin of particulate organic carbon in the marine atmosphere as indicated by its stable carbon isotopic composition, *Geophys. Res. Lett.*, 8, 345–348.

Chylek, P., V. Srivastava, L. Cahenzli, R. G. Pinnick, R. L. Dod, T. Novakov, T. L. Cook, and B. D. Hinds. 1987. Aerosol and graphitic carbon content of snow, *J. Geophys, Res.*, 92, 9801–9809.

Clain, M. P. 1995. Etude de la composante organique de l'aérosol atmosphérique, Ph.D. dissertation, Université de Savoie, 158 pp.

Clarke, A. D., and K. J. Noone. 1988. Soot in the Arctic snowpack: a cause for perturbations in radiative transfer. *Atmos. Env.* 19, 2045–2053.

Clarke, A. D. 1989. Aerosol light absorption by soot in remote environments. *Aerosol Sci. Technol.* 10, 161–171.

Cooke, W. F., and J. J. N. Wilson. 1996. A global black carbon aerosol model, Paper submitted to *J. of Geophys. Res.*

Crutzen, P. J., and M. O. Andreae. 1990. Biomass burning in the Tropics: impact on atmospheric chemical and biogeochemical cycles. *Science* 250, 1669–1678.

Darley, E. F., F. R. Burleson, E. H. Mateer, J. T. Middleton, and V. P. Osterli. 1979. Contribution of burning of agricultural wastes to photochemical air pollution, *J. Air Pollut. Contr. Assoc.*, 11, 685–690.

Delmas, R., P. Loudjani, and A. Podaire. 1991. Biomass burning in Africa: an assessment of annualy burnt biomass, in *Global Biomass Burning*, J. S. Levine ed., MIT Press, Cambridge, Mass., 126–132.

Delmas, R., J. P. Lacaux, and D. Brocard. 1995. Determination of biomass burning emission factors: methods and results *Environmental Monitoring and Assessment*, 38, 181–204.

Dinh, P. V., J. P. Lacaux, and R. Serpolay. 1994. Cloud active particles from African savanna combustion experiments, *Atmos. Res.* 31, 41–58.

Ducret, J., and H. Cacher. 1992. Particulate carbon in continental region rains and snows. *J. Atmos. Chem.* 15, 55–67.

Echalar, F., A. Gaudichet, H. Cachier, and P. Artaxo. 1995. Aerosol emissions by tropical forest and savanna biomass burning: characteristic trace elements and fluxes, *Geophys. Res. Lett.*, 22, 3039–3042.

Food and Agriculture, *FAO Production Yearbook*, 1989.

Food and Agriculture, *FAO Production Yearbook*, vol. 45, 265 pp, 1991.

Fischer, R. A. 1983. in *Potential Productivity of Field Crops on Different Environments*, International Rice Research Institute, Philippines, 526 pp.

Fishcer, K. S., and A. F. E. Palmer. 1983 in *Potential Productivity of Field Crops on Different Environments*, International Rice Research Institute, Philippines, 526 pp.

Franca, J. R. A., J. M. Brustent, J. Fontan, J. M. Grégoire, and J. P. Malingreau. 1993. A multi-spectral remote sensing of biomass burning in West Africa during 90/91 season, Paper presented at the E.G.S meeting, Wiesbaden, 3–7 May.

Gaudichet, A., F. Echalar, B. Chatenet, J. P. Quisefit, G. Malingre, H. Cachier, P. Buat-Menard, P. Artaxo, and W. Maenhaut. 1995. Trace element in tropical african savanna biomass burning aerosols, *J. Atmos. Chem.*, 22, 19–39.

Garstang, M., P. D. Tyson, H. Cachier, J. P. Lacaux, and L. Radke. 1996. Atmospheric transport of particulate and gaseous products by fires, NATO ASI series (J. Clarke, ed.), Springler Verlag, In *Sediment Records of Biomass Burning and Global Change*.

Gerstle, R. W., and D. A. Kemnitz. 1967. Atmospheric emissions from open burning, *J. Air Pollut. Contr. Assoc.*, 17, 324–327.

Hansen, A. D. A., B. A. Bodhaine, E. G. Dutton, and R. C. Schnell. 1988. Aerosol black carbon measurement at the South Pole: initial results, 1986–1987. *Geophys. Res. Lett.*, 15, 1193–1196.

Hao W. M., M. H. Liu, and P. J. Crutzen. 1990. Estimates of annual and regional releases of CO_2 and other trace gazes to the atmosphere from fires in the tropics, based on the FAO Statistics for the period 1975–1980, in *Fire in the tropical Biota*, J. C. Goldammer ed., 440–462.

Hao W. M., and M.-H. Liu 1995. Spatial and temporal distribution of tropical biomass burning, *Global Biogeochemical Cycles*, 8, 495–503.

Haywood, J. M., and K. P. Shine. 1995. The effect of anthropogenic sulfate and soot aerosol on the clear sky planetary radiation budget, *Geophys. Res. Lett.*, 22, 603–606.

Heintzenberg, J., and E. K. Bigg. 1990. Tropospheric transport of trace substances in the southern hemisphere, *Tellus*, 42B, 355–363.

Héry, J. L. 1993. Caractérisation des émissions particulaires par combustion de biomasse végétale: exemple d'une savane sèche d'Afrique du Sud. Fabrication artisanale de charbon de bois, Master, Univ. Paris VII, 26 pp.

Irvine J. E. 1983. In *Potential Productivity of Field Crops on Different Environments*, International Rice Research Institute, Philippines, 526 pp.

Jaffrezo J. L., R. E. Hillamo, C. I. Davidson, and W. Maenhaut. 1993. Size distribution of atmospheric trace elements at Dye 3, Greenland-II. Sources and Transport, *Atmos. Env.*, 27A, 2803–2814.

Jenkins, B. M., S. Q. Turn, R. B. Williams, D. P. Y. Chang, O. G. Raabe, J. Paskind, and S. Teague. 1991. in *Global Biomass Burning*, J. S. Levine ed., MIT Press, Cambridge, Mass., 305–317.

Jenkins, B. M., I. M. Kennedy, S. Q. Turn, R. B. Williams, S. G. Hall, S. V. Teague, D. P. Y. Chang, and O. G. Raabe. 1993. Wind tunnel modeling of atmospheric emissions from agricultural burning: influence of operating configuration on flame structure and particle emission factor for a spreading-type fire, *Environ. Sci. Technol.*, 27, 1763–1775.

Kaufman, T. J., A. Setzer, D. Ward, D. Tanré, B. N. Holben, P. Menzel, M. C. Pereira, and R. Ramussen. 1992. Biomass burning airborne and space borne experiment in the Amazonas (Base-A), *J. Gephys. Res.* 97, 14 581–14 599.

Ketsederis, G., J. Hahn, R. Jaeniche, and C. Junge. 1976. The organic constituents of atmospheric particulate matter, *Atmos. Env.*, 10, 603–610.

Lacaux, J. P., J. M. Brustet, R. Delmas, J. C. Menaut, L. Abbadie, B. Bonsang, H. Cachier, J. G. R. Baudet, M. O. Andreae, and G. Helas. 1995. Biomass burning in the tropical savannas of Ivory Coast: an overview of the field experiment Fire of Savanna (FOS-DECAFE 91), *J. Atmos. Chem.*, 22, 195–216.

Lambert, G., G. Polian, J. Sanak, B. Ardouin, A. Buisson, A. Jegou, and J. C. Le Roulley. 1982. Cycle du radon et de ses descendants: application a l'etude des echanges troposphere-stratosphere, *Ann. Geophys.*, 4, 497–531.

Le Canut, P., M. O. Andreae, G. W. Harris, F. G. Wienhold, and T. Zenker. 1996. Airborne studies of emissions from savanna fires in southern Africa: I. Aerosol emissions measured with a laser-optical particle counter, submitted to *Journal of Geophys. Res.*

Levine, J. S., W. R. Cofer III, D. R. Cahoon Jr., and E. L. Winstead. 1995. Biomass burning: a driver for global change, *Env. Sci. Technol.*, 29, 120–125.

Liousse, C., H. Cachier, and S. J. Jennings. 1993. Optical and thermal measurements of black carbon aerosol content if different environments: variation of the specific attenuation cross-section, sigma (s), *Atmos, Env.*, 27A, 1203–1211.

Liousse, C., J. R. A. Franca, H. Cachier, and F. Dulac. 1994. Monitoring of carbonaceous aerosols at Lamto, Ivory Coast and comparison with fire pixel numbers retrieved from AVHRR satellite data, Paper presented to the Fifth Conference on Carbonaceous Aerosols, Berkeley, August 23–26.

Liousse, C., J. E. Penner, C. CHuang, J. J. Walton, H. Eddleman, and H. Cachier. 1996. A global three-dimensional modeling of carbonaceous aerosols, Paper accepted at *J. Geophys. Res.*

Mahtab, F. U., and M. N. Islam. 1984. Biomass availability from field crops, Paper presented for Bangladesh Energy Planning Project Rural Energy Course, Government of Bangladesh, Dhaka.

Meland B., and R. Boubel. 1966. A study of field burning under varying environmental conditions, *J. Air Pollut. Contr. Assoc.*, 16, 481–484.

Menaut, J. C., L. Abbadie, F. Lavenu, P. Loudjani, and A. Podaire. 1991. Biomass burning in west African savannas. In *Global Biomass Burning* J. S. Levine ed., MIT Press, Cambridge, Mass., 133–142.

Middleton, J. T., and E. F. Darley. 1973. Control of air pollution affecting or caused by agriculture, in *Pollution: Engineering and Scientific Solutions*, E. S. Barrekette, ed., Plenum Press, New York, pp. 148–157.

Novakov, T., and J. E. Penner. 1993. Large contribution of organic aerosols to cloud-condensation-nuclei concentrations, *Nature*, 365, 823–826.

Ogren, J. A., P. J. Groblicki, and R. J. Charlson. 1984. Measurement of the removal rate of elemental carbon from the atmosphere, *Sci. Total. Environ.*, 36, 329–338.

Openshaw, K., Wood fuels the developing world, *New Scientist*, January, 31th, 1974.

Parungo, F., C. Nagamoto, M. Y. Zhou, A. D. A. Hansen, and J. Harris. 1994. Aeolian transport of aerosol black carbon from China to the ocean, Paper presented to the Fifth Conference on carbonaceous aerosols, Berkeley, August 23–26.

Penner, J. E., S. J. Ghan, and J. J. Walton. 1991. The role of biomass burning in the budget and cycle of carbonaceous soot aerosols and their climate impact, in *Global Biomass Burning*, J. S. Levine ed., MIT Press, Cambridge, Mass., 387–393.

Penner, J. E., R. E. Dickinson, and C. A. O'Neill. 1992. Effects of Aerosol from biomass burning on the global radiation budget, *Science*, 256, 1432–1433.

Penner, J. E., H. Eddleman, and T. Novakov. 1993. Towards the development of a global inventory for black carbon emissions, *Atmos. Env.*, 27A, 1277–1295.

Penner, J. E., C. A. Atherton, and T. E. Graedel. 1994. Global emissions and models of photochemically active compounds, in *Global Atmospheric-Biospheric Chemistry*, ed. R. Prinn, Plenum Publishing, New York, 223–248.

Penner, J. E., C. Liousse, C. Chuang, K. Taylor, K. Grant, and A. Grossman. 1996. Climate forcing by carbonaceous aerosols, in preparation, to be submitted to *Science*.

Radke, L. F., D. A. Hegg, J. H. Lyons, C. A. Brock, P. V. Hobbs, R. E. Weiss, and R. Rasmussen. 1988. Airborne measurements on smokes from biomass burning. In *Aerosols and Climate*, P. Hobbs and P. McCormick eds., A Deepack, Hampton, 411–421.

Robinson, J. M. 1989. On uncertainty in the computation of global emissions from biomass burning, *Climatic Change*, 14, 243–262.

Seiler, W., and P. J. Crutzen. 1980. Estimates of gross and net fluxes of carbon between the biosphere and the atmosphere from biomass burning, *Climatic Change*, 2, 207–247.

Smith K. R., A. L. Aggarwal, and R. M. Dave. 1983. Air pollution and rural biomass fuels in developing countries: a pilot village study in India and implications for research and policy, *Atmos. Env.*, 17, 2343–2362.

Strehler, A., and W. Stutzle. 1987. Biomass Residues, in *Biomass*, D. O. Hall and R. P. Overend eds., 75–102.

Susott, R. A., D. E. Ward, R. E. Babbit, and D. J. Latham. 1991. The measurement of trace emissions and combustion characteristics for a mass fire, in *Global Biomass Burning*, J. S. Levine ed., MIT Press, Cambridge, Mass., 245–258.

Turn, S. Q., B. M. Jenkins, J. C. Chow, L. C. Pritchett, D. Campbell, T. Cahill, and S. A. Whalen. 1993. Elemental characterization of particulate matter emitted from biomass burning: wind tunnel derived source profiles for herbaceous and wood fuels, Paper presented to the AGU Fall Meeting, San Francisco, December 6–10.

Walton, J. J., M. C. MacCracken, and S. J. Ghan. 1988. A global-scale lagrangian trace species model of transport, transformation and removal process, *J. Geophys. Res.*, 93, 8339–8354.

Ward, D. E., A. Setzer, Y. J. Kaufman, and R. A. Rasmussen. 1991. Characteristics of smoke emissions from biomass fires of the Amazon region-Base-A experiment, In "*Global Biomass Burning*," J. S. Levine ed., MIT Press, Cambridge, Mass., 394–402.

Warren, S. G., and A. D. Clarke. 1990. Soot in the atmosphere and snow surface of Antarctica, *J. Geophys. Res*, 95, 1811–1816.

Wolff, G. T., M. S. Ruthkosky, D. P. Stroup, P. E. Korsog, and M. A. Ferman. 1986. Measurements of SO$_x$, NO$_x$, and aerosol species on Bermuda, *Atmos. Env.*, 1229–1239.

World Resources 1992–1993, *A Guide to the Global Environment. Towards, Sustainable Development*. Oxford University Press, New York, 385 pp. 1992.

Yevich, R. and J. A. Logan. 1995. Assessment of the spatial distribution of biomass fuel consumption and burning of agricultural wastes. Paper presented at the Chapman Conference, March 1995.

Natural Organic Compounds as Tracers for Biomass Combustion in Aerosols

Bernd R. T. Simoneit, Mohammed Radzi bin Abas, Glen R. Cass,
Wolfgang F. Rogge, Monica A. Mazurek, Laurel J. Standley,
and Lynn M. Hildemann

Biogenic organic matter, consisting predominantly of lipids, biopolymers (lignin), and humic and fulvic acids, is now firmly established as a major carbonaceous fraction in atmospheric particles found in urban, rural, and remote locales (i.e., Arpino et al. 1972; Broddin et al. 1980; Cox et al. 1982; Duce et al. 1983; Eichmann et al. 1979, 1980; Gagosian et al. 1981, 1982, 1987; Marty and Saliot 1982; Matsumoto and Hanya 1980; Mazurek and Simoneit 1984; Mazurek et al. 1989; Ohta and Handa 1985; Rogge et al. 1993a; Simoneit 1977, 1979, 1980, 1984a,b, 1985, 1986, 1989; Simoneit and Mazurek 1982, 1989; Simoneit et al. 1977, 1980, 1983, 1988, 1990, 1991a,b,c). However, in comparison to the relatively extensive studies that have been carried out on nonpolar constituents of both biogenic and anthropogenic origins in these aerosols, only limited molecular information is available on polar compounds (e.g., Gagosian et al. 1987; Hawthorne et al. 1988, 1989, 1992; Mazurek et al. 1989; Rogge et al. 1993a,b; Schneider et al. 1983; Simoneit 1985, 1989; Simoneit and Mazurek 1982; Simoneit et al. 1983, 1988, 1993).

Biomass combustion is an important primary source of many trace substances that are reactants in atmospheric chemistry and of soot particulate matter that decreases visibility and absorbs incident radiation (e.g. Crutzen et al. 1985; Levine, 1991; Lobert et al. 1990; Seiler and Crutzen 1980). Not all sources of biomass have been characterized for combustion tracers. Thus, there is a need to develop additional specific tracers for this input process because only a limited number of molecular markers have been proposed.

This additional input of biogenic organic matter to the troposphere (urban, rural, and remote) from biomass combustion occurs by natural, as well as manmade, fires. Thermally altered (pyrolysis) and directly emitted molecular markers may be used as indicators. This concept has been applied preliminarily to tracing biomass burning in Oregon (Standley and Simoneit 1987, 1990, 1994), in China (Simoneit et al. 1991b), and Amazonia, Brazil (Abas et al. 1995). For example, retene, a thermal alteration product from resin diterpenoids (e.g., abietic acid), has been found in aerosols in Elverum, Norway, and in Oregon, and at trace levels in Los Angeles, California (Ramdahl 1983; Mazurek et al. 1989; Standley and Simoneit 1987, 1990, 1994), and China (Simoneit et al. 1991b). Retene was not detectable in aerosols of the Harmattan or in urban samples of Nigeria, nor in rural aerosols of southeast Australia, because conifer wood is not used for fuel in those areas (Simoneit et al. 1988, 1991c). Since there is no major terrestrial noncombustion source for retene, it is useful as an indicator of wood burning, especially conifer, but is not always concentrated enough for detection. Thus, additional tracers of thermally altered and directly emitted natural products need to be characterized to assign input sources of organic matter from biomass combustion to aerosols.

Experimental Methods

Samples

Large samples of smoke particulate matter were acquired by high volume filtration on prebaked quartz fiber filters, and for some the complementary fine particle fractions ($d_p < 2\,\mu m$) were collected after appropriate dilution on small prebaked quartz fiber filters (47 mm diam.) (e.g., Hildemann et al. 1991a; Standley and Simoneit 1987). Data are discussed from various types of smoke samples consisting of (1) the fine fraction from pine and oak wood combustion in a fireplace (Hildemann et al. 1991b; Rogge et al. 1993a; unpublished data), (2) controlled burns of forest litter in Oregon (Standley and Simoneit 1987, 1994), (3) alder wood combustion in a wood-stove (Standley and Simoneit 1990), (4) various grass and chaparral fires (Simoneit et al. unpublished data), and (5) a controlled fire of tropical forest litter in Amazonia, Brazil (Abas et al. 1995).

Lipid Isolation and Separation

The filters with particulate matter were extracted with methylene chloride (CH_2Cl_2) or other polar solvent mixtures (e.g., benzene/methanol) within the filter storage jars (Simoneit and Mazurek 1982). The solvent extracts were filtered for the removal of insoluble particles. Filtrates were first concentrated on a rotary evaporator and then by using a stream of filtered nitrogen gas to a volume of approximately 5 mL. Volumes were then adjusted to 5.0 mL exactly by the addition of CH_2Cl_2.

Typically, an aliquot (2.5 mL) of the filtrate was transferred to a 5 mL heavy-walled, conical vial, and 30 μL perdeutero-n-tetracosane (90 μg/mL) was added as internal standard. The mixture was reduced in volume to 500 μL and 3.0 mL of BF_3-methanol (Pierce) was added. The vial was capped and heated to approximately 60°C for 10 min. to esterify carboxylic acids. After cooling, the mixture was transferred to a separatory funnel with 30 mL of hexane and washed twice with saturated NaCl solution, discarding the aqueous layer. The organic layer was dried over Na_2SO_4. The solvent was evaporated under nitrogen to almost dryness and reconstituted with a known volume of hexane (final volume 500 μL). The derivatized extract was subjected to thin layer chromatography (TLC) using silica-gel plates (0.25 mm, Alltech) and eluted with a mixture of hexane and diethyl ether (9:1). The TLC plates had been cleaned prior to use by repetitive elutions with methanol and CH_2Cl_2, and before sample application they were activated in an oven at 120°C for 45 min. The weight of the total extract was quantified by evaporating an aliquot to dryness and weighing, and the fractions were quantified by gas chromatographic analyses.

The TLC fractions and the total and methylated extract fractions were subjected to gas chromatographic (GC) and gas chromatography-mass spectrometric (GC-MS) analyses. Alcohol fractions were converted to the trimethylsilyl ethers prior to GC and GC-MS analysis by reaction with N,O-bis-(trimethylsilyl)-trifluoroacetamide (BSTFA) for approximately 30 min at ~70°C under a nitrogen atmosphere.

Lipid Analyses

The GC analyses were carried out on a Hewlett-Packard Model 5840A gas chromatograph using a 25 m × 0.20 mm i.d. fused silica capillary column coated with DB-5 (J and W, Inc.). The GC-MS analyses were conducted on a Finnigan Model 4021 quadrupole mass spectrometer, interfaced directly with a Finnigan Model 9610 gas chromatograph, and equipped with a fused silica capillary column coated with DB-5 (30 m × 0.25 mm i.d.). The GC and GC-MS

operating conditions were as follows: isothermal at 65°C for 2 min (GC only), temperature programmed from 65 to 310°C at 4°C per min, and held isothermal at 310°C for 60 min, using helium as carrier gas. Mass spectrometric data were acquired and processed using a Finnigan-Incos Model 2300 data system. Molecular markers were identified by GC and GC-MS comparison with authentic standards and characterized mixtures. Unknown compounds were characterized by interpretation of their mass spectrometric fragmentation pattern.

Results and Discussion

Combustion Process

Combustion by burning other than incineration at ultrahigh temperatures is not completely destructive to the organic matter in the fuels. Thus, the injection of organic tracer compounds into smoke occurs primarily by direct volatilization/steam stripping and by thermal alteration based on combustion temperature. The degree of alteration of organic matter increases as the burn temperature rises and the moisture content of the fuel decreases (Simoneit et al. unpublished data). Although the molecular composition of organic matter in smoke particles is highly variable, the molecular tracers are generally still source specific (e.g., Hawthorne et al. 1988, 1989, 1992; Standley and Simoneit 1987, 1990, 1994, Abas et al. 1995). For example, retene, a thermal alteration product from resin diterpenoids (e.g., abietic acid) as described above, has been found in aerosols in Norway and Oregon, and at trace levels in some urban areas (Ramdahl 1983; Mazurek et al. 1991; Standley and Simoneit 1987, 1994; Simoneit et al. 1991a). Retene was not detectable in aerosols of geographic regions where conifers are not grown or used as fuel (Abas et al. 1995; Simoneit et al. 1988, 1991b). Since there is no known terrestrial noncombustion source for retene, it is useful as an indicator of conifer wood burning. However, retene does not always have a concentration adequate for detection in ambient aerosols. Thus, definition of additional tracers of thermally altered and directly emitted natural products (e.g., resin acids) will aid the assessment of the impact of biomass combustion on aerosols.

Organic Tracers

Smoke particles contain homologous compound series and biomarkers, which are derived directly from plant wax, gum, and resin by volatilization, and secondarily from pyrolysis of biopolymers (e.g., cellulose, lignin, cutin, suberin), wax, gum, and resin. Biomarkers or molecular tracers are the indicator compounds best utilized for confirmation of genetic sources of carbonaceous fractions in smoke emissions. As applied here, biomarkers are utilized as indicators of origins from natural product compounds of vegetation and their altered derivatives (residues) by partial combustion. The complexity of the organic components of smoke aerosol particles will be illustrated with three examples.

Conifer Wood Smoke The first example demonstrates that diterpenoids are excellent indicators for smoke from burning of gymnosperm wood (e.g., conifers, Ramdahl 1983; Standley and Simoneit 1987, 1994). Resin acids such as abietic or pimaric acids are produced by conifers. These compounds and their derivatives at various stages of thermal alteration have been found in ambient aerosols (Simoneit and Mazurek 1982; Simoneit 1989) and in smoke from slash and wood burning (Standley and Simoneit 1987, 1994; Simoneit et al. 1993). This can be illustrated with the mass fragmentograms for the extract fractions (acids as methylated derivatives) of fine aerosol particles from a fireplace where pine wood was burned (figure 48.1a, b; Rogge et al., unpublished data; Simoneit et al. 1993). The major diterpenoid compounds in the smoke are pimaric (2), [chemical structures are shown in Appendix 48.1], sandaracopimaric (3), isopimaric (4), dehydroabietic (5), and abietic (6) acids (figure 48.1a, b). Dehydroabietic acid can be regarded as an altered product from resin acids and the other diterpenoid acids are directly volatilized unaltered marker compounds. An aerosol sample taken during winter in Los Angeles is also shown for comparison (figure 48.1c, d). Dehydroabietic acid is the dominant diterpenoid marker and the other resin acids are present at reduced levels. $\Delta^{8,15}$-Pimaradien-18-oic acid (1, figure 48.1) appears to be an alteration product due to double bond migration, and retene and other hydrocarbon diterpenoid derivatives occur at trace levels only. This demonstrates that diterpenoid acids are recognizable tracers in the urban atmosphere for wood smoke.

The second example is smoke from a controlled slash burn of litter from conifers and shrubs illustrating the ketones formed by combustion (Standley and Simoneit 1987). The ketone fraction consists primarily of n-alkan-2-ones, n-alkanals, and triterpenones (figure 48.2). The n-alkan-2-ones range from C_{16} to C_{33}, with a $C_{max} = 27$ and strong odd-to-even carbon number preference. The n-alkanals range from C_{19} to C_{31}, with

a $C_{max} = 23$ and also a strong odd-to-even carbon number preference. Both n-alkanones and n-alkanals are partial oxidation products from alkanols, alkanes, and other aliphatic moieties, analogous as is observed in laboratory hydrous pyrolysis experiments (Leif and Simoneit 1995). The triterpenones are mainly the amyrones, taraxerone, lupenone, and a friedelin derivative (figure 48.2b). They are either natural products or low-temperature oxidation products from the respective triterpenol precursors and are characteristic of their higher plant sources.

Amazon Biomass Smoke The third example is a smoke sample from the burning of composited vegetation in Amazonia (Abas et al. 1995), and the various lipid fractions extracted from the particulate matter are shown in figure 48.3. The total extract is comprised mainly of n-alkanoic acids, n-alkanes, polynuclear aromatic hydrocarbons (PAH), and triterpenoids. The n-alkanes range from C_{17} to C_{37} with a $C_{max} = 29/31$ and high odd carbon number preference (figure 48.3b), which is similar as reported for aerosols from the Amazon region, indicating that alkanes from burning are indistinguishable from plant wax alkanes in the ambient aerosol (Simoneit et al. 1990). The same is the case for the n-alkanoic acids and n-alkanols (figure 48.3c, f).

More specific homologous aliphatic tracers for combustion are the n-alk-1-enes, n-alkan-2-ones and α,ω-alkanedioic acids (figure 48.3b, d, e). For example, the major series of n-alk-1-enes present ranges from C_{17} to C_{35}, with $C_{max} = 22$ (minor at 29/31) and a slight even carbon number predominance (CPI = 0.8, figure 48.4b). Alkenes are not dominant components in aerosols or plant waxes. The origin of these compounds is inferred to be from the biomass fuel, and based on their carbon number distribution primarily from n-alkanols by dehydration (n-alkan-1-ols are easily dehydrated to n-alk-1-enes by high temperatures), and to a minor degree from the n-alkanes by oxidation during incomplete combustion (compare figure 48.4b with c and a, respectively). The even carbon number dominance and $C_{max} = 22$ are derived from the n-alkanol distribution, and the minor dominance of n-C_{29} and n-C_{31} reflects the dehydrogenation of the n-alkanes.

The aliphatic hydrocarbon fraction contains a group of derivatives from the amyrins (e.g., β-7) (peaks 1–6, figure 48.3b, structures also shown in figure 48.5), which are various triterpadienes (e.g., ursa-2, 12-diene), noroleanene, norursene, and diaromatic A-noroleananes and A-norursanes (Abas et al. 1995). These compounds are not known as natural products

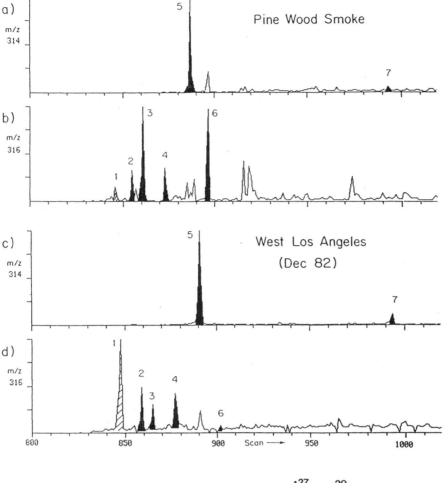

Figure 48.1 (*a, b*) Mass fragmentograms for the diterpenoid acids in pine wood smoke. (*c, d*) Fine particles of the West Los Angeles atmosphere in winter. Plots are molecular ions of methyl esters, m/z 314 and 316, numbers refer to compounds in text.

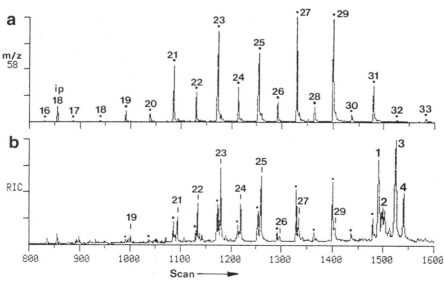

Figure 48.2 Salient features of a GC-MS analysis of a ketone fraction from a smoke aerosol in Oregon: (*a*) m/z 58 fragmentogram, numerals refer to carbon chain length of *n*-alkan-2-ones (ip 18 = 6,10,14-trimethylpentadecan-2-one); (*b*) total ion current trace (equivalent to GC trace of total fraction), peaks with dots are *n*-alkan-2-ones, peaks with dashes are *n*-alkanals, and the triterpenones are (1) taraxerone, (2) β-amyrone, (3) lupenone, and (4) a friedelenone

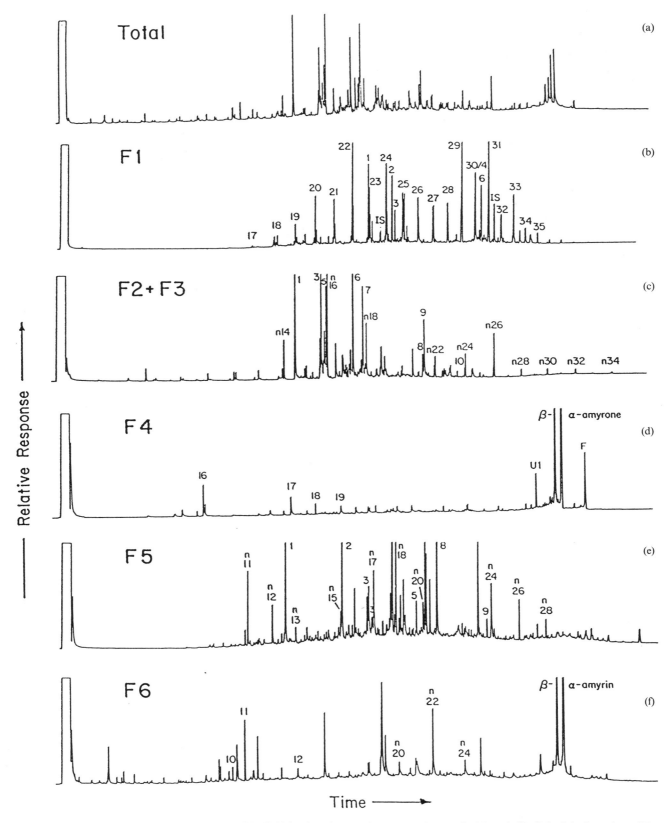

Figure 48.3 Representative gas chromatograms of the lipid fractions from an Amazon smoke sample: (*a*) total, (*b*) aliphatic hydrocarbons (F1, early eluting homolog of doublets is the *n*-alk-1-ene), (*c*) PAH and esters (F2 + 3), (*d*) ketones (F4), (*e*) dicarboxylic acids and oxy-PAH (F5), and (*f*) alcohols (F6) [ni = carbon numbers of homologous series, additional numbers (i.e., b: 1–6, c: 1–10, e, f: 1–12) see text, U1 = unknown triterpenoid, F = friedelin, and IS = internal standard, n-$C_{24}D_{50}$]

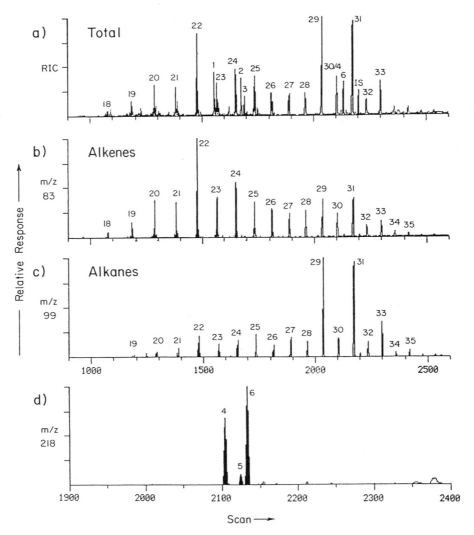

Figure 48.4 Salient features of the GC-MS data for the aliphatic hydrocarbons (F1) from an Amazon smoke sample: (*a*) total ion current trace, (*b*) alkenes, key ion m/z 83, (*c*) alkanes, key ion m/z 99, (*d*) triterpenes, key ion m/z 218 (numbers 18–35 refer to carbon chain length of homologous series and 1–6 are biomarkers, discussed in the text)

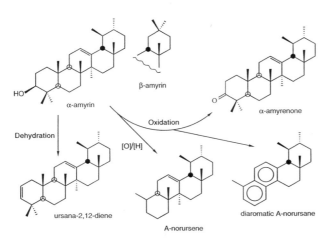

Figure 48.5 Thermal alteration products from amyrins in smoke

and are therefore indicators for combustion of biomass containing amyrin precursors. The α- and β-amyrins (7) are the predominant biomarkers in the total smoke lipids (figure 48.3*f*), and α- and β-amyrones (8) (mild oxidation products of amyrins) and friedelin (9) are also significant (figure 48.3*d*, *f*). Phytosterols from plant waxes are trace components in this smoke sample and consist mainly of β-sitosterol (10) with lesser amounts of other C_{29} and C_{28} isomers. This is characteristic as observed for other smoke emissions from vegetation (e.g., Simoneit et al. 1983, 1993).

The PAH are pyrogenic molecular markers in smoke, where the unsubstituted analogs are usually the characteristic compounds of higher temperature combustive processes. Major amounts of PAH are found in this sample (figures 48.3*c* and 48.6*a*, peaks *1–15*) and

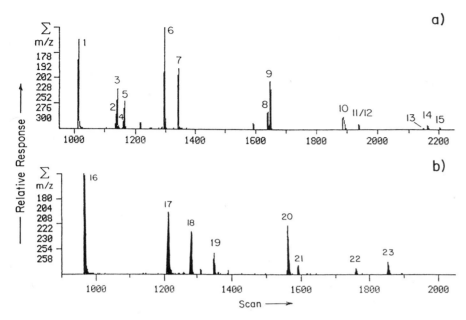

Figure 48.6 Salient features of the GC-MS data for the PAH and oxy-PAH in the Amazon smoke sample: (*a*) PAH in fraction 2 + 3 − (1) phenanthrene, (2) 3-methylphenanthrene, (3) 2-methylphenanthrene, (4) 9-methylphenanthrene, (5) 1-methylphenanthrene, (6) fluoranthene, (7) pyrene, (8), (9) chrysene (triphenylene), (10) benzofluoranthenes, (11) benzo(e)pyrene, (12) benzo(a)pyrene, (13) anthanthrene, (14) indeno(1,2,3-cd)pyrene, (15) benzo(ghi)perylene; (*b*) oxy-PAH in fraction 4 − (16) 9*H*-fluoren-9-one, (17) anthra-9,10-quinone, (18) 4*H*-cyclopenta(def)phenanthren-4-one, (19) methylanthra-9,10-quinone, (20) 11*H*-benzo(a)fluoren-11-one, (21) 7*H*-benzo(c)fluoren-7-one, (22) benz(a)anthra-7,12-quinone, (23) naphthanthrone (6*H*-benzo(cd)pyren-6-one)

consist primarily of phenanthrene (*1*), methylenephenanthrene, methylphenanthrenes (*2–5*), fluoranthene (*6*), pyrene (*7*), and chrysene (*9*), and traces of benzofluoranthenes (*10*), benzo (e&a) pyrenes, anthanthrene, indenopyrene, benzo(ghi)perylene, and coronene. Oxy-PAH are also significant (figures 48.3*e* and 48.6*b*, peaks *16–23*) and the predominant compounds are fluorenone (*16*), anthra-9, 10-quinone (*17*), cyclopenta-(def)phenanthren-4-one (*18*, 11), benzo(a)fluoren-11-one (*20*, 12), benzanthrone (*13*), and naphthanthrone (*23*, 14), with minor amounts of methylanthra-9, 10-quinones, benzo(c)fluoren-7-one and benz(a)anthra-7,12-quinone. The chemical structures of the PAH and oxy-PAH cannot be correlated to specific sources; they are resynthesis products from any high temperature combustive process using organic matter as fuel.

Phenolic products from the pyrolytic breakdown of lignin in vegetation have been proposed as tracers specific for plant classes (taxa) (e.g., Hawthorne et al. 1988, 1989, 1992; Simoneit et al. 1993). The dominant markers for lignin in this smoke sample found in fraction 6 (figure 48.3*f*) are vanillin (15), vanillic acid (16), syringaldehyde (17), syringic acid (18), and guaiacylacetone (19), with traces of various other phenolic products. This group of tracers confirms the relative

contribution to the smoke from each biomass taxon in the fuel (Abas et al. 1995).

Conclusions

The major organic components of smoke particles from tropical biomass are straight-chain, aliphatic and oxygenated compounds, and triterpenoids from vegetation waxes, resins/gums, and biopolymers. Several compounds (e.g., amyrones, friedelin, aromatic A-noroleananes, and other thermal derivatives from triterpenoids and from lignin—syringaldehyde, vanillin, syringic acid, vanillic acid) are potential key indicators for smoke components from combustion of such biomass. Burning of biomass from temperate regions (i.e., conifers) yields characteristic tracers from diterpenoids as well as phenolics and other oxygenated species, which are recognizable in urban airsheds. Biomass combustion smoke from tropical areas (i.e., hardwoods) also yields characteristic tracers. The precursor to product approach of organic geochemistry can be applied successfully to delineate tracers (biomarkers) for the environmental process of biomass combustion.

Acknowledgments

This research was supported by the U.S. Environmental Protection Agency under agreement R-819714-01-0 (to G.R.C) and R-823990-01 (to B.R.T.S) and by the South Coast Air Quality Management District. This manuscript has not been subject to the EPA's peer and policy review, and hence does not necessarily reflect the views of the EPA.

References

Abas, M. R., B. R. T. Simoneit, V. Elias, J. A. Cabral, and J. N. Cardos. 1995. Composition of higher molecular weight organic matter in smoke aerosol from biomass combutstion in Amazonia. *Chemosphere* 30, 995–1015.

Arpino, P., A. van Dorsselaer, K. D. Sevier, and G. Ourisson. 1972. Cires aeriènnes dans une forêt de Pins. *C. R. Acad. Sc. (Paris)* 275D, 2837–2840.

Broddin, G., W. Cautreels, and D. van Cauwenberghe. 1980. On the aliphatic and polyaromatic hydrocarbon levels in urban and background aerosols from Belgium and the Netherlands. *Atmos. Environ.* 14, 895–910.

Cox, R. E., M. A. Mazurek, and B. R. T. Simoneit. 1982. Lipids in Harmattan aerosols of Nigeria. *Nature* 296, 848–849.

Crutzen, P. J., A. C. Delany, J. Greenberg, P. Haagenson, L. Heidt, R. Lueb, W. Pollock, W. Seiler, A. Wartburg, and P. Zimmerman. 1985. Tropospheric chemical composition measurements in Brazil during the dry season. *J. Atmospher. Chem.* 2, 233–256.

Duce, R. A., V. A. Mohnen, P. R. Zimmerman, D. Grosjean, W. Cautreels, R. Chatfield, R. Jaenicke, J. A. Ogren, E. D. Pellizzari, and G. T. Wallace. 1983. Organic material in the global troposphere. *Rev. Geophys. Space Phys.* 21, 921–952.

Eichmann, R., P. Neuling, G. Ketseridis, J. Hahn, R. Jaenicke, and C. Junge. 1979. *n*-Alkane studies in the troposphere–I: Gas and particulate concentrations in North Atlantic air. *Atmosph. Environ.* 13, 587–599.

Eichmann, R., G. Ketseridis, G. Schebeske, R. Jaenicke, J. Hahn, P. Warneck, and C. Junge. 1980. *n*-Alkane studies in the troposphere–II: Gas and particulate concentrations in Indian Ocean air. *Atmosph. Environ.* 14, 695–703.

Gagosian, R. B., E. T. Peltzer, and J. T. Merrill. 1987. Long-range transport of terrestrially derived lipids in aerosols from the South Pacific. *Nature* 325, 800–803.

Gagosian, R. B., E. T. Peltzer, and O. C. Zafiriou. 1981. Atmospheric transport of continentally derived lipids to the tropical North Pacific. *Nature* 291, 321–324.

Gagosian, R. B., O. C. Zafiriou, E. T. Peltzer, and J. B. Alford. 1982. Lipids in aerosols from the tropical North Pacific: temporal variability. *J. Geophys. Res.* 87, 11 133–11 144.

Hawthorne, S. B., D. J. Miller, R. M. Barkley, and M. S. Krieger. 1988. Identification of methoxylated phenols as candidate tracers for atmospheric wood smoke pollution. *Environ. Sci. Technol.* 22, 1191–1196.

Hawthorne, S. B., M. S. Krieger, D. J. Miller, and M. B. Mathiason. 1989. Collection and quntitation of methoxylated phenol tracers for atmospheric pollution from residential wood stoves. *Environ. Sci. Technol.* 23, 470–475.

Hawthorne, S. B., D. J. Miller, J. J. Langenfeld, and M. S. Krieger. 1992. PM-10 high-volume collection and quantitation of semi- and nonvolatile phenols, methoxylated phenols, alkanes and polycyclic aromatic hydrocarbons from winter urban air and their relationship to wood smoke emissions. *Environ. Sci. Technol.* 26, 2251–2262.

Hildemann, L. M., G. R. Markowski, and G. R. Cass. 1991a. Chemical composition of emissions from urban sources of fine organic aerosol. *Environ. Sci. Technol.* 25, 744–759.

Hildemann, L. M., M. A. Mazurek, G. R. Cass, and B. R. T. Simoneit. 1991b. Quantitative characterization of urban sources of organic aerosol by high-resolution gas chromatography. *Environ. Sci. Technol.* 25, 1311–1325.

Leif, R. N., and B. R. T. Simoneit. 1995. Ketones in hydrothermal petroleums and sediment extracts from Guaymas Basin, Gulf of California. *Org. Geochem.*, 23, 889–904.

Levine, J. S. 1991. Introduction: Global biomass burning: Atmospheric, climatic, and biospheric implications. In: *Global Biomass Burning: Atmospheric, Climatic, and Biopheric Implications*, J. S. Levine (ed.), MIT Press, Cambridge, Mass., pp. xxv–xxx.

Lobert, J. M., D. H. Scharffe, W. M. Hao, and P. J. Crutzen. 1990. Importance of biomass burning in the atmospheric budgets of nitrogen-containing gases. *Nature* 346, 552–554.

Marty, J.-C., and A. Saliot. 1982. Aerosols in equatorial Atlantic air: *n*-alkanes as a function of particle size. *Nature* 298, 144–147.

Matsumoto, G., and T. Hanya. 1980. Organic constituents in atmospheric fallout in the Tokyo area. *Atmosph. Environ.* 14, 1409–1419.

Mazurek, M. A., and B. R. T. Simoneit. 1984. Characterization of biogenic and petroleum-derived organic matter in aerosols over remote, rural and urban areas. In: *Identification and Analysis of Organic Pollutants in Air*, ACS Symp., L. H. Keith (ed.), Ann Arbor Science/Butterworth Publishers, Woburn, Mass., pp. 353–370.

Mazurek, M. A., B. R. T. Simoneit, and G. R. Cass. 1989. Interpretation of high-resolution gas chromatography and high resolution gas chromatography/mass spectrometry data acquired from atmospheric organic aerosol samples. In: *Proc. 3rd Int. Conf. Carbonaceous Particles in the Atmosphere, Aerosol Sci. Technol.* 10, 408–420.

Mazurek, M. A., G. R. Cass, and B. R. T. Simoneit. 1991. Biological input to visibility-reducing aerosol particles in the remote aird southwestern United States. *Environ. Sci. Technol.* 25, 684–694.

Ohta, K., and N. Handa. 1985. Organic components in size-separated aerosols from the western North Pacific. *J. Oceanogr. Soc. Jap.* 41, 25–32.

Ramdahl, T. 1983. Retene–a molecular marker wood combustion in ambient air. *Nature* 306, 580–582.

Rogge, W. F., M. A. Mazurek, L. M. Hildemann, G. R. Cass, and B. R. T. Simoneit. 1993a. Quantification of organic aerosols on a molecular level: Identification, abundance and seasonal variation. In: *Proc. Fourth Int. Conf. on Carbonaceous Paricles in the Atmosphere. Atmosph. Environ.* 27A, 1309–1330.

Rogge, W. F., L. M. Hildmann, M. A. Mazurek, G. R. Cass, and B. R. T. Simoneit. 1993b. Sources of fine organic aerosol: 4. Particulate abrasion products from leaf surfaces of urban plants. *Environ. Sci. Technol.* 27, 2700–2711.

Schneider, J. K., R. B. Gagosian, J. K. Cochran, and T. W. Trull. 1983. Particle size distributions of *n*-alkanes and ^{210}Pb in aerosols off the coast of Peru. *Nature* 304, 429–432.

Seiler, W., and P. J. Crutzen 1980. Estimates of gross and net fluxes of carbon between the biosphere and the atmosphere from biomass burning. *Climatic Change* 2, 207–247.

Simoneit, B. R. T. 1977. Organic matter in eolian dusts over the Atlantic Ocean. *Proceedings of Symposium on Concepts in Marine Organic Chemistry. Mar. Chem.* 5, 443–464.

Simoneit, B. R. T. 1979. Biogenic lipids in eolian particulates collected over the ocean. In *Proceedings Carbonaceous Particles in the Atmosphere*, T. Novakov (ed.), National Science Foundation and Lawrence Berkeley Laboratory, Berkeley, Calif. 233–244.

Simoneit, B. R. T. 1980. Eolian particulates from oceanic and rural areas–their lipids, fulvic and humic acids and residual carbon. In: *Advances in Organic Geochemistry 1979*, A. G. Douglas and J. R. Maxwell (eds.), Pergamon Press, Oxford, pp. 343–352.

Simoneit, B. R. T. 1984a. Organic matter of the troposphere: III–characterization and sources of petroleum and pyrogenic residues in aerosols over the western United States. *Atmosph. Environ.* 18, 51–67.

Simoneit, B. R. T. 1984b. Application of molecular marker analysis to reconcile sources of carbonaceous particulates in tropospheric aerosols. In: *Proc. 2nd. Int. Conf. Carbonaceous Particles in the Atmosphere. Science of the Total Environ.* 36, 61–72.

Simoneit, B. R. T. 1985. Application of molecular marker analysis to vehicular exhaust for source reconciliations. *Int. J. Environ. Anal. Chem.* 22, 203–233.

Simoneit, B. R. T. 1986. Characterization of organic constituents in aerosols in relation to their origin and transport (a review). *Int. J. Environ. Anal. Chem.* 23, 207–237.

Simoneit, B. R. T. 1989. Organic matter of the troposphere-V: Application of molecular marker analysis to biogenic emissions into the troposphere for source reconciliations. *J. Atmosph. Chem.* 8, 251–275.

Simoneit, B. R. T., and M. A. Mazurek. 1982. Organic matter of the troposphere–II. Natural background of biogenic lipid matter in aerosols over the rural western United States. *Atmosph. Environ.* 16, 2139–2159.

Simoneit, B. R. T., and M. A. Mazurek. 1989. Organic tracers in ambient aerosols and rain. In *Proc. 3rd Int. Conf. Carbonaceous Particles in the Atmosphere. Aerosol Sci. Technol.* 10, 267–291.

Simoneit, B. R. T., R. Chester, and G. Eglinton. 1977. Biogenic lipids in particulates from the lower atmosphere over the eastern Atlantic. *Nature* 267, 682–685.

Simoneit, B. R. T., M. A. Mazurek, and T. A. Cahill. 1980. Contamination of the Lake Tahoe air basin by high molecular weight petroleum residues. *J. Air Poll. Contr. Assoc.* 30, 387–390.

Simoneit, B. R. T., M. A. Mazurek, and W. E. Reed. 1983. Characterization of organic matter in aerosols over rural sites: Phytosterols. In: *Advances in Organic Geochemistry 1981*, M. Bjorøy et al. (eds.), J. Wiley and Sons Ltd., Chichester, pp. 355–361.

Simoneit, B. R. T., R. E. Cox, and L. J. Standley. 1988. Organic matter of the troposphere-IV: lipids in Harmattan aerosol particles of Nigeria. *Atmosph. Environ.* 22, 983–1004.

Simoneit, B. R. T., J. N. Cardoso, and N. Robinson. 1990. An assessment of the origin and composition of higher molecular weight organic matter in aerosols over Amazonia. *Chemosphere* 21, 1285–1301.

Simoneit, B. R. T., J. N. Cardoso, and N. Robinson. 1991a. An assessment of terrestrial higher molecular weight lipid compounds in air particulate matter over the South Atlantic from about 30–70°S. *Chemosphere* 23, 447–465.

Simoneit, B. R. T., G. Sheng, X. Chen, J. Fu, J. Zhang, and Y. Xu. 1991b. Molecular marker study of extractable organic matter in aerosols form urban areas of China. *Atmosph. Environ.* 25A, 2111–2129.

Simoneit, B. R. T., P. T. Crisp, M. A. Mazurek, and L. J. Standley. 1991c. Composition of extractable organic matter of aerosols from the Blue Mountains and southeast coast of Australia. *Environ. Internat.* 17, 405–419.

Simoneit, B. R. T., W. F. Rogge, M. A. Mazurek, L. J. Standley, L. M. Hildemann, and G. R. Cass. 1993. Lignin pyrolysis products, lignans and resin acids as specific tracers of plant classes in emissions from biomass combustion. *Environ. Sci. Technol.* 27, 2533–2541.

Standley, L. J., and B. R. T. Simoneit. 1987. Composition of extractable organic matter in smoke particles from prescribed burns. *Environ. Sci. Technol.* 21, 163–169.

Standley, L. J., and B. R. T. Simoneit. 1990. Preliminary correlation of organic molecular tracers in residential wood smoke with the source of fuel. *Atmosph. Environ.* 24B, 67–73.

Standley, L. J., and B. R. T. Simoneit. 1994. Resin diterpenoids as tracers for biomass combustion aerosols. *J. Atmosph. Chem.* 18, 1–15.

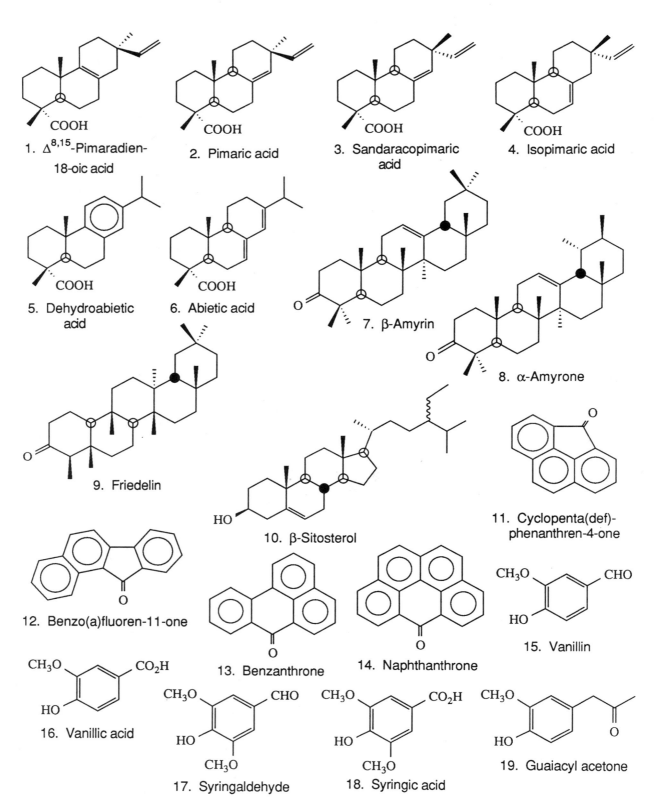

1. Δ^{8,15}-Pimaradien-18-oic acid

2. Pimaric acid

3. Sandaracopimaric acid

4. Isopimaric acid

5. Dehydroabietic acid

6. Abietic acid

7. β-Amyrin

8. α-Amyrone

9. Friedelin

10. β-Sitosterol

11. Cyclopenta(def)-phenanthren-4-one

12. Benzo(a)fluoren-11-one

13. Benzanthrone

14. Naphthanthrone

15. Vanillin

16. Vanillic acid

17. Syringaldehyde

18. Syringic acid

19. Guaiacyl acetone

Appendix 48.1 Chemical structures cited

The Size Distribution of Ambient Aerosol Particles: Smoke versus Urban/Industrial Aerosol

Lorraine A. Remer, Yoram J. Kaufman, and Brent N. Holben

Accurately determining the optical properties of atmospheric aerosol is significant in terms of both climate change (Penner et al. 1992; Charlson et al. 1991, 1992; Kaufman et al. 1991) and remote sensing (Fraser et al. 1984; Tanré et al. 1988; King et al. 1992). For example, in order to estimate quantitatively the direct effect of aerosols on climate, the fraction of aerosol backscatter must be known (Charlson et al. 1992). This requires knowing the aerosol phase function for different aerosol types such as sulfate aerosol from urban/industrial sources or smoke aerosol from biomass burning. Furthermore, estimating the fraction of backscatter requires an understanding of the way backscattering depends on atmospheric conditions (relative humidity, presence of clouds, aerosol concentration, etc.) or illumination direction (Kaufman and Holben 1997). Understanding aerosol optical properties is important in order to properly incorporate aerosols into model simulations of the earth's radiative budget and general circulation (IPCC 1995). Regional differences in aerosol type leading to regional differences in the optical properties and radiative forcing are apt to influence the general circulation of a regional and perhaps global scale. In remote sensing, applications correctly modeling the optical properties of the aerosol are vital (Tanré et al. 1992). Again, the aerosol model must correctly represent the aerosol type of the region being observed, and it also must represent the correct optical properties of that aerosol, because aerosol conditions change even within the same aerosol type. Failure to use the correct aerosol optical properties will contribute to errors in retrieval of the aerosol and of the land surface parameters under observation (Kaufman and Sendra 1988).

In the past, much of our knowledge of long-term aerosol optical properties has been based on ground-level measurements (WMO 1983). Aircraft measurements were limited in their temporal extent. The models resulting from these measurement techniques represent mean aerosol conditions for general aerosol types. The models may be modified as a function of relative humidity (Shettle and Fenn 1979), but, unfortunately, relative humidity varies significantly with height and is not a variable readily retrieved from remote sensing. Therefore, remote sensing applications and studies concerned with the total atmospheric column will find difficulty adapting the humidity variations to their needs. Ideally, aerosol optical properties should be determined from measurements of the total atmospheric column taken several times daily over an extended period of time. The analysis of the optical properties should be described not as a function of relative humidity, but as a function of a variable such as aerosol optical thickness, which describes the total atmospheric column and is more easily retrieved from satellite remote sensing. Manual sun/sky scanning spectral radiometers have been used successfully to retrieve spectral aerosol optical thickness and volume size distribution (Nakajima et al. 1989; Kaufman et al. 1994). A network of automatic instruments measuring direct solar and sky radiance at several stations in a region several times a day (Holben et al. 1995) will provide the information necessary to construct dynamical aerosol models representative of the total atmospheric column (Kaufman and Holben 1997). In this chapter, we analyze the data from two of these networks in two different aerosol regimes: smoke from biomass burning, and urban/industrial pollution. From this analysis, we can understand the unique features of the optical properties of each aerosol type, measured for the ambient undisturbed aerosol integrated on the vertical column.

Data

In the summer of 1993, two networks of automatic sun/sky scanning spectral radiometers were deployed. The first was set up in Brazil along the edge of the deforestation zone and in the Cerrado where both regions experience significant biomass burning during the dry season—late July through September (see Holben et al., chapter 60 this volume). Figure 49.1*a* gives the location of the Brazil instrument network. In this chapter, we concentrate on the analysis of the data

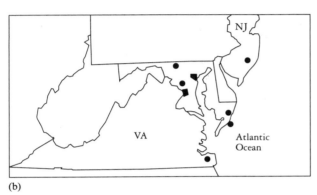

Figure 49.1 (*a*) Map of the partial 1993 sun/sky radiometer network in Brazil. This study uses data only from Brasília and Cuiaba located in the Cerrado region of Brazil. (*b*) Map showing the locations of the sun/sky radiometer network of the SCAR-A experiment operating in 1993.

measured at two stations in the Brazil network—Cuiaba and Brasília, both located in the Cerrado region in Brazil—and consider only data collected during August and September. A comparison between Cerrado and deforestation smoke in earlier and later parts of the dry season is described by Holben et al., chapter 60 this volume, and Kaufman and Holben (1997). The other network was installed in the mid-Atlantic region of the eastern United States seaboard (figure 49.1*b*) as part of the Sulfates, Clouds, And Radiation–Atlantic (SCAR-A) experiment. Data were collected from this second network from July through September at five stations. During the summer, the eastern seaboard typically experiences periods of extremely hazy conditions associated with urban/industrial pollution interspersed with some clearer periods. Often, the hazy episodes coincide with a moist

boundary layer and afternoon boundary layer cumulus clouds. These conditions occurred during SCAR-A. In situ measurements of the SCAR-A aerosol show that the aerosol is typically composed of sulfates, marine influence of salt, and also nitrates and organics.

The radiometers measure sky radiance in the sun's almucantar once per hour in the morning and late afternoon when the sun's zenith angle is greater than 60°. This procedure provides at most six almucantar measurements per station per day. However, the almucantars are checked for symmetry of the sky radiance on either side of the sun and asymmetrical almucantars are discarded. Asymmetry occurs due to nonhomogeneous atmospheric conditions or the presence of clouds. Altogether over 125 symmetrical almucantars were retrieved by each of the networks during their period of deployment.

The sky radiances from each symmetrical almucantar are inverted to achieve aerosol volume distribution (Nakajima et al. 1983). The 125-volume size distributions in each network were then sorted and ordered according to increasing aerosol optical thickness at 670 nm. All 5–10 size distributions were averaged to give the mean size distribution for a particular aerosol optical thickness. These curves are plotted in figure 49.2*a* for the biomass burning aerosol of the Brazil network and in figure 49.2*b* for the industrial aerosol of the SCAR-A network. In these figures we see several identifiable modes. The biomass burning aerosol (figure 49.2*a*) is composed of three modes corresponding to the accumulation mode (radii < 0.3 μm), stratospheric aerosols (0.3 μm $<$ radii < 0.8 μm), and a coarse mode (radii > 0.8 μm). The industrial aerosol (figure 49.2*b*) consists of the same modes except that the coarse mode can be further divided into a salt mode from the marine influence at radii near 1 μm and another coarse mode of larger particles (radii > 2 μm). The presence of the accumulation mode and salt mode was identified from simultaneous aircraft in situ measurements (Hegg et al. 1995). The presence of the stratospheric mode due to the eruption of Mount Pinatubo was identified from similar almucantar inversions that showed the mode a short time after the eruption, but not before the eruption (Kaufman et al. 1994). Detailed properties of the Pinatubo aerosol were measured by Pueschel et al. (1994) and Deshler et al. (1993).

Comparison between figure 49.2*a* and *b* shows that in the biomass burning regime the accumulation and coarse modes increase in volume as aerosol optical thickness increases, while in the industrial aerosol regime the accumulation mode and salt mode, but not

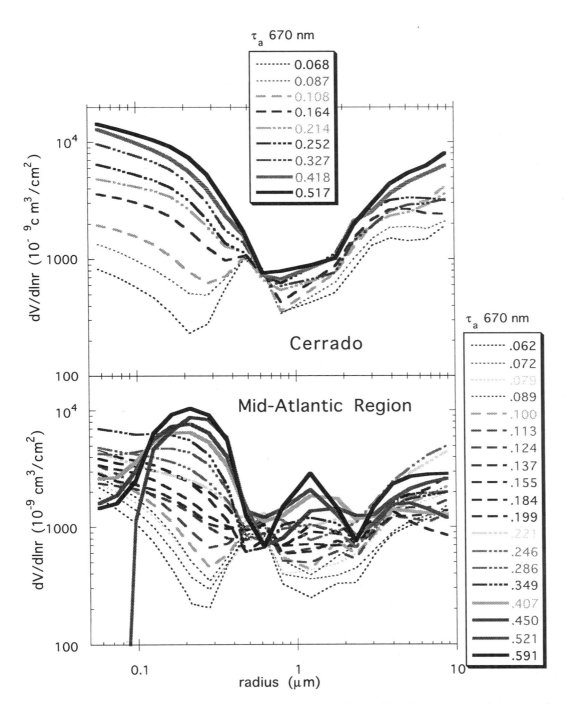

Figure 49.2 Volume size distributions derived from the spectral sky radiance with each curve representing a mean of approximately 10 observations and plotted as a function of particle radius and aerosol optical thickness (τ_a) at 670 nm. (*a*) Biomass burning aerosol from the Cerrado region of Brazil. (*b*) Urban/industrial aerosol from the SCAR-A network of the mid-Atlantic region of the eastern United States.

the coarse mode, increase in volume with increasing optical thickness. The stratospheric aerosol mode remains constant in both regimes, as expected. Further examination of the accumulation mode in the industrial regime reveals that not only is the mode increasing in volume with increasing aerosol optical thickness, but the size of the particles is also increasing. This is very different from the accumulation mode in biomass burning, which appears to maintain a constant modal radii for all optical thicknesses. These differences, especially the differences in the accumulation modes, will make significant differences in the optical and physical properties of the two different aerosol types.

Correction to the Inversion Scheme

Figure 49.2 presents the data as they were inverted from the sky radiances and averaged. The inversion algorithm assumes that beyond the size limits of the inversion (0.07 μm and 9 μm) the particle volume is zero. This assumption affects the resulting volume distribution by artificially increasing volume at the size limits of the inversion in order to compensate for the arbitrary and unphysically low volume just beyond the limits. At low optical thickness, the volume distributions of figure 49.2 suggest that the modal radii are much smaller than 0.05 μm. There is no evidence to support such a small modal radii of the volume distribution, and, therefore, the inverted distribution is incorrect. Likewise, a similar problem may exist at the large particle end of the distribution. These problems are isolated to the radius ranges near the edges of the figures.

Even though the inverted volume distributions misrepresent volume at the small and large particle ends, the inversion algorithm guarantees that these distributions do correctly represent most of the *optical properties* of the aerosol. The inversion algorithm fits the sky radiance data from the first 40° of the sun. This implies that a model constructed from these distributions will correctly represent the extinction and scattering as well as the phase function in the first 40°. Representation of the phase function beyond 40° is not guaranteed. Because the optical properties are well represented by the inverted volume size distributions, we can input the given size distributions into a Mie computation code (Dave and Gazdag 1970) to recalculate the aerosol single scattering path radiance $L_p(\theta, \lambda)$ in the first 40°. A more physical assumption is that the small particle accumulation mode is represented by a single log normal distribution (Hoppel et al. 1985, 1990). Therefore, the calculated path radiance can then be matched in a look-up table to a single log

normal that corresponds to the same single scattering path radiance. Thus, we can find a lognormal not only with the same *optical properties* of the inverted size distribution, but with what we assume are the correct *physical properties*. This correction to the accumulation mode and comparison to in situ measurements from aircraft (Hegg et al. 1995) is discussed in Remer et al. (1997).

Aerosol Model

We assume that each of the modes in figure 49.2, after correcting the accumulation mode, can be represented by a lognormal given by

$$\frac{dV}{d\ln r} = V_o \exp\left(-\frac{\left[\ln\left(\frac{r}{r_m}\right)\right]^2}{2\,\sigma^2} \right) \tag{49.1}$$

where $dV/d\ln r$ is the volume distribution, V_o is the amplitude of the lognormal, r is the radius, r_m is the modal radius and σ is the width of the mode. The sum of the modes will reproduce the original, corrected volume distribution curve for each particular optical thickness.

Stratospheric Aerosol

In both aerosol regimes we use the curves of lowest aerosol optical thickness to fit a lognormal function to the mode representing stratospheric aerosol (0.3 μm < radii < 0.8 μm). The fit parameters r_m, σ and V_o are listed in table 49.1. The differences between the Cerrado and the mid-Atlantic regions can be attributed to differences in latitude (McCormick et al. 1995). Kaufman et al. (1994), using an inversion technique similar to that used in this study, but with a handheld instrument, found the best fit to the stratospheric aerosol mode to be $r_m = 0.5$ μm, $\sigma = 0.6 - 0.8$ and $V_o = 1.8 \times 10^{-6}$ cm^3/cm^2. The mode size found in this study is very similar to that of Kaufman et al. (1994), with a much smaller σ and half as much volume of the stratospheric aerosol. Their data were collected during 1992, while our data were collected a year later, perhaps after much of the stratospheric aerosol had been removed by gravitational settlement. Deshler et al. (1993) also made in situ measurements from stratospheric balloons a year earlier, and found the data best fit by two modes. The larger of their modes ($r_m = 0.50 - 0.60$ μm, $\sigma = 0.20 - 0.30$) corresponds extremely well with the stratospheric mode of this study but the smaller mode ($r_m = 0.30 - 0.35$ μm, $\sigma = 0.30 - 0.50$) is not resolved by the inversion tech-

Table 49.1 Summary of lognormal parameters represent best fit to the different modes of the two aerosol regimes

Biomass burning	r_g (mm)	r_m (mm)	σ	V_o (10^{-9} cm^3/cm^2)
Accumulation	0.061	0.130	0.50	$f_4(\tau_{670})$
Stratospheric	0.382	0.51	0.31	984
Coarse	$f_1(\tau_{670})$	$f_2(\tau_{670})$	$f_3(\tau_{670})$	$f_5(\tau_{670})$

$f_1(\tau_{670}) = 1.003 - 1.30\tau_{670}$
$f_2(\tau_{670}) = 6.03 - 11.3\tau_{670} + 61\tau_{670}^2$
$f_3(\tau_{670}) = 0.695 + 0.812\tau_{670}$
$f_4(\tau_{670}) = -2419 + 45007\tau_{670}$
$f_5(\tau_{670}) = 2426 - 6282\tau_{670} + 36969\tau_{670}^2$

Industrial	r_g (μm)	r_m (μm)	σ	Vo (10^{-9} cm^3/cm^2)
Accumulation 1	0.036	0.106	0.60	$f_6(\tau_{670})$
Accumulation 2	0.114	0.21	0.45	$f_7(\tau_{670})$
Stratospheric	0.427	0.55	0.29	728
Salt	0.992	1.30	0.30	$f_8(\tau_{670})$
Coarse	0.671	9.50	0.94	1920

$f_6(\tau_{670}) = 2072 + 69828\tau_{670} - 196251\tau_{670}^2 + 150504\tau_{670}^3$
$f_7(\tau_{670}) = 342 - 7591\tau_{670} + 80040\tau_{670}^2 - 63072\tau_{670}^3$
$f_8(\tau_{670}) = -158 + 4129\tau_{670}$

nique of this study. However, Deshler et al. (1993) report that at the end of 1992 the smaller mode represents less than 30% of the total measured stratospheric aerosol volume, and that the trend over 1992 was for the volume in the smaller mode to decrease faster than the volume in the larger mode. By the summer of 1993, the smaller mode may have become insignificant in a volume size distribution.

Coarse Mode

For each of the curves in figure 49.2, we fit a lognormal function. Figure 49.3 shows V_o for each of the lognormal fits as a function of aerosol optical thickness. V_o increases with increasing optical thickness for the biomass burning aerosol regime, but does not exhibit a similar trend for the industrial aerosol regime. For the biomass burning aerosol, we can establish relationships between all the fit parameters (V_o, r_m and σ) and the aerosol optical thickness. For the industrial aerosol, the fit parameters are not functions of the aerosol optical thickness and so the coarse mode parameters are fixed at their mean value. Values are given in table 49.1.

Salt Mode

The lognormal fit to the salt mode found only in the industrial aerosol gives a mean modal radius of 1.3 μm with a mean σ of 0.30, with little variation of these mean values for differing aerosol optical thicknesses.

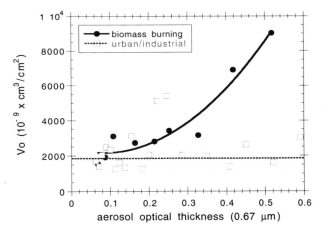

Figure 49.3 The volume of each coarse mode, V_o, plotted as a function of aerosol optical thickness for biomass burning (circles) and industrial aerosol (squares). Also plotted are the least squares quadratic fit to the biomass burning data (solid line) and the mean of the industrial data (dotted line).

Other studies also find a coarse mode with particle radii near 1 μm in the marine environment (Hegg et al. 1995; Hoppel et al. 1990). The salt mode volume of our data does grow with increasing optical thickness.

Accumulation Mode

Optically, the accumulation mode is the most important, and therefore, it will be analyzed in greater detail. The lognormal fit to the biomass burning accumulation mode gives a mean modal radius, r_m, of 0.13 $\pm$ 0.004 μm, with a mean σ of 0.50 $\pm$ 0.00 when averaged over all measured aerosol optical thicknesses. In situ measurements of biomass burning aerosol collected at the ground in Brazil, from aircraft over the Atlantic (Andreae, et al. 1994), and in Africa (Le Canut et al. 1994) also record aerosol sizes in the 0.12–0.16 μm range. Similar size distributions were measured in smoke from fires in North America (Radke et al. 1991). This indicates a remarkable stability in the size of organic smoke particles. All these measurements, including our data from the Cerrado, represent the mixed aged aerosol particle size. However, close to the fire source there is an evolution of particle sizes (see Martins et al., chapter 67 this volume), and the size evolution may continue through the first 24 hours after emission (Radke et al. 1991).

Up to this point we have successfully found a single lognormal function to fit each of the modes in figure 49.2. For stratospheric aerosol and industrial coarse mode, the best fit is a single lognormal with fixed r_m, σ, and volume. The salt mode and the biomass burning accumulation mode are best fit with a single lognormal with fixed r_m and σ, but with a variable V_o that increases with increasing optical thickness. The biomass burning coarse mode is best fit with all three parameters variable but still a single lognormal. However, the industrial accumulation mode is not well represented by a single lognormal, because the aerosol in the accumulation mode is the result of *two* physical processes: gas-to-particle conversion generating particles around $r_g = 0.03$, and cloud processing generating particles around $r_g = 0.1$ (Hoppel et al. 1985, 1990; Hegg et al. 1993; Kaufman and Tanré 1994). r_g is the modal radius of the number size distribution, dN/dr, represented by a lognormal

$$\frac{dN}{dr} = \frac{N_o}{r\sigma\sqrt{2\,\pi}} \exp\left(-\frac{\left[\ln\left(\frac{r}{r_g}\right)\right]^2}{2\,\sigma^2}\right) \tag{49.2}$$

where No is the number of particles and the other variables are defined as before in equation. (49.1). We

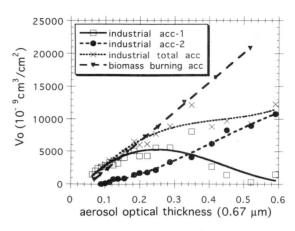

Figure 49.4 The volume of each accumulation mode, V_o, plotted as a function of aerosol optical thickness for the biomass burning aerosol (triangles) and the urban/industrial aerosol. For the industrial aerosol, squares represent the volume of particles created by gas-to-particle conversion (acc-1) and circles represent the volume of particles created by cloud processes (acc-2). Also shown is the sum of the two industrial aerosol accumulation modes (x's).

achieve a better fit to the industrial accumulation mode by assuming that the mode can be represented by *two* lognormals: acc-1 for gas-to-particle conversion and acc-2 for cloud processes.

Acc-1 dominates the aerosol for low optical thickness. At the lowest optical thicknesses, we fit a single lognormal to the data curves and correct the accumulation mode as described before. The fit results in a volume modal radius of 0.106 μm and σ of 0.60. We fit acc-2 to the data curves of figure 49.2b at the highest optical thickness where acc-2 dominates. The resulting modal radius is 0.21 μm with a σ of 0.45. The data curves of intermediate optical thickness are represented as a linear combination of the two lognormals found at the extremes of optical thickness. The calculated values of V_o as a function of aerosol optical thickness are shown in figure 49.4. The volume of acc-1 increases with increasing optical thickness until aerosol optical thickness reaches 0.30. Then the acc-1 volume experiences a sharp drop-off as the small particles are combined into bigger ones through cloud processing. Acc-2 volume exhibits a steady increase as a function of aerosol optical thickness. Figure 49.4 also shows 3rd-order polynomial fits through the data and the total volume of the accumulation mode found by adding the volumes of acc-1 and acc-2. The total volume increases rapidly at lower optical thicknesses and then levels off as the increase in optical thickness is dominated by the shift of particles from acc-1 to the more optically effective acc-2.

The volume size distributions resulting from the aerosol model for each aerosol regime are shown in figure 49.5 as a function of aerosol optical thickness at 0.67 μm. Figure 49.5 is a result of summing the contributions of each of the modes discussed above, weighted by the number of particles in that mode for the particular optical thickness. The accumulation modes appear to be different from the measured data in figure 49.2 due to the correction of the inversion scheme at the small particle end. The process of developing an aerosol model for each of the two types of aerosols reveals an important difference between them. The biomass burning accumulation mode results from one major physical process—emission from fires—and thus is well represented by a single lognormal. The industrial accumulation mode results from two separate processes—gas to particle conversion *and* processing through nonprecipitating clouds—and thus must be modeled by two lognormals. The inversion of the sky spectral radiance data measured these modes for the undisturbed aerosol integrated on the vertical column.

Testing the Models

Before the aerosol models can be used with the Mie calculation (Dave and Gazdag 1970) to study the optical properties of the two different aerosol regimes, they must be validated. The models are derived from the sky radiance measurements that were inverted, averaged, corrected, fit to lognormals, and averaged again. Naturally, to be correct, the models must be able to predict the original sky radiance from which they were derived, which is the sky radiance for scattering angles 0–40° from the sun. An independent test of the model and the correction for small particles is to determine if the model also fits the backscattering sky radiance at scattering angles near 120°. Scattering at 120° is more sensitive to the effect of the size distribution of small particles than is the radiance <40°. Figure 49.6 shows the sky radiance predicted by our models plotted with the original sky radiance observations as a function of optical thickness for 670 nm and 120° scattering angle. Single scattering albedo is 0.90 for the biomass burning aerosol (Kaufman et al. 1992) and 0.96 for the industrial aerosol (Fraser et al. 1984). Each curve represents a different assumption of surface reflectance. Actual surface reflectance at this wavelength is estimated to be 0.05 to 0.10, but varies depending on the location of each particular instrument. There is some scatter in the measurements, but we do see very good agreement between the model and the original observations.

Optical Properties of the Aerosol Regimes

The relative contribution of each mode toward total extinction of the aerosol at 670 nm is given in figure 49.7. For the biomass burning aerosol, we see that at very low optical thicknesses the coarse mode and the stratospheric component make substantial contributions to the total aerosol extinction, but at higher optical thicknesses the total extinction is completely dominated by the accumulation mode. For the industrial aerosol, the picture is similar except that we see an interesting relationship between the accumulation modes. At low optical thicknesses acc-1, the smallest particles, resulting from gas to particle conversion, dominate the accumulation mode contribution to extinction, but at higher optical thicknesses acc-2, the larger sized particles, resulting from cloud processes, contribute the most to extinction. Therefore, at moderate to high optical thickness, smoke aerosol is more dominated by smaller particles (0.13 μm) than is the industrial aerosol (with particles at 0.21 μm).

The wavelength dependence of the two aerosol regimes is shown in figure 49.8 for two different optical thicknesses. Figure 49.8a includes the stratospheric mode; figure 49.8b does not. As aerosol optical thickness increases, the slopes of the curves increase, indicating greater wavelength dependence. This effect is greater for biomass burning aerosol where the dominance of the relatively large sized stratospheric or coarse mode aerosol ($r_m > 0.50$ μm) gives way to the dominance of very fine accumulation mode particles ($r_m = 0.13$ μm). In the industrial aerosol, at larger optical thicknesses, the accumulation mode particles are larger ($r_m = 0.21$ μm) and thus mitigate the change in wavelength dependence.

The phase functions of the two aerosol regimes for wavelength of 670 nm are shown in figure 49.9. The biomass burning aerosol, in general, has a greater backscattering component than does the industrial aerosol. Figure 49.10 displays this phenomenon as a function of optical thickness. The phase function at 120° and 150° for wavelength of 670 nm for biomass burning remains higher than do the same quantities from the industrial aerosol regime for all values of optical thickness. There is a transition at low optical thickness from the larger sized particles of the stratospheric aerosol (which decrease the backscattering) to the smaller sized particles of the accumulation mode (which increase the backscattering). The industrial aerosol also shows the shift in backscattering to lower values again as acc-2 replaces acc-1 and the size of the

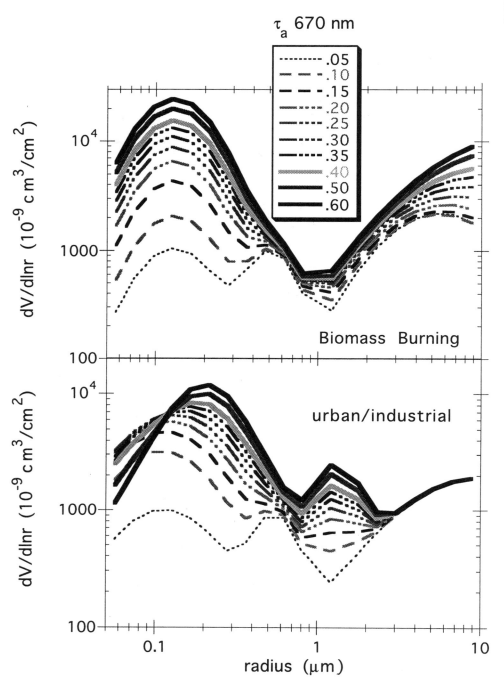

Figure 49.5 Volume size distributions resulting from the aerosol model of the urban/industrial aerosol (*bottom*) and biomass burning aerosol (*top*) as a function of aerosol optical thickness at 670 nm

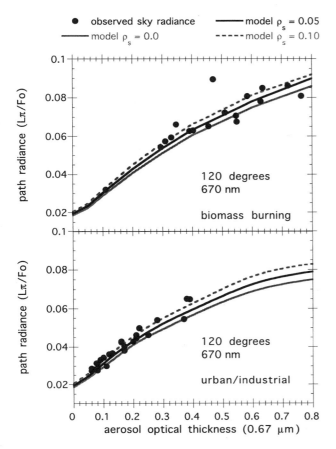

Figure 49.6 The sky radiance from original sun/sky radiometer measurements (points) and as predicted by aerosol models composed of the lognormal fits to the data (curves) for the biomass burning (*top*) and the urban industrial aerosol (*bottom*). Data and models are for 670 nm, solar zenith angle of 60° and a scattering angle of 120°.

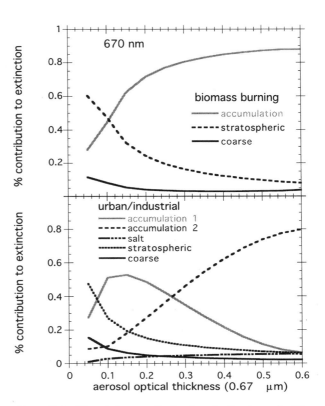

Figure 49.7 The relative contribution to extinction by each of the modes for biomass burning (*top*) and industrial aerosol (*bottom*)

particles increase again. Also shown in figure 49.10 are the values of the phase function at 120° and 150° for the continental model of the 6s code (Vermote et al. 1997). We see that the standard continental model does represent the mean characteristics of the industrial aerosol, but not that of the biomass burning aerosol. However, the continental model misses completely the dynamic character of industrial aerosol.

Discussion and Conclusions

We have used a data set consisting of sun/sky radiometer measurements to analyze the optical properties of smoke aerosol from biomass burning of Brazil's Cerrado region, and those of the mostly sulfate aerosol from urban/industrial pollution of the eastern United States. This data set is unprecedented in that it pro-

vides a regional climatology of the total column aerosol characteristics over several months of data collection. Over 125 measurements were used in the data set for each region. The sky radiance data are inverted to obtain volume size distributions that are then adjusted after the inversion to correct for a flaw that affects the resulting physical properties, but not the optical properties. Corrected volume size distributions from the SCAR-A experiment in the eastern United States have been compared with size distributions measured in situ by aircraft in situations where the aircraft and sun/sky radiometer measurements are spatially and temporally close. The remotely sensed/inverted size distributions agree favorably with the in situ measurements. No validation has been performed in a biomass burning environment, but such activity is planned using 1995 data from the Smoke, Clouds, And Radiation-Brazil (SCAR-B) experiment.

Understanding the optical properties of the two regimes is dependent on understanding the physical properties of the aerosol. For this reason, we created models to describe each aerosol type as a function of aerosol optical thickness. In both regimes, the aerosol volume increases with increasing optical thickness. In

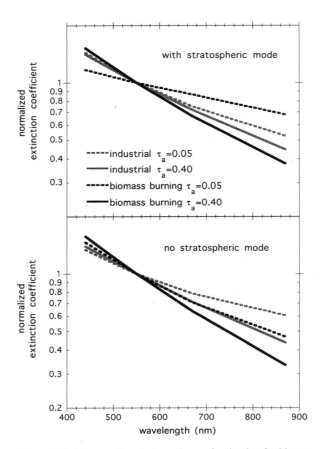

Figure 49.8 The wavelength dependence of extinction for biomass burning and industrial aerosol for aerosol optical thickness (τ_a) of 0.05 and 0.40 for both biomass burning and industrial aerosol. The top figure includes the stratospheric mode. The bottom figure is the same but with the stratospheric component removed.

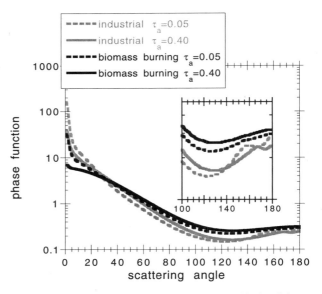

Figure 49.9 Phase function of biomass burning and industrial aerosol for aerosol optical thickness of 0.05 and 0.40

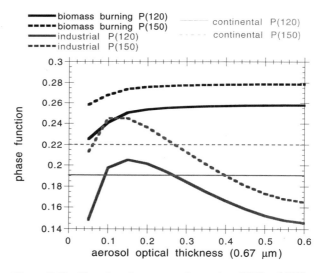

Figure 49.10 Phase function at scattering angles of 120° and 150° as a function of aerosol optical thickness for the biomass burning and industrial aerosol. Also plotted are the values for the continental model of the 6s radiative transfer code (Vermote et al. 1996).

the biomass burning environment, the accumulation mode and the coarse mode volume are both functions of the optical thickness. This suggests that these two modes have a common source. The source of the accumulation mode is obviously the smoke from the fires. The coarse mode may also result from the biomass burning. However, compositional analysis of the smoke in Brazil indicates that the coarse mode consists of ash and minerals (Artaxo et al. 1993). Even so, the coarse mode may be the indirect result of the biomass burning. Burning creates strong local winds and leaves bare soil, and the combination may act to stir up mineral dust into the air. In contrast, the coarse mode of the industrial aerosol is not dependent on aerosol optical thickness. It does not seem to share a common source with the accumulation mode, which is strongly dependent on optical thickness.

The industrial aerosol does have a mode at 1.3 μm, the salt mode, that is dependent on aerosol optical thickness. In this case, we do not suggest that the salt mode and accumulation mode share a common source, but instead that they share a common growth factor. Accumulation mode particles congregate and are transformed into larger particles by processing through nonprecipitating clouds. These processes occur in the mid-Atlantic region during stagnant conditions with no precipitation and little change of air mass. The stagnant conditions favorable to an increase in accumulation mode volume are also favorable to an increase in salt mode volume, due to a reduction in the amount of salt removed from the air by deposition.

The accumulation modes are the dominant factor in determining the optical properties of each aerosol regime. For biomass burning, the accumulation mode results from one physical process—the gas to particle conversion from fire emissions. Therefore, this mode stays a constant size for the entire range of optical thicknesses. Thus, except for very low optical thicknesses where the larger stratospheric and coarse modes make a substantial contribution, the wavelength dependence and phase function remain constant with optical thickness. In contrast, the industrial aerosol accumulation mode results from two distinct processes—gas-to-particle conversion from pollution emissions and cloud processing. Each process creates particles of different sizes with different optical properties. Aerosol optical thickness is correlated to the rapid increase of the larger accumulation mode. The result is that the wavelength dependence and phase function of the industrial aerosol are not constant with optical thickness. The reason that the optical thickness is closely correlated to the product of the cloud processing (acc-2) is not simply a matter of there being *more* accumulation mode particles in the air creating higher optical thickness, although that certainly is a factor. It is also a matter of the accumulation mode particles shifting to a larger size and a greater extinction, thus increasing the optical thickness.

The consequences of these differences are significant in terms of both climate change and remote sensing. The amount of backscatter is different for each aerosol regime, and in the case of the industrial aerosol, it varies as a function of aerosol optical thickness even within the same regime. Estimates of the direct radiative forcing of aerosols on the climate must use estimates of the backscatter. These estimates must then take into account the differences between and within aerosol regimes. In remote sensing of aerosol or the atmospheric correction of the aerosol for remote sensing studies of the land surface, the aerosol model and its optical properties must be assumed. Again, caution must be exercised in choosing an aerosol model. The model choice needs to be specific for the different types of aerosol and for the dynamic changes occurring within each aerosol regime. For example, blindly using the standard continental model in either of the regimes examined in this chapter would underpredict backscattering for the biomass burning aerosol, while completely missing the dynamic nature of the industrial aerosol.

Acknowledgments

We wish to thank I. Slutsker, the creator of the data base management software that made the data accessible for this analysis. We would also like to thank R. Kleidman, B-C. Gao, and E. Vermote who deployed the SCAR-A instrument network, and T. Eck and the many others owh deployed and maintained the Brazilian instrument network. Thanks also to A. Mattoo for her patient editing of this manuscript and figures.

References

Andreae, M. O., Anderson, B. E., Blake, D. R., Bradshaw, J. D., Collins, J. E., Gregory, G. L., Sachse, G. W., and Shipham, M. C. 1994. Influence of plumes from biomass burning on atmospheric chemistry over the equatorial and tropical South Atlantic during CITE 3 *J. Geophys. Res.*, 99D, 12 793–12 808.

Artaxo, P., M. A. Yamasoe, J. V. Martins, S. Kocinas, S. Carvalho, W. Maenhaut. 1993. Case study of atmospheric measurements in Brazil: aerosol emissions from Amazon Basin fires, in *Fire in the Environment: The Ecological, Atmospheric and Climatic Impact*, 139–158. John Wiley and Sons.

Charlson, R. J., J. Langer, H. Rodhe, C. B. Leovy, and S. G. Warren. 1991. Perturbation of the northern hemispheric radiative balance by backscattering from anthropogenic sulfate aerosol. *Tellus*, 43, 152–163.

Charlson, R. J., S. E. Schwartz, J. M. Hales, R. D. Cess, J. A. Coakley Jr., J. E. Hansen, and D. J. Hofmann. 1992. Climate forcing by anthropogenic aerosol. *Science*, 255, 423–430.

Dave, J. V., and J. Gazday. 1970. A modified Fourier transform method for multiple scattering calculations in a plane parallel Mie atmosphere. *Applied Optics*, 9, 1457–1466.

Deshler, T., B. J. Johnson, W. R. Rozier. 1993. Balloon borne measurements of Pinatubo aerosol during 1991 and 1992 at 41°N: Vertical profiles, size distribution and volatility. *Geophys. Research Letters*, 20, 1435–1438.

Fraser, R. S., Y. J. Kaufman, and R. L. Mahoney. 1984. Satellite measurements of aerosol mass and transport. *Atmos. Environ.*, 18, 2577–2584.

Hegg, D. A., R. J. Ferek, and P. V Hobbs. 1993. Aerosol size distributions in the cloudy atmospheric boundary layer of the North Atlantic Ocean. *J. Geophys. Res.*, 98, 8841–8846.

Hegg, D. A., P. V Hobbs, R. J Ferek, and A. P. Waggoner. 1995. Measurements of some aerosol properties relevant to radiative forcing on the East Coast of the United States. *J. Applied Meteor.*, 34, 2306–2315.

Holben, B. N., T. F. Eck, I. Slutsker, A. Setzer, A. Pereira, E. Vermote, J. A. Reagan, Y. J. Kaufman, and D. Tanré. 1995. *Sunphotometer Network Measurement of Aerosol Properties in the Brazilian Amazon*, Val d'Isere Conference, 17–21 January.

Hoppel, W. A., J. W. Fitzgerald, G. M. Frick, R. E. Larson, and E. J. Mack. 1990. Aerosol size distributions and optical properties found in the marine boundary layer over the Atlantic Ocean. *J. Geophys. Res.*, 95, 3659–3686.

Hoppel, W. A., J. W. Fitzgerald, and R. E. Larson. 1985. Aerosol size distributions in air masses advecting off the East Coast of the United States. *J. Geophys. Res.*, 90, 2365–2379.

IPCC Climate Change 1994,. 1995. Eds: J. T. Houghton, L. G. Meira Filhno, J. B. Hoesung Lee, B. A. Callander, E. Haites, N. Harris, and K. Maskell, Cambridge University Press, Cambridge.

Kaufman, Y. J., R. S. Fraser, and R. L. Mahoney. 1991. Fossil fuel and biomass burning effect on climate—heating or cooling? *J. Climate*, 4, 578–588.

Kaufman, Y. J., A. Gitelson, A. Karnieli, E. Ganor, R. S. Fraser, T. Nakajima, S. Mattoo, and B. N. Holben. 1994. Size distribution and scattering phase function of aerosol particles retrieved from sky brightness measurements. *J. Geophys. Res.*, 99D, 10 341–10 356.

Kaufman, Y. J., and B. N. Holben. 1997. Hemispherical backscattering by biomass burning and sulfate particles derived from sky measurements, special issue of *JGR-Atmospheres on Carbonaceous Aerosol.*, in press.

Kaufman, Y. J., and C. Sendra. 1988. Algorithms for atmospheric corrections of visible and near IR satellite imagery, *Int. J. Rem. Sens.*, 9, 1357–1381.

Kaufman, Y. J., A. Setzer, D. Ward, D. Tanré, B. N. Holben, P. Menzel, M. C. Pereira, and R. Rasmussen. 1992. Biomass burning airborne and spaceborne experiment in the amazonias (BASE-A). *J. Geophys. Res.*, 97D, 14 581–14 599.

Kaufman, Y. J., and D. Tanré. 1994. Effect of variations in supersaturation on the formation of cloud condensation nuclei. *Nature*, 369, 45–48.

King, M. D., Y. J. Kaufman, W. P. Menzel, and D. Tanré. 1992. Remote sensing of cloud, aerosol, and water vapor properties from the Moderate Resolution Imaging Spectrometer (MODIS), *IEEE Trans. Geosci. Remote Sensing*, 30, 2–27.

Le Canut, P., M. O. Andreae, G. W. Harris, F. G. Wienhold, and T. Zenker. 1994. Airborne studies of emissions form savanna fires in southern Africa: I. Aerosol emissions measured with a laser-optical particle counter. *J. Geophysical Res.*, submitted.

McCormick, M. P., L. W. Thomason, and C. R. Trepte. 1995. Atmospheric effects of the Mt. Pinatubo eruption. *Nature*, 373, 399–404.

Nakajima, T., M. Tanaka, M. Yamano, M. Shiobara, K. Arao, and Y. Nakanishi. 1989. Aerosol optical characteristics in the yellow sand events observed in May 1982 at Nagasaki, 2, Models. *J. Meteorol. Soc. Jpn.*, 67, 279–291.

Nakajima, T., M. Tanaka, and T. Yamauchi. 1983. Retrieval of the optical properties of aerosols from aereole and extinction data. *Appl. Opt.*, 22, 2951–2959.

Penner, J. E., R. E. Dickinson, and C. A. O'Neill 1992. Effects of aerosol from biomass burning on the global radiation budget. *Science*, 256, 1432–1434.

Pueschel, R. F., P. B. Russell, D. A. Allen, G. V. Ferry, and K. G. Snetsinger. 1994. Physical and optical properties of the Pinatubo volcanic aerosol: aircraft observations with impactros and suntracking photomete. *J. Geophys. Res.*, 99, 12 915–12 922.

Radke, L. F., D. A. Hegg, P. V. Hobbs, J. D. Nance, J. H. Lyons, K. K. Laursen, P. J. Reagan, and D. E. Ward. 1991. Particulate and trace gas emission from large biomass fires in North America, in *Global Biomass Burning*, The MIT Press, Cambridge, Mass. 209–224.

Remer, L. A., S. Gassó, D. A. Hegg, Y. J. Kaufman, B. N. Holben. 1997. Urban/Industrial Aerosol: Ground-based sun/sky radiometer and airborne in situ measurements. *J. Geophys. Res.*, in press.

Shettle, E. P., and R. W. Fenn. 1979. *Models for the Aerosols of the lower Atmosphere and the Effects of Humidity Variations on Their Optical Properties*, Air Force Geophysics Laboratory Report AFGL-TR-79-0214, 94 pp.,

Tanré, D., P. Y. Deschamps, C. Devaux, and M. Herman. 1988. Estimation of Saharan aerosol optical thickness from blurring effects in Thematic Mapper data. *J. Geophys. Res.*, 92, 15 955–15 964.

Tanré D., B. N. Holben, and Y. J. Kaufman. 1992. Atmospheric correction algorithm for NOAA-AVHRR products, theory and application. *IEEE J. Geosc. Rem. Sens.*, 30, 231–248.

Vermote, E. F., D. Tanré, J. L. Deuzé, M. Herman, and J. Morcrette. 1996. Second simulation of the satellite signal in the solar spectrum: an overview. *IEEE Trans. Geoscience Rem. Sensing*, accepted.

World Meteorological Organization (WMO)/Radiation Commission of IAMAP. 1983. Meeting of experts on aerosols and their climatic effects, WCP 55, Williamsburg, Va., 28–30 March.

Influence of Sample Composition on Aerosol Organic and Black Carbon Determinations

Tihomir Novakov and Craig E. Corrigan

Determining mass concentrations of major chemical species of biomass aerosols is a prerequisite for assessing the effects of biomass burning on the composition of the atmosphere and its influence on global and regional climate change. Accurate methods for chemical analyses of inorganic species are extensively used in most aerosol characterization studies. In contrast, techniques for characterizing carbonaceous material are much less accurate, and the results obtained by these methods are in many instances highly uncertain. Reasons for this situation are the chemical and physical complexity of carbonaceous (and other) aerosol material, sampling artifacts, and lack of appropriate standard materials.

At the simplest level, chemical characterization of carbonaceous aerosols involves determination of the concentration of classes of materials commonly referred to as total (TC), organic (OC), black (BC) (also known as elemental or graphitic), and carbonate carbon. Commonly used methods can be divided into two categories—thermal for total, organic, and black carbon; optical for black carbon determinations. The definitions of organic and black carbon used in conjunction with these methods are based on presumed volatilization, combustion, and optical properties of these classes of carbonaceous materials and are, therefore, operational and method dependent. These analytical problems are compounded by a lack of accepted standard materials and terminology; for example, terms such as "black," "elemental," and "graphitic" carbon are sometimes used interchangeably. For simplicity, here we will use the term "black carbon" as a synonym for both elemental and graphitic carbon.

Methods for total aerosol carbon determination rely on combusting the sample in oxygen, thereby converting all the carbon content to CO_2. The CO_2 is quantitatively determined, either directly as CO_2 or after methanation over a suitable catalyst to CH_4. Such a method was first applied to urban aerosols by Mueller et al. (1972) and different variants of this method are still in use.

Separation of organic and black carbon by thermal methods relies on the assumption that these components can be distinguished by their volatilization and combustion properties. These methods usually involve progressively heating a sample in an oxidizing or inert carrier gas and measuring the concentrations of gases evolved from the sample as a function of sample temperature. One method of evolved gas analysis used in this study is described below (Novakov 1981, 1982). Huntzicker et al. (1982) employs step-wise ramping of the sample temperature and two carrier gases—inert (pure He) and oxidizing (2% O_2 + 98% He mixture). This method defines OC as the sum of concentrations of species that volatilize at 350°C in a He/O_2 mixture and the fraction that volatilizes at 600°C in He. BC is defined as carbon oxidized in three temperature steps at 400, 500, and 600°C in a He/O_2 mixture. A modified version of the Huntzicker et al. apparatus was constructed by Chow et al. (1993). In this method, organic carbon is defined as the sum of the carbon component evolving in a He atmosphere at four discrete temperature steps (from ~25–120°C, 120–250°C, 250–450°C, and 450–550°C). The BC component is taken as the integrated carbon concentrations released in a 2% O_2 + 98% He atmosphere at 550, 700, and 800°C. Both of these methods monitor the changes in sample darkness to correct for sample charring.

In addition to the above methods that use either continuous or step-wise increasing of the sample temperature, several simplified two-step methods have been developed. Mueller et al. (1982) have used a method in which a sample is placed in contact with MnO_2 which serves as the oxidizing agent. OC is defined as the carbon evolved from the sample at 550°C. BC is determined by the difference between total carbon (measured on a second sample aliquot) and organic carbon. Another two-step thermal analysis procedure was developed by Cachier et al. (1989). In this method, a sample is first exposed to 340°C for 2 hr in pure oxygen to remove the organic material. BC is defined as the total carbon remaining on the sample after the precombustion treatment. Total carbon is determined by combustion of another aliquot of the untreated sample and the OC concentration derived by the difference. Wolff et al. (1982) used a two-step volatilization/oxidation procedure at a constant (650°C) temperature. During the first step, the concentration of carbon-containing gases evolved in a He atmosphere is defined as OC. In the second step, the carrier gas is changed from He to O_2, and the detected CO_2 is attributed to BC.

Black carbon concentrations are routinely monitored by optical methods based on measuring the light intensities transmitted through either precollected filters (Rosen et al. 1980) or in a continuous mode through the accumulating filter deposit by an *aethalometer* (Hansen et al. 1984). When BC is the principal light-absorbing species, its mass concentration (per unit filter area, in $\mu g\ cm^{-2}$) is given by $ATN = \sigma_{BC}BC$, where $ATN = -100\ \ln\ (I/I_0)$, I and I_0 are the light intensities transmitted through the loaded and blank filters, and σ_{BC} is the mass absorption coefficient or cross section for BC (in $m^2\ g^{-1}$). In practice, however, σ_{BC} should be viewed as an empirically derived proportionality constant rather than the absolute value of the BC absorption cross section. As discussed below, the empirical σ_{BC} values were found to be highly variable, depending on factors such as aerosol chemical composition, presence of other light-absorbing species, and, to a large extent, the accuracy of analytical BC determinations.

In this chapter, we present results on the characterization of filter-collected redwood (*Sequoia sempervirens*) needle and eucalyptus smoke particles by thermal, optical, and solvent extraction methods, with particular emphasis on determinations of BC and OC concentrations. As we will show, BC and OC concentrations of biomass smoke particles determined solely by any thermal method may significantly underestimate or overestimate the actual BC and OC concentrations. The derived concentrations are not only method dependent, but they are also critically influenced by the chemical composition of the samples, particularly the amounts of Na and K and certain organic materials present in these samples.

We first describe sample generation and collection and the methods by which these samples were analyzed. We then summarize our results that demonstrate the effects of sample chemical composition on the derived BC and OC concentrations. Finally, we discuss the implications of our results on the application of commonly used methods for the characterization of carbonaceous material in biomass smoke samples.

Table 50.1 Fuels and combustion conditions

Sample	Fuel	Amount (g)	Burn time (min)	Flame[a]
	redwood needles			
2	dry	5	1	F
3	dry + green	2.5	1	S
4	dry + green	2.5	1.2	S
5	green	5	1	S + F
	eucalyptus			
6		12	1	F
9		25	1.5	S
10		50	0.5	F
11		25	1	S

a. F—flaming; S—smoldering.

Experiments

Smoke particles used in this work were generated by burning redwood needles and eucalyptus bark. Approximately 10 g of redwood needles and 25 g of bark supported on a metal screen were burned in air under both flaming and smoldering conditions. Smoke particles from these fires were sampled after dilution with ambient air through a water-cooled duct, and collected on 47-mm diameter prefired quartz filters. The details of smoke generation are shown in table 50.1.

Smoke samples were characterized by a thermal and an optical method. The thermal method used in this study—Evolved Gas Analysis (EGA) (Novakov 1981, 1982)—is based on an approach originally introduced by Malissa et al. (1976). EGA involves progressive heating of a sample (a punch of 1 cm^2 or less taken from the quartz filter) from room temperature to about 600°C at a rate of 12.5°C min^{-1} in an oxidizing (O_2) or a neutral (N_2) atmosphere. In the oxidizing mode, the carbon-containing gases and vapors evolving from the sample as a result of volatilization, decomposition, and combustion of the carbonaceous material are converted to CO_2 over a MnO_2 catalyst maintained at $\sim 800°C$ and are detected by a nondispersive infrared (NDIR) analyzer. When the analysis is done with purified, oxygen-free N_2 as the carrier gas, the hot MnO_2 acts as the oxygen donor, converting the volatilized gases and vapors to CO_2, thus enabling use of the same NDIR detector in both modes of operation. The plot of the rate of evolution of CO_2 vs temperature gives the "thermogram" of carbonaceous material evolved from the sample. The area under the thermogram is proportional to the TC content of the sample. Carbon thermograms show a structure, often in the form of well-defined peaks, indicative of the classes of sample materials with different volatilization, decomposition, and combustion properties. The carbon concentrations corresponding to these classes of materials can be obtained from the areas under these peaks.

Samples were analyzed by EGA as collected (original state) and after removal of their organic and water-soluble content. For this purpose, replicate aliquots in the form of circular punches with areas ≤ 1 cm^2 were taken from exposed 47-mm diameter quartz filters and extracted with acetone and deionized water. To remove the organic material, the sample punches were immersed in acetone for about 4 hr. Water exposures were done in the same manner for time periods ranging from 20 min to several days. After extraction, the sample punches were dried in air at room temperature and analyzed.

Both the original and the extracted samples were also characterized by an optical transmission method similar to that described by Rosen et al. (1980) and Gundel et al. (1984). The purpose of these measurements was not to accurately determine the absorption coefficients, but rather to monitor the possible physical loss of BC from the samples during the extraction and to provide an estimate of BC concentration. Aliquots of collected filter samples extracted in 20 ml deionized water after 30-min sonication were analyzed for anions and cations by ion chromatography (Dionex 2020 I).

Results

The main features of the EGA thermograms of several smoke particle samples (recorded with O_2 as the carrier gas) are illustrated by the examples shown in figure 50.1a. Significant differences among these thermograms are obvious. The thermogram of sample #10 consists of a single peak at about 380°C. In contrast, the thermograms of samples #6 and #3 are more complex, showing features extending to about 100°C, in addition to the pronounced high temperature peaks. Thermal analyses of urban and source samples combined with simultaneous measurement of optical transmission through the sample collected on a quartz filter showed that the loss of optical absorption (or sample blackness) occurred simultaneously with the appearance of the high temperature thermogram peak appearing at 450°C to almost 600°C (Rosen et al. 1982). The aerosol carbon component giving rise to this peak was therefore termed "black carbon" (Novakov 1981, 1982). This high temperature peak could be used to derive the BC concentration, assuming that this peak results from the combustion of BC alone. The carbon

concentrations corresponding to the remainder of the thermogram could be assigned to organic carbon.

For the purpose of this discussion, we refer to the carbon concentration derived from the high temperature EGA peak of untreated (as collected) filter samples as the "apparent black carbon" (BC$_{app}$). A more accurate measure of the actual black carbon (BC$_{act}$) concentration can be obtained by analyzing samples from which most of the organic material has been removed by solvent (e.g., acetone) extraction. As BC is not soluble in organic solvents, the removal of OC from the sample should minimize its possible interference with BC peak assignment.

Thermograms of acetone-extracted samples are shown in figure 50.1b. The thermograms of the original (untreated) and the acetone-extracted sample #10 appear to be similar, indicating that the BC$_{app}$ and BC$_{act}$ concentrations are equivalent. That this sample is composed almost entirely of BC and has minimal OC is evident from the fact that the TC content of the acetone-extracted sample is only about 4% lower than that of the original sample (table 50.2). A different situation is seen for samples with a larger organic content, such as samples #6 and #3. The high temperature thermogram peak (BC$_{act}$) of the acetone-extracted sample #6 is considerably smaller than the BC$_{app}$ peak of the untreated sample. Even more drastic differences between BC$_{app}$ and BC$_{act}$ are seen for samples composed predominantly of organic material: one such example is sample #3, for which the BC$_{app}$ is about six times greater than the BC$_{act}$ concentration. In table 50.2 we also show the ATN values determined from untreated and acetone-extracted samples. The fact that the ATN values before and after extraction are similar (average ATN ratio 0.98 ± 0.13) shows that essentially no BC was removed from the samples during this treatment. (The ATN value of sample #10 is anomalously low because of the uneven sample coverage.) The results described so far indicate that a fraction of soluble organic material contributes to the BC$_{app}$ peak in the untreated smoke particle samples.

The approximate constancy of absorption cross sections (average value = 19.8 ± 5.0 m^2 g^{-1}) estimated from ATN values of untreated samples and BC$_{act}$ concentrations (table 50.2) demonstrates that black carbon (defined as BC$_{act}$) is the species principally responsible for light absorption of the filter-collected material. The contribution of organic species in biomass smoke to light absorption can be roughly estimated from the data shown in table 50.2. The ATN values and OC concentrations for the two samples (yellow and brown in color) with 98.5% organic

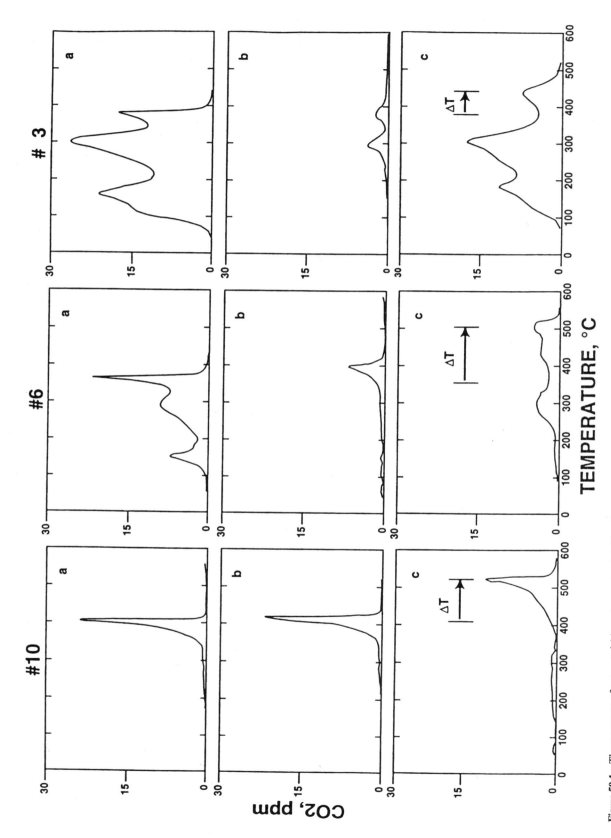

Figure 50.1 Thermograms of untreated (*a*), acetone-extracted (*b*) and water-extracted (*c*) samples. ΔT denotes the temperature shifts between untreated and water-extracted samples.

Table 50.2 Total carbon (TC), apparent black carbon (BC_{app}) and actual black carbon (BC_{act}) concentrations (all in $\mu g\ cm^{-2}$). Also shown, acetone extractable carbon mass fraction (C_{extr}), ATN, and σ_{BC} values (in $m^2 g^{-1}$) estimated from BC_{act} concentrations.

Sample	TC	C_{extr} (%)	BC_{app}	BC_{act}	ATN[a]	ATN ratio[b]	σ_{BC}
10	58.3	4.2	40.0	36.2	≥376	0.95	≥10.4
12	30.8	54.0	14.2	9.2	203	0.77	22.1
6	130.8	63.3	48.0	22.1	460	1.16	20.8
2	619	75.3	153	20.0	408	1.08	20.4
4	5580	80.9	1060	21.7	356	1.03	16.4
3	658	86.2	91.0	14.8	390	1.00	26.3
9	118	98.4	15.8	bd[c]	23.8	3.90	—
11	443	98.5	81.2	bd	55.3	1.86	—

a. Measured on untreated filters.
b. Ratio to ATN measured on acetone-extracted filters.
c. bd—below detection limit.

content (acetone soluble) are 23.8 and 55.3, and 118 and 443 $\mu g\ cm^{-2}$, respectively, resulting in apparent mass absorption cross sections of the total organic material in these samples of 0.20 and 0.12 $m^2\ g^{-1}$. The contribution of this material to the light absorption per unit mass of carbon is therefore about two orders of magnitude lower than that of black carbon.

A further insight into the thermal properties of biomass smoke particles is provided by the thermograms of water-exposed samples, examples of which are shown in figure 50.1c. These thermograms show that removal of water-soluble material results in a substantial increase in the BC_{app} peak temperatures over those seen in untreated samples. Water exposure of sample #10 (4.2% acetone-extractable organic content) resulted in a 130°C increase of the BC peak temperature above that of the untreated sample. Water treatment also resulted in a noticeable change in the BC peak shape but not in its area. Water treatment of sample #6 (63% acetone extractable) reduced the total carbon content from 131 $\mu g\ cm^{-2}$ to 83 $\mu g\ cm^{-2}$ (or by 37%) and caused the single BC_{app} peak seen in the thermogram of the untreated sample to separate into two distinct peaks at ~430°C and ~515°C. The carbon concentration corresponding to these two peaks accounts for 77% of the BC_{app} concentration of the untreated sample. We note that the carbon concentration represented by the 515°C peak ($\approx 26\ \mu g\ cm^{-2}$) agrees with the BC concentration determined from the acetone-extracted sample, suggesting that this component is due to the combustion of black carbon, while the peak at 430°C appears to be due to the water-insoluble component of a relatively oxidation-resistant organic component. Water treatment of sample #3

Table 50.3 Water-soluble mass fractions relative to total carbon concentration in untreated samples

Sample	C	K^+	Na^+	Cl^-	SO_4^{2-}
2	0.17	0.01	0.002	0.02	0.009
3	0.35	0.06	0.009	0.08	0.026
4	0.33	0.02	0.001	0.03	0.002
5	0.30	0.08	0.007	0.10	0.004
6	0.37	0.10	0.031	0.13	0.038
9	0.79	0.04	0.019	0.18	0.010
10	0.17	0.30	0.100	0.29	0.113
11	0.76	0.01	0.001	0.03	0.001

(86% organic) reduced the total carbon content by 35% (from 658 $\mu g\ cm^{-2}$ to 233 $\mu g\ cm^{-2}$). Exposure to water reduced the apparent BC concentration by ~20%, increased the BC_{app} temperature by ~80°C, and changed the shape of this peak.

We attribute the changes in the thermograms to removal of water-soluble organic and inorganic species from the smoke samples. (Concentrations of water-soluble carbon, K^+, Na^+, Cl^-, and SO_4^{2-} are listed in table 50.3.) The most abundant water-soluble inorganic species present in our samples were found to be K^+ and, to a lesser degree, Na^+. These metals are common constituents of biomass smoke particles and are also known combustion catalysts. Consequently, when these metals are present in a sample, the BC_{app} temperature is lower than in the water-extracted sample, in which these species are absent. We note that Na was shown to lower the combustion temperatures of laboratory-generated (Lin and Friedlander 1988) and urban (Grosjean et al. 1994) carbonaceous particles.

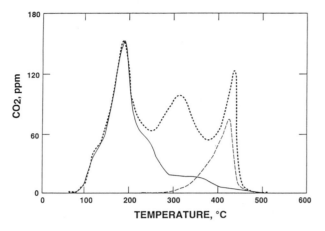

Figure 50.2 Thermograms of sample #11 measured with O_2 (dotted line) and N_2 (solid line) as the carrier gas. Thermogram (measured with O_2) of the residue remaining after the N_2 run (dashed line).

When catalytic species are present, all combustible material (including both the actual BC and the organic species that appear as a part of BC_{app}) will be rapidly oxidized when the temperature reaches a threshold value until it is consumed. This will result in a narrow, well-defined single thermogram peak. Conversely, in the absence of these catalysts, the high temperature thermogram should give an indication of the presence of species with differing combustion temperatures as indicated from the difference in BC_{app} peak shapes seen in water-exposed and untreated samples.

The EGA analyses desribed so far were all done by using O_2 as the carrier gas. Additional information about the nature of the organic material was obtained from the thermograms measured with both O_2 and N_2 as the carrier gas. The results for a predominantly organic (98.5% extractable) sample (#11) are shown in figure 50.2. A comparison of the N_2 thermogram with the one obtained in an O_2 atmosphere shows that the low temperature regions of both thermograms are identical. In contrast, the peak at about 320°C is only partially reproduced in the N_2 mode. The BC_{app} prominent in the oxygen thermogram is virtually absent in the nitrogen thermogram. This comparison clearly shows that the lowest temperature peak is derived entirely from volatile species, while the species responsible for BC_{app} and, to a lesser degree for the 320°C peak, are nonvolatile under our experimental conditions. The nonvolatile carbonaceous material can be recovered by reanalyzing in O_2 the residue remaining in the sample after the run in N_2. The oxygen thermogram of this residue (shown by dashed line in

figure 50.2) consists of a peak at the same temperature, with about the same concentration as the BC_{app} in the oxygen thermogram of the original sample.

The distinct BC_{app} peak at ~ 425°C ($BC_{app} = 81.2\ \mu g\ cm^{-2}$) seen in the O_2 thermogram shown in figure 50.2 must be derived from organic material because this sample is composed of $\approx 99\%$ acetone-soluble organic material, with BC_{act} below detection (table 50.2). Organic carbon can contribute to BC_{app} in two ways. The first possibility is that the sample contains an organic component having a combustion temperature corresponding to the observed BC_{app} peak. The second possibility is that the material giving rise to the BC_{app} peak is a product of pyrolysis (or charring) of lower-volatility organic species that combust at temperatures similar to BC. We estimated the contribution of the pyrolyzed organic carbon to BC_{app} from the ATN values of the original sample and the sample preheated (in O_2) to 345°C, corresponding to the onset of the BC_{app} peak. The ATN values before and after heating were found to be 55.3 and 130, respectively. This increase in ATN values shows that some charring of the sample in an O_2 atmosphere did occur. The contribution of this material to BC_{app} is small, approximately equal to ATN/$\sigma \approx 130/20 \approx 6.5$ $\mu g\ cm^{-2}$, or about 8% of BC_{app}. The ATN value at the end of the nitrogen run was found to be 286, which corresponds to an estimated pyrolysis contribution of $\approx 14\ \mu g\ cm^{-2}$, or about 18% of the total residual carbon concentration (80.0 $\mu m\ cm^{-2}$). The results of these analyses demonstrate that most of the carbon corresponding to BC_{app} in predominantly organic smoke samples is not derived from pyrolysis, but from relatively nonvolatile organic species.

Discussion

The results we have presented have several implications for determining black and organic aerosol carbon by both thermal and optical methods. Before discussing these implications, it is instructive to review the degree of disagreement in OC and BC determinations by commonly used methods. These disagreements are illustrated by the results of the recent Carbonaceous Species Methods Comparison Study (Shah and Rau 1990). In this intercomparison study, ambient urban samples, source samples, and, of particular relevance to biomass burning aerosols, two ambient samples heavily influences by residential wood burning were prepared and distributed to a number of laboratories for analyses by a variety of thermal methods.

Table 50.4 Methods and results of OC and BC concentration ($\mu g\ cm^{-2}$) determinations in wood smoke influenced ambient samples, compiled from Shah and Rau (1990)

Method[a]	Operational definitions		Sample #1			Sample #2		
	OC	BC	OC	BC	OC/BC	OC	BC	OC/BC
1	600°C, He	750°C, 2% O_2	88.3	10.4	8.5	374.5	21.7	17.2
2	600°C, He	TC–OC	47	16	2.9	195	33	5.9
3	600°C, He	650°C, 2% O_2	86.1	5.3	16.2	257.6	77.6	3.3
4	950°C, He	950°C, 2% O_2	82.4	21.8	3.8	390.5	78.7	5.0
5	610°C, He	750°C, 10% O_2	71.8	26.6	2.7	299.2	94.7	3.2
6	550°C, He	650°–800°C, 2% O_2	62.5	35.8	1.7	253.5	74.8	3.4
7	700°C, He	900°C, 2% O_2	82.7	8.4	9.8	320.5	11.7	27.4
8	525°C, He(MnO_2)	850°C, He(MnO_2)	86.1	3.8	22.6	364.7	1.6	228
9	400°C, O_2	700°C, O_2	92.9	6.5	14.3	392.1	11.9	32.9

a. Laboratories: (1) Environmental Monitoring Services, Inc., (2) Oregon Department of Environmental Quality, (3) US EPA, (4) General Motors Research Laboratory, (5) Oregon Graduate Institute, (6) Desert Research Institute, (7) Sunset Laboratory, (8) Environmental Research and Technology, (9) Global Geochemistry Corporation.

The intercomparison results show that most methods can accurately measure total carbon (within 5% of the common average). Organic and black carbon determinations of urban ambient samples showed, respectively, 6–9% and 28–31% variations from common averages (% variation taken as 100 × avg./std.dev.).

The discrepancies between OC and BC concentrations for the two wood smoke samples obtained by nine thermal methods were significantly greater. The operational definitions used by different methods and the analytical results for these samples, compiled from Shah and Rau (1990), are summarized in table 50.4. These data show that the total carbon concentrations (the sum of OC and BC concentrations) range from 63 to 101 $\mu g\ cm^{-2}$ and from 228 to 469 $\mu g\ cm^{-2}$ for the two sample sets. If the TC concentrations obtained by method 2 are excluded as being obviously too low, the remaining TC values agree within about 13–17% (concentration ranges 86–101 $\mu g\ TC\ cm^2$ and 328–469 $\mu g\ TC\ cm^2$)

Differences in OC and especially in BC determinations are, however, much larger. OC concentrations differ by about 50% (concentration ranges 62–93 μg OC cm^{-2} and 253–392 μg OC cm^{-2}). The reported BC concentrations range from 3.8–36 μg BC cm^{-2} and from 79–1.6 μg BC cm^{-2} for the two samples, resulting in 10- to 50-fold differences among individual determinations. Consequently, the OC/BC ratios differ by up to two orders of magnitude (OC/BC ranges 1.7–23 and 3.2–228). Furthermore, OC/BC ratios differ significantly, even when obtained by seemingly similar analytical procedures. For example, the OC/BC ratios obtained by two-step procedures (methods 1 through

4) differ by a factor of about 5, and methods that account for sample charring (methods 5 through 7) by factors of 5.7 to 8.5.

Implicit in methods 1–8 is the assumption that all organic material is removed from samples by volatilization in an inert atmosphere (He) at temperatures ranging from 525 to 950°C. Our results, however, show that a substantial organic fraction of biomass smoke particles is nonvolatile when heated in an inert atmosphere to about 550°C in a manner similar to those used in methods 5 and 6. Consequently, OC would be underestimated if this fraction were not included in the reported OC concentration. Furthermore, our results indicate that only a part of this nonvolatile material is due to pyrolysis and therefore not necessarily accounted for in methods that rely on monitoring the sample darkness during analysis. The residual material remaining after heating in an inert atmosphere could be oxidized in the subsequent combustion step and thus erroneously assigned to BC. The magnitude of the resulting error in BC will depend on the relative concentrations of OC (and its relatively nonvolatile component) and actual BC. Obviously, even a relatively small underestimation of OC would result in a large overestimation of BC for samples with the high OC/BC ratios common in biomass smoke particles.

Because the combustion temperatures of both apparent and actual BC are strongly influenced by catalytically active metals such as Na and K, no single temperature can be defined that distinguishes black and organic carbon in samples that contain these species. Methods that rely on thermal pretreatment of samples in oxygen at a fixed temperature to remove the organic material (for example, method 9 in table 50.4) could be more accurate when applied to samples that contain an appreciable fraction of black carbon and are relatively free of these catalytic materials. However, when K and Na are part of the sample, the black carbon will readily be oxidized at this temperature because of the lowered combustion threshold. This could result in underestimating the black carbon concentration not only in biomass smoke samples, but also (because of the catalytic effect of Na) in marine aerosol samples.

BC concentrations are routinely estimated from the empirical relationship $ATN = \sigma_{BC} BC$. However, the accuracy of such measurements using a single σ_{BC} value could be questionable because of the large variability in σ_{BC} values derived by different investigators (see Liousse et al. 1993). There is evidence that the ATN vs BC proportionality may depend on the kind of aerosol composition (i.e., chemical and physical) even when the same transmission measurements and procedures for BC determination are used. For example, Niessner and Petzold (1994) obtained σ_{BC} that varied from about 5 m^2 g^{-1} for rural aerosols to 18 m^2 g^{-1} for urban aerosols. Liousse et al. (1993) derived σ_{BC} values that were similarly variable depending on the sampling location. The lowest value of 5 m^2 g^{-1} was obtained in remote areas, and the highest value of 20 m^2 g^{-1} was measured in the African savannah regions. A significant part of this variability, particularly for samples with high organic content, is undoubtedly caused by errors in BC determinations, as illustrated by the large differences in apparent σ_{BC} that would be obtained if BC_{app} concentrations instead of BC_{act} were used (table 50.2).

Conclusions

The results of our experiments can be summarized as follows: (1) the organic fraction of biomass smoke particles analyzed includes a component ($BC_{app} - BC_{act}$), ranging in concentration from about 6–20% of total carbon or from 16–30% of acetone-extractable organic carbon that is relatively nonvolatile and has a combustion temperature close to that of black carbon; (2) combustion temperatures of this organic component and of black carbon in biomass smoke samples are significantly lowered when these contain K or Na; (3) when these metals are present in a sample, the combustion temperatures of both the black carbon and this organic material are indistinguishable; (4) about 20% of total organic material is nonvolatile when heated to 550°C in an inert atmosphere; (5) the contribution of colored organic material to the light absorption cross section of filter deposits is about two orders of magnitude smaller than that of black carbon; (6) the water-soluble fraction of biomass smoke samples includes a substantial organic component.

Because the combustion temperatures of both apparent and actual BC are strongly influenced by catalytically active metals such as Na and K, no single temperature can be defined that distinguishes black and organic carbon in samples that contain these species, even when aerosols with similar OC and BC content are sampled. Using a single temperature to separate BC from OC could result in an underestimation of the BC concentration not only in biomass smoke samples, but also (because of the catalytic effect of Na) in marine aerosol samples.

The assumption that all organic material is removed from samples by volatilization in an inert atmosphere may not be valid. The residual material remaining after heating in an inert atmosphere could erroneously be assigned to BC. The magnitude of the resulting error in BC will depend on the relative concentrations of OC (and its relatively nonvolatile component) and actual BC. Even a relatively small underestimation of OC would result in a large overestimation of BC for samples with high OC/BC ratios common in biomass smoke particles.

Estimating BC concentrations from the empirical relationship $ATN = \sigma_{BC} BC$ could be questionable because of the large variability in σ_{BC} values derived by different investigators (Liousse et al. 1993; Niessner and Petzold 1994)). A significant part of this variability, particularly for samples with high organic content, is undoubtedly caused by errors in BC determinations by thermal methods. Contribution of BC to light absorption per unit mass is about two orders of magnitude larger than combustion-generated (colored) organic material. The light attenuation caused by these species could by erroneously attributed to BC for biomass smoke samples with high OC/BC ratios. For comparison, the contribution of mineral dust to light absorption cross sections has been estimated to be 2–3 orders of magnitude smaller than BC (Schnell et al. 1994).

In summary, our results and discussion demonstrate that OC and BC concentrations determined by thermal and optical methods are not only method dependent, but also critically influenced by the overall chemical composition of the samples. Therefore, only a combined approach using different methods can give actual black carbon and organic concentrations.

Acknowledgment

This work was supported by the U.S. Department of Energy, Office of Health and Environmental Research, Environmental Sciences Division under contract DE-AC03-76SF00098.

References

Cachier, H., M.-P. Bremond, and P. Buat-Ménard. 1989. Determination of atmospheric soot carbon with a simple thermal method, *Tellus*, 41B, 379–390.

Chow, J. C., J. G. Watson, L. C. Pritchett, W. R. Pierson, C. A. Frazier, and R. G. Purcell. 1993. The DRI thermal/optical reflectance carbon analysis system: Description, evaluation and application in US. air quality studies, *Atmos. Environment*, 27A, 1185–1201.

Grosjean, D., E. L. Williams, E. Grosjean, and T. Novakov. 1994. Evolved gas analysis of secondary organic aerosols, *Aerosol Sci. and Technol.*, 21, 306–324.

Gundel, L. A., R. L. Dod, H. Rosen, and T. Novakov. 1984. The relationship between optical attenuation and black carbon concentration for ambient and source particles, *Sci. Total Environ.*, 36, 271–276.

Hansen, A. D. A., H. Rosen, and T. Novakov. 1984. The aethalometer—an instrument for the real-time measurement of optical absorption by aerosol particles, *Sci. Total Environ.*, 36, 191–196.

Huntzicker, J. J., R. L. Johnson, J. J. Shah, and R. A. Cary. 1982. Analysis of organic and elemental carbon in ambient aerosols by a thermal-optical method, in *Particulate Carbon: Atmospheric Life Cycle*, edited by G. T. Wolff and R. L. Climish, pp 79–88, Plenum, New York.

Lin, C., and S. K. Friedlander. (1988). A note on the use of glass fiber filters in the thermal analysis of carbon containing aerosols, *Atmos. Environment*, 22, 605–607.

Liousse, C., H. Cachier, and S. G. Jennings. 1993. Optical and thermal measurements of black carbon aerosol content in different environments: Variation of the specific attenuation cross-section, sigma (σ). *Atmos. Environment*, 27A, 1203–1211.

Malissa, H., H. Puxbaum, and E. Pell. 1976. Simultane Kohlenstoff- und Schwefelbestimmung in Stäuben, *Fresenius Z. Anal. Chem.*, 273, 109–113.

Mueller, P. K., R. W. Mosley, and L. B. Pierce. 1972. Chemical composition of Pasadena aerosol by particle size and time of day: Carbonate and noncarbonate carbon content, *J. Coll. Interface Sci.* 39, 235–240.

Mueller, P. K., K. K. Fung, S. L. Heisler, D. Grosjean, and G. M. Hidy. 1982. Atmospheric particulate carbon observations in urban and rural areas of the United States, in *Partuculate Carbon: Atmospheric Life Cycle*, edited by G. T. Wolff and R. L. Climish, pp. 343–370, Plenum, New York.

Niessner, R., and A. Petzold. 1994. Comparison study on thermal, optical, and photoelectrical methods for elemental carbon analysis, paper presented at the *Fifth International Conference on Carbonaceous Particles in the Atmosphere*, Berkeley, Calif., 23–26 August.

Novakov, T. 1981. Microchemical characterization of aerosols, in *Nature, Aim and Methods of Microchemistry*, edited by H. Malissa, M. Grasserbauer, and R. Belcher, pp. 141–165, Springer, Vienna.

Novakov, T. 1982. Soot in the atmosphere, in *Particulate Carbon: Atmospheric Life Cycle*, edited by G. T. Wolff and R. L. Climish, pp. 19–41, Plenum, New York.

Rosen, H., A. D. A. Hansen, R. L. Dod, and T. Novakov. 1980. Soot in urban atmospheres: Determination by an optical absorption technique, *Science*, 208, 741–744.

Rosen, H., A. D. A. Hansen, R. L. Dod, L. A. Gundel, and T. Novakov. 1982. Graphitic carbon in urban environments and the Arctic, in *Particulte Carbon: Atmospheric Life Cycle*, edited by G. T. Wolff and R. L. Climish, pp. 273–294, Plenum, New York.

Schnell, R. C., D. T. Kuniyuki, B. A. Bodhaine, and A. D. A. Hansen. 1994. The dust component of aerosol light absorption measured at Mauna Loa Observatory. Paper presented at the Fifth International Conference on Carbonaceous Particles in the Atmosphere, Berkeley, Calif., 23–26 August.

Shah, J. J., and J. A. Rau. 1990. Carbonaceous species methods comparison study: Interlaboratory round robin interpretation of results, Final report to Research Division, California Air Resources Board, Contract No. A832-154.

Wolff, G. T., P. J. Groblicki, S. H. Cadle, and R. J. Countess. 1982. Particulate carbon at various locations in the United States, in *Particulate Carbon: Atmospheric Life Cycle*, edited by G. T. Wolff and R. L. Climish, pp. 297–315, Plenum, New York.

Activation of Carbon Aerosol by Deposition of Sulfuric Acid

Wilson T. Rawlins, Shin G. Kang, David M. Sonnenfroh,
Karen L. Carleton, and Barbara E. Wyslouzil

Carbonaceous soot particles, injected into the atmosphere by biomass and fossil fuel combustion, can have a significant influence on radiative climate forcing. Black-carbon particles are well known to be hydrophobic, and thus are not expected to act as cloud condensation nuclei (CCN) in the atmosphere. However, adsorption of water-soluble species on particle surfaces can increase particle activity toward uptake of H_2O, potentially altering the role of soot particles in the atmosphere. Effluent streams from the combustion of biomass and fossil fuels contain substantial levels of NO_x, formed from combustion in air, and SO_2, formed from oxidation of fuel-bound sulfur. Continued oxidation of these species in the effluent stream generates lesser levels of HNO_x and H_2SO_4. Some or all of these species may play a role in the activation of soot particles for H_2O uptake. This chapter focuses on the effects of sulfuric acid, H_2SO_4.

In previous research (Wyslouzil et al. 1994), we demonstrated the H_2O activation of individual black carbon particles by adsorption of sub-monolayer amounts of H_2SO_4 vapor, and found H_2SO_4 solute activities that were consistent with soluble mass fractions observed in soot particles, collected from the exhaust stream of an aircraft jet engine (Hagen et al. 1992). Those initial results suggested that at least some of the soluble material on the engine-exhausted soot particles was indeed H_2SO_4, and raised the question of whether this activation mechanism could lead to previously unanticipated CCN behavior by soot particles. The chapter reports additional laboratory measurements of black-carbon hydration behavior and presents a discussion of the results in terms of critical supersaturation behavior as described by the Köhler theory.

Experiments

The experimental apparatus and measurement methods have been described in detail previously (Wyslouzil et al. 1994). The experimental measurements are conducted in a single-particle quadrupole trap, in which individual carbon particles are suspended electrodynamically and are exposed to well-controlled gas phase environments. Single, lightly-charged particles are confined radially by and AC field and are supported vertically by a DC balancing field. The DC balance voltage required to maintain the particle position is proportional to the particle's mass/charge ratio, and thus can be used to monitor relative changes in particle mass during the course of an experiment.

The particle trap is housed in a flow cell at controlled temperature and relative humidity. The relative humidity of the gas entering the cell is controlled by mixing flows of dry N_2 and N_2 that has passed through a water saturator, at typical total flow rates of ≈ 100 sccm. The relative humidity in the trap is determined from the measured flow rates and temperatures to a relative accuracy of $\pm 10\%$, as described in detail by Wyslouzil et al. (1994).

As in the previous work, synthetic Spherocarb particles were used to represent the behavior of black-carbon aerosol. True soot particles are highly uncertain and variable in both shape and surface area and are typically too small (<1 μm) to be studied as individual particles in our apparatus. In contrast, Spherocarb particles are well characterized in size, shape, purity, and surface area. They have a high C/H ratio (≥ 10), characteristic of black-carbon particles produced in high-temperature, fuel-rich diffusion flames. The particles used in these experiments were highly porous spheres with diameters of 125 to 150 μm, densities of ≈ 0.5 g/cm^3, and mass-specific surface areas of 800 to 1200 m^2/g. Virtually all of the available surface area is contained within internal pores. The mass-specific surface area of a single Spherocarb particle is equivalent to that of a nonporous, spherical particle of diameter ≈ 4 nm.

The experimental procedures were similar to those described previously (Wyslouzil et al. 1994). Small batches of particles were treated by overnight vapor deposition of 97 wt% H_2SO_4 in an oven at 120 to 140°C. Prior to H_2SO_4 treatment, some particle samples were dried at high temperatures for prolonged

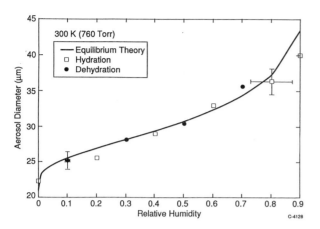

Figure 51.1 Observed and computed hydration and dehydration behavior of a single sulfuric acid aerosol

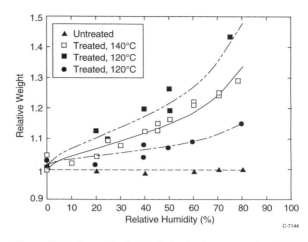

Figure 51.2 Prompt hydration behavior for untreated and H_2SO_4-treated carbon particles. The curves represent fits to the data points using equilibrium theory as described in the text.

periods. Some individual particles were also dried in the electrodynamic trap by exposure to flows of heated dry nitrogen overnight. For each experiment, a single untreated or H_2SO_4-treated Spherocarb particle was captured and balanced in the trap at room temperature and atmospheric pressure. The particle was then exposed to a flow of dry nitrogen until the DC balance voltage readings showed that the particle was equilibrated to its "dry" mass. The relative flow rates of dry and H_2O-saturated nitrogen were then adjusted to provide step changes in relative humidity at constant total flow rate. At each step change, the particle mass was allowed to equilibrate, and the particle mass change relative to the "dry" mass was determined from the DC voltage setting required to maintain the particle position at the saddle point of the AC trapping field.

Results and Analysis

The hydration experiments consisted of measurements of relative mass size change as a function of increasing and decreasing relative humidity step changes. An example hydration curve for liquid H_2SO_4 is illustrated in figure 51.1. The solid curve is an equilibrium calculation using the thermodynamic properties given by Zeleznik (1991). The close agreement between the hydration data and the equilibrium theory demonstrates the effectiveness of the experimental method for studies of equilibrium H_2O uptake measurements. The shape of the hydration curve illustrates the behavior that can be expected for H_2SO_4-coated carbon particles.

The hydration results for treated and untreated carbon particles fall into two different categories: (1)

treated particles for which only prompt reversible hydration due to H_2SO_4 coverage could be observed, and untreated particles that did not take up any H_2O; and (2) both treated and untreated particles for which slow, quasi-irreversible H_2O uptake could be observed, in addition to reversible hydration of the H_2SO_4-treated particles. The prompt hydration behavior, previously reported by Wyslouzil et al. (1994), is illustrated in figure 51.2. Particle masses were allowed to equilibrate for 15 to 30 min. at each relative humidity step. Clearly, the untreated carbon particles did not gain or lose any mass on this time scale for relative humidities up to 80%, and thus appeared to be immune to hydration as expected. In contrast, the H_2SO_4-treated carbon particles readily and promptly hydrated under subsaturated conditions. The hydration curve for a given particle was reproducible for both increasing and decreasing relative humidity, indicating a reversible process. There are significant differences in the observed hydration activities of the treated particles, which may be due in part to differences in the H_2SO_4 batch treatment temperatures and exposure geometries. However, it is also likely that substantial variation in apparent activity can be caused by differences in the accessible surface area for the individual particles. The data in figure 51.2 represent the median and upper and lower extremes of the observed hydration activities. The fitted curves in figure 51.2 were computed from an equilibrium treatment for H_2SO_4 hydration on an inert carbon substrate (Wyslouzil et al. 1994), and correspond to soluble mass fractions of 21%, 15%, and 6%.

Examples of longtime, compound hydration behavior are illustrated for untreated and H_2SO_4-treated

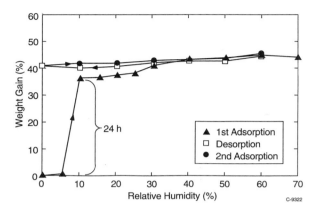

Figure 51.3 Long-term hydration behavior of untreated carbon particle

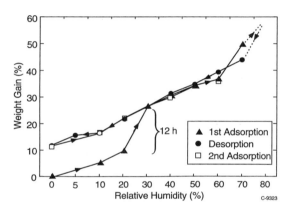

Figure 51.4 Compound hydration behavior of H₂SO₄-treated carbon particle, showing both the prompt and long-term components

particles in figures 51.3 and 51.4, respectively. This behavior was not previously observed by Wyslouzil et al. (1994), but became apparent when the particles were extensively dried and the hydration behavior was observed over time scales of several hours. As shown in figure 51.3, untreated Spherocarb particles exhibited a slow but substantial H₂O uptake that reached a saturation level over a period of several hours. The particle in figure 51.3 was dried in an oven at 70°C before use, and was left overnight ($\approx$ 24 hr) in the trap at a relative humidity of 10%, during which time it underwent a $\approx$ 40% mass increase due to H₂O uptake. As shown in the figure, continued exposure to increasing and decreasing relative humidity at intervals of 1 to 4 hr between steps resulted in little or no additional change in the particle's mass, signifying no further H₂O uptake or removal. Particles exposed to room temperature and heated flows of dry nitrogen over periods of many hours showed very slow mass loss, indicating that the H₂O uptake process may be reversible over very long

time scales. However, on the time scale of reversible H₂SO₄ hydration and dehydration (15 to 30 min per relative humidity step), the phenomenon is in effect irreversible.

H₂SO₄-treated particles showed a similar but somewhat lesser effect. For the experiment shown in figure 51.4, the particle was left in the trap overnight ($\approx$ 10 hr) at a relative humidity of 30%, resulting in an H₂O uptake of $\approx$ 10% of the particle's mass. Continued exposure to increasing and decreasing relative humidity, in 0.5-to-1 hr steps, produced reproducible and reversible hydration and dehydration behavior as observed in figure 51.2. The reversible H₂SO₄ hydration behavior in figure 51.4, when corrected for the irreversible H₂O uptake, agrees well with the most active hydration curve in figure 51.2 (21% soluble mass fraction). The long-term, quasi-irreversible, physical uptake of water by the Spherocarb material was observed consistently for some dozen treated and untreated particles tested in the trap; however, the amount of mass increase due to this process depended on exposure time at a given relative humidity, and varied significantly from particle to particle. The irreversible mass increase was 40 to 100% for untreated particles and 10 to 30% for H₂SO₄-treated particles.

The hysteresis behavior shown in figures 51.3 and 51.4 is characteristic of the effects of capillary condensation of H₂O in small pores of the Spherocarb material due to the inverse Kelvin effect (Adamson 1976). For sufficiently small pores and small contact angles (i.e., small radius of curvature), H₂O can condense at partial pressures considerably below the equilibrium vapor pressure over a flat surface. The cylindrical capillary condensation criteria are given by

$$N_A \, kT \ln(RH) = -\frac{2M\gamma \cos\theta}{\rho r} \tag{51.1}$$

where N_A is Avogadro's number, k is Boltzmann's constant, RH is fractional relative humidity, M is molecular weight, γ is surface tension, θ is contact angle, ρ is density, and r is the critical pore radius giving condensation at relative humidity RH. Evaluation of equation (51.1) in the limit of small contact angle gives the maximum critical pore radius as a function of relative humidity. For sufficiently long exposure times, we observe quasi-irreversible H₂O uptake for relative humidities of 10 to 40%, corresponding to pore radii of about 0.5 to 1 nm. The Spherocarb supplier quotes a volume-averaged mean pore radius of $\approx$ 0.75 nm, consistent with these observations.

The 40% mass increase for the untreated particle in figure 51.3 corresponds to filling of about half of the available pore volume, signifying that the surface area available for H_2SO_4 coverage must be considerably smaller than the nominal Spherocarb surface area. Particle-to-particle variations in the physical H_2O uptake occur due to variations in the individual particle sizes, porosities, and pore volume distributions. H_2SO_4-treated particles show smaller physical H_2O uptake occurring at somewhat higher relative humidities, perhaps due to obstruction of the smallest pores by the presence of H_2SO_4 adsorbate.

We conclude from the experimental data that two mechanisms for H_2O uptake by carbon particles can be identified: (1) prompt, solubility-induced equilibrium physisorption of H_2O by H_2SO_4 adsorbed on carbon surfaces; and (2) slow, physical trapping of H_2O by capillary condensation in ≈ 1 nm-sized pores. The H_2SO_4-induced process is consistent with adsorbed H_2SO_4 soluble mass fractions of $(14 \pm 6)\%$ resulting from the vapor deposition treatment of bulk powder samples. Using the reported bulk-averaged N_2 BET surface area of 864 m^2/g for Spherocarb (Waters et al. 1988) as an upper bound to the available surface area for H_2SO_4 adsorption, we estimate lower-bound equivalent surface coverage of approximately 0.2 to 0.4 monolayer of H_2SO_4 for soluble mass fractions of 10 to 20%.

Discussion And Conclusions

Atmospheric soot particles are typically in the 10-to-100-nm size range. Sampling of laboratory-simulated and aircraft-generated jet exhaust aerosols by an electrostatic analyzer particle counting method (Hagen et al. 1992; Whitefield et al. 1993) gives broad and sometimes bimodal size distributions over 10 to 500 nm, with maxima in the number distributions at diameters of 20 to 50 nm. Smoke particle size distributions measured above a flaming biomass fire are similarly broad, with maxima at diameters near 80 to 100 nm (e.g., Einfeld et al. 1991). The aircraft exhaust measurements also show that the small soot particles exhibit hydration behavior consistent with soluble mass fractions on the order of $\approx 10\%$ (Hagen et al. 1992). This result is in good agreement with our laboratory observations of 10 to 20% soluble H_2SO_4 mass fractions on Spherocarb, suggesting that H_2SO_4 may indeed be an activating agent for hydration of black carbon particulate in the atmosphere.

The effects of H_2SO_4 activation on the primary soot particles can be considered in terms of the Köhler theory, which describes critical supersaturation behavior for spherical aerosols (Pruppacher and Klett 1980). This theory applies to the limit of dilute solute/

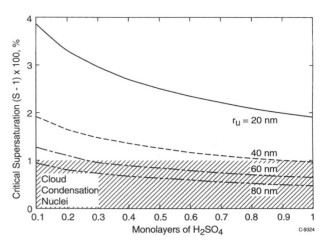

Figure 51.5 Köhler critical supersaturations as functions of surface coverage for H_2SO_4-coated carbon particles with initial radii of 20, 40, 60, and 80 nm

H_2O mixtures and provides useful guidelines for relating the laboratory results to atmospheric aerosol growth phenomena. Submonolayer coverage of an inert core by a water-soluble substance affects the subsaturation hydration behavior of the compound aerosol, as shown in our experiments, and also lowers the critical supersaturation level at which initially small aerosols will grow without bound to form cloud droplets. Using the equations given in Pruppacher and Klett (1980), we can calculate Köhler growth curves (saturation S vs. particle diameter) for different inert core diameters and solute coverages; these curves pass through maxima that define the critical supersaturations required for particle growth without bound. For a given initial particle size, increasing amounts of solute on the particle reduce the critical supersaturation level, eventually enabling CCN behavior (critical supersaturation $S_c < 1.01$). For smaller initial particle sizes, the critical supersaturation for a given coverage is higher, and correspondingly larger solute coverages are required to enable CCN behavior. The dependence of critical supersaturation on solute coverage for selected initial particle sizes is shown in figure 51.5. Clearly, particles with initial radii smaller than about 40 nm require solute coverages in excess of one monolayer to become CCN. The experimental measurements reported here indicate a typical H_2SO_4 coverage of at least 0.2 to 0.4 monolayer; the soluble mass fractions determined by Hagen et al. (1992) for ≈ 35-nm (diameter) soot particles correspond to a maximum solute coverage of 0.7 monolayer (assuming nonporous spherical surface area). These results taken together suggest that soot particles with initial radii larger than about 40 nm may contain enough H_2SO_4 to become CCN, given sufficient levels of H_2SO_4 in the plume. This limit is somewhat larger than the mean radii typcally observed for atmospheric soot particles as discussed above. Thus, it is likely that the larger black-carbon particles generated in fires will act as CCN. The fraction of the total black carbon that is affected is clearly a sensitive function of the actual size distribution of black-carbon particles produced by the fire.

These conclusions are consistent with phenomenological observations of the hydration characteristics of soot particls sampled from hydrocarbon combustion products in the laboratory and in the atmosphere. In laboratory combustion of liquid hydrocarbon fuels, Hagen et al. (1989) observed critical supersaturation spectra that extrapolate to $S_c \leq 1.01$ for particle radii greater than ≈ 40 nm, and Whitefield et al. (1993) observed $S_c \leq 1.01$ for particle radii greater than about 25 to 30 nm. Pitchford et al. (1991) performed airborne sampling of jet aircraft exhaust and observed $S_c \approx 1.04$ to 1.05 for soot particles of radii near ≈ 50 nm; their observed values extrapolate to a CCN threshold, $S_c \leq 1.01$, for particle radii greater than ≈ 30 nm.

The activation of black carbon as CCN by the H_2SO_4 solvation mechanism could significantly alter our view of the role of black carbon in the atmospheric radiation budget. Black carbon included in an H_2SO_4/H_2O aerosol should be less absorbing of solar radiation than dry black-carbon particles, and the compound aerosol should also have a larger scattering cross section because of its larger size. This leads to two interesting speculations: (1) black-carbon emissions may increase the local sulfate aerosol generation and/or cloud formation near the source, and (2) a significant mass fraction of black-carbon particulate emissions may end up in the form of modified sulfate aerosols, giving an unexpected radiative cooling effect rather than the expected warming effect.

Other combustion effluent species, such as SO_2, NO_2, and HNO_3, may also contribute to activation of soot particles. Furthermore, the mechanism of H_2SO_4 deposition on the particles in the combustion effluent stream is not known, and we have investigated only vapor deposition to date. We are continuing our investigations to examine these processes.

Acknowledgments

This work was supported by the NASA Atmospheric Effects of Aviation Program.

References

Adamson, A. W. 1976. *Physical Chemistry of Surfaces*, 3rd edition, John Wiley and Sons, New York, pp. 618 ff.

Einfeld, W., D. E. Ward, and C. Hardy. 1991. Effects of fire behavior on prescribed fire smoke characteristics, in *Global Biomass Burning*, J. S. Levine, Ed., MIT Press, Cambridge, Mass., pp. 412–419.

Hagen, D. E., M. B. Trueblood, and D. R. White. 1989. Hydration properties of combustion aerosols, *Aerosol Sci. Technol.*, 10, 63.

Hagen, D. E., M. B. Trueblood, and P. D. Whitefield. 1992. A field sampling of jet exhaust aerosols, *Particulate Sci. and Tech.*, 10, 53.

Pitchford, M., J. G. Hudson, and J. Hallett. 1991. Size and critical supersaturation for condensation of jet engine exhaust particles, *J. Geophys. Res.*, 96, 20 787.

Pruppacher, H. R., and J. D. Klett. 1980. *Microphysics of Clouds and Precipitation*, Reidel, Boston.

Waters, B. J., R. G. Squires, N. M. Laurendeau, and R. E. Mitchell. 1988. Evidence for formation of CO_2 in the vicinity of burning pulverized carbon particles, *Comb. and Flame*, 74, 91.

Whitefield, P. D., M. B. Trueblood, and D. E. Hagen. 1993. Size and hydration characteristics of laboratory simulated jet engine combustion aerosols, *Particulate Sci. and Tech.*, 11, 25.

Wyslouzil, B. E., K. L. Carleton, D. M. Sonnenfroh, W. T. Rawlins, and S. Arnold. 1994. Observation of hydration of single, modified carbon aerosols, *Geophys. Res. Lett.*, 21, 2107.

Zeleznik, F. J. 1991. Thermodynamic properties of the aqueous sulfuric acid system to 350 K, *J. Phys. Chem. Ref. Data*, 20, 1157.

Reconstruction of Prehistoric Fire Regimes in East Africa by Lake Sediment Analysis

Karl-Friedrich Weiss, Johann G. Goldammer, James S. Clark,
Dan A. Livingstone, and Meinrat O. Andreae

A few years ago, analysis of data from the Total Ozone Mapping Spectrometer (TOMS) instrument aboard Nimbus 7 suggested seasonally enhanced concentrations of tropospheric ozone over the southern tropical Atlantic between South America and southern Africa from August until October (Fishman et al. 1990). For a long time, southern Africa and parts of South America had been considered to be clean air regions because of low industrial emissions. Thus, the causes of these seasonally elevated tropospheric ozone concentrations needed to be investigated. Every year during the dry season, large vegetation fires take place in the savannas, producing significant emissions of trace gases to the atmosphere (Andreae 1991). Simultaneous occurrences of high ozone values and vegetation fires in both continents were observed and a connection between both phenomena was proposed. The SAFARI-92 experiment (Southern African Fire-Atmosphere Research Initiative) confirmed that high tropospheric ozone concentrations are the consequence of seasonal vegetation fires in the tropics (Andreae et al. 1994).

The reconstruction of Quaternary fire regimes will be helpful in clarifying whether high ozone concentrations were also present in the past. Our understanding of the role of vegetation fires in ecosystems has improved over the year, but little is yet known about past fires, especially in the tropics and subtropics. The 1994 NATO Advanced Research Workshop "Sediment Records of Biomass Burning and Global Change" also made clear that there are still many uncertainties (Clark et al. 1996).

Vegetation fires are documented since the Devonian (Francis 1961). In the further development of the earth's history, fire was established as a component of the ecosystems in many regions of the world. The use of fire by hominids has been confirmed for the last 1.5 million years (Brain and Sillen 1988).

Cores of lake sediment deposits contain highly useful stratigraphic sequences. Sedimentary charcoal is a product of vegetation fires indicating former fire occurrences. Details such as the number and the intra-annual timing of fires, however, are generally unknown (Clark and Robinson 1993). For the reconstruction of late-Quaternary fire regimes in East Africa, one lake sediment (F-core) recovered from Lake Mobutu Sese Seko, Uganda, was analysed to determine its content of charcoal particles. Because fire regimes are characterized by a combination of climate features, fuel types, and anthropogenic influences, additional information on environmental conditions during the last 30 000 years was necessary to interpret the changing frequencies of charcoal particles. Therefore, in order to outline the state of knowledge of environmental conditions during the late Quaternary, the second part of this work provides an extensive review and interpretation of the historic and prehistoric climate and vegetation, as well as the anthropogenic effects. Our results suggest a correlation between changes in climate and vegetation in savanna biomes and the fire activities in East Africa.

Technical Description

Collection Site

Between 1960 and 1970 several sediment cores were recovered from East African lakes by D. A. Livingstone (Duke University, Durham, North Carolina, United States). Lake Mobutu Sese Seko (formerly Lake Albert) is located at the West Rift Valley at the border of Uganda and Zaire between 1°03' to 2°15'N and 30°22' to 31°24'E. Today, the surface of the lake is 690 m above sea level, and the maximum water depth of 58 m is near the western shore of the lake (Harvey 1976). The tributaries feeding the lake are the Semliki River discharging form the south and the Victoria Nile discharging from the east side of Lake Victoria. The most important outlet is the White Nile at the northern end of the lake.

In 1971, the F-core investigated was taken by T. J. Harvey, under the guidance of D. A. Livingstone, from the northeastern part of the lake at 01°50.1'N and

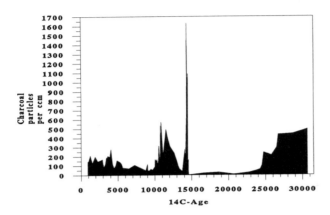

Figure 52.1 Number of charcoal particles per cm³ within the F-core from Lake Mobutu Sese Seko, Uganda, during the last 30 000 years (from *Weiss*, 1995)

Table 52.1 Depths of charcoal samples within the F-core, Lake Mobutu Sese Seko (Uganda)

Sampling step	Depth of the samples in cm
1	17, 51, 179, 306, 324, 358, 486, 613, 631, 665, 793, 917
2	11, 34, 115, 242, 396, 550, 648, 729, 824, 855, 886
3	10, 14, 21, 25, 30, 42, 83, 210, 351, 441, 518, 590, 635, 640, 644, 652, 656, 661, 697, 761, 808, 840
4	32, 38, 40, 59, 67, 75, 99, 195, 315, 377, 419, 570, 611, 612, 612.5, 632, 642, 646, 650, 654, 659, 816, 848
5	33, 36, 47, 55, 63, 79, 147, 187, 274, 320, 386.5, 425, 430, 436, 463.5, 502, 580, 622, 645, 651, 653, 655, 660, 820, 851.5

31°10.2′E. The core has a length of 917 cm and was recovered with a "Kullenberg Sampler" at ca. 46 m water depth. In 1991 this core was investigated by Sowunmi for pollen. In 1995 we investigated the F-core for charcoal particles. A separate core (G-core) was studied paleolimnologically by Harvey (1976).

Sampling
Samples were taken in five successive steps. During the first step, samples were taken arbitrarily from the core. After treating the samples of the first step in the laboratory, the charcoal particles were counted. The samples for the next steps were chosen at peaks of distinct changes in particle frequencies found in the previous step. This procedure was repeated four times, and a total of 93 samples were taken along the entire length of the core. Finally the results of all five steps were evaluated (figure 52.1). Table 52.1 shows the depths of all samples.

Table 52.2 ¹⁴C-data of the sediment core (F-core) from Lake Mobutu Sese Seko (Uganda)

Depth of ^{14}C samples (cm)	^{14}C-age (year BP)	Laboratory no.
35–39	3370 ± 250	Gif-6694
140–150	8750 ± 400	Gif-6695
450–460	10 980 ± 260	Gif-6696
658–673	14 700 ± 260	Beta-9901
902–907	29 900 ± 750	Gif-6698

Age Determination
Five radiometric (^{14}C) data were available to determine sedimentation rates and to allocate charcoal samples accordingly (table 52.2).

Sample Treatment
Sediment samples were heated in a water bath with distilled water, potassium hydroxide (KOH), hydrochloric acid (HCl), and hydrogen fluoride (HF), and centrifuged after each step.

Samples were then sieved through a 125 μm-mesh sieve to remove particles <125 μm. The residue contained quartz, organic pieces, and charcoal particles ≥125 μm. The particles were recognized clearly and counted with the help of a binocular.

Results

Charcoal Concentrations
Charcoal counts and age determination of the sediment samples provided a record of charcoal concentration in the sediment column for the last 30 000 years (figure 52.1). Five epochs are distinguished:

1. High charcoal concentrations occur between 30 000 and 26 000 B.P. followed by a subsequent rapid decrease.

2. Between 23 000 and 14 800 B.P. concentrations drop to minimal values.

3. Several oscillations with sometimes very high charcoal abundances occur in the late glacial period from 14 800 to 10 000 B.P.

4. The first half of the Holocene, up to 4000 B.P., is characterized by moderate concentrations.

5. A slight increase occurs over the last 4000 years.

Fire in African Savanna Biomes
The interpretation of the charcoal concentrations is based on the ecological interconnectedness of fire and

vegetation in African savannas. Long before hominids were able to control fire for their own use, vegetation fires were a natural component of the savanna ecosystems in southern African and had considerable influence on vegetation dynamics and composition. Fire return intervals in African savannas average one to three years. The fire return interval depends on the amount of precipitation and the duration of the dry season. Areas with relatively high rainfall burn nearly every year, whereas those with moderate to low moisture are influenced by fire at longer intervals (Menaut et al. 1991).

Most of the biomass consumed by savanna fires consists of grasses. Adult trees are rarely affected by fires, because most of the tree species are fire resistant. Only occasionally will the cambium be killed and result in mortality of the whole tree. The younger trees are not yet fire resistant, and most tree seedlings will be destroyed completely during each fire. They require a longer regeneration period than grasses, which regrow within a few weeks after a fire.

Interpretation by Comparison with Other Paleoenvironmental Information

Apart from astronomical forcing, changes in the atmospheric content of CO_2 and other greenhouse gases played an important climatic role during the Quaternary (Barnola et al. 1987). The climate changes of glacial/interglacial cycles altered vegetation composition, which in turn, influenced past fire regimes. The coldest epochs in Africa were at 420 000 B.P. and 16 000 B.P. The warmest period was during the last interglacial, approximately 125 000 years ago, when temperatures were slightly higher than they are today (Prell et al. 1979; Jouzel et al. 1987). In the tropics the temperatures were ca. 2°C higher between 140 000 and 116 000 B.P. than in the Holocene (Delmas 1992). Temperature fluctuations as manifested in the Vostock temperature record (Jouzel et al. 1987), as well as CO_2 and CH_4 concentrations (Barnola et al. 1987; Chappellaz et al. 1990), and the $\delta^{18}O$-values of the European Greenland Ice-Core Project (GRIP) (Dansgaard et al. 1993), correspond to each other and show significant agreements with other environmental oscillations (Delmas 1992). The $\delta^{18}O$-values of GRIP (Dansgaard et al. 1993) correlate very well with the results of the $^{18}O/^{16}O$ ratio of an equatorial Pacific sediment core from Shakleton and Opdyke (1973). All parameters mentioned reflect distinct temperature changes during the last glacial/interglacial transition at ca. 15 000 B.P.

In eastern Africa a cooler climate also prevailed during the last glacial maximum. The temperatures in the lowland tropics were reduced between 2°C to 11°C (Hamilton 1972; van der Hammen, 1974; Flenley, 1979b), and the former glaciation of Mt. Kilimanjaro and Mt. Kenya can be reconstructed by the traces of firnlines, that is, the limits of net accumulation and net ablation of the former glaciers. Even today there remain small ice caps on these mountains (Hamilton 1982).

The altitudinal shifting of the ice caps of tropical mountains during the glacial/interglacial periods led to repeated vegetation changes. Several lines of evidence show that these changes took place also in the lowland tropics (Flenley, 1979a). Livingstone (1971; 1975; 1982), Coetzee (1967), Agwu and Beug (1984), Maley (1991), Schulz and Pomel (1992), and Neumann and Ballouche (1992) reconstructed the ancient vegetation of equatorial and southern Africa mostly by pollen analysis. Expansion and shrinking of savanna and forest areas accompanied temperature and moisture changes between the glacial and interglacial periods. As a result, different fire activities established themselves over time.

The earliest natural vegetation fires are documented around 350×10^6 years ago (Francis 1961; Clark and Robinson 1993). They became an important part of the ecosystem development in many regions of the world. Savannas—defined as a vegetation type with more or less continuous grass layer and with various components of trees and bushes—have existed since the Eocene, as evidenced by the first occurrence of pollen of tropical *Poaceae* during this epoch (van der Hammen 1983). Also, fire scars on stems of savanna tree species show that savanna fires have occurred for at least 10×10^6 years (Dechamps and Maes 1990).

Pollen analysis from southwestern Uganda suggests that it was cool and dry from 38 000 to 32 000 B.P., slightly warmer and more moist between 32 000 and ca. 21 000 B.P., and very dry and cold during the last glacial maximum (Taylor 1988). A sharp decline in temperature began at 28 000 B.P. attaining the lowest temperatures between 20 000 and 12 000 B.P. (Hamilton 1982; van der Hammer 1983; Hamilton and Taylor 1991).

Between 29 000 and 25 000 B.P. the pollen spectra indicate moist and dry forest species Sowunmi (1991). High- and medium-altitude forests shifted down the slopes during glacial periods. The high percentage of pollen from lowland savannas, thicket and bushland speaks for a well-developed dry savanna and rather poor forest cover. The high percentage of aquatic and swamp pollen indicates well-established swamp vegetation at the shore of the lake (Sowunmi 1991). Between 30 000 and 25 000 B.P., the high number of charcoal particles may reflect more frequent vegetation fires due to extended grasslands. The lake level of Mobutu Sese Seko was as high as it is today. There was a slow reduction of the lake level from 25 000 to 18 000 B.P. causing it to fall below the outlet, and the lake became closed due to decreasing rainfall (Harvey 1976). Between ca. 25 000 to 22 000 B.P., *Poaceae* and charcoal decreased dramatically (Sowunmi 1991). There was also a reduction in the variety and amount of pollen from aquatic and swamp vegetation and other plant species between ca. 22 000 to 15 000 B. P. The most striking changes were in the pollen quantity of *Poaceae* and *Cyperaceae*, which declined continuously until 14 700 B.P. (Sowunmi 1991). The seasonal shift of the Intertropical Convergence Zone (ITCZ) was less distinct at 18 000 B.P. The subtropical high-pressure areas were diminished so that colder sea water could enter the equatorial zones. This led to reduced oceanic evaporation and, consequently, lowered precipitation (Nicholson and Flohn 1980; Shaw 1985). Decreased alkalinity from 18 000 to 14 000 B.P. suggests the level of Mobutu Sese Seko rose again (Harvey 1976). This is in contradiction to the reduced vegetation in the catchment of the lake, which is reflected by the grass pollen distribution in the F-core (Sowunmi 1991), and by the lowest values of charcoal particles between ca. 22 000 and 14 000 B.P. The period of open lake conditions during the glacial maximum is questionable because, in general, lake levels in equatorial Africa dropped until 15 000 B.P., and in East Africa until 13 000 B.P. (Kendall 1968; Street and Grove 1979; Street-Perrott et al. 1989). Stager et al. (1986) have not found evidence for an intensive humid phase during the last glacial maximum.

The period between 14 000 and 10 000 B.P. is characterized by several climatic and vegetational changes. At ca. 14 800 B.P., the climate shifted, and a highly-diversified vegetation developed where *Poaceae* were favored compared to other plant families (Weiss 1995). This is indicated by the strong peak of *Poaceae* pollen that occurred at 14 800 to ca. 14 000 B.P. Charcoal particle distribution shows a simultaneous strong peak. In the following millennia, the number of charcoal particles shifted several times from higher to lower levels. This corresponds with two more pollen analyses that show a similar fluctuating distribution of *Poaceae* pollen in the late glacial period, based on the P-2-core from Lake Victoria (Kendall 1968) and the 3PC-core

from the southern part of Lake Mobutu Sese Seko (Ssemmanda and Vincens 1993). *Poaceae* pollen in the F-core decreased continuously through the end of the Pleistocene.

The late glacial period from 12 000 to 10 000 B.P., slowly turned warmer (Livingstone 1967; van der Hammen 1983), interspersed by many cooler periods and a short glacial epoch from ca. 11 500 to 10 500 B.P. (Hamilton 1982; Mahaney 1987; Delmas 1992). The lake levels in Africa rose during this time (*COHMAP* 1988; Street-Perrott et al. 1989). The level of Lake Mobutu Sese Seko was low between 14 000 and 12 500 B.P. From 12 500 to 5000 B.P., the lake level was similar to its level 28 000 years ago due to higher inflow (Harvey 1976).

Without taking into account the breif fluctuations of the vegetational composition in late glacial time, the general tendency is for *Poaceae* pollen to decrease. To the same degree that *Poaceae* declined, the woody vegetation expanded (Kendall 1968; Sowunmi 1991; Ssemmanda and Vincens 1993).

Precipitation increased and average temperatures rose at the beginning of the Holocene (Hamilton 1982; Taylor 1988) resulting in the expansion of forest cover in East Africa (Kendall 1968; Sowunmi 1991). Between 10 000 and 8000 B.P. the seasonality became more prominent and the ITCZ migrated further from the equator during summer and winter. Rainfalls increased due to increasing oceanic evaporation (Nicholson and Flohn 1980; Shaw 1985). The warmest and wettest episode was between 8000 and 7000 B.P. (Kutzbach and Street-Perrott 1985). Below the closed canopy of tropical forests, little or no continuous layer of grasses and herbs exist. Tropical evergreen rain forests are rarely combustible, because of the lack of surface fuels and the moist microclimate. Tropical semideciduous lowland rainforests and tropical deciduous moist forests also burn less frequently than do savannas, depending on crown coverage and the seasonality of the climate. These climatic influences critically shaped the vegetational conditions until 4000 B.P. and may have caused reduced fire activity and fewer charcoal particles.

Since 5000 B.P. a reduction in precipitation occurred (Harvey 1976) without decreasing temperatures (Hamilton 1982). The level of Lake Mobutu Sese Seko dropped at ca. 5000 B.P. to its modern state (Harvey 1976). Increased grassy and decreased woody vegetation, prevailed in East Africa in the last 2000 to 3000 years (Kendall 1968; Livingstone 1975, Hamilton et al. 1989). The drier climate in the last millennia may have led to higher fire activity, which is reflected by the slight increase in the number of charcoal particles. Increased use of fire by more settled human populations may also have caused a higher occurence of fires.

At the transition from Pliocene to Pleistocene, savanna areas expanded due mainly to a drier climate and to a lesser degree, human influences (Kershaw et al. 1996). At ca. 1.5×10^6 B.P. hominids left their forest habitat and settled in the open savanna landscapes (Clark 1970; 1976). As the populations of hominids grew, they rapidly reduced the number of megaherbivores in their African savanna habitats. Hominids, therefore, had to shift their hunting activities to the smaller, but quicker ungulates of the savannas (Schüle 1990a). The earliest evidence of fire use by hominids is from the Swartkrans Cave in South Africa, 1.5×10^6 B.P. It is not known whether hominids were able to produce fire or were only tending fire (Brain and Sillen 1988). However, it is assumed that hominids used fire as a hunting tool. At about 30 000 B.P., nearly every biome was occupied by *Homo sapiens sapiens* (Clark 1970). In the early Holocene, intense settlement took place in Ethiopia and in the southern part of the Sahara where the first agricultural systems were established (Hamilton 1982; Sutton 1974). Before ca. 5000 B.P., when the climate became drier, farmers of the southern Sahara moved to equatorial Africa, where moister conditions made agriculture easier (Clark 1976; Neumann 1989). Pollen of secondary tree species in the southwest of Uganda, which in former times was limited to occasional natural disturbances, appeared more frequently (Hamilton et al. 1989). The forests became more open as more agricultural and pastural land was needed. The repeated peaks of *Poaceae* found by Hamilton et al. (1989) are interpreted by the authors to be caused by secondary grasslands. In the lower elevation around Lake Mobutu Sese Seko and Lake Victoria, the vegetation is dominated by lowland savanna formations and evergreen and semievergreen thicket and bushland. More than 80% of the modern flora in this area has been altered by anthropogenic influences (Sowunmi 1991). The mountain ranges of nearly all elevations in East Africa were burned periodically over a long time for pasture management (Flenley 1979b). Hunters and gatherers in Central Africa became agriculturalists in the second half of the Holocene and used fire as an efficient tool for agriculture, hunting, and many other purposes, and they still do today (Flenley 1979b; Hamilton 1982; Hamilton et al. 1986; 1989; Schüle 1990b).

Ultimately, it cannot be shown whether climate change or human activities has had more pronounced effects on East Africa's vegetation. Palynologists believe mainly that anthropogenic influences had a dominating influence on the ecological development during the last millenia (Sowunmi 1991; Schulz and Pomel 1992; Pyne 1993; Kershaw et al. 1996).

Conclusions

The changes in charcoal frequencies during the last 30 000 years found in the F-core are the result of changing environmental conditions. The comparison of charcoal particles distribution with data from the paleoecological literature shows ecological coherences between the recurrence of vegetation fires and the corresponding climatic and vegetational conditions. The interpretation is based primarily on the antagonism between grass and woody vegetation and the alternating extension of grasslands and forests in the African savanna biome. Varying abundances of charcoal particles do not allow any quantitative conclusions; however, a rough approximation of former fire regimes in East Africa is possible by interpreting the charcoal particle depositional history.

During the last glacial maximum when temperatures were low and precipitation reduced, the vegetation cover was sparse and fewer charcoal particles were found in the core. This may be because vegetation fires did not spread over large areas due to the lack of fuels. Periods with moderate temperatures and rainfall allowed the development of large, extended savannas and marginal forest cover. At these times, the highest number of charcoal particles appeared, possibly reflecting higher frequencies of vegetation fires. The climatic conditions with high temperatures and precipitation that existed in the early Holocene led to the shift of the tree line to higher altitudes and to a reduction of the grass cover under closed canopies. Moderate numbers of charcoal particles reflect fewer vegetation fires.

The charcoal concentrations in the F-core show that the number of charcoal particles entering Lake Mobutu Sese Seko fluctuated considerably between 30 000 and 900 B.P. Thus, the level of fire activities was temporarily higher, but also lower when compared with the level at the beginning of the last millennium. We propose that the concentrations of tropospheric ozone from vegetation fires during the last 30 000 years may have fluctuated accordingly. More research is needed to verify the results of this case study.

Acknowledgements

This work was supported by the Max Planck Society (Germany), and an NSF grant to J. S. Clark and the

Department of Zoology of Duke University, Durham, North Carolina, United States).

References

Agwu, C. D. C., and H.-J. Beug. 1984. Palynologische Untersuchungen in marinen Sedimenten vor der west-afrikanischen Küste, *Palaeoecology of Africa*, 16, 37–52.

Andreae, M. O. 1991. Biomass burning: its history, use, and distribution and its impact in environmental quality and global climate, in *Global Biomass Burning*, edited by J. S. Levine, pp. 3–21, MIT Press, Cambridge, Mass.

Andreae, M. O., J. Fishman, M. Garstang, J. G. Goldammer, C. O. Justice, J. S. Levine, R. J. Scholes, B. J. Stocks,. A. M. Thompson, B. van Wilgen, and the STARE/TRACE-A/SAFARI Science team. 1993. Biomass burning in the global environment: First results from IGAC/BIBEX field campaign STARE/TRACE-A/SAFARI-92, in *Global Atmospheric-Biospheric Chemistry* edited by R. G. Prinn, pp. 83–101. Plenum Press, New York.

Andreae, M. O. et. al., Biomass burning in the global environment: First results from the IGAC/BIBEX field campaign STARE/TRACE-A/SAFARI-92, in *Global Atmospheric-Biospheric Chemistry*, edited by R. G. Prinn, pp. 83–101, Plenum, New York.

Barnola, J. M., D. Raynaud, Y. S. Korotkevich, and C. Lorius. 1987. Vostock ice core provides 160 000-year record of atmospheric CO_2, *Nature*, 329, 408–414.

Brain, C. K., and A. Sillen. 1988. Evidence from the Swartkrans cave for the earliest use of fire. *Nature* 336, 464–466.

Chappellaz, J., M. Barnola, D. Raynaud, Y. S. Korotkevich, and C. Lorius. 1990. Ice-core record of atmospheric methane over the past 160 000 years, *Nature*, 345, 127–131.

Clark, J. D. 1970. *The Prehistory of Africa*, Thames and Hudson, London, 302 pp.

Clark, J. D. 1976. African origins of man the toolmaker, in *Human Origins*, edited by G. L. Isaac, and E. R. McCown, pp. 1–53, W. A. Benjamin, Calif.

Clark, J. S., and J. Robinson. 1993. Paleoecology of fire, in *Fire in the Environment. The Ecological Atmospheric and Climatic Importance of Vegetation Fires*, edited by P. J. Crutzen, and J. G. Goldammer, pp. 193–214, Dahlem Workshop Reports ES 13 Berlin, John Wiley and Sons, Chichester.

Clark, J. S., H. Cachier, J. G. Goldammer, and B. J. Stocks, (eds.). 1996. *Sediment Records of Biomass Burning and Global Change*, Springer-Verlag, Berlin, Heidelberg, New York, in press.

Coetzee, J. A. 1967. Pollen analytical studies in East and Southern Africa, *Palaeoecology of Africa*, 3, 1–146.

COHMAP Members. 1988. Climatic changes of the last 18 000 years: Observation and model simulations, *Science*, 241, 1043–1052.

Dansgaard, W., S. J. Johnson, H. B. Clausen, D. Dahl-Jensen, N. S. Gundestrup, C. U. Hammer, C. S. Hvidberg, J. P. Steffensen, A. E. Sveinbjörnsdottir, J. Jouzel, and G. Bond. 1993. Evidence for general instability of the past climate from a 250-kyr ice-core record, *Nature*, 364, 218–220.

Dechamps, R., and F. Maes. 1990. Woody plant communities and climate in the Pliocene of the Semliki Valley, Zaire, in *Evolution of Environments and Hominidae in the African Western Rift Valley*, edited by N. T. Boaz, pp. 71–94, Virginia Museum of Natural History, Martinsville.

Delmas, R. J. 1992. Environmental informations from ice cores, *Rev. Geophysics.*, 30, *1*, 1–21.

Fishman, J., C. E. Watson, J. C. Larsen, and J. A. Logan. 1990. Distribution of tropospheric ozone determined from satellite data, *J. Geophys. Res.*, 95, 3599–3617.

Flenley, J. R. 1979a. The late Quaternary vegetational history of the equatorial mountains, *Prog. Phys. Geogr.*, 3, 488–509.

Flenley, J. R. 1979b. *The Equatorial Rain Forest: A Geological History*, Butterworths, London, Boston, 162 pp.

Francis, W. 1961. *Coal, Its Formation and Composition*, Arnold Publ. Ltd., London, 806 pp.

Hamilton, A. C. 1982. The interpretation of pollen diagrams from Higland Uganda, *Palaeoecology of Africa*, 7, 45–149.

Hamilton, A. C. 1982. *Environmental History of East Africa. A Study of Quarternary*, Academic Press, London, New York, Paris, 328 pp.

Hamilton, A. C., and D. Taylor. 1991. History of climate and forests in tropical Africa during the last 8 million years. *Climatic Change* 19, 65–78.

Hamilton, A. C., D. Taylor, and J. C. Vogel. 1986. Early forest clearance and environmental degradation in south-west Uganda, *Nature*, 320, 164–167.

Hamilton, A. C., D. Taylor, and J. C. Vogel. 1989. Neolithic forest clearance at Ahakagyezi, western Uganda, in *Quaternary and Environmental Research on East African Mountains*, edited by W. C. Mahaney, pp. 435–463, A. A. Balkema, Rotterdam, Brookfield.

Hammen van der, T. 1974. The Pleistocene changes of vegetation and climate in tropical South America, *J. Biogeography*, 1, 3–36.

Hammen van der, T. 1983. The palaeoecology and the palaeogeography of savannas, in *Tropical Savannas*, edited by F. Bourlière, pp. 19–35, Elsevier Sci. Publ. Co., Amsterdam.

Harvery, T. J. 1976. The paleolimnology of Lake Mobutu Sese Seko, Uganda-Zaire: The last 28 000 year, Ph.D. dissertation, Department of Zoology, Duke University, Durham, North Carolina, 103 pp.

Jouzel, J., C. Lorius, J. R. Petit, C. Genthon, N. I. Barkov, V. M. Kotlyakov, and V. M. Petrov. 1987. Vostok ice core: a continuous isotope temperature record over the last climatic cycle (160 000 years), *Nature*, 329, 403–408.

Kendall, R. L. 1968. An ecological history of the Lake Victoria basin, Ph.D. dissertation, Duke University, Durham, North Carolina, 193 pp.

Kershaw, A. P., M. B. Bush, G. S. Hope, J. G. Goldammer, and K.-F. Weiss. 1996. The contribution of humans to past biomass burning in the tropics, in *Sediment Records of Biomass Burning and Global Change*, edited by J. S. Clark, H. Cachier, J. G. Goldammer, and B. J. Stocks, Springer-Verlag, Berlin, Heidelberg, New York, in press.

Kutzbach, J. E., and F. A. Street-Perrott. 1985. Milankovitch forcing of fluctuations in the level of tropical lakes from 18 to 0 kyr BP, *Nature*, 317, 130–134.

Livingstone, D. A. 1967. Postglacial vegetation on the Ruwenzori Mountains in equatorial Africa, *Ecological Monography*, 37, 25–52.

Livingstone, D. A. 1971. A 22 000-year pollen record from the plateau of Zambia, *Limmology and Oceanography* 16, No. 2, 349–356.

Livingstone, D. A. 1975. Late Quaternary climatic change in Africa, *Annual Review of Ecology and Systematics* 6, 249–280.

Livingstone, D. A. 1982. Quaternary geography of Afrika and the refuge theory, in *Biological Diversification in the Tropics*, edited by G. T. Prance, pp. 523–536, Columbia University Press, New York.

Mahaney, W. C. 1987. Paleoenvironmental problms in the East African mountains, *Palaeoecolology of Africa* 18, 245–255.

Maley, J. 1991. The African rain forest vegetation and palaeo-environments during late Quaternary, *Climate Change* 19, 79–98.

Menaut, J-C., L. Abbadie, F. Lavenu, P. Loudjani, and A. Podaire. 1991. Biomass burning in West African Savannas, in *Global Biomass Burning*, edited by J. S. Levine, pp. 133–142, MIT Press, Cambridge, Mass.

Neumann, K. 1989. Holocene vegetation of the Eastern Sahara: charcoal from prehistoric sites, *The African Archeological Review* 7, 97–116.

Neumann, K., and A. Ballouche. 1992. Die Chaine de Gobnangou in SE Burkina Faso—Ein Beitrag zur Vegetationsgeschichte der Sudanzone W-Afrikas, Frankfurt, *Geobot. Kolloq.* 8, 53–68.

Nicholson, S., and H. Flohn. 1980. African environmental and climatic changes and the general circulation in the late Pleistocene and Holocene, *Climatic Change* 2, 313–348.

Prell, W. L., W. H. Hutson, and D. F. Williams. 1979. The suptropical convergence and Late Quaternary circulation in the Southern Indian ocean, *Marine Micropaleontology* 4, 225–234.

Pyne, S. J. 1993. Keeper of the flame: A survey of anthropogenic fire, in *Fire in the Environment. The Ecological Atmospheric, and Climatic Importance of Vegetation Fires*, edited by P. J. Crutzen, and J. G. Goldammer, pp. 245–266, Dahlem Workshop Reports ES 13 Berlin, John Wiley and Sons, Chichester.

Schüle, W. 1990a. Human evolution, animal behaviour, and quaternary extinctions: A paleo-ecology of hunting, *Homo* 41/3, 228–250.

Schüle, W. 1990b. Landscapes and climate in prehistory: Interaction of wildlife, man and fire, in *Fire in the Tropical Biota: Ecosystem Processes and Global Challenges*, edited by J. G. Goldammer, pp. 273–318, Ecological Studies 84, Springer-Verlag, Berlin, Heidelberg, New York.

Schulz, E. and S. Pomel. 1992. Die anthropogene Entstehung des Sahels, *Würzburger Geographische Arbeiten* 84, 263–288.

Shackleton, N. J., and N. D. Opdyke. 1973. Oxygen isotope and palaeomagnetic stratigraphy of Equatorial Pacific Core V28-238: oxygen isotope temperatures and ice volumes on a 10^5 and 10^6 year scale, *Quaternary Research* 3, 39–55.

Shaw, T. 1985. Past climates and vegetation changes, in *Historical atlas of Africa*, edited by J. F. Ade Ajayi, and M. Crowder, Chapt. 6, Longman, London.

Sowunmi, M. A. 1991. Late Quaternary environments in equatorial Africa: Palynological evidence, *Palaeoecology of Africa* 22, 213–238.

Ssemmanda, I., and A. Vincens. 1993. Vegetation and climate in the Lake Albert basin during the last 13 000 years B. P.: palynological contribution, *C. R. Académie des Sciences Paris*, t. 316, Série II, 561–567.

Stager, J. C., P. N. Reinthal, and D. A. Livingstone. 1986. A 25 000-year history of Lake Victoria, East Africa, and some comments on its significance for the evolution of cichlid fishes, *Freshwater Biology* 16, 15–19.

Street, F. A., and A. T. Grove. 1979. Global maps of lake-level fluctuations since 30 000 B. P., *Quaternary Research* 12, 83–118.

Street-Perrott, F. A., D. S. Marchand, N. Roberts, and S. P. Harrison. 1989. *Global Lake-Levels Variations from 18 000 to 0 Years Ago: A Palaeoclimatic Analysis*. Oxford University, United Kingdom. Prepared for United States Department of Energy, 213 pp.

Sutton, J. E. G. 1974. The aquatic civilization of middle Africa, *Journal of African History* 15, 527–546.

Taylor, D. M. 1988. The environmental history of the Rukiga Highlands, South-West Uganda, during the last 40 000 to 50 000 year, unpublished D. Phil. thesis, University of Ulster, United Kingdom.

Weiss, K.-F. 1995. Rekonstruktion prähistorischer Feuerregime in Ostafrika durch Seesedimentkerne, unpublished dissertation, University of Gießen, Germany, 109 p.

Index